008409130

U. Förstner · G. T. W. Wittmann

Metal Pollution in the Aquatic Environment

With Contributions by
F. Prosi and J. H. van Lierde

Foreword by Edward D. Goldberg

With 102 Figures and 94 Tables

Springer-Verlag
Berlin Heidelberg New York 1979

Professor Dr. ULRICH FÖRSTNER
Institut für Sedimentforschung
Universität Heidelberg
Im Neuenheimer Feld 236
6900 Heidelberg 1/FRG

Dr. GOTTFRIED T. W. WITTMANN
Department of Chemistry
University of Pretoria
Pretoria 0002/South Africa

TD
427
H45
F62

ISBN 3-540-09307-9 Springer-Verlag Berlin Heidelberg New York
ISBN 0-387-09307-9 Springer-Verlag New York Heidelberg Berlin

Library of Congress Cataloging in Publication Data. Förstner, Ulrich. Metal pollution in the aquatic environment. Bibliography: p. Includes index. 1. Heavy metals – Environmental aspects. 2. Water – Pollution. I. Wittmann, Gottfried, 1929–. joint author. II. Title. TD427.H45F62.553'.7. 79-11122.

This work is subject to copyright. All rights are reserved, whether the whole or part of the material is concerned, specifically those of translation, reprinting, re-use of illustrations, broadcasting, reproduction by photocopying machine or similar means, and storage in data banks. Under §54 of the German Copyright Law, where copies are made for other than private use, a fee is payable to the publisher, the amount of the fee to be determined by agreement with the publisher.

© by Springer-Verlag Berlin Heidelberg 1979
Printed in Germany.

The use of registered names, trademarks, etc. in this publication does not imply, even in the absence of a specific statement, that such names are exempt from the relevant protective laws and regulations and therefore free for general use.

Cover design: Dietrich Bogumil, Weinheim.

Typesetting, offsetprinting and bookbinding: Brühlsche Universitätsdruckerei, Lahn-Gießen.
2132/3130-543210

Foreword

Aquatic chemistry is becoming both a rewarding and substantial area of inquiry and is drawing many prominent scientists to its fold. Its literature has changed from a compilation of compositional tables to studies of the chemical reactions occurring within the aquatic environments. But more than this is the recognition that human society in part is determining the nature of aquatic systems. Since rivers deliver to the world ocean most of its dissolved and particulate components, the interactions of these two sets of waters determine the vitality of our coastal waters. This significant volume provides not only an introduction to the dynamics of aquatic chemistries but also identifies those materials that jeopardize the resources of both the marine and fluvial domains. Its very title provides its emphasis but clearly not its breadth in considering natural processes.

The book will be of great value to those environmental scientists who are dedicated to keeping the resources of the hydrosphere renewable. As the size of the world population becomes larger in the near future and as the uses of materials and energy show parallel increases, the rivers and oceans must be considered as a resource to accept some of the wastes of society. The ability of these waters and the sediments below them to accommodate wastes must be assessed continually.

The key questions relate to the capacities of aqueous systems to carry one or more pollutants These may be based upon impacts on public health, the communities of organisms, navigation, recreation, aesthetics, or economics of aquatic environments. Of concern is the amount of material that the water body can carry before the impact becomes unacceptable. The amount, essentially determined by a titration of the pollutant with the water body, becomes evident at an endpoint. One of the goals of environmental managers is the identification of these endpoints.

Perhaps the most extensive set of endpoints evolves from the entries of radionuclides into the oceans from facilities of the nuclear fuel cycle. Human health is to be protected from the consumption of artificial radioactivities in fish, shellfish, or edible algae or through exposure to radiation on beaches. The acceptable amounts of radionuclides in natural waters are related to potential damage to the human body from exposure to ionizing radiation. The endpoints are found in the levels in organisms or beach solids.

Sometimes a catastrophe initiates a concern about a pollutant. The Minimata Bay episode identified methyl mercury as a contaminant in the marine system which could endanger human health. Endpoints for this material are found in methyl mercury levels of commercial fish.

The impact of DDT, a biocide, upon nontarget organisms, such as fish-eating birds, could have provided endpoints for its entry into natural waters. Its ingestion results in egg-shell thinning, and populations decline as a result of reproductive failures in the birds. Either of these phenomena might have been translated into an endpoint.

In this volume we have background information for managing our environment. Whether we seek endpoints for metal pollutant loads or merely understanding of natural phenomena, the relevant concepts are to be found herein.

Edward D. Goldberg
Scripps Institution of Oceanography

Preface

This work represents an attempt by the authors to compile and evaluate the present state of affairs with regard to heavy metal pollution in streams, lakes and near-shore marine areas. The considerable interest in, and apprehension about the role and fate of heavy metals in aquatic systems, as displayed by scientists and authoritative bodies, is warranted on the grounds of several disasters which are related to mercury and cadmium poisoning. However, in order to determine the degree of heavy metal enrichment it is essential to establish suitable norms and criteria.

In their book, *Schwermetalle in Flüssen und Seen als Ausdruck der Umweltverschmutzung* (Springer-Verlag Berlin/Heidelberg/New York, 1974), Ulrich Förstner and German Müller adopted a geochemical and sedimentological approach for assessing heavy metal pollution in aquatic systems. The practice of separating the chemically active fine-grained sediment fraction and subsequent analysis of the metal content thereof has proved to serve as a reliable pollution indicator. Further basic research as well as the interest displayed by colleagues in the above mentioned book, and in several papers published by the present authors, both jointly and individually, have been incentives to compile this work.

Although an evaluation of different analytical techniques—especially in connection with water analysis—lies beyond the scope of this study, we hope to have stimulated further research relating to heavy metal pollution. In particular, attention is drawn to hitherto unresolved problems—such as those associated with chemical speciation, the availability and toxicological properties of different species, the uptake mechanisms of metals by organisms, and the remobilization of heavy metals from sediments.

We wish to record our appreciation for such special information which was graciously made available to us by unnamed colleagues and which would otherwise have been well-nigh unobtainable. In particular, we are specially indebted to Dr. Prosi for the chapter on bio-accumulation of metals and for the assistance of Mr. J. H. van Lierde on the subject of water purification. We further gratefully acknowledge the support of those colleagues with whom we have had fruitful discussions and who have been of invaluable assistance to us, especially the efforts of Professor German Müller, Bernhard Schaule and Dr. S. R. Patchineelam. Both latter colleagues, as well as Dr. I. Thornton (Imperial College, London) and Dr. U. Schöttler (Dortmund) kindly also gave access to some of their unpublished information. Furthermore, our appreciation includes the contributions of J. Broadbridge,

S. Knapton, and D. Godfrey for typing and translation assistance as well as of U. Kästner for his technical drawings.

Finally, we gratefully acknowledge the continued financial support of our research projects by various organizations, in particular the German Research Society, the Universities of Heidelberg and Pretoria and the South African Water Research Commission.

Heidelberg/Pretoria
May, 1979

U. Förstner
G. T. W. Wittmann

Contents

Chapter A
Introduction 1

1 Aquatic Ecosystems 1
2 Pollutants in Water 1

Chapter B
Toxic Metals (G. Wittmann) 3

1 Metals 3
  1.1 Classification of the Elements 3
  1.2 Classification of Metals 6
  1.3 Trace Metal Species in Aquatic Systems 7
2 Trace Metals and Organic Life 8
  2.1 Trace Elements Essential to Human Life 9
  2.2 Deficiency and Oversupply 11
  2.3 Metal Toxicity 12
  2.4 Health Hazard Due to Certain Trace Elements 14
  2.5 Accumulation of Toxic Substances in the Aquatic Food Chain 17
  2.6 Catastrophic Episodes of Metal Poisonings 18
    2.6.1 Mercury Poisoning 18
    2.6.2 Cadmium Poisoning 21
    2.6.3 Lead Poisoning 23
    2.6.4 Copper Poisoning 24
    2.6.5 Chromium Poisoning 25
3 Water Quality Criteria: Standards 26
  3.1 Introduction 26
  3.2 Criteria Development 26
  3.3 Water Quality Criteria 29
4 The Sources of Metal Pollution 30
  4.1 Geologic Weathering 31
  4.2 Mining Effluents 33
  4.3 Industrial Effluents 38

	4.4 Domestic Effluents and Urban Stormwater Runoff	43
	4.4.1 Domestic Effluents	43
	4.4.2 Urban Storm Water Runoff	44
	4.4.3 Spoil Heaps	47
	4.5 Metal Inputs from Rural Areas	48
	4.6 Atmospheric Sources	49
	4.7 Special Sources	55
	4.8 Multi-Source Effects	57
5	Metal Analysis	61
	5.1 Media of Pollution Assessment	62
	5.2 Sampling and Analytic Methods	68
	5.2.1 Sampling	68
	5.2.2 Analytic Methods	68

Chapter C
Metal Concentrations in River, Lake, and Ocean Waters (U. Förstner)   71

1	Distribution of Major Ions	71
	1.1 Natural Salt Concentrations	71
	1.2 Man-Made Contamination	73
2	Chemical Conditions for Trace Metals in Natural Waters	73
	2.1 Chemical Speciation in Freshwater and Seawater	75
	2.1.1 Analysis of Trace Metal Speciation	75
	2.1.2 Freshwater/Seawater Model	78
	2.2 Redox Conditions in Natural Waters	79
3	Trace Metals in Seawater	82
	3.1 Natural Distribution	83
	3.2 Man-Made Effects	86
	3.2.1 Atmospheric Input of Metals	88
	3.2.2 Metal Input from Sewage Effluents	89
4	Trace Metals in Inland Waters	90
	4.1 Natural Contents	90
	4.2 Metal Pollution in River Water: Regional Examples	93
	4.2.1 Heavy Metal Pollution in United States Water Systems	93
	4.2.2 Metal Pollution in Inland and Coastal Waters of Great Britain	96
	4.2.3 Heavy Metals in River Water of the Federal Republic of Germany	98
	4.2.4 Heavy Metals in River Water of the U.S.S.R.	102
	4.2.5 Heavy Metals in Waters of Japan	104
	4.3 Metal Transport in Freshwater Systems	104
	4.3.1 Water Discharge and Metal Transport	104
	4.3.2 Annual Cycles of Metal Transport	107

Chapter D
Metal Pollution Assessment from Sediment Analysis (U. Förstner) .. 110

1 Introduction 110
  1.1 Soluble/Solid Equilibrium 110
  1.2 Surface Samples and Sediment Cores 112
2 Metal Investigations on Aquatic Sediments 113
  2.1 Sampling and Storage 114
    2.1.1 Soils and Sediments 114
    2.1.2 Grab and Sore Samplers 114
    2.1.3 Bottom Sediment Traps 115
    2.1.4 Suspended Materials 115
    2.1.5 Recovery of Pore Waters 115
    2.1.6 Storage 115
  2.2 The Mechanical Sediment Analysis 116
  2.3 Mineralogical Analysis 116
  2.4 Chemical Analysis of Nutrient Components (C−N−P) 117
    2.4.1 Determinations of Oxidizable Matter (Organic
          Carbon) by the Chromic Acid Method 117
    2.4.2 Determination of Kjeldahl Nitrogen 117
    2.4.3 Determination of Total Phosphorus 117
  2.5 Sediment Digestion in Metal Analysis 118
    2.5.1 Hydrofluoric Acid Decomposition 118
    2.5.2 Hydrochloric-Nitric Acid (Aqua Regia)
          Decomposition or Digestion by Nitric Acid 118
    2.5.3 Lithium Metaborate Fusion (with Simultaneous
          Determination of Silica) 118
    2.5.4 Transfer of Solid Suspensions into Graphite
          Cuvettes 118
3 Geochemical Reconnaissance of Aquatic Sediments 119
4 Grain-Size Effects 121
  4.1 Grain-Size Dependencies of Trace Metal Concentrations 122
  4.2 Reduction of Grain-Size Effects 124
    4.2.1 Extrapolation from Grain-Size Distribution 124
    4.2.2 Metal Concentrations vs Surface Area 126
    4.2.3 Separation of Clay/Silt and Fine Sand Fractions by
          Sieving 126
    4.2.4 Separation of the Pelitic Fraction (< 2 μm) in Settling
          Tubes 127
    4.2.5 Treatment with Dilute Acids (Hydrochloric Acid,
          Nitric Acid) 128
    4.2.6 Mineral Separation: Quartz Correction Method 128
    4.2.7 Comparison with "Conservative" Elements 129
    4.2.8 The Relative Atomic Variations of Elements 130
5 Factors Controlling the Distribution of Metals in Aquatic
  Sediments 131

6 Natural Metal Content—Civilizational Accumulation 133
6.1 Average Shale: Global Standard Value 133
6.2 Fossil Lake Sediments: Standards Regarding Environmental Data 135
6.3 Fossil Fluviatile Deposits: Regional Influences 135
6.4 Short, Dated Sediment Cores: 200 Years of Industrial Development 136
6.5 Recent Lake Deposits in Relatively Unpolluted Areas 137
6.6 Metals in Suspended Matter: Background Values in Storm Water 137
6.7 Background Values and Nonpoint Sources 138

7 Lake Sediments as Indicators of Heavy Metal Pollution 140
7.1 Interference: Geochemical Background and Man's Impact ... 140
7.2 Metal Pollution in Lake Sediments (Examples) 143
7.3 Metal Contamination Recorded in Dated Sedimentary Cores 146
7.4 Mercury Poisoning of Lakes 151
   7.4.1 Sources of Mercury Pollution 151
   7.4.2 Swedish Lakes 152
   7.4.3 Canadian Lakes: Clay Lake 153
   7.4.4 Laurentian Great Lakes 153

8 Metal Pollution in River Sediments 157
8.1 Geochemical Reconnaissance of Mercury 158
8.2 Stream Sediments: a Response to Environmental Contamination 158
8.3 Heavy Metal Enrichment of River Sediments by Man-Made Influences 163

9 Assessing Metal Pollution in the Sea by Sediment Study 171
9.1 Mercury Contamination—Forms of Metal Enrichment in Coastal Sediments 172
   9.1.1 Minamata—Industrial Metal Contamination in Japanese Coastal Waters 172
   9.1.2 Firth of Clyde—Sewage Sludge Disposal 174
   9.1.3 Southern Californian Coast—Sewer Outfalls and Atmospheric Influences 176
   9.1.4 New Haven—Unregulated Effluent Discharge 178
9.2 Marine Waste Deposits in the New York Metropolitan Region . 178
9.3 Industrial Effluents in New Bedford Harbor, Mass. 182
9.4 Heavy Metal Enrichment in the North Sea, Baltic Sea, and Mediterranean Sea 184
   9.4.1 Metal Pollution in the Mediterranean Sea 184
   9.4.2 Metal Pollution in the North Sea 185
   9.4.3 Metal Contamination of the Baltic Sea 188
   9.4.4 Other Restricted Basins—Fjords 189
9.5 Heavy Metal in Estuarine Sediments 191
   9.5.1 The Estuarine Environment 191

Contents   XIII

    9.5.2 The Rhine Estuary 193
    9.5.3 The Elbe Estuary 194
    9.5.4 Mixing Processes 194

Chapter E
Metal Transfer Between Solid and Aqueous Phases (U. Förstner) ... 197

1 Residence Times of Metals in Aquatic Systems 197
2 Types of Metal Association in Sediments 200
  2.1 Classification of Chemical Phases in Sediment 201
  2.2 Heavy Metals in Detrital Minerals 201
  2.3 Heavy Metal Precipitation 203
    2.3.1 Hydroxides 203
    2.3.2 Sulfides 205
    2.3.3 Carbonates 205
  2.4 Cation Exchange and Adsorption 207
  2.5 Sorption onto Clay Minerals 210
  2.6 Sorption and Coprecipitation on Hydrous Fe/Mn-Oxides and Fe-Sulfides 213
    2.6.1 Formation of Hydrous Mn and Fe Oxides 214
    2.6.2 Sorption of Heavy Metals onto Fe/Mn Oxides 216
    2.6.3 Coprecipitation of Trace Elements with Iron Sulfides .. 218
  2.7 Metal Associations with Organic Substances 220
    2.7.1 Organic Substances in Natural Waters 221
    2.7.2 Sorption and Complexation of Metals by Humic Substances 222
    2.7.3 Coagulation and Flocculation of Metal-Organic Matter . 223
    2.7.4 Associations of Metal-Organic Compounds to Sediments 224
  2.8 Sorption of Trace Elements on Carbonates and Phosphates .. 227
3 Metal Accumulation in Aquatic Sediments—Interactions and Effects of Various Processes and Sinks 230
  3.1 Hydroxidic Coatings on Clay Minerals 230
  3.2 Organic Coatings on Clay Minerals 232
  3.3 Interactions Between Hydrous Metal Oxides, Organic Substances, Carbonate, and Phosphate 232
  3.4 Significance of the Different Sinks in Natural Systems 234
  3.5 Non-Conservative Effects of Trace Metals in Estuaries 236
4 Determination of Chemical Phases in Natural and Polluted Sediments 238
  4.1 Proportion of the Individual Types of Metal Associations in Natural and Polluted Aquatic Sediments 239
  4.2 Chronological and Grain-Size Variation of Trace Metal Bonding 242
5 Mobilization of Heavy Metals from Sediments 247

- 5.1 Saltwater/Sediment Interactions ... 247
  - 5.1.1 Desorption Experiments ... 248
  - 5.1.2 Estuary-Sediment Boundary ... 249
- 5.2 Redox Changes and Metal Release ... 250
  - 5.2.1 Chemical Factors Affecting Metal Distribution in Interstitial Water ... 252
  - 5.2.2 Physical Processes Affecting Metal Release from Pore Water ... 254
- 5.3 Metal Release by Acidic Water ... 258
  - 5.3.1 Acid Mine Drainage ... 259
  - 5.3.2 Acid Precipitation ... 260
- 5.4 Mobilization of Metals by Organic Complexing Agents ... 262
- 5.5 Mobilization of Heavy Metals by Microbial Activity ... 265
  - 5.5.1 Microbial Interactions in Natural Environments ... 265
  - 5.5.2 Bacterial Leaching of Metals ... 266
  - 5.5.3 Microbial Action in the Mercury Cycle ... 267
  - 5.5.4 Bacterial Methylation of Arsenic, Lead, and Selenium ... 269

Chapter F
Heavy Metals in Aquatic Organisms (F. Prosi) ... 271

1 Physico-Chemical Influences on the Toxicity and the Uptake of Heavy Metals with Respect to Organisms ... 273
  - 1.1 Temperature and Oxygen Content ... 273
  - 1.2 Water Hardness ... 274
  - 1.3 Organic Compounds ... 276
  - 1.4 pH Values ... 279
  - 1.5 Salinity ... 280
2 Biologic Factors Affecting Heavy Metal Concentrations in Aquatic Organisms ... 281
  - 2.1 General Physiologic Behavior ... 282
  - 2.2 Life Cycle and Life History of the Organism ... 282
  - 2.3 Seasonal Variations of Metal Content in Organisms ... 284
  - 2.4 Species-Specific and Individual Variability ... 285
  - 2.5 Contamination by Food and Intestine Content ... 285
3 Heavy Metal Enrichment in Limnic and Marine Organisms at Different Trophic Levels ... 286
  - 3.1 Autotrophic Organisms ... 286
    - 3.1.1 Phytoplankton ... 286
    - 3.1.2 Macroalgae ... 288
    - 3.1.3 Freshwater Algae ... 290
    - 3.1.4 Mosses ... 291
    - 3.1.5 Higher Plants ... 292

3.2 Heterotrophic Organisms 294
    3.2.1 Zooplankton 295
    3.2.2 Bivalves 297
    3.2.3 Higher Marine Crustaceans 304
    3.2.4 Freshwater Crustaceans 305
    3.2.5 Marine and Freshwater Fish 306
3.3 The Mobilization of Heavy Metals from Sediment by
    Aquatic Biota 313
3.4 Food Chain Enrichment in Aquatic Life 318

Chapter G
Trace Metals in Water Purification Processes
(U. Förstner and J. H. van Lierde) 324

1 Heavy Metal Removal for the Production of Drinking Water 324
  1.1 Obtaining Water by Bank Filtration 324
  1.2 Artificial Recharge of Groundwater by Land Spreading
     and Injection 327
  1.3 Direct Water Purification by Traditional Physico-Chemical
     Treatment (PCT) and Related Advanced Methods 330
     1.3.1 Traditional Removal of Trace Metals by Pre-Clarification,
         Chlorination, Flocculation, and Filtration 331
     1.3.2 Heavy Metal Removal by Chemical Precipitation 333
     1.3.3 Activated Carbon Filtration in Drinking Water
         Purification 334
     1.3.4 Heavy Metal Removal by Ion Exchange 336
     1.3.5 Potential Metal Enrichments in the Water Distribution
         System 337
2 Heavy Metals in Industrial and Domestic Effluents 340
  2.1 Effluents from the Electroplating Industry 340
  2.2 Mercury Removal from Chlor-Alkali Plant Effluents 343
  2.3 Prevention and Control of Acidic Mine Drainage 344
  2.4 Heavy Metals in Urban Drainage Systems–Biologic
     Treatment (BT) 345
     2.4.1 Metal Extraction in the Mechanical (Primary)
         Sedimentation Unit 346
     2.4.2 Reduction of Metal Loads in the Biologic Stage 346
  2.5 Tertiary Physico-Chemical Treatment of Wastewater 347
3 Heavy Metals in Sewage Sludges 352
  3.1 Land Application of Sewage Sludges 352
  3.2 Impact of Heavy Metals on Groundwater Quality 355
  3.3 Sewage Sludge Disposal to the Sea 358

Chapter H
Concluding Remarks 360

1 Disposal Versus Reuse 363
2 Alternative Materials 364

Appendix 367

References 399

Subject Index 475

The heavy metal concentrations were given mainly in
$\mu g/l$ (water) and *ppm* (sediment, biological materials).

1 $\mu g/l$ (1 $\mu g/dm^3$) equals
(for water) $\approx$ 1 μg/kg *or* 1 mg/t *or* $10^{-7}\%$ *or*
1 ppb (parts per billion)

1 ppm (parts per million) equals
1 mg/kg *or* 1 g/t *or* $10^{-4}\%$

Chapter A
Introduction

Since the Industrial Revolution, the efforts of removing pollutants from the natural environment have not been able to keep pace with the increasing amount of waste materials and a growing population that further aggravates the situation. This has often resulted in the transformation of lakes, rivers, and coastal waters into sewage depots where the natural biologic balance is severely upset and in some cases totally disrupted.

1 Aquatic Ecosystems

The major reason for the particular sensitivity of aquatic systems to pollution influences may lie in the structure of their food chains (Stumm, 1976; 1977). Compared with land systems, the relatively small biomass in aquatic environments generally occurs in a greater variety of trophic levels, whereby accumulation of xenobiotic and poisonous substances can be enhanced.

The adverse effects of waste materials became acute in inland water systems due to their traditional role as receiving bodies for effluents. Simultaneously, more areas have become dependent on surface water for their water supply because of the depletion of natural groundwater reserves and the difficulty in exploiting new sources. At the same time, this precarious situation is not limited to inland waters since rivers carry their load of pollutants—either in dissolved, colloidal or particulate form—to oceans. In many cases harmful substances enter the food chain and are concentrated in fish and other edible organisms particular in nearshore areas. This development is all the more cause for concern at a time when the oceans are being increasingly considered as future suppliers of protein for the growing world population.

2 Pollutants in Water

With the growth of technology, two groups of substances in particular have a lasting effect on the natural balance in aquatic systems: *nutrients,* which promote unrestricted biologic growth and, in turn, oxygen depletion, and *sparingly degradable* ("refractory") *synthetic chemicals* and other waste substances which often imply multiple effects on the aquatic ecosystem.

There are various sources of pollutants in the aquatic ecosystem. Atmospheric emissions from industry and households and runoff from agriculture both contribute to water pollution. The most important source is, however, the waste water fed directly into the aquatic system. Experts estimate that industrial and domestic waste water introduces up to a million different pollutants into natural waters. These include substances that are not considered dangerous; many of them add a disagreeable odour or taste to the water and others significantly upset the ecosystem without being directly harmful to humans. Other groups do, however, have direct and indirect influences on the human organism and can cause grave damage. Substances such as polycyclic aromatics, pesticides, radioactive matter, and trace metals can directly endanger human life.

The latter group of pollutants—which are dealt with in this book—are of note in two respects: firstly, trace metals are not usually eliminated from the aquatic ecosystems by natural processes, in contrast to most organic pollutants, and secondly, most metal pollutants are enriched in mineral and organic substances. Toxic metals such as mercury, cadmium, arsenic, copper, and many other species tend to accumulate in bottom sediments from which they may be released by various processes of remobilization, and—in changing form—can move up the biologic chain, thereby reaching human beings where they produce chronic and acute ailments.

Chapter B
Toxic Metals

1 Metals

Of the chemical elements, metals make up the largest group; their characteristics, however, differ greatly within the biosphere. In the last few years, it has become especially clear that it is less the total concentration of a certain element that produces a negative or positive effect on the organism, but rather the specific compound form that decisively influences the toxic effect of an element.

1.1 Classification of the Elements

The chemical elements are conveniently arranged in the long form of the Periodic Table (PT), (Fig. 1), which consists of seven horizontal rows known as *periods* (or series) and sixteen vertical columns called *groups* (or families). Inspection of Figure 1 reveals that

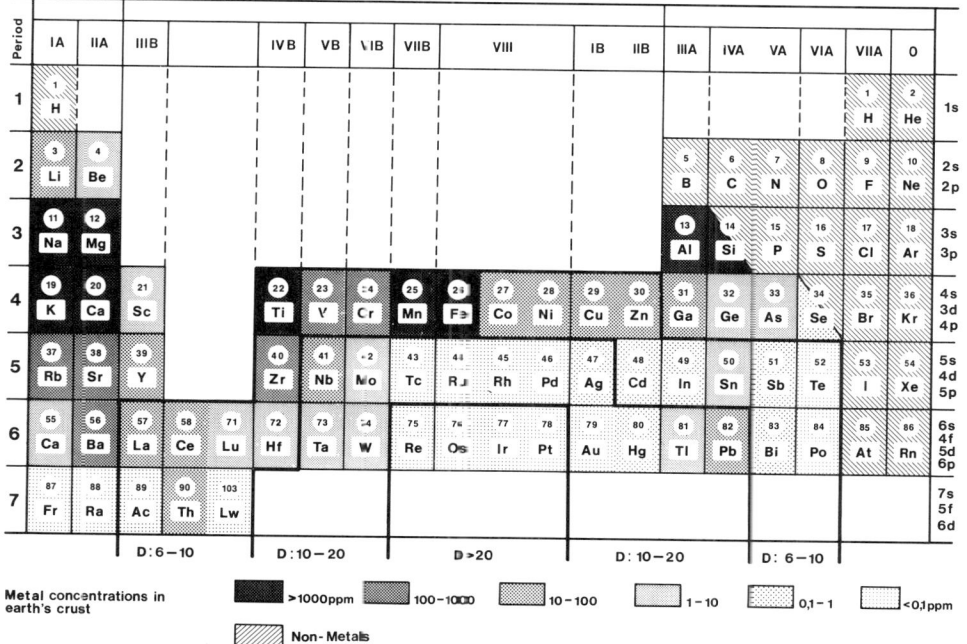

Fig. 1. Periodic table of elements

there are eight main groups designated IA to VIIA and group 0, containing the inert gases. From the fourth period onward a gap occurs between the groups IIA and IIIA as a result of electrons occupying d-orbitals, thus constituting a block of *transition* elements. Electron occupation of f-energy levels results in the lanthanide series (4f-level) and the actanides of trans-uranium series (5f-level).

A natural consequence of the long form of the PT is that elements with similar electron configurations—which are chiefly responsible for chemical behavior—are ordered according to increasing atomic number. In general, members within the same group resemble one another closely. Within a given period the properties of the elements vary gradually from a highly electropositive (metallic) character at the left-hand side of the series, to a highly electronegative (nonmetallic) character at the end of the series.

Some 65 of the known elements—excluding the trans-uranium series—are metallic in character. The term "metal" designates an element which is a good conductor of electricity and whose electric resistance is directly proportional to the absolute temperature. In addition to this distinctive characteristic, metals share several other typical physical properties such as: high thermal conductivity; high density; malleability and ductility—the ability to be drawn into sheets and wires. Several nonmetallic elements exhibit one or more of these properties, so that the only feature that defines a metal unambiguously is the electric conductivity which decreases with increasing temperature. The elements boron, silicon, germanium, arsenic and tellurium conversely have a low electric conductivity, which increases with a rise in temperature. In consequence, they are regarded as metalloids (or halfmetals) situated in the PT, between metals and nonmetals, the latter being nonconductors of electricity. Thus a diagonal passing through boron and tellurium in the PT subdivides the elements into metals, metalloids and nonmetals. This distinction is apparent from an inspection of the main groups IA to VIIA. Group VA, for example, reveals a marked increase in metallic character with increasing atomic number (N and Bi!). However, this classification should not be rigorously applied since there are numerous exceptions where elements placed near the borderline occur in more than one allotrophic modification. For example, carbon in the form of diamond is extremely hard and has no electric conductivity. Graphite, on the other hand, is very soft and is an extremely good conductor of electricity and can therefore be used as electrodes in many electrochemical processes. It is evident that the properties of elements do not vary haphazardly. Most elements have closely related properties, and these *similar properties recur in a periodic manner with increase in atomic number.* This rule is a statement of the *periodic law.* Some of the physical and chemical properties which reveal periodicity are:

conductivity (electrical and thermal)
density
atomic and ionic radii
electronegativity
oxidation numbers

It is of interest to note that of the ten most abundant elements in the earth's crust, seven are metals: aluminum at 7.5 weight percent, iron at 4.7, calcium 3.4, sodium 2.6, potassium 2.4, magnesium 1.9, and titanium at 0.6 weight percent (Giddings, 1973).

Classification of the Elements

Furthermore, the abundance of elements in the lithosphere generally decreases with increasing atomic mass. With the exception of titanium and manganese (0.1%) one finds the *trace elements* at the other end of the distribution pattern i.e. lacking in natural occurrence (< 0.1%). Abundance, availability and the significance of trace metals are discussed in Section B 2.

Element densities reveal the recurring periodicity involved to be a function of increasing atomic number. It is not surprising that the alkali metals (group IA) occupy the lowest positions for each successive period, since the addition of an outer electron to their inert gas configuration leads to a relatively large increase in atomic radius. Characteristically, metals generally have higher densities than nonmetals; the value in excess of 5.0 g/cm^3 has often been quoted to distinguish heavy metals.

Elements which readily release electrons are generally referred to as being *electropositive*. This is a generalized qualitative concept which applies to metals in contrast to nonmetals, which do not easily yield electrons. On the other hand, an element that generally tends to accept electrons in compound formation is termed electronegative. The term electronegativity (EN)[1] is a measure of the power of an atom to attract electrons to itself in a covalent bond (Pauling[2]). Although a precise definition is elusive and no unambiguous experimental method exists for establishing numerical values, the EN concept is widely used in chemical literature. When applied judiciously, electronegativity values serve a most useful qualitative (and semi-quantitative) purpose. Pauling used bond energies to calculate relative electronegativity values for the various elements, arbitrarily defining the EN of fluorine as the most electronegative element at 4,0. The assigned values have been listed in Table 1.

Table 1. Electronegativity values of the elements (Pauling)

H 2,1																	He —
Li 1,0	Be 1,5											B 2,0	C 2,5	N 3,0	O 3,5	F 4,0	Ne —
Na 0,9	Mg 1,2											Al 1,5	Si 1,8	P 2,1	S 2,5	Cl 3,0	Ar —
K 0,8	Ca 1,0	Sc 1,3	Ti 1,5	V 1,6	Cr 1,6	Mn 1,5	Fe 1,8	Co 1,8	Ni 1,8	Cu 1,8	Zn 1,6	Ga 1,6	Ge 1,8	As 2,0	Se 2,4	Br 2,8	Kr —
Rb 0,8	Sr 1,0	Y 1,2	Zr 1,4	Nb 1,6	Mo 1,8	Tc 1,9	Ru 2,2	Rh 2,2	Pd 2,2	Ag 1,9	Cd 1,7	In 1,7	Sn 1,8	Sb 1,9	Te 2,1	I 2,5	Xe —
Cs 0,7	Ba 0,9	57–71 1,1–1,2	Hf 1,3	Ta 1,5	W 1,7	Re 1,9	Os 2,2	Ir 2,2	Pt 2,2	Au 2,4	Hg 1,9	Tl 1,8	Pb 1,8	Bi 1,9	Po 2,0	At 2,2	Rn —

[1] Electronegativity must not be confused with electronaffinity, which refers to the energy released when gaseous atoms of the electronegative halogens undergo the following general reaction:

$$X_{(g)} + e^- \rightarrow X^-_{(g)} \quad \Delta H = -q \text{ kJ/mol } X_{(g)} \text{ atom.}$$

[2] Paulings Method to estimate electronegativities is based on bond energy data, and is described in detail in many inorganic chemistry texts (e.g., Cotton, F. A., Wilkinson, G.: Advanced inorganic chemistry, a comprehensive text. 2nd ed. New York: Interscience Publishers, 1966).

An inspection of this table reveals that the pattern displayed is in accordance with the periodic law: within a given PT group of the main group elements there is a general trend for the elements to display lower electronegativities with increasing atomic number, the lowest values occurring toward the bottom left-hand corner of the PT. Conversely, the EN increases gradually from left to right in each period of the main groups, pointing to the most EN elements (nonmetal character) in the top right-hand corner of the PT.

The *oxidation number* of an element is a direct consequence of the electronegativity concept. Whereas oxidation numbers coincide numerically with the ionic charges of ions, they are derived on the basis of assigning the shared electron pair to the more EN atom forming a covalent bond. The combining capacity of an element expressed as valence is thus related to its oxidation number or oxidation state; but valence and oxidation number are not necessarily synonymous concepts. For example, in vitamin B_{12}, cobalt is the central metal. The cobalt cannot be referred to as a Co^{3+} ion since it is covalently bonded. In modern terminology it is preferable to refer to cobalt as being in a +III oxidation state, i.e. Co(III).

1.2 Classification of Metals

Metals react as electron-pair acceptors (Lewis acids) toward electron-pair donors (Lewis bases) in possibly the most important type of chemical reaction, which may be generalized by the equation:

$$A + :B \rightarrow A:B$$

The resulting species may be termed an ion pair, a metal complex, a coordination compound, or a donor-acceptor complex.

On the grounds of experimental evidence, Pearson has arrived at a classification of acceptors and donors into "hard" and "soft" categories to explain differences in the stability of the species AB. As a general rule it was found that hard acceptors prefer to bind to hard donors and soft acceptors prefer to bind to soft donors to form stable compounds (Pearson, 1968 a, b). Since this principle serves to systematize a great deal of chemical knowledge and is usefully employed to explain various phenomena, an abbreviated list of metal acceptors and ligand donors is included in Table 2. Electron

Table 2. Classification of acceptors and donors (adapted from Pearson, 1968a) R = alkyl or aryl

Hard acceptor	Intermediate	Soft acceptor
H^+, Na^+, K^+, Be^{2+}, Mg^{2+}	Fe^{2+}, Co^{2+}, Ni^{2+}	Cu^+, Ag^+, Au^+, Tl^+
Ca^{2+}, Mn^{2+}, Al^{3+}, Cr^{3+}	Cu^{2+}, Zn^{2+}, Pb^{2+}	Hg_2^{2+}, Pd^{2+}, Cd^{2+}, Pt^{2+}
Co^{3+}, Fe^{3+}, As^{3+}		Hg^{2+}, CH_3Hg^+
Hard donor	Intermediate	Soft donor
H_2O, OH^-, F^-, Cl^-	Br^-, NO_2^-, SO_3^{2-}	SH^-, S^{2-}, RS^-
PO_4^{3-}, SO_4^{2-}, CO_3^{2-}, O^{2-}		CN^-, SCN^-, CO,
		R_2S, RSH, RS^-

mobility or polarizability and electronegativity are the chief criteria for classification as hard or soft. A hard acceptor is characterized by low polarizability, low electronegativity, large positive charge density (e.g., high oxidation state and small radius); the converse is true for a soft acceptor. Similarly, a hard donor is characterized by low polarizability, but high electronegativity and a high negative charge density, whereas the opposite holds true for a soft donor atom.

Since electron pair acceptors are broadly classified as acids (Lewis), and electron pair donors are referred to as bases (Lewis), the preferential bonding between hard species in contrast to soft species, has become known as the HSAB principle (hard and soft acids and bases).

One of the first observations of the HSAB principle at work occurs in nature: Some metals occur in the earth's crust as ores of oxide or carbonate, whereas other metals occur as sulfides. The explanation is that the hard acids, for example, Mg^{2+}, Ca^{2+}, Al^{3+}, form strong bonds with the hard bases, e.g. O^{2-} or CO_3^{2-}. Conversely, softer acids, Hg_2^{2+} or Hg^{2+}, or Pb^{2+} prefer soft bases such as S^{2-}. Hard acid–soft base, or soft acid–hard base combinations according to HSAB, do not form strong bonds; their ores will have been leached away by water prior to any ore recovery.

The number of d-electrons is of great importance for the transition metal ions. It has been pointed out that no really good acceptor ions i.e., soft acids exist which do not have at least a half-filled d-shell. This accounts for an anomaly in going across a transition series, for example, from Ca to Zn. The ionization potentials of these metals increase due to increasing nuclear charge. In consequence, one would expect an increase in electronegativity and a corresponding increase in hardness. Chemically, however, the elements become *softer* as one goes across from Ca to Zn. This must be ascribed to the increasing number of d-electrons, a factor which outweighs increasing electronegativity.

For cations of the 3d-series, a reasonably well-established rule on the sequence of complex stability, the Irving-Williams series, is valid. According to this sequence, the stability of complexes increases in the series $Mn^{2+} < Fe^{2+} < Co^{2+} < Ni^{2+} < Cu^{2+} > Zn^{2+}$. Since this sequence is valid for almost every ligand, it has become customary to explain the Irving-Williams series in terms of the crystal field theory. The ions Mn^{2+} and Zn^{2+}, with respective d^5 and d^{10} outer electron configurations, have no crystal field stabilization energy, whereas this steadily increases in the series from d^6 to d^8.

1.3 Trace Metal Species in Aquatic Systems

The term speciation refers to the particular physical and chemical forms in which an element occurs. Until recently most environmental research on trace metals was based on an assessment of the total metal concentration. It has become increasingly evident that the environmental impact of a particular metal species may be more important than the total metal concentration (Sibley and Morgan, 1975).

A general scheme of metal speciation—mainly based on the particle size fractions—has been given by Stumm and Bilinski (1972), reproduced here in Table 3. In practice, the first step applied in separation of particulate from soluble metals involves filtration through a 0.45 μm pore size membrane filter. Further differentiation of the solid phases is described in Chap. E.4. The group of soluble *metal ions* can be subdivided as follows (Guy and Chakrabarti, 1975):

Table 3. Types of metal species in waters (after Stumm and Bilinski, 1972)

Metal species	Range of diameters (μm)	Examples
Free aquated ions		$Fe(H_2O)_6^{3+}$; $Cu(H_2O)_6^{2+}$
Complex ionic entities		AsO_4^{3-}, UO_2^{2+}, VO_3^-
Inorganic ion-pairs and complexes		$CuOH^+$, $CuCO_3^0$, $Pb(CO_3)_2^{2-}$
		$AgSH^0$, $CdCl^+$, $Zn(OH)^-$
Organic complexes, chelates and compounds	0.001	$Me - OOCR^{n+}$, HgR_2
		$\begin{array}{c} CH_2 - C = O \\ / \quad \backslash \\ H_2N \quad\quad O \\ \backslash \quad / \\ Cu \\ / \quad \backslash \\ O \quad\quad NH_2 \\ \backslash \quad / \\ O = C - CH_2 \end{array}$
Metals bound to high molecular weight organic materials	0.01	Me-humic/fulvic acid polymers
Highly-dispersed colloids	↓	FeOOH, Mn(IV) hydrous oxides
Metals sorbed on colloids	0.1	$Me.aq^{n+}$, $Me_n(OH)_y$, $MeCO_3$, etc. on clays, FeOOH, organics
Precipitates, mineral particles, organic particles		$ZnSiO_3$, $CuCO_3$, CdS in FeS, PbS
Metals present in live and dead biota		Metals in algae

(Arrows in middle column indicate: In true solution; Dialysable; Membrane filterable; Filterable)

(Me = metal; R = alkyl)

1. simple aquated metal ions,
2. metal ions complexed by inorganic anions, such as $CuCO_3$,
3. metal ions complexed by organic ligands such as amino, fulvic and humic acids.

The basic approaches to identify the chemical species of trace metals in soluble form are discussed in Chapter C.

2 Trace Metals and Organic Life

It has been known for several decades that trace quantities of certain elements exert a positive or negative influence on plant, animal, and human life. However, more recently, greater interest has been taken with regard to the specific role of these elements. Generally, the term trace element is rather loosely used in current literature to designate the elements which occur in small concentrations in natural biologic systems. The growing public concern over the deteriorating quality of the environment has led to a generalized usage when referring to trace elements. Thus, for practical purposes, other terms such as "trace metals", "trace inorganics", "heavy metals", "microelements" and "micronutrients" will be treated as synonymous with the term trace elements.

2.1 Trace Elements Essential to Human Life

The fact that metal ions have a biologic significance is contradictory to the classical concept that inorganic chemistry is restricted to nonliving chemical systems, whereas the living world falls within the realm of organic and biochemistry. Modern research has led to a broader understanding of the inextricability of overlapping concepts in the field of applied chemistry, such as occur in nature, and stresses the need to diverge from artificial compartmentation. It has been borne out by experimental evidence that the role of heavy metal ions in living systems follows the pattern of natural availability and abundance of the same metals occurring in nature. (Vahrenkamp, 1973; Williams, 1967; Wood, 1974, 1975).

No organic life can develop and survive without the participation of metal ions. Current research has revealed that life is as much inorganic as organic. Due to the inorganic chemists's lack of consideration for living systems, a totally false impression of the chemistry of living matter has been created.

The basis of bioinorganic chemistry was laid by Richard Willstätter, who received the Nobel Prize for chemistry in 1915 for his contribution toward unravelling the chemistry of chlorophyll, a magnesium complex present in all plants, which is capable of synthesizing carbohydrates from carbon dioxide and water. The significance of the role of magnesium in the metabolism of carbohydrates was not understood at that time, since complex chemistry (often referred to as coordination chemistry), had only been established shortly before by Alfred Werner, who received the Nobel Prize for chemistry in 1914. However, when it became known that iron(II) in hemoglobin is an essential constituent of red blood cells for the respiratory process, it could hardly be denied that metal ions could be excluded from an active participation in the living cell.

An element is essential when (1) it is consistently determined to be present in all healthy living tissues within a zoological family, whereby tissue concentrations from species to species should not vary by a wide range, (2) deficiency symptoms are noted with depletion or removal, which disappear when the elements are returned to the tissue, (3) the deficiency symptoms should be attributed to a distinct biochemical defect (on the molecular level). (Overhcff and Forth, 1978).

It is well known that the major ions such as sodium and potassium, also magnesium and calcium, are essential to sustain biologic life. However, it is less known that at least some further six metals, chiefly transition metals, are essential for optimal human growth, development, achievement and reproduction. The essential elements which the human body requires have been listed in Table 4 in accordance with the PT, including approximate concentration levels based on an average 70 kg body weight (Vahrenkamp, 1973).

Inspection of Table 4 reveals marked differences in the concentrations of the listed ten elements, currently known to be essential to human life. In accordance with these levels, a distinction must be drawn with regard to the alkali and alkaline earth metals on the one hand, and the heavier metals exceeding the atomic mass of calcium—often referred to as heavy metals—on the other.

It is of interest to classify the essential metals according to their biologic role.

Sodium and Potassium. Both metals are highly concentrated in bodily fluids and are widely distributed throughout the human body. Accordingly, they are not considered

Table 4. Essential metals and respective concentrations for human beings (expressed in mg/70 kg body weight; Vahrenkamp, 1973)

Group / Period	IA	IIA	VIB	VIIB	VIII			IB	IIB
3	11 Na 70,000	12 Mg 40,000							
4	19 K 250,000	20 Ca 1,700,000		25 Mn 30	26 Fe 7,000	27 Co 1		29 Cu 150	30 Zn 3,000
5		42 Mo 5							

as trace elements, but rather as "macroelements". Being the most mobile cations, it is not surprising that apart from an involvement in metabolic processes these ions participate in nerve impulse conduction via the brain. The exact functional capacity in this respect has not been elucidated as yet (Williams, 1971).

Magnesium and Calcium. These are the next most mobile metal ions which are widely distributed throughout the human body and thus cannot be considered as being trace metals. Their unequal distribution pattern is similar to their alkali metal counterparts. Magnesium ions chiefly participate in functions within the cell; they are found complexed to nucleic acids and are necessary for nerve impulse transmissions, for muscle contractions and metabolic functions. Calcium ions having a greater affinity for oxygen-containing ligands are less mobile than magnesium ions. This results in the presence of crystalline calcium salts, e.g., phosphate and oxalate, in the circulatory blood system. The insolubility of many calcium salts is reflected by the formation of bones and teeth in which calcium is deposited as hydroxyl-apatite, $Ca_5(PO_4)_3(OH)$ (Vahrenkamp, 1973). Thus calcium is by far the most abundant metal in the human body.

Trace Metals. The essential transition metals including zinc are capable of forming stable coordinate bonds to fixed positions of immobile protein molecules, where they function mainly as catalysts i.e., they induce or enhance enzymatic activity. In the highly specific metalloenzymes the metal is firmly associated with the protein and often constitutes the active centre of the living cell, catalyzing only one specific reaction or type of reaction. This explains why in some cases trace concentrations have such a powerful, directive influence on the biologic functions within the human body.

Two separate functions performed by essential metals may be distinguished:
1. Redox reactions. Since only transition metals with partially filled d-orbitals have stable oxidation states differing by only one unit, they are involved in electron transfer processes. This applies specifically to the three systems $Fe(II)/Fe(III)$, $Cu(I)/Cu(II)$ and $Mo(V)/Mo(VI)$.

Since these metals become firmly incorporated into proteins, their redox properties are controlled by the coordinated ligands.

2. Direct participation of the metal in governing the reaction mechanism. This applies especially to cobalt and zinc. A brief summary is given with regard to the individual role of each essential trace metal listed in Table 4.

Manganese. The second most abundant metal in nature (exceeded only by iron). The chemistry of Mn^{2+} resembles that of Mg^{2-}, both ionic species preferring weaker donors such as phosphate and carboxylate groupings to form stable bonds. Due to the ready interchangeability of manganese and magnesium in biologic systems, the specific biochemical role of manganese has proved extremely elusive for many years. It was known that many enzymes are activated by manganese in vitro, a property notably shared with magnesium. Unequivocal evidence of manganese specificity was established in 1962, the importance of which was enhanced by the discovery that this element is involved in glucose utilization (Underwood, 1971).

Iron. The most abundant transition element, is also probably the most well-known metal in biologic systems [hemoglobin in blood—the oxygen-carrying protein molecule of blood regarded as the most important iron(II) complex consisting of the globin protein with four heme units attached to it].

Cobalt. Relatively scarce in the earth's crust, but the human body requires vitamin B_{12}, which is a cobalt(III) complex, to form hemoglobin. In fact, cobalt is widely distributed throughout the human body, without, however, excessive concentration in any particular organ or tissue. Having the ability to occupy low symmetry sites in enzymes, cobalt(II) is an enzyme activator.

Copper. Long before copper was recognized as an essential element, it was shown to exist in combination with the blood protein of snails. Today it is known that Cu(I) is found in enzymes capable of carrying oxygen as hemoglobin does, and that it is actually required in the formation of this substance.

Zinc. One of the most abundant of the essential elements required by the human body and approximately 100 times as abundant as copper (Vahrenkamp, 1973). Zinc appears to be present in all mammals. As with cobalt(II), zinc has the ability to occupy low symmetry sites in enzymes, and can therefore function as an essential constituent of several of them.

Molybdenum. This metal is involved in electron transfer processes as in xanthine and purine oxidations of milk. Nitrogen fixation is also coupled to a molybdenum process.

2.2 Deficiency and Oversupply

Essential trace metals such as zinc become toxic when the nutritional supply becomes excessive. A metal in trace amounts (smaller than 0.01% of the mass of the organism) is essential when an organism fails to grow or complete its life cycle in the absence of that metal. However, the same trace metal is toxic when concentration levels exceed those required for correct nutritional response by factors varying between 40- and 200-fold. (Venugopal and Luckey, 1975).

Studies pertaining to the toxicity of trace metals follow the general trend that an undersupply leads to a deficiency, sufficient supply results in optimum conditions, but an oversupply results in toxic effects and lethality in the end.

These facts are graphically displayed in Figure 2a, which illustrates the essentiality of trace metals as a dose-response curve ranging from deficiency to oversupply. The plateau region reflects concentrations for optimum growth, health, and reproduction, and indicates that a deficiency as well as an oversupply have detrimental effects. A wide plateau corresponds to a low inherent toxicity of the metal, whereas a narrow plateau reveals a small range between required and harmful doses. Furthermore, it is of interest to note that all metals essential to life are toxic when supplied in concentrations in excess of the optimum concentration levels.

Metals which may not be as yet identifiable as serving a beneficial biologic function are referred to as nonessential (Fig. 2b). However, the concept of essentiality is under constant review as research makes progress; an unequivocal classification is therefore impossible. Nonetheless, there is no doubt that all metals are potentially hazardous to living organisms, and not necessarily at large exposure levels. This topic shall be dealt with in greater detail in a subsequent section of this chapter.

2.3 Metal Toxicity

The various factors influencing the toxicity of trace metals in aquatic organisms compiled by Bryan (1976) are presented in Table 5. The metal species formed in water is handled in Chapter C (dissolved components) and Chapter E (particulate-associated metals); the physiologic factors (such as temperature, pH and salinity) are described in Chapter F. The main emphasis is here given to the presentation of some basic relations between potentially toxic metals and water organisms and their consumers.

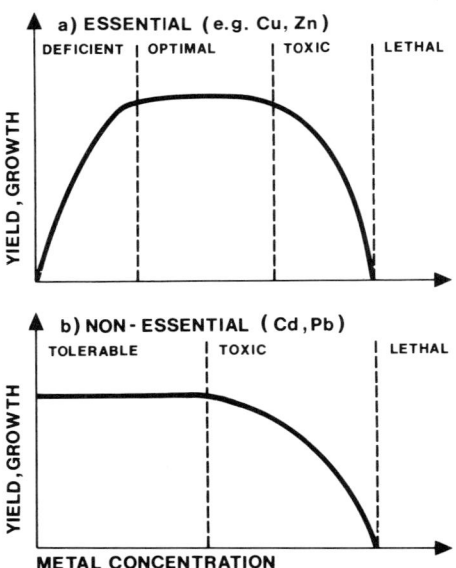

Fig. 2 a and b. Deficiency and oversupply of essential and nonessential trace elements (modified after Baccini and Roberts, 1976)

Table 5. Factors influencing the toxicity of heavy metals in solution (Bryan, 1976)

Form of metal in water	inorganic / organic	soluble	ion / complex ion / chelate ion / molecule
		particulate	colloidal / precipitated / adsorbed
Presence of other metals or poisons	joint action	more-than-additive / additive / less-than-additive	
	no interaction		
	antagonism		
Factors influencing physiology of organisms and possibly form of metal in water	temperature / pH / dissolved oxygen / light / salinity		
Condition of organism	stage in life history (egg, larva, etc.) / changes in life cycle (e.g., moulting, reproduction) / age and size / sex / starvation / activity / additional protection (e.g., shell) / adaptation to metals		
Behavioral response	altered behavior		

The metal ions used by biologic systems must be both abundant in nature and readily available as soluble species. Abundance generally restricts the available metals to those of atomic number below 40, some of which are virtually unavailable due to the low solubility of their hydroxides e.g., aluminum and titanium. Viewed from the standpoint of environmental pollution, metals may be classified according to three criteria: (1) noncritical, (2) toxic but very insoluble or very rare, and (3) very toxic and relatively accessible. Such a classification has been made by Wood (1974) and is listed in Table 6.

Table 6. Classification of elements according to toxicity and availability

Noncritical			Toxic but very insoluble or very rare		Very toxic and relatively accessible		
Na	C	F	Ti	Ga	Be	As	Au
K	P	Li	Hf	La	Co	Se	Hg
Mg	Fe	Rb	Zr	Os	Ni	Te	Tl
Ca	S	Sr	W	Rh	Cu	Pd	Pb
H	Cl	Al	Nb	Ir	Zn	Ag	Sb
O	Br	Si	Ta	Ru	Sn	Cd	Bi
N			Re	Ba		Pt	

Several elements not listed fit more than one category, but should not be neglected in an environmental sense, as for example, manganese. The list of lanthanides and actinides has also been omitted, primarily due to their exceptionally rare natural occurrence.

From a comparison of the classification of acceptors and donors (Table 2) with the classification of elements according to toxicity and availability (Table 6) it is possible to establish a relationship which reveals that (1) noncritical elements such as Na, K, Mg, and Ca are *hard* acceptors (Lewis acids), which in accordance with the HSAB theory only form stable bonds to *hard* donors (Lewis bases) such as H_2O, OH^-, Cl^-, etc., (2) most of the very toxic and relatively accessible metals also appear to qualify as *soft* acceptors. These metals form particularly stable bonds to *soft* donors, such as $-SH$ groups, being amongst others active sites of many proteins. At this stage, explaining toxicity in terms of stable soft-soft bonds, requires a measure of caution and further research.

The principle of preferential bonding as explained by the HSAB principle of Pearson goes a long way toward predicting and explaining encountered phenomena which lead to health hazards. Although the microbial methylation of methyl mercury of vitamin B_{12} poses a known health hazard, one is tempted to forecast similar methylation of tin, palladium, platinum, gold, and thallium, which may be found to undergo similar methylation in the environment by virtue of periodic similarities. It is known on the other hand, that cadmium alkyls are unstable in aqueous medium; it is therefore fortuitous to make generalizations. Although arsenic compounds are known to be methylated under anaerobic conditions to form toxic organoarsenial compounds, and have been found on analysis to accumulate in shellfish from Norway (Wood, 1974), predictions that similar metabolic reactions occur for selenium, tellurium, and sulfur do not appear to be of practical significance at the present state of research.

2.4 Health Hazard Due to Certain Trace Elements

Arsenic. although widespread in plant and animal tissues, has become synonymous with "poison" in the public mind. In spite of its toxicity, it has been employed for its medical virtues in the form of organic arsenicals, and in partial prevention of selenosis.

It appears that the stable, soluble inorganic arsenites and arsenates are readily absorbed by the digestion tract, abdominal cavity, and muscle tissue. Excretion of arsenate is faster than of arsenite, mostly in urine. Arsenate has a low order of toxicity and does not inhibit any enzyme system due to its lack of affinity to hydroxo and thiol groups; but ATP synthesis is inhibited by a AsO_4^{3-} by uncoupling oxidative phosphorylation and replacing the stable phosphoryl group. In contrast, arsenite inhibits thiol-dependent enzymes, binds to tissue protein as keratin disulfides in hair, nails, and skin and is retained in the body for a prolonged period.

Long-term ingestion of arsenic-contaminated drinking water from wells has produced gastrointestinal, skin, liver, and nerve tissue injuries. In Taiwan, widespread occurrence of Black-foot Disease resulted from high levels of arsenic in ground waters–the maximum arsenic value amounted to 2.5 mg/l (Haberer and Normann, 1971). Chronic arsenic poisoning appears to be regional, but certainly not limited to Taiwan. For example, in the Mexican village, Toreon, 60% of the inhabitants showed varying degrees of chronic intoxication from drinking water which contained 4–6 ppm As (Zaldivar, 1974).

The importance of arsenic as a health hazard is well known; examples of chronic poisoning have been reported for water with arsenic concentrations of 210 µg/l to 1000 µg/l (Aston et al., 1975).

Beryllium is permanently retained in mammalian tissues. The phosphates and sulfates are stable and insoluble. Be links more easily to O than to N or S atoms in ligands for complex formations. The sites of deposition depend mainly on complexing; the non-diffusable form is mainly encountered as insoluble phosphates. Be is toxic due to its high retention, leading to pneumonites, cardiac strain, and heart disease in humans.

Cadmium being the middle member of the periodic sub-group consisting of Zn, Cd, and Hg reveals intermediate properties. All three elements display a profound capacity of combining with $-SH$ (and imidazole-containing ligands); the stability of such complexes increases in the order of $Zn < Cd < Hg$. Cd and Hg compete with and displace Zn in a number of Zn-containing metalloenzymes by irreversibly binding to active sites, thereby destroying normal metabolism. This, of course, leads to the above stability sequence which is best explained on the basis that sulfhydryl is a soft donor, cadmium and mercury are soft acceptors, whereas zinc is a borderline case (see Table 2). On the premise that soft acceptors prefer to bind to soft donors, the displacement of zinc is readily explained.

Chromium is one of the least toxic of the trace elements on the basis of its over-supply and essentiality. Generally, the mammalian body can tolerate 100–200 times its total body content of Cr without harmful effects. Chromium (VI) compounds are approximately 100 times more toxic than Cr(III) salts. The stomach acidity leads to reduction of Cr(VI) to Cr(III) of which gastrointestinal absorption is less than 1%.

Copper. The presence of copper in plant and animal tissues was recognized more than 150 years ago. Long before it was recognized as an essential element in the diets of birds and mammals, it was detected as a component of blood proteins of snails. The first indication that copper deficiency occurs naturally in livestock dates back to 1931 with regard to "salt-sick" cattle in Florida. Later it was established that copper deficiency is a casual factor in a disease of sheep and cattle occurring in parts of Holland where it was termed *lechsucht*. The basic biochemical defects were subsequently related to enzymatic activity of copper.

Chronic copper poisoning of sheep in Australia revealed that Cu retention was dependent on the molybdenum status of the diet, which in turn depended upon the inorganic sulfate status of the diet and the animal (Underwood, 1971). This impressive discovery was followed by further revelations that there are interactions among trace elements such as a reciprocal copper-zinc antagonism. In other words, copper toxicity or deficiency in animals is not merely dependent upon copper intake, but also upon dietary levels of Zn, Fe, and Ca.

Copper is an essential metal in a number of enzymes. Excessive intake of copper results in its accumulating in the liver. Generally, copper toxicity is increased by low Mo, Zn, and SO_4^{2-} intake.

Lead resembles the divalent alkaline earth group metals in chemical behavior more than its own Group IVA metals. It differs from the Group IIA metals in the poor solubility of lead salts such as hydroxides, sulfates, halides, and phosphates. Metabolism of Pb and Ca are similar both in their deposition in and mobilization from bone. Since Pb can remain immobilized for years, metabolic disturbance can remain undetected. Under normal conditions more than 90% of the lead retained in the body is in the skeleton. Although lead is a nonessential element it is present in all tissues and organs of mammals.

The large affinity of Pb^{2+} for thiol and phosphate-containing ligands inhibits the biosynthesis of heme and thereby affects membrane permeability of kidney, liver and brain cells. This results in either reduced functioning or complete breakdown of these tissues, since lead is a cumulative poison.

Mercury is considered a nonessential but highly toxic element for living organisms. Even at low concentrations, mercury and its compounds present potential hazards due to enrichment in the food chain. Poisoning by methylmercury compounds presents a bizarre neurological picture as observed in large-scale outbreaks in Japan and Iraq. Damage is chiefly in the cerebellum and sensory pathways with lesions in the cerebral cortex of man.

The higher toxicity of mercury than cadmium cannot be attributed to the smaller ionic radius and greater penetration of the Hg^{2+} ion (Venugopal and Luckey, 1975). The profound capacity of the soft acid (acceptor) CH_3Hg^+ to bind soft ligands such as —SH groups of proteins is a more plausible explanation for the high toxicity of methylmercury compounds.

Molybdenum is an essential nutrient for all nitrogen-fixing organisms and occurs regularly in all plant and animal tissues. High dietary concentrations of molybdenum are harmful to several animal species; all cattle are susceptible to molybdenosis, a scouring disease known as tearts which arises from ingestion of excessive amounts of this element from the herbage of affected areas. Acute or chronic toxic effects have not been reported for humans.

Selenium in trace amounts is essential for growth and fertility in animals and for the prevention of various diseases. Seleniferous soils and vegetation may lead to an oversupply and result in poisoning. All degrees of selenosis exist ranging from a mild, chronic condition to an acute form resulting in the death of the animal. Chronic selenosis is characterized by dullness and a lack of vitality; stiffness and lameness, due to erosion of the joints of the long bones. In acute selenium poisoning the animals suffer from blindness, abdominal pain, and death may result from respiration failure.

The high general toxicity of selenium toward animals and man is possibly due to the replacement of thiol groups by SeH groups.

Silver generally occurs in the form of salts having a low solubility and is consequently excreted in feces. Most of the soluble Ag^+ (aq) ingested remains impregnated in tissues forming a stable bond to —SH or —SR groupings. When larger quantities are assimilated by skin tissue, silver sulfide leads to permanent discoloration of both skin and eye tissue, a condition known as argyria. Silver salts are exceptionally toxic toward freshwater fish.

Thallium is discharged in effluents from processing of base metal sulfide ores, such as galena and chalcophyrite (Robinson, 1973). Most of the thallous compounds are relatively soluble and high removals were not obtained either with conventional or physicochemical waste water treatment (Hannah et al., 1977).

The limited use of thallium will not lead to a global contamination of the aquatic environment, but localized problems may exist or develop in the future, mainly as a result of mineral processing. Investigations performed by Zitko et al. (1975) indicate that the acute toxicity to some species of fish, such as Atlantic salmon, is approximately equal to that of copper; the toxicity of copper-thallium and zinc-thallium mixtures is not additive.

Zinc is one of the most abundant essential trace elements in the human body. It is a constituent of all cells, and several enzymes depend upon it as a cofactor. Concern has arisen because of the intimate connection of zinc with cadmium in the geosphere and biosphere. In soils, the ratio of cadmium varies from approximately 1:300 to 1:2900 by mol; in seawater, the molar ratio is 1:35; in the adult human kidney, the molar abundance of cadmium can reach or exceed three-fifths that of zinc, varying widely from individual to individual and from one geographic area to another (Schroeder et al., 1967). Moderately increased zinc concentrations in water stemming from the release of zinc from drainage pipes due to corrosion do not induce any clinical manifestations. The results of laboratory tests on animals indicate, however, that the metabolism of humans may be affected (for example their mineral and enzyme budget), especially of children and patients already suffering from irregular metabolism.

2.5 Accumulation of Toxic Substances in the Aquatic Food Chain

The Minamata Bay disaster during the early 1950s (discussed later in Sect. B.2.6 in more detail) concerns the mysterious neurologic illness and human fatalities among fisherfolk who subsisted mainly on fish. Since this disease also prevailed among local sea birds and household cats, investigations led to the discovery that the consumption of high concentrations of mercury compounds accumulated in fish and shellfish had evoked disastrous end effects in the nutritional food chain (Goldwater, 1971). After unravelling the cause of the mysterious "Minamata Illness", society suddenly became aware of the existence of toxic metals in the environment. Moreover, this awareness has led to further research—which is still in progress at various centers—with regard to the accumulation of toxic substances in the food chain.

A typical biologic food chain for mercury has been depicted in Figure 3 (from Hartung, 1972). Decay of organic material in the aquatic environment—possibly enriched by the disposal of sewage and industrial effluents—together with detritus formed by natural weathering processes, provides a rich source of nutrients in both the bottom sediments and the overlying water body. Microorganisms and microflora are capable of incorporating and accumulating metal species into their living cells from these supply sources. Subsequently, small fish become enriched with the accumulated substances. Predatory fish again, generally display higher levels than their prey. Eventually man, consuming the fish, inevitably suffers from the results of an enrichment having taken place at each trophic level i.e., where less is excreted than ingested.

The high toxicity of mercury(II) compounds has long been known. The conversion of inorganic mercury to the more toxic monomethyl and dimethyl mercury was first detected in aquarium sediments. Subsequently, it was discovered that microorganisms are capable of this transformation (Jensen and Jernelöv, 1969). Thus a pathway was

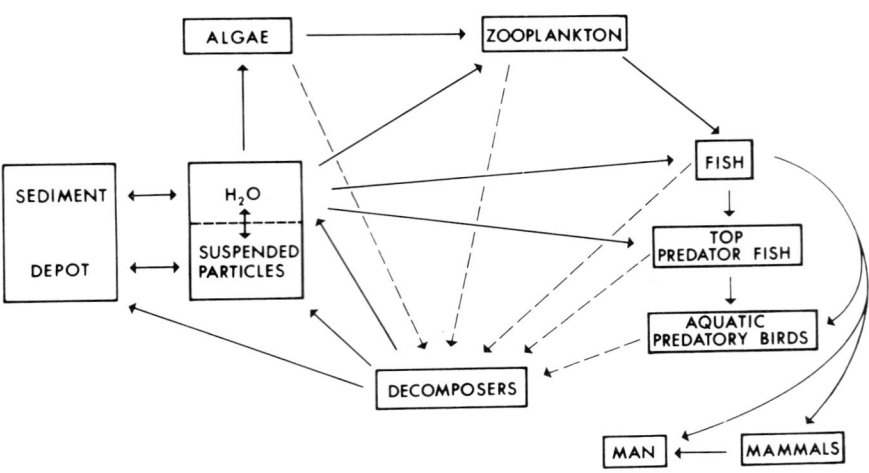

Fig. 3. Food chain model for mercury (Hartung, 1972; reproduced with permission of Ann Arbor Science Publishers)

uncovered by which mercury could enter the biologic food chain. Monomethylmercury, being the most toxic mercury compound known until recently, is not tightly bound to sediments, and is somewhat water-soluble and volatile. It is rapidly assimilated by living organisms and then retained relatively stable (Vallee and Ulmer, 1972).

With regard to the element mercury, it is generally accepted that large predatory species such as swordfish and tuna usually have higher levels of mercury in their tissue than lower species in the food chain (Ratkowsky et al., 1975). A recent study conducted by these authors revealed that the position of the fish in the food chain appears to be an important factor in determining its mercury content. Approximately 51% of individual fish, of species whose diet predominantly consists of other fish, had mercury concentrations in excess of 0.5 mg/kg. In contrast, only 24% of invertebrate predators and 7% of individuals of herbivorous habits had mercury concentrations in excess of 0.5 mg/kg.

In a study performed in Japan it was shown that local inhabitants had relatively higher concentrations of Cd and Hg in their blood, and much higher levels of both these toxic elements in body tissue than American and European counterparts (Sumino et al., 1975). Such differences may be due to diets, geographic conditions or to man-made pollution.

In Chapter F metal enrichments in the aquatic food chain will be further investigated. It will become clear that the model for mercury presented here cannot be generalized for most other metals.

2.6 Catastrophic Episodes of Metal Poisonings

2.6.1 Mercury Poisoning

The outbreak of a hitherto unknown and mysterious, noninfectious neurologic illness amongst inhabitants—especially fishermen and their families who mainly subsist on sea-foods—living around Minamata Bay in southwestern Kyushu, Japan, was recognized late in 1953 (Fig. 4). The patients who had consumed fish and shellfish from that region progressively suffered from a weakening of muscles, loss of vision, impairment of cerebral functions, and eventual paralysis which in numerous cases resulted in coma and death (Fujiki, 1972). This disease became known as the Minamata Disease and after extensive investigations it was revealed in 1959 that the deaths were caused by the consumption of fish and other foodstuffs contaminated with methylmercury.

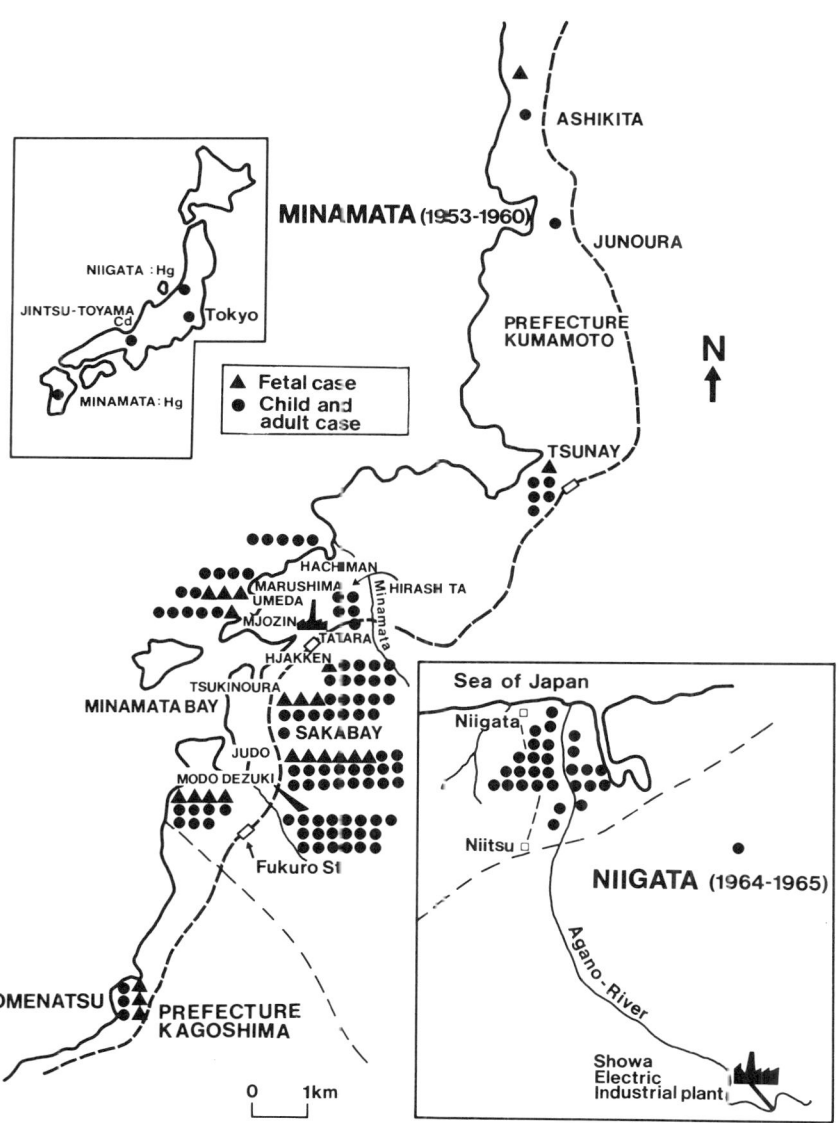

Fig. 4. Mercury poisoning in the Minamata region (Takeuchi, 1972; reproduced with permission of Ann Arbor Science Publishers)

The source of methylmercury contamination was traced to effluents from various plants of the chemical firm Chisso Co., manufacturers of plastics (PVC). For several years before 1953, these plants had discharged methyl-mercury formed from acetaldehyde and inorganic mercury (used as catalyst) into the drainage channel which leads into the Bay. The Minamata Bay incident was the first known modern case of mercury pollution in the aquatic environment.

A second outbreak of Minamata Disease occurred in Japan during 1964–1965 among inhabitants—again mainly fishermen and their families—living near Niigata, who regularly eat seafish and shellfish from this area and also river fish from the inflowing Agano River carrying effluents from the Showa electrical industrial plant (Fig. 4).

A third case of mass-poisoning by mercurial effluents in Japan was reported during May 1973 by a research team of the Kumaoto University. They established that 59 inhabitants from the city of Goshonoura on the island Amakusu, opposite of Minamata, had become afflicted with Minamata Disease.

Several other alarming incidents of mercurial poisoning have been reported from other continents after it became appreciated that mercury can produce hazards in aquatic systems. Investigations of the aquatic environment in Sweden revealed abnormally high concentrations of mercury compounds in fresh- and salt-water fish and other aquatic organisms, which led to a ban in 1967 by the Swedish Medical Board on the sale of fish from approximately 40 lakes and rivers (Goldwater, 1971). In one particular case the source of mercury pollution was traced to discharges of mercurial fungicides by a paper mill. However, even after the use of the fungicide had been discontinued, high mercury concentrations prevailed in fish from the vicinity of the plant. By 1964 Swedish ornithologists had noted that a significant decrease in wild bird populations had occurred (e.g., some birds of prey were almost extinct). The most severe incident of man-made mercury poisoning occurred during 1972 in Iraq. The first outbreaks were reported from northern Iraq where farmers had received wheat seed treated with mercurial fungicides (ethylmercury p-toluene sulfonanilide) from Mexico, and ate the seed instead of planting it. When local government authorities began to note this self-inflicted plague, it was announced that any farmer possessing treated seeds was liable to prosecution involving the death sentence. Subsequently, the peasants disposed of these hazardous seeds into nearby rivers and lakes. The combined effects of this blunder have caused the loss of an estimated 5,000 to 50,000 lives and the permanent disability of more than 100,000 people (possibly more than 500,000) (Bakir et al., 1973).

Poisoning by ingestion of alkyl-mercury-treated maize was reported by Derban (1974) from investigations conducted in Ghana. Due to a lack of safety precautions, inhabitants of the village Yalovi in the Volta region simply washed the maize thoroughly with warm water to remove the mercury-based pesticide, as they had done previously to DDT-treated seed. Instead of planting the maize the inhabitants of the named village used the seed to supplement their dietary requirements. Out of 250 persons living in the village, 144 reported sick after having eaten the maize.

Environmental mercury poses a human health hazard to native communities in Canada. In some instances their traditional diet, e.g. of fish, is contaminated, and mild symptoms of methyl mercury intoxication have been identified (C. T. Charlebois, *Ambio* 7, p. 204–210, 1978).

2.6.2 Cadmium Poisoning

Mercury is regarded as the most toxic metal, followed by cadmium, lead, and others although there is no rigid order of toxicity (Bryan, 1971). Contamination of the aqueous environment by cadmium appears to be less widespread than by mercury but has nonetheless hazardous effects on humans. During 1947 an unusual and painful disease of a "rheumatic nature" was recorded in the case of 44 patients from villages (e.g., Fuchu) on the banks of the Jintsu River, Toyama Prefecture, Japan (Friberg et al., 1974). During subsequent years, it became known as the "itai-itai" disease (meaning "ouch-ouch") in accordance with the patients' shrieks resulting from painful skeletal deformities (Kobayashi, 1971). It is impossible to state a precise figure on the incidence of itai-itai disease over the years; it is estimated that approx. 100 deaths occurred due to the disease until the end of 1965.

However, the cause of this disease was completely unknown until 1961 when sufficient evidence led to the postulation that cadmium played a role in its development (Hagino and Yoshioka, 1961). Based on the findings of further government-supported studies, the Japanese Ministry of Health and Welfare declared in 1968: "The itai-itai disease is caused by chronic cadmium poisoning, on condition of the existence of such inducing factors as pregnancy, lactation, imbalance in internal secretion, aging, defiency of calcium, etc."

The incubation period for chronic cadmium intoxication varies considerably, usually between 5 and 10 years but in some cases up to 30 years! During the first phase of poisoning a yellow discoloration of the teeth ("cadmium ring') is formed, the sense of smell is lost and the mouth becomes dry. Subsequently, the number of red blood cells is diminished which results in impairment of bone marrow. The most characteristic features of the disease are lumbar pains and leg myalgia, these conditions continuing for several years until the patient becomes bed-ridden and the clinical conditions progress rapidly. Urinary excretion of albuminous substances result from severe kidney damage. Cadmium-induced disturbances in calcium metabolism accompanied by softening of bones, fractures, and skeletal deformation takes place with a marked decrease in body height—up to 30 cm! (Friberg et al., 1974).

The source of cadmium pollution of the Jintsu River was a zinc mine owned by the Makioko Co. and situated some 50 km upstream from the afflicted villages (Fig. 5). Cadmium is always associated with the natural occurrence of zinc. During World War II production of zinc and lead from the mine was increased without sufficient accompanying treatment of the plant effluents and flotation sludge which were discharged directly into the upper stream of the Jintsu River. The sludge became deposited downstream and caused great damage to the rice crop, the rice fields having been irrigated or flooded by water from the Jintsu River. It is of interest to note that no itai-itai cases were found outside the region where the crop had been damaged; all patients were found to reside within 3 km of the river bank and in the low-lying rice field areas which had been flooded by the polluted river water. After the mine constructed a retaining dam in 1955, pollution of the Jintsu River ceased and the number of itai-itai cases rapidly declined. The first cadmium analyses of rice were performed in 1960 and revealed that in 20 samples of rice from the endemic area the average cadmium concentration (wet weight) was more than ten times higher (0.68 $\mu g/g$) than in approxi-

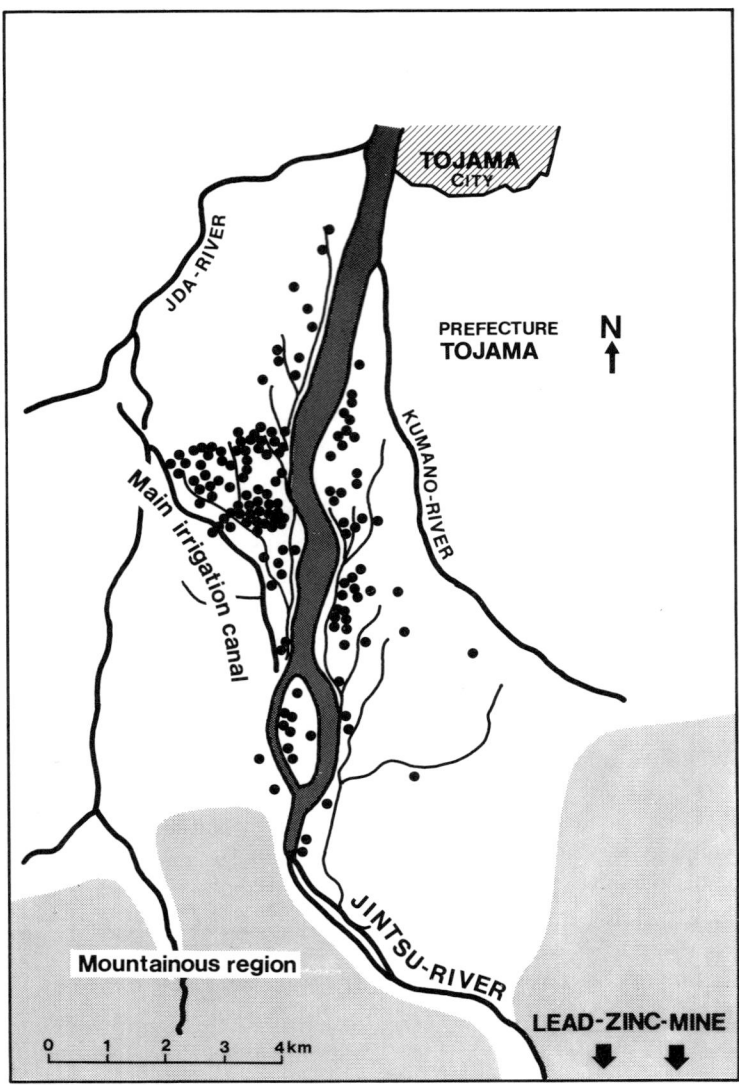

Fig. 5. Cadmium poisoning in the lower Jintsu River area (Kobayashi, 1971)

mately 200 samples from other areas of Japan (0.066 $\mu g/g$) (Friberg et al., 1974; Kobayashi, 1971).

After the first alarming discovery and correct diagnosis of the itai-itai disease, global contamination of the aquatic ecosystem—similar to mercury pollution—could no longer be ruled out. Numerous investigations have, however, led to the conclusion that cadmium is—at least in the aquatic environment—"not a second mercury" (Anon, 1971a). Nevertheless, a local enrichment of cadmium, as in the Jintsu River area, cannot be excluded for other aquatic environments.

2.6.3 Lead Poisoning

The toxicity of lead has, in contrast to mercury and cadmium, been known to mankind for many centuries. The Greek poet-physician Nicander described the disease known as plumbism, which is caused by acute lead poisoning, over 2000 years ago (Stöfen, 1974a).

Especially young children who live in dilapidated buildings are exposed to acute lead poisoning because of the habit of eating nonfood substances such as lead containing peeling paint, putty, and plaster. A chip of paint, the size of an adult's thumbnail contains between 50 and 100 mg of lead. Childhood lead poisoning is a real problem in many of the older urban areas (especially slums) and is a major source of brain damage, mental deficiency, and serious behavioral problems. Yet plumbism is an insidious disease and difficult to diagnose. Until recently it often remained unrecognized and was largely ignored by physicians and public health officers (Chisolm, 1971).

Poisoning from lead-glazed ceramics has been forgotten and rediscovered periodically since antiquity. The Greeks knew about the danger but the Romans did not; they stored food supplies, cider, and wine in such earthenware vessels (Stöfen, 1974a). Although lemon or lime juice was recommended in 1753 as a preventative for scurvy, it was warned against storage of such juices in earthenware jugs. In this context it is of interest to take note of a recent appalling case of lead poisoning reported from Canada. Two boys drank grapefruit juice that had been stored in an earthenware jug for 3 hours and were taken ill. One boy died of lead poisoning—the juice contained 175 ppm lead! (Anon, 1972a).

It is evident from the procedure of glazing, that the initial usage of such colored porcelain or ceramic vessels is unavoidably accompanied by release of lead, especially if they contain acidic liquids such as fruit juices (Anon, 1972a). However, if glazing has been conducted correctly, this should not present a health hazard, i.e., if lead-glazing has been conducted sufficiently long at sufficiently high baking temperature.

The widespread and general use of lead due especially to its exceptional properties such as a high degree of ductility and low corrosiveness, has resulted in lead being concentrated in the environment. As early as during the days of the Roman Empire of Julius Caesar the ruling upper-class made extensive use of lead water pipelines, lead-coated wine vats, lead pigmented cooking and drinking utensils, etc., having acquired the necessary manifacturing skills from their Greek conquerors. The Greeks had also gathered rainwater from roofs protected by a lead covering, via lead-covered guttering and drain-pipes and stored the water in lead-lined cisterns. It has been maintained that the resulting lead poisoning in all probability contributed towards the decline and downfall of the ruling classes of both the Roman and Greek Empires (Stöfen, 1974a).

That the lead content of drinking water is enhanced by lead piping especially if the water is "soft" (a low calcium concentration) has been repeatedly documented for so-called lead epidemics. For example, 250 cases of lead intoxication were recorded in Leipzig (Germany) during the summer of 1930. Investigation of drinking water revealed lead enrichment; the highest value of approximately 25 mg/l was recorded for samples taken during the early morning hours which then dropped to approximately 1 mg/l for samples collected in the course of the day (Coenen et al. 1972a).

Many of America's largest cities use lead pipe extensively for water service connections. Approximately 10% of the total lead pipe used in the United States may be accounted for by the various public utility services, which include water service (Patterson, 1965).

Cases of acute lead poisoning are not frequently encountered but do occasionally occur, as illustrated by the following example from 1969. A British military unit stationed in Hongkong was overtaken by acute lead poisoning and suffered from severe vomiting, intestinal cramps, and circulatory disorder. The source was traced to lead chromate, which had been used to improve the coloration in curry powder. Analysis revealed a lead content of 1.08%! Since this spice is imported to many countries it should be subject to governmental control (Coenen et al., 1972a).

One of the most common consequences of long-term lead poisoning is chronic kidney infection, known as nephritis. During 1929 a physician in Australia became aware of the high incidence of nephritis and early death in Queensland. Investigation revealed that especially children drank rainwater collected from house roofs protected by a lead-pigmented paint. In 1954 it was subsequently established that of 352 adults in Queensland who had had childhood lead poisoning 15 to 40 years earlier, 165 had died, 94 of chronic nephritis. Similarly, nephritis has been reported for consumers of home-stilled (moonshine) whiskey. In the distillation unit lead-soldered tubing is used and in some instances a discarded automobile radiator (containing more lead solder) serves as condensor. It is not surprising therefore, that most samples of confiscated moonshine contain lead (Chisolm, 1971).

2.6.4 Copper Poisoning

An interesting case study of marine copper pollution resulting in serious poisoning of fish off the coast of Holland has been recorded (Roskam, 1972). Innumerable dead fish of widely diversified species and size were found between Scheveningen and Ijmuiden. This led to the conclusion that the phenomenon was of regional nature and had to be ascribed to acute toxification, the source of which was discovered accidentally: several kilograms of copper sulfate crystals were discovered buried under sand near Noordwijk. From subsequent copper analyses, it was established that wherever dead fish were encountered, the seawater contained a Cu^{2+} concentration of several hundred $\mu g/l$ (normal concentration is $1-3$ $\mu g/l$). Groundwater samples taken inland from Noordwijk contained 2.5 $\mu g/l$!

This investigation was followed by a study of the pattern of dilution along the coastline in the direction of the current flow (south to north), which is illustrated in Fig. 6. It was found that mixing of the available billions of cubic meters of unpolluted seawater with the original "copper bell" required two weeks to attain a fivefold dilution, as indicated by the dates in Figure 6 which apply to the period March to April 1965.

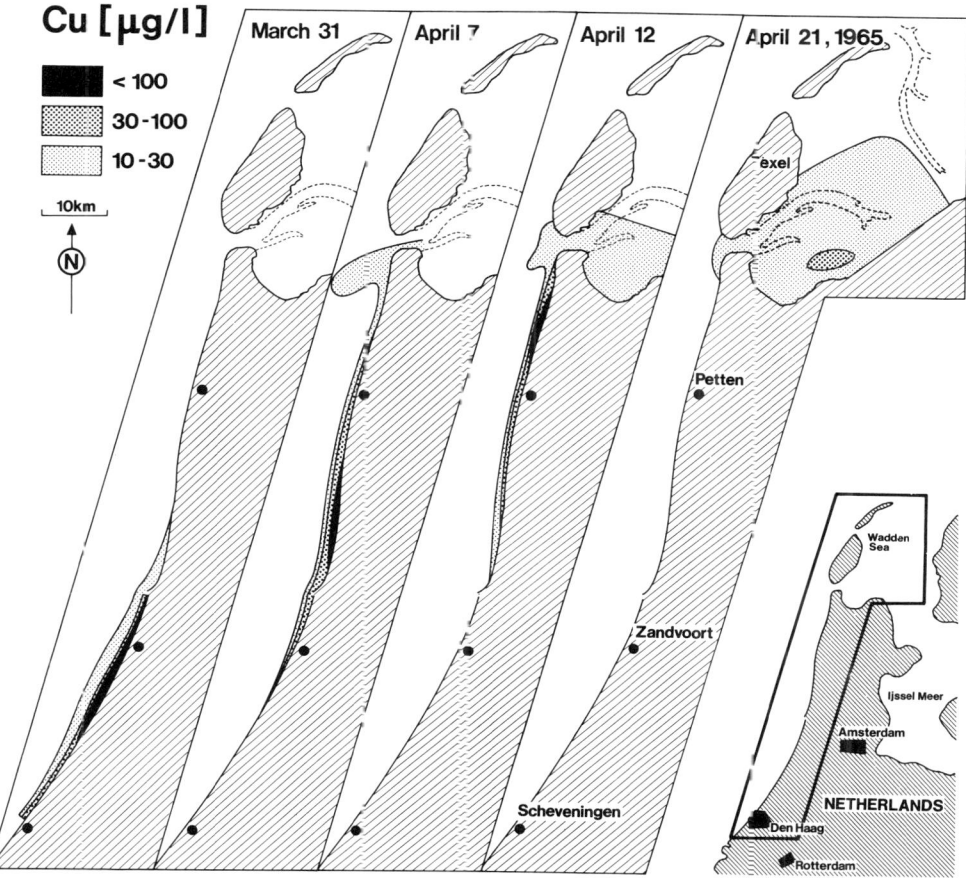

Fig. 6. Copper pollution in coastal waters of the Netherlands (after Roskam, 1972)

2.6.5 Chromium Poisoning

An incidence of catastrophic heavy metal poisoning was reported from highly toxic Cr(VI) contained in untreated slimes and factory wastes. The largest chromium consumer in Japan, the Nippon Chemical Industrial Co., has deposited approximately 530,000 tons of unreduced slimes and wastes containing hexavalent chromium around Tokyo and in the neighboring Chiba Prefecture. Due to its hardening properties, this waste material has gained extensive use for constructional purposes, such as fortification of reclaimed land and building sites, school and recreational grounds, etc. Complete housing blocks along the Bay of Tokyo have even been erected on spoil heaps containing highly toxic Cr(VI) compounds.

This alarming state of affairs must be viewed against an awe-inspiring background. Already during 1960 fatal incidents of lung cancer were reported from the Kiryama factory of the Nippon-Denko Concern on the island of Hokkaido; medical warnings were issued that exposure by inhalation to dust containing chromium in high oxidation states (IV) and (VI) was associated with malignant growth in the respiratory tract and painless perforation of the nasal septum. However, no particular heed was taken by the authorities, because it is a "tragic irony" that the factory workers themselves regarded a perforated, otherwise undeformed nose as a status symbol, displaying "seniority" and loyalty toward a particular employment.

The extent of the present catastrophe cannot as yet be judged by the few official figures—30 dead; over 200 incurables—which only appear to herald the beginning of an unparalleled, scandalous disaster. During August 1975 it was found that drinking water in Tokyo obtained from groundwater near the Cr(VI)-containing spoil heaps, contained more than 2000 times the official threshold limit. In Nagoya the municipal effluent water had turned yellow: extremely toxic chromate ions had revealed their presence! To combat this precarious situation the authorities are considering covering all spoil heaps with a protective layer of tarmac (SPIEGEL Magazine 37/1975).

3 Water Quality Criteria: Standards

3.1 Introduction

The importance of determining adverse effects of chemicals upon human health has gained momentum during the past decade, both on scientific and emotional grounds. The primary reason for setting standards, whether recommended or mandatory, is to protect human health. Standards for drinking water, air, and food are intended to limit the concentration of chemicals below levels that produce harmful effects. However, this approach presumes that a sound scientific data base exists to define the maximum exposure levels for a specific chemical compound (Bull, 1974). With regard to inorganic constituents, professional groups have developed acceptable criteria for most constituents, although some limits have a rather unscientific base. This is not to be regarded as an analytic inability in establishing enforceable standards, but should rather be seen as a lack of quick reliable means of quantifying some metals at low concentrations (Robeck, 1974). Furthermore, in setting drinking water standards, an extensive knowledge of both the qualitative and quantitative nature of the effects of a particular chemical on human health is required (Bull, 1974). The actual problem thus arises from the fact that for ethical reasons humans cannot be deliberately exposed to potentially harmful chemicals so as to study their effects on human health.

3.2 Criteria Development

Although it has been well established that many inorganic constituents enter inland waters from natural or man-made sources (see Chap. B.4), their significance to surface water quality depends on many interdependent factors. Not only is the abundance and widespread occurrence of a particular constituent of importance, but also its availability in the form of solubilized species. Table 7 lists some of the most well-characterized toxic inorganic species. This list is not meant to be comprehensive. Chemical speciation is indeed a complex problem when applied to real systems due to interactions between

Table 7. Sources and aqueous ionic species of toxic metals (after O'Connor, 1974)

Toxic metal	Principal aqueous ionic forms	Sources	
		Mineral	Man-made
As	AsO_2^- (AsIII), arsenite	$FeAsS$	Herbicides,
	AsO_4^{3-} (AsV), arsenate	As_2S_3	Fertilizers,
		AsO_2	Detergent presoaks
		$FeAs_2$	
		As_4S_4	
Cd	Cd^{2+}	$CdCO_3$	Electroplating, pigmentation
		CdS	Photography
		CdO	
Cr	Cr^{3+} (CrIII)		Metal plating
	CrO_4^{2-} (CrVI), chromate	$Fe_2Cr_2O_4$	Industrial dyes
	$Cr_2O_7^{2-}$ (CrVI), dichromate		Ink
Pb	Pb^{2+}	PbO, PbS	Auto and boat fuel,
		$PbCO_3$	Ammunition
		$PbSO_4$	
Hg	Hg_2^{2+}, mercurious	Hg_2Cl_2	Manufacturing of chlorine
	Hg^{2+}, mercuric	HgS, HgO	Electronics, pesticides, fungicides
Se	SeO_3^{2-} (SeIV), selenite	trace constituents	Smelting of copper
	SeO_4^{2-} (SeVI), selenate	of sulfide ores	
Ag	Ag^+	Ag_2O, $AgCl$	Electroplating, food and
		AgS, AgF	beverage processing

various ionic species. In natural waters the formation constants of hydroxo complexes, especially of hard cations are large when compared to most other inorganic ligands (Stumm and Morgan, 1970).

From an inspection of Table 7 it is apparent that in the case of bivalent anions such as CO_3^{2-}, S^{2-}, S_2^{2-}, and PO_4^{3-}, which may successfully compete with OH^- for coordination sites to form insoluble precipitates, such species have not been listed. This is in accordance with a fundamental principle of pharmacology and toxicology that inherent toxicity is related to bioavailability (Tardiff, 1972). Accordingly, the availability of a particular chemical depends on the species present in the aquatic environment available for uptake by aquatic plant organisms, fish and man.

When setting permissible limits or ultimate goals for drinking water standards, cognizance must be taken of the bioaccumulation via the food chain; moreover, it is imperative to impose limits which not only protect man's health on the basis of trace metal quantities in surface water from which potable water is extracted, but to consider the environmental impact of waters discharged to the environment.

Such considerations involve the ecosystem as a whole, self-purification of river systems, biologic treatment plants, the effects of trace metal enrichment on biologic purification treatment, the effects on crustaceans, fish, and ultimately on man. Table 8 represents an attempt to compile data relevant to the aforementioned categories at the present (published) levels concerned.

The factors important for setting drinking water standards for chemicals have been listed (Tardiff, 1972) as being: (1) exposure (type and degree); (2) population

Table 8. Toxicological tolerance levels of some metals and compounds in mg/l

Element/Compound	Biologic Purification[a]	Self-purification[a]	Crustacean[b]	Fish[b]	Mammal[b]	Man[b]
$Al(SO_4)_3$			136	1.8 (trout)	12 g/kg	
Ag-compounds	> 0.7		0.01 – 0.03	0.003 – 0.1		(0.05 mg/l chron.)
As-compounds			4 – 9.1	1 – 23	2 – 15 mg/kg	2 mg/kg (0.05 mg/l chron.)
$BaCl_2$	1000		83 – 1500	0.15 – 8.3 g/l	30 – 500 mg/kg	3.3 – 8.3 mg/kg
$BeSO_4$ (Be)				1.3 (sunfishes) 0.2 (others)		
H_3BO_3				1 – 10 g/l	1 – 5.1 g/kg	5 – 20 g
$CdCl_2$ (Cd)	1–5	0.1	0.03 – 0.4	3.0 (trout)	0.07 – 0.15 mg/kg	50 –500 mg/kg
Co-compound (Co)		5	(0.5) protozoa	0.01 – 0.1	0.7 – 1.5 g/kg	50 –500 mg/kg
Chromate [Cr(VI)]			0.3 – 0.7	0.015 – 0.195	(0.45 – 11 mg/l chron.)	0.5 – 5 g/kg
$Cr_2(SO_4)_3$ (Cr)	2–5	0.3	0.03 – 0.1	1.2 – 200		
$CuSO_4$ (Cu)	1	0.01	0.08 – 0.8	0.03 – 0.8 (trout)	(8 g)	(8 g)
$FeSO_4$ (Fe)	> 35	Fe-deposits	1.62 – 152	0.9 – 152	0.5 – 5 g/kg	0.5 – 5 g/kg
$HgCl_2$ (Hg)		0.018	0.03 – 0.1	0.15 – 0.25 (trout)	0.1 – 1 g)	
Mn-compounds (Mn)			0.5 – 1 g/l	0.05 – 1.2 g/l		0.5 – 5 g/kg
Ni-compounds (Ni)	6	0.1	0.0055 – 1 g/l	0.8 – 55		50 –500 mg/kg
$Pb(NO_3)_2$ (Pb)	5		3 – 170	0.33 – 200 (trout)	2 g/kg	
Se-compounds			(183) protozoa	2 – 10.5	(5 – 10 mg/l chron.)	(0.01 mg/l chron.)
Sn-compounds				2		
$ZnSO_4$ (Zn)	1–3	0.1	19.4	1 – 5 (trout)	1.9 – 2.2 mg/kg	

[a] Data after Liebmann, 1958.
[b] Data after Jung, 1973, and the Hygiene-Institut des Ruhrgebiets, Gelsenkirchen.

exposure; (3) physical state of chemical, and (4) toxicity towards man and experimental animals. By inference, account is taken of environmental exposure via other media such as food and air, when determining a safe intake of drinking water. The question arises whether this is generally accomplished. For example, population exposure to lead by ingestion of water, foodstuffs, and the inhalation of atmospheric particulates has been extensively studied. It was found that urban residents and those residing in areas close to highly travelled motorways are clearly exposed to higher amounts of Pb than the average citizen (see Chap. B.4). In consequence, a realistic appraisal should lead to a lowering of generally applicable Pb limits in drinking water in such areas. However, no literature evidence has been found to substantiate this contention.

Usually, the case of mercury is cited, water standards having been adjusted downward on the assumption that all fish in the daily diet contain 0.5 ppm of Hg in the form of methylmercury. Whether this is realistic for the weekend angler who enjoys his sport and catch, is debatable.

With regard to the different types of drinking water contaminants, the trace metals have received considerable attention in terms of their toxic effects during the past few years. Unfortunately, it has to be admitted that many basic questions regarding this group of chemicals still remain unanswered (Bull, 1974). As pointed out before, the question of chemical speciation poses one of the most difficult problems to be resolved by the chemist, pharmacologist, and toxicologist, especially in the context of synergic effects as encountered in natural waters. For example, the interaction of trace elements (mercury and cadmium) in detoxification has not been unravelled; nor has multiplication of toxicity effects on metal ion combinations—such as nickel and zinc, copper and zinc, or copper and cadmium—which may lead to a fivefold increase in toxicity (Haberer and Normann, 1971), been either explained or rejected on a scientific basis.

With regard to the fourth factor, namely the toxicity toward man and experimental animals, recourse taken to experimental animals has provided much helpful data, especially in respect of fish tests. The classical method for determining the toxicity of a given chemical toward an aquatic organism involves the LD_{50} (or LC_{50}) test. This implies an evaluation of the tested organism of which 50% is killed within a given time-interval at the applied dose for concentrations (usually 96 h, sometimes 48 h). Although criticized for lack of endurance, randomness, follow-up of surviving members of the species, and the employment of single chemicals (Gray and Ventilla, 1973), such tests are widely employed. Advances are being made, for instance, with regard to the real toxicity of copper in relatively simple chemical systems. Tests showed that copper toxicity was apparently related to the soluble form of copper ion occurring in the presence of carbonate ion (Chap. F).

3.3 Water Quality Criteria

In accordance with toxicity data obtained from human clinical investigations, and various other studies such as animal experiments, drinking water standards have been proposed by various governmental bodies. A brief summary is given in Table 9 compiled by Hattingh (1977).

Table 9. Drinking water quality criteria for trace metals which might affect public health[a]

Parameter	USPHS (1962)	Japan (1968)	USSR (1970)	WHO European (1970)	WHO Intern. (1971)	SABS (1971)	NAS (1972)	Australia (1973)	US EPA (1975)	FRG (1975)
Arsenic	10	50	50	50	50	50	100	50	50	40
Barium	1,000	–	4,000	1,000	–	–	1,000	1,000	1,000	–
Cadmium	10	–	10	10	10	50	10	10	10	6
Chromium	50	50	100	50	–	50	50	50	50	50
Copper	1,000	10,000	100	50	50	1,000	1,000	10,000	–	–
Lead	50	100	100	100	100	50	50	50	50	40
Mercury	–	1	5	–	1	–	2	–	2	4
Selenium	10	–	1	10	10	–	10	10	10	8
Silver	50	–	–	–	–	–	–	50	50	–
Zinc	5,000	100	1,000	5,000	5,000	5,000	5,000	5,000	–	2,000

[a] As proposed by the World Health Organization (WHO), US Public Health Service (USPHS), South African Bureau of Standards (SABS), Russia (USSR), USA National Academy of Sciences (NAS), Australia, Japan and Environmental Protection Agency (EPA) of the USA. All concentrations in $\mu g/l$. Compiled by Hattingh (1977), except for F.R.G. data (Schöttler, 1977).

Finally, it is of importance to note that maximum permissible concentrations (U.S.S.R.) and threshold limit values (U.S.) have been established within the field of occupational hygiene (Roschin and Timofeevskaya, 1975). These values pertain to the control of occupational exposure with regard to airborne particulates. In consequence, they are of no relevant importance in the present context.

4 The Sources of Metal Pollution

In general, it is possible to distinguish between five different sources from which metal pollution of the environment originates: (1) geologic weathering, (2) industrial processing of ores and metals, (3) the use of metals and metal components, (4) leaching of metals from garbage and solid waste dumps, and (5) animal and human excretions which contain heavy metals.

Upon attempting to locate the source of metal input of receiving water bodies, a distinction is often made between diffused *nonpoint* and *point sources.* Essentially, rural areas are regarded as nonpoint sources, since the metal supply originates from vast regional areas.

However, nonpoint sources also include a substantial fraction of urbanized areas. Nonetheless, in highly industrialized zones it is often possible to pinpoint the source of localized effluent discharges responsible for metal contamination. As discussed in Chap. D sediment analysis in particular is a unique technique to trace such point sources of metal pollution.

4.1 Geologic Weathering

This is the source of baseline or background levels. It is to be expected that in areas characterized by metal-bearing formations, these metals will also occur at elevated levels in the water and bottom sediments of the particular area. Obviously, mineralized zones, when economically viable, are explored to retrieve and process the ore. This in turn leads to disposal of tailings, discharge of effluents and possibly smelting operations which result in atmospheric pollution. In consequence, the general problem arises of how to distinguish between natural geologic weathering and metal enrichment attributable to human activities.

Not many examples are known for which the interactions between natural weathering processes and mineralized zones are completely devoid of a human contribution. For example, there is evidence that the high mercury content in rocks encountered in the catchment of the La Grande River, Canada, may be responsible for high mercury levels in organisms (Boyle and Jonasson, 1973). It was found that the Aphebian shale in central and northern Quebec—near the headwaters of the La Grande—contained mercury levels averaging 0.5 ppm, which these authors regard as being high.

Regional geochemical maps have been compiled for some areas and, in exceptional cases, countrywide. These maps are based on computerized data of multi-element analysis and reflect the natural composition of both soils and rocks, thus providing baseline information on the occurrence and distribution of trace elements.

Such regional geochemical maps for England and Wales have revealed that the background level of arsenic in stream sediments lies in the concentration range 0–7 ppm, whereas the higher concentrations in anomalous areas can be in excess of 150 ppm (Aston and Thornton, 1975). High concentrations are connected with sulfide mineralization, for instance in South-West England, North Wales, and the Lake District of North-West England.

A study conducted by Colbourne et al. (1975) confirmed that the stream sediment patterns for arsenic and copper in the Dartmoor area of South-West England may be correlated with significant enrichment of these elements in soils derived from rocks within the metamorphic aureole around the Dartmoor granitic intrusion. Previously it had been concluded that the source of arsenic within the metamorphosed country rocks was the result of hydrothermal activity during phases of granitic intrusion. Arsenic is in fact an ubiquitous element and has been reported to be present in solids at an average concentration of 5 mg/kg.

Arsenic-rich hot springs arising from geothermal activity feed the Waikato River of North Island, New Zealand. Submerged aquatic plants from this river were found to contain a maximum of 650 mg/kg dry mass as compared to arsenic levels below 12 mg/kg in plants growing in natural soils (Reay, 1972).

Similarly, geothermal sources in North Island are a natural source for mercury enrichment. A comparative study (Weissberg and Zobel, 1973) between the effects of mercury-containing effluents from a pulp and paper mill into Lake Maraetai and geothermal discharges in Lake Rotorua, revealed that mercury accumulation in sediments from the industrial source did not exceed the enrichment in sediments from the geothermal source. It was thus concluded that mercury contamination of Lake Rotorua is solely due to natural enrichment, especially since sediments from both lakes contained no apparent variation with increasing depth (i.e., age).

Release rates for unmined molybdenum-rich deposits have been determined for a number of locations in Colorado and New Mexico. One particular mineralized study site is situated in the undisturbed alpine environment of Mt. Aetna, Central Colorado. The concentration of molybdenum is low in waters draining the Mt. Aetna area and it has been concluded (Runnells, 1972) that the mineralized deposit could have been detected during geochemical prospecting by sampling the soils and sediments of the area instead of analyzing the waters and plants.

When confronted by a problem such as the impact of a particular trace metal in a broad context, it is inevitable that some form of general model with regard to sources, transport and effects will be resorted to.

Several theoretical models have been presented for the natural cycle of mercury, which has received much attention during the past decade as a result of catastrophic mercury poisonings due to man's impact. A noteworthy contribution (Garrels et al., 1975) related to the differences between pre-man cycle and man-included cycles of mercury. Inspection of Figure 7 reveals that mass transfer in the pre-man sedimentary cycle of Hg significantly resulted between the earth's surface and the atmosphere. The estimated total flux to the atmosphere amounting to 250×10^8 kg/yr exceeds the land-ocean-sediment transfer by an approximate 20-fold factor. This is not surprising because of the well-known volatility of Hg. On account of the low boiling points and vapor pressures of mercury compounds, one can expect a higher mercury content in the vapor state than is normally detected in atmospheric "particulates". Furthermore, the model reveals that vapor transport from land and sea surfaces to the atmosphere prior to man's activities is balanced by equal mass transfers of mercury by rain in the opposite direction.

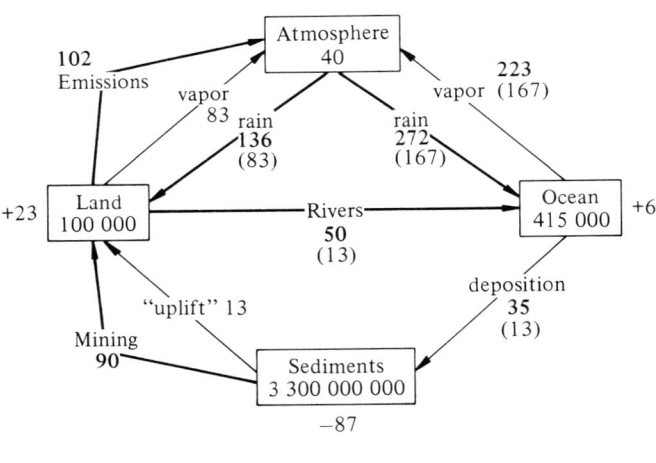

Fig. 7. Present-day cycle of mercury in the environment. Reservoir masses, gains (+) and losses (−) in units of 10^8 g (after Garrels et al., 1975; reproduced by permission of the authors from "Chemical Cycles and the Global Environment"; Copyright by William Kaufmann, Inc., Los Altos, California 94022. All rights reserved)

This pre-civilizational mercury cycle has been disturbed by an increased rate of Hg input resulting from mining activities, emissions from chlor-alkali production, combustion of fossil fuels, roasting of sulfide ores and, in general, the increased utilization of Hg by man. Precipitations such as rain probably contain large amounts of Hg^o, which may again be easily recycled from soils to the atmosphere. Bacterial methylation of inorganic mercury certainly represents an important intermediary process for the release of mercury from soils, sediments and surface waters. This topic will be dealt with in greater detail at a later stage.

4.2 Mining Effluents

Exposure of pyrite (FeS_2) and of other sulfide minerals to atmospheric oxygen and moisture results in one of the most acidic of all known weathering reactions. The sulfidic components (S_2^{2-}) in pyrite is oxidized to sulphate (SO_4^{2-}) whereby acidity (H^+) is generated and Fe^{2+} ions are released. Once this reaction has been initiated by atmospheric oxidation, a cycle is established whereby Fe^{2+} is oxidized to Fe^{3+}; the latter ion is capable of oxidizing pyrite—thereby taking over the initial role of oxygen—to produce additional Fe^{2+} and acidity. During recent years it has been emphasized that bacterial action, for example by *Thiobacillus ferroxidans*, can assist the oxidation of Fe^{2+} (aq) in the presence of dissolved oxygen.

The serious effects of mine effluents on the water quality in rivers and lakes, as well as on the biotopes, particularly on the fish population, have been known for many years. One of the very first descriptions of this problem is the fifth report of the 1868 River Pollution Commission (Anon. 1874) in Britain, where especially grave damage was caused by the dispersal of toxic metals from lead, zinc, and arsenic mines in mid-Wales (Lewin et al., 1977):

"All these streams are turbid, whitened by the waste of the lead mines in their course; and flood waters in the case of all of them bring down poisonous slimes which, spreading over the adjoining flats, either befoul or destroy the grass, and thus injure cattle and horses grazing on the dirtied herbage, or, by killing the plants whose roots have held the land together, render the shores more liable to abrasion and destruction on the next occasion of high water."

Agricultural problems arising from metal contaminations influenced by past and present mining activities were investigated by Griffith (1918), who explained the unproductiveness of certain fields in north Cardiganshire as due to toxic levels of lead and zinc in the soils. In a study published in 1924 on the fauna of rivers in the Aberystwyth District of Cardiganshire, Carpenter showed that in spite of the laws existing in Britain (Pollution of Rivers Act, 1876; Fisheries Act, 1891), several rivers in this area were completely lacking in fish, probably as a result of pollution by the local lead mines there, some of which had remained in operation. Whatever the deeper causes for the absence of fish, there could be no doubt that a certain analogy existed to the findings of Thresh (1922; cf. Carpenter, 1924) of the Counties Health Laboratory who conducted investigations on the effect of natural waters on lead foil.

Generally, *Welsh* rivers and lakes are remarkably devoid of urban and industrial metal pollutants since little or no wastes are discharged into the feeder streams. However, significant concentrations of heavy metals have caused a severe deterioration of the water quality. For example, the presence of lead, copper, and zinc has caused a high mortality rate amongst fish and other living organisms in some Welsh streams (Abdullah and Royle, 1972). It has thus been concluded that heavy metals are leached from the outcrops of mineralized zones and spoil heaps of disused mines drained by such streams. (Alloway and Davies, 1971a; Davies, 1972, 1976, 1977; Davis and Lewin, 1974; Brown, 1977; Davies and Roberts, 1978).

In contrast to modern chemical engineering and technology employed for the development of new mining areas, past mining operations are today being scrutinized as an important source of pollution. For example, in the highly mineralized areas of *South-West England*, copper and tin have been mined for centuries. Since arsenic was regarded as an undesirable constituent until the mid-nineteenth century, the associated arsenic-rich wastes were dumped alongside the copper and tin mines. The origin of high arsenic values in waters and sediments (500 to > 5000 ppm in the sediments within the Tamar River catchment) has been attributed to four supply sources (Aston, et al., 1975). These are: (1) natural geologic weathering of mineralized zones; (2) erosion and dissolution of mine spoil heaps; (3) surface runoff from soils; and (4) dispersion of arsenic fumes from smelters.

Studies related to the chemical composition of present mine drainage from the Erzgebirge — one of the worlds's richest ore mining districts, which has been exploited for centuries in *Saxony* (East Germany) — have revealed that there is practically no plant growth in the vicinity of the age-old Mulden mine. The soil surrounding this area has been markedly enriched with arsenic and lead from effluents and emissions which resulted from alchemistic ore-processing dating back to the Middle Ages. Water sample analyses conducted between 1958 and 1962 (Leutwein and Weise, 1962) have shown that mine drainage resulting from percolation of this widespread mining area contains high levels of Pb, Zn and, in many cases, Bi and As.

In *Poland* Pasternak (1973) investigated waste waters from the region of Bolesaw where large deposits of lead and zinc are mined and processed. The results indicate variable concentrations of lead and zinc in flotation effluents, the maximum reported concentrations being 10.3 mg/l for Pb and 1.7 mg/l for Zn. Surprisingly, the underground water from the mine at Bolesaw, which discharges directly into the Sztola River, is considerably more enriched with Pb (maximum value 300 mg/l) and Zn (maximum value 1800 mg/l).

In the *United States* the increasing consciousness of the environment since the 1960s has led to the recognition that metal accumulation in mine effluents is one of the main problems of water protection (Reppert, 1964; Smith and Frey, 1971; Parsons, 1977). In Colorado alone, 450 miles of surface streams are classified as affected by mill tailings and metal drainages (Wentz, 1974); Hill (1973) reports Fe levels of over 200 mg/l in mine drainage. In groundwater of the mineral belt of the Front Range, Colorado, the limit set by the U.S. Public Health Service is exceeded in 14% of the samples for Cd, 51% for Fe and 74% for Mn (Klusman and Edwards, 1977). In Idaho, it is mainly the Kellogg Smelterville Area in Silver Valley (Miller et al., 1975) and the Cataldo Mission Flats in the catchment area of the Coeur d'Alène River (Galbraith et al., 1972) as well as other areas in the Coeur d'Alène Basin (Ellis, 1932; Mink et al., 1971; Rabe and Bauer, 1977), which are most affected. Pollution by mine drainage in the Cheyenne River System, Western South Dakota, and in the Upper Clark Fork River and selected tributaries is described by Mink et al. (1972). Water, sediment, and fish from the North Anna River in Virginia affected by acidic mine drainage from abandoned pyrite mines were analyzed by Blood and Reed (1975) and characteristic enrichment of toxic trace metals were found.

The *New Lead Belt of South East Missouri* is the world's largest lead-mining district, having produced an unprecedented 495,090 short tons of lead in 1972 (Jennett et al., 1973). Moreover, the development of these mineral resources has the unique distinction of being conducted with a commitment on the part of industry to protect the surrounding environment by controlling waste disposal. Prior to mining activities the concentrations of lead and zinc in local streams were found to be in the 4–6 μg/l range. Subsequently, mining, milling and smelter wastes were channelled into separate streams—also a unique situation, governed by the regional topography—thus providing ideal conditions for studies on environmental pollution. During the early 1970s an interdisciplinary research team from the University of Missouri chose 23 sampling sites and analyzed more than 10,000 water samples (Bolter et al., 1972, 1975; Gale et al. 1972, 1973, 1976; Jennett and Wixson, 1971). It was found that the dissolved heavy metal content of a control stream, unpolluted by industrial wastes, was minimal due to the characteristic slightly basic pH (7.8–8.3) pertaining to this region. However, pond dams contained an appreciable amount of unrecovered lead. The general assessment of the situation is as follows:

Although efficient and modern engineering designs have been incorporated into the mines and mills within the New Lead Belt—where galena (Lead) is the principal ore mined, sphalerite (zinc), chalcopyrite (copper), and silver are recovered as by-products—lead, copper, zinc, cadmium, and other trace metals have been released into a formerly unaffected ecosystem (Wixson et al., 1973). Thus, based on the premise of pollution abatement by the lead mining industry, modern technology has hopelessly failed to meet its own commitments to society. The admission that improved techniques must be found to prevent fine particulate material from being blown and washed into regional streams, substantiates this point.

Comparable developments were also observed in many mining areas of *Canada,* especially in those with sulfidic mineral occurrences and particularly in places where, due to the lack of carbonates, the water has a low buffer capacity. In the 1950s and 60s with the aid of improved trace analytic methods, these connections between the degree of acidity and the dissolved metal loads became more and more apparent. Examples can be seen in the studies by Boyle et al. (1955) on the heavy metal content of stream and spring waters in the Keno Hill-Galena Hill area, Yukon Territory. Studies were carried out on the same problem by Hawley (1972) with investigations in the province of Ontario, and by Hoos and Holman (1972) in New Brunswick and Rose Creek, Yukon.

Harvey (1976) reviews several examples of acid pollution problems from Brunswick mines: one of the worst cases of acid mine waste destroying valuable fish resources is the Brunswick No. 6 mine, where the wastewater had the composition of zinc 389 ppm, copper 31 ppm and iron 131 ppm, with a pH of 3.0 (Montreal Engineering).

The Little River, where the old Brunswick No. 12 mine was located, plus the still-operating mill, is just as heavily polluted and is reported to have no fish life downstream from the site.

An ecological survey conducted to evaluate the effects of a disused copper-lead-zinc mine in *Australia* (Weatherley and Dawson, 1973) revealed that the area remains disfigured by slime dumps, although the mine closed down in 1962. The mining area situated at Captains Flat, 50 km from Canberra, is drained by the Molonglo River. Some 35 years ago a flood occurred which conveyed an enormous amount of tailing deposits from the tailings area of the mine to the fairly productive adjacent flats. Today, 15 km downstream along the Molonglo River, the area is a virtual wasteland. Since fauna and flora are particularly sensitive to an oversupply of zinc, the release of this metal has been regarded as the chief detrimental influent. In consequence of this disaster, the Australian Government has introduced a pollution monitoring system based on zinc concentrations as a pollution indicator (Anon, 1974).

At Rum Jungle in Northern Australia, 64 km south of Darwin, uranium and copper has been mined since the mid-1960s. An estimated 1300 tons of copper had been released and dispersed on the River Finniss floodplain; in addition, 90 curies of radium, whose fate is still uncertain, had been leached from the tailing dump (Watson, 1975). Biologic evidence of severe pollution has been found at the mine site.

Despite a relatively short history of tin and tungsten mining in *Tasmania,* the sparsely populated island has suffered from mining activities. The coarse tailings and supernatants have effected lead, zinc, copper, cadmium, manganese, iron, and sulfuric acid enrichment far beyond the borders of the mining areas (Tyler and Buckney, 1973). It was found that bordering farmers were unable to utilize the water resulting from the creeks that contain these mining effluents; fish and normal biota were found to be absent in these areas.

In *New Zealand* investigations were performed by Ward et al. (1976) on the distributions of copper, cadmium, lead, and zinc in waters and natural vegetation around the Tui Mine, Te Aroha, and on the silver content of soils, stream sediments, waters and vegetation near a silver mine and treatment plant at Maratoto, Coromandel Region (Ward et al., 1977). Metal levels in soils increased near the treatment plant (due to aerial fall-out) and also in natural vegetation growing over the ore deposits; the metal content of stream waters and stream sediments, though anomalous near the deposits, decreased progressively with increasing distance from the source.

The *Republic of South Africa* is renowned for its rich endowment of exploitable mineral resources, and especially for containing the world's largest platinum and gold repositories.

Since the initial discovery of a gold-bearing conglomeritic outcrop some 5 km to the west of the present city of Johannesburg toward the end of the 19th century, further discoveries have led to seven major goldfields being developed along a gold-strike of over 380 km in length. Now gold and uranium are recovered at considerable depths below the surface, sometimes exceeding 3 km. It is acknowledged that these goldfields are situated along the periphery of the geologic Witwatersrand Basin, as depicted in Figure 8, the largest known sedimentary repository of gold (Pretorius, 1975).

The initial amalgamation process to extract gold was soon supplemented by cyanidation since sulfide-containing ores inhibited the process of amalgamation. Furthermore, since the exploitation of uraninite following World War II, several slimes

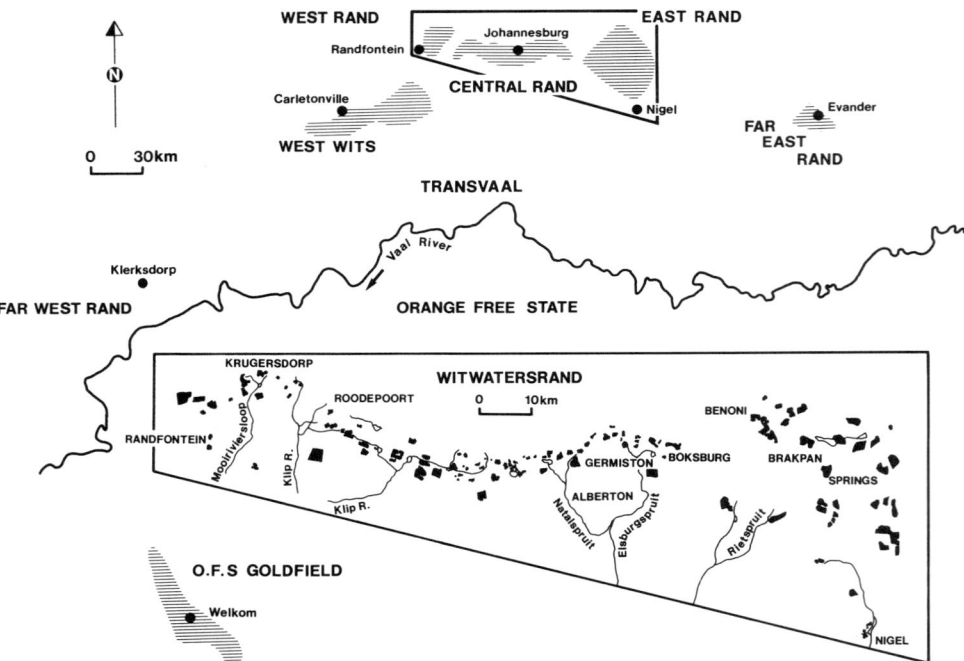

Fig. 8. South African gold mining areas and the distribution of tailings in the Witwatersrand region

dams also serve as disposal sites for the extraction of uranium by pyrolusite (MnO_2) oxidation in sulfuric acid medium.

The slimes dams, resulting from the exploitation of gold- and uranium-bearing ores, are an impressive source of metal supply. The magnitude of the slimes disposal areas may be appreciated from an inspection of the graph given by Förstner and Wittmann (1976), which depicts some of the 247 slimes dams situated along the Witwatersrand proper, i.e., a belt approximately 120 km long stretching from Randfontein in the west to Nigel in the east (Fig. 8).

Mine drainage does not occur only from the mine itself but also from waste rock dumps and tailings areas. The latter two sources often contain a high concentration of sulfides and/or sulfo salts which are associated with most ore and coal bodies. The most commonly occurring sulfides are those of iron, namely the minerals pyrite, pyrrhotite, and marcasite.

It stands to reason that conditions leading to acid drainage of coal mines similarly pertain to gold mining operations in South Africa. Pyrite is the predominant ore mineral in the Witwatersrand reefs—it may exceed 90% of the ore minerals present—while pyrrhotite occurs sporadically and marcasite is a very rare constituent. Lietenberg (1972), in discussing the refractory properties of Witwatersrand ores, mentions that "the presence of the sulfate radical suggests that the leach solutions consist of dilute sulfuric acid derived from decomposed sulfides". A qualitative spectrographic analysis revealed the presence of the following elements: Au, Ag, B, Cu, Co, Cr, Ti, Fe, U, Th, Pb, Hg, Mn, V, Ni, Zr, Al, Mg, Ca, Ba, Na, K, Li, and P and the platinum-group-metals (by the same author). "This analysis shows that a number of mineral constituents, including chromite, zircon, gold, and the platinum group metals that are normally highly refractory to acid attack, had been leached in situ".

In fact, it has been established that insidious seepage results from the gold/uranium slimes dams and is characterized by extremely low pH values, high sulfate content, and high metal concentrations. Table 10 contains a brief summary of values obtained for the seven goldfields (Wittmann and Förstner, 1976a, b; 1977a, b).

Table 10. Hydrochemistry of effluents from mining wastes. (The major ions, Mn and Fe are given as mg/l, and trace elements in μg/l)

Goldfield	pH	SO_4^{2-}	Mn	Fe	Cr	Co	Ni	Cu	Zn	Cd	Pb
Klerksdorp[a]	2.7	2,700	14	110	700	3,900	6,800	1,600	9,000	23	20
West Wits[b]	1.7	11,130	206	550	4,000	3,300	6,400	5,400	26,000	52	290
West Rand[c]	3.7	2,950	18	3	20	2,250	1,540	1,220	4,680	7.0	81
Central Rand[d]	3.0	4,500	41	124	72	2,610	15,900	540	13,200	6.1	32
East Rand[e]	4.0	340	4	–	–	270	1,920	60	1,150	0.2	–
Evander[f]	3.2	2,960	10	29	61	1,500	3,800	400	3,000	2.5	2
O.F.S.[g]	2.6	6,530	44	274	60	2,060	8,900	2,320	5,220	5	10
Normal river water[h]		11	0.007	0.1	1	0.2	1	3	10	0.5	0.5
Enrichment (max)		1,000	30,000	5,500	4,000	19,500	15,900	1,800	2,600	105	580

[a] Furrow from Vaal Reefs G.M. [b] Furrow from West Driefontein G.M. under Carletonville-Fochville Highway. [c] Stream from Krugersdorp at Randfontein-Roodepoort Road. [d] Eastern inflow to Elsburg Dam at Germiston-Boksburg Road. [e] Outflow from Nigel Dam. [f] Furrow from Winkelhaak G.M. [g] Furrow from Loraine G.M. [h] see also Table 28.

The levels of dissolved manganese, cobalt, and nickel exceed the normal surface water values by a factor > 10,000 for each individual metal; iron, chromium, zinc, and sulphate concentrations are at least increased 1000-fold; whereas lead and cadmium in many instances are encountered at values exceeding a 100-fold enrichment. A comparison of these metal concentrations with drinking water standards (see Chap. 8.3) clearly reveals that the maximum values of all metals determined in gold/uranium mining effluents significantly surpass the permissible levels.

The high zinc and lead values are attributable to the cyanidation process for the recovery of gold, whereas high manganese values result from the oxidation of uraninite by pyrolusite (MnO_2) in sulfuric acid medium. However, thucholite, described as an "enigmatic hydrocarbon" (Feather and Koen, 1973) is a common cause of uraninite losses. It is therefore not surprising that the environmental impact from uranium recovery does not rest with heavy metal toxicity and pollution as such, but with the hazards associated with radioactivity. In the United States it was established that only approximately 15% of the radioactivity in uranium ore is removed during processing, the remaining 85% being deposited as abandoned tailings. Radioactive decay leads to the formation of two distinct types of potentially hazardous radiation conditions— gamma radiation and the emission of gaseous radon—a process which continues for thousands of years because of the long half-life of radium, the main radioactive waste product (Controlling the radiation hazard from uranium mill tailings. Report to Congress, [Anon, 1975a]). These results imply that the mining industry is suffering considerable economical losses due to valuable and recoverable metals being discharged to the environment.

4.3 Industrial Effluents

The disposal of industrial wastes is often conducted without critical appraisal of the losses incurred. Usually no consideration is taken with regard to the deleterious environmental impact upon the receiving water body.

There are numerous sources of industrial effluents leading to heavy metal enrichment of the aquatic environment. The classic example is the discharge of the catalyst methylated mercury chloride into Minamata Bay from a factory manufacturing plastics. Contrary to expectations, microorganisms converted the sedimented compound to monomethyl-mercury, which led to an enrichment of this most toxic compound in fish consumed by the local fisher-folk. The Minamata Bay disaster has been discussed elsewhere.

The major industrial uses of various economically important heavy metals have been compiled in Table 11 (after Dean et al., 1972). An inspection of this table reveals that most heavy metals under consideration are employed in widely diversified fields such as petroleum refining, steel and fertilizer production, etc. On the other hand, several industries function on a basis where only one specific heavy metal is involved, for example, the use of chromium in the tanning industry. However, in general, the multipurpose usage of numerous heavy metals may lead to difficulties in tracing the source of origin of water pollution conclusively.

Table 11. Heavy metals employed in major industries (after Dean et al., 1972)

	Cd	Cr	Cu	Fe	Hg	Mn	Pb	Ni	Sn	Zn
Pulp, papermills, paperboard, building paper, board mills		X	X		X		X	X		X
Organic chemicals, petrochemicals	X	X		X	X		X		X	X
Alkalis, chlorine, inorganic chemicals	X	X		X	X		X		X	X
Fertilizers	X	X	X	X	X	X	X	X		X
Petroleum refining	X	X	X	X			X	X		X
Basic steel works foundries	X	X	X	X	X		X	X	X	X
Basic nonferrous metal-works, foundries	X	X	X		X		X			X
Motor vehicles, aircraft-plating, finishing	X	X	X		X			X		
Flat glass, cement, asbestos products, etc.		X								
Textile mill products		X								
Leather tanning, finishing		X								
Steam generation power plants		X								X

In order to appreciate the vast quantities of metal-containing discharges and the relative contributions from various sources of supply, the New York wastewater may serve as an illustrative example. From an inspection of Table 12 it follows that the *electroplating industry* is the major contributor of chromium and nickel to the city's wastewater treatment plants; similarly, relatively high values have been recorded for copper, zinc, and especially cadmium which resulted from electroplating wastes.

Table 12. Relative contributions of metals to New York City wastewater treatment plants (after Klein et al., 1974)

	Percentage of total weights received				
Source	Cu	Cr	Ni	Zn	Cd
Water supply	20	0	0	7	0
Electroplaters	12	43	62	13	33
Other industrial	7	9	3	7	6
Runoff	14	9	10	31	12
Residential	47	28	25	42	49
Unknown	0	11	0	0	0
Total (kg/day)	1160	674	509	1780	73

With regard to the other sources (runoff, residential wastes, water supply), it was previously noted by Klein et al. (1974) that (1) except for nickel at 62%, the electroplating industry does not contribute the major portion of the metals; (2) industry excluding platers contributes less than 9% of the metals; (3) the contribution from residential wastes varies from 25% to 49%; in all instances for copper, zinc, and cadmium contamination, while residential wastes contribute considerably greater amounts than do the electroplaters; (4) the water supply contributes 67% more copper than do the electroplaters; and (5) stormwater runoff is the source of more copper and zinc than are the electroplaters.

Chemical and electrochemical methods are employed in the metal finishing and allied industries for the purpose of protection and/or the decoration of a variety of metal surfaces (e.g., Lowe, 1970). Most of the processes are followed by rinsing operations to remove the excess chemicals and other waste material from the treated surfaces, thus giving rise to effluents. Notably, pickling and electroplating give rise to high waste metal concentrations.

After machining operations, degreasing is often employed to remove excess oil, usually by immersion in trichloroethylene. The cleaning-out operation may contain a significant amount of finely divided metal particles (Lowe, 1970). Since it is necessary for the surfaces to be free from oil and grease, degreasing is often employed as a preliminary stage prior to pickling, a process used extensively for cleaning metal surfaces of rust and scale or other undesirable surface conditions. This treatment usually consists of reaction with concentrated mineral acid, particularly sulfuric acid—although other acids such as HCl, HF, H_3PO_4 and mixtures of these acids are sometimes used—to obtain a clean surface for further operations.

Obviously, most effluents from pickling and dipping operations are strongly acidic and contain an appreciable amount of dissolved metals. A typical example from sulfuric acid pickling of copper and brass is quoted by Lowe (1970) as containing 250–300 mg H_2SO_4/l, 60–90 mg/l soluble Cu, 20–30 mg/l soluble Zn, and 10–15 mg/l suspended solids.

The mechanical processes designed to remove surface irregularities and other blemishes, and to impart a smooth lustrous surface are generally referred to as polishing. Wastes produced from polishing are in the form of dusts and flocs which are retained in the dust extraction system. However, an increasing amount of surface finishing is nowadays being done *en masse* by wet processes, by rotational or vibrational processes. In general, these effluents contain high values of suspended solids, from 2000 to 4000 mg/l, which can include a significant amount of metal. In addition, metals may become solubilized; for example, the copper content has been reported to attain a maximum value of 130 mg/l (Lowe, 1970).

Metal finishing is achieved by chemical and electrochemical processes in special cases, e.g., to produce highly lustrous surfaces on intricately shaped objects made of brass, steel, aluminum, etc. Generally, solutions consisting of mixtures of sulfuric and phosphoric acids with additions of nitric or chromic acids are used to achieve polishing and brightening to passivate the electrochemically and chemically brightened surface. It is well known that such surfaces otherwise tarnish rapidly upon exposure to atmospheric oxygen. Effluents from these processes are strongly acidic and may contain the toxic hexavalent chromium or the less toxic trivalent form.

The *electrolytic production of chlorine and sodium hydroxide* from brine solutions using mercury electrodes has led to considerable mercury pollution as a result of mercury losses which have entered the aquatic system. It has been estimated that the Dow Chemical chlor-alkali plant at Sarnia has discharged some 91,000 kg of mercury compounds into the St. Clair River system during the period from 1949 to 1972 (Wood, 1972). Such discharges have subsequently been curtailed in the United States and a subsidiary of the particular company operates the chlor-alkali manufacture on the basis of the diaphragm process, without the utilization of mercury, near Stade, West

Germany (pers. comm.). Nonetheless, the production of chlorine continues unchecked elsewhere, by means of the mercury electrode process.

The principal source of chromium results from discharges of industries using large amounts of chromates or dichromate as in the *textile industry and the leather tanning industry*. In the tanning industry the preliminary treatment is by means of alkali solutions used to remove the outer skin and dirt from hides. The second stage known as tanning is directed toward conditioning the hide to keep it from becoming rough or brittle. Chromium(III) compounds are frequently employed for this process.
Polish tannery wastes from chrome tanneries have been reported to vary between 9 and 140 mg/l (Koziorowski and Kucharski, 1972). When considering effluents with inorganic pigments content resulting from the *paint industry*, the great variety of raw materials makes it impossible to approach the problem in a general manner. One possible source is illustrated by an intensive investigation into the cause of cadmium enrichment of sediments from the Neckar River flowing through Heidelberg, West Germany. Water analyses conducted upstream and downstream from this city revealed a maximum cadmium concentration of 220 $\mu g/l$ at the confluence of the Enz and Neckar Rivers. Subsequently, the investigators were able to trace the source of cadmium pollution of the Enz River. In this particular case the discharges contained high enough cadmium levels so that the source of pollution could be traced by means of water analyses (Förstner and Müller, 1974a).

In another study (Kneip et al., 1974), cadmium and nickel enrichment was found in an aquatic system in the marshes and coves along the eastern side of the Hudson River, N. Y. Discharges of these metals were attributed to waste waters from a nickel-cadmium *battery plant*—which again illustrates the wide diversification of industrial sources of toxic heavy metals.

The environmental pollution from Cd and Zn discharged from a Braun tube (used in T.V. sets) factory in Japan was investigated by Asami (1974): The slag at the wastewater outlets and settling tank was 4820–4500 and 15,500–37,500 ppm respectively for Cd and Zn concentrations.

A typical example of pollution caused by the *iron and steel industry* is presented by the steelworks near the Tees Estuary, Scotland. The so-called Clevelent effluent from the Teesside Steel Works (British Steel Corporation) was monitored during June and July 1973 (Prater, 1975). It was found that iron and manganese had the highest mean concentration, the blast furnaces and the ferro-manganese plant respectively being the major contributors. Zinc and lead occured at similar concentration levels. Although there are a variety of sources for zinc and lead within the works, the iron and ferro-manganese blast furnaces are considered to be the most important. These four elements were also found to be the most abundant in the final effluent which discharges the following approx. quantities to the Tees Estuary (in kg/day): Fe 7500, Mn 2300, Zn 850, Pb 310, Cu 25.

A comparison of these quantities of metal discharges to other estuarine locations in the United Kingdom, which is given in Table 13, clearly indicates that the effluent from the Teeside Steel Works contribute considerable loadings of zinc and, in particular, lead to the Tees Estuary.

It is often overlooked that heavy metal pollution results from the industrial usage of *organic compounds containing metal additives*. Apart from the well-known case of

Table 13. Discharges of copper, zinc and lead in various coastal areas of the United Kingdom. Loadings expressed in terms of kg/d (after Prater, 1975)

Area	Copper	Zinc	Lead
Liverpool Bay[a]	820	4400	10
Firth of Clyde[b]	40– 100	160– 260	70–80
Thames Estuary[b]	1600–4000	350–1700	–[c]
Tees Estuary[d]	25	850	310

[a] Figures are minimum values as they do not include discharges from local authority outfalls.
[b] Figures are minimum and maximum values and only refer to metals dumped in sewage sludge.
[c] No data available.
[d] Discharges from S. Teesside steelworks only.

gasoline (containing tetraethyl lead as additive), there are numerous other examples to support this contention. Thus, heavy duty oil often contains lead as an additive, whereas lubricating oil is usually supplimented by molybdenum sulfide. Heavy metals are also added to various stearates: For example, Zn, Sn, Pb, and Cd are employed as stabilizers and additives in the manufacture of synthetic rubber and PVC; lead stearate as softener in the manufacture of nitrocellulose; copper stearate for mineral flotation; chrome stearate as anti-corrosion agent, etc. (A detailed list has been compiled by Herrig, 1969).

The above mentioned examples serve to illustrate the wide spectrum of heavy metal contaminants released to the environment by the utilization of metal enriched organic substances. Consequently, an inherent difficulty arises when attempting to trace the source of metal pollution related to these diffuse sources of origin. Nonetheless, a comprehensive investigation conducted by the New York City wastewater authorities (Klein et al., 1974) has revealed many sources of metal pollution resulting from industrial branches, which hitherto had received little attention. Disregarding

Table 14. Metals in industrial wastewaters (after Klein et al., 1974)

Industry	Average concentrations in $\mu g/l$				
	Cu	Cr	Ni	Zn	Cd
Meat processing	150	150	70	460	11
Fat rendering	220	210	280	3,890	6
Fish processing	240	230	140	1,590	14
Bakery	150	330	430	280	2
Miscellaneous foods	350	150	110	1,100	6
Brewery	410	60	40	470	5
Soft drinks and flavorings	2,040	180	220	2,990	3
Ice cream	2,700	50	110	780	31
Textile dyeing	37	820	250	500	30
Fur dressing and dyeing	7,040	20,140	740	1,730	115
Miscellaneous chemicals	160	280	100	800	27
Laundry	1,700	1,220	100	1,750	134
Car wash	180	140	190	920	18

Domestic Effluents

metal-containing discharges by electroplaters, 13 industrial activities were examined, as listed in Table 14.

An inspection of Table 14 reveals that laundries, as well as ice cream and soft drink manufacturers discharge wastes rich in copper; textile dyeing and laundry wastes have high chromium contents; bakery wastewaters contain high levels of nickel; several industries discharge high concentrations of zinc; and finally, fur dressers and dyers discharge exceptionally high concentrations of all five listed metals.

4.4 Domestic Effluents and Urban Stormwater Runoff

Metal enrichment which results from residential areas is treated in accordance with its source of origin. Thus, on the one hand there are domestic effluents which are usually discharged from a relatively well-defined point source. On the other hand, urban stormwater runoff is characterized by a diffuse drainage pattern—only partially contributary towards the metal content of domestic effluents—and together with rural areas (see Sect. B.4.6) belongs to the most important nonpoint sources of metal loads in inland waters.

4.4.1 Domestic Effluents

Generally speaking, these wastewaters probably constitute the largest single source of elevated metal values in rivers and lakes.

Domestic effluents may consist of (1) untreated or solely mechanically treated wastewaters, (2) substances which have passed through the filters of biologic treatment plants, either solubilized or as finely divided particulates, and (3) waste substances passed over sewage outfalls and discharged to receiving water bodies—often the sea in coastal residential areas.

Examples for (1) and (2) may be taken from the Federal Republic of Germany. According to official figures pertaining to the early 1970s, only 38% of approximately 7×10^9 m^3 domestic effluents received full biologic treatments with the required oxygen replenishment; 52% was discharged to the receiving waters either untreated or after mechanical treatment only; 10% received partial biologic treatment. Furthermore, it must be noted that soluble species containing phosphorus, nitrogen, or metals are only retained by a further "third" treatment stage (see Chap. G.2). Insufficiently purified domestic effluents also transport an appreciable amount of suspended material to receiving waters and may thus lead to a substantial increase of the natural sediment load. For example, Hellmann (1972a) has shown that between one-third to one-half of the suspended material load of the Neckar River near Stuttgart (FRG) consists of waste particulates from the treatment of domestic effluents. After deposition, such material, being rich in organic substances, leads to a high oxygen demand, thus causing a deterioration of water quality. In addition (in the cited example), the navigation of canalized river sections may be hampered.

Solid wastewater particles may cause appreciable metal enrichment of the suspended load in waters. A clue as to the magnitude of these effects is obtained by comparing the normal values for sewage sludges with the corresponding metal levels in crustal rocks—the latter data serving as natural occurring metal levels in solid substances—in Table 15.

From an inspection of these values it is evident that hardly any enrichment can be expected for nickel and chromium. By contrast, the concentrations of copper, lead, zinc,

Table 15. Comparison of metal concentrations in sewage sludges

	Average sewage sludge (mg/kg)[a]	Average crustal rocks (mg/kg)[b]	Ratio (a)/(b)
Nickel	60	80	0.8
Chromium	240	200	1.2
Copper	700	45	15
Lead	450	15	30
Zinc	2600	65	40
Cadmium	10	0.2	50
Silver	20	0.1	200

[a] Sweden: Berggren and Odén, 1972; England and Wales: Berrow and Webber, 1972; Michigan: Blakesley, 1973; Median-values from Page (1974), and average crustal rocks. [b] Mason, 1958.

cadmium, and silver reveal a marked influence of domestic effluents. These elevated levels of Cu, Pb, Zn, and to a lesser degree Cd, are due to *corrosion* within the urban water supply network; the high values of silver mainly result from its photochemical use i.e., partly an industrial source (Preuss and Kollmann, 1974).

Similar metal enrichment may also be expected to result from metal species contained in solution. However, due to the localized nature of the hydrochemical conditions, it is more complicated to define the situation than in the case of solid substances. A typical example stems from the presence of detergents in domestic effluents.

The use of *detergents* also creates a possible pollution hazard, since common household detergent products can affect the water quality. Angino et al. (1970) found that most enzyme detergents contained trace amounts of the elements Fe, Mn, Cr, No, Co, Zn, Sr, and B. Moreover, spectrographic evidence for the presence of arsenic was found in three enzyme presoaks, three heavy duty enzyme detergents, one heavy duty detergent and one detergent aid. The presence of high arsenic levels for these detergents was found to vary between 31 and 45 ppm As, with the enzyme presoak containing between 51 and 73 ppm. Since waste waters containing detergents enter the urban water drainage system and many sewage or waste effluent plants do not remove arsenic. the maximum concentration of 8 ppb, found in the Kansas River at Topeka, is very close to the permissible level of 10 ppb, and may thus be regarded as a potential health hazard.

4.4.2 Urban Storm Water Runoff

With regard to pollution resulting from urbanized areas, there is an increasing awareness that urban runoff presents a serious problem of heavy metal contamination. Heavy rainfall in urban areas is no longer regarded as only a downpour of "rainwater" (Sartor et al., 1974) since they often contain shock loads of contaminants, The runoff frequently presents insurmountable problems especially with regard to the local sewage works, which may become overloaded and thus overflow (Field and Lager, 1975). A statistical summary (Bradford, 1977) revealed that urban stormwater runoff has long been recognized as a major source of pollutants to surface waters.

However, it appears that the early studies conducted during the 1950s, have not received much attention. Bradford reports that in Sweden, the Soviet Union and in the United States it was found that the 5-day biochemical oxygen demand (BOD_5) values in stormwater runoff from urban areas exceeded the concentration in domestic wastewaters (having received secondary treatment) by an approximate tenfold factor in some instances. Moreover, it was concluded from these studies that the shock load contained by urban stormwater runoff could be 100 to 1000 times greater than that for sanitary wastewater. This implies that urban stormwater is a significant source of pollution and may indeed have a dominating effect on the existing quality of the receiving water body.

During the past decade, efforts were made to identify and quantify this form of pollution from urban watersheds (Bradford, 1977). Street dust, dirt and various solids were among the pollutants considered as dependent variables, whereas climate, season, and land usage are primarily variable but independent parameters. In this statistical summary it was revealed that there are few significant trends in the occurrence of dust and dirt loading on streets and pollutant concentrations as a function of independent parameters "which are expected to influence their occurrence" (Bradford, 1977). Furthermore, this statistical summary is of great interest since it counters many prejudices. A brief summary of the findings is given below, but the interested reader is referred to the original text:

1. In urbanized areas, loading rates are below average despite heavy traffic. However, BOD_5 values and lead concentrations are above average, in accordance with expectations.
2. In residential areas it was expected that the use of fertilizers would lead to high phosphate concentrations. There was, unexpectedly, no significant phosphate increase in street-born solids.
3. It was thought that tree foliage could be regarded as a major indicator of street pollution. Analysis revealed, unexpectedly, that tree-landscaped areas reflected loads below the average.
4. There is no consistent relationship between the automobile traffic—discharging volatile "antiknock" tetraethyl lead—and lead pollution.

A logarithmic decrease of runoff containing lead from paved highways was found with regard to progressive rainfall (Sylvester and DeWalle, 1972). This is in accordance with the general finding that the first downpour contains the highest Pb content and most of the accumulated lead in paved areas is washed out within a short period of time.

The important feature is that potential contamination may occur during periods of storm runoff, whereby trace elements resulting from atmospheric emissions and subsequently deposited on various surface material may be transported to the nearby drainage system. Furthermore, studies by Bolter et al. (1974) indicate that lead is leached by humic and other acids, thus increasing its availability for runoff rather than seepage into the upper soil layer.

The problem of urban stormwater and snow has been most aptly formulated with: "when a city takes a bath, what do you do with the dirty water?" (Field and Lager, 1975). The stormwater discharges and combined sewage overflows from urban areas present problems of ever increasing importance. Whipple and Hunter (1977) have recommended treatment of urban runoff and sewage overflows in combination with detention storage. In their investigations they found that the problem has not really been seriously considered with respect to heavy metals. After monitoring heavy metals from two heavily urbanized watersheds at Lodi, New Jersey, they establihsed high concentrations of lead, zinc, and copper after a storm event. The highest concentrations usually occurred within the first 30 min of storm runoff which reveals that the first flush carries exceptionally high loads. Inspection of Figure 9 reveals that the highest dissolved lead concentrations occur within approximately ten minutes prior to the peak discharge. However, in another study Whipple and Hunter (1977) established

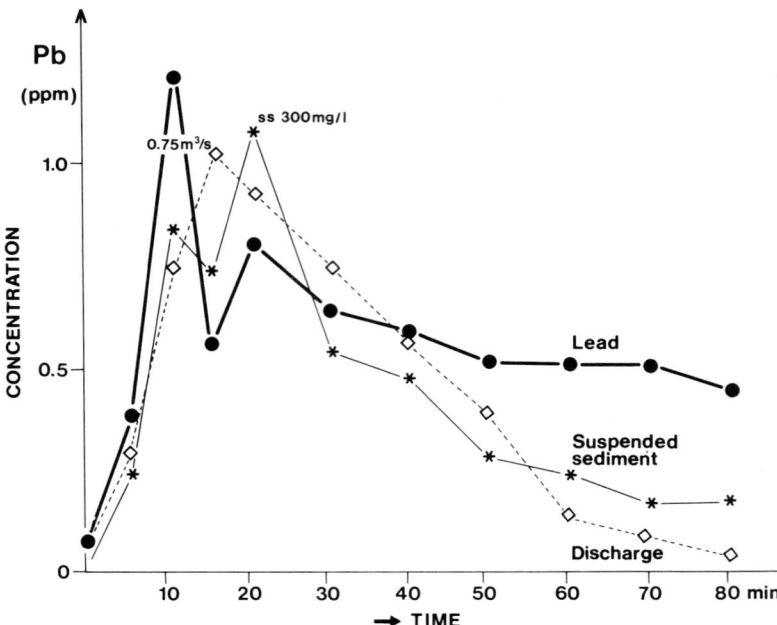

Fig. 9. Behavior of lead during urban storm water events (Whipple and Hunter, 1977; permission of Water Pollution Control Federation)

that it should not be assumed that the observed metal amounts move downstream uninterruptedly. On the contrary, these authors showed that most of the lead and zinc at Lodi were in the form of particulates. As might be expected, such material is deposited as sediment upon arrival at a receiving body not subjected to flood conditions.

From many investigations it is known that urban stormwater contains metal concentrations which vary extremely with regard to locality and time of sampling (Malmquist, 1975). This is not surprising in the light of the aforementioned findings, which are, of course, valid for all pollutants in stormwater. No significant predictions as to the quality of stormwater can be made without a knowledge of the pollutant sources. These have been summarized diagramatically by Malmquist (1975) as illustrated in Figure 10.

In accordance with Bradford's statistical results, there is no direct correlation between the dependent variables and the independent parameters. However, it is of interest to note that polluted air (treated in detail in Sect. 4.6. under "atmospheric pollution") has been incorporated as a source of *corrosion* which leads to metal enrichment of stormwaters.

This is, of course, an important source of heavy metal enrichment resulting from residential areas as has been shown by an investigation conducted in Gothenburg, Sweden. Although presented as a preliminary study in which the stormwater runoff from a total watershed (0.16 km^2) and from a house roof (900 m^2) were continuously monitored during the autumn of 1974, the comparison of data obtained shows that more copper and zinc is released from the roof but less lead than from the watershed per unit area. It may thus be concluded that corrosion of copper and zinc roof fittings are an important contributory source toward enrichment of these metals in urban stormwater runoff (Malmquist, 1975).

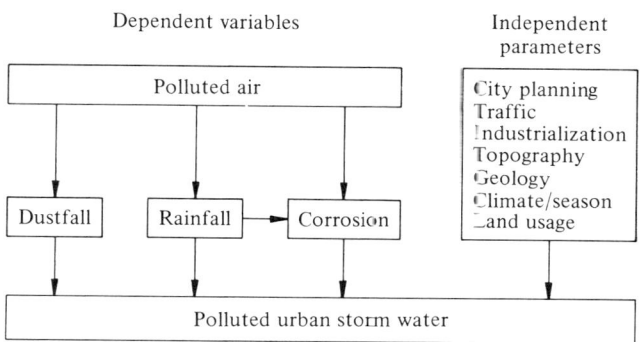

Fig. 10. Sources of metal pollution in urban stormwater (after Malmquist, 1975)

In the northern hemisphere urban drainage presents another problem because of low winter temperatures, the resultant snowfall, and formation of ice on tarmac. It is well known that *salts* (notably NaCl and $CaCl_2$) are used to depress the freezing point of water and accordingly de-ice traffic pathways. Oliver et al. (1974) have pointed out that chloride contamination is presenting an ever-increasing problem in Canadian and American cities where salt is used to keep roads clear of snow during winter. According to an estimate (Whipple and Hunter, 1977), nine million tons of salt are spread annually for de-icing highways in the United States.

It stands to reason that if massive quantities of salt are employed for purposes of de-icing, the removal of urban snow, which is also contaminated with automobile exhaust gases, must result in regional lead and chloride contamination. If the disposal sites are judiciously chosen instead of dumping the snow directly into water courses, it has been demonstrated that this source of lead pollution is significantly reduced (LaBarre, 1973). The significance of pH and chloride ion concentration on the behavior of heavy metal pollutants was studied by Hahne and Kroontje (1973). Their results indicate that both chloride and hydroxy complexes may contribute to the mobilization of Hg^{2+}, Cd^{2+}, Zn^{2+}, and Pb^{2+} ions in the environment.

4.4.3 Spoil Heaps

Heavy metal pollution resulting from spoil heaps may be treated in the context of domestic sources, bearing in mind that industrial sources likewise contribute to the formation of such deposits.

For example, disposal of neutralized spent pickling liquors into an abandoned strip mine in eastern Ohio has been proven to degrade both surface- and groundwater, resulting in acid mine drainage (Pettyjohn, 1975). Although percolation from spoil heaps often has to penetrate a formidable layer of filter material before reaching the groundwater table, it has been shown that traces of arsenic, cyanides, phenoles, etc., render the water unsuitable for drinking purposes (Flögl, 1958).

Investigations conducted in the vicinity of 10 domestic refuse heaps near Aachen, West Germany (F.R.G.) have revealed elevated concentrations of cadmium and copper (Heitfeld and Schöttler, 1973). These authors concluded that this was a potential hazard for drinking water quality (Sect. G.3.2.2).

4.5 Metal Inputs from Rural Areas

In the United States almost 97% of the total land area is essentially rural in nature (McElroy et al., 1975). Commercial forest, grassland and cultivated agricultural areas (cropland) are, on the basis of land area, potentially large contributors of nonpoint sources leading to environmental pollution in aquatic systems.

It is of special interest to note that soil cultivation has been estimated to be responsible for 95–99% of soil erosion (McElroy et al., 1975). The sediment resulting from soil erosion is today recognized as being the largest single pollutant affecting water quality. Robinson (1973) correctly points out that "... sediment is our *greatest pollutant*". This author emphasizes the seriousness of water pollution by quoting the President of the United States, who stated that "water pollution has three principal sources: municipal, industrial, and agricultural wastes. Of these three, the most troublesome to control are those from agricultural sources: animal wastes, fertilizers, pesticides, and in particular *eroded soil*." Not only is the storage capacity of man-made impoundments reduced by silting, agricultural runoff provides sufficient nutrients (usually only regarded in terms of nitrogen and phosphorus-containing constituents) to cause excessive algal growth (Toerien et al., 1975).

In contrast to being eroded, soil may become *enriched* with hazardous heavy metals by the application of plant nutrients and crop protective measures. Rock phosphates and *phosphatic fertilizers* often contain high levels of trace elements, especially cadmium.

The analysis of 21 of the most common Swedish fertilizers revealed that in four instances the Cd concentrations exceed the tolerance limit of 15 mg/kg set by the Swedish National Board of Health and Welfare for agricultural use of sewage sludge. The highest value found was 30 mg/kg Cd (Sternström and Vahter, 1974). Prior to this investigation apparently little attention was paid to fertilizers being a source of cadmium pollution. That this problem is not restricted to Sweden is borne out by the Cd concentrations in Australian commercial fertilizers, which have been shown to vary between 18 and 91 mg/kg (Williams and David, 1973).

It has been found that the uranium concentration in Mississippi River water exceeds that of the Amazon by more than 20 times. Although many chemical plants and municipalities dump wastes into the river, the increased and widespread usage of phosphate fertilizers containing uranium must be taken into consideration when evaluating the elevated uranium-238 concentration in the Mississippi.

The concentrations of uranium in sediments from most of the investigated rivers which flow into the Gulf of Mexico have increased considerably during the course of two decades. Spalding and Sackett (1972) have attributed the high values to uranium in phosphate fertilizers and pointed out that rivers (by way of example, the Rio Grande) which drain uranium-bearing strata contain approximately 1 μg/l. Similar concentrations were found by these investigators for the Colorado, Brazos, and Trinity Rivers, which drain highly developed agricultural areas. Moreover, a 44% increase in fertilizer consumption took place in the 5-year period from 1962 to 1967. On the basis of an average uranium content of 150 μg/g in phosphate fertilizers, these investigators estimated that during 1967, 2.6×10^5 kg (or 285 short tons) of U_3O_8 were applied to cultivated lands in the U.S. portion of the Gulf of Mexico.

Land application of *dredged sediment* and other waste materials from construction and demolition processes can be beneficial for specific materials and sites. Lee et al. (1976) cautioned, however, that not all dredged material is suitable for land application. Accordingly, the dilemma is to distinguish between dredged materials and application sites which are compatible with regard to the environmental impact. *Sewage sludge* is often used as a plant nutrient source on cultivated soils. In many cases the sludge contains trace elements with toxic properties. Upon decomposition of the sludge, these elements are released and become available for plant uptake (Chap. G).

The use of *herbicides* as grass control along highways and other areas has given rise to concern. Since the introduction of arsenicals such as MSMA (monosodium methanearsonate) to control the growth of Johnson grass, the estimated U.S. production of arsenical herbicides during 1971 has risen to 16.8×10^6 kg (Isensee et al., 1973). Although three species of fish were tested and showed good tolerance toward MSMA, tolerance of other species is not known (Andersson et al., 1975). It appears that in this particular case the concern has been overrated. MSMA is a contact herbicide which is rapidly inactivated upon contact with soil and probably degraded to carbon dioxide and arsenate; the latter, precipitated as insoluble salt, becomes unavailable for uptake in the food chain (Edwards and Davis, 1975).

4.6 Atmospheric Sources

Natural and man-made processes have been shown to result in metal-containing airborne particulates. Depending on prevailing climatic conditions, these particulates may become wind-blown over great distances; nonetheless they are subjected to the fate that they are ultimately returned to the lithosphere as precipitations by rain- or snowfall. The results of several intensive studies conducted to establish the extent of atmospheric metal enrichment have been summarized in Table 16.

For a given element, X, an EF value close to unity indicates crustal weathering as atmospheric source for that particular element. Thus, the respective EF values indicated that

Table 16. Enrichment of metals in atmospheric particles relative to the earth's crust

	Enrichment factor (EF = $(X/St)_{atm}/(X/St)_{crust}$)			
	United States urban air[a]	North Atlantic Westerlies[b]	South Pole[c]	Low level marine atmosph. particulates[d]
Iron	1.0 (St)	1.1	2.1	1.0 (St)
Aluminum	0.5	1.0 (St)	1.0 (St)	— —
Cobalt	2	3.5	4.7	0.45
Chromium	11	—	—	1.2
Nickel	12	—	—	1.5
Vanadium	42	13	1.4	1.4
Copper	83	84	93	3.4
Zinc	270	40	69	13
Arsenic	310	—	—	—
Silver	830	—	—	—
Mercury	1,100	—	—	—
Cadmium	1,900	300	—	—
Lead	2,300	2,300	2,500	29
Selenium	2,500	16,000	18,000	—
Antimony	2,800	3,000	1,300	—

[a] Anon, 1972; [b] Zoller et al., 1974; [c] Duce et al., 1972; [d] Chester and Stoner, 1973a.

cobalt, and to a lesser extent chromium and nickel, in the atmosphere result from normal crustal erosion. The high vanadium enrichment found in the North Atlantic, but not in Antarctica, is probably due to vanadium produced by the burning of heavy fuel oil (containing high concentrations of vanadium porphyrin complexes) along the east coast of North America (Duce et al., 1974a). In contrast, the elements copper, zinc, arsenic, silver, mercury, cadmium, lead, selenium, and antimony are enriched in the aeolian particulates relative to average crustal material by one to four orders of magnitude. Some of these effects are of an anthropogenic origin, but it has also to be taken into account that many of the enriched elements are relatively volatile in their inorganic forms compared to the other metal compounds (Duce et al., 1974b).

A systematic investigation of lead contained in low level marine atmospheric particulates—collected some 15 m above sea level from both the Norther and Southern Hemisphere in the Eastern Atlantic—had led to the results contained in Table 17.

Table 17. Pb concentrations in atmospheric particulates of the Eastern Atlantic (after Chester and Stoner, 1973b)

Wind system	Pb concentration (ppm)	Enrichment factor (Fe = 1)	Potential particulate source area
Westerlies 50°N–30°N	1452 (720–2000)	113	Western Europe
NE-trade/Westerlies 30°N–20°N	384 (185– 685)	46	Western Europe, North Africa
NE-trades 20°N–5°N	140 (59– 238)	12	North Africa, West Africa
Intertrop. Converg. Zone 5°N–0°N	116 (86– 142)	10	Central Africa
SE-trades 0°S–30°S	269 (139– 490)	23	Angola, Southwest Africa
Off South African coast (variable w.)	695 (162–1450)	56	South Africa

It can be seen from an inspection of Table 17 that there is a marked decrease in the Pb content in the particulates southwards from the westerlies to the Inter-Tropical Convergence Zone (ITCZ) in the Northern Hemisphere, and an increase from the ITCZ to the zone of variable winds of the South African coast. This profile offers evidence of anthropogenic effects on the Pb concentrations of particulates from oceanic regions (Chester and Stoner, 1973b). The classic proof that airborne lead particulates may become transported over vast distances has been provided from concentrations of lead in the ice sheets from the North Pole and Greenland by Murozumi et al. (1969). Their graphical representation (Fig. 11) indicates that the concentrations have increased from less than 0.001 μg Pb/kg ice in 800 B.C. to more than 0.200 μg Pb/kg ice in modern times.

This finding is paralleled by analyses of mosses which are capable of concentrating airborne lead. In Sweden it was found that local mosses contained an average level of between 80 and 90 mg Pb/kg

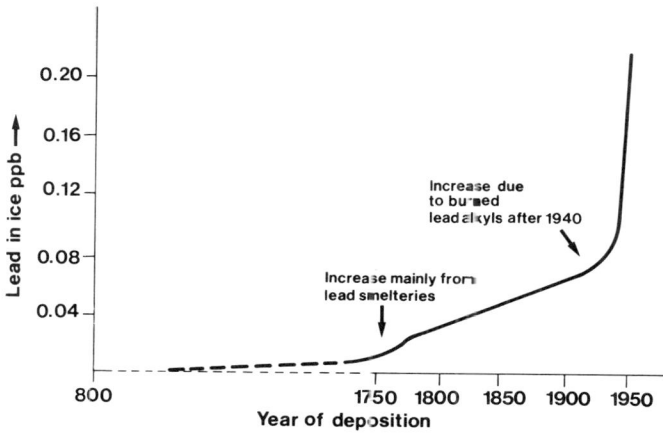

Fig. 11. Lead concentrations in Greenland ice (Murozumi et al., 1969)

in comparison to values in the region of 20 mg/kg for samples collected during the years 1860–1875 (Rühling and Tyler, 1968).

Piotrowicz et al. (1973) found a characteristic lead enrichment in the surface microlayer (~150 μ) of seawater samples, which they attributed to atmospheric fallout. If the observation of Garrett (1967) concerning organic enrichment in surface mono-molecular layers ($\sim 2 \times 10^{-3}$ μm) is also applicable to trace metals, the concentration of lead would then be approximately 10^5 to 10^6 times higher in this layer than in the sub-surface layer (Piotrowicz et al., 1973). The potential significance of the resulting high concentrations of heavy metals and other pollutants to biologic life in this restricted zone of the sea has been pointed out by Duce et al. (1972).

That atmospheric precipitations may lead to a considerable enrichment of heavy metals in the environment, cannot be disputed. A recent study revealed that 15%–36% of the Pb entering Lake Ontario from the Niagara River was attributable to atmospheric precipitations (Shiomi, 1973). Since the beginning of the industrial era there has been a considerable increase in the use and redistribution of lead in the environment. After the discovery in 1920 that alkylated lead compounds increase the antiknock qualities of gasoline (petrol) in spark ignition engines, the amount of lead additives today varies from 0.39 g/l (1.5 g/U.S. gal.) for regular to 0.55 g/l for premium (Engel et al., 1971). During combustion the lead additives undergo thermal and oxidative breakdown to form inorganic compounds such as PbClBr and PbO (Pierrard, 1969; Heichel and Hankin, 1972), most of which are discharged as exhaust gases. It has been estimated that two-thirds of these emissions have a particle size of less than 5 μm and readily form an aerosol which can be inhaled; approximately one-third is deposited in the surface close to the roads. In the Netherlands alone over one million kilograms of lead is exhausted to the atmosphere and redeposited annually (Rameau, 1972).

In the United States it was found that *lead in gasoline is the largest single source of air pollution*. The enormous increase in lead consumption by the gasoline industry is revealed by the figures pertaining to 1941 and 1970, viz. 45,350 and 252,600 metric tons respectively. Aerosols account for about one-third of the industrial lead added to the oceans, as shown by the estimates listed in Table 18 (from Patterson et al., 1976b). About 40,000 tons of these aerosols are added annually by dry deposition and washout from the atmosphere, while about 60,000 tons are added by rivers and sewers as storm runoff from paved surfaces which collect the aerosols on land (Patterson et al., 1976b).

Table 18. Approximate lead input for total oceans (in tons/year) (from Patterson et al., 1976b)

Industrial inputs	
Aerosols (gasoline)	37,000
Aerosols (smelters and forest fires)	3,000
Rivers and sewers (soluble, mainly from aerosols)	60,000
Rivers and sewers (solids)	200,000
Neolithic inputs	
Aerosols	1,000
Rivers (soluble)	13,000
Rivers (solids)	100,000

Approximately 75% of the lead issues from exhausts of engines, the remainder being incorporated in the oil, oil filters, engines, exhaust systems and silencers (Huntzicker et al. 1975; Goldberg, 1975a).

As early as 1963, Tatsumoto and Patterson (1963a) attributed high concentrations of lead in surface seawater of the Los Angeles Basin, as compared to deeper waters, to automotive aerosol fallout. Patterson (in Duce et al., 1974a) found that the atmospheric deposition of Pb into a 12,000 km^2 area of the Southern California Bight accounted for approximately 45% of the pollutant lead input, the remaining 55% arising from wastewater, stormwater runoff, and river input. In a study of trace metals on the New York Bight, Duce et al. (1975) estimated that approximately 13% of the input of Pb to a 10,000 km^2 area of the bight was from atmospheric fallout; the rest originated from barge dumping, runoff, sewage, and river input (Winchester and Duce, 1977).

Additional sources of atmospheric metal enrichment, such as the *high temperature anthropogenic sources,* are of special importance on a global scale. Examples to illustrate the annual metal emissions from the *combustion of fossil fuels* and the *production of cement* have been compiled in Table 19 (from Bertine and Goldberg, 1971; Goldberg, 1976).

By a simplified group assignment in the last column of Table 19, the emission characteristics of the listed elements become apparent. The elements of Group A cause atmospheric loads which are either irrelevant (Co, Cr) or of little importance (Cu) as compared to their natural occurrences; emissions from coal by far exceed those from oil-burning. Group B, consisting of nickel and vanadium, is representative of those elements that also do not cause high metal values in airborne particles; however, in contrast to Group A, exceptionally high metal constituents result from oil-burning (Ni and V porphyrins). Similarly, combustion of oil leads to exceptionally high mercury values (Group C), but the total anthropogenic contribution toward mercury enrichment observed in airborne particles is controversial. The elements (Cd, Se, As) corresponding to Group D are emitted in appreciable quantities by the combustion of coal and the production of cement, whereas oil-burning is responsible for a negligible proportion. It is noteworthy that the high temperature processes involved in cement production leads to atmospheric release of considerable amounts of arsenic and zinc—definitely

Table 19. Metal emissions from fossil-fuel burning

	Fossil-fuel mobilization[a]		Cement production[b]	
	Coal Tons/year	Oil Tons/year	Tons/year	Group
Cobalt	700	20	–	A
Chromium	1,400	50	–	A
Copper	2,100	23	–	A
Nickel	2,100	1,600	–	B
Vanadium	3,500	8,200	–	B
Mercury	400	1,600	100	C
Cadmium	140[c]	2	80	D
Selenium	420	30	700	D
Arsenic	5,000	10	3,200	D
Zinc	7,000	40	30,000	Burnt as alkyls:
Lead	3,500	50	30,000	300,000[d]

[a] Data from Bertine and Goldberg, 1971.
[b] Goldberg, 1976.
[c] Calculated for Illinois coal (Ruch et al., 1973).
[d] Estimate of Murozumi et al., 1969.

expected to enrich adjacent waters as fallout. Finally, in comparison to the emission (and subsequent re-entry) of lead alkyls, all other atmospheric sources of pollution are of less importance.

Despite the high efficiency of modern electrostatic precipitation, *lignite-burning power plants* are a typical source of trace metals. Heinrichs (1977) has shown that this is locally of particular interest in the lower Rhine district near Cologne, where about 90% of the total brown coal tonnage of the F.R.G. is fired in as few as six power plants.

Another source of atmospheric pollution has been described by Brumsack (1977). Here comparison of trace metal contents in plants and soils downwind and upwind from *brickworks* indicate characteristic anomalies for Tl, Bi, Hg, Pb, Cd, and Zn.

Fallout from fossil fuels and geothermal emissions are important sources of atmospheric pollution. Especially *coal-fired electric power plants* have recently come under scrutiny as potential sources of mercury and other heavy metal pollution. Since the amount of mercury in coal was not well known, 36 American coals were analyzed some years ago to establish their mercury content; it was found that the values were generally < 1 ppm. However, 11 samples contained mercury concentrations ranging from 1.5 to 33 ppm. Although the concentration of mercury in most coals appears small, coal is consumed at such an enormous rate that an estimated 3000 metric tons of mercury are released annually in the U.S.—a figure which is comparable to that emitted from industrial processes—far exceeding the upper limit of mercury released by weathering processes where 230 metric tons/annum are released (Joensuu, 1971).

Analyses of 23 samples of Illinois coal indicate a cadmium content ranging from less than 0.3 to 28 ppm; the maximum value of Zn was 3,100 ppm, whereas the average level of Hg was 3.3 ppm

(Gluskoter and Lindahl, 1973). The fallout of a major coal burner on the eastern shore of Lake Michigan was investigated (Klein and Russell, 1973). Soils in the surroundings of this plant are enriched in Ag, Cd, Co, Cr, Cu, Fe, Hg, Ni, Ti, and Zn. The enriched area covers some 300 km^2, although a precipitator having 90% efficiency for all metals–excluding Hg–is in operation.

Billings and Matson (1972) have estimated that one-fifth of the mercury discharged from the power plant enters nearby Lake Michigan.

In another investigation conducted at the Kineaid Power Plant–a part of the Lake Sangchris complex in central Illinois–it was revealed that 2.7 million metric tons of coal were burned by the power plant between September 1973 and August 1974. An estimated 97% of the mercury content, i.e., 54 kg, was vaporized and emitted to the atmosphere (Anderson and Smith, 1977).

Geothermal sources, such as volcanic eruptions, have caused significant atmospheric pollution. A report by Eshleman et al. (1971) on the volcanism on the island of Hawaii has substantiated that significant amounts of mercury are emitted during volcanic activity on the island. These results revealed that some 98% of the mercury issuing from the Hawaiian fumaroles is either in the form of gas or particulates < 0.3 μm in diameter. Despite military uses and industrial consumption of mercury, these authors suspect that volcanic activity is responsible for large quantities of mercury supplied to the environment.

Geothermal sources have been harnessed to supply electrical energy, by way of example the Warakei hydroelectric plant situated near the volcanic belt of North Island, New Zealand. The only significant environmental effects produced are due to arsenic and mercury enrichment. Sediments from the nearby Lake Aratiatia contain between 20 to 30 mg As/kg–the average for soils in nonthermal areas is 5 mg kg^{-1}–which points to a high level of arsenic being discharged by the geothermal plant. Similarly, evidence suggests that the Wairakei plant makes a contribution to the mercury contamination of the river (Axtmann, 1975).

Based on sample analyses from New Zealand's North Island, Weissberg and Zobel (1973) showed that discharges from natural hot springs or from drill holes producing hot water or steam for geothermal power may result in mercury pollution. Although extremely low concentrations of mercury have been discharged from natural geochemical sources over centuries, these sources are less localized and more difficult to control than industrialized sources of geothermal power.

Atmospheric pollution occurs during the metallurgical process known as *smelting,* a process whereby an ore or ore-concentrate is fused with suitable fluxes (i.e., material added to a furnace charge to combine with the gangue and form a fusible slag). Obviously, smelting operations cause the emission of particulates which often contain toxic constituents that precipitate in the environment. Despite the installation of high-intensity electrostatic precipitators and high refractory-lined stacks in accordance with modern technological advances, the problem of pollution and smelting operations has not yet been overcome.

This fact can be illustrated by several examples, such as the pollution caused by sulfur dioxide emissions in smelting operations conducted at the Sudbury Basin of Ontario, Canada. Here, on the site of one of the world's major ore deposits of nickel and copper, three smelters are currently in operation. One of these smelters alone (the Copper Cliff smelter) discharged approximately 2.45×10^9 kg of sulfur dioxide during 1970. Steps were taken to curtail the inevitable discharge of sulfur dioxide pollution by increasing the height of stacks, and by building production plants for sulfuric acid and sulfur. However, it was found that damage to vegetation continues to spread over greater areas (Hutchinson and Whitby, 1974).

Special Sources

Although the company concerned (INCO) and the Federal and Provincial Governments involved have individually determined the composition of stack emissions from the Copper Cliff smelter, extreme secrecy surrounds these analysis reports (Harvey, 1976). The best available data pertains to the 1971 emissions: 2000 tons nickel, 1800 tons copper, 155 tons zinc, 12 tons cadmium, 240 tons lead, 110 tons arsenic and 17 tons selenium were registered in that year (Steel Labour, Oct. 1973; lit. cit. Harvey, 1976). These substantial air-borne quantities are subsequently precipitated as atmospheric fallout onto the surrounding area and eventually become transported to the aquatic environment.

Smelting operations have long been known to contain toxic substances. During the late 18th and early 19th centuries industrial workers suffered from arsenic poisoning caused by dust emitted from smelting operations (Luh et al., 1973). Despite modern technological advances, smelting operations continue to contaminate the environment. This is illustrated by numerous examples, such as the large copper smelter at Tacoma, Washington, which has been revealed as a major anthropogenic source of arsenic and antimony to Puget Sound (Crecelius et al., 1975).

These authors have shown that Puget Sound, a remote estuary in north-western Washington State, receives considerable amounts of arsenic and antimony which are produced as by-products in the local smelter. Since almost all of the arsenic trioxide sold in the United States originates at the Tacoma smelter, the fact that it is produced merely as a by-product should not be misleading. Data compiled by Crecelius (cit. Crecelius et al., 1975) have revealed that arsenic and antimony are released to the environment in three different ways. The major emissions, however, are in the form of stack dust which enters the atmosphere at an estimated rate of 2×10^5 kg/yr of As_2O_3 and 2×10^4 kg/yr of antimony oxides.

Although there are no cadmium ores as such, cadmium is ubiquitous in the environment. Primarily, cadmium is closely associated with zinc ores. Industrial production of cadmium compounds in the United States began in 1907 (McCaull, 1971). The processing and refining of cadmium-bearing ores in zinc, lead, copper and cadmium industries released an estimated 0.95 million kg (2.1 million pounds) of cadmium into the air in 1968, constituting 45% of the total 2.1 million kg (4.6 million pounds) cadmium discharged during that year. Cadmium was lost to the atmosphere chiefly from smelting zinc concentrates to remove impurities. Emissions containing cadmium sulfate, cadmium sulfide and cadmium oxide are condensed to be collected as dust. However, the efficiency of dust collection varies from plant to plant and losses depend on the number of times the cadmium-containing fumes are recycled. Accordingly, although the cadmium content of concentrates processed in lead and copper smelter is low, losses are relatively high due to the number of times the fumes have to be recycled to increase the cadmium concentration to the stage where recovery is economical.

Sources of heavy metal pollution resulting from smelting operations are widely diversified. For instance, lead pollution has been localized to a battery smelter in western Canada (John, 1971); and a zinc smelting plant near Odda, Norway was found to discharge large quantities of solids and solutions to the nearby fjord; the respective amounts of copper, lead, and zinc are 300, 4500 and 6000 kg/day (Skei et al., 1973).

4.7 Special Sources

There are numerous other sources of heavy metal enrichment which cannot be explicitly subdivided into point and nonpoint sources. Such sources of "mixed" origin will simply be treated as "special" sources. For instance, corrosion of iron or steel is widespread and occurs when atmospheric oxidation leads to anodic oxidation i.e., the dissolution of iron in the presence of moisture, which acts as an electrolyte. Thus corrosion is increased in coastal areas where wind-blown salts (e.g., sodium chloride) enhance the conductivity of the electrolyte and facilitate the current flow between —what may be regarded as miniature voltaic cells—anodic dissolution (corrosion) and cathodic deposition (rust). The release of acidic gases such as SO_2, NO_2, HCl, etc., to the atmosphere results in acidic precipitations which subsequently act as electrolytes in electrochemical corrosion.

Corrosion may also result from chemical interaction with the materials concerned. It is known that the Cl^- ion has a catalytic corrosive effect on heat exchangers and cooling towers due to the formation of soluble iron complexes. In the past, chromate salts were used to control corrosion in cooling water systems and wastes from chrome-plating baths. These salts pose a particular problem as Cr(VI) is suspected of having carcinogenic effects, and it certainly interferes with waste water treatment. Onstott et al. (1973) treated cooling water blown down with electronically generated $Fe(OH)_2$ and quantitatively reduced Cr(VI) to Cr(III), and then precipitated the latter without pH adjustment. The effect of chromium from cooling tower drift onto vegetation was assessed at the Oak Ridge Gaseous Diffusion Plant using tobacco plants. Plants were harvested at 1-week intervals and it was found that an inverse relationship exists between folio chromium concentrations and distance from the towers. Associated with high chromium concentrations, it was established that a 75% inhibition of leaf growth had taken place (Parr et al., 1976). Recently it was found that 20 ppm sodium molybdate acts as an effective corrosion inhibitor for iron in typical cooling towers (Robitaille, 1976), providing the concentration of competing anions such as Cl^- and SO_4^{2-} is not high.

A study performed by Bellinger and Benham (1978) on the levels of metals in dockyard sediments and water deals particularly with the contributions from *ship-bottom paints*. Elevated amounts of copper and zinc in the harbor deposits taken from Brocklebank Dock at Liverpool, from Tilbury Dock, and from the Manchester Ship Canal adjacent to the dry docks of Manchester Marine Ltd., probably stem from antifouling paints; lead may be partly introduced into the harbor areas from anticorrosive and primer paints. Direct toxic effects are not thought to be important, since many of the dockyard organisms are components of the ship-fouling ecosystem, indicating "a toxic resistance of considerable economic importance".

Oil-drilling operations require chemicals which contain various trace elements and may thus present a potential pollution problem (Montalvo and McKown, 1975). However, this is a localized source of contamination, which is confined to the various chemical forms of trace elements and their location in sediments. Unstable species may become bioavailable, whereas stable forms, which become incorporated in the inner layers of clay materials, are essentially unavailable to biota. Further investigation is needed before ultimate conclusions can be drawn.

Investigations into the possible cause of mercury concentrations in livers of arctic and sub-arctic wading birds showing a 10- to 20-fold increase during winter i.e., during the period they are feeding in the Wash in eastern England, led to investigations of mercury contents of rivers draining this predominantly rural catchment area. Sites in the south-west of the Wash bulb-growing area were tested for possible mercury enrichment, since *bulb-dipping in mercurial fungicides* was thought to result in mercury pollution. Although no large-scale contamination was found, appreciable sources of mercury enrichment were found to occur in ditches from two bulb-dipping sites; the highest mercury residue in a sediment sample, wet mass, was found to contain 223 ppm (Moriarty and French, 1977).

Similarly, the *treatment of seed potatoes* with mercurial solution, to control a wide range of latent diseases and rots, has been carried out in Scotland for over two decades. In a study by Caines and Holden (1976), total mercury concentrations of up to 20 ppm and 12 ppm wet mass were found in the muscle tissue of brown trout and grayling, respectively. These high levels, indicative of a polluting source, were traced to a potato treatment plant, which operated by dipping washed potatoes into a bath containing methoxyethylmercuric chloride at a concentration of 120 ppm mercury. After three to four days, when the concentration fell to 80 ppm, the liquor was discarded. Detoxification attempts by sodium sulfide treatment did not meet with success, as appears from the high mercury level (40 ppm) in the plant effluent and a maximum concentration of 100 ppm dry mass in stream sediments. The revised treatment with ferrous sulfate, releasing only 1 ppm to the effluent, was adopted during 1973 and led to a striking reduction in the mercury levels in fish; in 1974 no trout or grayling was found to contain more than 0.4 ppm in muscle tissue. The mean values, however, were still significantly higher than the levels normally expected for these species (< 0.1 ppm). This is in accordance with the observation that large proportions of mercury lost by effluents are retained by sediments and released over a period of years (Grimstone, 1972).

4.8 Multi-Source Effects

In conclusion, it appears appropriate to take note of various authors who have expressed their view on the origin of metal pollution. With regard to *air pollution*, Friedländer (1973) maintains that irrespective of the source of pollution, whether natural or manmade, it can be established that a characteristic set of chemical elements are emitted in approximately fixed proportions. If the sources in a polluted region are known, it is sufficient to determine elemental concentrations at a given point and solving a set of simultaneous linear algebraic equations. The particulate matter was determined in the air of Pasadena and it was found that approximately 15% resulted from primary natural sources and 25% from primary man-made sources. Some 40% of the total results from atmospheric reactions with hydrocarbons which are converted from the gas phase to the particulate form. Collectively, approximately 70% of the total particulate burden can therefore be accounted for; the other 30% must be ascribed to particulate contamination of the aqueous environment.

Many of the elements included in this survey are associated with the natural background aerosol, for instance sodium, chlorine, silicon, and aluminum. Other species, termed "exotic", include lead, zinc, and barium, and are chiefly a result of man's activities.

A different approach toward the problem of metal enrichment has resulted in a comparison of heavy metal with the natural concentration of these elements in the different spheres (lithosphere, pedosphere, hydrosphere, and atmosphere). The ratio *metal consumption (in tons/a)* avg. metal content in a specific sphere (g/ton) is introduced as a measure of the *relative pollution potential* of each element in a certain sphere.

Inspection of Table 20 reveals that the most toxic elements, i.e., mercury, cadmium lead, and copper have become significantly enriched.

Table 20. World's consumption of heavy metals in 1968; metal contents in soils. Index of relative pollution potential for the pedosphere; and Technophility Index

	Consumption × 1000 t/y (Sames, 1971)	Soils (ppm) (Bowen, 1966)	Index of relative pollution potential (Förstner and Müller, 1973a)	Technophility Index (Nikiforova and Smirnova, 1975)
Iron	400,000	38,000	1	5.3×10^7
Manganese	9,200	850	1	5.0×10^7
Copper	6,400	20	30	1.1×10^9
Zinc	4,600	50	10	5.4×10^8
Lead	3,500	10	35	1.6×10^9
Chromium	1,700	100	2	2.0×10^8
Nickel	493	40	1	9.0×10^7
Tin	232	10	2	–
Cadmium	15	0.06	25	–
Mercury	10	0.03	30	1.5×10^9

A similar approach has been used by Nikiforova and Smirnova (1975), where, of the technologically active elements, the greatest danger for living organisms are metals with maximum technophility, which are capable of forming highly contrasting anomalies in the environment. Technogenic migration of a metal and the degree of its utilization in the noosphere are estimated through its *technophility index (TP)*, which represents a ratio of the annual output of a metal to its "Clarke" (mean concentration in the earth's crust).

The higher the TP of a metal the more intensively it is involved in the technogenic migration. Mercury and lead are characterized by a high TP value of 1.5 and 1.6×10^9 respectively (Table 20, last column). It has been shown by Nikiforova and Smirnova (1975) that the TP index is unstable with time and that each metal can further be characterized by its own rate of TP growth. For example, the TP of lead has grown two and a half times from the beginning of the century up to the present time and will have increased in the year 2000 by four and a half times.

A different approach is based on the *utilization of the individual metals* on a broad scale, thus eliminating localized conditions. Figure 12 serves as an example of two most toxic metals—mercury and cadmium—and is a reflection of United States consumption (1968; D'Itri, 1972; Anon. 1971a).

Estimates of environmental contamination resulting from the manufacture and use of the associated products have also been incorporated into this table and been subdivided into three categories: the largest single source of mercury pollution results from

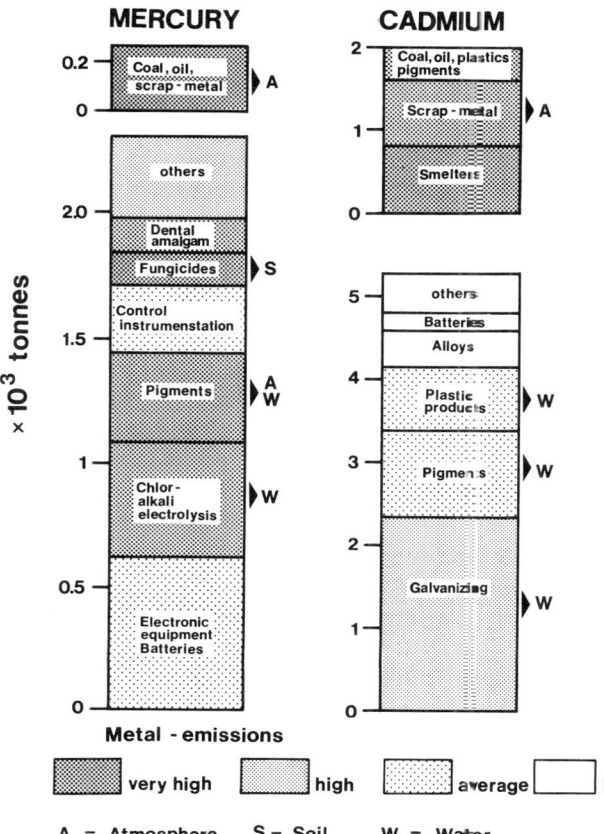

Fig. 12. Sources of mercury and cadmium emissions in the United States. Hg-values from d'Itri (1972), Cd-values from Environmental Science and Technology (Anon. 1971a)

the Hg-electrode process in the chlor-alkali industry; mercury-based fungicides formerly (1968) contributed extensively toward mercury in the environment, but the use of such products has since been curtailed; cadmium enrichment in waters mainly results from the galvanizing industry.

The production figures reflected in Figure 12 do not include metal contamination resulting from the recovery from ore and wastes: for instance, one-third of the cadmium utilized in such processes—augmented by combustion of cadmium-stabilized plastics and pigments—is emitted to the atmosphere.

A large percentage of metal enrichment in airborne particles, which eventually contaminate waters as fallout, stems from the combustion of fossil fuels and the production of cement (Sect. B.4.6)

A *regional investigation* concerning the origin of elevated zinc levels in surface waters was conducted by Schleichert and Hellmann (1973) for the *lower Rhine catchment*. Concentrations of solubilized zinc varied between 100 to 300 $\mu g/l$ at a flow rate of 1000–2000 m^3/s. These authors maintain that:
1. Zinc values of *uncontaminated surface waters* are in the vicinity of 10–15 $\mu g/l$.
2. Elevated Zn concentrations from 11–28 $\mu g/l$ result from *domestic effluents*. The

Zn content due to corrosion of the water network has been estimated to exceed that of human excretions by a threefold factor.
3. The influence of *contaminating rainwater* was not accurately determined in this study. Nevertheless, the average value of 100–200 μg/l Zn in rainwater indicates an important contributary factor.
4. Zinc enrichment above 50 μg/l may usually be traced to *industrial effluents* with a large measure of certainty. The main sources of supply are the galvanizing industry, viscose plants, candle and soap factories.

The different sources of metal pollution have also been investigated in a more confined regional area and in greater detail. These studies concern *New York City* and the adjacent coastal zone, and were conducted by numerous investigators, such as Gross et al., (1971), Klein et al. (1974), and Mueller et al. (1976). Prior to these investigations, city officials knew that the existing 250 electroplating firms discharged the following amounts of metal wastes to sewers daily: 227 kg Cu, 154 kg Cr(VI), 477 kg Ni, 304 kg Zn, and 30 kg Cd. Approximately 85% of these totals entered the sewage treatment plants, whereas the remainder was discharged to the harbor as wastewater. Although the electroplating industry is an acknowledged contributor of several metals, considerably larger quantities of metals are introduced from other sources (Klein et al., 1974).

The New York Bight Project was begun in 1973 to collect, compile, and eventually assess the data on contaminant discharges into the Bight, which is a natural resource of great economic and environmental value (Mueller et al., 1976). These authors have given a comparative evaluation of the four sources of metal inputs considered in this project: barge dumps, atmospheric fallout, runoff, and wastewater. The Bight proper receives direct discharges from the first two sources, whereas wastes from the latter two originate in the coastal zones (i.e., Long Island and New Jersey).

The metal input into the New York Bight has been estimated to result from the sources shown in Table 21.

In Figure 13, lead has been singled out to depict the relevant importance of the individual contributary sources. As illustrated, all four sources considered in the New York Bight Project are of importance: barge discharges (mainly dredge spoils, also

Table 21. Percentages of New York Bight metal loads by source (Mueller et al., 1976)

	Direct Bight		*Coastal zone*				
	Barge	Atmospheric	Wastewater		Runoff		
			Municipal	Industrial	Gauged	Urban	Groundwater
Cadmium	82	2	5	0.6	5	5	< 1
Chromium	50	1	22	0.8	10	16	< 1
Copper	51	3	11	9	10	16	< 1
Iron	79	3	5	0.5	6	6	< 1
Mercury	9	–	71	2	13	5	–
Lead	44	9	19	3	6	19	< 1
Zinc	29	18	8	2	21	22	< 1

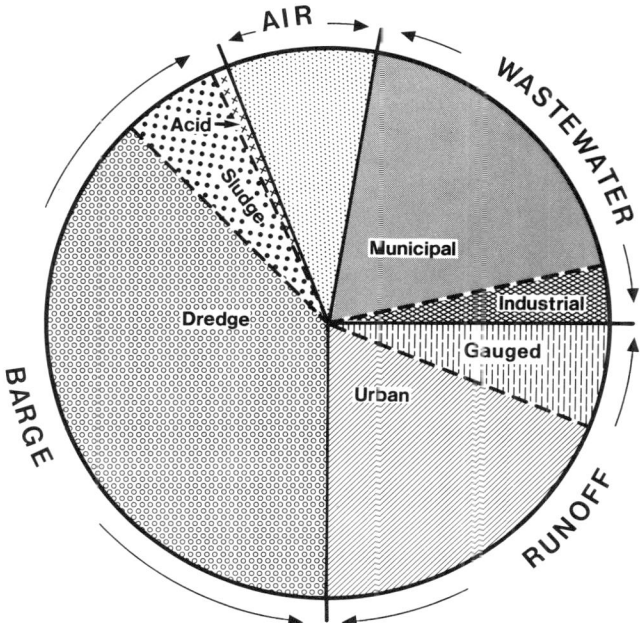

Fig. 13. Sources of lead in the New York Bight (Mueller et al., 1976; with permission of Water Pollution Control Federation)

containing sewage sludge and acid wastes) constitute the major source of lead contamination; both wastewater and runoff contribute significant fractions in accordance with the characteristics of domestic effluents and urban runoff (see Sect. B.4.4); the lowest supply source of lead results from atmospheric fallout, as may be expected (see Section B.4.6).

It has been estimated that annually some 4500 tons of lead are introduced into the Bight (Mueller et al., 1976). In view of this alarming quantity, it is of considerable interest to evaluate the possible ecological impact, especially since the largest fraction of lead results from dredged sediment disposal (see Table 21, column under "Barge"). In this context, Lee (1977a) has pointed out that the loads of contaminants entering the Bight cannot serve as a basis for judging potential water quality effects. A correct assessment of the effects of contaminants on the water quality must be based on the fraction of the pollutant from each source which becomes available within the receiving water body. This aspect is dealt with in greater detail elsewhere (Chaps. E and F).

5 Metal Analysis

The discovery of ubiquitous environmental contamination by mercury has caused much concern about the possible presence of other toxic elements in water, sediment, and biota. Concurrent development of new and improved analytic techniques have fortunately provided the means to study and monitor these elements at extremely low concentrations. In consequence, sophisticated methods of analysis are being employed to an ever-increasing extent in fields which appear to be widely diversified. For example, regional geochemical reconnaissance surveys are principally conducted in the search for metalliferous deposits. However, the data obtained also serve as background levels for the assessment of anthropogenic influences (Thornton and Webb, 1973).

5.1 Media of Pollution Assessment

Various media are analyzed to assess, monitor, and control metal pollution. The most obvious medium is *surface water*. However, it has been established that for a given sampling station monitored over a long period of time the metal values of the collected samples tend to vary by several orders of magnitude, although the samples may have been collected at short-time intervals. Such fluctuations are attributable to a large number of variables, such as daily and seasonal variations in water flow, surreptitious local discharges of effluent, changing pH and redox conditions, the input of treated secondary sewage, detergent levels, salinity and temperature.

Pollutant concentrations in particulate matter often provide a more stable and convenient means of obtaining an indication of the state of associated waters. In a number of polluted rivers in the United States (Kopp and Kroner, 1968) and Central Europe (De Groot et al., 1973b; Heinrichs, 1975), the amounts of heavy metals in water and particulates were determined for a certain specified period of time. From the ratios of heavy metals in water to those adsorbed to particulates, a sequence of "mobility" may be deduced, which is characterized by the examples in Table 22. Alkali and alkaline earth metals are predominantly present in a dissolved form and, therefore, highly mobile;

Table 22. Percentages in particulate-associated metals of total metal discharge (solid + aqueous phases) in polluted rivers in the U.S.A. and Europe

Metal (example)	U.S. Rivers[a]	F.R.G. Rivers[b]	Rhine (Neth.)[c]
Sodium	–	0.5%	–
Calcium	–	2.5%	–
Strontium	21%	–	–
Boron	30%	–	–
Cadmium	–	30%	45%
Zinc	40%	45%	37%
Copper	63%	55%	64%
Mercury	–	59%	56%
Chromium	76%	72%	70%
Lead	84%	79%	73%
Aluminum	98%	98%	–
Iron	98%	98%	–

[a] Kopp and Kroner, 1968, [b] Heinrichs, 1975, [c] De Groot et al., 1973.

trace metals like boron, zinc, and cadmium have ratios of dissolved species to particulate species of between 2:1 and 1:1; copper, mercury, chromium, and lead exhibit ratios of the aqueous phases to the solid phases between 1:2 and 1:4; iron, aluminum (and manganese under normal Eh conditions in rivers) are almost totally transported as solid particles. Sediment analyses, therefore, are particularly useful with respect to the less mobile elements, i.e., most of the heavy metals.

Although *sediment analyses* do not furnish quantitative data on the absolute degree of pollution, they can play a key role in ascertaining relative factors of enrichment whereby sources of pollution in the aquatic environment may be traced and monitored. These methods will be discussed in detail in Chap. D.

At this stage it is of interest to take note of the contributions made by the Applied Geochemistry Research Group, Imperial College, London. This body has intensively applied geochemical reconnaissance by analyzing more than 50,000 sediment samples taken from tributary drainage at road/stream intersections to compile geochemical atlases. These maps reveal the broad-scale distribution of some 20 elements for England, Wales, and Northern Ireland on an average of one sample per square mile (Thornton and Webb, 1973; Aston and Thornton, 1977). It has been shown by this group that regional geochemical mapping is of immediate practical value in the application to various environmental problems, such as those associated with agriculture, fisheries, pollution, public health, and sludge dumping.

The following two figures, depicting molybdenum and arsenic distribution patterns in the UK, serve to illustrate the aforementioned statements. Widespread patterns of excess molybdenum have been discerned and correlated with clinical hypocuprosis in cattle in some of these areas (Thornton and Webb, 1973).

Figure 14a shows the distribution of molybdenum in stream sediment in England and Wales. Seventy-seven percent of the animals tested in one of these areas were found to have low copper values and responded favorably to dietary copper supplementation, gaining 30–70 lbs (14–32 kg) per animal over a 6-month period (Thornton and Webb, 1973; Thornton, 1977a, b).

Figure 14b is an illustration of the arsenic distribution pattern and reflects the contamination from mining, urban and industrial sources (Aston and Thornton, 1975; Aston et al., 1975). On the basis of this and similar maps the authors have been able to highlight potential problem areas where anomalous high concentrations of metal enrichment may result in the deterioration of potable water quality.

Recent investigations by Aston and Thornton (1977) on the composition of stream sediments have been shown to provide a useful, stable indicator of associated waters. Bottom sediments collected from streams of both mineralized (mining) and unmineralized areas were sieved through mesh to obtain the < 200 μm fraction for analysis. Water samples collected from the sediment sampling sites were acidified to pH less than 2 and analyzed unfiltered. These authors have found temporal variations in stream sediments, which they ascribed to seasonal changes in bed loads resulting from (1) weathering fluctuations, and (2) associated changes in grain-size distribution. It is interesting to note that average metal concentrations in sediment fraction < 200 μm have been contrasted with maximum seasonal concentrations.

Inspection of Table 23 reveals that in waters of the mineralized tributary (A) of the Carnon River, the highest desirable levels (HDL) as recommended by the WHO (1971–see Sect. B.3) are exceeded for some elements. For instance, Fe (195 μg/l) is present in excess of the recommended value (100 μg/l); the Cu level shows a tenfold increase (515 μg/l as compared to 50 μg/l. On the basis of these and previous findings, the above-mentioned investigators have employed stream sediment composition to delineate the problem of potentially contaminated associated waters. The salient factors are the degree of "contrast" and the assignment of "threshold" values, contrast being defined as the highest anomalous concentration divided by the average background concentration. It was found that high contrast values are readily identified for contaminated areas but that "the most useful contrast value is that which relates to the HDL value for the particular trace element in water". This is closely related to the problem of threshold values for stream sediment

Fig. 14. Map showing the distribution of molybdenum (**a**) and arsenic (**b**) in stream sediments in England and Wales (compiled by the Applied Geochemistry Research Group, Imperial College, London, as part of the Geochemical Atlas of England and Wales, financed by the Wolfson Foundation)

composition i.e., "the concentration level in sediments at which it is likely that associated waters may, on occasion, exceed the HDL value" (Aston and Thornton, 1977). It is evident that due to considerable variations in the physical and chemical environments of the catchments, it is an extremely difficult problem to establish universally acceptable contrast and threshold levels for potentially toxic elements in bottom sediments of inland waters. However, a notable suggestion of tentative values for the thresholds of selected elements in stream sediments has been made and is given in Table 24.

Media of Pollution Assessment 65

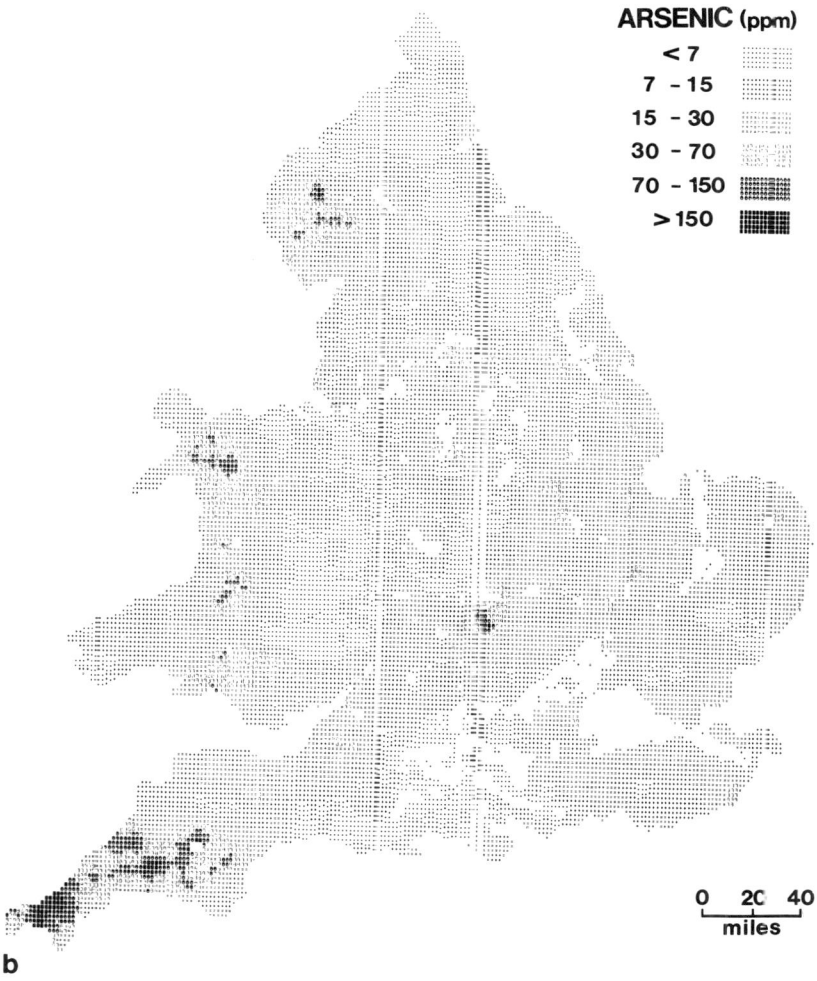

Fig. 14b

The fine-grained sedimentary deposits derived from suspended material in which the sorbed, co-precipitated, incorporated or otherwise bonded metal contents tend to accumulate, are particularly suited to these investigations. However, it is often overlooked that it is *imperative* to base such finding on a standardized procedure with regard to particle size, since "there is a marked decrease in the content of metals as sediment particle size increases" (Drifmeyer and Odum, 1975). Special emphasis on this aspect will be given in Chapter D.4.

Bio-monitoring of heavy metal pollution has been given ever-increasing attention due to the ability of various marine organisms to accumulate trace elements. On the basis

Table 23. Average stream sediment compositions and seasonal maximum concentrations of elements in associated water (after Aston and Thornton, 1977)

Element	Tributary A (mineralized/mined)		Tributary B (unmineralized)	
	Average conc. in sediment (ppm)	Max. conc. in water (μg/l)	Average conc. in sediment (ppm)	Max. conc. in water (μg/l)
Fe	6.7%	195	3.0%	26
Mn	727	280	490	9
Cu	3245	515	89	8
Pb	291	4	89	1.5
Zn	2750	2000	320	12
Cd	3	5	2	1

Table 24. Tentative threshold values in stream sediments in comparison to the highest desirable levels in associated soft water bodies (after Aston and Thornton, 1977)

Element	HDL value (μg/l)	Tentative threshold value (ppm)
Fe	100	6% (w/w)
Mn	500	1000
Cu	50	1000
Pb	50	500
Zn	5000	2000
Cd	10	10
As	50	100

that molluscs are well-known concentrators of heavy metals, Goldberg (1975a) proposed a global mussel watch utilizing *Mytilus edulis* and other similar species as a method of assessing marine pollution. The International Environmental Program Commitee (IEPC) has endorsed the practical program of "Mussel Watch", which would entail the analysis of specimens from some 100 open ocean and coastal sites worldwide. Although the general merit of this proposal has been acknowledged, the immediate task of getting some global monitoring has not yet been tackled.

Bio-monitoring has of late been strongly recommended in connection with *algae*. Since analysis of *algae* can only reflect accumulations that occurred during a limited period of growth—less than one year—it would appear that they are potentially suitable indicators of localized, short-term pollution effects i.e., secondary effects in contrast to sediments, since the latter are long-term indicators of more primary effects and water analysis, which may be subjected to large fluctuations.

Several investigators have proposed the use of *aquatic microphytes* to monitor heavy metal pollution in freshwaters. Mayes and McIntosh (1975) found that concentrations of Cd and Zn in coontail (Ceratophyllum demersum) reach a steady state at six to eight weeks. Maximum concentrations of 20 ppm Cd (dry mass) and 260 ppm Pb were observed in a highly contaminated lake. Again the question arises whether such results can be regarded as a reliable pollution indicator. Shoals of coontail from an uncontaminated control pond were placed in enclosures and Cd and Pb accumulations were monitored over a period of 12 weeks. The accumulation of Cd and Pb was found to reach a maximum after six to nine weeks and subsequently found to decrease by the 12th week.

Since elemental mercury has a low solubility, the levels found in surface waters seldom exceed 1.0 μg/l, irrespective of the mercury input. In consequence, water analysis is a poor indicator of mercury pollution. Although a large percentage of mercury is associated particulate matter, bottom sediments contain the largest quantities of mercury in water bodies, as might be expected (Williams and Coffee, 1975). The largest concentrations occur at the sediment/water interface in the uppermost layers of ooze. Under conditions of high velocities, there are no ooze deposits, whereas in areas of reduced velocities they occur but are subjected to variations in current velocities along the bottom, leading to scour deposits. Thus ooze sampling is impractical as an indicator of mercury pollution because of irregularities in ooze accumulations. This has been clearly demonstrated by Williams and Coffee (1975) using the technique of packaging organic material in perforated polythene bags for determining trace levels of radionuclides. Generally, this technique is used for the determination of the amount of mercury and tends to simulate its natural concentration by the metabolism of microorganisms. In consequence, different microbial communities known to concentrate mercury from the aquatic environment were chosen from three different aquatic habitats. Commercial dogfood containing mercury background levels of 24 μg/kg (on a wet basis) was used in a dry, ground form. This substance was chosen as the organic substrate for concentrating mercury, due to its high organic content (93%), ease of packaging, and low cost.

Trees have also been used in an attempt to monitor the environmental pollution of lead and to distinguish between the different sources or origins by determining lead isotopic ratios. These studies are based on the fact that lead-204 is not known to have been formed by a radioactive decay process, whereas lead-206 and other lead isotopes are formed by the decay of uranium and thorium.

Following the above-mentioned studies, pine trees growing along the banks of the Spokane River, Idaho have been used to monitor the past history of pollution with regard to Hg, Cr, Ag, Rb, Zn, Co, and Fe. The metal content of tree rings and sediment core analysis roughly agree (Sheppard and Funk, 1975).

It appears that in various modifications the method of isotopic ratios can be used to solve environmental problems with regard to lead pollution (Ault et al., 1970; Gast, 1970; Holtzmann, 1970).

Chow et al. (1973) have determined the rates of lead accumulation in the deposits off the coast of California by radiometric means. Increased rates of lead accumulation became evident in the late 1940s, approximately 25 years after the introduction of lead alkyls into gasoline. Much higher values of lead were found in sediments immediately off a sewer outfall (Whites Point), which suggests that there is an additional input of lead from stormwater runoff and from industrial wastes introduced by rivers. The lead in gasoline additives is usually derived from tertiary or older lead ores and has a quite distinctive isotopic composition, provided it is derived from the same source of supply, which depends on localized conditions. The Whites Point sediment adjacent to a sewer outfall appear to contain lead primarily derived from gasolines;

similarly, recently deposited sediments of the San Pedro Basin appear to have been influenced by gasoline lead.

5.2 Sampling and Analytic Methods

5.2.1 Sampling

Sampling of water, sediment or biota is a difficult and complex problem. The experimental approach involves field studies which have to be based on prior evidence to justify the desirability of incurring financial expense and expenditure of time and effort. It is imperative that experimental studies are carefully designed in order to evaluate the resulting data objectively. Accordingly, the choice of sampling sites is of utmost importance in order to avoid trivial results. Sampling methods and techniques have been described in detail in many texts (e.g. Maienthal and Becker, 1976: „A survey on current literature on sampling, sample handling for environmental materials and long-term storage." Interface 5, 49–62; 196 references).

The technique of sampling sediments by means of a grab or corer and the subsequent separation of the pelitic fraction < 2 μm or other grain-size fractions, digestion and analysis have been explained in Chapter D.

For metal analysis of river water it is convenient to investigate combined samples (daily, weekly, monthly) composed of aliquots in proportion to the water discharge or to dissolved salt mass transport (e.g., Gibbs, 1977). For the calculation of mass transport of trace elements in rivers, consideration of the respective water discharge is absolutely necessary.

5.2.2 Analytic Methods

Reliable and sensitive analytic methods have an important role in determining and combatting the environmental impact of metal pollution. The elements monitored are not only those that are toxic towards humans in trace amounts—such as mercury, cadmium, lead, and arsenic—but include the wider spectrum which are toxic toward certain animal species or which become enriched in the food chain.

During the past years the need for determining trace amounts of metals has resulted in an increased commercial availability of analytic instrumentation; the current trend is displayed by an emphasis on both diversification of instrumental techniques and the degree of sophistication of the equipment. [The latest literature is reviewed in the July, 1977 issue of Journal Water Pollution Control Federation, in the section entitled Nature and Analysis of Chemical Species under the heading Trace Organics (up to 1974) or Inorganics (from 1975)].

It is realized that no single analytic technique can be used for all analyses. Some of the requirements for an analytic method to be acceptable are (1) sensitivity, specificity, and accuracy; (2) rapidity of analysis and ease of operation; (3) the possibility of automation; (4) low cost of the equipment, and (5) reliability of results, i.e., freedom of interface effects (Norval and Butler, 1974). Figure 15 gives a summary (from Tölg, 1973) of some important analytic techniques and their respective detection limits under optimal conditions.

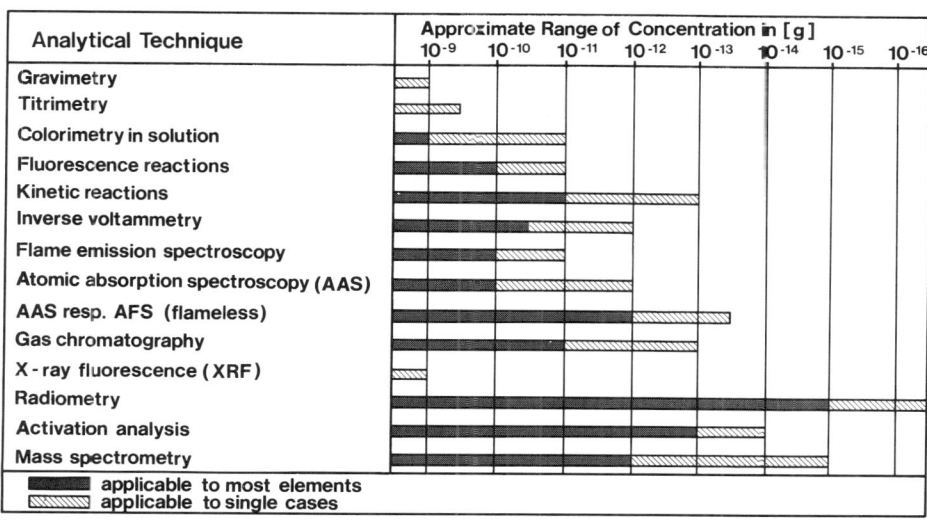

Fig. 15. Methods of trace element analysis (after Tölg, 1973)

The atomic absorption method (AA) enables metallic elements to be determined with remarkable sensitivity and accuracy and is currently the most prevalent method of metal analysis of water and effluents (Kopp, 1977). Although AA is a precise technique and easy to operate, this does not imply a guarantee for the validity of the data obtained. Especially when the concentration of metals in complex wastes is high, the absorbance of the trace metals being determined is affected. To some extent compensation can be achieved by preparing standards to match the matrix of the samples.

Although AA is a very sensitive technique, with detection limits in the $\mu g/dm^3$ range, flame interference at extremely low metal concentrations has been the major source of problems. Most manufacturers have recently developed systems to overcome this problem by the introduction of nonflame atomizers, such as the graphite furnace. Atomization is effected by very rapidly heating each sample aliquot to a high temperature, necessitating small volume samples, which is accomplished in situ Burrell, 1972). When the flameless technique is used, the detection limits are usually enhanced by a factor of 10^3. However, this technique does not replace the conventional flame method, but should be regarded as a supplement for special studies. Being quicker and more precise, the conventional flame technique is given preference for routine analyses and automation (Norval and Butler, 1974).

Emission spectroscopy has fallen into a state of neglect because of the overwhelming success of AA. Whereas in AA the signal depends on the number of unexcited atoms in the flame at a given moment, in flame emission the signal arises from excited atoms. The most promising development in the field of emission spectroscopy is the new source of excitation known as a plasma i.e., an electrically generated luminous gas containing a significant fraction of ionized atoms or molecules in inert gas. The ionized particles may interact with magnetic fields and thus acquire sufficient energy to excite the atomic or molecular species under investigation.

Recently, commercial high frequency plasma sources coupled with direct reading spectrometers have been introduced for the purpose of multi-element analysis (Kopp, 1977). The inductively coupled plasma (ICP) derives its sustaining power by induction from high-frequency magnetic fields. It has been reported that the ICP has virtually ideal characteristics for simultaneous multi-element determinations, because one set of experimental parameters is essentially optimal for the determination of all metals and metalloids. However, from a commercial standpoint, the costs involved in acquiring the equipment are prohibitive as compared to AA.

X-ray fluorescence spectrometry has undergone a dramatic change in interest due to the introduction of energy dispersive X-ray fluorescence into instrumentation. Efficiency is based on the resolution of the detector. Solid state lithium–silicon detectors provide a sufficient resolution provided they are maintained below $-160\,°C$. An interesting discussion on X-ray methods for environmental water analysis has been presented by Leyden (1974).

Neutron activation analysis provides a powerful analytic technique for the detection of trace elements. Although one of the most sensitive methods available for the accurate and precise determination of many trace elements under optimum conditions (Meinke, 1973), the costs, availability, and range of elements to be analyzed places a severe restriction on the usefulness of this technique.

Electroanalytic methods such as polarography have been important analytic tools but are usually effective only within a limited range of concentrations. To improve the usefulness of conventional polarography, several modifications have been introduced, such as pulse polarography. However, the usefulness of voltammetry lies in its ability to differentiate oxidation states rather than as a tool for the determination of metal concentrations (Kopp, 1977).

Mass spectroscopy applied in the field of water analysis has inherent drawbacks. Spark source mass spectroscopy will perhaps find a greater field of application when applied to solid samples such as sediments, soils, and sludges (Kopp, 1977).

Chapter C
Metal Concentrations in River, Lake, and Ocean Waters

According to rough estimates 875 km^3 of water evaporate daily from the oceans. After condensation, approx. 775 km^3 of this total is transported from the oceans to the continents by wind action where is becomes deposited as rain, snow, mist or dew together with approx. 160 km^3 of condensate from continental sources. The most important part of this cycle occurs during transport from land to sea via ground and river waters: these 100 km^3 per day constitute the source of our available freshwater supply. However, freshwater from drainage runoff fluctuates greatly according to region. The Amazon and Orinoco Rivers, for example, carry nearly one-quarter of the total surface waters practically unused to the sea, while in other parts of the world water shortages due to climatic and geologic conditions or factors such as overpopulation, concentration of industries, etc., may prevail. Of groundwater reserves, which constitute 0.6% of the total global water reserves, little more than one-tenth to one-fifth can be exonomically exploited. Thus, at the utmost, only one-third of the total fresh water reserves can be utilized at the present stage of development; of this, only a limited amount is potable due to the heavy salt load from natural or industrial sources.

1 Distribution of Major Ions

1.1 Natural Salt Concentrations

The chemical composition of inland waters results from different environmental factors operating simultaneously but with differing influences and efficiency (Gorham, 1961): the soil and rock composition (chemical behavior, solubility and secondary factors), climatic conditions (rainfall, temperature), morphology, fauna and flora, the time factor —and to an increasing extent—anthropogenic influences.

The most important determining factor for the natural development of natural inland waters is the amount of rainfall ("precipitation") and the chemical composition of the rainwater.

Characteristic developmental trends of natural waters can be deduced from the analytic data for inland waters which were gathered by Livingstone (1963) from various different climatic zones around the world over a period of several decades (see Fig. 16).

The salt content in rivers varies between 10 mg/l and approximately 2000 mg/l; levels between 100 mg/l and 200 mg/l are about average. Lakes, however, show wider distribution of salinity than rivers: freshwater lakes, generally with an outflow and thus similar to dams, contain up to approximately 300 mg/l of dissolved salts. Higher salt concentrations—as high as 700 grams per litre of water—are found in basin lakes with low precipitation but high rates of evaporation. The inter-relationship of calcium,

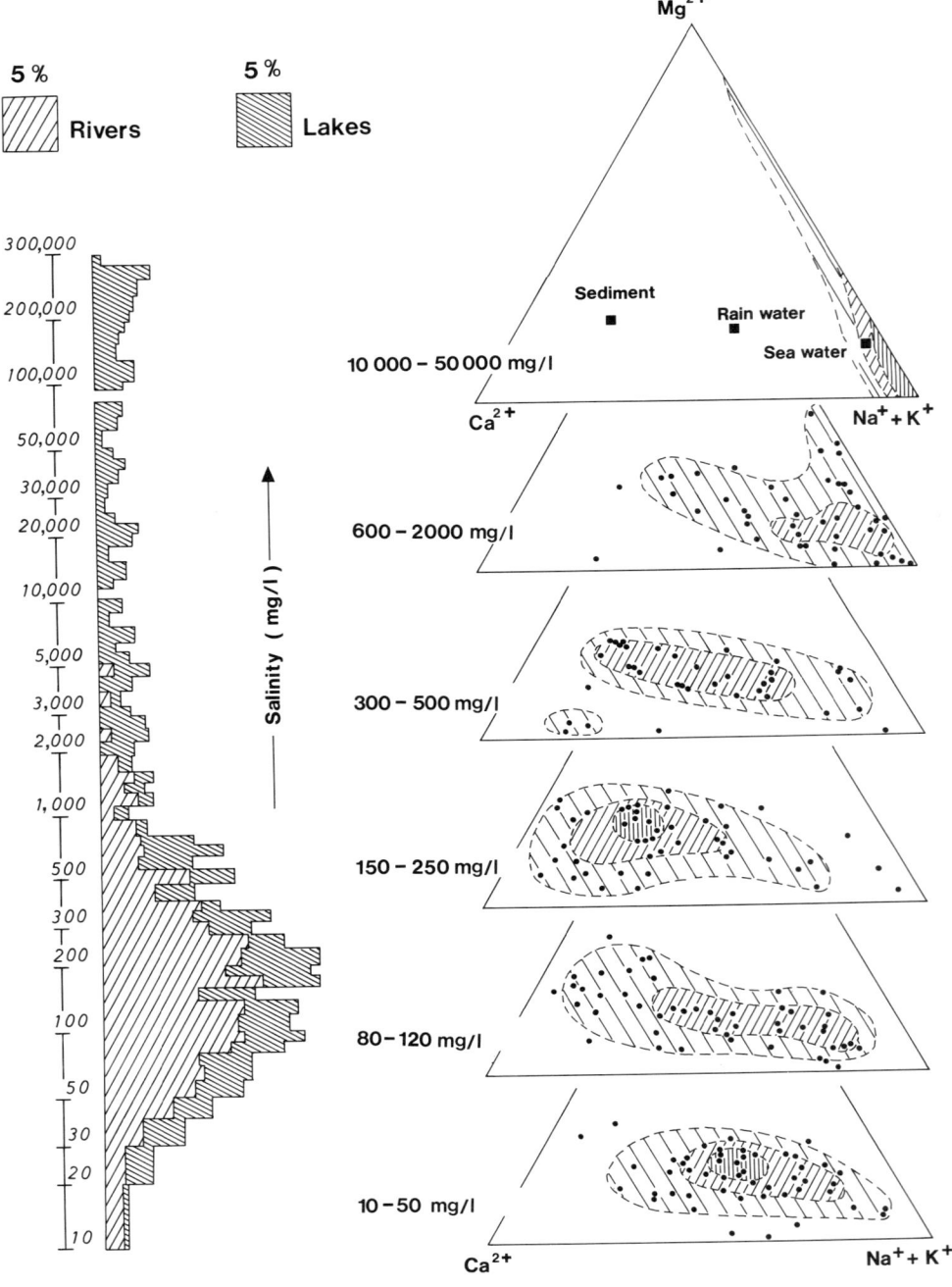

Fig. 16. Salt concentrations and distribution of major cations in continental waters (Förstner, 1973, after data from Livingstone, 1963)

magnesium and alkali metal ions in inland waters for different stages of salinity is illustrated in Figure 16.

In river waters with an extremely low salt content—encountered only in areas of high rainfall—the distribution maximums of the chief cations correspond exactly to the average rainwater composition (Wedepohl, 1969). With increasing salinity, the cation ratios change in the direction of the average chemical properties displayed by sedimentary rocks (Poldervaart, 1955)—a result of intensive interaction between water and underlying formations. In the more arid areas, the process of salt enrichment in inland waters is ultimately carried further by the high rate of evaporation; precipitation reactions and sorption processes, especially carbonate precipitation (Gibbs, 1970a, 1971), play an increasingly important role and lead to an advanced differentiation of the solution. Finally, the extremely saline waters of arid lakes become predominantly enriched by sodium chloride.

1.2 Man-Made Contamination

The natural development of salt contents in freshwater is increasingly affected by anthropogenic influences—in some areas even in such a way that the natural chemical form is no longer recognizable. The following examples point out this phenomenon.

Of the largest lakes in North America, Lake Erie and Lake Ontario appear to be most affected by extensive changes in the salt load. On comparing concentrations of different chemicals reported by various investigators, Beeton (In: Vallentyne, 1974) was able to show that concentrations of sodium, chloride, and sulfate in Lake Erie and Lake Ontario had more than doubled since 1910. In Lake Erie the changes were more pronounced than in the other Great Lakes because the water in this lake is relatively shallow and more than 12 million people live within the lake's drainage basin.

Figure 17 shows that these changes in chemistry were to a considerable extent passed on to Lake Ontario. Still more pronounced are the changes occurring in the salt content of several rivers in central Europe. Water analyses made in the past century were compared with newer test results and indicate an unhealthy development for our rivers and lakes (Table 25). In the example of the Danube, chloride, sulfate, bicarbonate, sodium, calcium and magnesium ions have practically not changed in the duration of the investigations. In contrast, the salt content of the Rhine has increased approximately three times, that of the Weser as much as eight times in the period from 1887/1893 to 1971. The changes occurred in the sodium chloride components, and originate in both cases from the waste water output of potash mills. The salt load in the Rhine River—about 10 million tons yearly (Hinrich, 1972)—derived to almost 50% from Alsatia potash mills. In the case of the Thuringian influents, the water quality in the lower section of the Weser has decreased to such an extent that an attempt was made to establish marine fish in colonies to replace the normal freshwater fish.

2 Chemical Conditions for Trace Metals in Natural Waters

The field of trace element analysis in natural waters is still very much in a stage of development. As simple and inexpensive equipment became available, the measuring

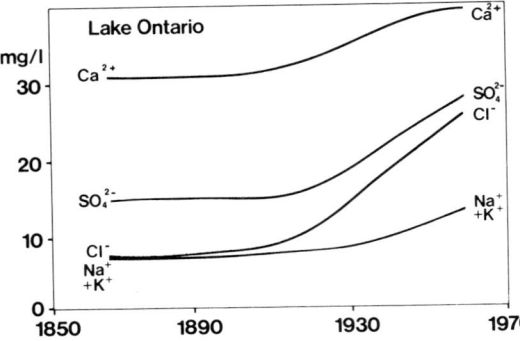

Fig. 17. Changes of the chemistry of Lake Ontario waters from 1850 to 1967 (after Vallentyne, 1974: *The Algal Bowl*, reproduced with permission of the Department of the Environment, Fisheries and Marine Service, Ottawa, Canada)

Table 25. Concentration of major ions in Danube, Rhine and Weser river waters

	Danube (Regensburg)		Rhine (Mainz)		Weser (Bremen)	
	1893[a]	1971	1887[b]	1971	1893[c]	1971
HCO_3^-	232	204	146	153	124	168
SO_4^{2-}	15	35	20	146	64	235
Cl^-	3	18	7	159	49	1233
Ca^{2+}	58	43	57	65	52	56
Mg^{2+}	14	14	7	16	9	151
Na^+	3	8	4	94	29	574
K^+	2	3	2	11	5	42
Total mg/l	327	325	243	644	332	2459

[a] Schwager. [b] Egger. [c] Seyfert from Livingstone (1963).

of trace metals became popular in the scientific world. However, it soon became apparent that because of contamination of samples incurred during handling and sampling, the results must be regarded as of doubtful validity. It is therefore highly probable that over the next years, new revised data on most of the trace metals will be made public. Especially in regard to metal concentrations in seawater, the first promising investigations are now available showing the general development trends and the important factors of influence (Sect. 3). The case of inland waterways is more difficult, but even here decisive improvements can be expected (Sect. 4). Concerning the complex problem of seawater analysis, Boyle et al. (1977) validly summarized that "a priori arguments on the extent of precautions in sampling and analysis are not in themselves sufficient, nor is the fact that the numbers are lower than previously reported. The primary criteria must be interlaboratory agreement and the oceanographic consistency of the data themselves. Regional variations should be compatible with what is known of the large-scale physical and chemical circulation of the oceans."

2.1 Chemical Speciation in Freshwater and Seawater

The knowledge that not the concentration of a metal, but its chemical behavior in specific surroundings—especially chemical speciation and the reactions involved in the transformation of species—is often the main factor by which pollutants affect the aquatic environment, is very important for the analyses of trace metals in respect to environmental questions.

Studies on the toxicity of copper (Steeman-Nielsen and Wium Anderson, 1970; Davey et al., 1974) have revealed that the toxicity of this trace element is largely dependent on the formation of organic copper complexes. Barber (1972, cit. Morel et al., 1973) noted a strong response of the system to rather small additions of chelators or metals in phytoplankton experiments. Hutchinson and Stokes (1975) found in lake waters that the toxicity of copper for algae is not dependent on the overall copper concentration, but rather on its mere presence in even a small quantity. This is confirmed by experiments of Jackson and Morgan (1975), which also indicated that free ion activity is a good, but not necessarily the only, indicator of copper toxicity to marine phytoplankton (Sibley and Morgan, 1975). It has been demonstrated that organic ligands can inhibit the uptake of essential metals (Manahan and Smith, 1973; Brown et al., 1974) or may raise the toxic threshold (Andrew, 1976, Davies et al., 1975; Zitko, 1976; Mancy and Allen, 1977). The presence of organic compounds, such as fulvic acids, NTA, and EDTA in the water generally reduces the toxicity of heavy metals, presumably by complexing the free metal ions (Jenne and Luoma, 1975; Chynoweth et al., 1976, Hart and Davies, 1978).

Special problems arise in analysis of chemical speciation: *natural samples* are often contaminated—or original conditions change, *laboratory experiments* are not conducted at natural levels of trace metals, and *theoretical models* have no sufficient basis for influencing parameters.

2.1.1 Analysis of Trace Metal Speciation

According to Guy and Chakrabarti (1977) three major questions must be answered in regard to the speciation of metal ions:
1. Is the soluble metal present as a complexed species or as a simple aquated ion?
2. Is the species charged?
3. What is the size of the metal species?

Two analytic techniques have in particular been applied in metal ion speciation: anodic stripping voltametry (ASV) and ultrafiltration. The first procedure divides the metal species into two categories: electroactive (aquo ions and "labile" complexes) and electroinactive (organic complexes and colloidal species). Ultrafiltration and dialysis is used to divide the metal species into different sized fractions. The species that pass through the smallest pore size are generally taken to be free metal ions or small complexes.

Analytic work has been performed with ASV, for example, by Matson (1968), Stiff (1971), Zirino and Healy (1972), Fukai (1973), Chau and Lum-Shue Chan (1974), Gardiner and Stiff (1975) and Bilinski et al. (1976). Ultrafiltration for differentiation of metal species has been employed by Gjessing (1970) and Schindler et al. (1972); dialysis procedures have been used, for example, by Benes and Steinnes (1974).

An analytic scheme for chemical speciation in natural waters was presented by Florence and Batley (1977). Of the eight measurements (Fig. 18), seven groups of species can be quantified and are described by the authors as follows:

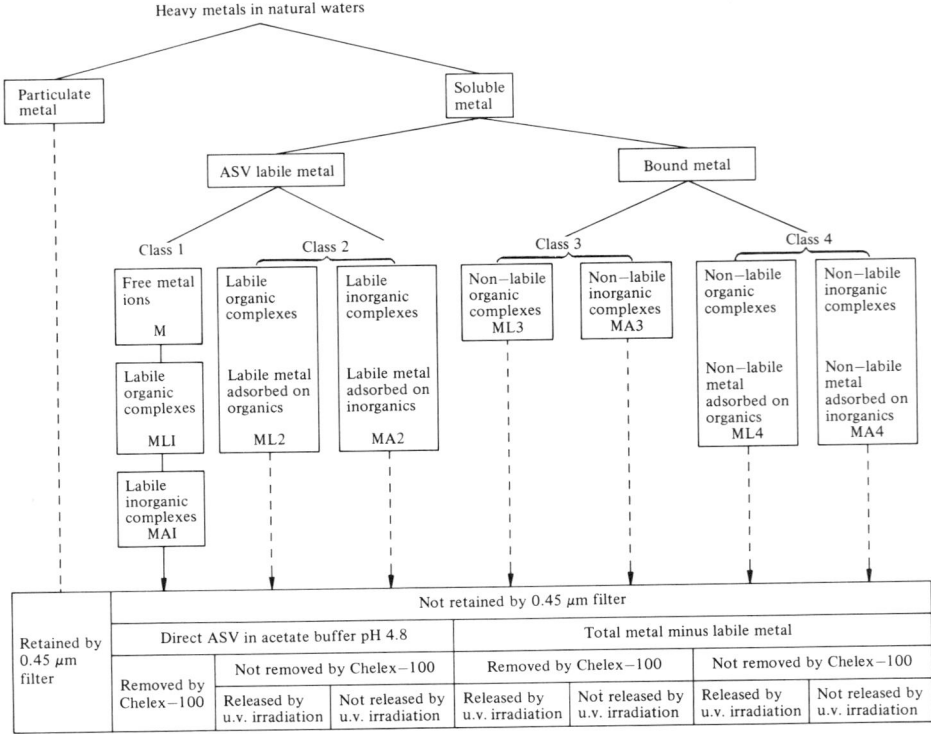

Fig. 18. Analytical scheme for the differentiation of chemical species of trace metals in natural waters (Florence and Batley, 1977; Talanta 24, p. 155; reproduced with permission of Pergamon Press, Oxford)

"$M + ML1 + MA1$. In seawater at neutral pH, these groups will consist principally of simple inorganic complexes such as chloride, sulphate, carbonate, and hydroxide. In freshwater, citrate and amino acid complexes may be present.

$MA2, MA4, ML2, ML4$. Since significant concentrations of metal complexes, stronger than the corresponding metal-Chelex-100 complexes, are unlikely to be present in seawater, these species would be mainly metal adsorbed on, or occluded in, organic and inorganic colloidal particles. The situation may be different in freshwater, which generally has a higher organic content, and hence possibly higher concentrations of strongly chelating organic ligands such as humic and fulvic acids.

$MA3, ML3$. These species may include some metal dissociated from colloids and retained by the resin. Also included are some of the humic and fulvic acid complexes, which should be at least partially dissociated by Chelex-100 resin."

Hart and Davies (1977) have proposed a pre-analytic scheme to determine (1) the total filtrable metal fraction, (2) the ultrafiltrable fractions, and (3) the ion-exchangeable fraction. The total filtrable fraction (M_{tf}) includes trace metals in a free ionic form as inorganic or organic complexes as in, or associated with, colloidal forms. The "ultrafiltrable" fraction (M_{uf}) includes all trace metals existing as free ions or as complexes, provided they are smaller than ~2 nm in diameter, and excludes the greater proportion of the colloidal iron and manganese oxides and perhaps metal-humic acid complexes as well. The ion-exchangeable metal fraction (M_{ie}) retained by the Chelex-100 resin includes the free ionic forms of the trace metal, and that trace metal associated with any labile complexes whose stability constant is less than that resulting from the combination of the trace metal with the Chelex resin.

According to Hart and Davies (1977) three physicochemical forms of trace metal associations in the filtrable fraction can be broadly characterized:

A—labile forms, determined by M_{ie}/M_{tf}

B—strongly bound organic or inorganic forms, determined by $\dfrac{M_{tf}-M_{ie}}{M_{tf}}$

C—trace metals as, or associated with, colloidal species: $\dfrac{M_{tf}-M_{uf}}{M_{tf}}$

The following results were reported for the chemical forms of Fe, Cd, Cu, Pb, and Zn in a sediment interstitial water sample and a sample of the overlying water from an urban creek in Melbourne, Australia:

Iron was found to be considerably more concentrated in the interstitial water (8,800 µg/l) than in the bulk water (60 µg/l); 97% of this metal was in colloidal form. *Cadmium* in the bulk sample was predominantly in labile forms (~70%), while the major proportion of Cd in the interstitial water was associated with the colloidal fraction (~80%). Similarly, ~90% of the *copper* in the bulk water was present in labile form, while in the interstitial water copper was almost evenly distributed between labile species (~40%) and more strongly bound complexes (~45%). *Zinc* and *lead* in the bulk samples were mainly present in a labile form; in the interstitial water the major proportion (~60%) was associated with the colloidal phase, 20% to 25% in labile forms and 10% to 15% in more tightly bound forms. These differences in the interstititial and bulk waters were probably due to the very great amount of colloidal iron oxide present in the interstitial water (Hart and Davies, 1977).

It should be noted that the usefulness of some of these differentiations is doubtful at this time, as the sum of the concentrations of the individual species often exceeds the total metal concentration determined for similar samples by standardization laboratories, which use ultra-clean contamination-controlled analytical techniques (Schaule, pers. comm.). At the same time, however, the separation of more or less labile metal species (e.g., by filtration, ion exchange and subsequent determination with AAS) will provide significantly more information on physico-chemical behavior, bioavailability, and toxicity of a metal than is obtained from bulk analysis of water samples.

Another approach to the question of metal speciation is the computation of equilibrium models on the basis of available thermodynamic information. In his classic paper on *The Physical Chemistry of Seawater,* Sillen (1961) calculated the principal solid phases and aqueous species of major cations, ligands, and several trace metals. The effect of pH changes on the speciation of zinc, copper, cadmium, and lead in seawater was computed from equilibrium programs by Zirino and Yamamoto (1972). Morel and Morgan (1972) described a numerical method for computing equilibria in aqueous systems and applied it to compute the equilibrium distribution for a hypothetical model system of 20 metals and 31 ligands; the system involves over 700 complexes and over 80 possible solids. Morel et al. (1973), Dyrrsen and Wedborg (1974), Ahrland (1975), and Stumm and Brauner (1975) have reviewed the available information of metal speciation in seawater. From the work of Morel et al. (1973) some common features of the interactions between trace metals and inorganic ligands in the pH-range of natural water between 5 and 9.5, as computed from model equilibrium conditions, are summarized in Table 26. The species are listed in accordance with increasing pH values.

Table 26. Predominant trace metal species under all conditions of computations in the models of Morel et al. (1973). The species of each metal are listed according to pH

	Species accounting for more than 90%	... for a few percent
Ag	Ag^+, $AgCl$, $Ag_2S_{(s)}$	
Cd	Cd^{2+}, $CdCO_{3(s)}$, $Cd(OH)_{2(s)}$, $CdS_{(s)}$	$CdSO_4$, $CdCl^+$
Co	Co^{2+}, $CoCO_{3(s)}$, $Co(OH)_{3(s)}$, $CoS_{(s)}$	$CoSO_4$, $CoCl^+$
Cu	Cu^{2+}, $Cu_2CO_3(OH)_2$, $CuCO_3$; $Cu(OH)_{2(s)}$, $CuS_{(s)}$	$CuSO_4$
Fe	$Fe(OH)_2^+$, $FePO_{4(s)}$, $Fe(OH)_{3(s)}$, $FeCO_{3(s)}$, $FeS_{(s)}$, $FeSiO_{3(s)}$	
Hg	$HgCl_2$, $Hg(OH)_{2(s)}$, $HgS_{(s)}$, $Hg_{(liq)}$, HgS_2^{2-}, $Hg(SH)_2$	
Mn	Mn^{2+}, $MnCO_{3(s)}$, $MnO_{2(s)}$, $MnS_{(s)}$	$MnHCO_3^+$, $MnSO_4$, $MnCl^+$
Ni	Ni^{2+}, $Ni(OH)_{2(s)}$, $NiS_{(s)}$	$NiSO_4$
Pb	Pb^{2+}, $PbCO_{3(s)}$, $PbO_{2(s)}$, $PbS_{(s)}$	$PbSO_4$, $PbCl^+$
Zn	Zn^{2+}, $ZnCO_{3(s)}$, $ZnSiO_{3(s)}$, $ZnS_{(s)}$	$ZnSO_4$, $ZnCl^+$

A general pattern is evident for well-aerated conditions: The free ions are found mainly at low pH, the carbonate and then the oxide, hydroxide or even silicate solids precipitate at higher pH. In most cases, using the computer model (Morel et al., 1973), the trace metal ions are either free or controlled by solids with superabundant ligands. Only four important soluble complexes were found in the computations for the pH range under consideration: copper carbonate, mercuric chloride, mercuric sulfide, and silver chloride. In considering the stability constants for organic metal complexes which may be expected to occur in seawater, one may conclude that only a few metals are likely to be significantly affected by organic ligands.

Exceptions are Cu(II) and Fe(II) where according to model calculations (Stumm and Brauner, 1975) and chemical data (Theis and Singer, 1974), significant amounts occur as organic complexes in seawater. With regard to many other metals, inorganic ligands either compete with organic ones, e.g., OH^- and Cl^-, or abundant cations such as Mg^{2+} and Ca^{2+} compete for the organic functional groups. Kester et al. (1975), however, emphasized, that significant contrasts in chemical behavior of organic material are likely to be found in the biochemically active upper layer of the oceans and the refractory substances which persist in deep regions of the oceans.

2.1.2 Freshwater/Seawater Model

Sibley and Morgan (1975) pointed out that experimental data has led to conclusive evidence with regard to a change from a *freshwater to a seawater milieu*. The parameters causing the most significant changes (at constant pH and Eh) are (1) different ionic strengths, (2) lower content of adsorbing surfaces in seawater, (3) different concentrations of trace metals, (4) different concentrations of major cations and anions, and (5) usually higher concentrations of organic ligands in freshwater systems.

Characteristic developments of the trace element series are shown in Figure 19 (data from Sibley and Morgan, 1975). For most metals in the freshwater model the particulate components, especially the adsorbed metals, dominate. Since no single ligand controls speciation as chloride does in seawater, the dissolved complexes are

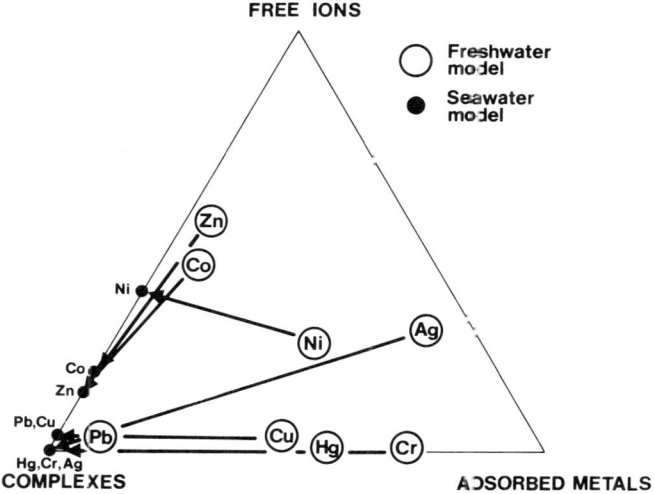

Fig. 19. Evaluation of the changes in chemical species of selected trace elements from a seawater/freshwater model (data from Sibley and Morgan, 1977)

considerably more variable in freshwater than in seawater (Sibley and Morgan, 1975). When the freshwater reaches the seawater environment, chloro-complexes become the dominant species for Cu, Zn, Hg, Co. Only Ni has the free ion as a main species, while Cr forms complexes with hydroxide. Calculations by Sibley and Morgan (1975) indicate that in this system, dominated by seawater, adsorption is negligible for all these metals. This is mainly due to the fact that with an increase in ionic strength the adsorption density of metal ions on particle surfaces decreases on account of the competitive exchange in the electric double layer (see Sect. E.3.4).

In addition to the species listed in Table 26, many metals are capable of existing in different oxidation states, which markedly govern metal speciation. Whether the metal species exist in a cationic, neutral or anionic form, often depends on the oxidation state, e.g., $Mn(H_2O)_6^{2+}$, MnO_2, MnO_4^-. Furthermore, solubility and complex formation are usually a direct consequence of the oxidation state, as displayed by differences pertaining to Fe(II)/Fe(III) and Mn(II)/Mn(IV).

2.2 Redox Conditions in Natural Waters

Only a few elements—C, N, O, S, Fe, and Mn—are predominant participants in aquatic redox processes. Water in solubility equilibrium with atmospheric oxygen has a well-defined pE of 13.6 (for P_{O_2} = 0.21 atm, Eh = 800 mV at pH 7 and 25 °C). Calculations based on $pE°$ values (Stumm and Morgan, 1970) of other elements show that they almost exist completely in their highest naturally occurring oxidation states: C as CO_2, HCO_3^- or CO_3^{2-} (with reduced forms less than 10^{-35} M); N as NO_3^- (with NO_2^- less than 10^{-7} M); S as SO_4^{2-} (with SO_3^{2-} or HS^- less than 10^{-20} M); Mn as MnO_2 (with

Mn^{2+} less than 10^{-10} M); and Fe as Fe_2O_3 or FeO(OH) (with Fe^{2+} less than 10^{-18} M). Even NO_3^- should result from the oxidation of atmospheric N_2.

The pE range in which certain redox reactions are possible can be estimated by calculating equilibrium composition as a function of pE. This has been done, for example, for iron and manganese and is depicted in Figure 20 (from Stumm and Morgan, 1970). Under highly oxygenated conditions (pE > 11) iron and manganese occur only as Fe(III) and Mn(IV) hydroxides or hydrated oxides. Biologic catalyzed oxidation of organic matter causes a downward depletion of oxygen in sediments and pE values are lowered. In consequence, concentrations of soluble Fe(II) and Mn(II) increase at the expense of Fe(III) and Mn(IV) with depth. In sediment interstitial waters the respective concentrations are controlled by two factors: (1) by the solubility of the respective carbonates and sulfides—the latter of less importance to Mn; and (2) by upward diffusion of ions out of the reduced zone leading to reprecipitation as Fe(III) and Mn(IV) at or near the sediment/water interface.

Dissolution and reprecipitation of Fe(III) as hydroxide tends to occur at pE values (~4) which are considerably lower than the equivalent reaction involving Mn(IV) (pE ~7.5), as may be seen from an inspection of Figure 20. Furthermore, Fe(II) acts as a reducing agent towards higher oxides or hydroxides of manganese and can incorporate appreciable amounts of Mn(II) into their structures (Stumm and Morgan, 1970). The overall effect of such reactions is that Mn(II) appears prior to Fe(II) upon progressive lowering of pE in interstitial waters and is also reoxidized to Mn(IV) phases closer to the sediment/water interface than Fe(III). In consequence, the vertical movement of Mn in sediments is more pronounced than that of iron.

The graphic representation of equilibria between chemical species as a function of pE (or Eh) at a particular pH is but one of several possible graphic treatments. Often equilibria between chemical species in a particular oxidation state are represented as

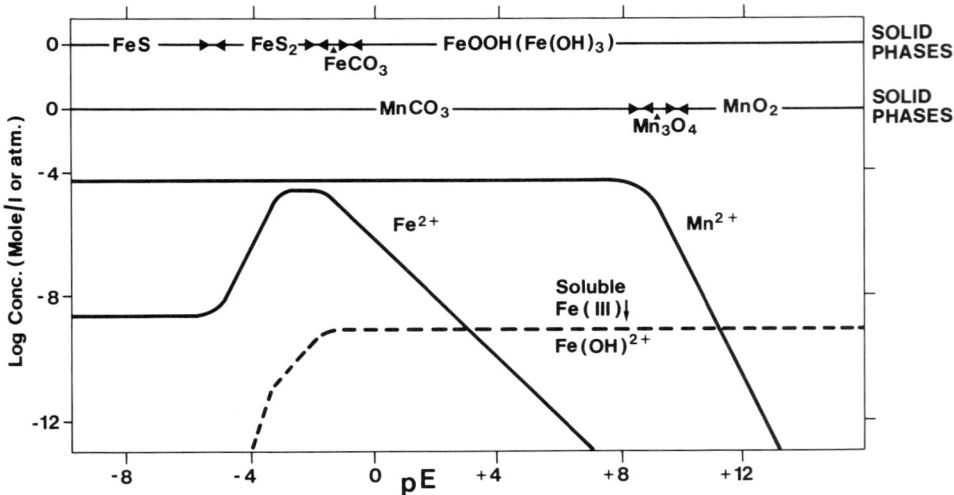

Fig. 20. pE-dependencies for iron and manganese in natural waters (Stumm and Morgan, 1970. *Aquatic Chemistry*. Wiley Interscience, fig. 7-9c, p. 331)

a function of pH and solution composition. These two representations are often combined into Eh-pH (or pE-pH) stability field diagrams.

An example is presented in Figure 21 (from Hem, 1970) indicating both the solubilities of *iron* in mol per liter and the fields of stability for the solid forms of iron (shaded area).

Stability regions for the solids $Fe(OH)_2$, FeS_2 and FeS are given, assuming activities of sulfur species as 96 mg/l (SO_4^{2-}), carbon species as 61 mg/l (HCO_3^-), and dissolved iron as 5.6 µg/l; these seem to be realistic values for normal freshwater systems. At concentrations of HCO_3^- higher than in the present model, a fourth solid phase, $FeCO_3$ (siderite), may occur at the respective Eh and pH values; the region shown in Figure 21 between the solubility lines of 10^{-7} for $Fe(OH)_3$ under oxidizing conditions and FeS_2 under reducing conditions is then filled by $FeCO_3$-precipitates.

From an inspection of Figure 21 it is clear that iron solubility is very low under two different Eh-pH conditions. First, there is the strongly reducing condition which covers a wide pH range and falls within the stability field of pyrite (FeS_2). Secondly, a moderate oxidative condition above pH 5 coincides with the $Fe(OH)_3(s)$ stability area. Between these two regions iron is relatively soluble, especially at low pH values.

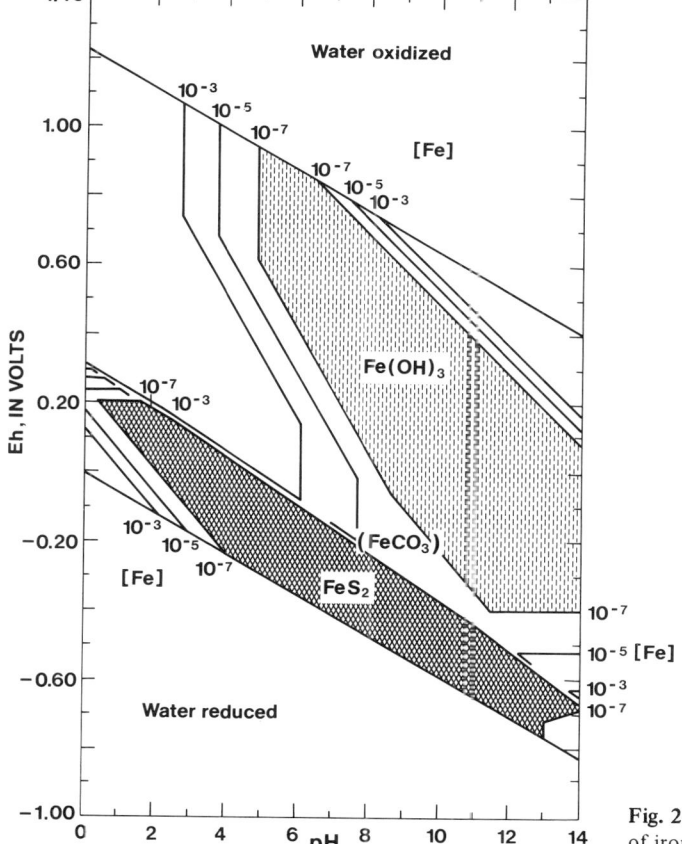

Fig. 21. pH and Eh fields of stability of iron (after Hem, 1970)

It is also evident from the solubility diagram that relatively small shifts in Eh or pH can have a large effect on the solubility of iron. Thus, when pyrite is exposed to oxygenated water, iron will be solubilized. This fact is of importance to the formation of acid mine drainage from spoil heaps containing pyrite.

In more recent investigations, theoretical data and actual measured results of iron concentration in water are increasingly in agreement. Turekian (1969) has pointed out that the global average of 670 μg/l in river water (Livingstone, 1963) is probably too high; "it is doubtful that the iron listed as Fe really exists in true solution in the amount listed." In the compilation of Head (1971), the values of dissolved iron in seawater range between 81 μg/l (inshore Atlantic off the mouth of the Amazon River; Ryther et al., 1967) and 0.25 μg/l (Northwest Atlantic; Spencer and Brewer, 1969).

According to the equilibrium values with Fe(OH)$_3$, dissolved iron in aerated waters, such as seawater, should only occur in values of 0.003 μg/l (Berner, 1970). Even the influence of river water with pH-values of 7 (seawater: pH 8) would only increase solubility by a maximum factor of 10. It is considered likely that one of the reasons for the discrepancies observed in the data lies in the intricate separation of the phases, as iron compounds of less than 5000 Å (separation by millipore filters) can be largely colloidal and incorrectly reported as monomeric dissolved iron. Concentrations of iron were found to decrease drastically when filters of a pore size less than 0.45–0.50 μm or an ultra-centrifuge were used (Lengweiler et al. 1961). Hem et al. (1973) have shown that some naturally occurring compounds of Fe, Al, and Ti in stream sediments may be caught onto 0.1 μm membrane filters after passing a 0.45 μm filter. It was therefore suggested by Kennedy et al. (1974) that the ratio of concentrations in the filtrates of both pore sizes may be a convenient way of determining the relative amount of particulate removed during filtration. In addition, the presence of Ti concentrations significantly exceeding 1 μg/l seems to be a warning that clay minerals may have passed through the filter. It has been seen that the concentrations of Al and Fe in particular, will vary with the degree of filter clogging, the pH after acidification, and the standing time before analysis. Thus they propose that the most reliable data for these elements in solution are obtained by filtration with 0.1 μm membranes and analyzed immediately after collection. Determinations on samples from some California coast range streams by Jones et al. (1974) using these methods indicated distinctly lower values of 1.3 μg/l for Al and 0.7–6.6 μg/l for Fe than in previous studies on the concentrations of these elements.

3 Trace Metals in Seawater

The difficulties encountered in separating the dissolved and the solid phases of iron by filtration necessarily lead to a degree of uncertainty in determining the dissolved concentrations of other metals in the water. Iron particulates (and manganese particulates) are capable of coprecipitating and sorbing other metals and thus act as scavengers. Still greater difficulties arise as a result of inadequate sampling, storage and analysis procedures. Especially in the taking of samples contamination is considerable, due to the proximity of ships and unprotected metal sampling devices. It is assumed that many of the earlier data on dissolved heavy metal contents in seawater are on the whole too high. In the recent past, however, decisive breakthroughs in analytic and handling techniques of seawater samples have been recorded, leading to our greater understanding of trace metal geochemistry in the marine environment.

Cooperative intercalibration studies by which many samples from one location are analyzed in different laboratories have been performed. These data enable the evaluation of the various aspects controlling trace metal concentration in seawater profiles both from the open ocean and near-coast environments and seemingly begin to reveal the major processes and mechanisms controlling trace metal distribution in these areas.

Voltammetric determination of toxic metals from water matrices, using Hg-amalgam techniques, prove to have conclusive advantages over other methods of trace analysis in aquatic systems (Barendrecht, 1956; Neeb, 1969; Brainina, 1974). A practical example is selected here from the work of Nürnberg and colleagues (Nürnberg, 1977; Valenta et al., 1977).

During a recent extended field study involving 225 sampling stations along the Ligurian and Thyrrhenian coast, concentration levels of dissolved Cd, Pb, and Cu were studied; the reported data classification is shown in Table 27.

Table 27. Levels of dissolved heavy metals in seawater (examples from the Ligurian and Tyrrhenian Coast; Nürnberg, 1977). Values in µg/l

Level	Cadmium	(n)	Lead	(n)	Copper	(n)
Low	0.005–0.009	(59)	0.018–0.09	(86)	0.13–0.19	(7)
Elevated	0.021–0.050	(33)	0.21–0.50	(34)	0.55–1.00	(58)
High	0.052–0.452	(9)	0.51–2.42	(9)	1.1–3.6	(45)

It can be seen that most of the examples of cadmium and lead fall into the category "low metal content", whereas copper dominates in the categories "elevated" and "high". In areas where the water is rich in algae and/or suspended particulate matter, concentrations of the category "low" are usually encountered. This is related to the trace metal uptake by organisms and/or metal removal due to chemisorption of the suspended inorganic and dead organic matter. On the other hand, it has been shown that levels of dissolved Cd, Pb, and Cu in the category "high" of Table 27 are usually observed near the main shipping routes to the ports.

3.1 Natural Distribution

The distribution of minor and trace elements in seawater was first compared and discussed by Fabricand et al. (1962) and Schutz and Turekian (1965a, b). They showed that some trace elements such as Sr, Cs, Rb, U, and Mo are very constant with depth and locality, whereas others such as Co, Ag, and Ni vary according to locality. Apart from man's influence, which greatly affects coastal waters, the considerable fluctuations observed in the case of some metals can be attributed mainly to their close relationship to biologic cycles. Only in the past few years, however, have reliable experimental data become available.

One important process in these areas is the removal of phosphorus and nitrogen (nutrient for water organisms) and calcium and silicon (which compose the skeletal parts of plankton) from surface water. The rate of decrease depends on local conditions such as increased biologic productivity in the marine food chain and in coastal areas of cold, upwelling water. These conditions are characterized by circulatory processes which determine the occurrence of biologically active compounds. According to Johnston (1964) the supply of chelating substances is frequently the most crucial aspect of phytoplankton nutrition in seawater. Barber and Ryther (1969) confirmed this hypothesis and demonstrated that in recently upwelled water phytoplankton growth is initially limited not by inorganic nutrients, trace metals or vitamin deficiencies, but rather by the absence of certain chelating substances. As the upwelled water ages, the organisms gradually enrich the water with organic compounds, some of which may be effective chelating agents.

In recent years, the metal enrichments in these upwelling zones have been of special interest. Calvert and Price (1971) found that concentrations of nickel and zinc in the fine-grained, organic-rich sediments of the shelf are in general relatively high in comparison to recent sediments in shallow water or other shelf sediments. This enrichment appears to confirm the suggestion of Brongersma-Saunders (1965) that sediments deposited in areas of upwelling are rich in metals, presumably because of direct contributions of metals by planktonic material.

In this respect, the relationship between the phosphorus content and the uranium enrichments observed in many areas is of special interest (Koczy et al., 1967; Baturin et al., 1971; Veeh et al., 1973, 1974).

Trace element removal from seawater by planktonic organisms can take place (1) through assimilation by organisms with subsequent transport as fecal material, (2) by body absorption on sinking, or (3) by adsorption onto organic detritus (Brewer, 1975). Calculations by Spencer and Brewer (1969) on the distribution of Cu, Zn, and Ni in the Sargasso Sea and the Gulf of Maine indicated that a simple metal uptake in planktonic species does not necessarily lead to large variations in the dissolved element concentrations. The biologic flux of trace elements, however, includes other mechanisms such as grazing and excretion of the phytoplankton crop by zooplankton. Brewer (1975) cites Polikarpov's suggestion that primary consumers take only small amounts of trace substances from food and that adsorption from the surrounding water is the principal mechanism by which minor elements are accumulated in body tissue. Early investigations by Martin (1970) on the vertical transport of trace metals by zooplankton suggested that the greater availability of food in the surface layers resulted in a more rapid breakdown of zooplankton, and thus less time was available for adsorption of metals from the water by the exoskeleton.

The above processes decisively influence the chemical composition of marine ferromanganese nodules, which are economically valuable due to their high trace metal content (Calvert and Price, 1977). The areas of manganese-rich nodules, containing high concentrations of nickel, copper, molybdenum, and zinc, coincide with regions of relatively high organic production in the northern equatorial region of the Pacific. The ultimate source of these metals is considered to be the biologic debris which accumulates at a high rate in such regions. Below the carbonate compensation depth, the bulk of the calcareous debris dissolves, thereby releasing incorporated metals. The metals then undergo diagenetic remobilization (see Chap. E.5.2) and are incorporated into ferromanganese oxides. Under special circumstances there might be direct removal from overlying seawater.

A typical depth profile for *cadmium* in mid-ocean areas is given in Figure 22 (Boyle et al., 1976) along with profiles of the nutrients phosphate and silicate. It is readily apparent that cadmium, phosphate and silicate are depleted in the surface water (approx. 0.01 μg Cd/kg seawater) relative to the deeper ocean water ($\sim$0.07 μg Cd/kg); the concentration increases with depth by a factor of at least 10. Such a distribution indicates uptake by organisms at the surface and regeneration from sinking biologic debris deeper in the water column. The particular high covariance of Cd with phosphate suggests that cadmium occurs in a shallow cycle like the labile nutrients, rather than deeper in the ocean in silicates.

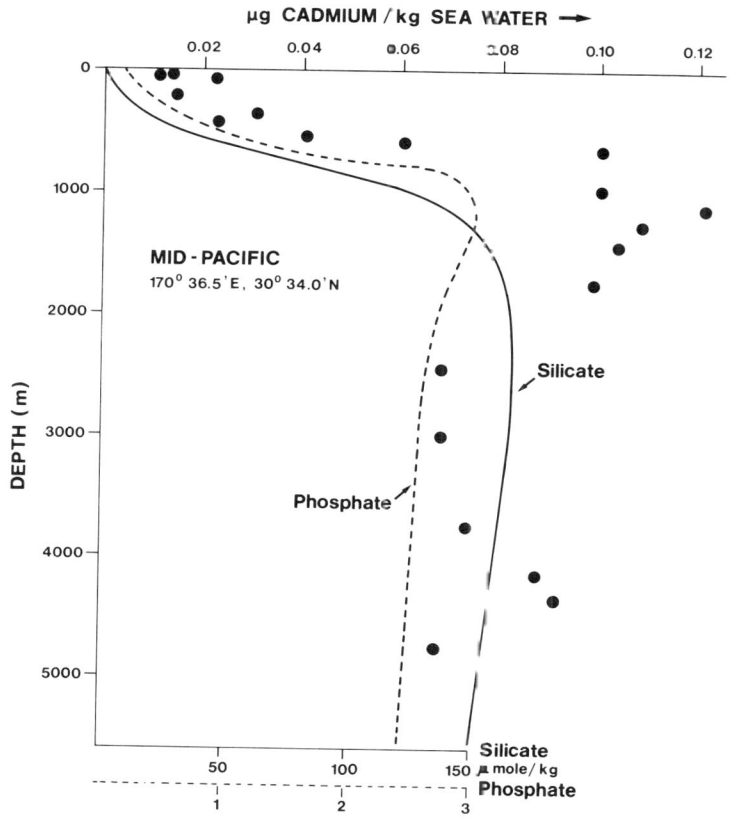

Fig. 22. Cadmium, phosphate and silicate profiles for a mid-Pacific sampling station (after Boyle et al., 1976; Nature (London), 263, p. 42)

Martin et al. (1976) found cadmium data from the surface waters off Baja California varying from 0.0004 µg/l to roughly 0.070 µg/l; by extrapolation, maximum values should approach 0.100 µg/l in the deep Pacific. This means an approximate 25-fold increase from deep to surface water. The strong correlation of Cd to P in both seawater and plankton suggests that phosphorus and cadmium are taken up together, either actively during phytoplankton growth (with an approximate ratio of 1 mg P for each 1 µg of Cd), or—after growth ceases—passively by adsorption (Martin et al., 1976). The authors believe the second process or some approximation of it to be more likely.

From studies in the Sargasso Sea, Bender and Gagner (1976) concluded that cadmium concentrations in surface waters are less than 0.01 µg/kg and probably less than 0.005 µg/kg; cadmium concentrations in deep water samples range from 0.016 to 0.055 µg/kg at a mean value of 0.025 µg Cd/kg seawater.

Boyle and Edmond (1975) found *copper* variations greater than a factor of three (from 0.56 µg/l to 0.185 µg/l) in surface waters across the Antarctic Circumpolar currents south of New Zealand. Three detailed copper profiles from the Antarctic, the Bering Sea and the northeast Pacific show systematic depth variations from 0.06 µg/l to 0.34 µg/l copper (Boyle et al. 1976). There was a significant correlation with nitrate with a molar ratio of 1:9200. It was suggested that the copper concentrations in the nutrient-poor surface waters of low and middle latitudes should be less than 0.006 μg/l Cu and the maximum values in the deep Pacific should be close to 0.23 μg/l Cu, i.e., a 1:40 depletion of copper in the surface water compared to values from deep water. Copper values from the Sargasso Sea (Bender and Gagner, 1976) range from a mean of approx. 0.12 μg/kg in surface water to 0.15 μg/kg in deep water. Similarly, copper data determined by Moore and Burton (1976) from the eastern Atlantic Ocean profiles range from 0.10 μg Cu/l in the surface to approximately 0.20 μg/l in the deeper water, and thus do not indicate such strong vertical gradients as has been observed by Boyle and co-workers in some areas of the Pacific. In the latter cases, where a close correlation with nitrate exists, copper can be classified as a limiting nutrient (Boyle and Edmonds, 1975). In North Pacific surface waters Bruland (pers. comm.) observed a systematic decrease in Cu concentrations towards the open ocean (from 0.100 to 0.4 μg/kg).

Nickel has been measured by Sclater et al. (1976) on four GEOSECS profiles from the Atlantic and Pacific. Values range from 0.16 μg Ni/kg in surface waters to 0.75 μg Ni/l in the deeper North Pacific. The shape of the Pacific profiles indicates that nickel is involved in the biogeochemical cycle, as it is incorporated in both the soft and hard parts of organisms. While ferromanganese phases (see above) may be the ultimate sink for nickel, they do not control its distribution in the water column. Bender and Gagner (1976) from studies in the Sargasso Sea found Ni concentrations of 0.10 ± 0.02 μg/kg in surface waters and 0.20 ± 0.04 μg/kg in deep water.

Recent *zinc* data obtained by Bruland et al. (1978) from samples taken under conditions of strict contamination control exhibit strong depletion at the surface and enrichment at greater depths (the samples were taken with a modified GoFlow system and were checked by a specially protected seawater sampler for lead analysis designed and constructed by Schaule and Patterson of the California Institute of Technology). The deep maximum value of 0.622 μg/l and the average surface concentration of 0.0085 μg/l yield a high concentration ratio of about 70. The same study revealed a highly significant correlation between Zn and Si, which suggests that diatoms play an important role in the biogeochemical cycling of zinc.

Presently known concentration levels of metals in open ocean seawater are compiled in Table 28. In the open ocean *total concentrations* of trace metals can be considered as nearly equivalent to *dissolved concentrations* (analytic advantage); these values should be less affected by anthropogenic influences than it would be the case in near-shore marine areas. However, it should be taken into account that there are very large local variations arising from the various degrees of productivity of surface waters. By way of example, Bruland (pers. comm.) found in North Pacific surface waters minimal cadmium values of 0.0002 μg/kg–that is 1/50 the value of that shown in Table 28.

3.2 Man-Made Effects

The often investigated metals mercury and lead will here serve as examples for anthropogenic influenced metal concentrations in seawater. For both elements, however, the natural effects are also to be considered. The enrichment of mercury at deeper parts of water profiles has been ascribed to mercury discharge from natural hot springs in tectonically mobile zones (Boström and Fischer, 1969; Carr et al., 1975). It was suggested that these accumulations originated either from degassing of the mantle by volcanic activity, or from being adsorbed onto colloidal precipitates which are a common constituent of such crestal sediments (Chester and Aston, 1976). Klein and Goldberg (1970), in contrast, have attributed the depletion of mercury in surface waters to mercury uptake by plankton and subsequent conveyance to greater depths by the biologic activities of the main food web, leading to an enrichment. Brewer (1975) suggested that although living phytoplankton does not take up mercury in large quantities, adsorption onto dead cells is rapid. It appears likely to Brewer (1975) that oceanic concentrations are buffered by

Table 28. Background of trace metals in seawater and freshwater

Element	Seawater μg/l	Author(s)	Freshwater μg/l	Author(s)
Aluminum	1	Sackett and Arrhenius (1962)	< 30	Kennedy et al. (1974)
Antimony	0.21	Brewer et al. (1972)	0.1[a]	–
Arsenic	2.1	Johnson and Pilson (1972)	2	Kanamori and Sugawara (1972)
Barium	20	Turekian and Johnson (1966)	10	Turekian (1966)
Beryllium	0.006	Merill et al. (1960)	0.01	Reichert (1973)
Boron	4450	Culkin (1965)	10	Konovalov (1969)
Cadmium	0.01 (s) 0.07 (d)	Boyle et al. (1976)	0.07	Boyle et al. (1976)
Chromium	0.08 (s) 0.15 (d)	Cranston and Murray (1978) [Cr(VI)]	0.5	Trefry and Presley (1976)
Cobalt	0.04	Robertson (1970)	0.05	Turekian et al. (1967)
Copper	0.1 –0.04 (s) 0.04 (d)	Bruland (pers. comm.) Boyle et al. (1977)	1.8	Gibbs (1977); Boyle (1978)
Gold	0.01	Schutz and Turekian (1965)	0.01	Crocket (1974)
Iron	1.3	Chester and Stoner (1974)	< 30	Kennedy et al. (1974)
Lead	0.005–0.015 (s) 0.001 (d)	Schaule and Patterson (1978)	0.2	Trefry and Presley (1976)
Lithium	173	Chow and Goldberg (1962)	1	Heier and Billings (1970)
Manganese	0.2	Brewer (1975)	< 5	Kennedy et al. (1974)
Mercury	0.011	Gardner (1975)	0.01[a]	–
Molybdenum	10	Sugawara and Okabe (1960)	1	Kharkar et al. (1968)
Nickel	0.2 (s) 0.7 (d)	Sclater et al. (1976)	0.3	Gibbs (1977)
Selenium	0.04 (s) 0.13 (d)	Measures and Burton (1978)	0.1[a]	–
Silver	0.01	Robertson (1971)	0.3	Kharkar et al. (1968)
Strontium	8100	Chow and Thompson (1955)	50	Turekian (1966)
Tin	0.01	Smith and Burton (1972)	0.03	Hamaguchi and Kuroda (1970)
Titanium	1	Griel and Robinson (1952)	< 1	Kennedy et al. (1974)
Uranium	3.3	Rona et al. (1956)	0.5	Bertine et al. (1970)
Vanadium	1.9	Sugawara et al. (1956)	0.9	Sugawara et al. (1956)
Zinc	0.01 (s) 0.62 (d)	Bruland et al. (1978)	10	Wedepohl (1972)

[a] Estimate according to references in Appendix A
s Surface water.
d Deep water.

scavenging in this manner, and that much of the Hg brought into the coastal zone by runoff is immobilized by adsorption.

On the basis of analytic data from 75 stations around the world, ten of which were depth profiles, Gardner (1975) found average Hg concentrations ranging from 0.0112 μg/l in the Southern Hemisphere to 0.0335 μg/l for the Northern Hemisphere.

This significant increase is suspected to be mainly the result of the high levels of mercury in industrial pollution injected into the Northern Hemisphere by the United States, Europe and Japan. Other factors influencing the distribution of dissolved mercury on a local or regional scale include upwelling, land runoff, high loads of suspended matter, and volcanic action. It has been suggested that the average values for the Southern Hemisphere might be considered as a natural background level for dissolved mercury in the ocean.

Recent investigations by Baker (1977) on water samples from the Irish Sea and North Sea indicate that elevated concentrations of mercury in this area are mainly caused by sewage sludge dumping. "High spots" of mercury contamination are apparent in the Thames Estuary, Liverpool Bay, the area off the Humber Estuary, and to a lesser extent the Bristol Channel; samples from the North Sea show elevated values at the mouths of the Rhine and Elbe Rivers.

Until 1975, information on the speciation and concentration of *lead in seawater* was not reliable despite decades of measurement (Patterson et al., 1976b) because during collection, handling and analysis, contamination of seawater with artifact lead has been widespread. Even the work of Tatsumoto and Patterson (1963a, b) and Chow and Patterson (1966), which gave the proper perspective of the occurrence of lead in seawater, has to be regarded as adversely affected by artifact contamination during collection. True concentrations of lead in open North Pacific surface waters were found to range from 0.005 to 0.015 g/kg, with the lower values typical for productive near-coast areas relatively unaffected by industrial pollution (Fig. 23; Schaule and Patterson, 1978). In coastal waters near urban regions, lead values may be significantly elevated and reach more than 0.150 μg/l in waters highly polluted with sewage (Patterson et al., 1976a). It is believed that in areas of high sewage pollution most of the particle lead in the water is associated with the sewage and not with plankton; total lead concentrations appear to decline to 0.03 μg/l before contributions of lead in plankton become significant (Patterson et al., 1976b). The vertical profile of Pb in the North Pacific as compared to the deep water values in the upper part of the water column shows a distinct enrichment, which is explained by atmospheric inputs derived from automobile exhausts and emissions from smelters (Schaule and Patterson, 1978). This general structure is modified depending on local conditions of productivity.

3.2.1 Atmospheric Input of Metals

The concentration of trace metals in the surface layers of the sea is influenced by several mechanisms (Wallace et al., 1977): (1) advective transport from regions of high source strength, such as rivers, (2) advective transport by gas bubbles (Wallace and Duce, 1975), (3) vertical turbulence, (4) aeolean transport of land-derived, mainly inorganic particles (Hoffman et al., 1974) and (5) biologic production. Examination of the relative variations in particulate trace metals from a series of surface bucket samples collected between the New England coast of the United States and Bermuda, led to the conclusion that *organic matter* most probably regulates the amounts of particulate trace metals in open-ocean surface waters and is important for continental shelf and slope waters as well (Wallace et al., 1977). The ratios of the metal concentrations in the surface water particulates and those in the pelagic sediments have been found

Metal Input from Sewage Effluents 89

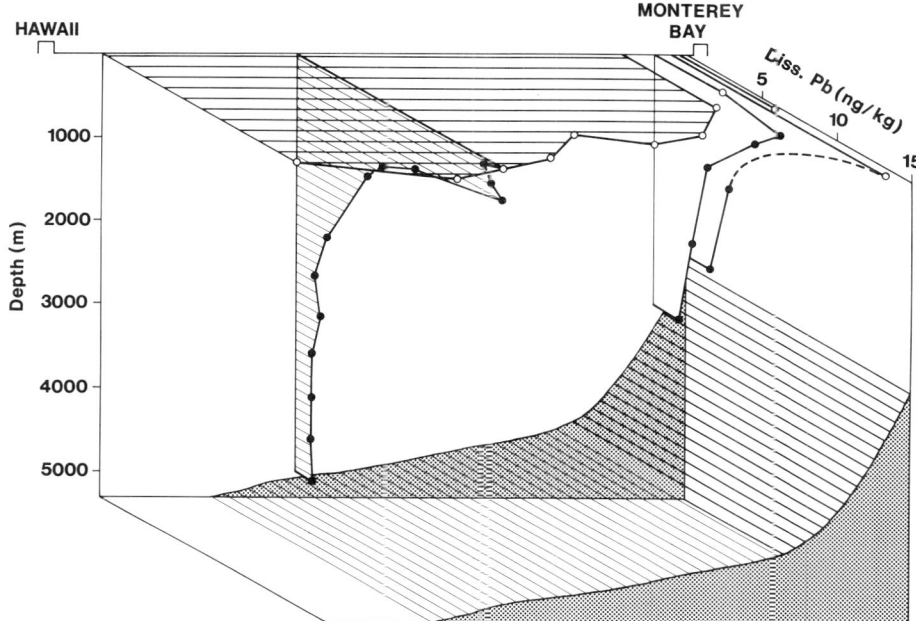

Fig. 23. Distribution of dissolved lead in northeast Pacific between Hawaii and California (Schaule and Patterson, 1979; reproduced with permission of the authors)

particularly high for copper, zinc, lead, and cadmium; i.e., metals with a high index of the "relative pollution potential" (Table 20). It would seem that associations between digested trace metals and particulate organic carbon take place in situ and that the trace metals introduced from the atmosphere are a direct source of these enrichments in the mixed layer of open-ocean water.

3.2.2 Metal Input from Sewage Effluents

A characteristic example of the effects of sewage effluents on the concentrations of dissolved metals in coastal seawater has been given by Martin et al. (1976) from investigations conducted in the immediate vicinity of the Los Angeles County outfall at White's Point.

The beam transmissiometer-temperature record showed three turbid layers (solid line in Fig. 24): (1) the near-surface layer at about 3 m water depth indicated high clorophyll concentrations from phytoplankton, (2) well-defined, 8 m thick lens centered at 15 m was most likely enriched by material from the sewage effluent; the mid-depth lens was trapped below a strong thermocline centered at 12 m, (3) a near-bottom effluent plume is stratified below a weaker thermocline at a water depth of 26 m.

Metal concentrations from water samples collected by a tube pump in each of the water layers clearly show increased values, which are influenced by sewage material (Fig. 24). Maximum factors of enrichment of dissolved trace metals in the main outfall

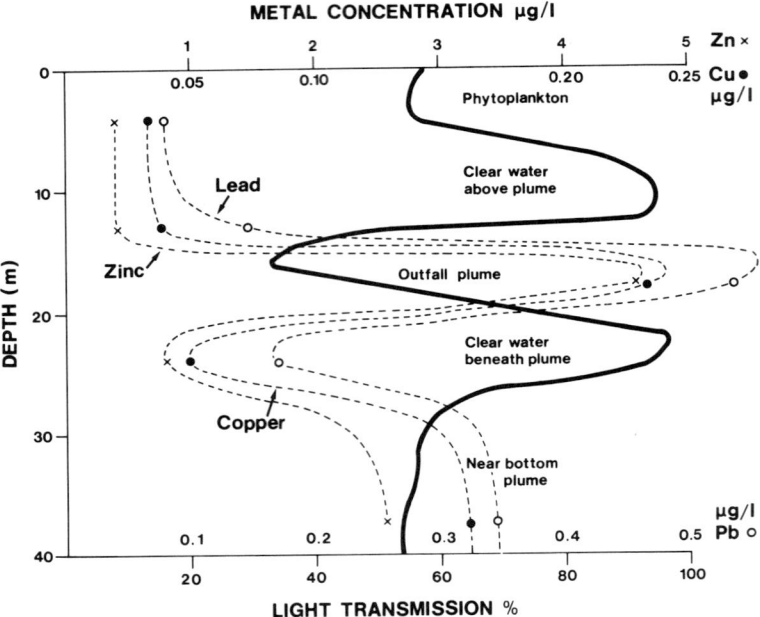

Fig. 24. Depth distribution of light transmission (*solid line*) and concentrations of zinc (*crosses*), copper (*closed circles*) and lead (*open circles*) in a water column near Los Angeles County wastewater outfall (after Martin et al., 1976. In: *Marine Pollution Transfer*. Lexington Books, D.C. Heath and Co., Mass.)

plume, as compared with the surface water composition, are 13 for lead and zinc, 8 for copper and 3–4 for cadmium and nickel.

Since due to an increased nutrient supply and elevated temperatures such effluent plumes offer especially favorable conditions for fish, it can be expected that in just such zones, a considerable uptake and enrichment of toxic trace elements can also occur (see Chap. F).

4 Trace Metals in Inland Waters

4.1 Natural Contents

It is even more problematic to establish global mean values for the individual trace elements in inland waters than in seawater. This is due in the first place to the variety of rock formations and the fluctuations in water transport (especially in rivers). The accompanying changes in metal transport will be treated in greater detail later. At this stage, however, attention must be drawn to the large contribution made by the contamination from the "civilizational" factor, i.e., man-made influences which until now have generally caused greater pollution in continental water than in coastal water.

In order to determine the extent of this pollution, background values are most desirable which, although not "absolute", serve as *guidelines* for water quality control and as indicators of the extent and possible consequences of contamination whenever they are exceeded. An attempt has been made (see Table 28) to compile such a list of normal background values.

Many published results obtained from geochemical exploration (trace metal determination in this context were first conducted in waters draining mineralized bed zones) were discarded; yet the remaining results still fluctuate within several orders of magnitude. Likewise, levels from heavily polluted regions were not included in the list of possible background values, even though most analytic data do, in fact, stem from such areas.

Preference has been given to metal values near the lower end of the distribution pattern with special significance attached to those values which were reported in at least several examples. However, a larger number of examples with equally low values is available for only a few metals, e.g., zinc and copper; in most cases it was necessary to estimate the values.

A major source of reference for the choice of trace metal concentrations in inland water is the *Handbook of Geochemistry* (edited by K. H. Wedepohl, 1969–1978), section I "Abundance in Natural Waters and in the Atmosphere". Bibliographies on trace metals in water have been edited by the U.S. Office of Water Resources Research, NTIS, Springfield (Va.) for arsenic, chromium, strontium, lead, copper, zinc, mercury, and manganese (Anon., 1970a; Anon., 1971b–g; Anon., 1972c). The series *Water Quality Criteria Data Book*, edited by Arthur D. Little Inc. contains a compilation on inorganic chemical pollution of freshwater in vol. 2 (Anon., 1971h). Review *Concentration of Trace Metals in River Waters* was published by the Water Research Centre, Medmenham and Stevenage Laboratories, United Kingdom (Wilson, 1976). Here, Appendix A gives a summary of data found in current literature, including examples of both normal values obtained for contaminated waters.

It should be noted that there is still a good deal of controversy as to the origin of the sometimes strong divergences in the metal contents of inland waters, which are even greater than those for seawater. For example, Valenta et al. (1977), using differential pulse stripping voltammetry, measured cadmium values of approximately 0.005 μg/l in water samples from Lake Constance (Überlingen). This is nearly two orders of magnitude under the "normal" level as determined by atomic adsorption spectrometry in unpolluted water bodies. Similarly, lead and copper register only 1/1000th or 1/20th of the respective "normal" values registered in Lake Constance near Lindau using atomic absorption spectroscopy (Quentin and Winkler, 1974). For some data, there might be a characteristic influence of nonlabile metal complexes, which are not accessible to anodic stripping voltammetry if there is no step of irradiation prior to the metal analysis which breaks down the complexes (Gardiner and Stiff, 1975). It is fairly clear that the analytic problems encountered in trace metal analysis in water are still considerable, that problems are also confronted with the storage of the samples, and—what is decisive for rivers—the rate of water discharge can be a strong influence (see Sect. 5.1).

The investigations, with improved sample extraction, storage and handling procedures, now carried out on the large rivers of relatively unpolluted areas, such as the Amazon, Congo, and Mekong Rivers, should contribute considerably to our knowledge of trace metal geochemistry in freshwater systems. For the interpretation of the concentrations and the behavior of the major ions in such waters, investigations on the river system of the Amazon play a major role (e.g., from Gibbs, 1970b, 1972). For trace metals, investigations on the Amazon and Yukon Rivers allowed new interpretations of earlier statements (Gibbs, 1977): "The two river systems were selected in order to observe the maximum difference in their environments—one being a tropical basin, the other a subarctic system. The striking similarity in the concentrations (dissolved concentra-

Table 29. Concentrations of dissolved transition metals in Amazon River (Macapa, Brazil) and Yukon River (Alakanuk, Alaska) determined from combined monthly samples in proportion to the dissolved salt transport

	Amazon (µg/l)	% of total[a]	Yukon (µg/l)	% of total[a]	World river % of total[a, b]
Iron	34	0.6	50	0.05	0.13
Manganese	19	17	20	10	2.9
Chromium	2.02	17	2.33	13	2.5
Copper	1.77	6.9	2.00	3.3	0.98
Nickel	0.27	2.7	0.43	2.2	0.5
Cobalt	0.06	1.6	0.1	1.7	0.29

[a] Transport of transition metals in solution as percentage of the total mass of metal discharged in the system (dissolved +particulate phases).
[b] Estimate based on the concentration averages of the Amazon and Yukon Rivers multiplied by world discharge values. All data from Gibbs (1977).

tions of trace metals are listed in Table 29) and percentage distributions in the various transporting phases for the Amazon and Yukon Rivers leads to the conclusion that the concentration and percentage distributions are fundamental process characteristics of natural—unpolluted—river systems". Estimates on the world river transport of transition metals to oceans indicate that less than 3% is associated with dissolved species (Gibbs, 1977; Table 29, last column).

It should be noted that the various subsystems of the large river systems may differ widely both with respect to concentrations and percentage distributions. Gibbs (1977) gives the example of the Amazon system where the contributary Rio Negro carries very little sediment and therefore almost the entire load of transition metals is transported in the dissolved-complexed phase. When the Rio Negro reaches the main channel of the Amazon River, the sediment-related metal transport overwhelms the minor load carried in solution. Irion and Förstner (1975) found from studies on Amazon "lake" sediments that deposits contributed by rivers from regions with deeply leached soils—"blackwater rivers" (e.g., Rio Negro)—are significantly depleted of iron, manganese, zinc, copper, and cobalt, when compared to the "whitewater" river deposits, e.g., from Rio Solimoes. Recent analyses of copper in the Amazon River system by Boyle (1978) indicate similar distributions for the dissolved species; the values of the black- or clearwater examples center around 0.5 µg Cu/l, whereas the Cu-concentrations of the second group of rivers, originating in the Andes, pre-Andes and southwestern Amazon lowlands, range between 1.5 and 2.0 µg/l.

At this point, an attempt should be made to compare the individual "background values" estimated for each element with the corresponding metal analyses from inland waters of areas which are more affected by pollution. In Table 30 (1) *river water* from the relatively unpolluted Danube River in Austria, (2) *lake water* from Lake Michigan, and (3) *spring water* from the upper Neckar River area are compared.

This comparison shows that the composition of spring water is particularly valuable in providing a standard for assessing the contamination of certain waters. By including regional and local factors, and especially by taking the lithogenic influences into consideration, the anthropogenic effects can, in most cases, be determined with reasonable accuracy. On the other hand, the example of the Danube brought the surprising fact to light that in water systems which flow through relatively highly popu-

Table 30. Comparison of metal values (filtered samples) from Table 28 with data from river water (Schroll et al., 1975), inland lake water (Copeland and Ayers, 1975), and spring water (Lodemann and Bukenberger, 1973)

Element	Typical values Table 28 µg/l	River Water Danube/Austria µg/l	Inland Lake Water (Lake Michigan) µg/l	Spring Water (upper Neckar River) µg/l
Antimony	0.1	–	0.23	–
Arsenic	2	–	1	–
Barium	10	20 – 67	37	–
Boron	10	–	–	10[b]
Cadmium	0.07	0.07 – 2.6	0.3	0.1
Chromium	0.5	0.4 – 1	1.7	1 – 5
Cobalt	0.05	–	0.18	0.5
Copper	2	3 – 7	5	2 – 5
Lead	0.2	2 – 4	1.5[a]	0.1 – 1
Manganese	< 5	2 – 9	1	2 – 7
Mercury	0.01	0.1 – 1	0.027	–
Nickel	0.3	1 – 7	3.0	0.5 – 2
Selenium	0.1	–	0.083	–
Silver	0.3	–	0.3	–
Strontium	50	160 – 310	97	–
Titanium	< 1	2 – 9	–	–
Vanadium	0.9	–	0.2	–
Zinc	10	7 – 28	16	4 – 10

[a] Edgington and Robbins, 1976.
[b] Dietz, 1975.

lated areas and are thus more heavily contaminated with communal effluents, most trace elements exhibit only light, toxicologically insignificant enrichments. To a certain extent, it is the physico-chemical conditions which effect a rapid elimination of dissolved components by transferring them into the solid phase. The size of the water body also plays an important role in the "digestion" of a specific quantity of effluent. In the following section examples are given of cases where this self-purification no longer functions, either due to too large an effluent quantity, to a water system which is too small, or to certain hydrochemical conditions which increase the solubility of heavy metals.

4.2 Metal Pollution in River Water: Regional Examples

4.2.1 Heavy Metal Pollution in United States Water Systems

Two comprehensive investigations on dissolved metal loads in water systems have been carried out in the United States until 1970:
1. A five-year summary (October 1962–September 1967) of trace metals in rivers and lakes, compiled by Kopp and Kroner (1968), U.S. Department of the Interior,

Federal Water Pollution Control Administration (FWPCA), Division of Pollution Surveillance, Cincinnati, Ohio.

2. A review of selected minor elements in surface waters of the United States, October 1970, prepared by Durum et al., (1971), U.S. Geological Survey (USGS), in cooperation with the U.S. Bureau of Sport Fisheries and Wildlife; reports on several river systems have been presented in the Water Supply Papers of USGS.

The FWPCA Report (1) investigated 130 stations, which were selected according to the criteria of (a) human and/or animal concentrations; (b) industrial activity including agricultural and heat emission sources; (c) recreational use areas; (d) state and national boundaries; and (e) potential problem areas. A total of 1500 samples were analysed for 18 trace elements. In general, weekly samples received from the various stations were collected for three-month periods. In the investigation program of the USGS (2), 720 individual samples were taken from October to November 1970 from rivers and lakes and analyzed for arsenic, cadmium, chromium (hexavalent), cobalt, lead, mercury, and zinc. Sampling stations fell within three categories: (a) surface water sources of public water supplies for cities with a population exceeding 100,000, (b) water courses downstream from major municipal and/or industrial complexes, and (c) USGS hydrologic bench-mark stations. These stations are located in undeveloped drainage basins in the major physiographic regions of the country. Table 31 includes a selection from these two data collections and shows (1) the number of violations of the 1968 Water Quality Criteria (permissible) and (2) the maximum values for the individual elements.

A comparison of the naturally occurring metal content, as presented in Table 28, shows that the maximum values are enriched in each case by factors between 100 (B, Cr, Cu, Ag, Zn) and 1000 (As, Cd, Hg, V, Mo, Pb). The extreme rate of increase for cobalt and zinc (factor of 10^5 and 10^4, respectively), observed in Mineral Creek near Big Dome, Arizona, can probably be attributed to the influence of acidic mine effluents. Past mining activities have left their scar on the arid land of the *Colorado River area*. Mine drainage and tailing piles still exert an adverse influence on the quality of the water. The elevated metal contents in Lima, the station farthest upstream on the Colorado, stem partly from active mines and uranium plants at Rifle, Grand Junction and Gunnison. The extremely high vanadium content in the river is probably due to the presence of an oil shale extraction plant at Rifle. A possible explanation for the high boron concentrations at the Yuma station may be the influence of irrigation return drainage, as the water is used for extensive agricultural development. The metal enrichments of the *Allegheny River* (Co, Mn) originate from oilfield brines, acid mine drainage and mill wastes, partly from the tributary of the Kiskiminetas River. The quality of the water taken from the intake of the Pittsburg filtration plant is affected by discharges from industries including steel, fabricated steel products, clay, glass, paper, petroleum, food products, and stone products. The *Monongahela River* and its principle tributary, the *Youghiogheny River,* are polluted by acid mine drainage where pH-values below 3.0 have been observed. The Cuyahoga River flows into the eastern end of Lake Erie and is largely influenced by Cleveland (1.6 million population) and its complex of chemical, automobile, paper, and metal plating industries (Kopp and Kroner, 1968). The investigations of Hem et al. (1971) also indicate potential problems in a few areas. A very high cadmium value of 90 ppm found in the *Tennessee River* at Whitesburg, was possibly caused by industrial emissions. Generally, the higher concentrations of cadmium in water occur in areas of high population density. The same applies to lead which was detected less frequently in samples collected at bench-mark sites, than in those from

Table 31. Metal contents in US waters (samples passed through 0.45 μm filter)

	Federal Water Pollution Control Administration Kopp and Kroner (1968, 1500 samples)				U.S. Geological Survey Durum et al. (1971, 720 samples)		
	Limit μg/l	Violation	Max. value μg/l	River	Violation	Max. value μg/l	River
Arsenic	50	3%	336	Maumee River at Toledo, Ohio	2%	1,100	Sugar Creek near Ft. Mill, S.C.
Beryllium	–	–	1.2	Monongahela River at Pittsburg, Pa.	–	–	
Boron	–	–	1,800	Colorado River at Yuma, Ariz.	–	–	
Cadmium	10	0.4%	120	Cuyahoga River at Cleveland, Ohio	4%	130	Mineral Creek at Big Dome, Ariz.
Chromium	50	0.3%	112	St. Lawrence River at Massena, N.Y.	0%	17	Schuylkill River below Reading, Pa.
Cobalt	–	–	48	Allegheny River at Pittsburg, Pa.	–	–	
Copper	1,000	0%	280	Monongahela River at Pittsburg, Pa.	–	4,500	Mineral Creek near Big Dome, Ariz.
Iron	300	2%	952	Sabine River near Ruliff, Texas	–	–	
Lead	50	2%	140	Ohio River at Evansville, Ind.	1%	890	St. Croix River at Baring, Maine
Manganese	50	5%	3,230	Allegheny River at Pittsburg, Pa.	–	–	
Mercury	5	–	–		0%	4.3	James River near Boaz, Missouri
Molybdenum	–	–	1,100	Arkansas River at Coolidge, Kansas	–	–	
Nickel	–	–	130	Cuyahoga River at Cleveland, Ohio	–	–	
Silver	50	0%	38	Colorado River at Loma, Colorado	–	–	
Vanadium	–	–	300	Colorado River at Loma, Colorado	–	–	
Zinc	5,000	0%	1,182	Cuyahoga River at Cleveland, Ohio	0.5%	12,000	Mineral Creek near Big Dome, Ariz.

Table 32. Metal concentrations (filtered samples) in Welsh rivers (Abdullah and Royle, 1972; 1974a)

	Manganese	Zinc	Copper	Nickel	Lead	Cadmium
Minimum concentration	0.8	11.0	0.7	0.5	0.7	0.1
River Ystwyth	7–15	200–270	–	–	2.0–6.0	0.4–3.8
River Rheidol	9–28	50–130	–	–	1.3–2.4	0.2–4.7
River Dovey (R. Twymyn)	–	153–438	1.3–3.7	–	3.7–17.6	0.2–4.9
River Mawddach	3–25	14–50	1.9–3.8	–	–	0.1–1.0
River Dwyryd	–	55	3.8	3.5	1.4	2.6

public water supply sources and from streams below metropolitan-industrial areas (Durum et al., 1971).

4.2.2 Metal Pollution in Inland and Coastal Waters of Great Britain

Rivers in Great Britain have been protected from high levels of metal contamination by strict legislative control of discharges: trace metal pollution is principally restricted to estuaries and a narrow coastal margin (Jaffe and Walters, 1975). Nevertheless, there are locally elevated metal concentrations from past and present mining activities which may attain quite considerable dimensions. Examples are given, particularly from areas in Wales and Cornwall. The metal contents in rivers and lakes of Wales gave rise to great concern and led to intensive investigations carried out in the 1920's by Carpenter (1924) on the effects of metal poisons on water organisms. More recent investigations on metal contents in Welsh rivers and lakes were carried out by Abdullah and Royle (1972) from September 1970 to October 1971. These results, together with the results of a detailed study on the cadmium contents in rivers of Wales by Abdullah and Royle (1974a) are summarized in Table 32 and Figure 25.

The rivers that originate from the mineralized areas are the River Ystwyth, River Rheidol, and River Twymyn; the waters of these rivers are characterized by high zinc, lead, and copper levels. The largest total amount of metal is found in the River Twymyn of the River Dovey system; the lead, zinc and copper levels are among the highest encountered in the entire region. The River Mawddach's course runs through the Dolgelly mineral zone and its water generally contains high levels of Cu and Mn. The unusually high average manganese concentrations found at the head of the Mawddach, Dwyryd, and Dysynni estuaries were probably caused by occasional interactions between saline water and the suspended hydrated manganese oxide carried by the streams (Abdullah and Royle, 1972). Collectively, the average cadmium levels from monthly samples, as shown in Figure 25 from 31 stations, clearly reflect the influence of mineralization zones (shaded areas). The rivers and lakes in regions where no mineral deposits are known show cadmium levels ranging between 0.1 and 0.6 $\mu g/l$. The annual average cadmium level in the rivers of the mineralized regions are found to range between 1.2 and 4.7 $\mu g/l$, with the highest recorded concentrations being 20 $\mu g/l$. The annual average values for these rivers is approximately ten times greater than that found in streams free from mineral contamination (Abdullah and Royle, 1974).

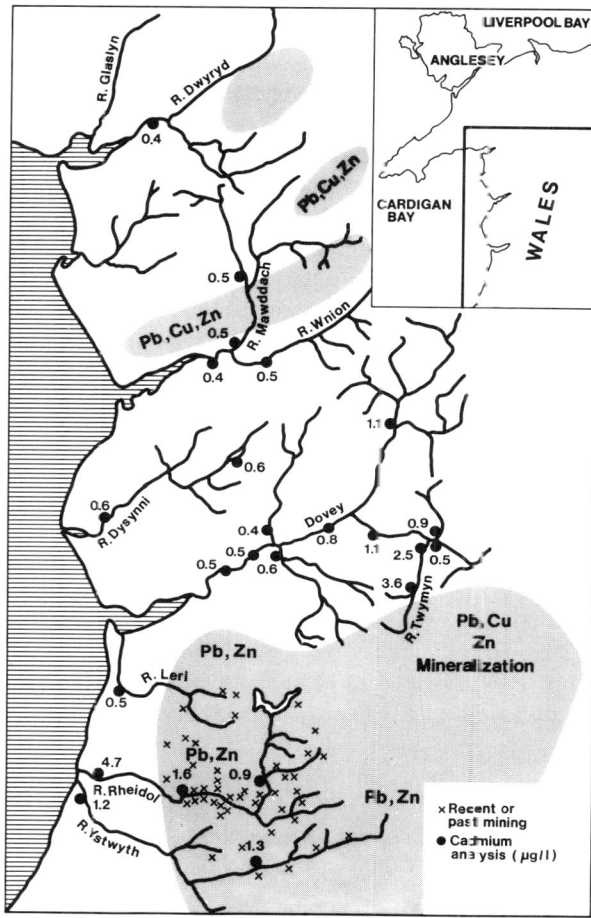

Fig. 25. Concentrations of cadmium in Welsh rivers (after Abdullah and Royle, 1974a)

Coastal waters are also affected by metal contamination: In the *Bristol Channel* industrial and domestic effluent carried by the Severn River and the Avon River together with the runoff from the mineralized zone in north Devon contribute to the trace metal occurrence and distribution (Abdullah et al., 1972). In a detailed study, Butterworth et al. (1972) demonstrated that the contents of zinc, cadmium, and lead regularly decrease from the Severn Estuary to the Bristol Channel: zinc from 52 μg/l near Porthishead to 12 μg/l off Hartland (open sea 10 μg/l); cadmium from 5.8 μg/l to 0.3 μg/l (0.1 μg/l); lead from 2.5 μg/l to 0.4 μg/l (0.3 μg/l).

The highest metal pollution in the coastal regions of the United Kingdom is encountered in Southern Cornwall. Aston et al. (1974) found strong metal enrichments chiefly in the 50 km^2 catchment areas of the *Carnon River*, which receives effluents from working metalliferous mines, (refer to Chap. B.4). These enrichments are probably connected with the relatively low pH-values of 4 to 6. The following average values were determined: iron 5120 μg/l (3700–6050); manganese 500 μg/l (200–1000); zinc 4950 μg/l (1520–10,000); copper 1080 μg/l (960–1150). Bryan and Hummerstone (1973a) found similar data for the Carnon River and also determined very high metal contents in Restronguet Creek, which drains into the Fal Estuary.

4.2.3 Heavy Metals in River Water of the Federal Republic of Germany

The routine control of the water systems in the Federal Republic of Germany is primarily a task for the authorities of the individual states. In addition, however, a number of other institutions, such as health authorities, water works and industries, are all concerned with the question of water quality control, though sometimes only on local basis. In 1969, the German Research Society (DFG) initiated a research program, having as central theme *Water Research–Pollutants in Water*. Within the framework of the program, a "Metals" research team began looking into the occurrence and extent of metal contamination in inland waters of the Federal Republic of Germany and attempted to coordinate the various activites being carried out in this field. The results of a preliminary investigation extending over several years were summarized in 1975 in an interim report (Anon 1975b); a conclusive report is planned for 1979.

In order to illustrate the general situation, data are listed below of trace element investigations which were carried out by the Institute for Sediment Research on important rivers of the Federal Republic of Germany as part of this specialized program. Individual samples were used in each case, which were collected between October 1971 and February 1973 from the Rhine, Elbe, Weser, Ems, and Danube Rivers as well as from their most important tributaries. The varying water levels were, however, not taken into account (see following section). The sampling sites are indicated in Figure 26a; Fig. 26 b shows the concentrations of zinc, copper, and lead in the water of the main rivers. The maximum frequency distribution (67 individual values) is approx. 20 μg/l for the zinc content, between 4 and 5 μg/l for copper, and approximately 2.5 μg/l for lead. The data for nickel and cadmium reached frequency maxima at 4 μg/l and 0.5 μg/l respectively. The corresponding background data of these metals coincide with the positive slopes of the represented distribution curves.

Deviations from these average values to higher heavy metal contents are generally indicative of the anthropogenic contamination of water systems. Such deviations are particularly frequent and extensive in the zinc content of the river samples analyzed in this investigation. Lead and copper values however, also reach high levels in some cases. Elevated zinc, lead, and copper concentrations are found mainly in the *upper Elbe and lower Rhine;* zinc is generally present in relatively high amounts in the *Weser.* Copper in the *Danube* above Ulm, nickel in the *upper Elbe,* and cadmium in the *lower Neckar* (a tributary of the Rhine) and in the *lower Rhine* itself, are enriched at factors of up to 20 above the average values of relatively unpolluted water samples.

The analysis results for metals in central and lower Rhine sections indicate that lead, iron, and manganese clearly exceed the acceptable maximum values for raw or drinking water, World Health Organization (WHO). In some cases, cadmium, chromium, mercury, and selenium are present in critical concentrations. Approximately 50% of the nickel and arsenic values are already approaching a critical point (Interim Report, "Metals" research group, Anon, 1975b). The development of heavy metal loads in the Rhine was compared by Heinrichs (1975) with the population density in catchment areas:

> The Rhine River is one of the most heavily polluted rivers on earth and had been given the appropriate designation of the "majestic cesspool of Europe". Approx. 70% to 80% of the pollutant loads which flow into the Rhine originate from emittors in the Federal Republic of Germany. It appears that since 1970 a slight improvement has occurred in some regions especially in the upper

Rhine and Lake Constance, in spite of recent increases in certain pollutants. In this way, the content of organo-chloro-compounds such as chloroform, tetrachlorcarbon, chlorophenol, or pesticides containing chlorine, particularly in the lower Rhine, has risen to a significant level and has reached with nearly 10% of all dissolved organic substances very serious concentrations with respect to health and the recovery of drinking water (Authority Board *Environment Problems of the Rhine* Anon, 1976a). In its report for 1977 the *IAWR* (International Association of Waterworks in the Rhine catchment area) stated that pollution by organo-chlorine compounds, nitrates, and phosphates continued to increase despite a higher water level in this period than during the previous years. The problem of chloride pollution (see p. 73) has not been solved, although 30 mill. dollars have been spent for underground pressing. Limiting values of trace metals for drinking water processing (see Chap. G) were still exceeded for mercury, chromium, iron, and lead during 1977; there is no significant decrease of the high cadmium level from 1975–1977.

Figure 27 shows a comparison of the contents of cadmium in Rhine water with the population figures per square kilometer. The first distinct increase in cadmium content can be seen in the catchment area of Basel. The concentrations then decrease slightly as far as Karlsruhe. In the entire area stretching from Mannheim to Mainz, which includes the heavily populated and industrialized areas of the Rhine-Neckar and Rhine-Main, the cadmium values reach higher concentration maxima. After another decrease in the cadmium values, the greatest increase in cadmium concentrations then occurs in the catchment area of Cologne. The values increase continuously as far as the Dutch border, although after leaving the Ruhr area the population and industrial densities decrease sharply again. In this section the Rhine water is no longer regenerated.

The increase in metal contents in the Rhine between Lake Constance and the Dutch/German border can be seen in Table 33 (data of Heinrichs, 1975), in which data from the most important Rhine tributaries are also included. The concentrations of zinc and cadmium increase along this stretch by factors of 45 and 35 respectively. Mercury, iron, manganese, and chromium contents are between 12 and 20 times higher. Only lead and silver show an increase of less than ten, probably in connection with the state of equilibrium between solution and solid phases. The water samples from the Rhine tributaries exhibit mainly high quantities of zinc and iron in the Main and Mosel Rivers. The relatively low cadmium content of Neckar River water was probably determined from samples taken after the spring of 1973, at which time an improved purification system at the pigment manufacturing plant at Besigheim (Enz/Neckar) commenced operations (see Chap. D.8.3).

Table 33. Metal contents (μg/l) in waters of the Rhine and several tributaries after data by Heinrichs (1975). The numbers refer to the measuring stations in Figure 27

	Cr	Mn	Fe	Cu	Zn	Ag	Cd	Hg	Pb
1. Rhine (Stein)	0.5	4.7	4.7	1.1	4.3	0.25	0.1	0.04	0.6
2. Neckar (Heidelberg)	2.7	3.8	21.	4.6	18.	0.23	1.6	0.40	2.6
3. Main (Hochheim)	3.8	131.	84.	8.3	222.	0.24	0.5	0.14	0.7
4. Mosel (Alken)	3.8	3.8	54.	3.0	146.	0.12	0.4	0.04	1.3
5. Ruhr (Essen)	2.7	3.4	9.0	8.3	81.	0.25	0.6	0.08	0.2
6. Rhine (Hamborn)	11.	62.	67.	17.	201.	0.51	3.7	0.49	4.2

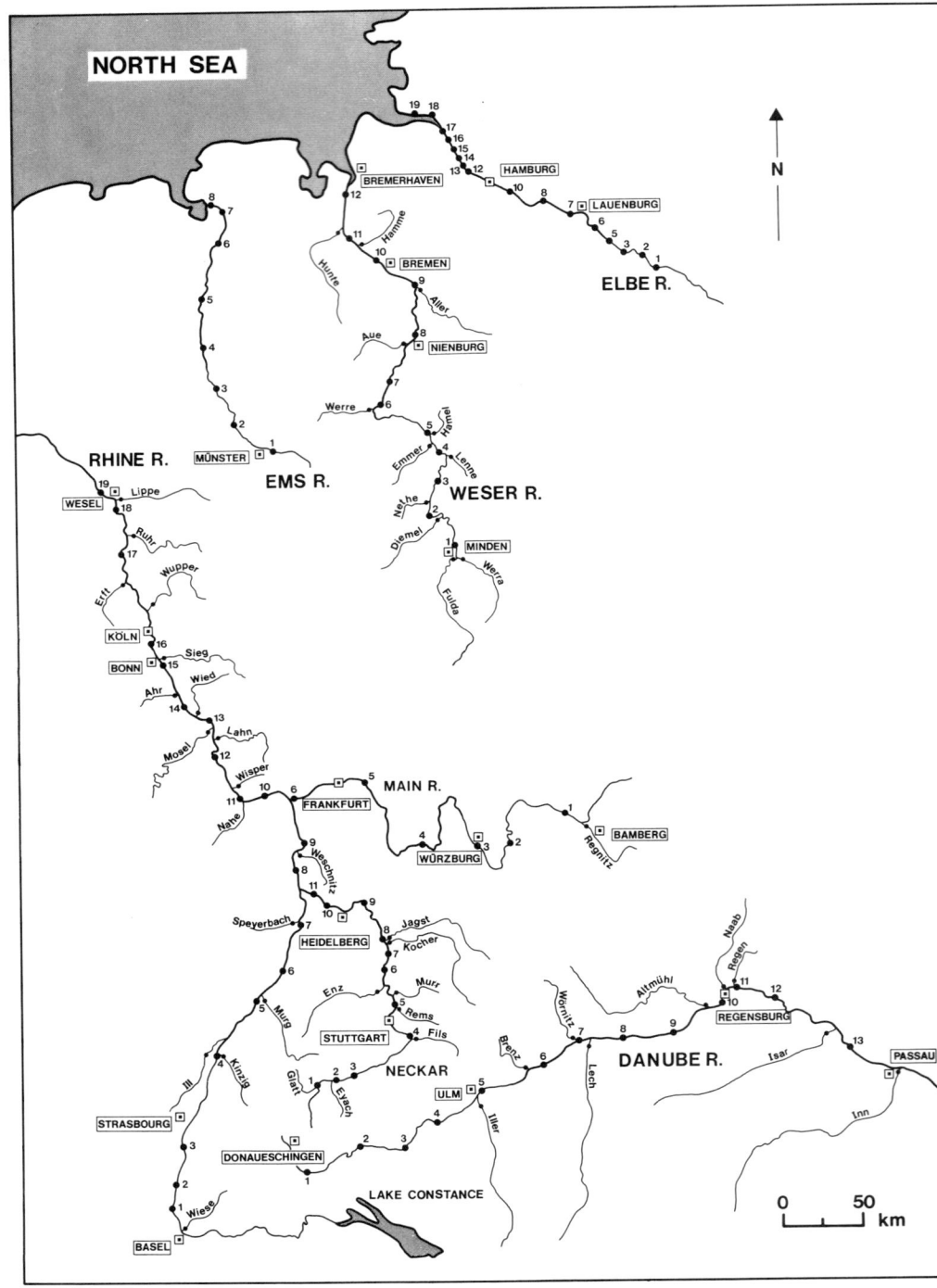

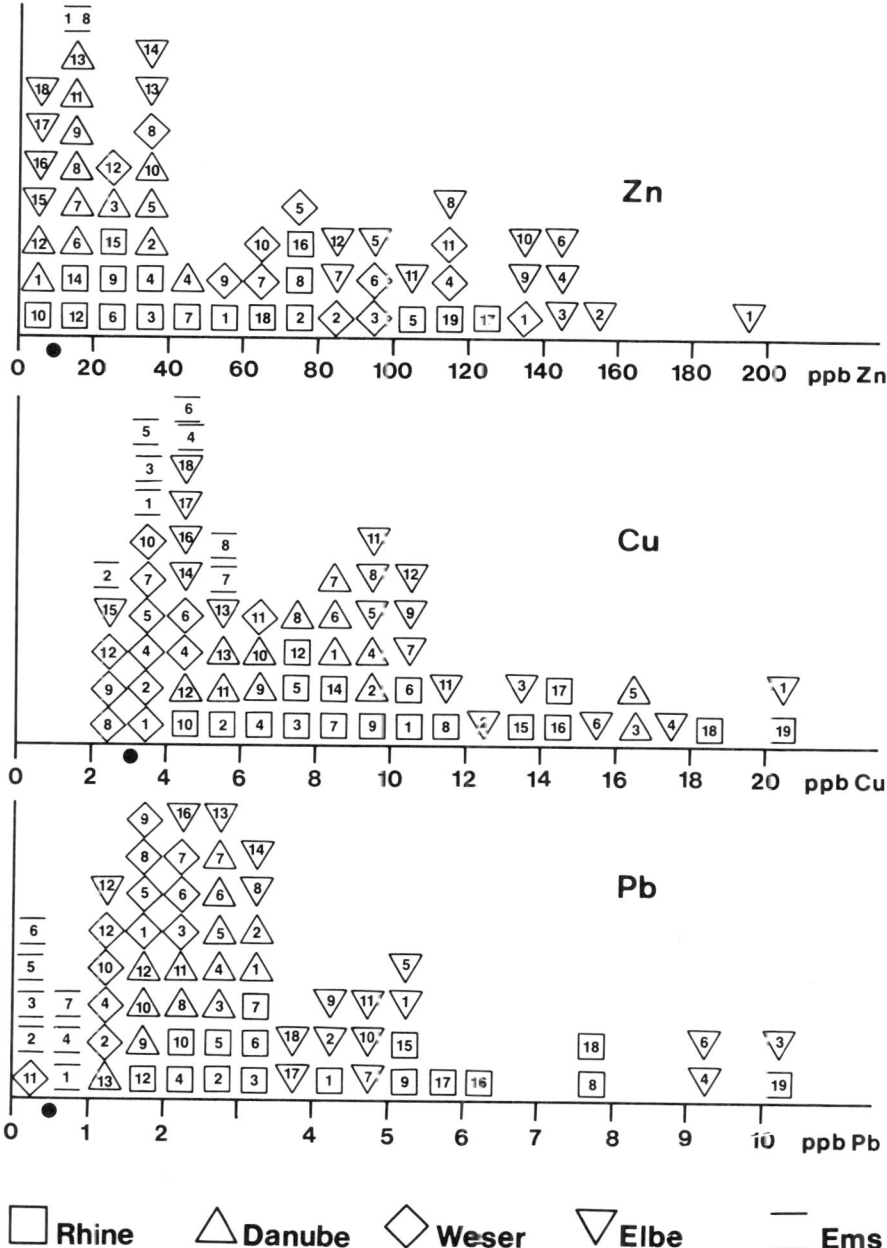

Fig. 26. Metal studies in major rivers of West Germany: *left* sampling locations, *right* frequency distribution of Zn, Cu, and Pb data

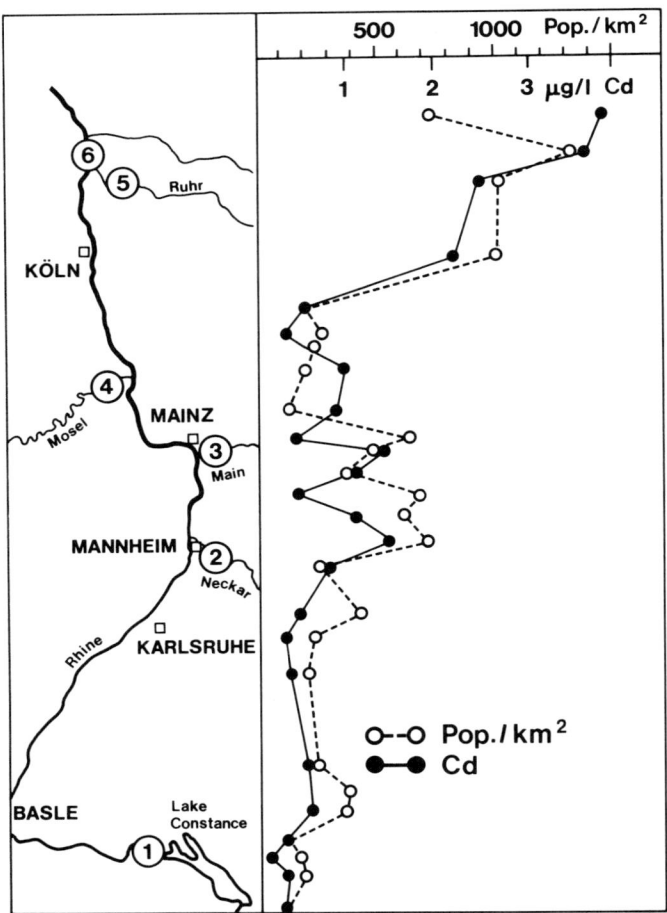

Fig. 27. Comparison of population densities and cadmium concentrations in the catchment area of the river Rhine (Federal Republic of Germany; from Heinrichs, Dissertation Göttingen 1975; reproduced with permission of the author)

4.2.4 Heavy Metals in River Water of the U.S.S.R.

Investigations of heavy metals in surface- and groundwaters of the U.S.S.R. have a long tradition. An early goal was to prospect the economic mineral deposits, exemplified by the investigations of Ududov and Parilov (1961) on 3490 samples from rivers and lakes in Siberia (see Appendix A). In more recent studies toxicologic and environment-hygienic aspects are generally considered.

An example is the study of Konovalov and Nazarova (1975) on the heavy metal load of the European U.S.S.R. (west of the Ural Mountains). A distribution map for copper found in this study is reproduced in Figure 28.

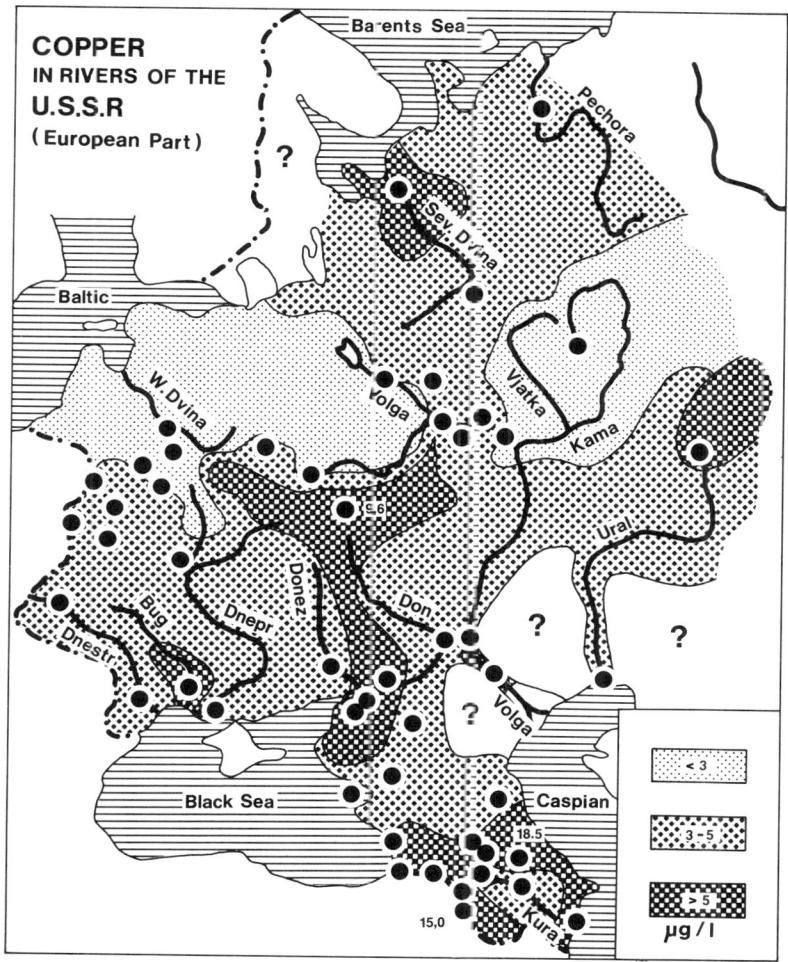

Fig. 28. Concentrations of dissolved copper in major rivers of the western U.S.S.R. (after Konovalov and Nazarova, Gidrokhim. Mater 62, p. 41, 1975)

During typical hydrologic phases between 1971 and 1972, samples from 45 sites on twenty larger rivers were investigated at: (1) normal water levels in winter, (2) beginning of spring highwater, (3) maximum spring highwater, (4) falling spring highwater, (5) and normal water level of summer. From each investigation site at least three samples and no more than 35 (an average of 10 per site) were measured for the concentrations of B, F, I, V, Ni, Mn, Cu, Zn, and Mo. The results are shown by three distinct levels in the map: especially low contents of copper (< 3 µg/l) were determined in the river water of the western Dvina and its tributaries, the upper course of the Volga and Vjatka Rivers, and large areas of the Kama River catchment basin. Greater Cu-concentrations (> 5 µg/l) were found in the drainage area of the Dnjepr and Don Rivers

(9.6 µg/l) and especially in the upper course of the Kura River (mean value up to 18.5 µg/l). This enrichment appears to be partially due to anthropogenic influences, but this aspect of the study was not discussed in detail.

4.2.5 Heavy Metals in Waters of Japan

Minamata disease–methylmercury poisoning–was first reported late in 1953 (Fujuki, 1972). However, the *mercury* content was not determined until 1959, when waste water from the vinylchloride and acetaldehyde plants was found to contain 0.1 µg/l and 20 µg/l Hg, respectively. Larger quantities of mercury were discharged with drainage waste when the plants were shut down and cleaned (four times annually). According to analyses from July and October 1966, the mercury concentrations of the waste from the acetaldehyde plant were 366 µg/l and 336 µg/l. In the investigations of May 1968 the concentration of mercury in the waste water from the vinyl-chloride plant was 40 µg/l; of that, methylmercury was 0.3 µg/l (Fujuki, 1972). In 1974 the Environmental Agency of Japan released the results of a national survey on mercury pollution in water, sediments, and fish (Anon. 1974, cit.: Buffa, 1976): approximately 3800 water samples from more than 600 sites were analyzed and found to be generally below the environmental quality standard of 0.5 µg/l total mercury. Two percent of the samples collected in areas where sediments have shown high mercury levels (Chap. D.8.3) exceeded the standard.

In a review on pollution with inorganic chemicals in Japan, Goto (1973) reports that 35 areas are polluted with *cadmium*. Particularly high concentrations were found in some parts of the Jintsu River, where the so-called Itai-Itai disease occurred in Toyama Prefecture. The analytic values of cadmium in this area were as follows: well water: 1 µg/l; mine waste water: 5–61 µg/l (average 17 µg/l); river water: 1–9 µg/l (Goto, 1973).

The effects of the Togane *arsenic* mine north of Nagoya (Gifu-Prefecture) on the water of the Wada River (the mine ceased operations in 1957) was studied by Kato et al. (1973) from March 1972 to February 1973. The As concentration in the river water was usually less than 30 µg/l. However, during the high water period, a maximum value of 1.44 mg/l was found near the closed mine, whereas 2 km downstream the As value was 40 µg/l.

4.3 Metal Transport in Freshwater Systems

Two factors are important for the transport of trace metals within freshwater systems, both of which show strong temporal variations, water discharge and biologic productivity. The first factor is especially significant in rivers, the second in lakes.

4.3.1 Water Discharge and Metal Transport

It has been emphasized by Colby (1963) that "difference in source of sediments at the time of the peak flow, as compared with that at the time of the peak sediment concentration for certain storms in a river system might be significant in studies of waste dispersal". Different dispersal behavior of the water constituents, and especially of the sediment fractions, must be considered when evaluating the sources and distribution of pollutants in rivers. A general discussion of the possible relationship between metal concentrations and discharge was presented by Hellmann (1970a) from the example of the zinc loads in the River Rhine (Fig. 29 is a reproduction of his Figure 4, including some simplifications made by Wilson, 1976). The heavy lines represent the major developments of the non-filtrable (~ solids) and filtrable (~ dissolved) fractions of the metal load with increasing discharge: (1) a decrease in the filtrable fraction by dilution, and (2) an increase of the non-filtrable fraction with discharge as a result of re-suspension of particles from the river bed and banks. The shaded areas on the upper section of both strong curves indicate minor effects of sorption, as in the case of the solid fractions, and remobilization from the particulates, which increases the dissolved metal load, i.e., the filtrable fraction, to some extent. The decrease of the amount of sorbed cations with increasing discharge can be explained by the higher percentage of relatively coarse material, which usually exhibits lower exchange capacities, by a lesser amount of dissolved cations, due to dilution, and the shorter residence times of both solids and dissolved ions in the river channel, which in turn influence the attainment of equilibrium between both phases. On the other hand, metal cations are increasingly released from solid substances into the aqueous phase at higher water discharges owing to desorption or dissolution processes. There does not appear to be any significant variation in the total metal load with changing water discharges; the decrease of this curve, for intermediate water flows (Fig. 29) should be considered as being rather hypothetical. It will be shown in Chap. D and from data of Hellmann and Griffatong (1972) that deeper erosion of the river channel—for example after a storm—will effect a significant lowering of the concentrations of solid pollutants, and may thus considerably influence the transport characteristics of the metals.

Investigations into the dependency of trace metal contents from water discharge were carried out on the Rhine River by Schleichert (1975). At the sampling site at Koblenz on the middle section of the Rhine, water samples were taken every working day between March 1973 and March 1974. The particles larger than 0.45 μm were then separated using a membrane filter and the heavy metal contents were determined by X-ray fluorescence analysis.

In spite of the considerable fluctuations, which are confusing at first sight, important conclusions can be made by taking the discharge data into consideration: (1) each discharge maximum is associated with a concentration minimum, (2) the concentration changes in trace elements occur more or less in the same form, and (3) extreme concentration peaks, independent of discharge, are rare. The particular differences in the development of the summer and winter values will be discussed in Sect. D.6.6, which deals with the problem in various catchment areas. Chromium was used as an example of the development of metal contents in suspended material in the presence of variable discharge quantities, as it does not follow the pattern of other metals in the Rhine catchment area, e.g., cadmium which reveals differences between their summer and winter values (Fig. 30a). Unfortunately, there is still little information regarding these phenomena, probably due to the relatively large expenditure pertaining to this particular type of study. According to data by Schleichert (1975), the corresponding chromium contents in dissolved state of the Rhine River are also shown in dependency on

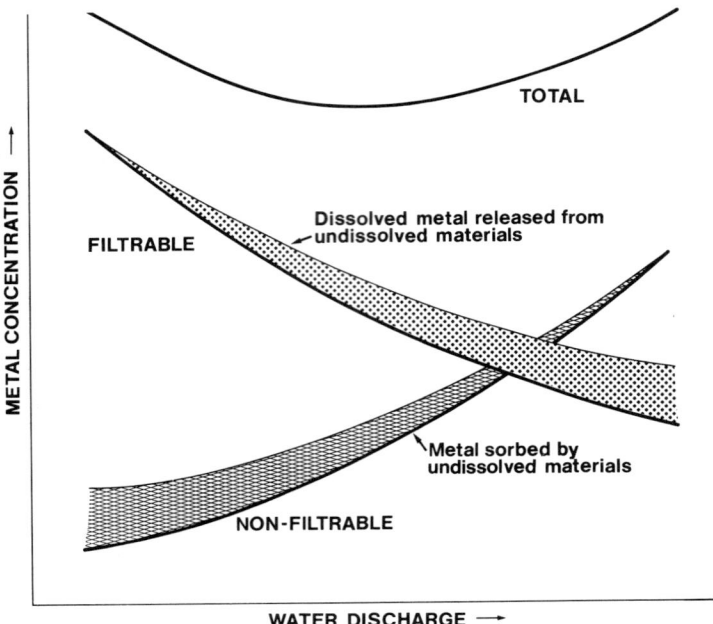

Fig. 29. Schematic presentation of the transport modes of trace metals (example: zinc) in rivers (after Hellmann, 1970a; modified by Wilson, 1976; with permission of Water Research Centre, England)

the water discharges (Fig. 30b). This indisputable relationship was determined here only in the case of chromium.

It should be noted that other investigations on heavy metal transport by suspended sediments (e.g., Turekian and Scott, 1967; Angino et al., 1974) — to be treated in greater detail in Sect. D.8 — have not given such strong indications of a relationship between trace metal content of the suspended phase and water discharge as is the case with the Rhine River, which is probably due to a lesser degree of contamination. In this context the findings of Aston and Thornton (1977) in their study from Cornish catchments are of interest, as they found a significantly smaller variation of heavy metals in both sediments and water of unmineralized areas than in the tributaries of the Carnon, Red, and Gannel Rivers which are influenced by past and present mining and smelting industries.

A clear dependency of the metal concentrations in suspended sediments on the varying loads of the suspended matter was found by Wagner (1976) both for more strongly and for less contaminated rivers feeding into Lake Constance. The examples of the cadmium and lead concentrations in the Rivers Rhine (less polluted in this part of its course) and Schussen (polluted from both domestic and industrial sources) indicate that these effects are particularly pronounced at higher pollution intensities: as the concentrations of suspended matter increase from approximately 20 mg/l to more than 700 mg/l, the concentrations of cadmium decrease from 15 ppm to 1 ppm in the Schussen and from 2 to approximately 0.2 in the Rhine. The lead concentrations decrease from 220 ppm to 120 ppm in the suspended matter of the Schussen and from 120 ppm to approximately 50 ppm in Rhine particulates. The latter example draws attention to the concentrations of lead which are influenced most by diffuse sources such as surface runoff from rural areas and storm water runoff from municipalities, both of which are contaminated above all by atmospheric lead emissions; the strong increase of cadmium in the suspended sediments of the Schussen is however probably due to industrial point emissions.

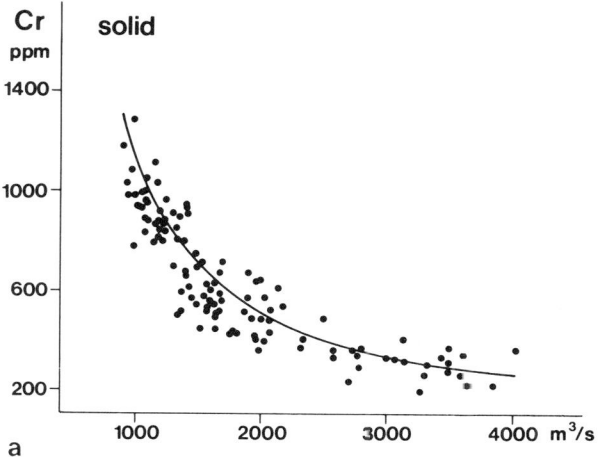

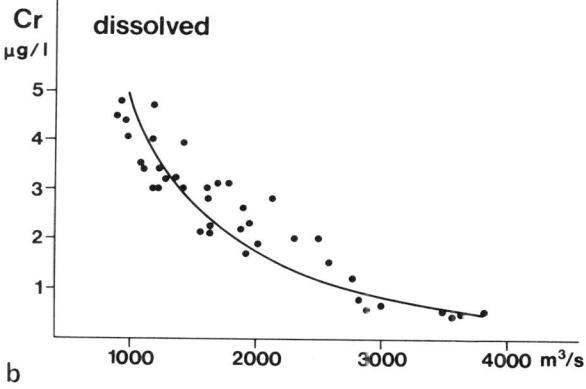

Fig. 30. Metal transport vs. water discharge. Chromium concentrations in suspended solids (a) and in dissolved phases (b) of the Rhine River (Schleichert, 1975; with permission of Bundesanstalt für Gewässerkunde, Koblenz)

4.3.2 Annual Cycles of Metal Transport

Recent studies on the transport and distribution of trace metals in freshwater systems have revealed characteristic developments in annual cycles. Carpenter et al. (1975) and Troup and Bricker (1975) investigated the temporal variability of trace metals transported by the *Susquehanna River* to the Chesapeake Bay. At first they found that the trace metal concentration correlated well with the amounts of solids discharged: the concentrations seemed to be highest in the spring and lowest in the summer and fall. Upon closer inspection, however, Mn, Ni, Zn, and Co exhibit large concentrations in January, and Cu, Cr, and Mn have concentration peaks in the late spring or early summer. When data are calculated for weight concentrations of metals in the solid fraction, it is found that all metals generally peak during December and January and secondary peaks occur for Co, Cr, Ni, Cu, and Mn in July. Troup and Bricker (1975) suggest that since decaying organic matter is abundant in the Susquehanna River during these two periods, the high concentrations may be the result of metal bonding to particulate matter.

Grimshaw et al. (1976) investigated the seasonal variations in the concentration and supply of dissolved zinc to polluted aquatic environments in mining areas of mid-Wales. They found that a "flushing" effect occurs for a short period at the onset of increasing runoff, during which a rise in metal concentration takes place. It is suggested that summer convection storms in mining areas are particularly significant for river biota, which may also explain observed variations in metal concentration in coastal waters, in addition to changes arising from varying rates of metabolic activity.

An important factor in controlling the trace metal content in natural waters is the ability of planktonic material to adsorb some metals from solution. These effects have been studied by Abdullah and Royle (1974b) in two surveys carried out in April and June, 1971 on the *Bristol Channel* (where metal enrichment originates from both mineralization zones and waste disposal; see p. 97). The plot of the amount of acid extractable metal present in the suspended matter against the weight of solid in suspension shows a first-order relationship for Zn, Cd, and Cu during *April*. This uniformity suggests that the distribution of the particulates is controlled by mixing and turbulence and that little or no fractionation by settling takes place in the area studied. For the *June* data zinc, copper, and cadmium plots show positive anomalies at stations situated in the outer part of the Channel indicating that the additional metal is due to agencies other than runoff. Similar seasonal effects have been found by Knauer and Martin (1973) from studies on the uptake of Cu, Zn, Mn, Cd, and Pb by plankton in Monterey Bay, California (see Chap. F).

Baccini (1976) studied the concentrations and sedimentation of the mesotrophic *Lake of the Four Cantons* (Bay of Horw) in Switzerland. Here the input of metals is

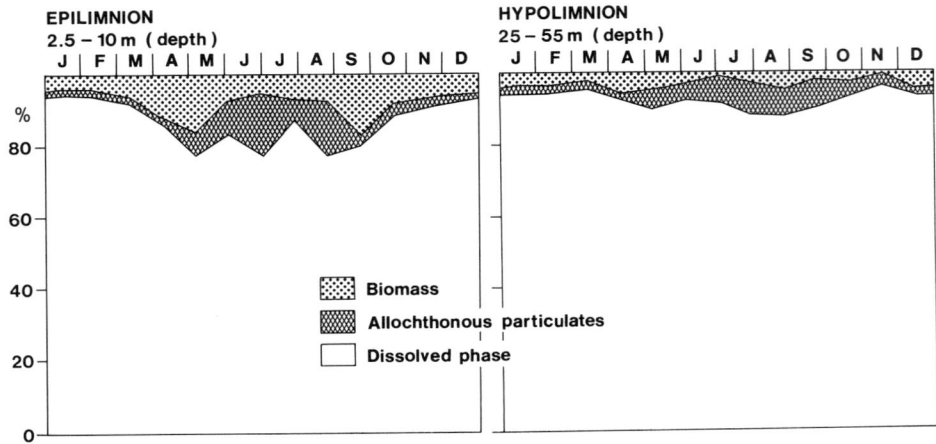

Fig. 31. Seasonal variations of copper concentrations in biomass, allochthonous particulates and water of the Bay of Horw (Lake of the Four Cantons; Baccini, 1976)

predominantly from rivers. While iron is retained almost completely, nearly 50% of the copper, zinc, and cadmium is transported further. A comparison of the retention of the dissolved parts shows that copper is only retained to a small extent (0%–25%). Temporal variations in metal distribution in the three phases—biomass, allochthonous particulate, and dissolved phase—were calculated in a model in which particulate iron serves as an indicator for the allochthonous particulate material phase and particulate organic nitrogen as an indicator for the biomass phase. Figure 31 shows the example of copper from the data of Baccini (1976): a comparison of the three phases of copper reveals that the biogenic portion of the total particulate phase from the epilimnion is approximately the same as the portion of the allochthonous phase and can even be higher during periods of high production. The distribution in the hypolimnion, however, indicates that the allochthonous portion dominates during periods of greatest sedimentation (May to September). Three important conclusions were drawn (Baccini, 1976):

1. Copper and zinc introduced in soluble form are transported into particulate through the agency of plankton. The biogenic portion of the total concentration varies between 5% and 15% for copper and 5% and 20% for zinc.
2. The decomposition of sedimentary plankton permits the partial return to solution of the metals and the amount in the solution phase rises in the hypolimnion.
3. The allochthonous particles are deposited on the lake bottom relatively unchanged.

From investigations on the partitioning and transport of lead in Lake Washington it was shown by Baier and Healy (1977) that the path of Pb to the sediments (70% of the total input) is not direct but involves cycling through both liquid and solid phases. Differences in Pb content of seston between winter and summer arise predominantly from dilution by fixed carbon resulting from the higher biologic productivity in summer. The relative large amount of Pb sorbed on particulates potentially can produce a toxic benthic environment. Data from waters of the English Lake District, which are affected by lead mining, indicate that littoral macrophytic vegetation plays a key role in the recycling of heavy metals within a lake system, that sometimes—at higher metal accumulation rates—gives rise to faunistic abnormalities (Welsh and Denny, 1976).

Chapter D
Metal Pollution Assessment from Sediment Analysis

The tracing of pollution sources by means of water analysis frequently gives rise to difficulties which may usually be ascribed to sampling procedure and the physical-chemical condition pertaining to the investigated species rather than to the accuracy and precision of analytic techniques. Determination of heavy metal concentrations in surface water samples collected from rivers at short intervals reveals fluctuations of several orders of magnitude. These may be partially explained by variations resulting from water discharge and the predominance of certain source areas leading to irregular effluent emissions. This phenomenon has been discussed in Chap. C. Exchange processes between interstitial and surface water effect fluctuations along the river course. These processes are influenced by pH and Eh conditions, the type and quantity of complexing agents, their biologic activity and by salinity and temperature. Therefore, in addition to having limited experimental data available, the extremely complex system of influencing factors generally makes it impossible to arrive at any definite conclusion regarding the source of pollution.

1 Introduction

1.1 Soluble/Solid Equilibrium

A special difficulty arises from contaminants which are not readily soluble but become rapidly fixed to particulate matter in the receiving water body. This applies, in particular, to the heavy metals chromium, mercury, and lead. Even close to the point of input, the metal content in water decreases to its normal level making detection difficult, except by means of a very closely knit system of water sampling.

These interrelationships are best illustrated by means of a practical example. Figure 32 shows the distribution of mercury in Lake Ontario, in surface and in bottom water (Chau and Saitoh, 1973) as well as in sediments (Thomas, 1972). This proves that no relationship exists between the mercury distribution in *surface and bottom water;* it is also impossible to pinpoint a specific pollution center from the lateral distribution of the water data (with the exception of elevated mercury concentration on the northern shore of the lake, which is only substantiated, however, by a single analysis). In contrast, metal contents in the *sediments* reveal quite different distribution patterns, having a distinct grouping of elevated mercury levels near the southern bank of the lake, especially close to the mouth of the Niagara River. Detailed investigations carried out by Fitchko and Hutchinson (1975) have in fact shown that the dispersion pathways point to the Niagara River as the prime source of mercury input to Lake

Soluble/Solid Equilibrium

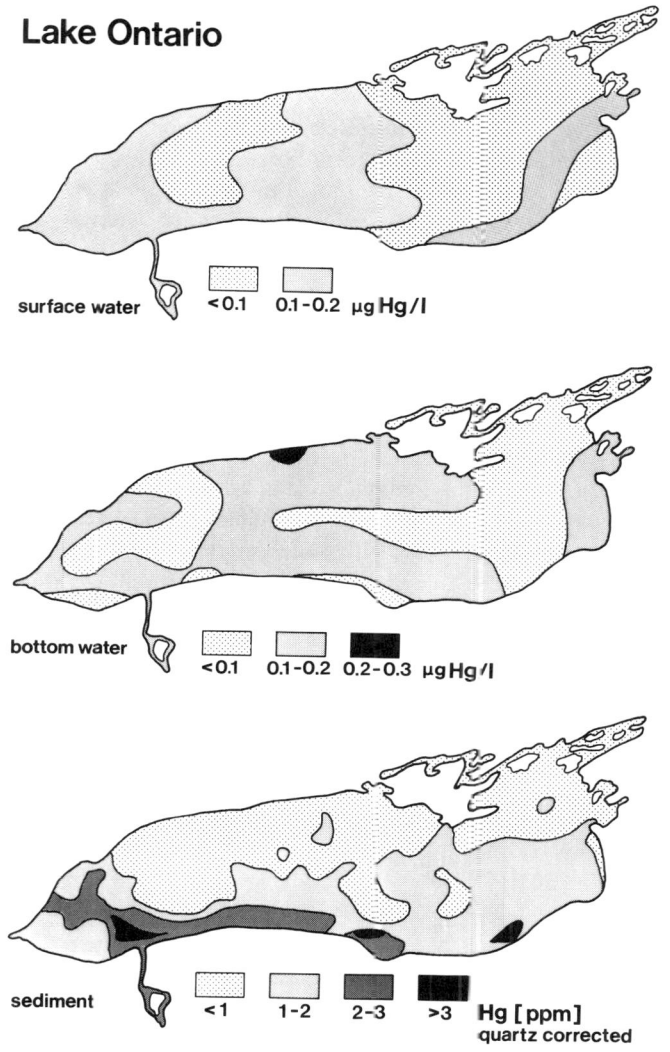

Fig. 32. Concentrations of mercury in surface and bottom waters (Chau and Saitoh, 1973) and in sediments (Thomas, 1972) from Lake Ontario

Ontario. The eastward extension of the Niagara mercury "plume" is probably due to coastal currents (Simons, 1972), whereas the enrichment of mercury at the mouth of the Genesee and Oswego Rivers appears to result from the contamination within the catchment areas.

The establishment of metal levels in sediments can therefore play a key role in detecting sources of pollution in aquatic systems. Although sediment analyses do not represent the extent of intoxication, they can be employed on a semiquantitative basis in comparative studies to trace the sources of pollution, such as surreptitious discharges by factories. Under favorable conditions, pollution sources can even be detected long after input has taken place. Furthermore, it is possible to determine their development of pollution intensity from dated sediment cores provided they contain fine-grained depositions, in which the sorbed, precipitated, organically bonded metal concentrations are accumulated. Finally, an investigation of sediment particles is likely to be useful from an analytic point fo view since trace element content in particulate matter is 1,000 to 100,000 times higher than the corresponding metal content in the associated aqueous phase.

1.2 Surface Samples and Sediment Cores

The example from Lake Ontario has provided evidence that lateral variations in the chemical composition of surface sediments act as a guide to local pollution centers. Similarly, qualitative profiles of sediment data along a river's course can be used to evaluate characteristic influences from industrial, municipal and agricultural sources, provided that the grain-size effects—which strongly influence the metal values—are taken into account. One field of sedimentary investigations, which is particularly useful in the present context, is marked by the study of vertical profiles from fine-grained deposits in lakes, impoundments, estuaries and coastal basins. Sediment cores provide a historical record of events occurring in the watershed of a particular lake or bay and enable a reasonable estimate of the background level and changes in input over an extended period of time.

This approach is especially valuable if the rate of sedimentation is known. Modern dating methods of recent sediments entail the determination of lead-210 contents, a radioactive isotope produced in the U-238 series. Furthermore, cesium-137 contents resulting from radioactive fallout achieved maximum values during the period from 1959 to 1960 and again in 1964. Examples from lacustrine sediment cores have been given by Krishnaswamy et al. (1971); Ravera and Premazzi (1971); Ritchie et al. (1973); and Robbins and Edgington (1975). Further attempts to date sediment cores are discussed in the later sections.

Variations in the stratigraphy of Recent lake sediments were first performed to trace cultural activities on the rate of eutrophication by Hutchinson and Wollack (1940) in Linsley Pond, Conn., by Murray (1956) from Wisconsin lakes, and by Ohle (1956) from lakes in northern Germany. Man-made effects have been evaluated from the distribution of phosphorus and other nutrients (Livingstone and Boykin, 1962; Whiteside, 1965; Frink, 1967; Wentz and Lee, 1969; Duthie and Sreenivasa, 1971), from the distribution of iron monosulfide (Müller, 1967) and in changes of diatom assemblages (Stockner and Benson, 1967). Pollen variations in Recent lake sediments reflect historical changes in land use (Vuorela, 1970; Solomon and Kroener, 1971; Anderson, 1973; Kemp et al. (1974). Poon and Sheih (1976) found a close relationship between phosphorus and iron from sedimentary cores both in the lacustrine and marine coastal environments. Kemp et al. (1976) made a thorough study on the simultaneous increase of these nutrient elements and of heavy metal from short, dated core profiles. Sedimentary cores studies are particularly valuable for applications in mining areas where anomalous metal distribution patterns due to geochemical and man-made influences may overlap. We will give examples of the distinction of both influencing factors by comparison between the natural background data in the deeper parts of the cores profiles and the sometimes extreme enrichments within the top layers of the sedimentary sequences.

2 Metal Investigations on Aquatic Sediments

Investigations on metal pollution in water by means of sediments commence with the determination of general hydrologic data such as: water temperature, pH-Eh values and conductivity and, in the case of river systems, water flow rate at the time of sampling. For routine investigations on sediments, an analytic procedure, represented here in graphic form, was conceived. The individual steps of this procedure are summarized in Figure 33.

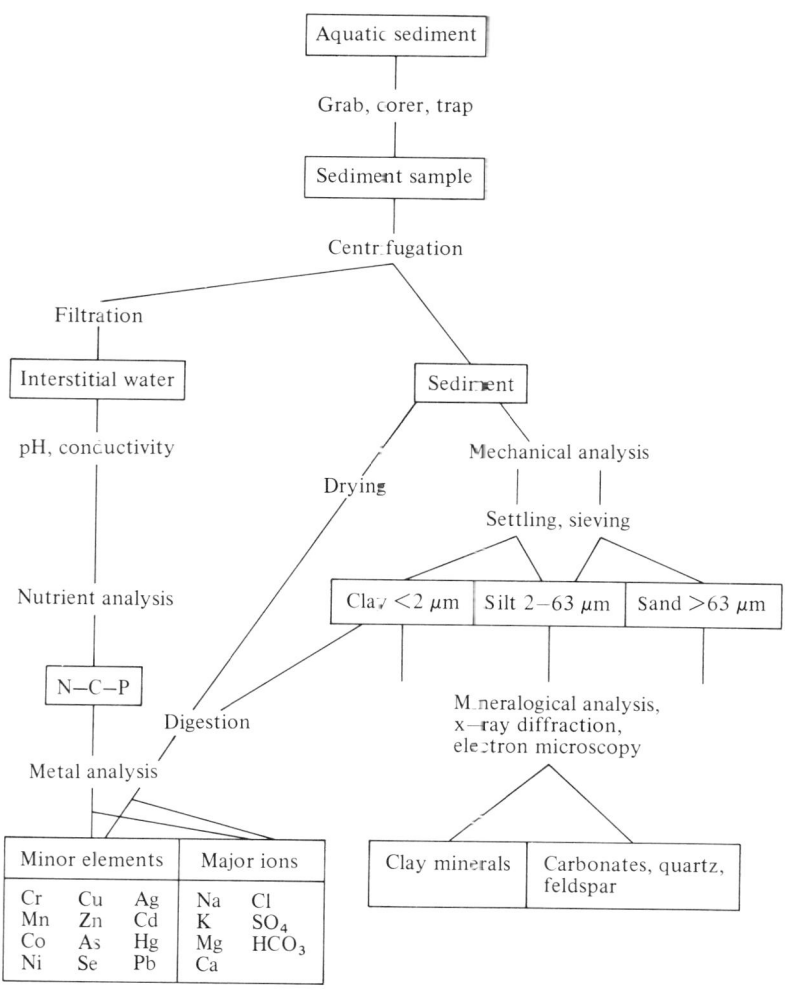

Fig. 33. Schematic sequence for analytic procedures in metal pollution assessment from sediments

2.1. Sampling and Storage

Several devices have been developed and marketed for the extraction of surface sediments, sediment cores, interstitial water and suspended material. In most cases the particular apparatus serves only one purpose, as briefly discussed below.

2.1.1 Soils and Sediments

Sampling and storage of soils, rocks, minerals and dry sediments present fewer problems than those encountered with water and biologic sampling material (see Chap. B.5). Nonetheless, it is advisable to observe stricter precautionary measures than those incorporated into sampling procedures generally employed (Maienthal and Becker, 1976). For instance, the use of an ordinary spade for soil sampling is preferable to the use of an auger (Morrison and Pierce, 1974). Tools which are encased in Teflon have been especially recommended (Patterson and Settle, 1975); in all cases the material selected for analysis should be collected from the inner part of the sample material which has not been in direct contact with the metal of the sampling device. Another important requirement is to observe the textural features of the sample and make a preliminary classification according to grain-size. When sampling from beneath water coverage, great care must be exercised to leave the top layers undisturbed (Hakanson and Uhrberg, 1973).

2.1.2 Grab and Core Samplers

Sampling from sediments below water coverage depends on the type of sample required. Coarse-grained consolidated material may be recovered—even from beneath overlying ice—by means of a mud-grab. This spring-loaded device is constructed from noncorrosive material and capable of extracting samples of 15 cm in diameter.

Normally, fine-grained bottom sediments are collected with the aid of a sediment grab of the Van Veen type. The catch that keeps the two bowl-shaped sections of this dredger apart is released upon making contact with bottom sediment; withdrawal subsequently leads to a closure of the half sections and capture of some 2 kg of sediment material with a penetration depth exceeding 20 cm.

If special importance is attached to collecting undisturbed sediment samples, the Ekman-Birge grab is most suitable. The box is furnished with two flaps which are spring-loaded and enclose the sample material upon withdrawal. A modification of this grab after Lenz enables the subdivision of the sample into five profiles of 20 cm thickness by means of inserting horizontal plates.

The increased interest displayed in the reconstruction of historical developments of pollution influences has led to widespread sampling of sediment cores. Such profiles may cover the last two hundred years of industrial development and in accordance with an average sedimentation rate for lacustrine and marine coastal environments of approximately 1 to 5 mm per year, which is applicable to moderately humid climates, the procedure entails sampling of a core of 20 cm to 1 m in length. Obviously, in areas where the annual sedimentation rate is much higher due to greater erosion, longer cores have to be extracted.

Gravity corers are used particularly in the study of sediment textures and evaluations of paleoclimatic changes for which cores of 2–3 m of fine-grained sediment material are required (under favorable conditions cores of 10 m can be recovered). A major disadvantage arises from the fact that the top section of the core—which is of particular interest in pollution studies—is often lost.

The Kullenberg piston corer (Kullenberg, 1947) has been widely applied in marine and lacustrine studies. This device has been occasionally modified: one such adaption for environmental studies allows the recovery of 1 m cores in a plastic liner (Poon and Sheih, 1976).

Box-corers are employed when undisturbed cores with a large cross-section are desired, such as needed for the investigation of sedimentary structures, soil mechanics, or pore-water chemistry of marine sediments. These devices usually recover a core having a surface area of 30 × 40 cm and a depth of 30–40 cm from the bottom sediment material.

2.1.3 Bottom Sediment Traps

These have been used during the past 50 years for the determination of sedimentation rates. When first introduced in Winona Lake, Indiana (Scott and Miner, 1936), 56 men were needed to handle these traps. Since then, many techniques have been developed which include bottles, funnels, and pots suspended from cables. Bottom sediment traps have been constructed even for sandy river deposits (for example Aurand and Behrens, 1966). Recent developments have been described by Håkanson (1976), a new bottom trap having been introduced. The bottom plate consists of Plexiglas which is roughened to increase friction; brass arms with lead weights increase the stability and eliminate disturbances by means of a sophisticated anchor system.

2.1.4 Suspended Materials

For their recovery methods are necessary which can register suspended material contents as low as approximately 1 mg/l. Müller et al. (1976) described a portable high-pressure filtration system (12 atm) in which 8 liters of river water are filtered through a 0.45 μm membrane filter within approx. 20 min. These filters are constructed from stainless steel and are connected in pairs; the second filter is silver-coated to permit direct analyses by X-ray diffraction techniques. Other methods for the recovery of particulate matter are continuous-flow centrifugation and suspended sediment traps, which are immersed into the flow medium.

2.1.5 Recovery of Pore Waters

Investigations of pore solution are of particular interest with regard to post-depositional reactions and diagenesis. Pore waters have been extracted from sediments by leaching (Emery and Rittenberg, 1952; Kullenberg, 1952), by centrifugation (Powers, 1957), and by squeezing (Manheim, 1966). Instruments were developed in which inert gases were utilized to prevent oxidation during water extractions. Nonetheless, it has been shown that temperature and pressure exert a strong influence on the composition of interstitial waters (Fanning and Pilson, 1971) and that much of the difficulty attached to the study of dissolved species in natural aquatic systems arises from collecting samples without disturbing the physico-chemical conditions in the medium to be investigated. Sayles et al. (1976) constructed a sampler for the in situ collection of marine sedimentary pore waters. Filtered pore solutions from six depths in the upper two meters of sediment and bottom water were extracted by a hydraulic pump system and stored in capillary tubes (5–20 cm^3). The sampling device was successfully tested and used in a wide variety of sediments including calcareous oozes, terrigenous muds, and pelagic clays at depths up to 6000 m.

In situ separation of dissolved and particulate material has been successfully employed for sampling interstitial water from lake sediments (Mayer, 1976). The technique essentially consists in placing a dialysis bag with particle-free distilled water into the sediment and allowing for equilibration to take place between inserted and ambient water—some 7 to 8 h at 25 °C. The dialysis bag is then retrieved and the contents filed into a sample bottle for subsequent analysis. Hesslein (1976) has described an in situ sampler for pore water studies where dialysis membranes are incorporated to the aqueous phases at 1 cm intervals after equilibration with the surrounding water. Further analysis of interstitial waters is discussed in Chapter B.

2.1.6 Storage

The storage of the solid materials is subjected to stringent conditions. Usually air-drying should not have much effect on the total trace element content, but if speciation, organic extractable trace elements, etc. are of interest, any procedure of drying may lessen the validity of the sample analysis (Maienthal and Becker, 1976). For instance, air-drying of the sample markedly affects cation exchange capacity and iron speciation, and appears to decrease the pH slightly (Schalscha et al. 1965). Furthermore it was stated that the longer a soil or sediment sample is kept air-dried, the greater the amount of water soluble and organic material that can be extracted (Birch, 1960). Drying or freezing may result in irreversible change which will also affect the complexation state of the trace metals (Hesse, 1971). In the opinion of many authors freezing at $-20°$ to $-30\,°C$ is the best method of storage.

2.2 The Mechanical Sediment Analysis

Mechanical sediment analysis will be discussed here only in such detail as is of consequence for heavy metal studies of river, lake, and sea deposits. A detailed description of procedures and statistical evaluation of the data can be found in Müller (1964). Mechanical analysis normally begins with the sieve procedure, the separation of the sand fractions (> 63 μm) from the finer-grained sediment particles. These are then separated into silt and clay fractions by allowing the heavier particles to settle using water as suspension medium. This method, however, can cause a deceptive contamination of the sediment due to metal corrosion of the sieve walls and base. For this reason, it is preferable to place all samples irrespective of grain-size in Atterberg settling tubes and separate the < 2 μm fraction by repeated suspension and settlement cycles (five to ten times). This is followed by a separation of the increasingly coarser silt fractions.

[In our studies on aquatic sediments, evenly spaced intervals were selected for the subdivision of the silt and centrifuged clay fractions into distinct grain sizes. The silt fraction was divided into grain size intervals of 2–6.3 μm (fine silt), 6.3–20 μm (medium silt), and 20–63 μm (coarse silt). The intervals in the clay fraction were correspondingly fixed at < 0.06 μm, 0.06–0.2 μm, 0.2–0.63 μm, and 0.63–2 μm. Given a height of fall of 25 cm and a temperature of 22 $°C$, the rates of fall (according to Stokes's law) were as follows: for the fraction 2 μm = 17 h 30 min, for 6.3 μm = 2 h 45 min, for 20 μm = 8 min 20 s.] The final step is the separation of the sand content by sifting (through plastic sieves). The fraction < 2 μm can be separated further by centrifugation. This is particularly useful when a metal compound is enriched in a limited range of grain size fractions, e.g., caused by a specific industrial manufacturing process, from which it has to be determined more accurately by chemical or X-ray analysis. The coarser grained material is sieved at smaller intervals than silt and clay fractions in order to carry out a more accurate evaluation of the histrogram and summation curves. The terms "medium grain size" and "sorting" are particularly informative with regard to the transportation and deposition conditions of the sediment in question.

The above-mentioned procedure is ambiguous, especially with regard to substances which have abnormal specific gravity values, as in the case of sewage sludge or sediments with a high content of organic carbon. The settling rate of the fraction < 2 μm for silica, is approximately 10 mm/h according to Stokes's Law. A 2 μm diameter particle or organic solid with a specific gravity of 1.1 should settle approximately 20 times more slowly. This leads to completely incomparable results and it has therefore been proposed that in such cases the grain-size differentiation should be conducted by using filters and sieves (Chen and Lockwood, 1976). Modern filter membranes are capable of providing a base of closely controlled pore sizes, ranging from 0.2 μm (and possibly smaller) to 8 μm. Processes of filter clogging present particular problems in this procedure.

2.3 Mineralogical Analysis

The main aim of the mineralogical analysis is to determine the origin of the individual sediment components; this is generally done by an investigation of the distinct grain-size fractions.

The *clay minerals* are determined from the fraction < 2 μm previously freed of most interfering organic substances by H_2O_2 treatment (cf. Jackson et al. 1950). The sample is disintegrated with ultrasonic waves and suspended onto a glass slide. These clay preparations are then successively run through an X-ray diffractometer (Jenkins and De Vries, 1970) in an air-dried state, after treatment with ethylene-glycol and heating to 350° or 500 °C in the range of $2°–32°\ 2\theta$ (CuKα radiation). The individual mineral components are determined from the position of their basal reflexes, whereby kaolinite and chlorite can generally be distinguished by the (002)/(004) peak splitting (Brown, 1961). Since a quantitative method of determining clay minerals does not exist in the strictest sense, and even semi-quantitative determinations require a great deal of work, the contents of montmorillonite, illite, chlorite, and kaolinite were established in a simplified manner as arbitrary values derived from the intensities of the strongest basal reflexes.

The *carbonate* determinations were carried out in each fraction. The total carbonate content is generally gasometrically measured by the CO_2 development in treatment with 6 N hydrochloric acid (Scheibler method, cf. Müller, 1964; "Karbonate Bombe": Müller and Gastner, 1971); the individual components calcite and dolomite were quantitatively determined by X-ray analysis.

Microscopic investigations of *heavy minerals* were carried out on the fine and medium sand fractions after separating heavy from light floating mineral particles in liquids of intermediate density (2.90–2.96 g · cm^{-3}; tetrabromethane, bromoform). A summary of diagnostic characteristics of heavy minerals is given in Milner (1962) and Füchtbauer and Müller (1970).

2.4 Chemical Analysis of Nutrient Components (C–N–P)

In waters contaminated with metals the nutrient budget is generally also disturbed. In particular, the contents of organic carbon, nitrogen and phosphorus are of importance with regard to the condition of a water body. In what follows, the analytic procedure for the determination of carbon, nitrogen, and phosphorus has been mentioned, partly in accordance with the Manual of Methods in Aquatic Environment Research of the Food and Agricultural Organization (FAO) prepared by Olausson (1975a).

2.4.1 Determinations of Oxidizable Matter (Organic Carbon) by the Chromic Acid Method

This method (Walkley and Black, 1934) excluded 90%–95% of the elementary carbon present as graphite and charcoal, and therefore only leads to an evaluation of the organic matter present in the sediment sample. Since carbon constitutes an average 58% of the soft organic residue in sediments, the carbon content can be converted to organic matter content by multiplying the former by a factor of 1.7. The sediment sample is either frozen or dried in an oven (105 °C for 24 h). The dried sample is ground in an agate mortar and then passed through a 0.2 mm nonferrous sieve. A carefully weighed amount of the powdered sample containing approx. 10–25 mg carbon is placed in a 500 ml conical flask. Using a pipette add exactly 10 ml of the 1N dichromate solution and mix carefully by swirling. Then add 20 ml of concentrated sulfuric acid and mix by gently agitating the flask for 1 min. Let the mixture react for 20 to 30 min. Dilute sample to 200 ml with distilled water, add 10 ml of concentrated phosphoric acid, 0.2 g of sodium fluoride and 1 ml of diphenylamine indicator. The sample is now back-titrated with 0.4 N ferrous ammonium sulfate solution. In the beginning the color of the sample is dull green and then becomes a turbid blue as the titration proceeds. At the end, this color changes sharply to a brilliant green, resulting in a sharp end point.

2.4.2 Determination of Kjeldahl Nitrogen

The Kjeldahl method is most frequently used to determine organic nitrogen in the form of ammonia nitrogen which results from biochemical decomposition of nitrogenous matter. Knowledge of the Kjeldahl nitrogen (total org. –N and NH_3) is an important pollution control, especially with regard to wastewater. In the modified Kjeldahl method described by Scheiner (1976) the liberated NH_3 is reacted with indophenol in hypochlorite medium and the resulting indophenol blue dye is determined spectrophotometrically by measuring the absorbance at 635 nm (1 cm light path). Reproducible results were obtained in the range of 0.02 mg/l in a buffered system.

2.4.3 Determination of Total Phosphorus

The sample is treated with concentrated nitric acid and concentrated perchloric acid in order to release the organic phosphorus. An acid ammonium molybdate reagent is added to a suitable volume of the filtered, digested sample. The resulting phosphomolybdate complex is reduced to an intense blue coloration by ascorbic acid; the intensity of the color is a measure of the spectrophotometric determination at 882 nm. Samples with a phosphorus concentration ranging between 20 and 1,500 ppm can be determined in this manner.

2.5 Sediment Digestion in Metal Analysis

The chemical analytic procedures have already been described in Chapter B. Atomic absorption spectroscopy, which is commonly used for metal analysis, necessitates the disintegration of the sediment sample. Determinations can be carried out using a suitable method chosen from one of the following procedures (further information is available in the standard works on silicate and trace analysis, e.g. Bennett and Hawley, 1965; Maxwell, 1968; Koch and Koch-Dedic, 1974; Hermann, 1975):

2.5.1 Hydrofluoric Acid Decomposition

The procedure which is usually followed for dissolving sediments for spectroscopic analysis employs either *sulfuric acid* or a mixture of *nitric and perchloric acids* in combination with *hydrofluoric acid*. Decomposition of the material is effected by reaction of hydrofluoric acid with silica forming gaseous silicon tetrafluorides. As the evaporation proceeds, most of the metal fluorides are converted to sulfates or perchlorates which in turn can be dissolved by hydrochloric acid (Bennet and Hawley, 1965). The $HF-H_2SO_4-HNO_3$ decomposition permits the ready extraction of the hydrofluoric acid residue, and the acid mixture can then be handled more safely; a definite disadvantage is the possible formation of sulfates which can be dissolved only with great difficulty. The advantage of the $HF-HClO_4$ decomposition is that the perchlorates—with the exception of K, Rb, and Cs compounds—are more soluble (Hermann, 1975). If the substances to be analyzed contain bituminous or other organic components, they must first be oxidized by heating in order to reduce the danger of explosion; this can also be achieved by adding nitric acid to the substance. The following procedure for sediment samples is used: Weigh 250 mg of the sample material in a platinum crucible and evaporate to dryness with a mixture of 5 cm³ HF and 2 cm³ HNO_3 on a sand bath. Repeat evaporation with 2 cm³ HNO_3. Dissolve the material with 5 cm³ dilute HNO_3 (15%) by heating for 30 min and transfer to a 50-cm³ volumetric flask.

2.5.2 Hydrochloric-Nitric Acid (Aqua Regia) Decomposition or Digestion by Nitric Acid

Used for the determination of more volatile elements, e.g., mercury, arsenic, and cadmium: Transfer 50 mg of sample material to a 50 cm³ measuring flask and treat with 3 cm³ HNO_3–HCl (1:3) on a sand bath for 30 min at moderate temperature (ca. 60 °C). Allow for cooling to room temperature and dilute with distilled water to 50 cm³. Alternatively, PTFE-bombs are used for digestion of sediment samples to be analyzed for Cd, Hg, and As: 100 mg of powdered sediments are weighed into the Teflon-bomb and 5 cm³ of aqua regia is added. The sealed bomb is heated at 110 °C for 2 h in an oven.

2.5.3 Lithium Metaborate Fusion (with Simultaneous Determination of Silica)

Place 50 mg of sample material in a platinum crucible and mix with approx. 200 mg $LiBO_2$ before heating for 15 min at a temperature of 1100 °C; chill the molten matter. Add 25 cm³ of 10% hydrochloric acid and dissolve the entire mass using a magnetic stirrer. Transfer quantitatively and dilute to 50 cm³ in a volumetric flask.

2.5.4 Transfer of Solid Suspensions into Graphite Cuvettes

Trace metals in solid material can be determined without digestion provided the particles are sufficiently fine-grained to ensure homogeneous distribution in suspension. This prerequisite is always met by the < 2 μm fraction. The danger here of secondary metal contamination is negligible.

Numerous investigations using sediment analyses in pollution control have shown that it is unnecessary to obtain full digestion of all sediment components, including metals bound into the internal structures of silicates and other detrital minerals, since the pollution effects usually occur at the surface of the sediment particles and in the autochthonous precipitates. Anderson (1974) has therefore suggested a simple $HCl-HNO_3$ (1:1) digestion for the determination of heavy metals by atomic absorption spectroscopy. This procedure also circumvents the need to redissolve the lead precipitates of the filtration residue when determining for lead, compared for example with the methods involving H_2SO_4 reagents. Further digestion procedures pertaining to the selective extraction of distinct sedimentary phases will be discussed in Chapter E. In the following, sediment studies employing more or less complete digestion methods for the determination of metal concentrations will be dealt with.

3 Geochemical Reconnaissance of Aquatic Sediments

Geochemical investigations of stream sediments have long been standard practice in mineral exploration. The method of *stream sediment sampling* is preferentially employed in remote areas in order to obtain a preliminary idea of the possible mineralization zones (Boyle et al., 1955; Hawkes et al., 1957). By more extensive sampling and analysis of the metal content in water, soils, and plants, the probable enrichment zones can be narrowed down and, in some cases, localized as exploitable deposits (Hawkes and Webb, 1962).

In mineral exploration three categories are used to characterize the distribution of metals: background, threshold, and anomaly. If one takes the "log-normal distribution of the elements as a fundamental law of geochemistry" into account (Ahrens, 1957), it is possible to determine by statistical analysis the "threshold" and "anomaly" levels as deviations from this distribution. An example of this is given in Figure 34 (after Hilmer, 1972) from measurements of the cobalt content of 670 individual samples from river sediments taken from the area around the Meggen deposit in Nor-

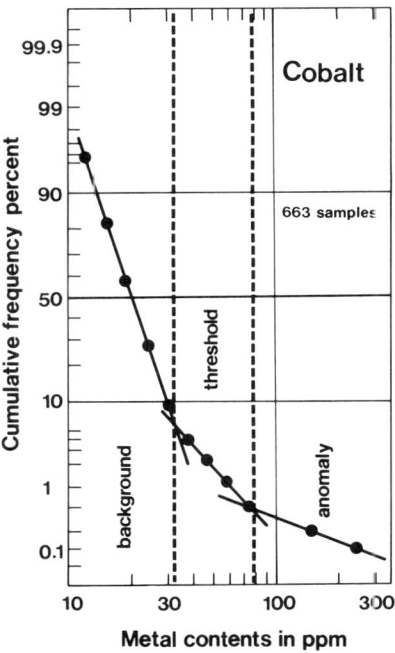

Fig. 34. Geochemical statistics: distribution of cobalt in 663 samples of fluviatile sediments from the lead-zinc district of Meggen (Federal Republic of Germany; after Hilmer, 1972)

thern Germany. From the breaks in the straight line, which result from the overlapping of each two logarithmic frequency distributions, it is possible to distinguish the three categories "background", "threshold" and "anomaly". For a detailed geochemical prospection, a classification of this kind should be developed for every region. This would necessarily entail a relatively closely knit system of sampling sites which at present is not available in the type of short-term operation commonly used in water control. The need arises therefore to work out simpler, more easily obtainable reference values, in particular for the "background". Several alternative possibilities will be discussed later (Sect. 6).

Within the framework of the present study it is not possible to give all the numerous methods for the graphic and statistic evaluation of metal analysis in water. In stream sediment sampling for geoprospecting, Nicol et al. (1969) have shown that with statistical and mathematical models it is possible to reduce the number of variables; examples of factor analysis of stream sediment geochemical data are given by Saager and Sinclair (1974) from the Mount Nansen area, Yukon Territory, Canada. For different mining environments of mid-Wales and Colorado, Wolfenden and Lewin (1978) have found systematic relationships between metal concentration and particle size and distance downstream. For environment-related investigations, Symader (1976) and Rump (1976) developed mathematical prediction models, which have shown that by means of a main-component analysis the number of the water quality parameters can be reduced in half without great loss of information. Examples of practical application have been given by Symader and Thomas (1978) and Neuland et al. (1978); the former used hierarchical grouping analysis for interpretation of average heavy metal pollution in flowing waters, the latter applied the means of the time-series analysis to differentiate the behavior of Mn and Fe and of metal pollutants such as Cu, Cd, and Pb in water and sediments of the Red Main River in the Federal Republic of Germany. Synoptic graphical presentations of correlation coefficient matrices were used by Davaud (1977) to demonstrate the differences in chemical associations in recent and older sediments from Swiss lakes, not only due to natural processes such as diagenesis, but also to anthropogenic influences such as eutrophication and excess loadings of heavy metals. Dahmen et al. (1976) introduced graphic multi-factor diagrams where the reproduction of more than two dimensions of value scales important in ecology on a single plane is made possible, also using appropriate systems of colors to make it easier to read the diagrams. Wolfenden and Lewin (1977) explain how to produce a three-dimensional view of lead values in sediment samples from a meander loop of River Rheidol in mid-Wales, which is polluted by mining wastes. Examples of more conventional graphic representations of lateral metal distributions in river sediments are shown in Figure 46 (Banat et al., 1972b).

It is often the case in geochemical exploration that the metal element under consideration is not directly investigated, but rather the investigation is carried out on a so-called indicator or guide element. Pathfinder elements, those associated with the ore but having dispersion characteristics superior to those of the economic minerals, play an important role in exploration geochemistry. [By reason of the complexity and diversity of geologic bodies, Schroll (1977) has proposed the term "indicator parameters", including both multi-element analysis and statistical methods of evaluation]. For example, as native mercury is associated with many sulfide ores of zinc, copper, lead, and native gold and silver, great success has been registered in the search for sulfide ores – in addition to improvement of analytic methods of the past years (Hawkes and Williston, 1962; Friedrich and Kulms, 1969).

For 20 years *lake sediment geochemistry has also been used as a guide to mineralization.* A center of recent activities in this respect is Canada which has a greater area covered by freshwater lakes than any other country. As early as 1956, Schmidt established anomalous metal distribution patterns in lake sediments bordering on areas of mineralization in New Brunswick and Quebec. From studies in Saskatchewan, Arnold (1970) and Dyck (1971) inferred that at low sampling density lake sediments reveal outline metalliferous areas as accurately as stream sediments. Lake sediments geochemistry, at the reconnaissance level, was first described by Allan (1971) in a survey covering an area of 3800 km² in the Coppermine Region. Since then, thousands of mountain lakes (Appalachia, Cordillera), Arctic lakes, Prairie lakes and, in particular, Shield lakes have been analyzed both for geochemical exploration and environmental management (summarized in Allan et al., 1972, 1974; Coker and Nichol, 1975; Allan, 1977).

The general sequence of events leading to the ultimate *dispersion of metals into lake sediments* was depicted by Allan et al. (1974) in the study "Mercury and arsenic levels in lake sediments from the Canadian Shield". According to the authors, the fate of metal ions, originally derived from chemical weathering and mechanical disintegration of host rock is controlled by many factors involving atmospheric precipitation, water movement, soil movement, changes in redox and pH conditions, absorption-desorption processes, chemical complexation, precipitation and hydrolysis, uptake by and decay of vegetation and biochemical-bacterial interactions. Whether or not a specific flush load of freshly leached metal ions eventually reaches a lake system intact or widely dispersed, depends on the relative interplay of these factors. A comparison of the predominance of metal in host rocks and in lake deposits is limited to a much narrower area than the corresponding levels for bedrock of the catchment area (Fig. 35). This means that a relatively small number of samples will generally suffice in the exploration of lake sediments as compared to the number needed for soil or rocks. It can be assumed that by careful selection of lake sediment samples the minimum number could even be further reduced. Apart from this, it would be possible to break down the wide range of the measurement data by dividing the lake sediment samples according to grain sizes to arrive at satisfactory conclusions as to the movement of metal in the catchment area of the lake with relatively few samples.

4 Grain-Size Effects

Grain size and grain-size distribution of clastic sediments are primarily influenced by physical processes in the transportation and deposition area.

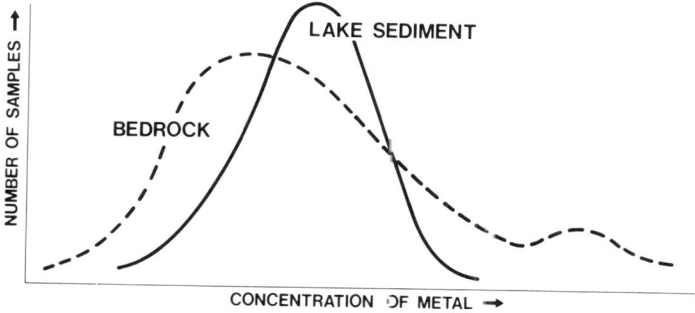

Fig. 35. Schematic presentation of the dispersion process of metals into lake sediments: the effect of dispersion on relative concentrations (Allan et al., 1974, *Mercury and arsenic levels in lake sediments from the Canadian Shield*)

A large proportion of the fluviatile transport of matter occurs in the form of suspended material with mean grain sizes of 2–63 μm (silt grain size; grain sizes of 10–20 μm are very common). During a decreased rate of flow, i.e., if the total load exceeds the available energy, the suspended material gradually sinks to the bottom and is deposited in the river bed where it is partially incorporated into the bottom sediment. The particles with larger mass (and correspondingly coarser grains) are the first to be deposited, followed by the increasingly lighter, finer-grained components. The fact that these deposits partly remain intact in spite of any subsequent increased rate of flow is due to specific cohesion characteristics of the fine-grained particles, which Hjulström (1934) empirically recorded in his well-known diagram. While the size of the grains transportable in suspension increases exponentially with the flow of the current, energy of a comparable magnitude is required to lift very small sediment particles (clay or fine silt) from the river bed and to bring them into a suspended state, as is the case for fine sand particles. The deposits of the river bank frequently consist of several "sedimentation units" beginning at the base with coarse-grained particles and progressing upwards into finer-grained material. Such a sedimentation sequence is generally formed under decreasing water flow conditions. Compared with rivers, where shifts between the processes of erosion and deposition can take place within short intervals of time and space, lakes and marine basins exhibit a more continuous accumulation of sediment materials particularly in deeper water. One can generally expect a decrease in the sediment grain sizes from the shore towards the center of the basin; here too, however, currents can give rise to special conditions (e.g., Thomas et al., 1972).

4.1 Grain-Size Dependencies of Trace Metal Concentrations

The distribution of metals in the grain-size spectrum is shown in Figure 36 from analyses of nickel and zinc in sediment samples from Lake Constance and the lower Rhine. Lake Constance is, generally speaking, relatively uncontaminated; this will be discussed more thoroughly in Section 7. The lower Rhine sample, on the other hand, exhibits very strong anthropogenic pollution influences (cf. Sect. 8). In both samples, a general increase of metal concentrations from coarse to finer-grained fractions can be observed. Different developments occur at either end of the grain-size spectrum: whereas the increase in metal concentrations in the grain size > 63 μm is determined by detrital minerals rich in heavy metals, the relatively low metal concentrations in the grain fractions < 0.2 μm may be due to a reduced adsorption potential of the partially less crystalline or amorphous substances. This development can still be seen in samples of the fraction < 0.06 μm although these results are not reproduced here owing to difficulties in the separating of the gypsum and salt contents which lead to inaccuracy in the data.

A comparison of the nickel and zinc concentrations in both samples examined here clearly shows that zinc is generally more subject to anthropogenic influences than nickel. Whereas the nickel concentrations in the Lower Rhine sediment are maximally twice as high as in the Lake Constance deposits, the corresponding zinc concentrations in the Lower Rhine deposits are five to ten times higher (than the comparative values from the Lake Constance samples). The heaviest enrichment occurs in the finer-grained fractions.

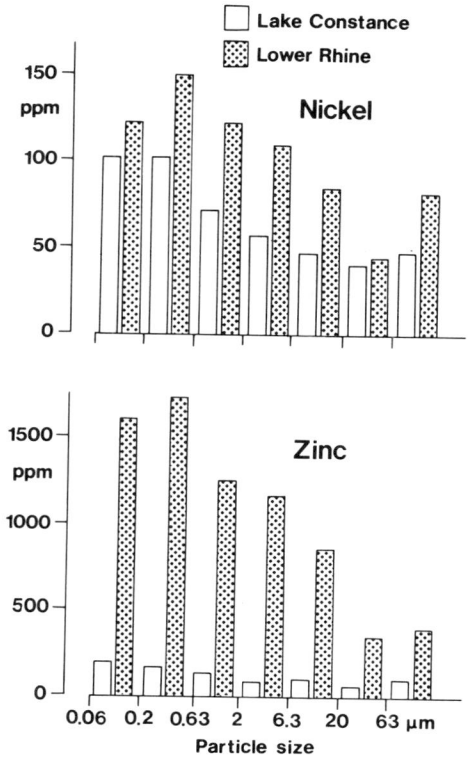

Fig. 36. Grain-size distribution of the concentrations of zinc and nickel in sediments from Lake Constance and the Lower Rhine

In a study on the transport phases of transition metals in the Amazon and Yukon Rivers, Gibbs (1977) formulated calculations on the mass transportation of various elements and characteristic metal associations with particulates in relation to the grain-size spectrum. Figure 37a shows the transported metals in suspended sediments from the Amazon River as a function of the concentration of the element under consideration, together with size distribution of the selected sample. The resulting evaluation shows that even though solid particles with small diameters have the highest concentrations of all the transition elements, there are simply many more particles of intermediate size. It can therefore be seen that more than 90% of the iron content in the Amazon River is transported in the grain-size interval of 0.2 μm to 20 μm. A particularly interesting insight into the forms of metal transport by river sediments is presented in Figure 37b (from Gibbs, 1977), from which it is obvious that the concentrations of the transition metals in the different phases (selective extraction procedures for the metal associations studied will be treated in Chap. E) do not have a similar distribution with size. The transportation mode of iron in the hydrous oxide state ("coatings", mainly on clay minerals) is approximately 1 to 2 μm, whereas the mode in crystalline (predominantly silicate) particles is located at approximately 10 μm.

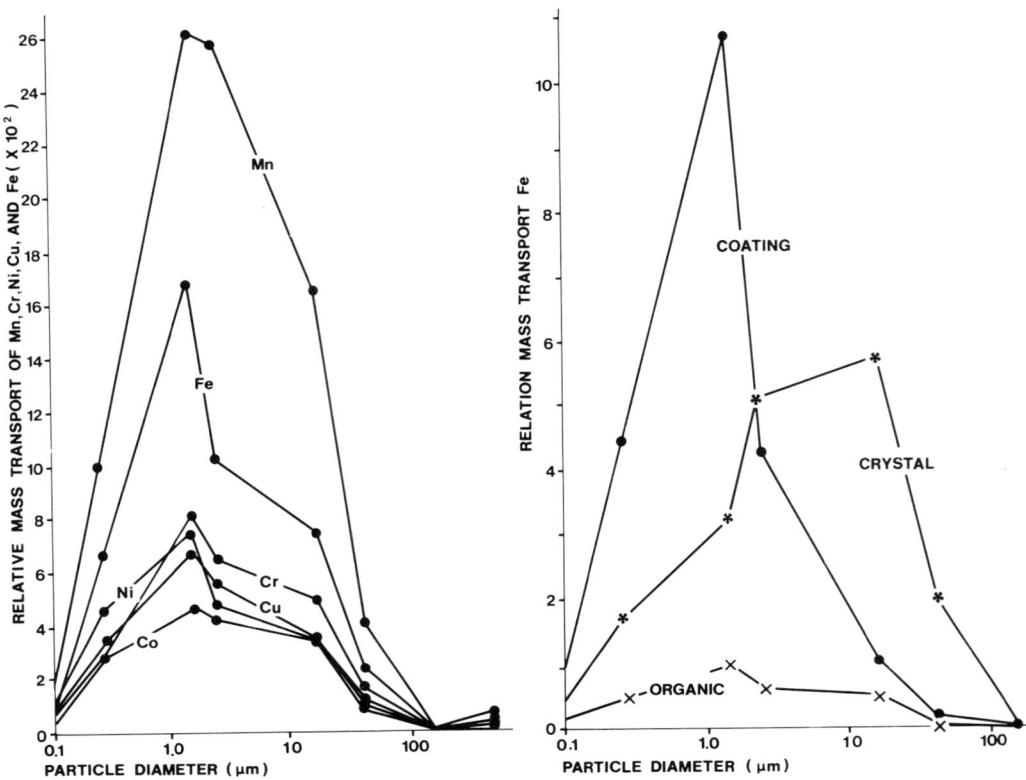

Fig. 37. Grain-size dependency of transport phases of transition metals in suspended sediments from the Amazon River (Gibbs, 1977; Geol Soc. Amer. Bull. *88*, pp. 840–841; reproduced with permission of the author, Copyright by The Geological Society of America)

4.2 Reduction of Grain-Size Effects

From the examples mentioned above, it is obvious that grain size exercises a determining influence on the metal concentrations, not only with regard to the samples selected for pollution control but also to the natural background data. This fact must be taken into consideration particularly with respect to studies of river sediments where variations in grain size are usually large. One can even go as far as to state that without a correction for grain-size effects, a mutual comparison of metal data in fluviatile deposits would be impossible. Several procedures to minimize grain-size effects on trace metal data are summarized below.

4.2.1 Extrapolation from Grain-Size Distribution

In a study of the amounts and behavior of mercury and other metals in the suspended matter of the highly polluted River Rhine in the Netherlands and—for comparison—in

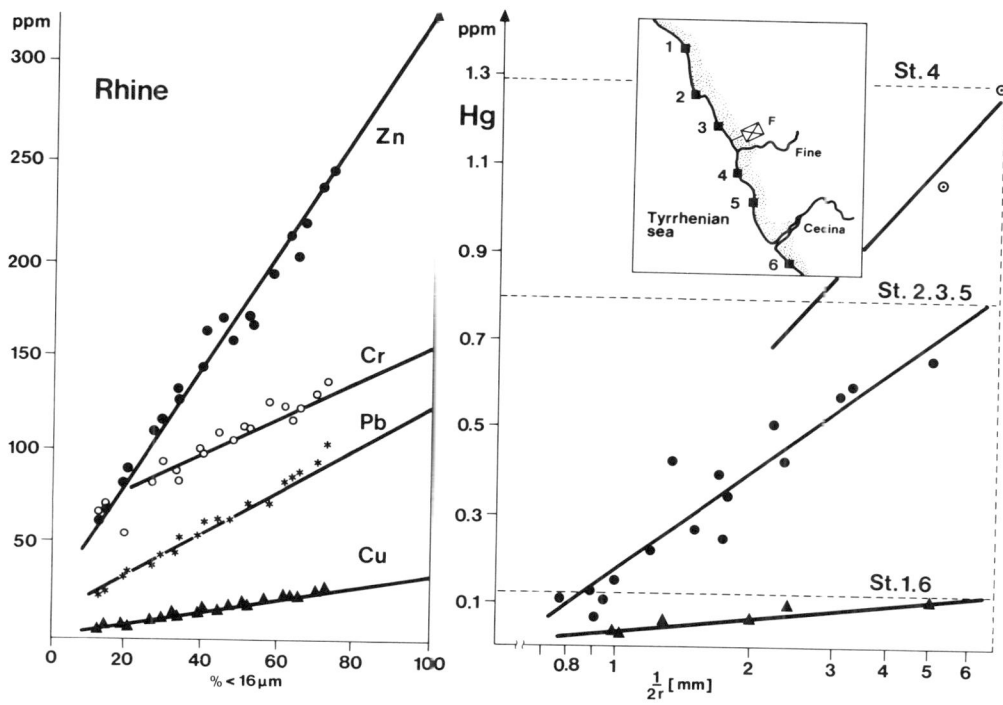

Fig. 38. Reduction of grain size effects by extrapolation from grain-size distribution (*left* Salomons and de Groot, 1977; *right* Renzoni et al., 1973)

the less polluted estuary of the River Ems, De Groot et al. (1971) used an extrapolation method for correcting the metal analysis data: "Due to a preferential occurrence of the heavy metals in the finest grain-size fractions, linear relationships are always found between the amounts of the heavy metals and the fraction of particles < 16 micrometers (expressed as a percentage of the $CaCO_3$-free mineral constituents in the oven-dry sediment) in samples from the same location." In Figure 38 (left), these relationships are shown for the elements zinc, chromium, lead, and copper in sediments from the Rhine River in the Netherlands (from Salomons and De Groot, 1977). According to De Groot (1964) "these linear relationships make it possible to characterize the content of a specific metal of a whole group of co-genetic sediments by a single value, the content obtained by extrapolation of the fraction < 16 micrometers to 100%". In the study *Pollution History of Trace Metals in Sediments, as Affected by the Rhine River* Salomons and De Groot (1977) used the concentrations of metals at 50% <16μm, since this value corresponds with the mean grain-size composition of estuarine sediments from Rotterdam Harbor, a major sedimentation area in the Netherlands. A similar approach was used by Renzoni et al. (1973) for the assessment of mercury pollution from a large chloralkali factory on the Tuscan coast of the Tyrrhenian Sea. Repeated analyses of mercury in the bottom sediments of stations 1-6, 2-3-5, and 4 showed that the mercury is strictly related to the average diameter of the sediment grains, and has the following formula:

$$y = bx + a \quad \text{where,}$$
$$y = \text{mg Hg/kg dry sediment}$$
$$x = \log \cdot \frac{1}{2\bar{r}}$$
$$2\bar{r} = \text{average diameter of the sediment grains}$$

The typical straight lines obtained by this pooling are shown in Figure 38 (right). It is suggested that the slope of regression of mercury concentration on the $\log \frac{1}{2\bar{r}}$ can replace the pollution index.

4.2.2 Metal Concentrations vs Surface Area

The familiar interdependency of grain size and specific surface area of sediments is used here for the indirect determination of relative grain-size values. This method is particularly well suited to serial determinations within a defined catchment area.

Both samples chosen here for copper and nickel stem from samples taken at two mile intervals from the Ottawa and Rideau Rivers in the Ottawa area (Oliver, 1973). Plots of the nickel and copper concentrations vs surface area are shown in Figure 39. The overall shape of these typical curves is roughly the same for all metals, but the actual contour of the curve varies from metal to metal and from area to area. It is now possible to make allowance for varying surface areas of the samples. Readings which are considerably above the concentrations vs surface area are termed anomalous. While no unusually large discharge of nickel in the river is indicated in these locations (Fig. 39a), several copper values show a clear deviation from the curve (Fig. 39b).

The main advantage of this method is that in comparison to separation it is possible to estimate the influence of the specific surface area and to identify those substances in the sample which are a determining factor in the enrichment of trace metals in recent aquatic sediments, i.e., clay minerals, Fe-hydroxides, fine-grained carbonate and organic material. The disadvantage of this method lies in the relatively large number of samples required for plotting a curve relating to the metal content and specific surface area. In addition, it is worthwhile noting that the grain-size spectrum (more accurately, the grades of a specific surface area) should be relatively evenly covered.

4.2.3 Separation of Clay/Silt and Fine Sand Fractions by Sieving

In many studies on metal pollution in river sediment, the clay/silt fraction (and in some cases the fine sand fraction as well) is separated with nylon sieves in order to reduce at least the strongest grain-size influences. Examples for the use of the < 200 μm fraction can be found in Copeland's (1972) study on *Mercury in the Lake Michigan Environment* and in Thornton et al.'s (1975) ambitious study of the geochemistry of sediments from rivers and estuaries in Great Britain. Others working on sediment analysis prefer to use the < 63 μm fraction (silt/sand boundary). This latter method was used in the exploration studies undertaken with the help of river and lake sediments by the Geological Survey of Canada (Cameron, 1974), which were described in greater detail in Section 3 of this chapter. In their study *Trace Elements in Sediment from the Lower Severn Estuary and Bristol Channel*, Chester and Stoner (1975a) used the < 61

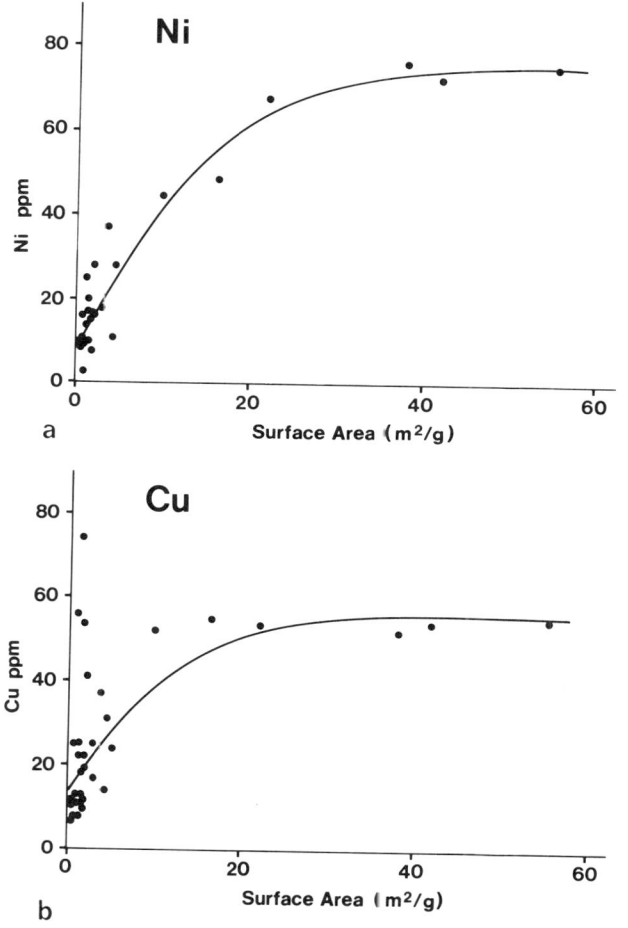

Fig. 39a and b. Reduction of grain-size effects by determination of the specific surface area. a nickel; b copper. Examples from the Ottawa and Rideau Rivers, Canada (after Oliver, 1973)

micrometer fractions which were separated by wet sieving with re-distilled water through nylon sieves.

4.2.4 Separation of the Pelitic Fraction (< 2 μm) in Settling Tubes

The Institut für Sedimentforschung of Heidelberg University (Banat et al., 1972a, b, c), followed by a number of others (e.g., Gadow and Schäfer, 1973; Helmke et al., 1975) introduced the method of separating the fine-grained sediment at the silt/clay interface. Since the fraction < 2 μm contains not only clay minerals but also hydrous oxides, sulfides, organic substances and other material, it is more appropriately termed the "pelitic fraction". The fine-grained sediment, in which most of the substances active in metal bonding are enriched, is separated from the coarser-grained, more or less inert (with regard to metal accumulations) sediment fractions.

The separation is carried out in Atterberg sedimentation cylinders, a modification of the procedure described in Section 2; the fraction < 2 μm is extracted only once or twice in the grain-size analysis in order to minimize the possibility of contamination and the remobilization of the metals from the sediment material. Consequently, the < 2 μm is evaporated at 60 °C in a porcelain dish. The disadvantage of this method is that it involves only the grain size and not the material properties. The advantage lies in its simplicity as a method to employ in serial determinations.

4.2.5 Treatment with Dilute Acids (Hydrochloric Acid, Nitric Acid)

When assessing the influence of trace elements in sediments on the environment, and in particular the question of remobilization of heavy metals, one must primarily consider the most mobile fraction of these elements, which are bound to the sediment in sorbed, precipitated or co-precipitated (on carbonates and hydrous Fe/Mn oxides) and complexed form. Goldberg (1954) proved that a large part of the "hydrogenous" metal fraction can be dissolved by selective attack with dilute acids; a residue of mainly inert bound metals in the natural rock detritus ("residual fraction") can be found predominantly in the coarser grain sizes. The metal data thus obtained—by means including the method of selective mineral separation to be discussed later—compare favorably with data from clay size fractions.

Studies by Cross et al. (1970) on the distribution of manganese, iron, and zinc in sediments of the Newport River Estuary, North Carolina, and by Gross et al. (1971) on the marine waste deposits of the New York Metropolitan Region were among the first to consider problems of environmental protection; Cross used 0.1 N hydrochlorid acid, whereas Gross used 1 N HCl for the extraction. In the most recent studies, other acids have come into use for the extraction of metal species associated with carbonates, hydrous oxides, sulfides and surface sorbed materials, in particular 0.1 N nitric acid (Jones, 1973), 0.1 N hydrochloric acid (Duinker et al., 1974), dilute acetic acid (Loring, 1975), and 25% acetic acid in a solution of hydroxylamine hydrochloride (originally introduced by Chester and Hughes, 1967; Bruland et al., 1974). The various processes used in selective chemical leaching of aquatic sediments will be discussed thoroughly in Chapter E.4.

4.2.6 Mineral Separation: Quartz Correction Method

Thomas (1969a) and Thomas et al. (1972) demonstrated that quartz is a minor constituent in the sub 4 μm grain-size fraction and that there is direct correlation between mean grain size and quartz content. As such, quartz-corrected values of trace metals give a realistic assessment of the degree of metal concentrations or adsorption by the fine particulate. This is especially true for those trace metals which occur in very small quantities in quartz and feldspars. The gravimetric determination of total quartz plus feldspar was carried out according to the Trostell and Wynne (1940) method. This method, based on fusion with potassium pyrosulfate, preferentially removes the layered silicates (clays), organic and inorganic carbon, and sulfides, leaving a residue of quartz plus feldspar with resistent heavy minerals such as zircon (Thomas et al., 1976). Apart from this, Thomas (1969b) differentiated between the quartz and feldspar portions in the sediment of Lake Ontario.

The metal concentration can thus be calculated in the following manner:

$$\text{Trace metal (quartz-corrected)} = \frac{\text{Trace metal (observed)} \times 100}{(100 - \text{quartz \%})}$$

Using this method, a number of mercury anomalies were traced in the sediments of the Laurentian Great Lakes, their occurrence being attributed partly to natural and partly to civilizational influences. (Lake Ontario—see Fig. 32: Thomas, 1972; Lake Huron: Thomas, 1973; Lake Superior, Lake St. Clair. Lake Erie: Thomas, 1974, and Thomas and Jaquet, 1976).

4.2.7 Comparison with "Conservative" Elements

Bruland et al. (1974) expressed the variation of the trace metal content in sediment cores from the Southern Californian basin as metal/aluminum ratios. Aluminum is assumed to have had a uniform flux to the sediments over the past century from crustal rock sources. Consequently, changes in the water, salt, $CaCO_3$ or organic matter content, especially in the upper layers, can be compensated.

A fundamental investigation into the question of element enrichment in Recent sediments and the possibility of internal comparison was performed by Kemp et al. (1976) on sediment cores from Lake Erie. The observations of Kemp et al. (1974) initially held true for most of the other examples from the Great Lakes, according to which the concentration of a number of trace elements is much greater near the sediment—water interface than at the *Ambrosia* pollen horizon or below this chronological marker of approx. 120 BP (corresponding to early agricultural development following forest clearance). To quantify the results, Kemp et al. (1976) introduced a ratio that is designated as the Sediment Enrichment Factor (SEF):

$$SEF = \frac{\frac{E_s}{Al_s} - \frac{E_a}{Al_a}}{\frac{E_a}{Al_a}}$$

where E_s = the observed elemental concentration in the surface cm of the sediment
E_a = the observed elemental concentration below the *Ambrosia* horizon.
Al_s = the aluminum concentration in the surface cm of the sediment
Al_a = the aluminum concentration below the *Ambrosia* horizon

Table 34 depicts the SEFs and the respective interpretation from Kemp et al. (1976) for three profiles from the western, central and eastern basin of Lake Erie.

Those elements labeled "conservative" in Table 34 are of particular value when trying to determine proportionate levels of metals in deposits. Further calculations by Kemp et al. (1976) show that Al enters into strong positive relationships with K and with clay. This is in agreement with the earlier observation that aluminum is indicative of the clay mineral content, while the relationship of Al to K conforms to the dominance of illite as determined by X-ray diffraction. With respect to other elements e.g., vanadium and beryllium, it is not certain whether these elements should be placed in the "conservative" or "enriched" grouping. Bruland et al. (1974) demonstrated the enrichment of V in southern California sediments to be of anthropogenic origin. Strontium should be added to the table as a typical carbonate element.

Table 34. Sediment enrichment factors in Lake Erie deposits

Element	Western Basin	Central Basin	Eastern Basin	Element Behavior
Silicon	0	0	0	
Potassium	0	0	0	
Titanium	0	0	0	Conservative
Sodium	0	−0.3	0	
Magnesium	0	0	0	
Mercury	13.2	9.7	7.3	
Lead	3.4	2.6	3.7	
Zinc	1.7	1.4	2.2	Enriched
Cadmium	4.0	0.8	2.7	
Copper	1.4	1.1	0.8	
Organic carbon	1.1	1.3	1.7	
Nitrogen	1.2	2.1	2.0	Nutrient
Phosphorus	0.5	0.3	0.8	elements
Carbonate-C	0.4	−0.4	0	Carbonate
Calcium	−0.3	−0.4	−0.3	elements
Iron	0	0.3	0	Mobile
Manganese	0.4	0.4	1.1	elements
Sulfur	0.3	0	1.0	

(Values between −0.2 and 0.2 are designated as zero).

4.2.8 The Relative Atomic Variations of Elements

Many efforts have been made to gain a more general and geographically independent index of element geochemistry or contamination than is at present provided by the distribution patterns of single elements. The first was done by Cowgill and Hutchinson (1966) in a study on the geochemistry of Laguna de Petenxil sediments. A further example has been given in the previous section as the "sediment enrichment factor", represented by the ratio of the element under consideration to another element of little variablitiy e.g., an element with "conservative behavior".

As proposed by Allan and Brunskill (1977), an even more comprehensive system of element correlation is encountered in the index of relative atomic variations (RAV). Linear regression analyses were performed on metal data from Lake Winnipeg to determine correlation coefficients for all possible pairs of elements, e.g., the heavy metals Ni, Cu, Co, Cr, Zn, Fe, and Mn. Where significant correlations existed, the slopes of the linear regression were calculated (from the metal concentration in μmoles/g), representing the index of the relative atomic variation. It has been suggested that the analogy in RAV-values for different sites implied a similarity or possible homogeneity in the large number of processes present in the geochemical cycle including weathering, transport, deposition, and diagenesis. This hypothesis, tested by Allan and Brunskill (1977) for many diverse limnological sites in Canada, revealed that the correlations disappear or alter in magnitude at sites of cultural and natural contamination. Although no data is presented on rivers, the determination of RAV-values for their sediments is considered to be of greater significance than from lake deposits, since metal data are greatly influenced by hydrologic and textural (grain-size) factors.

5 Factors Controlling the Distribution of Metals in Aquatic Sediments

The interpretation of the enrichment of chemical elements in sediments by anthropogenic pollution must take into account a number of geologic, mineralogic, hydrologic and biologic processes controlled both by internal and external factors:

a) Allochthonous influences which can be subdivided into "natural" and "civilizational" effects;

b) Autochthonous influences, comprising the mechanisms of precipitation, sorption, enrichment in organisms and organc-metallic complexing during sedimentation, as well as the postdepositional effects of diagenesis.

In order to evaluate the major influences controlling the distribution of trace metals in aquatic sediment samples, approx. 100 lakes situated in south and western Australia, South America, East and South Africa, central and eastern Europe, and western Asia were investigated (Förstner, 1973; Förstner, 1977a–c).

Analytic data of Pb, Cu, Ni, Cr, Zn, Sr, and Mn in the $< 2~\mu$m-fraction of lake sediments is summarized in Figure 40. The outer solid lines represent the frequency distribution for all 87 samples of each element under consideration. Average shale values are indicated by triangles.

A general inspection of the frequency curves in Figure 40 indicates largely log-normal distributions of the elements lead, copper, nickel, chromium, and zinc, which are most likely to be present when the element concentrations in the various rock sources are virtually similar *and* when the subsequent change in the metal content due to internal factors (e.g., precipitation and diagenesis) is relatively small. On the other hand, there are characteristic deviations from log-normal values for the examples of *strontium* and *manganese.* With regard to manganese, the distribution curve exhibits a distinct assymetry to low metal values. The behavior from *diagenetic effects,* where reducing conditions remobilize manganese, leading to a depletion of manganese in the sediments, will be explained later. In the case of strontium, the higher concentrations (up to 50% in some samples), result from the incorporation into *carbonate minerals,* particularly in aragonite.

A more detailed review of the data in Figure 40 indicates some anomalous high nickel and, in particular, chromium values. These deviations originate from the typical influence from gabbroid rocks. In the European lakes concentrations of lead, zinc, and copper exhibit on the average higher values than is the case for the overall distribution of these elements in the 87 samples (Förstner, 1978a). This apparently points to a civilizational influence, in particular for the enrichment of lead and zinc.

To validate the above mentioned element associations, further evidence can be derived from the statistical evaluations of the analytic data:

a) There is a high degree of positive correlation between the elements Fe, Cr, Ni, Co, and Mn, and to a lesser degree Cu, indicating the influences from the lithology of

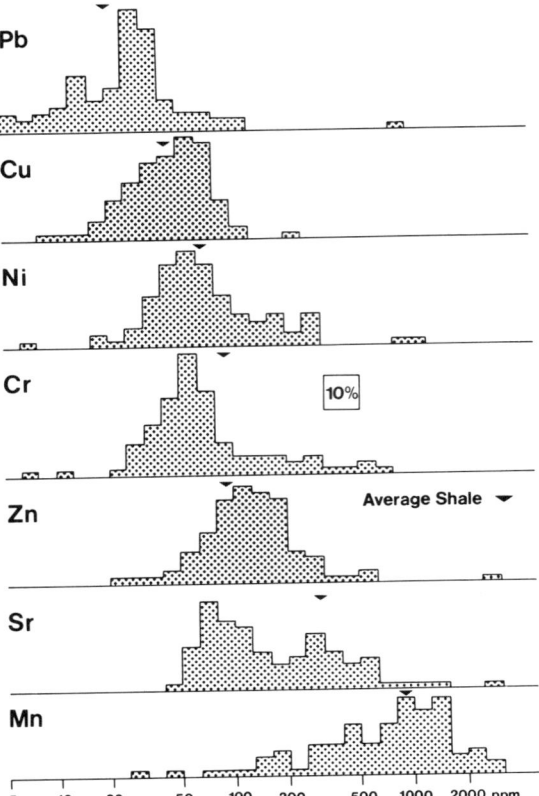

Fig. 40. Frequency distribution of metal concentrations in the pelitic fractions (< 2 μm) of sediment samples from 87 lakes

the lake's catchment area. b) A strong correlation between the strontium values and total carbonate contents could partly be associated with allochtonous influences; authigenic formations of carbonate minerals should in addition be considered as a possible mechanism of strontium enrichments in some lake sediments. c) Obviously, elevated concentrations of carbonate, strontium, and organic carbon effect the decrease of the iron concentrations in lake sediments, as is indicated by the negative correlations between these components. d) The carbon content also appears to be a diluting factor for the chromium concentrations. e) A high degree of positive correlation particularly exists between the metals zinc, lead, cadmium, and mercury; organic carbon, and to a lesser extent copper indicate association with cultural activities. It is of special note that the "anthropogenic" elements of the evaluations fully coincide with the group of "enriched" elements, as derived by Kemp et al. (1976) from sedimentary core investigations in Lake Erie (Table 34), and Hakanson's (1977) "contaminating elements" from studies on the four largest Swedish Lakes.

6 Natural Metal Content—Civilizational Accumulation

Trace metals in Recent sedimentary deposits can generally be divided into two categories: in accordance with their predominant source of origin, either as "lithogenic" or "anthropogenic" (Hellmann 1970a), often simply referred to as "geochemical" and "man-made" ("civilizational") respectively. Metals such as zirconium, rubidium, and strontium, which are derived from rock material by natural weathering processes, constitute the first group. The second group is made up of metals which have become enriched chiefly as a result of man's activities, and includes among others chromium, cobalt, nickel, copper, zinc, cadmium, mercury, and lead.

Between these two groups there are combinations; for example, the enrichment of mobile elements such as manganese and iron, which may well have had civilizational origins such as extreme eutrophication. By their own accumulation, these metals can cause other elements to accumulate. Even the precipitation of carbonates can be influenced by "civilizational" means and this can in turn influence metal levels. The interrelationship between lithogenic and anthropogenic influences on heavy metal enrichment of sediments will be discussed and illustrated later in greater detail.

For the determination of characteristic metal concentrations of a sediment—for example, when prospecting for ore deposits—it is sometimes possible to dispense with the differentiation between lithogenic and anthropogenic components. Nonetheless, when attempting to determine the "extent of pollution" in a lake or river by means of the heavy metal load in sediments, it is of primary importance to establish the natural level of these substances, i.e., the "pre-civilizational" level (Shimp et al. 1971) and then substract it from existing values for metal concentrations in order to derive the total enrichment caused by anthropogenic influences.

In order to obtain an ideal comparative basis for environmental studies, the following criteria should be fulfilled so as to achieve representative values for metal concentrations: a large number of sediment samples must be analyzed, which correspond with recent deposits in their (a) grain-size distribution, (b) material composition and (c) conditions of origin, and (d) which are uncontaminated by civilizational influences. In practice all these criteria cannot be fulfilled simultaneously. Several attempts have been made to solve this problem which is briefly called the "background question". As Hellmann (1972a) observed in a basic study on the "definition and significance of the background in hydrologic investigations pertaining to environmental hygiene", it is not merely a question of measuring the "background value", but also of determining the more general underlying factors e.g., parameters such as the quantity of effluent, season, temperature, comparison with other waters, and hydrologic interrelation before these data can be used with optimal effect in an environmental discussion.

6.1 Average Shale: Global Standard Value

The rock standard is a world-wide standard in general use and satisfies the basic requirement of being uncontaminated and based, for most elements, on a large number of sediment samples. Extending this standard by incorporating the grain size (and, to a certain extent, the medium of formation), fossil argillaceous sediments seem to be the best medium for comparison with recent aquatic sediments.

In order to assess the influence of different magmatic rocks on the *composition of aquatic sediments,* the relative percentage of individual rock types must be taken into consideration. It is chiefly *granitic and granodioritic* rocks forming up to 86% of the magmatites that exercise the greatest influence on the composition of the weathering products (Wedepohl, 1969). The data in Table 35 (Turekian and Wedepohl, 1961 and other authors) indicate that—with the exception of strontium—all trace elements increase with the transition from *granite* to *shale*. Whereas there is only a light increase in the average concentrations of barium, zinc, copper, lead, uranium, and silver, the enrichment for boron, arsenic, antimony, selenium, and mercury is between factors of 5 and 10. The higher co-precipitating amount of hydrous manganese and iron oxides in the *deep-sea clays* provide an explanation for the particularly strong enrichments of Mo, Ba, Cu, Pb, Co, and Ni in these deposits.

Compared to the shale values, the trace metal content in *sandstone* is more or less "diluted" (Table 35). This is mainly the result of the extremely low trace metal concentrations encountered in quartz or feldspar components. A relatively small decrease is observed in the content of Mn, Ba, B, Cu, Pb, W, and Ag, which have up to a third of the shale levels. In contrast, the trace metal content of Sr, Ni, Co, As, Be, Mo,

Table 35. Minor and trace elements in granitic rocks and argillaceous, sandy and calcareous sediments, ordered according to the metal content in shales (after Turekian and Wedepohl, 1961)

	Granitic rocks	Shales	Deep-sea clays	Sandstone	Carbonates
Manganese	540	850	6700	390[a]	1100
Barium	420	580	2300	190[a]	10
Strontium	440	300	180	20	610
Vanadium	88	130	120	20	20
Boron	9	100	230	35	20
Zinc	60	95	165	16	20
Chromium	22	90	90	35	11
Nickel	15	68	225	2	20
Lithium	24	66	57	15	5
Copper	30	45	250	15[a]	4
Lead	15	20	80	7	9
Cobalt	7	19	74	0.3	0.1
Arsenic	1.9	13	13	1	1
Uranium	3.0	3.7	1.3	0.45	2.2
Beryllium	2.0	3.0	2.6	0.3[a]	0.2[a]
Molybdenum	1.0	2.6	27	0.2	0.4
Tungsten	1.3	1.8	4.5[b]	1.6	0.6
Antimony	0.2	1.5	1.0	0.04[a]	0.2
Selenium	0.05	0.6	0.17	0.05	0.08
Mercury	0.08	0.4	0.001–0.4[c]	0.03	0.04
Cadmium	0.13	0.3	0.43	0.02[a]	0.035
Silver	0.15[a]	0.27[a]	0.11	0.12[a]	0.19[a]

[a] Horn and Adams, 1966. [b] Krauskopf, 1967. [c] Boström and Fisher, 1969.

Sb, Se, Hg, and Cd in sandstone exhibits only a tenth or less of their respective values in shales.

Most *carbonate rocks* also show a "dilution" of trace metals compared with the corresponding shale levels (Table 35). Only manganese and strontium are enriched; there is little difference in the Pb, U, W, and Ag values. The concentrations of Cr, Li, Cu, As, Be, Hg, and Cd, in contrast, are greatly reduced, having approximately one-tenth of the amount in shale. A very strong reduction can be observed in Ba and Co, the content of which is only 1/50 and 1/100 of their respective amounts in argillaceous rock. A correction for carbonate-free substance, such as is often carried out for comparison purposes, may lead to inadequate results (see Chap. E.4) The samples used for setting a rock standard such as the *shale standard* stem from different environments which, in some cases—such as under reducing conditions or at higher carbonate contents —are not comparable. Nevertheless, a comparison with a "shale standard" is a quick and practical means of tracing high metal enrichments which may constitute a source of dangerous environmental pollution or an economically interesting mineral deposit. Once the sources responsible for these accumulations in sediments have been traced, they can subsequently be evaluated more precisely by a more refined—but time consuming—procedure which will be described in a later section.

6.2 Fossil Lake Sediments: Standards Regarding Environmental Data

An improvement of the results on the use of a global shale standard can only be obtained if the values of rock samples from defined formation environments are used for comparison with the actual data. The first example stems from a fossil lake (Förstner, 1977d):

The Ries Crater in Southern Germany, which was formed by a meteorite around 14 million years ago, was filled at first with predominantly clastic and later by increasingly chemical or biogenic lake deposits approx. 320 meters thick. From the deposits of this crater, 25 core samples were selected, the pelitic fraction was separated and ten trace metals were measured. Average values are listed in the first column of Table 36. The comparison with the shale standard values indicates that the concentrations of iron, manganese, chromium, and copper are relatively low; this can be explained by the elevated carbonate contents (iron) and the absence of stronger influences from basic rocks (for chromium and to a lesser extent copper). Diagenetic effects have been discussed with regard to the manganese distribution (Förstner, 1977). On the other hand, there is a distinct increase of the lithium concentrations, particularly within the deeper part of the lacustrine sediment sequence. This appears mainly to be due to high evaporation during the earlier stages of the lake's development.

6.3 Fossil Fluviatile Deposits: Regional Influences

Whereas the metal content in lacustrine deposits is to some extent influenced by autochthonous factors, particularly in the case of elevated amounts of chemical and biogenic constituents, the composition of fluviatile deposits is predominantly controlled by lithogenic influences from the source areas. Deviations of the metal composition from the global standard in different localities are naturally more likely to occur in smaller catchment areas than, for example, in sediments of the lower course of large rivers. Comparative samples from earlier fluviatile deposits are therefore of special interest for smaller catchment areas.

Table 36. Concentrations of metals (values in ppm), organic carbon (%), and carbonate (%) in pelitic fractions (< 2 μm) of fossil and recent aquatic sediments. Shale standard values from Turekian and Wedepohl (1961)

	Fossil lake sediment (Ries-Lake)[a]	Fossil river sediment (Rhine)[b]	Recent lake sediments mostly from remote areas (n = 87)[c]					Shale standard
Iron	18,200	32,350	43,400	(11,500	−	67,300	)	46,700
Manganese	406	960	760	(100	−	1,800	)	850
Strontium	252	184	151	(60	−	750	)	300
Zinc	105	115	118	(50	−	250	)	95
Chromium	59	47	62	(20	−	190	)	90
Nickel	51	46	66	(30	−	250	)	68
Lithium	203	91	45	(15	−	200	)	66
Copper	25	51	45	(20	−	90	)	45
Lead	16	30	34	(10	−	100	)	20
Cobalt	15	16	16	(4	−	40	)	19
Mercury	0.5	0.2	0.35	(0.15−		1.50	)	0.4
Cadmium	0.2	0.3	0.40	(0.10−		1.50	)	0.3
Org. C	3.5%	1.8%	1.6%	(< 0.2%−		3.7	%)	−
Carbonate	36%	11%	16%	(0	−	70	%)	−

[a] Arithmetic mean of 25 values.
[b] Arithmetic mean of 4 values.
[c] Metal = median of 87 values; organic carbon = arithmetic mean of 87 values.

The example from the Rhine River is typical of the extent to which metals are enriched by cultural influences. The samples stem from two cores taken from boreholes approx. 200 m from the present-day river bank of the Rhine near Cologne, from a maximum depth of 25 m through predominantly gravelly and sandy river deposits. In each case, the clay fraction was separated and analyzed. The upper Tertiary deposits were reached at a depth of 23.6 m.

Considerable fluctuations in the heavy metal concentration were established: zinc, mercury, and lead were particularly abundant in the uppermost sediment layer, probably due to cultural influences; the copper and chromium content was more enriched in the deeper Rhine sediments. The amount of manganese, cadmium, and lead was also relatively high in the middle section. Most of the trace metals studied here, however, reflected lower concentration in the sedimentation units between 6–7 m and 16–17 m respectively, which according to their clay mineral composition can be considered as counterparts of the recent Rhine sediments. Four samples from these two layers, indicating particularly low heavy metal contents, were selected for the calculation of the mean values given in Table 36, second column.

Since the investigated river deposits have probably been exposed to secondary contamination by seepage of surface water and groundwater, minimum levels such as these can best be used for a background comparison. These values in fact correspond very closely to Turekian and Wedepohl's (1961) shale standard data. As in the previous example, low values for chromium and nickel can be explained by reduced influences form gabbroid rocks.

6.4 Short, Dated Sediment Cores: 200 Years of Industrial Development

The expense involved in drilling operations to obtain sediment cores from great depths can hardly be justified merely for the purpose of obtaining background values. Furthermore, it has been demonstrated that the most important effects of metal enrichment can be determined even by means of short cores from consolidated sediments. In lakes and marine coastal basins where these investigations are predominantly carried out, the average annual rate of deposition for relatively fine-grained sedimentation (clay and fine-to-medium silt fractions are predominant here) is from 1 mm to 5 mm. A core profile of approximately 1 m, which is relatively inexpensive to obtain, covers a historical period of at least 200 years and its development can be traced by virtue of the metal content in the individual sediment layers. This period of time corresponds to the phase of greatest industrial growth which is also characterized by an often exponential increase in the discharge of effluents. It is especially useful if the individual phases of development can be dated by means of certain granulometric or material characteristics in the sediments or by isotope measurements In Sections 7 and 9 several examples are given of this method of investigation into heavy metals in sediments from the lacustrine and marine environments.

6.5 Recent Lake Deposits in Relatively Unpolluted Areas

The practice of comparing the metal concentrations in environmentally polluted deposits with those of sediments from relatively uncontaminated regions also represents a usable possibility. This method is particularly suited for lake deposits which have been formed under a variety of influences (cf. Sect. 5) as it is often possible to find comparable examples in remote areas. The average composition of 87 samples of pelitic lake sediments was calculated for statistical mean values (medians); deviations from the mean are given as "confidence limits" (range for 90% of the data) for each element studied (Table 36, third column).

Variation coefficients are relatively low for iron (90% of the data fall between 12,000 and 69,000 ppm), zinc (45–220 ppm), and copper (20–80 ppm), indicating that these mean data may primarily be regarded as "background values" for pelitic lacustrine sediments. This is confirmed by the fact that these elements correspond closely to those of the average shale composition.

Manganese and strontium, on the other hand, cover particularly wide ranges of values due to the effects of diagenetic remobilization (for Mn) and coprecipitation with carbonate minerals (for Sr).

6.6 Metals in Suspended Matter: Background Values in Storm Water

In a study on suspended matter in waters, Hellmann and Griffatong (1972) investigated the „background" problem of heavy metals in suspended particulate and sediments. They introduced the possibility of using *storm water sediment* as a basis for geochemical comparison, since in these deposits "the influence of sewer input can be disregarded because of the favorable dilution and because the organic component is only weakly represented in favor of the inorganic."

Using this process, Hellmann and Griffatong (1972) determined the natural trace element concentrations of solid matter in the whole of the Rhine River system: the zinc concentrations were between 150 and 300 ppm; the copper as well as the lead content was approx. 70 ppm. A comparison with "background data" determined by other methods, i.e., particularly with the fossil Rhine core sediments, shows that the percentages of zinc and lead (both metals are very heavily enriched as a result of cultural influences) are two to three times higher than the real "background" levels even during the highest water levels. This suggests that the contaminated deposits and suspended material of the river are only partly eliminated even when the storm water levels persist for a long period. It could also imply that even during such times, or perhaps particularly during periods of storm water, considerable amounts of effluent suspended material are introduced into the water.

In a detailed study, Schleichert (1975) points out that even the varying influences of different watersheds on the composition of suspended material in the lower course of a river must be taken into consideration: Figure 41 shows the chromium and cadmium content in Rhine River suspended particulate which was collected every working day near Koblenz. The values for the spring/summer period are marked here by open circles, those for the autumn/winter period by dots. Both examples indicate a characteristic decrease of the metal concentrations with increasing water discharges (see Chap. C). Chromium data (Fig. 41 a) reflected no difference between summer and winter samples. In the case of cadmium (Fig. 41b), which is very strongly enriched in the Main and Neckar Rivers (see Sect. 8), the November and December values exceed the maximum values for the period May to October, despite the much higher water discharge. The reason for this peculiarity of cadmium transport in suspended material (data for zinc, copper, and lead can be similarly divided into spring/summer and autumn/winter curves) lies in the fact that the suspended material, rich in heavy metals, is held back in periods of poor flow (summer) in the lock-regulated Neckar and Main Rivers. In autumn/winter, on the other hand, it is carried into the Rhine in increased quantities by stronger currents (Schleichert, 1975).

6.7 Background Values and Nonpoint Sources

The practical experience of the last few years has shown that positive proof of pollution sources can best be obtained by means of sediment analysis near specific sources, particularly at points where industrial sewage is discharged into surface water systems. Additional difficulties arise in tracing specific influences of metal pollution from mixed sewage systems, particularly where municipal water predominates. There are other metal enrichments, which simply cannot be pinpointed to a specific supply source and which are consequently described as resulting from diffuse or nonpoint sources. They may be locally included in the background value in order to obtain easier evaluation of the definite or point sources in a particular area.

Typical nonpoint sources for elevated metal concentrations are represented by the runoff from agricultural land and by atmospheric precipitations. Sheet erosion or seepage may, for example, transport higher concentrations of cadmium and arsenic—resulting from fertilizers or sewage sludge—into adjacent waters, both in solution and as solid particles. Lead from atmospheric precipitations can also reach surface waters in a similar way: on the one hand by mechanical erosion of soil particles which are transported in the form of fluviatile particulates and deposited in inland water systems and marine basins, and on the other hand from the chemical processes of leaching and mobilization of the heavy metals from particulates in either a dissolved or complexed

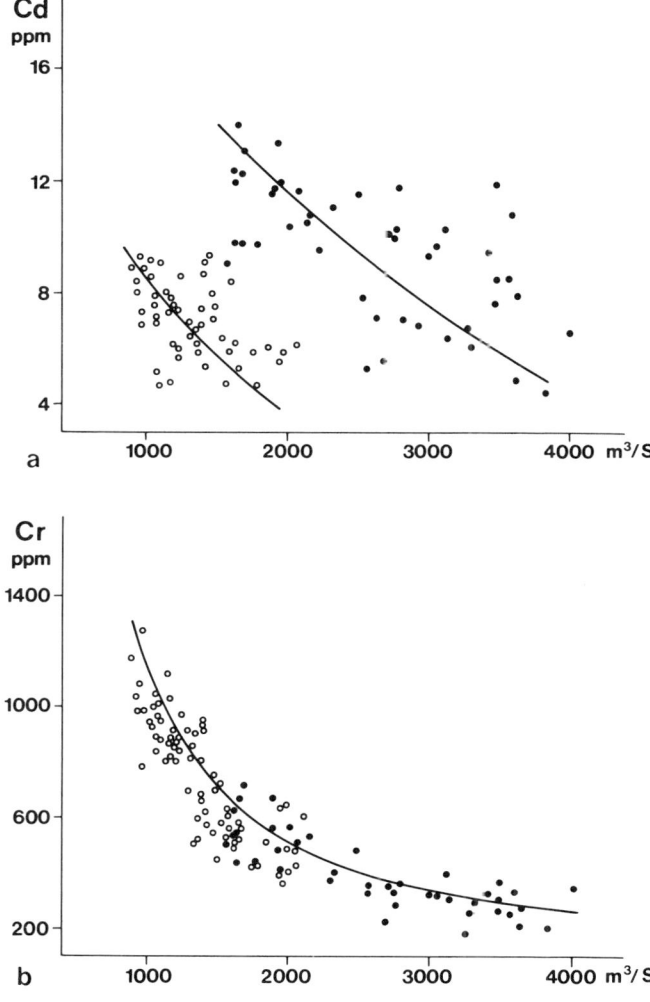

Fig. 41a and b. Water discharges vs. metal concentrations in sediments in the Rhine River (sampling station: Koblenz). From: Schleichert (1975), Deutsche Gewässerkundliche Mitteilungen, *19* (6), p. 155. With permission of Bundesanstalt für Gewässerkunde

form and a reprecipitation onto organic and inorganic sediment particles in a depositional basin. In densely populated areas a considerable amount of atmospheric precipitation with its metal load enters the surface waters via the public sewage system, thus rendering a clear differentiation between the influences of direct sewage and atmospheric heavy metal contamination impossible. These various interactions must be considered when pollution assessment is undertaken by sediment analysis. In the following three sections on heavy metal pollution in lakes, rivers and marine coastal zones, a number of examples will outline the possibilities, as well as the limits, of these methods.

7 Lake Sediments as Indicators of Heavy Metal Pollution

The significance of sediments in the assessment of the quality of aquatic systems is well illustrated by examples from the lacustrine environment. The reason for this lies in the fact that natural freshwater lakes have been the centers of important cultural developments since the earliest days of civilization. As a consequence of increased population and industrialization densities, the threat of pollution has become most acute in areas dependent on natural and man-made lakes as a source of potable water.

The first observations of changes in the lacustrine environment, as indicated in the variations in the sediment stratigraphic records, were made on Lake Zürich and described by Nipkow (1920). Hutchinson and his team (1943) and Züllig (1956) drew attention to a major application of geochemical research in the study of Recent lake sediments. Since "sediments may be regarded as a response of the conditions in an aquatic system" (Züllig), the study of freshwater deposits could play a key role in ascertaining the complex hydrochemical interactions which take place in lakes and rivers. Also approximately 20 years ago, the study of freshwater sediments was initiated in Canada as an aid to geochemical reconnaissance i.e., for tracing mineral deposits of economic value (see Sect. 3). During the last two decades sediment analysis has acquired a new dimension by being employed as a tool to trace man-made pollution influences in inland and coastal waters.

7.1 Interference: Geochemical Background and Man's Impact

Environmental as well as explorational geochemistry requires a knowledge of the naturally occurring metals in order to evaluate man's environmental impact: "Both the exploration and environmental geochemist can be looking for the same type of areas, those with high metal concentrations, but obviously from a different motivation" (Allan, 1974).

The latter aspect of pollution monitoring from lake sediments attracted much attention when mineral exploration was followed by large-scale *mining and processing* activites in certain areas.

Mining activities are often associated with higher metal levels in the environment. Difficulties arise, therefore, in distinguishing natural metal anomalies—which are the prerequisite for the economic interest in ore exploitation—from the subsequent effects of ore extraction procedures. This is especially valid for areas where mining operations have been carried out for some time and where environmental contamination by mining operations occurs in one or more of the following ways: (1) contamination by tailings introduced as solids into drainage systems, (2) by leachates from on-shore tailing piles, (3) by effluents from mines or mining plants, and (4) by airborne particulate from crushers and smelters.

One of the best examples of such an *exploration–environmental* geochemistry "overlap" was given by Allan (1974) from the large *nickel mining area at Sudbury*. Figure 42 shows the distribution of nickel in sediment cores in and around the Sudbury Basin. Approximately 150 cores of lake sediments were collected from sites of mining acitivites of the Superior Province of the Southern Canadian Shield. The mining region

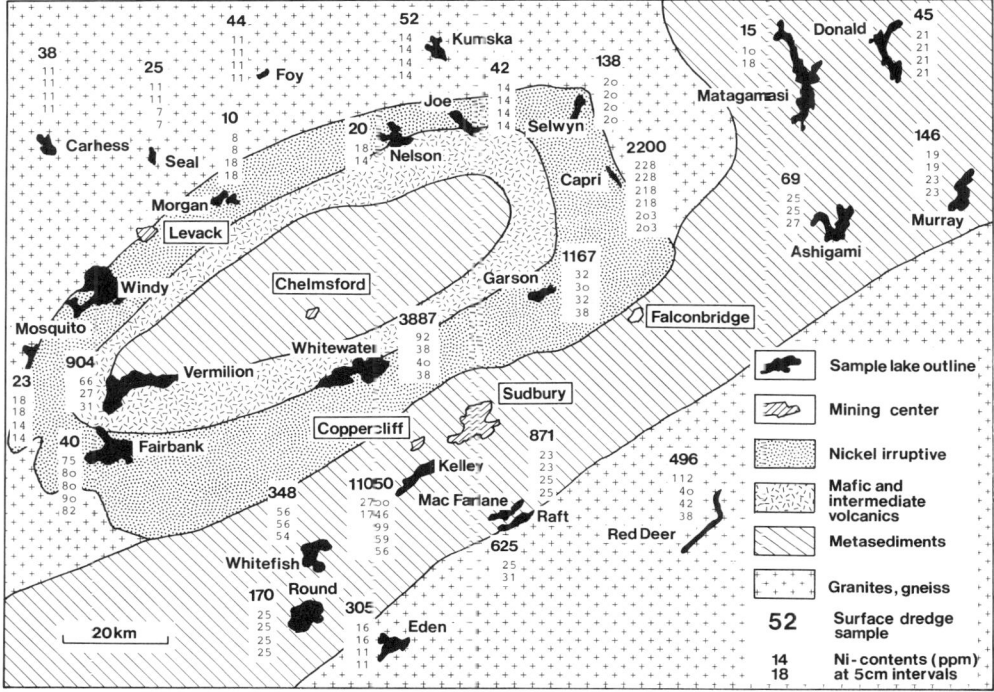

Fig. 42. Distribution of nickel in sediment cores from lakes in the Sudbury mining area of Canada (after Allan, 1974, Geological Survey of Canada, Paper 74−1/B, p. 45)

of Sudbury consists of a Proterozoic norite intrusion, surrounded by Archaean granites. The intrusion, referred to as Sudbury Nickel Irruptive, contains many Ni-Cu sulfide ore deposits. The oldest mines at Sudbury have been in operation for about 80 years. Apart from the nickel contamination around Sudbury, the effects of arsenic at Red Lake and of copper at Chibougamau were also studied in detail.

The cores were taken from the center of the lakes (as the center is most likely to reflect the overall variation within the lake in relation to man's activities.) Most of the lakes had accumulated less than 10 cm of sediment within the last hundred years. Thus samples of sediment cores taken from below this depth reflect the natural levels which occurred prior to ore exploitation, whereas samples within the upper 10 cm level can be expected to indicate the extent of subsequent mining acitivities.

This is substantiated by the fact that only surface samples collected with the aid of a grab contain abnormally high nickel levels. In general, the nickel levels in the different sections of core samples do not decrease significantly with increase in depth, neither are they consistently higher in lakes within or without the Sudbury Basin. Most nickel concentrations in the 10- to 15-cm section of core samples vary between 10 and 50 ppm; exceptions are the Lakes Capri (north of Falconbridge) with an average of 200 ppm Ni in the deeper part of the core, and Lake Fairbank, which contains elevated Ni concentrations of 80 ppm in that section. In Kelley Lake, adjacent to the Copper-cliff smelters, the nickel content, even at levels down to 15 cm of the sediment, was elevated to such an extent that "it may prove economically feasible to reclaim this or other lakes for their metal content" (Allan, 1974).

An example of the *interface of lithologic influences and man-made pollution* effects with regard to metal concentrations in lake sediments from South Africa is described (Wittmann and Förstner, 1975): The *Hartbeespoort Dam,* which represents a highly eutrophicated impoundment of South Africa (Toerien et al., 1975), is situated 40 km to the west of Pretoria between the Daspoort Hills on the southern side, and the Magaliesberg Mountains constituting the northern boundary (Fig. 43). From the western catchment area the dam is fed by the Magalies and Skeerpoort Rivers, which drain a predominantly agricultural basin. In contrast, the south-eastern Crocodile River inflow stems from the highly industrialized Kempton Park and Modderfontain areas north of Johannesburg. Graphic representation of these chemical results illustrate that heavy metal enrichment of the Hartbeespoort Dam sediments may be subdivided into two groups:

1. The western section of the dam has been enriched with chromium (Fig. 43), iron, cobalt, and nickel, which is in agreement with the higher levels of these metals in the sediments of the inflowing Skeerpoort and Leeuspruit Rivers.
2. The southeastern zone of the dam shows a marked enrichment of zinc (Fig. 43), lead, mercury, and cadmium, which again is in agreement with the higher levels of these metals in the sediments of the Crocodile River.

The similar increases in the chromium, nickel, cobalt and iron levels in the western zone indicate that the enrichment is largely due to the lithology of the catchment areas (basic intrusive rocks). On the other hand, the enriched mercury, lead, cadmium and zinc

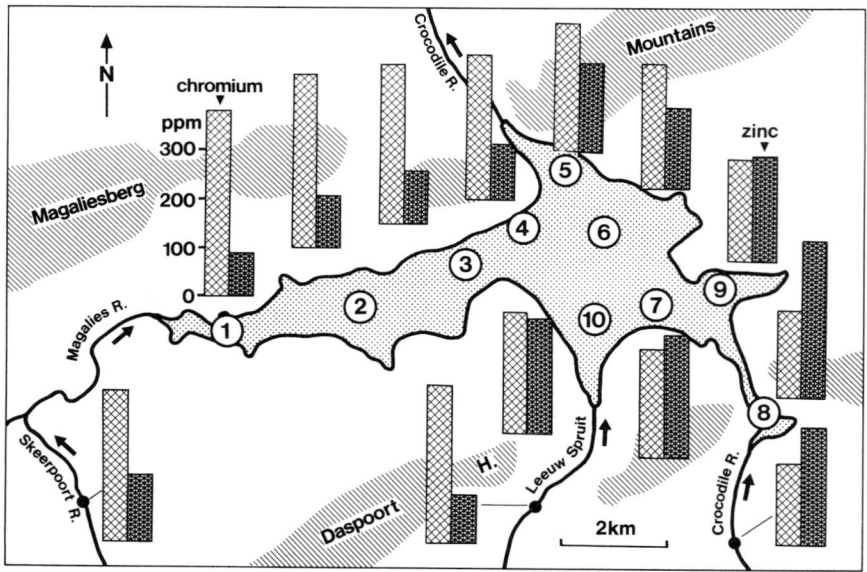

Fig. 43. Distribution of zinc and chromium in pelitic sediments from Hartbeespoort Dam (Republic of South Africa; data from Wittmann and Förstner, 1975)

levels encountered in the southeastern zone of the Hartbeespoort Dam are most probably due to cultural effects. The high zinc and lead values can possibly be ascribed to domestic waste effluents, while the cadmium and, in particular, the mercury enrichment have resulted from industrial discharges (Wittmann and Förstner, 1976a, b).

7.2 Metal Pollution in Lake Sediments (Examples)

Since the beginning of the 1970's, a large number of sediment research projects have been carried out in North America and Europe. These investigations had the goal of determining the temporal development and intensity of metal pollution. Table 37 presents a summary of such metal investigations completed in various lakes of North America. Some of these examples, along with a consideration of mercury pollution, will be examined in the next section.

Table 37. Heavy metal pollution studies in sediments of lakes in North America ("c" = core samples)

Example	Trace elements	Author(s)
I. Laurentian Great Lakes		
Lake Michigan	(c) As	Ruch et al. (1970),
	Se	Copeland (1971)
	(c) Zn, Pb, Cr, Cu, Be,	Shimp et al. (1971), Frye and
	Co, La, Sc, V, Ni	Shimp (1973). Leland et al.
		(1973), Leland (1977)
	(c) Hg	Kennedy et al (1971)
	As	Seydel (1972)
	Hg, Zn, Cu, Cr u. a.	Copeland and Ayers (1972),
		Klein (1975)
	(c) Pb	Edgington and Robbins (1974)
	(c) Cr, Zn, Cd, Ag u. a.	Cline and Chambers (1974)
	Hg	Copeland (1972)
	Zn, Cu u. a.	Baker-Blocker et al. (1975)
	Cr, Ni, Cu, Zn, Cd, Hg	Pezzetta and Iskandar (1975)
	Zn, Pb, Cu	Auer et al. (1976)
	Zn, Pb, Cu etc.	Cahill et al. (1977)
	(c) Pb, Zn, Cu, As	Leland (1977)
	(c) As, Cd, Cr, Cu etc.	Robbins and Edgington (1977)
Lake Erie	Hg	Sivisankara et al. (1971)
		Skoch and Turk (1972)
	(c) Hg	Kovacik and Walters (1973)
		Walters et al. (1974a, b)
		Wolery and Walters (1974)
	(c) Cr, Co, Cd, Zn, Ni,	Walters et al. (1974)
	Sb, As, Hg, Cu	
	Hg	Thomas and Jaquet (1976)
	(c) Hg, Pb, Zn, Cd, Cu	Kemp et al. (1976)
	Se	Adams and Johnson (1977)
	(c) Hg, Cr, Ni	Walters (1977)

Table 37 (continued)

Example	Trace elements	Author(s)
Lake St. Clair	(c) Pb, Cu, Zn, Cr, Co, Ag,	Thomas et al. (1977)
Lake Ontario	(c) Hg	Thomas (1972)
	(c) Zn, Pb, Cu, Hg etc.	Kemp and Thomas (1976a)
	(c) Pb	Farmer (1978)
Lake Huron	Hg	Thomas (1973)
	(c) Hg	Kemp et al. (1974)
		Kemp and Thomas (1976b)
	Cd, Hg, Pb, Cu	Brown and Chow (1977)
	Co, Cr, Zn	Owen and Ullman (1977)
Lake Superior	Hg	Thomas (1974)

II. Other lakes, dams and ponds

Example	Trace elements	Author(s)
Western Michigan Lakes	Hg	D'Itri et al. (1971)
	Hg, Cd, Pb	Mathis and Kevern (1973)
	(c) Cu, Cr, Zn	Dunning et al. (1975)
		Wheeler and Dunning (1976)
		Anderson et al. (1978)
Wisconsin Lakes	Se	Wiersma and Lee (1971)
	Hg	Iskandar et al. (1972)
	Hg	Konrad (1972)
	(c) As	Shukla et al. (1972), Kobayashi and Lee (1978)
	(c) Hg	Syers et al. (1973)
	(c) Cu, Zn, Cd, Pb, Cr	Iskandar and Keeney (1974)
Upper Peoria Lake	(c) Hg, Pb, Zn, As, Cd, Ni, Cu	Collinson and Shimp (1972)
Palestine (Indiana)	Cd, Pb	Wentsel and Berry (1974)
London (Tennessee)	Cu, Pb, Zn	Perhac (1974a)
Clearwater (Missouri)	Pb	Jennett and Wixson (1977)
Chautauguar (N. Y.)	As	Ruppert et al. (1974)
Lake George	(c) Cu, Zn, Cr	Schöttle and Friedman (1974)
Seneca (N. Y.)	Hg	Blackburn et al. (1977)
Reservoirs in Connecticut	Cd, Cr, Cn, Pb, Zn	Bertine and Mendeck (1978)
Reservoirs in N. Carolina	Hg	Abernathy and Cumbie (1977)
Lake Oahe (S. Dakota)	Hg	Walter et al. (1973)
Ft. Gibson, Grand (Okla.)	Zn, Pb, Cd	Pita and Hyne (1975)
Carl Blackwell (Okla.)	Cu, Pb, Zn	Sias and Wilhm (1975)
Reservoirs in Texas	Zn, Cu	Seagle and Ehlman (1974)
L. Coeur d'Alene (Idaho)	Hg, Zn, Cu	Maxfield et al. (1974)
Lake Washington	(c) Pb	Crecelius and Piper (1973)
	(c) Hg, Zn, Cu, Pb, Sb, As	Barnes and Schell (1973)
	(c) As	Crecelius (1975)
Wapato Lake (Washington)	Pb, Zn, Cr, Cd	Wisseman and Cook (1977)
Urban storm-runoff retention basin (Fresno, Calif.)	Pb, Zn, Cu	Nightingale (1975)

An especially noteworthy example is here referred to: Wentsel and Berry (1974) found the cadmium content in Palestine Lake, Palestine, Indiana, a state-owned lake of approx. 1 km² used for recreational fishing purposes, to be 2678 ppm (in dry-weight sediment). This proves to be an enrichment of about 7000 times that of unpolluted clay sediments. An electroplating plant is suspected of discharging cadmium via a small ditch into the western part of the lake, whereas the eastern part seems to be affected only to a small extent (mean 3.6 ppm Cd). In this context, it is pointed out that according to United States Environmental Protection Agency (EPA) proposals in 1973, no cadmium should be allowed to be discharged into lakes of less than 500 acres (2 km²).

The last example of Table 37 is from a lake type that, due to its metal contents, is increasingly becoming a problem. To control urban area flooding, more and more storm-drainage basins are being installed in which metals from communal runoff collect, heavily enriched with lead, zinc, cadmium, and other metals. Nightingale (1975) in his study of an example from California points out that the accumulation of lead, especially in the first few centimeters of soil in basins used for, among other things, recreation, could conceivably become an environmental health hazard. Another sort of lake basin is the sewage disposal pond. Investigation by Lund et al. (1976) concerning heavy metal content in soils beneath such sewage sludge and effluent disposal ponds shows that a movement of zinc, cadmium, copper, chromium, and nickel occur in various manners. The distribution of metals with depth was closely related to the changes in chemical oxygen demands, suggesting that the metals have moved as soluble metal-organic complexes (see Chap. G).

One of the crucial points of metal investigations in *Europe* is the sediment research in *Sweden* (Bengtson and Fleischer, 1971; Axelsson and Hakanson, 1971–75; Hakanson, 1972–76; Hakanson and Uhrberg, 1973, 1976; Hakanson and Ahl, 1975; Fredriksson and Qvarfort, 1973; Jernelöv and Lann, 1973; for a detailed compilation, see Hakanson, 1977); investigation of mercury pollution refers especially to these accounts. Other studies on lakes in *Finland* have been done, among others, by Hinneri (1974) and Särkkä et al. (1978). *Switzerland* has also been the scene for a large number of studies concerning the burden of heavy metals in sediments of lakes (Vernet and Thomas, 1972a, b; Blackburn, 1973; Baccini and Roberts, 1976; Davaud, 1977; Vernet et al., 1977a; Davaud et al., 1978); and in Northern Italy, in Lago Maggiore (Damiani and Thomas, 1974) and Lugano Lake (Premazzi, 1973). Metal pollution was reported by Stern and Förstner (1976) and Molnar et al. (1978) from lakes and dams in Slovenia, *Yugoslavia*. Contents of mercury and other trace elements in mud samples of Lake Neusiedl, *Austria*, were studied by Richter et al. (1974). For a long time *Poland* has been the focal point of metal investigations in Eastern Europa, through research by Pasternak and his team (Pasternak and Glinski, 1969; Pasternak and Antoniewicz, 1970, 1971; Pasternak, 1974a, b; Pasternak and Glinski, 1972), as well as other authors (Oporowska, 1976, Krasnicki and Szcepanski, 1976). Data have been given by Datsko et al. (1964); Datsko and Krasnov (1965); Krasnov and Kuz'menko (1967), Mun and Idrisova (1967), and Nakhshina (1974) on the distribution of trace elements in sediments of a reservoir in the *U.S.S.R.* The first investigations of trace metal contents in lake sediments from the *Federal Republic of Germany* were carried out by Groth (1971) on samples from Schleswig-Holstein. Studies of heavy metals in Southern German lakes were done on the Bodensee (Lake Constance) by Förstner and colleagues (Förstner et al., 1974; Förstner and Müller, 1974b; the combined development of heavy metals and benzopyrene was investigated by Müller (1977a, b) and Müller et al. (1977). Although research on drinking water reservoirs in West Germany done by members of the Deutscher Verein des Gas- und Wasserfachs (DVGW) uncovered no abnormal contents of trace metals in the sediments, the analyses by Mihm et al. (1976a, b) on samples taken from the pre-reservoir of Wahnbachtal dam at Siegburg showed very heavily increased contents of zinc and lead.

Many recent investigations carried out on lakes in areas of high mineral content (see Sect. B.4) have brought to light distinct enrichments of metals such as Zn, Cu, Pb, As. Increased metal concentrations also often arise in lake, reservoir, and pond sediments, originating from mine wastes – for example Lake Burley Griffin on the Molonglo River in *Australia* (Anon. 1974). In contrast, investigations on Lake Biwa in *Japan* (Nakamura et al., 1974; Satake et al., 1975, Tatekawa et al., 1975) have shown local heavy metal enrichments that are more probably attributable to rapid industrialization, urbanization and land development. Stronger contamination by zinc, lead, and cadmium was found by Goldberg et al. (1976) in sediment cores from the Palace Moat, Tokyo.

7.3 Metal Contamination Recorded in Dated Sedimentary Cores

Some of the previously mentioned studies utilized sediment cores (marked "c" in Table 37) which enable a determination of the geochemical background of certain regions and of cultural accumulation.

These investigations of the vertical distribution of trace elements in sedimentary cores clearly indicate the increase of metal pollution during the last 100 to 200 years. Apart from local effects which will be treated below, a general increase in the concentrations of zinc, lead, cadmium, and mercury was observed in the lakes studied. Figure 44 presents two sedimentary profiles from different areas, one from North America, the other from England, both showing the same evolution of the mercury pollution during the last centuries which culminates in a sharp increase at the onset of the Industrial Revolution: Profile 1 in Figure 44 stems from the exemplary study of the mercury contamination in sediments of Lake Ontario undertaken by Thomas (1972). The core profile was taken from the southeastern part of Lake Ontario which exhibited no characteristic local anomalies. High amounts of mercury occur in the top 6 cm of the core. Below 8 cm the mercury concentrations decrease significantly to a level ranging between 500 and 550 ppb. At approximately 25 cm, a relatively low

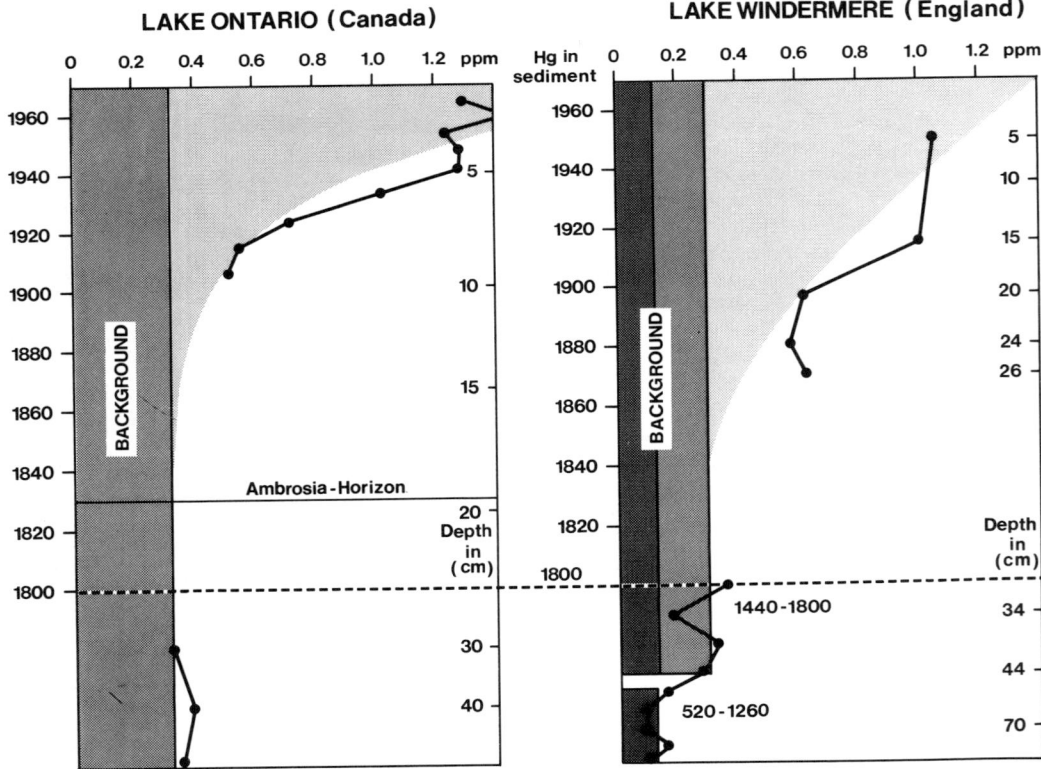

Fig. 44. Mercury in sedimentary core profiles from Lake Ontario (Thomas 1972) and Lake Windermere (data from Aston et al., 1973)

value of 140 ppb was found in the layers of decayed organic material known as the Ambrosia pollen which resulted from the deforestation carried out by the early settlers (during the eighteenth century).

At lower layers from 29 to 50 cm the samples have elevated mercury concentrations varying between 300 and 400 ppb, thus reflecting the background level. On the basis of this time scale, a chronology of mercury input into the Rochester Basin can be made. Input of industrial mercury occurs at a sediment depth of 9 cm, approximately coinciding with the turn of the century. In sediments from 1906 to ca. 1943, there is a spectacular and continuous rise in the mercury concentrations attributable to the industrial expansion of the region. From the early 1940s until the present, mercury values have fluctuated, but in general they show a consistent but less rapid increase in concentration, with present levels being approximately four times higher than the natural background level (Thomas, 1972).

Profile 2 in Figure 44 shows the distribution of mercury in a 1 m sediment core from Lake Windermere (England) investigated by Aston et al. (1973) using a ^{210}Pb radionuclide dating technique. The bottom section of the core (i.e., between 50 and 80 cm), representing the sedimentation period from 520 to 1250 A.D., apparently reflects a background value of about 120 ppb Hg. Since 1400 there has been a gradual increase in Hg concentrations in sediments which may be correlated with man's activities. These include: denudation of land surface, industrialization, mining and quarrying, burning of fossil fuels, and sewage disposal. Much of the Hg thus released initially enters the atmosphere, although some is introduced directly into natural waters (Aston et al., 1973).

In five lake examples, Lake Constance, Lake Michigan, the Wisconsin Lakes, Lake Washington and Lake Erie, the vertical distribution of a large number of elements in core profiles was determined simultaneously. Table 38 summarizes the results of these studies by listing (1) the background levels of minor elements in the deeper sections of the core, (2) the maximum values in the upper layers, and (3) the factors of enrichment as the quotient (2):(1). In making horizontal comparison, it should, however, be borne in mind that these samples differ widely in composition and texture; grain-size dilution has not been uniformly minimized by analyses of the clay fraction with particle sizes of < 2 μm in all instances. The core profile from Lake Monona, for example, exhibits higher portions of carbonate sediments which may be responsible for the lower background levels of some elements (Iskandar and Keeney, 1974). The data from Lake Michigan registered here originate from profiles of the central area of the southern part of the lake where most minor element concentrations are significantly higher than in the core profiles from near-shore areas, probably owing to the finer grain size in the sediments of the lake center.

Lake Constance and Lake Michigan: Mixed Sewage Inputs. Low concentrations of heavy metals are reflected by the concentrations of iron, cobalt and nickel in most of the lake sediment sequences. These findings agree with the results from highly polluted river sediments e.g., from the lower Rhine section where the Fe, Co, and Ni content is influenced, in particular, by geochemical factors (Section 8).

As has been demonstrated for the mercury concentration in the sedimentary cores from Lake Windermere, there are several environmental influences to be considered in

Table 38. Distribution of minor elements in sedimentary profiles from Lake Constance (Förstner and Müller, 1974), Lake Michigan (Ruch et al., 1970; Shimp et al., 1971; Kennedy et al., 1971; Frye and Shimp, 1973), Lake Monona/Wisconsin (Lake Minocqua) (Syers et al., 1973; Shukla et al., 1972), Lake Washington (Barnes and Schell, 1973; Crecelius and Piper, 1973; Schell, 1974; Crecelius, 1975), and Lake Erie (Walters et al., 1974). Data of background and maximum value in parts per million (ppm). F = factor of enrichment

	Lake Constance			Lake Michigan			Wisconsin Lakes			Lake Washington			Lake Erie		
	background	max. value	F	background	max. value	F	background	max. value	F	background	max. value	F	background	max. value	F
Zinc	124	380	3	120	317	2.5	15	92	6	60	230	4	7	42	6
Chromium	50	153	3	77	85	1	7	49	7		n.d.		13	42	4.5
Nickel	55	50	1	54	44	1	34	50	1.5	(iron:		1)	40	95	2.5
Copper	30	34	1	44	75	1.5	22	268	12	16	50	3	18	59	4
Lead	19	52	3	40	145	3.5	14	124	9	20	400	20		n.d.	
Arsenic		n.d.		11	22	2	(2	51	25)	10	200	20	0.6	3.2	5.5
Mercury	0.2	0.8	4	0.04	0.2	5	0.24	1.12	5	0.1	1.0	10	0.004	4.48	12
Cadmium	0.21	0.68	3		n.d.	2.5	2.5	4.6	2		n.d.		0.14	2.4	17

their complex interrelations. Generally, it is difficult to establish the association of enhanced metal concentrations in sediments with a definite source of pollution. As a rule, however, one can postulate that moderate factors of enrichment between approx. 2 to 10 originate mainly from mixed inputs of industrial effluents and sewage from domestic and agricultural sources. This appears to be especially valid for the accumulation of mercury in recent sedimentary sequences for which core profiles show an approximately fivefold concentration increase during the past 100 years as compared with the geochemical background value. The same statement is valid, to a greater or lesser degree, for the increased enrichment rates of zinc, lead, and cadmium in recent lacustrine sediments from industrialized areas.

Further specification of sewage discharge can be made by the use of additional indicators e.g., the content of nitrogen, phosphorus, and organic carbon within the same sedimentary profiles. In the example from Lake Constance, N and P concentrations are compared with those of the heavy metals found in the corresponding segments of a dated core from the central part of the lake. Similar trends of both zinc/lead and nitrogen/phosphorus concentration ratios indicate that these substances originate simultaneously in the public sewage system, whereas the increase of cadmium and chromium within the upper layers of the sediment sequence can be explained by industrial emissions, e.g. from electroplating and tanning industries, respectively. These findings are substantiated by analyses of surface sediments from the various streams emptying into Lake Constance, which reveal characteristic influences of both industrial and municipal sewage disposal (Förstner and Müller, 1974b).

Wisconsin Lakes: Algicides and Herbicides. Metal pollution resulting from the use of pesticides is demonstrated by examples from lakes in Wisconsin: *Copper* concentrations in the sediments from Lake Monona increase with depth to a maximum (434 ppm Cu at 15–20 cm sediment depth) and then decline sharply (Syers et al., 1973). These effects are a result of the intermittent treatment of the lakes with copper sulfate to control algal growth from 1918 to 1944. Since the rates of increase of copper concentrations in sediments from other lakes e.g., Lake Michigan and Lake Constance, are generally low, local copper accumulations may, as a rule, be taken as indicators of previous administrations of copper for algal control. The amount of *arsenic* is not available for Lake Monona and is therefore taken from a study of Lake Minoqua in northeastern Wisconsin (Shukla et al., 1972). The characteristic increase of arsenic in these sediments was explained by the fact that several Wisconsin lakes had been treated with sodium arsenite to reduce the population of noxious weeds. It is assumed that fertilizers (Kanamori and Sugawara, 1965) and, above all, household detergents (Anginc et al., 1970) are responsible for the general enrichment of arsenic concentrations, as observed for example in the sediments of Lake Michigan (Seydel, 1972).

Lake Washington: Airborne Particulate Fallout. Characteristic sources of metal pollution are constituted by smelters and by coal-fired power plants (Ruch et al., 1973), as well as by heavy city traffic as was shown by Chow et al. (1973) in lead isotope studies on marine sedimentary cores.

The Tacoma copper smelter, situated 50 km to the south of Seattle (Washington), has for a long time been considered one of the main sources of atmospheric pollution along the northwestern border of the United States. The smelter, which began operation in 1890 as a lead producer, at present releases stack dusts containing approx. 80,000 kg of lead and 150,000 kg of arsenic per annum. According to the distribution pattern of metal concentrations, which was evaluated by determining the copper content of present-day Douglas fir needles (Crecelius and Piper, 1973), the dominant direction of the transport of particles from the smelter is directly in line with Lake Washington.

Lead concentrations were analyzed in two sedimentary cores from Lake Washington. The results of one profile from the center of the lake show an increase from about 25 ppb Pb at the 30 to 50 cm depth to about 400 ppm near the surface.

With a sedimentation rate of 3 mm/year, the history of lead pollution in Lake Washington can be drawn as follows (Crecelius and Piper 1973): The lead content in sediment deeper than 25 cm presumably represents the background concentration of that metal. The first increase in lead at about 25 cm depth level corresponds to the time when the land along the western slope of Lake Washington was first developed in 1880. The population of Seattle increased rapidly during the next 20 years, rising from 7000 to more than 100,000 in 1900. During this period, operations began at the Tacoma smelter. Local input from the Seattle community and atmospheric releases from the smelter are considered responsible for the increase in lead observed in the sediments between depths of 25 cm and 18 cm. The sharp decrease in lead in the sediments deposited between 1916 and 1920 may be due to two factors: (1) the diversion of the Cedar River into the lake, which may have diluted the lake sediments with coarse sediments of lower lead concentrations and (2) the conversion of the smelters from lead to copper production. Above this minimum, lead increases strongly to its present value of 400 ppm. Most of this increase is probably due to the introduction of lead-containing gasoline additives during the 1920's.

In a recent study Crecelius (1975) examined the arsenic cycle in Lake Washington by determining the inputs from rivers, rain, dustfall, storm water and removal of As in the outlet water, and those due to sedimentation. Sedimentary cores from the lake center indicate a strong enrichment of arsenic in the surface materials exceeding 200 ppm As, decreasing with depth to the usual background concentration of about 10 ppm. These abnormally high concentrations in the surface sediments are largely attributed to atmospheric input of partially soluble arsenic-rich dust from the Tacoma smelter. An arsenic budget for Lake Washington shows equal supplies from the atmosphere and from rivers, and a removal by outflowing water (45%) and by accumulation in the sediments (55%).

Barnes and Schell (1973) estimated that 99% of the current lead input, 93% of the mercury, 68% of the zinc, and 56% of the copper input into Lake Washington sediments originate from aeolian processes. However, since trace metals from atmospheric fallout, as well as from fluviatile suspended load, often occupy inert positions within the particles, their direct effects on both water quality and aquatic life are less detrimental than those arising from solute metal discharges.

Lake Erie: Industrial Pollution. Walters et al. (1974a, b) determined the trace and minor element concentrations in sediment cores from eight stations in Lake Erie. The profile example 19-1 which is listed in Table 38 was taken from the central basin of Lake Erie near Cleveland at the mouth of the Cuyahoga River. It would appear that these values are representative of the moderate pollution effect in that area, but there are other examples from the harbors of Cleveland and Buffalo where enrichment of elements such as mercury or cadmium is found to be very high. Profile contours, in particular from the concentrations of mercury indicate (Wolery and Walters, 1974) that early cultural activity might have occurred sometime around 1835. Major increases in zinc, arsenic, and copper during 1939–1955 correspond to the general growth in industry during World War II and the Korean conflict. It is assumed that the establishment of the chemical plants in the Cleveland area in 1949 caused a characteristic increase in mercury concentrations; the major break in 1955 coincides with the opening of the Detrex chlor-alkali plant at nearby Ashtabula, Ohio. The chromium increase would then have occurred in the late forties, in reasonable agreement with the growth of the Cleveland electroplating industry (Walters et al., 1974).

7.4 Mercury Poisoning of Lakes

One of the most serious forms of environmental pollution is the contamination of lakes by mercury wastes. This is due to the specific toxic behavior of mercury, arising from the biosynthesis of mercury-alkyls (see Chap. E), and to a distinct residue-forming tendency of these substances in aquatic ecosystems. The problem of mercury pollution in lakes therefore merits a more detailed discussion.

7.4.1 Sources of Mercury Pollution

Mercury concentrations in sediments represent a particularly good indicator for various forms of cultural activities e.g., from agricultural use of fungicides, atmospheric fallout from power plants, influences from mining and smelting processes, sewage input, etc. Sediment cores from lakes, which had been investigated with respect to an assessment of historical changes in the pollution intensities, mostly exhibit a characteristic increase of the mercury concentrations during the last few centuries even if there is no significant variation of other metals (except in many cases, of lead). Two typical examples of such profile cores have been presented from North America and England in Figure 44; additional examples are listed in Table 39, in order to demonstrate the great variety of sources, which affect the enrichment of mercury in recent aquatic sediments.

Among the examples of Table 39, there are some which are situated far removed from direct industrial or municipal effluents, but nonetheless have been strongly enriched in mercury. One likely source of mercury accumulations in remote areas is particulate matter, either from industrial plants burning fossil fuel, or from the smelting of copper, lead, and zinc ores (Mathis and Kevern, 1973). For example, sediments from Lake Saoseo and Lake Tuma in southern Switzerland, both situated at high altitudes, are contaminated with mercury of atmospheric origin, probably derived from the industrial complexes of northern Italy (Vernet and Thomas, 1972a).

On the other hand, there is the example of Lake Vättern, the second largest lake in Sweden with a surface area of 1912 km^2. Sediments were investigated in its southerly part where most contamination was expected to have occurred, resulting mainly from the industrialized and densely populated region around Jönköping and Huskvarna. The mercury concentrations, however, did not exceed 520 ppb in the surface sediments and decreased to a background level of less than 100 ppb Hg.

Stronger increases in mercury concentrations from municipal sewage effluents often originate from plants accepting industrial wastes (Evans et al., 1973). Such discharges might be the cause of mercury accumulations in sediments from Lake Trummen

Table 39. Concentrations and sources of mercury in lake sediments (examples)

Lake/Dam	max. conc. ppm (background)	Source	Author
Orly Res. (Paris)	0.18 (0.15)	Municipal–background	Cumont and Montiel (1973)
Lake Winnipeg	0.23 (0.07)	Industrial–Winnipeg R.	Allan and Brunskill (1977)
Lake Lucerne (Switzerland)	0.49 (0.10)	Municipal–Steinibach	Blackburn (1973)
Rietvlei Dam (Rep. S. Africa)	0.52 (0.12)	Agricultural	Veres and Hasty (1976)
Lake Vättern (Sweden)	0.52 (0.10)	Municipal–Jönköping	Axelsson and Håkanson (1973)
Lake Sangchris (Illinois)	0.57 (0.28)	Coal-fired power plant	Anderson and Smith (1977)
Lake Oahe (South Dakota)	0.63 (0.03)[a]	Gold-mining (Historical)	Walter et al. (1973)
Lake Beldany (Poland)	1.77 (0.15)	Agricultural, industrial	Krasnicki and Szczepanski (1976)
Lake Ekoln (Sweden)	2.10 (0.10)	Industrial–Uppsala	Axelsson and Håkanson (1973)
Lake Saoseo (Switzerland)	2.20	Atmospheric fallout	Vernet and Thomas (1972)
Lake Trummen (Sweden)	3.30 (0.68)[a]	Municipal–Växjö	Björk (1970) in Jernelöv and Lann (1973)
Lake Coeur d'Alène	6–10	Mining–d'Alene River	Maxfield et al. (1974)
Moste Dam (Yugoslavia)	17.7 (0.50)	Smelter–Jesenice	Štern and Förstner (1976)
Lago Maggiore (Italy)	20.1 (0.24)[a]	Industrial–Toce River	Damiani and Thomas (1974)
Pickwick Res. (Tenn./U.S.A.)	32.6 (1.3)[a]	Industrial–Tennessee R.	McMullen (1973)

[a] Minimum value from surface samples.

in southern Sweden, an extremely eutrophic lake which has been polluted with waste water from the town of Växjö (Jernelöv and Lann, 1973).

A comprehensive survey of the effects of mercury pollution in lakes, rivers, and coastal regions was first undertaken in Sweden. After severe cases of bird poisoning resulting from seed dressings containing methylmercury dicyanidiamide in the early 1960's, the distribution of mercury was investigated in other parts of the environment. These studies led to the detection of industrial effluents as the major source of pollution.

7.4.2 Swedish Lakes

Sediments from lakes in Southern Sweden were examined for mercury pollution originating from pulp and paper mills and from chloralkali plants.

Värmlandssjön, the eastern main basin of *Lake Vänern*, is highly polluted by mercury. Contamination originates mainly from the chlor-alkali industry of Skogfall (Håkanson, 1975). Emissions have successively been reduced from 3000 kg/yr during the period 1920–1968, to 500 kg/yr during 1969–1971, and to approx. 30 kg/yr in 1974 (the latter figure is approximately the minimum loss for mercury-cell operations).

One of the smaller (1.8 km^2) but most polluted lakes in Central Sweden is Lake Björken. Maximum mercury concentrations of up to 11 ppm Hg were measured in the surface sediments by Axelsson and Håkanson (1973). Hasselrot (cit. A. & H) reports mercury levels of more than 50 ppm from neutron activation analyses of Lake Björken sediments. The principal source of contamination is wastes from a pulp mill discharged just upstream from Lake Björken. Between 1955 and 1965 about 352 kg phenylmercury was used (the equivalent of about 180 kg metallic Hg) and probably most of it reached the lake.

7.4.3 Canadian Lakes: Clay Lake

One of the areas in Canada which is most heavily polluted with mercury was found in the Wabigoon River System in northwestern Ontario. Fish from Clay Lake, about 80 km downstream from Dryden, contained up to 16 ppm of mercury. The source of toxification is believed to be a chlorine-alkali plant at Dryden which, from 1962 to 1969, before effluent control was introduced, discharged a total of 10,000 kg of mercury into the waste water. After stringent enforcement of legislation early in 1970, mercury discharges have been reduced to about 3% of the previous values.

Analyses of sediments, water and biota were performed by Armstrong et al. (1972) and Armstrong and Hamilton (1973), in particular on samples from Clay Lake. This lake has a surface area of about 6 km^2, a maximum depth of 24 m and an average depth of 8 m. Mercury distribution in sedimentary profiles from the central parts of the lake show concentrations of more than 8 ppm Hg in the upper 4 to 6 cm of the cores, whereas the Hg content below 6 cm was usually less than 1 ppm. The background level in the sediments of Clay Lake is estimated at 0.1 ppm Hg (as was also found in Swedish lakes and Laurentian Great Lakes). From the median values of 3.1 ppm, the total amount of pollutant mercury in the top 6 cm was calculated at about 2000 kg.

7.4.4 Laurentian Great Lakes

The mercury problem in the Great Lakes Region roused public interest at the beginning of 1970 when commercial fishing was banned in the waters of the St. Clair River –Lake Erie System. Mercury concentrations attaining values of 5 ppm, i.e. ten times the current permissible level laid down by the U.S. Food and Drug Administration Act, were found in walleye from Lake St. Clair in April and June 1970 by Bails (1972). In bottom sediments the highest value recorded was an alarming 86 ppm. After a rather extensive review of all the industries and communities, two main sources of mercury pollution were found in that area (Fimreite, 1970; Turney, 1972). One was the Wyandotte Chemical, Mich., which has a chlor-alkali plant located on the Detroit River; and the Dow Chemical chlor-alkali plant on the St. Clair River at Sarnia, Ontario. In Michigan, chlor-alkali manufacturing, utilizing continuous mercury cathode cells, began in 1939 at Wyandotte where approx. 5 to 10 kg of mercury were discharged daily. Chlor-alkali operations at Sarnia commenced in 1947 and had an average discharge of about 10,000 kg annually before the control system was introduced in 1970. Figure 45 shows the decrease of mercury in the river sediments downstream, starting with more than 60 ppm Hg near Sarnia at the sources of pollution. In the St. Clair River, the border between Michigan and Ontario, the Hg concentration declines exponentially with distance from the source. Since the highest concentrations of mercury are found in the sediments along the Canadian side of the St. Clair River, it has been suggested that the mercury is transported in particulate form as part of the sediment load rather than as soluble or methylated compounds. According to Cline et al. (1973),

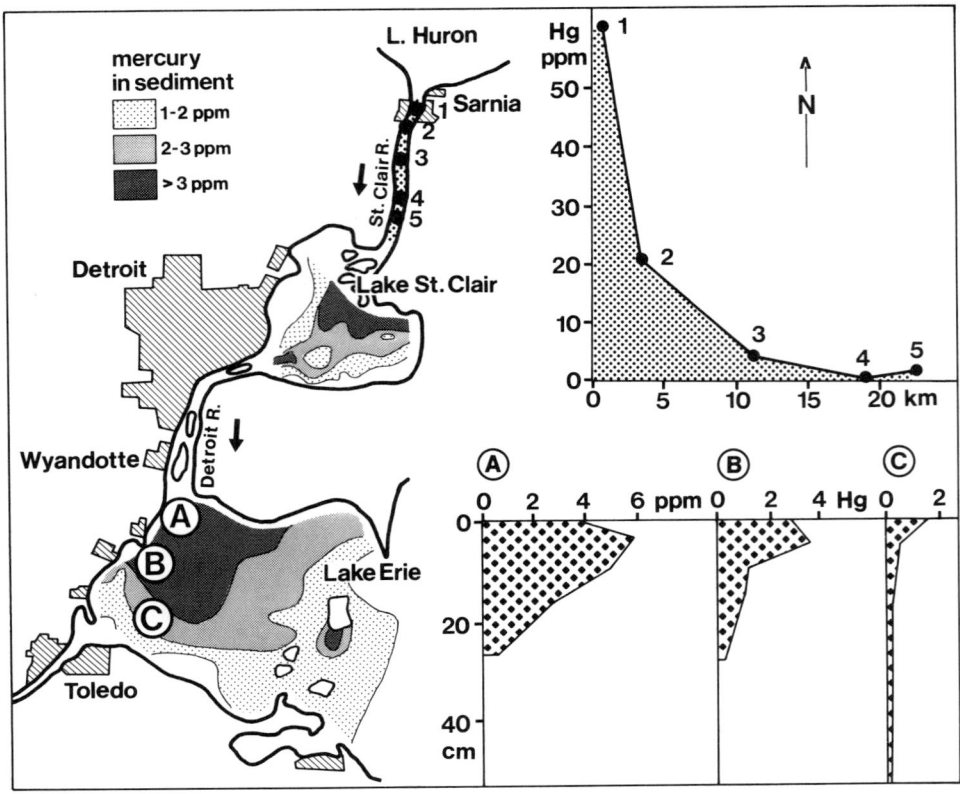

Fig. 45. Distribution of mercury in sediments from Lake St. Clair, St. Clair River, and western Lake Erie (after Cline et al., 1973; Thomas, 1974; Kovacik and Walters, 1973)

mercury concentrations in the sediments of the St. Clair River correlate not only to the distance from the pollution source, but also to the organic fraction of the sediment. It is hypothesized that mercury is often associated with particulate organic complexes. The authors thus consider the "possibilities that these particulate complexes enter the food chain through ingestion, with methylation then occurring in vivo".

Mercury concentrations in Lake St. Clair indicate a distinct gradient with increasing distance from the St. Clair delta and a remaining enrichment zone along the line between the south bend cut-off and the Detroit River (Thomas, 1974); the first feature is in agreement with an upstream source of Hg in the St. Clair River; the latter anomaly, closely associated with the shipment channel, suggests that this must be the prodominant flow of water with associated particulates from the St. Clair River to the Detroit River.

The effects of mercury pollution from the St. Clair/Detroit Rivers on the composition of the sediments in Western Lake Erie were studied by Kovacik and Walters (1973). Distribution of mercury was determined by analyses of sedimentary cores, examples of which are given in a north-south profile in Figure 45. The high level of mercury near the mouth of the Detroit River is due to rapid deposition by the western flow as it enters Lake Erie. As the contaminated water mass flows south along the Michigan-Ohio shore, it deposits progressively smaller amounts of mercury into the lake. The mercury contamination of 1 to 6 ppm in the top 10 cm segment of the sediment cores decreases exponentially with depth to a low background concentration of 0.04 to 0.09 ppm. (Frequency distribution of mercury data from over 4000 sediment core depth-intervals indicate an average baseline level of 0.04 ppm Hg [Walters, 1977]). A chief source of contamination resulting in values 100-fold in excess of the normal Hg content is the mercury discharge from the above mentioned chlor-alkali plants on the Detroit and St. Clair Rivers. In addition, the authors believe that the surface enrichment zone is also a result of mercury pollution from coal fly ash. This theory is substantiated by Turney's calculations (1972), according to which, for example, the Detroit Edison Company and Consumer's Power in Michigan emitted approx. 10,000 kg of mercury from burning coal. This is in the order of the former discharge from the Dow Chemical chlor-alkali plant at Sarnia.

Restoration of Mercury-Contaminated Lakes

Following legislation and enforcement to limit mercury pollution, the mercury concentrations in both sediments and organisms of Lake St. Clair decreased significantly during the period 1970–1974 (Thomas, Myslik, Toronto Conf. 1975). There is evidence that mercury-polluted sediments are rapidly eroded from the shallow basin of Lake St. Clair. More serious problems, however, may arise in cases of lower rates of either erosion or sedimentation. With respect to Clay Lake, Armstrong and Hamilton (1973) stated that "even with the present control, there is little hope of any rapid improvement in the degree of pollution because of the disturbance of sediments by animals".

In order to understand the full extent of the mercury problem, one has to consider the enormous loss rate during former years. From the total of 2865 tons of mercury purchased in the U.S., in 1968, 76% or 2160 tons were lost to the environment. Walters et al., (1974) have estimated the pollutant mercury accumulated in the western basin of Lake Erie at 228 metric tons. According to Kemp et al. (1974), the Lake Ontario reservoir contains a mass of 500 to 600 metric tons of mercury.

Several methods for the restoration of mercury-contaminated water bodies have been suggested (Jernelöv and Lann, 1973); (1) dredging of polluted sediments, (2) converting mercury to mercury sulfide with a low methylation rate (anaerobic conditions), (3) binding of Hg to silica or coprecipitation with hydrous Fe and Mn oxides (aerobic conditions), (4) increasing the pH to give volatile dimethyl mercury rather than monomethyl mercury, (5) isolating the polluted sediments from the water body by means of physical barriers, such as polymer film overlays (Widman and Epstein, 1972) blanket plugs of waste wool (Tratnyek, 1972), sand and gravel overlays (Bongers and Khattak, 1972, Fujino, 1977).

Dredging processes (which will be treated in more detail in Chap. E) are ambiguous with regard to the mercury problem. One main disadvantage has been described by Reimers et al. (1975): Since sulfur-bearing materials are particularly good absorbents of both inorganic and methylmercury (see below) any contact with oxygen will reduce the capacity of anaerobic bottom sediments to keep these substances fixed. In this context, the studies of Frimmel and Winkler (1975), on the mercury distribution in the Main River (Germany) are of particular interest and indicate that only 2% of the total mercury content in the sediment, as against 65% in the pore water, occurs in the form of methylmercury-chloride. Furthermore, dredging activities will disperse unrecovered sediment over large areas and would be likely to result in an increase in the mercury content over a period of time in the flora and fauna of the water course (Smith, 1972). Problems arising from the disposal of mercury-contaminated materials should also be considered. Finally, model studies by Jernelöv and Åsell (1974) suggest that even if all inorganic mercury in a heavily contaminated smaller area (constituting about 90% of all mercury in the system) could be removed, the methylation rate would still not be reduced by more than 40%.

The major advantage, on the other hand, of dredging mercury-contaminated sediments, lies in the possibility of mercury recovery especially at high concentrations. Several procedures have been investigated by Smith (1972) in an EPA-contracted study, which include methods of particle size fractionation, gravity separation, flotation and roasting. Chemical leaching was found to be the most effective method in recovering mercury: treatment with 1% hypochlorite at pH 6.0–6.5 proved to be rapid and effective (up to 98%–99% in many cases) at ambient temperature and no reactions occurred with carbonate. Mercury solubilized by this method could be recovered through the leachate method used to treat chlor-alkali plant effluents (e.g., metal reduction by iron, absorption on carbon, precipitation with aluminum, sulfide precipitation, etc.—see Chap. G), and subsequently the spent leachate could be returned to the water course. Hypochlorite, widely used in water treatment, is relatively innocuous to the aquatic environment and is inexpensive and readily available in the vicinity of chlor-alkali plants where many mercury-laden sediment deposits occur. In-place treatment might be particularly useful in mercury recovery from industrial holding ponds (Smith, 1972).

The feasibility of methods such as providing anaerobic conditions or raising the pH-values appear to be unfavorable, particularly since changes in lake chemistry will cause other adverse effects in the ecosystem. The alternatives to the dredging (and successive recovery) procedures will, in particular, lie in the *covering of mercury-contaminated sediments* with material that decreases the rate of release of methyl mercury to the water (Jernelöv et al., 1975).

In a comprehensive study Reimers et al. (1975) have analyzed the sorption characteristics of mercury in natural sediments, synthetic sediments, and synthetic scavengers, such as alloys, activated carbon and ion exchange, nylons, polyesters, polypropylene, agricultural products, wood, and chicken feathers. It has been shown that all sediments or scavengers containing sulfur possess high affinity to the mercury compounds. The greatest uptake of mercury has been observed in organic mercaptans; the sulfides are next in adsorption capacity, but when high sulfide contents occur, inorganic mercury forms a soluble mercury disulfide complex. A sequence of materials for mercury sorption is listed in Table 40.

Polyethylene-nylon films, sands and gravels—which are mostly used for covering contaminated sediments—indicate only a low sorption capacity for inorganic mercury and almost none for methylmercury; the inhibition mechanism must be due not to sorption or binding but rather to stabilization of the colloidal sediment by reduction of sediment agitation (Reimers et al., 1975). Nonetheless, this treatment seems to be particularly promising in areas where large quantities of sand- and gravel-sized materials are supplied by mining, dredging and demolition activities. Thomas (1974) reports the case of an unintentional but very effective covering of mercury-polluted sediments by

Table 40. Mercury-binding substances in the order of their sorption capacity (from Reimers et al., 1975)

Sulfur-containing sediments (low-oxygen contents) e.g., peat and mercaptans	Commercial scavengers (silicon alloys, act. carbon etc.).	Three-layer clay (illite, montmorillon.) Protein scav. (wool, chicken feathers) reduced sand	Fe/Mn-oxides natural org. sediments free of sulphur	Synthetic org. scavengers, not containing sulphur but amines. Two-layer clays (kaolinite), fine sand
>	>	>	>	

waste products from the mining industry. Sediment cores from the Silver Bay of Lake Superior, where tailings from a ore processing plant are disposed, indicate that mercury enrichment, which had occured at sediment depths of 2 to 3 cm, had been interrupted in the upper part of the profile; above this level, the Hg-content decreased to its background value in Silver Bay sediments.

An iron overlay using crushed automobiles has been proposed as an additionally useful approach for the treatment of mercury-contaminated sediments (Smith, 1972).

Iron will convert soluble mercury in water to elemental mercury; during this process, hydrated ferric oxide is formed, which is an effective coprecipitator of mercury ions. In addition, iron reduces toxic methylmercury to elemental mercury. It was suggested by Smith (1972), that iron scrap in combination with sand, clay, or other environmentally acceptable material should function as an effective barrier preventing any methylmercury found in the sediment from entering the overlaying water and contaminating the flora and fauna of the water body.

8 Metal Pollution in River Sediments

Mineral exploration based on the distribution of ore metals in stream sediments has become a generally accepted method of mineral reconnaissance. Hawkes (1976) has presented the following idealized formula relating the metal content of an anomalous stream sediment sample (Me_a) and the upstream area (A_a) with the grade (Me_m = metal content of material undergoing active erosion within the mineralized area) and the surface area (A_m) of the mineralized area, provided the background (Me_b) is known and constant:

$$Me_m A_m = A_a (Me_a - Me_b) + A_m Me_b.$$

The formula says that the grade-tonnage factor ($Me_m A_m$) is equal to the product of the drainage area above the sample site times the excess of the anomalous value over background. In the Hawkes (1976) formula there is no provision for the contamination of the drainage system by addition of metals from mine waters, dumps, tailing or industrial waste.

Generally, the variation in trace metal content of stream sediments can be characterized as a function of potential controlling factors by the following model (Dahlberg, 1968):

$$T = f(L, H, G, C, V, M, e)$$

in which T represents the resulting trace metal concentration, L the influence of lithologic units, H the hydrologic effects, G the geologic features, C the cultural (man-made) influences, V the type of vegetational cover, M the effects of mineralized zones and e for the error plus effects of additional factors not explicitly defined in the model. In mineral exploration, where the method of stream sediment sampling was initially applied successfully, the main problem was to maximize the factor M i.e., to eliminate other effects as far as possible. The effects of the cultural influences and the mineralized zone are found to be highly intrusive. As has been demonstrated by a large number of investigations in the field of mineral exploration, traditional mining regions are particularly susceptible to contamination by metals. It is usually difficult to distinguish between contamination of industrial or domestic origin, "natural" pollution, and pollution resulting from mining activites. Airborne dust and particulate material from smelter stacks, wind-blown particles and leachate from the tailing ponds all contribute to an increase in the metal concentrations in river sediments.

8.1 Geochemical Reconnaissance of Mercury

Mercury and its compounds play a particularly important role in geochemical reconnaissance in stream sediments. Since at normal pH values dissolved mercury is rapidly adsorbed onto sediments, hydrochemical sediment methods for the prospecting of mercury are required for the analysis of particulate material, such as clay, iron oxides or organic matter (Jonasson and Boyle, 1972). This element acts as a guide in prospecting for sulfide deposits (see Sect. 3.2), and is of special interest as a "natural source of environmental pollution". Examples of geothermal mercury, arsenic and antimony pollution are described from New Zealand and Long Valley, California (Ritchie, 1961; Reay, 1972; Weissberg and Zobel, 1973; Davey and Van Moort, 1974; Mariner and Willey, 1975).

Characteristic enrichments of toxic metals from mining operations and natural sources also occur in fluviatile sediments in other areas: Bombace et al. (1974) and Batti et al. (1975) found maximum mercury concentrations in river deposits from the Mt. Amiata region (Tuscany, Italy) of up to 290 ppm, approximately 500 times higher than the background values for this region. More than 6000 ppm Hg and approximately 2% As were registered in stream sediments from the Mt. Avala Cinnabar mine in Serbia (Maksimović and Dangić, 1973). In a review of the Hg concentration in rocks, soils, and stream sediments of the United States Pierce et al. (1972) indicated that mercury values greater than 1 ppm, as found for example in the Coeur d'Alène district (Gott et al., 1969) and in the Taylor Mountains, Alaska (Clark et al., 1970), are considered worthy of further investigation as a possible result of (a) mercury mineralization processes or (b) surface contamination by mercury-bearing wastes.

8.2 Stream Sediments: a Response to Environmental Contamination

Pollution reconnaissance can be carried out in stream sediments in the same manner as in mineral exploration. Two effects, however, must be considered when chemical data from active stream sampling are used for identification of pollution sources: (1) under conditions of high water discharge, erosion of the river bed takes place, and generally leads to a lower degree of contamination. Laszlo et al. (1977), for example, obtained data from the Sajo River, Hungary, which indicated that three months after a flood, the mercury concentration in the bottom sediment had been reduced to approximately one-quarter of the values found immediately before the flood. (2) It is usually imperative in stream sediment studies to base metal analyses on a standardized procedure with regard to particle size. Several methods to effect this procedure have already been described.

As an introductory example results from systematic studies are described, which had been carried out by the Institute for Sediment Research at Heidelberg University in the early 70s (Banat et al., 1972a, b, c; Förstner and Müller, 1973a, b, 1974a). Analyses of the pelitic fraction ($< 2\ \mu m$) have been used in assessing the most significant centres of heavy metal pollution of the major rivers—Danube, Rhine, Neckar, Main, Ems, Weser, and Elbe—draining West Germany. Figure 46a–d contains graphic representation of the Cd, Hg, Pb, and Zn values found for the respective sampling stations for these rivers, whereas values for the minor tributaries are indicated alongside each major river (small print). A total of 126 samples were analyzed. For the elements under consideration here, the following is a brief summary of results:

Cadmium. Generally the values range between 10 and 20 ppm; they are especially low in the upper Rhine. The highest cadmium levels were found in the lower Neckar River (88 ppm), in the Danube catchment drained by the Naab River (45 ppm), and in the upper Elbe River (38 ppm).

Mercury. Bottom sediments from the Danube, Neckar, and Weser Rivers contain less than 2 ppm Hg. Extremely high values were found in a tributary of the Rhine, the Wupper River (70 ppm); mercury contamination was found to be high in the upper Elbe river (20–35 ppm), in the Rhine-Main area (15–20 ppm), and in the middle Rhine (15–20 ppm).

Lead. Average values ranging between 100 to 200 ppm Pb were found in the sediments of the Danube, Neckar, and Ems Rivers. High values were recorded for several rivers such the Elbe (650 ppm) and Main (650 ppm); sediments from the Aller River, a tributary of the Weser River, contained the highest Pb levels, namely 1200 ppm

Zinc. Average zinc enrichment was found to pertain to the Danube and the upper Rhine and Main River drainage patterns (500 ppm). Zn contamination was more pronounced in several other drainage systems—generally exceeding 1000 ppm—especially in the Weser River (2800–3100 ppm) with its tributary streams, Aller (2200 ppm) and Werra (2250 ppm).

The distribution of heavy metals in the *Rhine system,* as reflected by the sediment composition, is separately depicted in Table 41, which summarizes the data of eight elements; it includes data on sediments from Lake Constance through which the Rhine flows, as well as of sediment samples from the Rhine Estuary in the North Sea. The Lake Constance data stem from the bottom section of a 50 cm sediment core taken from the central part of the lake basin (Förstner et al., 1974). As indicated by Table 41, the "background" values for Lake Constance correspond closely to the average shale values (Turekian and Wedepohl, 1961). Between stations 1 and 2, the Rhine River is already significantly contaminated by mercury, lead, and cadmium. These metals are enriched by a factor of 7 or more compared to the metal concentrations in the deeper sediment layers in Lake Constance. A considerable increase of chromium in the sedi-

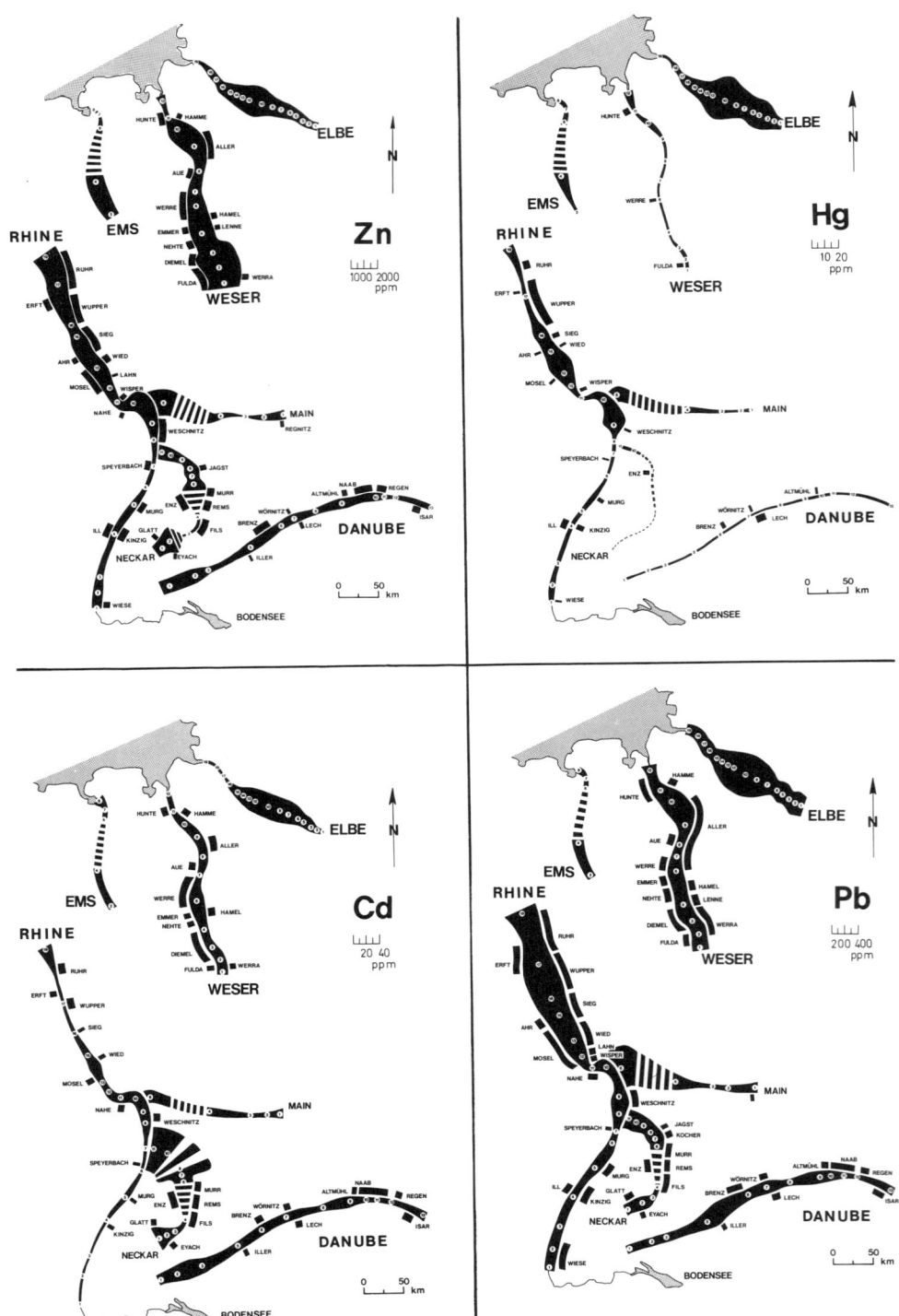

Fig. 46. Heavy metals in the pelitic fraction of sediments from major rivers in the Federal Republic of Germany (Förstner and Müller, 1974a)

9.5.2 The Rhine Estuary

During the past decade de Groot (1966), and de Groot et al. (1971, 1973) have studied in great detail the transport of heavy metals in suspended matter from the highly polluted Rhine River into the North Sea and Wadden Sea. Certain results from these investigations are summarized in Figure 52. Taking the total amount of any particular heavy metal in the sediment at the fresh tidal area of Biesbosch (chlorinity = $0^o/oo$) as 100%, one observes a steady decrease in percentage towards Harvinvliet (chlorinity = $2^o/oo$). Cadmium, mercury, copper, zinc, lead, and chromium undergo a loss of 95% to 70% as compared to the Biesbosch values; cobalt, nickel, and iron decrease by 60% to 45%. Changes in the levels of manganese, samarium, scandium, and lanthanium are insignificant.

The sharp decrease in the heavy metal concentrations in the sediment downstream from the fresh water tidal area of the Rhine was explained by *mobilization processes* which, according to de Groot and co-workers (de Groot, 1966; de Groot et al., 1968, 1971) should particularly be caused by intensive decomposition of organic matter, the products of which form soluble organometallic complexes with the consequence of a release of metals from the sediments into the water. Similar effects were described by Martin et al. (1970, 1971, 1973, 1976) from the Gironde estuary in France. The authors of these studies based their strongest arguments in support of the remobilization hypothesis on the obvious resemblances in the order of sequence for both the rates of decrease of the particulate-associated metals during the estuarine mixing and their respective position within the Irwing-Williams series (Irving and Williams, 1948) of the relative stability of metal chelates.

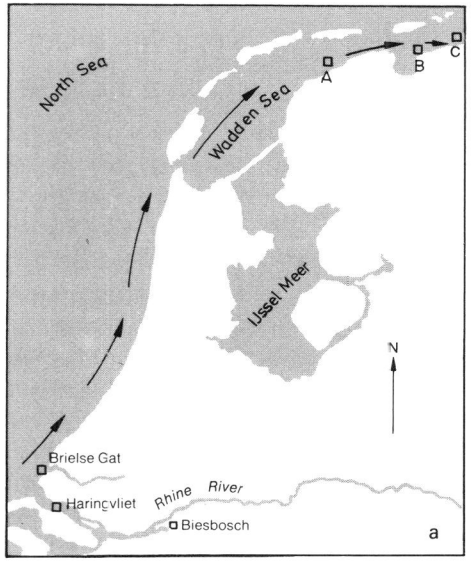 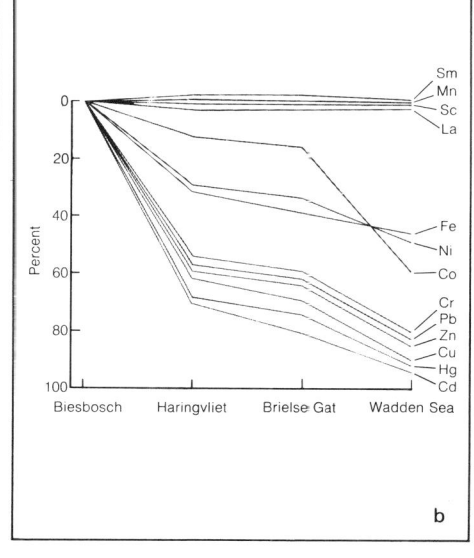

Fig. 52. a Direction of movement of sediments in the Rhine estuary, North Sea, and Wadden Sea. **b** Mobilization of metals in the Rhine Estuary, North Sea, and Wadden Sea, expressed as a percentage of the original contents. (Both parts after de Groot et al., 1971, 1973)

9.5.3 The Elbe Estuary

Förstner and Müller (1974a), and Müller and Förstner (1975) investigated a profile along the Elbe River between Gorleben and Brunsbüttelkoog from samples of 19 stations. Eight heavy metals were determined in the clay fraction of the sediments and six metals in the Elbe water. Data of cadmium, lead, copper, chromium, and zinc in both sediment and water are summarized in Figure 53.

The concentration of all heavy metals examined tended to increase considerably between station 1 and station 7 (or, for some metals, station 8). Between stations 7 (or 8) and 19 the opposite is observed—a very sharp decrease of concentration occurs between stations 7 and 12; between stations 13 and 19 only slight changes occur.

Taking the content of a specific heavy metal at station 7 (or 8), where maxima occur, as 100%, the cadmium, mercury, and copper values at station 19 have lost more than 92% of their maximum concentration, whereas the decrease of cobalt and nickel is only 34% and 64% respectively. This general tendency is approximately the same as is observed in de Groot's sediment studies (Fig. 52).

Surprisingly, however, the metal concentrations in the water did not increase in the estuarine mixing zone, as would be expected from the remobilization model of de Groot and associates. Conversely, there is a general decrease of most of the metals studied (limitations of the data arising from single sampling and large fluctuations have to be considered), which indicates the presence of other or additional mechanisms controlling the distribution of trace metals in the Elbe estuary.

9.5.4 Mixing Processes

Förstner and Müller (1974) proposed that the distribution of heavy metals in the clay fraction of sediments of the Elbe River tidal area can be explained by a mixing of North Sea sediment with Elbe River sediment. This interpretation is supported by investigations on the movement of sand grains in the Elbe estuary conducted by Simon (1953) (Fig. 54), which indicate that the sand of the river bed between Cuxhaven (North Sea) and St. Margarethen (between stations 18 and 17) is derived nearly exclusively from the North Sea ("Seesandzone" of Simon). Between St. Margarethen and Luhe (close to station 13) a mixing zone exists which is composed of both North Sea and Elbe sand ("bed load mixing zone"). Further upstream the zone consisting of sand particles is exclusively derived from transport by the Elbe River.

A further support to the "mixing" theory evolves directly from the trace metal data of the sediment samples: the different loss rates of a specific metal correlate well with the ratio of man-made pollution to the natural background value, that is, the higher the portion of a specific metal concentration induced by man, the greater the decrease. Cobalt and nickel have the lowest rate of enrichment whereas mercury and cadmium have the highest enrichment in sediments (Förstner and Müller, 1973). The simplest explanation for this development is the mixing of highly polluted river-induced material with sea-derived sediment particles, which are less contaminated.

The similar behavior of the heavy metals in the sediments of the Rhine estuary, and in the North Sea and Wadden Sea, may be explained by the same mixing mechanism. The mixing of fluviatile and marine sediments in the estuary area of the Rhine River

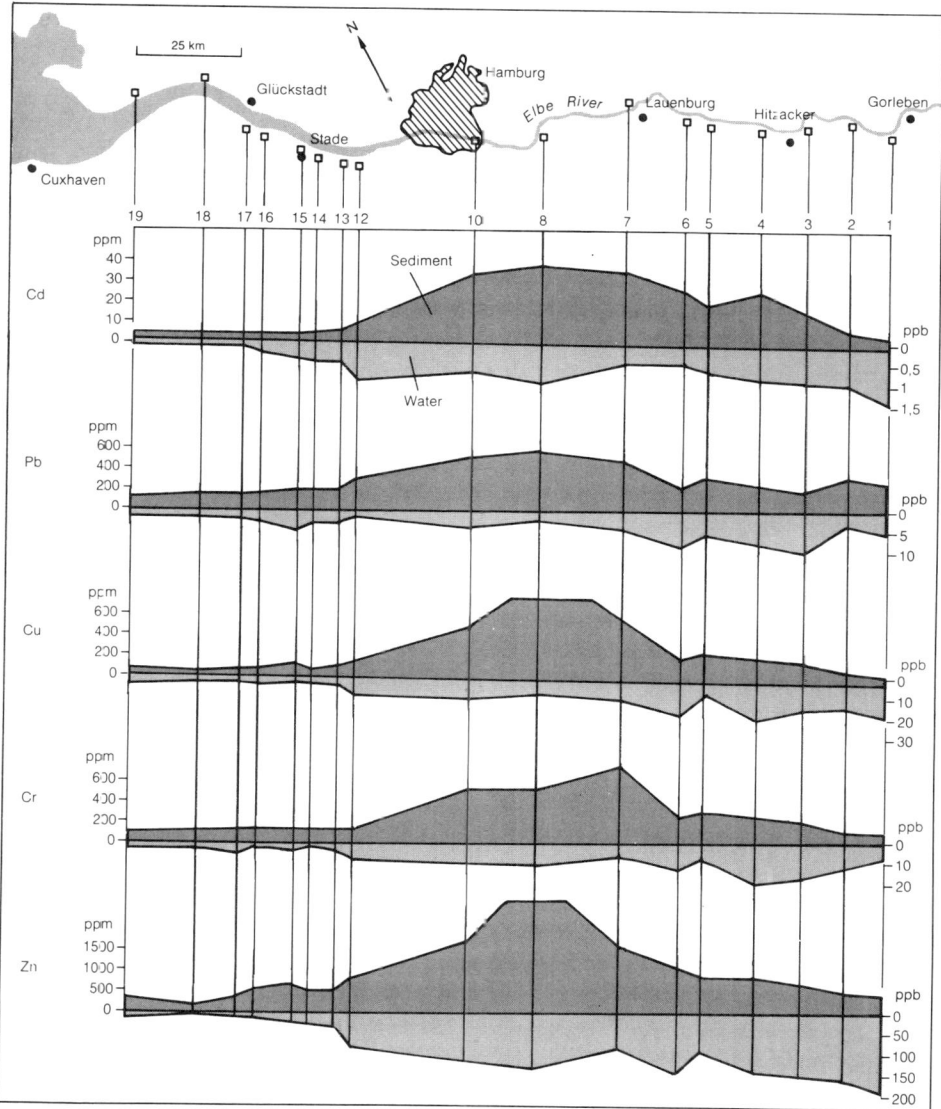

Fig. 53. Distribution of heavy metals in sediments (pelitic fraction) and water from the estuarine mixing zone of the Elbe (Müller and Förstner, 1975)

has been confirmed by isotope investigations of clay minerals (Salomons, et al., 1975), carbonate minerals and organic substances (de Groot, pers. comm.).

Other processes, such as the formation of soluble organometallic complexes with the metals from the sediments or partial desorption of heavy metals by competing cations, especially sodium, from the seawater could play an additional—but only

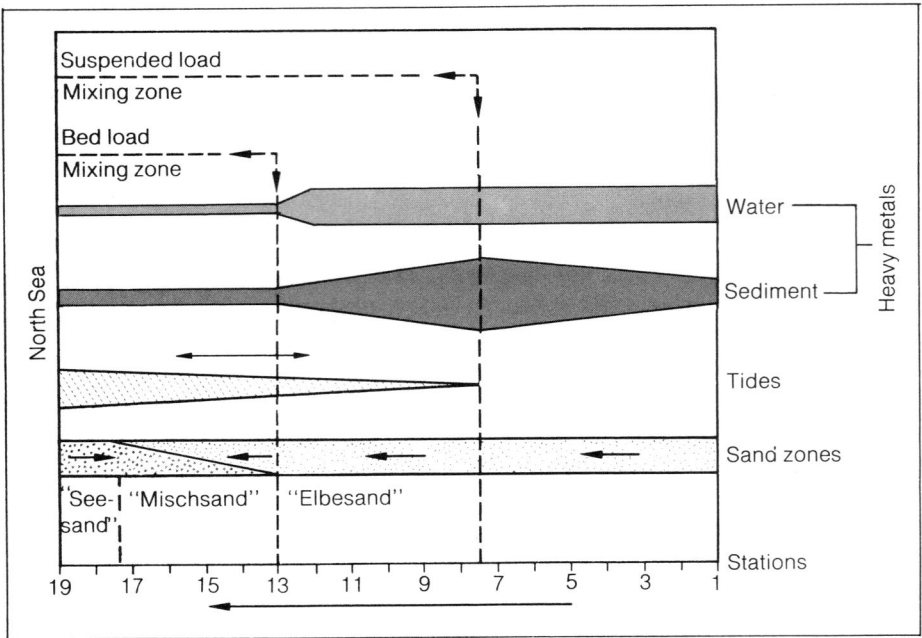

Fig. 54. Schematic presentation of the distribution of sediments and heavy metals in the estuarine mixing zone of the Elbe (Müller and Förstner, 1975, based on investigations of Simon, 1953)

minor—role in the depletion of heavy metals in estuarine sediments. In Chap. E the complex interactions of the various factors which influence the distribution of trace metals in water and sediments of estuaries will be further discussed.

Chapter E
Metal Transfer Between Solid and Aqueous Phases

The greater part of the dissolved heavy metals transported by natural water systems is, under normal physicochemical conditions, rapidly adsorbed onto particulate material. However, heavy metals immobilized in bottom sediments do not necessarily stay in that condition, but may be released as a result of chemical changes in the aquatic milieu. The following sections will pay special consideration to the interaction between water and sediment, and the possibilities of immobilization and remobilization of metals in unpolluted as well as in polluted systems.

1 Residence Times of Metals in Aquatic Systems

In order to gain insight into the adsorbing capacities which sediments display towards metal i.e., whether a certain metal tends to remain solubilized or to adhere to the particulate material, one could compare their sequence of *residence times* in the aqueous phase of a given system.

Barth (1952) has defined residence times by the following formula:

$$\tau = \frac{A}{dA/dt}$$

where A is the total amount of the element in suspension or solution in a specific water body and dA/dt is the amount introduced into the water body (river, lake, ocean) within a specific interval of time. The calculation of the residence times of elements in aquatic solution should include the following variables: (1) surface water and groundwater flow; (2) atmospheric fallout and evaporation; (3) sedimentation and remobilization (Bowen, 1977).

In 1958 Goldberg and Arrhenius determined the residence times of a number of metals by means of their rates of removal from the ocean by sedimentation processes. These values agree remarkably with those of Barth (1952), who performed similar calculations using statistics for the rate of supply of elements to the oceans by rivers. In Table 54 the most recent estimate for the oceanic residence times for metals are listed according to the data of Goldberg et al. (1971), and Brewer (1975).

It is evident that the elements Al, Ti, Cr, and Fe, which readily form insoluble hydroxides, have relatively short residence times. Silicon and aluminum are further reactants active in formation of clay minerals, and enter the oceans to a considerable extent as suspended solids. Iron and titanium, which also pass rapidly through the hydrosphere, probably prevail as solid phases during the major part of their residence times, even in the pelagic areas of the oceans (Goldberg and Arrhenius, 1958). The importance of the residence times in predicting the increases in concentration of

Table 54. Mean residence times (in years) of metals in oceans

	Ocean[a]	Lake Michigan[b]	Lake Washington[c]
Aluminum	1×10^2	3.1×10^{-1}	–
Iron	2×10^2	4.4×10^{-1}	3.0×10^{-2}
Lead	4×10^2	–	5.8×10^{-2}
Chromium	6×10^3	1.3×10^1	–
Antimony	7×10^3	1.4×10^2	–
Manganese	1×10^4	8.0×10^{-1}	–
Titanium	1.3×10^4	–	–
Silicon	1.8×10^4	–	–
Selenium	2×10^4	8.2×10^1	–
Copper	2×10^4	8.9×10^1	1.3×10^0
Zinc	2.1×10^4	3.7×10^1	1.8×10^0
Cobalt	3×10^4	1.0×10^1	–
Barium	4×10^4	4.3×10^1	–
Arsenic	5×10^4	4.5×10^1	–
Mercury	8×10^4	2.2×10^1	1.2×10^0
Vanadium	8×10^4	2.6×10^0	–
Tungsten	1.2×10^5	–	–
Molybdenum	2×10^5	–	–
Cesium	6×10^5	–	–
Cadmium	1×10^6	4.1×10^2	–
Lithium	2.3×10^6	–	–
Uranium	3×10^6	–	–
Strontium	4×10^6	4.4×10^2	–
Sodium	6.8×10^7	4.0×10^2	2.4×10^0

[a] Goldberg et al., 1971; Brewer, 1975.
[b] Klein, 1975.
[c] Barnes and Schell, 1973.

artificial radionuclides introduced into the oceans by nuclear detonations or as wastes from reactors was also pointed out. The radioisotopes of strontium and cesium, both hazardous for living organisms, have very high residence times in the oceans, and lead to radioactive decay in the aquatic environment. The alkali and alkaline earth metals have long residence times which decrease with increasing atomic number, following the decrease of the effective hydrated ionic radii and hydration energies (Brewer, 1975). It should be noted that quoted residence times are to be viewed with caution, since they are based on the assumption that complete mixing of the input takes place in time periods that are short compared to the residence times (Brewer, 1975). It is obvious, however, that this is not the case for a number of elements, such as silicon, phosphorus and barium, which exhibit variations of more than an entire order of magnitude with depth. On the other hand, substances on the ocean bed have particularly long residence times (10^7 to 10^8 years). This raises the possibility of utilizing the ocean beds for dumping waste materials.

Because nuclear waste disposal has become a pressing issue, many eyes are turned to the ocean bottom as a possible alternative disposal site. In a special issue (1977) of the journal Oceanus with the title *High-level nuclear wastes in the sea bed?*, various aspects of this question are discussed. The "multiple barrier concept" is based on the fact that the residence time during containment should be $= X \times 10^6$ years. In this respect, $T_{containment} = T_{waste\ form} + T_{canister} + T_{rock} + T_{sediment} + T_{ocean}$, where for the single *barriers* the following numerical values are set (Hollister, 1977): $T_{waste\ form} = 10^3$ to 10^X yr (where "x" = F(solubility)); T_{rock} is yet uncertain since bulk permeability due to thermal concentration fracturing is unknown; $T_{sediment} = 10^6$ yr/100 m for pure diffusion or 10^{13} yr/100 m for sorption + diffusion; T_{ocean} is considered to range from 10^2 to 10^3 yr, but may be less if there is a biologic short circuit. The relatively better possibilities seem to be the center of oceanic gyres (great circular currents) and of lithospheric plates, the so-called MPGS (mid-plate, mid-gyre areas). These regions are thousands of kilometers from areas of crustal destruction, and are covered with soft oxidized clay, which exhibits ion-retention and permeability characteristics adequate to contain chemically and physically the waste for the periods needed. Calculations of Duursma and Gross (1971), and Heath (1977) indicate that in ten million years the elements under consideration would diffuse over the following distances: strontium, 4–32 m; cesium, 8 meters; zinc, 2 to 6 m; cerium, 1 m; and thorium, less than a third of a meter. With that in mind, Turekian and Rona (1977) have suggested that fracture zones are potential disposal sites for radioactive waste, especially the flanks of the Mid-Atlantic Ridge, and represent effective physiographic, sedimentary, chemical, and oceanographic barriers in their aseismic deep canyons. In addition, the major producers of radioactive wastes are likely to be near the Atlantic Ocean.

Table 54 also shows two examples of lakes in which several data of residence times have been determined in order to formulate mass balances. The sequence of residence times agrees largely with oceanic data, although in Lake Michigan, for example, the water volume is 5×10^5 units less than the total volume of seawater, and the residence times are about four orders of magnitude lower than the oceanic residence times (Klein, 1975); in Lake Washington the residence times are once again one to two orders of magnitude lower than in Lake Michigan (Barnes and Schell, 1973). In the case of several metals in these two lakes, atmospheric input—also from "communal sources" (Klein)—plays a considerable role. In Lake Michigan deposition by airborne particulates account for at least one-fifth of the total input of As, Ag, Cr, Fe, Zn; for Hg, Sb, Se the major source is also of atmospheric nature (Klein, 1975). In Lake Washington 99% of the current input of lead and 93% of the mercury input are attributed to aeolian processes (Barnes and Schell, 1973).

It has previously been noted that characteristic differences exist for the behavior of a number of elements in oceanic and continental water. These differences can be demonstrated from the ratio of the background or average concentration in continental waters and in oceanic water. The values range from a 100-fold higher concentration of zinc in continental waters compared with the minimum content in seawater, to a sodium enrichment in oceans that is 10,000 times that of the freshwater average value. The order of sequence of these data is similar to the respective sequence of the residence time values for the oceans i.e., the elements exhibiting short residence times in the ocean are often depleted in the seawater relative to river water. It can be concluded that a group of metals consisting primarily of aluminum, iron, manganese and lead, then of zinc, copper, titanium, and chromium, is very effectively eliminated from the dissolved state in the river water upon reaching the marine environment. With regard to the above mentioned elements, a characteristic "nonconservative" behavior may be expected for the estuarine mixing zone. In contrast, a group of metals, such as sodium, strontium, lithium, and calcium, and to a lesser degree cesium, molybdenum, uranium, tungsten, vanadium, barium, and antimony tend to be more associated with the aqueous phase in oceans than is the case under the less saline conditions that prevail in rivers and lakes. This is partly due to the lack of chemical reactivity of solvated mono- and divalent cations (Na, Sr, Li, Ca, Cs, and Ba), and seems also to result from the formation of anionic complexes in seawater (for Mo, W, U, V, and Sb), which are not readily adsorbed or incorporated by clay minerals, hydrous iron oxides, and solid organics, and consequently accumulate in seawater.

Similar "nonconservative" effects have also been observed in freshwater systems, where the concentrations of dissolved heavy metals are significantly reduced along the water's course, beginning in the sewer system and continuing in the river or lake.

Figure 55 (from Koppe, 1973) demonstrates the immobilization of heavy metals in the water. For several weeks zinc and lead salts were emitted at a certain point in a river section in the Ruhr catchment area in measurable quantities. It was demonstrated that the lead content decreases rapidly and at a distance of 70 km is no longer present in measurable quantities. The zinc concentration decreases in a similar fashion, although to a somewhat lesser extent.

It has been noted by Bowen (1975) that in most rivers equilibrium cannot be completely achieved between the solution and the solid phases due to the short residence times: the Thames, for example, flows approximately 250 km in 14 h, whereby chemical equilibrium would require approximately 200 h. Nonetheless, the use of surface water from contaminated rivers and lakes for drinking water purposes is often only possible because of the tendency of most of the heavy metals to adhere to solid particles, in particular when the metal concentration is markedly high.

2 Types of Metal Association in Sediments

According to the classification by Goldberg (1954) natural metal enrichments in solid substances can originate from the following five sources:
a) Lithogenous formations: weathering products from the source areas or rock debris from the river bed. This material undergoes only slight change, regardless of long residence times.
b) Hydrogenous formations: particles, precipitation products, adsorbed substances, formed due to physico-chemical changes in the water.
c) Biogenous formations: biologic remains, decomposition products of organic substances as well as inorganic siliceous or calcareous shells.

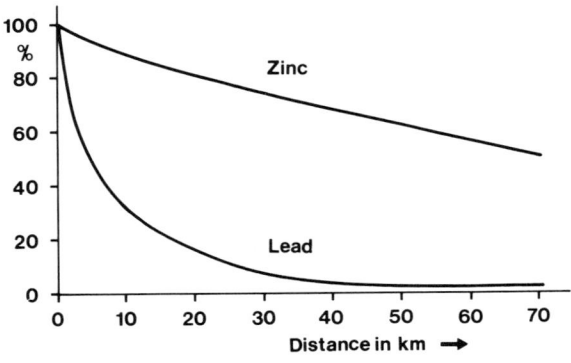

Fig. 55. Immobilization of lead and zinc from artificially induced lead and zinc solution in a section of the Ruhr River catchment area (from Koppe, 1973)

d) Atmogenous formations: metal enrichment resulting from atmospheric fallout.
e) Cosmogenous formations: extra-terrestrial particles.

One could include civilizational metal enrichments among the first four processes because the by-products from the erosion of refuse dumps and mine waste are an important source of lithogenous solids as well as of soil particles, which can be carried into the aquatic system via the surface runoff. Hydrogenous metal enrichment in solids may occur during the total course of a primary dissolved contaminant—i.e., as early as in the waste effluent, later in the sewage treatment plant, or ultimately in the river, lake, or sea. The same holds true for biogenous formations. Atmospheric contributions can enter water by several means e.g., as direct input in dissolved or solid form (particularly on large water surfaces), or by way of enrichment processes in the soil and plants. From the variety of influences between these individual processes it is clear that classification of enrichment can often prove to be difficult. This is particularly true when no single characteristic source of metal pollution is obvious i.e., when it originates from diffuse nonpoint sources (see Chap. B.4). When this is the case, the composition of the grain sizes and the material composition of the deposit (e.g., content of organic material) can be useful indicators. It is evident, however, that the key to understanding metal enrichment in aquatic systems lies in identifying the different types of metal associations in sediments.

2.1 Classification of Chemical Phases in Sediment

Despite their dissimilar originations, natural and environmentally related types of metal bonding have characteristics in common. Following his observations on particulate substances from the Amazon and Yukon Rivers, Gibbs (1973) suggested four groups of heavy metal associations in aquatic solid substances. They can be characterized by the following bonding processes:

(1) adsorptive bonding, (2) coprecipitation by hydrous iron and manganese oxides, (3) complexation by organic molecules, (4) incorporation in crystalline minerals. This categorization is expanded in Table 55 to include all main types of metal associations, such as occur in both natural and polluted water systems.

2.2 Heavy Metals in Detrital Minerals

Heavy metals as major, minor or trace components, frequently in inert lattice positions, can be transported and deposited in the mineral substances of natural rock debris. The silicate minerals feldspar and quartz usually have very low heavy metal contents.

The distribution of elements in *minerals* is determined by the physico-chemistry of the source medium (magma, lava, aqueous solution) and by crystal-chemical factors, i.e., ionic radii, valences and electron configuration. Characteristic examples of trace metal concentration in minerals of magmatic origin are given in Table 56 (from Wedepohl, 1972, 1974a, b): (1) the data do not indicate that *copper* has a strong tendency to be incorporated in any particular crystal structure of rock-forming minerals, but appears to be enriched more in the earlier counterparts (olivine, pyroxene) of the magmatic differentiation series (Bowen series) than in the later crystallization products (alkali feldspar, quartz); (2) *Zinc*, on the other hand, is preferentially incorporated into distinct structural positions of silicates and oxides, where it replaces ferrous iron and magnesium. Biotites

Table 55. Carrier substances and mechanisms of heavy metal bonding (from Förstner and Patchineelam, 1976)

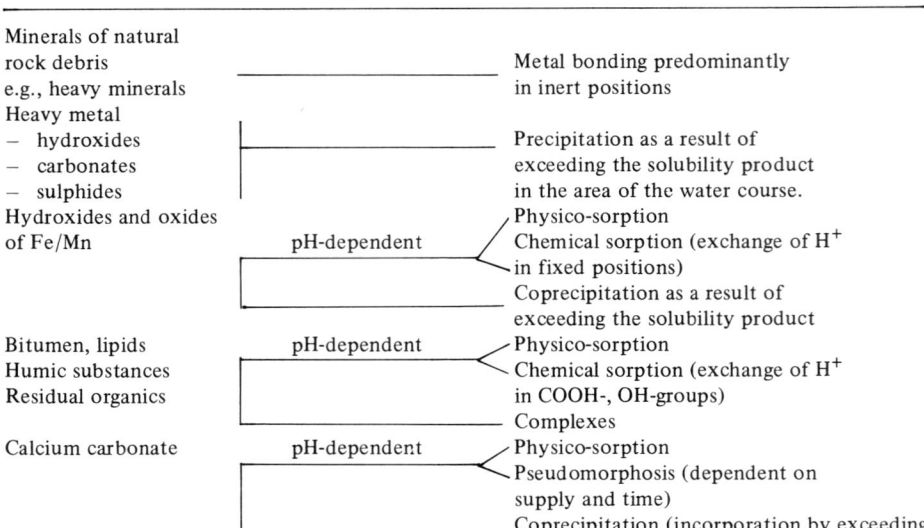

Minerals of natural rock debris e.g., heavy minerals		Metal bonding predominantly in inert positions
Heavy metal – hydroxides – carbonates – sulphides		Precipitation as a result of exceeding the solubility product in the area of the water course.
Hydroxides and oxides of Fe/Mn	pH-dependent	Physico-sorption Chemical sorption (exchange of H^+ in fixed positions) Coprecipitation as a result of exceeding the solubility product
Bitumen, lipids Humic substances Residual organics	pH-dependent	Physico-sorption Chemical sorption (exchange of H^+ in COOH-, OH-groups) Complexes
Calcium carbonate	pH-dependent	Physico-sorption Pseudomorphosis (dependent on supply and time) Coprecipitation (incorporation by exceeding the solubility product)

and amphiboles are often very rich in zinc. Some pure zinc minerals exist in both structural forms; (3) *Lead* in the silicate structures replaces the positions of large monovalent or divalent metals, especially potassium. This explains the enrichment in both muscovite and alkali feldspars.

Clay minerals, and particularly heavy minerals contain significantly higher metal concentrations and frequently exhibit a close relationship to a specific source material of a regional nature (rock, ore dike). In his study of the geochemistry of relatively uncontaminated sediments from the Gulf of Paria Hirst (1962) suggested that the elements Cr, V, Cu, Pb, and to a lesser extent Co and Ni, enter the deposition basin structurally combined in the lattices of degraded *clay minerals*. The differences between the ratios of these elements to Al, Mg, Fe, K, and (Na + K) indicate that Cr and Cu favor bonding to illite, whereas V, Co, and Ni favor montmorillonite.

Table 56. Concentrations of copper, zinc, and lead in rock-forming minerals (according to data from Wedepohl, 1972, 1974a, b)

	Copper			Zinc			Lead		
	(n)	range ppm	mean ppm	(n)	range ppm	mean ppm	(n)	range ppm	mean ppm
Olivine	(51)	6– 960	115	(7)	50– 82	63	(8)	0.2– 7.2	2
Pyroxene	(90)	4–1000	120	(25)	16– 200	97	(20)	0.3– 20	6
Amphibole	(40)	1– 300	78	(51)	34– 690	196	(75)	1 – 70	11
Biotite	(660)	1– 480	86	(666)	34–4000	527	(259)	7 – 95	21
Muscovite	(12)	5– 152	36	(17)	24– 200	59	(32)	6 – 70	20
Plagioclase	(180)	8– 700	62	(23)	1– 50	17	(16)	1 – 70	20
Alkali feldspar	(10)	1– 20	4	(4)	10– 24	15	(419)	2 –700	53
Quartz	(1)		2	(5)	4– 11	7	(12)	0.1– 3	1

2.3 Heavy Metal Precipitation

The solubility product of an electrolyte is a special type of equilibrium constant, referred to as K_{sp}. This concept is applied to a saturated solution in which a dynamic equilibrium exists between solid substance and its aquated ions, for example:

$$MX_2 \text{ (s)} \rightleftharpoons M^{2+} \text{ (aq)} + 2 X^- \text{ (aq)}$$

for which we can write

$$K_{sp} \text{ for } MX_2 = [M^{2+}] \times [X^-]^2$$

for example

$$K_{sp} \text{ for } PbCl_2 = [Pb^{2+}] \times [Cl^-]^2 = 1.7 \times 10^{-5}$$

For a compound such as As_2S_3, we have

$$As_2S_3 \text{ (s)} \rightleftharpoons 2 As^{3+} \text{ (aq)} + 3 S^{2-} \text{ (aq)}$$

$$K_{sp} \text{ of } As_2S_3 = [As^{3+}]^2 \times [S^{2-}]^3 = 5 \times 10^{-27}$$

In general, the solubility product principle has the following interpretation: In any aqueous solution in equilibrium with a slightly soluble ionic substance, the product of the molar ion concentrations, each raised to the power equal to the coefficient of the ion in the solubility equation, has a constant value (at a given temperature, usually 25 °C). If the solubility product of a particular substance is exceeded, precipitation of the compound occurs until the product of the ionic concentrations is $\leqslant$ to the K_{sp} value. This means, for instance, that under conditions of high S^{2-} ion concentration, the $[As^{3+}]$ will be exceedingly low because of the set condition that states when $[As^{3+}]^2 \times [S^{2-}]^3 > 5 \times 10^{-27}$, precipitation of As_2S_3 takes place.

The concentration of a metal ion in the initial stages of precipitation is primarily dependent upon the type and concentration (or more precisely the activity) of the anionic species in the solution, as well as the existing pH value. In surface water and in the pore solutions of sediments, there is a predominance of chloride, sulfate, bicarbonate, and (under reducing conditions) anionic species derived from hydrogen sulfide. The chlorides (with the exception of silver and mercury (I) chloride and $PbCl_2$ $K_{sp} = 1.7 \times 10^{-5}$) and sulfates of all heavy metals discussed here are readily soluble, whereas the carbonates, hydroxides, and sulfides dissolve only with difficulty. In the following discussion only the latter compounds with a limited degree of solubility in water are treated.

2.3.1 Hydroxides

According to the work of Feitknecht and Schindler (1963) the processes of precipitation of metal hydroxides result in several forms, which may behave quite differently in the aquatic environment with respect to the effects of coprecipitation or later, redissolution. The "active" form, probably in most cases equivalent to "amorphous"

(Krauskopf, 1967) or very fine crystalline precipitate with disordered lattice (Stumm and Morgan, 1970), is derived from strongly oversaturated solutions. These precipitates may persist in metastable equilibrium with the solution and may slowly convert—often for geologic time spans—into the "aged" forms, thereby becoming more stable and "inactive". The inactive solid phases with ordered crystals are also formed from solutions that are only slightly oversaturated (Stumm and Morgan, 1970).

Calculations of the solubilities of metal hydroxides will have to consider that the solid species are not only in equilibrium with simple, uncomplexed forms Me^{2+} and OH^-, but that there may occur several ionization steps, which usually exhibit lower solubility constants. An example is given by Krauskopf (1967);

$$Cd(OH)_2 \text{ (aged)} \rightleftharpoons Cd^{2+} + 2\,OH^-;\ K_T = [Cd^{2+}] \times [OH^{-2}] = 10^{-14.4}$$

$$Cd(OH)_2 \text{ (aged)} \rightleftharpoons CdOH^+ + OH^-;\ K_1 = [CdOH^+] \times [OH^-] = 10^{-9.5}$$

$$CdOH^+ \rightleftharpoons Cd^{2+} + OH^-;\ K_2 = \frac{[Cd^{2+}] \times [OH^-]}{[CdOH^+]} = 10^{-4.9}$$

The equilibrium constant K_T expressing the total ionization (activity) has also been referred to as K_{sO}; similarly, the constants K_1, K_2 etc., for the equilibrium with hydroxo metal-ion complexes, are found in literature as K_{s1}, K_{s2} etc. (Stumm and Morgan, 1970).

In Table 57 examples of metal hydroxides are arranged according to the sequence of increasing stabilities of the aged form, assuming complete ionization has taken place. The diagram presentation of Jenne (1968, p. 354) indicates that for each hydroxide the minimum solubility values lie within the pH range of 9 to 12. A lowering of pH

Table 57. Negative logarithms of solubility products of heavy metal hydroxides (total ionization constants), carbonates and sulfides (pH = 7 at 25°) (from Krauskopf, 1967; according to data from Sillen, 1964)

Hydroxides		$-\log K_T$	Sulfides		$-\log K_{sp}$	Carbonates	$-\log K_{sp}$
$Mn(OH)_2$	(aged)	10.9	MnS	(pink)	9.6	$MnCO_3$	10.2
$Mn(OH)_2$	(cryst.)	12.7	MnS	(green)	12.6		
$Cd(OH)_2$	(aged)	14.4	CdS		27.8	$CdCO_3$	11.3
$Fe(OH)_2$	(aged)	15.1	FeS		17.2	$FeCO_3$	10.5
$PbO + H_2O$	(red)	15.3	PbS	(galena)	27.5	$PbCO_3$	13.1
$Co(OH)_2$	(blue)	14.2	CoS	(alpha)	20.4	$CoCO_3$	12.8
$Co(OH)_2$	(pink, aged)	15.7	CoS	(beta)	24.7		
$Zn(OH)_2$	(amorphous)	15.5	ZnS	(wurtzite)	21.6	$ZnCO_3$	10.8
$ZnO + H_2O$	(aged)	16.8	ZnS	(sphalerite)	23.6		
$Ni(OH)_2$	(active)	14.7	NiS	(alpha)	18.5	$NiCO_3$	6.9
$Ni(OH)_2$	(aged)	17.2	NiS	(gamma)	25.7		
$CuO + H_2O$	(active)	19.7	CuS		36.1	$CuCO_3$	9.6
$CuO + H_2O$	(tenorite)	20.5	Cu_2S		48.0	$Cu_2(OH)_2CO_3$	33.8
$HgO + H_2O$	(red)	25.4	HgS		52.4		
$Cr(OH)_3$	(aged)	37.4					
$Fe(OH)_3$	(aged)	39.1					

effects a marked increase in solubility: in neutral solutions the solubilities are increased by several orders of magnitude, while at pH 4 complete dissolution is largely achieved (example in Fig. 56).

2.3.2 Sulfides

Heavy metal sulfides are practically insoluble at neutral pH (Table 57). Fe, Mn, and Cd sulfides are readily soluble in hydrochloric acid, whereas Ni and Co sulfides dissolve with more difficulty; Cu, Pb, and Hg sulfides are only soluble in oxidizing acids, the most effective being nitric acid.

2.3.3 Carbonates

The solubility of carbonates in aqueous solution (Table 57) is highly dependent on CO_2 partial pressure. For example, the low solubility of $PbCO_3$ (~ 2.1 mg/l) in distilled water can be increased severalfold in the presence of CO_2 since a dissolution (similar to that of $CaCO_3$) occurs by virtue of bicarbonate formation:

$$Me(II) CO_3 + H_2O + CO_2 \rightleftharpoons Me^{2+}(aq) + 2 (HCO_3)^-(aq).$$

Consequently all carbonates tend to be more soluble in the presence of CO_2.

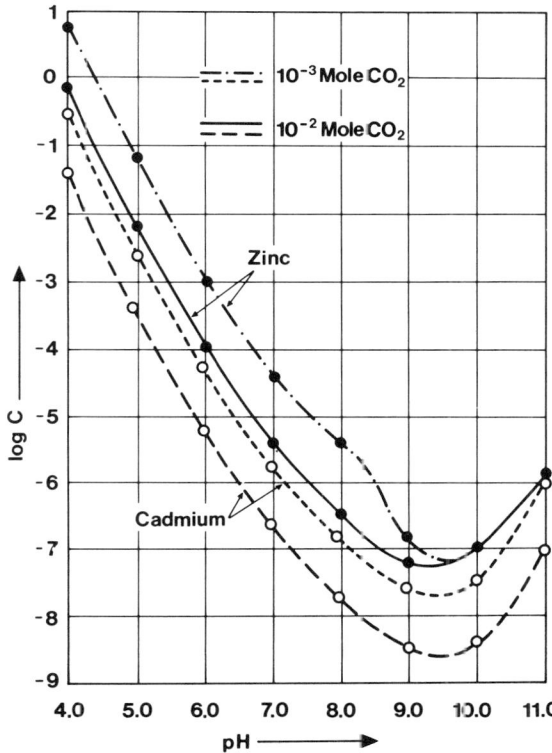

Fig. 56. Solubilities of zinc and cadmium in waters as a function of pH. Total dissolved carbon dioxide species concentrations 10^{-3} and 10^{-2} mol/l, respectively. Ionic strength of 0.0; log C in mol Zn or Cd (from Hem, Water Resources Research, 8 (3), 1972. Copyright by American Geophysical Union)

Precipitation, primarily of hydroxides, sulfides and carbonates, occurs within a water body (sewage system, purification plant, river or sea) when the corresponding solubility product is exceeded. The interactions of a variety of factors play an important role in this context with the result that the solubility data, obtained in pure individual systems in distilled water, only represent a guide to the conditions existing in natural water systems.

The solubility curves for cadmium and zinc, illustrated in Figure 56 (from Hem, 1972), are a more realistic representation since these have been computed by taking the bicarbonate content normally encountered in inland waters into account i.e., between 10^{-3} and 10^{-2} mol dissolved CO_2 per liter (equivalent to 61 and 610 mg HCO_3^-). The solubility curves for both zinc and cadmium show a minimum at pH = 9.3. Taking pH values of 7 to 8 normally encountered in river water (as well as oxidizing conditions) into account, it has been calculated that between 100 and 1000 μg/l cadmium may be present in the dissolved state (cf.: the maximum threshold limit for cadmium in drinking water is below 5 μg/l). The type of precipitation product can be determined from the stability relations in Eh-pH diagrams. Since the most important elements in controlling metal ion solubilities in natural systems are commonly sulfur and carbon, the diagram of zinc (Fig. 57, from Hem, 1972) includes the presence of S and C. For example, in the system $Zn + S + CO_2 + H_2O$ where the zinc concentration is 10^{-5} molar (~ 650 μg/l Zn together with 10^{-3} dissolved CO_2 and sulfur), three solid phases are possible—the sulfide, carbonate and hydroxide of zinc (Fig. 57). In aerated waters, zinc carbonate is the stable phase if the pH falls below 8.3; above pH 8.3 precipitation of zinc hydroxide occurs.

The Eh-pH diagrams of most other heavy metals are comparable to those of zinc. In the presence of free oxygen (a positive value for the standard reduction potential), Me^{2+} is stable at pH values of less than approx. 7 to 8. With increasing pH, first the

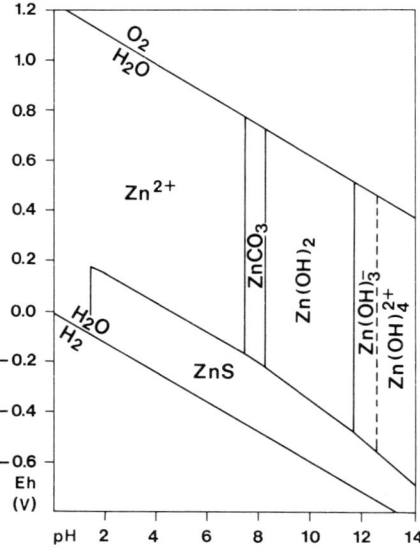

Fig. 57. Fields of stability of solids and predominating dissolved zinc species in system $Zn + CO_2 + S + H_2O$ at 25° and 1 atm. pressure in relation to Eh and pH. Dissolved zinc activity, 10^{-5} mol/l; dissolved carbon dioxide and sulfur species, 10^{-3} mol/l (from Hem, Water Resources Research, 8, 1972. Copyright by American Geophysical Union)

carbonate and then the hydroxide becomes the stable phase. For negative values of the reduction potential, the sulfide remains the stable phase over a wide pE range.

The inclusion of further soluble components in these systems e.g., of silicon and phosphorus, and the codetermination of complex species in the case of some metals, have reduced the discrepancy between the calculated solubility data and the actual conditions observed in nature.

Hem (1972) for example found that the solubilities of the carbonate and hydroxide species of zinc might be too high to explain its concentration in normal surface waters; it has been proposed therefore that the stable solid phase might be willemite, Zn_2SiO_4. The stable solid phases of *lead* in an oxygenated freshwater environment appear to be $PbCO_3$, $Pb_3(OH)_2(CO_3)_2$, and the phosphate compounds pyromorphite and plumbogummite (Nriagu, 1974). Hem and Durum (1973) found that the solubilities of the first two species generally determine the lead concentrations in surface water; in a small area close to pH 8 hydrocerussite is the least soluble compound in seawater.

2.4 Cation Exchange and Adsorption

A number of sediment-forming materials with a large surface area—particularly clay minerals, freshly precipitated iron hydroxides, amorphous silicic acids, as well as organic substances—are capable of sorbing cations from solution and releasing equivalent amounts of other cations into the solution—i.e., by cation exchange. The mechanism which results in cation exchange is based on the sorptive properties of negatively charged anionic sites—SiOH-, $AlOH_2$-, and AlOH-groups in clay minerals, FeOH-groups in iron hydroxides, carboxyl and phenolic OH-groups in organic substances—towards positively charged cations. The balancing of negative charges of the lattice is a selective process which accounts for preferential adsorption of specific cations and the release of equivalent charges associated with other species. Furthermore, all fine-grained materials with a large surface area are capable of accumulating heavy metal ions at the solid–liquid interface as a result of intermolecular forces. This phenomenon is termed *adsorption.*

Adsorption can arise from electrostatic attraction alone, in which case it is relatively nonspecific or nonselective, or it can arise from electrostatic attraction augmented by hydrogen bonding coordinate bonding, or London–van der Waals bonding (Parks, 1967). In this context it has been suggested that adsorption does not differ mechanically from precipitation, since both processes are in reality just different expressions for the same phenomenon—the concentration of solutes on the solid side of the solid/solution boundary (Jenne, 1976).

Surface phenomena of this kind can best be explained by the *electric double layer model* (Mysels, 1959, van Olphen, 1963). An example is described in Fig. 58 from the work of Parks (1967): The solid surface (of silica) contains ions which are not fully coordinated and are thus electrically charged. On exposure to water vapor, ionic and charged sites are converted to surface hydroxide or MOH-groups. Further adsorption of water vapor occurs at first spottily, then in coherent layers of hydrogen-bonded water. This water is probably more highly dissociated than bulk water (Fripiat et al., 1965). As the thickness of the adsorbed water increases, its properties approach those of the bulk liquid. In this state hydrogen ions which were mobile on the gas/solid interface are free to diffuse into the bulk aqueous phase. The result is a hydroxylated, negatively charged surface layer and a diffuse layer of positive ions. When the solid is immersed directly in water, all these steps occur simultaneously. Exchange processes on the surface of the particles involve the release of hydrogen ions or other cations and the adsorption or surface complexing of metal ions. Stumm and Morgan (1970) suggest for the sorption on microcrystalline iron and manganese hydroxide, that while these processes with Group I and II cations take place predominantly

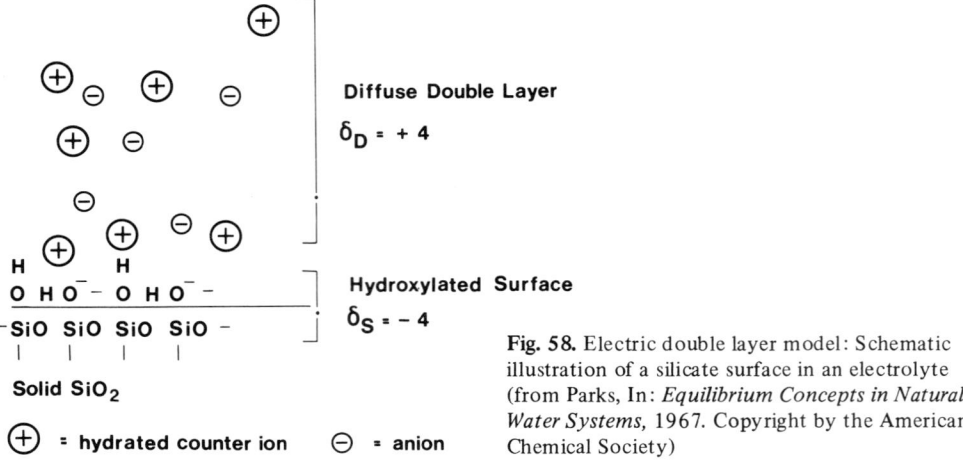

Fig. 58. Electric double layer model: Schematic illustration of a silicate surface in an electrolyte (from Parks, In: *Equilibrium Concepts in Natural Water Systems,* 1967. Copyright by the American Chemical Society)

in the diffuse part of the electric double layer, the transition and heavy metal ions become specifically attached to the surface.

It has been shown that if the surface area is large, the development of the electric double layer may lead to important changes in solution composition and pH (Parks, 1967). Because so much of the behavior of suspensions is determined or modified by the *charge associated with the solid phases,* Parks (1967) introduced the concept of the zero point of charge, ZPC, or isoelectric point of the solid, IEP$_{(s)}$, both terms involving pH as a master variable. The ZPC is the pH at which the solid surface charge from all sources is zero. If the charge is established only by H^+, OH^-, and species capable of interacting with H^+, OH^- or H_2O to form other species present in the solid lattice, then ZPC may conveniently be given the special name IEP$_{(s)}$ (Parks, 1967). The ZPC or IEP$_{(s)}$ values give indications on the behavior of the various sorbends in the aquatic environment. Hydrous oxides, for example, will form cation exchangers under basic conditions when the surface charge is negative; when the environment is acidic, the surface charge will be positive and therefore anion exchangers are formed. The presence of surface charges prevents coagulation in a low ionic strength solution, whereas by changes of the pH in the direction of ZPC the rate of coagulation and hence sedimentation can be increased. Examples of ZPC values of some common sorbends for heavy metal ions in natural water are given in Table 58 (Stumm and Morgan, 1970):

Table 58. Zero point of charge values of common adsorbents in natural waters (from Stumm and Morgan, 1970; calcite values from Somasundaram and Agar, 1967)

Material	pH$_{ZPC}$
α-Al$_2$O$_3$	9.1
α-Al(OH)$_3$	5.0
γ-Fe$_2$O$_3$	6.7
Fe(OH)$_3$ (amorphous)	8.5
MnO$_2$	2 −4.5
SiO$_2$	2.0
Kaolinite	4.6
Montmorillonite	2.5
Calcite	8 −9.5

A state of equilibrium exists between the number of cations accumulated by an adsorptive substance and their concentration. If the heavy metal content of the solution is increased, the quantity of adsorbed cations also increases; desorption occurs as a result of lowering the concentration, since a new equilibrium has to be achieved.

Sorption processes may be described by two alternative equations:

$$x/m = k \, C_{eq}^{1/n} \tag{1}$$

$$\frac{C_{eq}}{x/m} = \frac{C_{eq}}{S} + \frac{1}{kS} \tag{2}$$

where x/m = the quantity of metal adsorbed per unit weight of adsorbent (mol metal/100 g adsorbent)
 C_{eq} = the equilibrium concentration of metal remaining in solution (mol metal)
 k, n, S = constants

Equation (1) is called the Freundlich isotherm (Freundlich, 1926), while for Eq. (2) the term Langmuir adsorption isotherm is used.

At intermediate values of C_{eq}, the Langmuir equation reduces to the Freundlich equation (Knipling, 1965). Because the Freundlich equation is parabolic, adsorption should mathematically increase indefinitely with increasing concentration; that is, there should be no adsorption maximum. In contrast, the Langmuir equation describes a rectangular hyperbola passing through the origin and tends towards a definite adsorption maximum (Browman and Chesters, 1977). Guy and Chakrabarti (1975) demonstrated that sorption of heavy metals Cu, Pb, Cd, and Zn onto clay mineral follows Freundlich isotherms, whereas sorption onto humic acid and hydrous Fe/Mn oxides follows Langmuir isotherms.

The sum of exchangeable cations (including H^+) constitutes the "exchange capacity", expressed in milli-equivalents (meq)/100 g material. Exchange capacities of important sedimentary materials are presented in Table 59.

With clay minerals, which are particularly common in pelitic sediments and suspended material, the exchange capacity increases markedly in the order kaolinite < chlorite < illite < montmorillonite. This increase corresponds accordingly with the reduction of particle size and the related increase of surface area. The exchange capacity of organic substances is particularly high, especially that of humic acids. A low percent-

Table 59. Specific surface area and exchange capacities of several sorption active substances

Material	Specific surface area (m^2/g)	Exchange capacity (meq/100 g)
Calcite (< 2 μm)	12.5[a]	–
Clay minerals:		
Kaolinite	10 – 50[b]	3 – 15[e]
Illite	30 – 80[b]	10 – 40[e]
Chlorite	–	20 – 50[e]
Montmorillonite	50 – 150[b]	80 – 120[e]
Freshly precipitated Fe-hydroxide	300[c]	10 – 25[e]
Amorphous silicic acid	–	11 – 34[e]
Humic acids from soils	1900[d]	170 – 590[f]

[a] Suess, 1973. [b] Heling, pers. comm. [c] Fripiat and Gastuche, 1952. [d] Gapon, 1947. [e] Scheffer and Schachtschabel, 1966. [f] Marshall, 1964.

age of organic material can thus cause a marked increase in the exchange capacity of the total sediment. The complexing effect of organic components is discussed in Sect. 2.7.

The affinity of the cations towards exchangers is governed by the following factors (Scheffer and Schachtschabel, 1966):

a) Valence and Hydration Effects. The affinity increases with increasing oxidation number (valence effect)

$$Me^+ < Me^{2+} < Me^{3+} ...$$

and with a decrease in the diameter of the hydrated cations, thus producing higher charge densities in the alkali and alkaline earth series (hydration effect).

$$Ba < Sr < Ca < Mg < Cs < Rb < K < Na < Li$$

b) Concentration of Solution. As the concentration of a solution increases, the number of exchanged cations likewise increases; exchangers in equilibrium with cations of different valences show preference for species with higher charge densities, an effect which becomes more pronounced upon dilution (increased valence effect).

c) Cation Exchange on Organic and Inorganic Substances. Organic substances possess a high degree of selectivity for divalent ions as opposed to monovalent ones. The affinity of heavy metal ions is greater than that of alkaline earth and alkali ions, as illustrated by the following affinity series:

$$Pb > Cu > Ni > Co > Zn > Mn > Ba > Ca > Mg > NH_4 > K > Na$$

For synthetic sodium silicate exchangers (zeolites) Reynolds (1935) proposed the following sequence: $Cu > Pb > Ni > Ag > Zn > Cd$. Qualitative experiments show that a similar sequence exists for clay minerals (Weiss and Amstutz, 1966).

d) Reactions Involving Hydrolized Cations. An apparent increase in cation exchange capacity may result from increased hydrolysis of the exchanging cations, since the hydroxy complexes, i.e., $CuOH^+$, $FeOH^{2+}$ etc., are sorbed by most solids in preference to the uncomplexed cations, for example Cu^{2+}, Fe^{3+} (James and Healy, 1972; Jenne, 1976).

e) Specific Reaction Between Inorganic Exchangers and Cations. The different properties of the exchange sites in the lattice, the influence of electrostatic field strength as well as cation sorption mainly influence the affinity for K^+ and NH_4^+ ions. These observations confirm that a greater affinity can generally be attributed to the series of heavy metals than to the metals of the alkali and earth alkali groups. Since the "normal" cation content of limnic and marine sediments, characterized by their aqueous chemistry, consists almost exclusively of calcium, magnesium, sodium, and potassium, the heavy metal ions can be exchanged for these ions and in this way be sorbed onto the sediment particles.

As yet it is impossible to establish an order of affinities generally applicable to the individual heavy metals based solely on ionic charge and radius. Other factors, such as the tendency to form hydroxyl species or the tendency to adopt a particular geometric relation to the crystal lattice of the exchanger, also play a significant role.

The rate at which cation exchange takes place depends on the exchange capacity of the material as well as the cation species and their concentration. With clay minerals, a complete exchange takes place most rapidly with kaolinite, whereas illite is slow to effect complete exchange.

2.5 Sorption onto Clay Minerals

The clay mineral uptake capacity for ions, particularly for cations, is primarily governed (Grim, 1968) by:

a) Broken bonds around the edges of the silicon-aluminum units, which can be balanced by adsorbing cations. They occur predominantly on noncleavage surfaces and thus on vertical planes, in a position parallel to the c-axis of the layer. The number of broken bonds and hence the adsorption capacity for cations increases as the particle size decreases and moreover, should be even further increased by lattice distortions which are in turn reflected in an appropriately diminished degree of crystallinity. This form of cation adsorption occurs chiefly for kaolinite.

b) Substitution of Si^{4+} by Al^{3+} in the tetrahedral layer and of Al^{3+} by divalent cations in the octahedral layer of the structural units of some clay minerals. This process of substitution can leave behind unbalanced charges, which are then balanced by cations, in this instance predominantly on cleavage surfaces. Whereas substitution in a tetrahedral layer leads to an extremely strong bonding of cations and sometimes to their incorporation in fixed positions (e.g., potassium in mica), the substitution of Al^{3+} in the octahedral layers causes the balanced cations to be only loosely bonded. This results in their exchange under altered external conditions. Substitution of one of 400 Si atoms by Al in kaolinite creates 2 mEq/100 g cation exchange capacity.

The second type of cation bonding by substitution with the lattice is particularly significant for the minerals of the mica (illite) and chlorite groups.

The pH value may dominate in the adsorption processes of heavy metals onto clay minerals. The H^+ ions compete with heavy metal cations for exchange sites in the system, thereby partially releasing the latter. The heavy metal cations are completely released under circumstances of extreme acidic conditions. From infrared absorption spectra, Hildebrand and Blum (1974a) determined that at pH > 7, the Pb^{2+} adsorption process on clay minerals follows the same pattern as a chemical sorption, with the marginal surfaces participating with AlOH-, $AlOH_2$- and SiOH-groups. A large number of investigations have been performed on the adsorption of heavy metals onto clay minerals. The studies of Correns (1924) concerning the adsorption of copper from very dilute solutions onto kaolinite were continued by Heydemann (1959) with various clay mineral species. It was established that the adsorption of copper on clay minerals occurs according to the Freundlich adsorption isotherm. A more recent example of the adsorption characteristics of zinc is given in Figure 59 (Bourg and Filby, 1974).

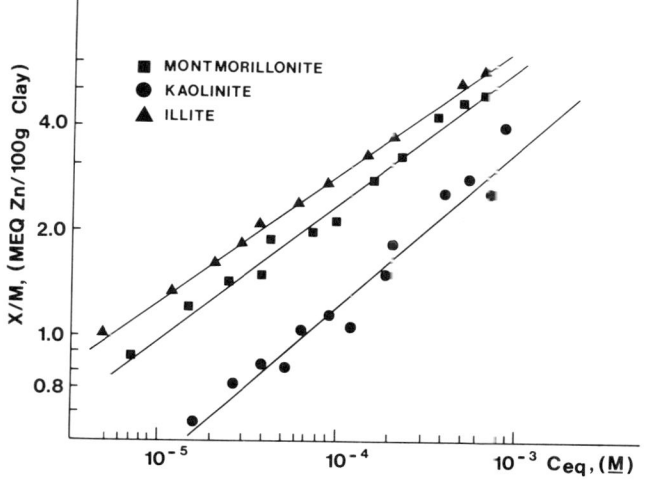

Fig. 59. Adsorption isotherms of zinc on different clay minerals. (Bourg and Filby 1974. Proc. Int. Conf. *Transport of Persistent Chemicals in Aquatic Ecosystems.* Ottawa)

Since the metal adsorption in the different clay species does not indicate any maximum value (Freundlich adsorption isotherm), it can be concluded that the zinc cations do not occupy specific ion exchange sites in the clay structure. On the other hand, the speed of the adsorption process—the reactions are completed after only 10 min—suggests an ion exchange mechanism as opposed to pure physical adsorption. Bourg and Filby (1974) explain that in addition to adsorption of Zn^{2+} between layers to neutralize excess negative charges, Zn^{2+} ions may also diffuse into the clay mineral lattice in comparatively unlimited amounts.

Particular insight into the dynamics of sorbing processes has been provided by studies with radioactive tracers. Ros Vicent et al. (1976) investigated the leachability of different trace metals and radionuclides which had been sorbed for long periods by marine sediments: Treatment with acetic acid/ammonium acetate solution at pH = 5.4 caused 100% of the cadmium, 50% of the manganese and lead, 33% of the zinc and cobalt to be leached, whereas iron, silver, and copper remained unresponsive. The experiments indicate that cadmium is adsorbed only in exchangeable positions without being immobilized onto fixed inter-crystalline positions. As Duursma (1976) demonstrated, the "past history" of the adsorption conditions of a heavy metal onto clay minerals may also play an important role. In contrast to the desorption of zinc-65 that can be realized in the laboratory directly after sorption, the zinc-65 in the Columbia River estuary essentially remains bonded to the sediment, thus acting as a tracer for the transport of river sediment to the sea (Cutshall et al., 1973). This may be ascribed to the tracer becoming more effectively fixed in the crystal structure of minerals during a longer period of transport.

The selective affinity of the clay minerals for certain heavy metals is an effect which has not yet been fully understood. Factors such as the valence of the ions, hydration behavior, electron negativity and ionization potential must be considered. Mitchell (1964) established the following—empirical—sequence for the affinity of heavy metals toward clay minerals:

Pb > Ni > Cu > Zn

A series of instructive experiments on the competitive sorption of heavy metals onto clay mineral have been performed by Soong (1974). Figure 60 gives the example of the adsorptive history of lead, copper, and zinc in a clay/sand mixture column containing 4% montmorillonite. Here, q is the concentration of the cations adsorbed and U is the amount of cation-feeding per 100 g of adsorbent. The competitive power of each cation species is illustrated by plotting the dq/dU derivative against U—a negative change implies desorption. At the beginning of the process, when there are sufficient available sites on the mineral, all the cations in the solution are sorbed at a comparable and fairly rapid rate. When the adsorption positions become densely covered by the adsorbates, the competition begins to grow stronger. The affinitive Pb, for instance, repels even the less affinitive Cu and Zn. This results in some of the already adsorbed Cu and Zn returning to the solution.

Soong (1974) accounted for these effects with the explanation that lead has a special affinity for the clay mineral structures due to its ionic radius, which is very similar to that of potassium (a metal primarily incorporated onto clay minerals). At the same time, lead is also capable of replacing potassium in the montmorillonite lattice (Marshall, 1949). Since the ionization potentials of copper and zinc are quite similar (Cu: 2.71; Zn: 2.70; Pb: 1.66 eV.), the different adsorption behavior of these two elements can only be attributed to polarization effects. Cu^{2+} does in fact possess an

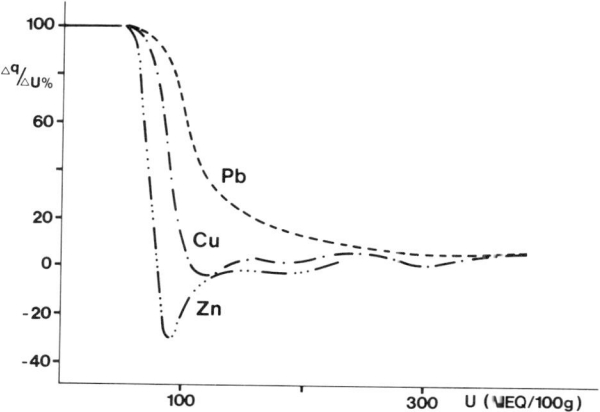

Fig. 60. Competitive adsorption/desorption (sorption history experiments) of lead, copper, and zinc in a clay/sand mixture column containing 4% montmorillonite (Soong, 1974; dissertation University Heidelberg)

unpaired electron in the 3d-orbitals and is therefore more strongly polarized than Zn^{2+} (with no unpaired 3d electrons) and exhibits a greater affinity to anionic sites.

Investigations carried out by Bittell and Miller (1974), and Lagerwerff and Browner (1972, 1973) on lead, cadmium, and calcium to establish selectivity coefficients for sorption on montmorillonite, illite, and kaolinite indicate that Pb^{2+} and Cd^{2+} may compete with common divalent ions in the soil e.g., with Ca^{2+}, for clay adsorption sites. Whereas Cd^{2+} competes more or less on an even basis with Ca^{2+} for such sites, the adsorption of Pb^{2+} is favored by a factor of 2 to 3 over Ca^{2+}. This suggests that there may be relatively more Cd^{2+} in a soil solution and thus available to plants.

Very little is known about the specific interaction of the wide range of ionic complexes that occur in natural aquatic systems. In river systems, for example, different hydroxo-, carbonato- and aquo-inorganic metal complexes occur. The adsorption of complexed ions by anionic sites is often more stable than the affinity for less hydrated cations of soil. Recent observations of natural aquatic systems have shown that adsorption of heavy metals onto clay minerals is, on the whole, relatively insignificant. Studies by Block and Schneider (1967–70) on the sorption of radionuclides by Rhine-suspended material show that a significant uptake of heavy metals onto clay minerals within the water does not occur. This is possibly due to other processes which, prior to clay adsorption, greatly reduce the heavy metal content in solution. On the other hand, Jenne (1976) proposed that the most significant role of clay-size minerals in trace element sorption by soils and sediment is that of a mechanical substrate for the precipitation and flocculation of organics and secondary minerals. Among the latter group of substances hydrous iron and manganese oxides have shown particularly strong affinities for trace elements. These affinities involve mechanisms of adsorption and coprecipitation.

2.6 Sorption and Coprecipitation on Hydrous Fe/Mn-Oxides and Fe-Sulfides

The hydrous oxides of aluminum, iron, and manganese, particularly the redox-sensitive Fe- and Mn-hydroxides and -oxides under oxidizing conditions, constitute significant *sinks* of heavy metals in aquatic systems. These hydroxides and oxides readily sorb or coprecipitate cations and anions; even a low percentage of $Fe(OH)_3$ and MnO_2 has a controlling influence on the heavy metal distribution in an aquatic system. Practical application of the sorption ability of hydrous metal oxides has been made in water and waste water treatment, particularly by the use of iron and aluminum salts for the removal of phosphorus compounds, organic contaminants and trace metals. Under reducing conditions the sorbed heavy metals are readily mobilized; accumulations of hydrous Fe/Mn oxides can therefore act as a major *source* of dissolved metals in natural waters. The latter mechanism is particularly effective in the presence of higher concentrations of dissolved organic material. Observations from heavily polluted waters indicate that both actions frequently coincide in these systems (Jenne, 1976).

2.6.1 Formation of Hydrous Mn and Fe Oxides

In soils and in particulate materials of the aquatic environment, Fe and Mn hydroxides or oxides commonly occur as coating on minerals and finely dispersed particles. They can be present in X-ray amorphous, micro-crystalline and in more "aged" crystalline forms (see Sect. E.3.3). The active forms exhibit high specific surface areas of up to 300 m^2/g for MnO_2 (Buser and Graf, 1955) and 230 to 320 m^2/g for FeOOH (Fripiat and Gastuche, 1952).

Forms of Hydrous Fe/Mn Oxides. With respect to the aquatic chemistry of iron oxides and hydroxides, $Fe(OH)_3$, Fe_3O_4 (magnetite), amorphous FeOOH, and α-FeOOH (goethite) are primarily to be considered. What are usually referred to as "hydrous ferric oxides" or "ferric hydroxides" are more likely to be poorly crystalline FeOOH (Stumm and Morgan, 1970). The mineralogy of Mn oxides is extremely varied and often taxonomically confusing (Jeffries and Stumm, 1976). A simplified scheme for polymorphism in the manganese dioxide has been given by Giovanoli, 1969). Among the more important Mn(IV) species "δ-MnO_2" two different forms have been found to exist: (a) manganese(III)manganate(IV), $Mn_7O_{13} \cdot 5\ H_2O$ and (b) sodium-manganese(III)manganate(IV), $Na_4Mn_{14}O_{27} \cdot 9\ H_2O$. MnOOH species usually consist of mixtures of Mn(II) and (IV) or Mn(III). Paths of reactions have been described by Bricker (1965).

Areas of Formation of Hydrous Fe/Mn Oxides. Iron and manganese ions enter the aquatic system from both ground and surface water. In groundwater, these ions generally occur in reduced oxidation states in the form manganese (II) and iron (II). However, in the presence of dissolved oxygen in water e.g., where groundwaters are discharged to the surface through *springs,* the Fe(II) and Mn(II) in the neutral or weakly alkaline pH range are oxidized to insoluble $Fe_2O_3 \cdot (H_2O)_x$ and MnO_x. Another important boundary for the formation of hydrous Fe/Mn oxides lies at the thermocline regions in *eutrophic lakes and fjords,* where the anoxic hypolimnion water is in contact with oxidized sediment.

The degree of acidity/alkalinity has a characteristic influence on the precipitation of hydrous Fe oxides, as demonstrated from a third boundary, where neutralization of *acidic waters* takes place. This is borne out by examples from acid mine drainage (see Sect. 5.1), the *junction of rivers* exhibiting different pH values (e.g., Deer Creek/Snake

River in Colorado: Theobold et al., 1963) and from *mixing zones of acid river water with seawater* (e.g., Mullica River/Great Bay in New Jersey; Coonley et al., 1971). A further boundary, particularly for the deposition of manganese oxide, is the surface of carbonate minerals such as calcite, where there is a *microzone of higher pH* (Lee, 1975).

Finally, many examples show that the manganese oxidation may be enhanced to a considerable extent by *microorganisms* (Schweisfurth, 1972).

Processes and Factors Influencing the Formation of Hydrous Fe/Mn-Oxides. Stumm and Lee (1960) have established that the oxidation *process* of Fe^{2+} to Fe^{3+} takes a few hours at pH = 5, while at pH = 7, only a few minutes are required.

$$2 Fe^{2+} + 1/2 O_2 \rightarrow 2 Fe^{3+} + H_2O$$

Hydroxide ions often possess a greater affinity to Fe^{3+} than to other organic and inorganic bases. Ferric iron hydrolyses to form insoluble ferric hydroxide (Stumm and Morgan, 1970):

$$Fe^{3+} + 3 H_2O \rightarrow Fe(OH)_3 (s) = 3 H^+$$

Under natural conditions in aquatic systems, however, Fe(III) combines with OH^- ions as well as with other ligands e.g., phosphates and organic substances, forming both insoluble and soluble complexes.

The oxidation of Mn(II) is autocatalytic (Stumm and Morgan, 1970) and may be represented as follows (reactions are not balanced with respect to water and protons):

$$Mn(II) + 1/2 O_2 \xrightarrow{slow} MnO_2(s)$$

$$Mn(II) + MnO_2(s) \xrightarrow{fast} Mn(II) \cdot MnO_2(s)$$

$$Mn(II) \cdot MnO_2(s) + 1/2 O_2 \xrightarrow{slow} 2 MnO_2(s)$$

Morgan (1967) reported that the oxidation of manganese (II) by dissolved oxygen yields a stoichiometric ratio $MnO_{1.9}$, but only under highly alkaline conditions; oxidation under other conditions led to considerable adsorption of manganese(II) from solution (Lee, 1975).

There are several *influencing factors* for the formation of hydrous iron and manganese oxides:

a) Iron is readily oxidized by dissolved oxygen to the ferric form in the alkaline-neutral to lightly-acid *pH range*. On the other hand, manganese requires a much higher pH for equivalent rates of oxidation (Stumm and Lee, 1960). The rate of oxidation of both iron and manganese increased by a factor of 100 for each pH unit increase (Morgan and Stumm, 1964).

b) The chemical oxidation of manganese is influenced by *inorganic ions* such as HCO_3^- or SO_4^{2-} (Hem, 1964). Experiments by Schweisfurth (1972) indicate that orthophosphates in concentrations above 50 mg/l inhibit the oxidation of manganese, whereas pyrophosphates up to 50 mg/l catalyze such processes. Polyphosphates in small amounts also have a positive effect on the manganese oxidation. When oxides and hydroxides sink to the bed of a water system, they may stabilize by complexing with organic or inorganic ligands (Lee, 1975).

c) *Organic substances,* especially those containing hydroxo- and/or carboxylic functional groups, are capable of reducing Fe(III) and MnO_2. The same substances, however, can also catalyze the oxygenation rate. Stumm and Morgan (1970) have shown that when this happens, the ferrous-ferric system acts merely as a catalyst for the oxidation of organic material by oxygen. Sorbed or coprecipitated organics may inhibit crystallizations (Schwertmann, 1966). In addition, the sorption capacity of hydrous metal oxides may be significantly changed with aging due to the sorption of other materials in solution (Lee, 1975). Ferric iron tends to form complexes with natural water organics. By adding ferric iron to the system, these organics and the iron become associated to form larger particles and become colloidal in character (Hall and Lee, 1974).

d) Hydrous metal oxides are part of a reciprocal relationship with heavy metals. Not only do the hydrous Fe/Mn metal oxides exert a significant influence on the heavy metals, but *heavy metals* such as copper may have an influence on these hydrous metal oxides as well (Lee, 1975). By way of example, Stumm and Lee (1961) have found that the presence of copper *catalyzes* the oxidation of ferrous sulfate by dissolved oxygen. This has also been confirmed for the oxidation of manganese during water treatment (Jenne, 1968).

e) $Fe(OH)_3$ colloids, partly peptized by organic substances, can also be sorbed by *clay minerals* if the ionic product Fe^{3+} and 3 OH^- exceeds the solubility product (Follet, 1965). Subramanian and Gibbs (1972) found that the smaller the grain size of the minerals, the more Fe-hydroxides become bonded to clay minerals as coatings. Pravdic (1970), and Neihoff and Loeb (1972) have shown that sediment particles in seawater carry a negative charge and may thus provide centers for the nucleation of hydrolysis products of iron.

2.6.2 Sorption of Heavy Metals onto Fe/Mn-Oxides

The mechanism of heavy metal sorption on hydrous oxides has been demonstrated by Hildebrand and Blum (1974) from the example of the Pb^{2+} interactions with Fe hydroxides. Pb is incorporated in the Stern layer of the hydroxide structure and hydrogen ions are exchanged, so that lead exhibits a strong affinity for the hydroxyl group of the FeO(OH) crystal. Gadde and Laitinen (1973) observed in their investigation that the sorption of Pb^{2+} at pH values between 5 and 7 steadily decreases. In this pH range FeOOH is positively charged: lead therefore appears specifically sorbed. Similar results were obtained by Lockwood and Chen (1973, 1974) for the sorption of Hg^{2+} by hydrous manganese oxides and ferric hydroxide. Jeffries and Stumm (1976) studied the sorption characteristics of a defined Mn mineral species (buserite) and found that data did not agree with a simple 1:1 (Cu:H) exchange model of adsorption; this discrepancy may be due to some degree of 1:2 surface exchange and/or specific adsorption. Subramanian (1976) found that a combination of exchange reactions and solid solution can satisfactorily explain the observed chemistry of natural ferro-manganese material. Studies of Hohl and Stumm (1976), Schindler et al. (1976) and James and McNaughton (1977) have formulated the adsorption of hydrolysable metals as a surface complex with electrostatic interactions.

In a review on the interaction of such metal ions as SiO_2, Al_2O_3, and MnO_2, Stumm et al. (1976) favored the latter model above other possible mechanisms (Gouy-Chapman-Stern-Graham model accounting for specific and electrostatic adsorption; adsorption-hydrolysis model; ion-solvent interaction model; cation exchange model involving replacement of protons). In their proposed model, the hydrous oxide surface groups ≡Me–OH or =Me$\langle^{OH}_{HO}$, which are similar to amphoteric functional groups in polyelectrolytes, are treated as complex forming species. In solution of monomeric metal species pH-dependence is explained with the basicity of the MeO$^-$ group and the affinity of this group to the metal ions. Strong adsorption of polymeric or colloidal species, on the other hand, seems to be influenced by the nature of the substrate only to a small degree, as long as the surface charge is opposite to the charge of the adsorbing species.

Sorption of Transition Series Elements onto Hydrous Iron and Manganese Oxides in Rivers. Since the mid-sixties the economic value of the deep-sea ferromanganese deposits, particularly based on the characteristic enrichment of trace elements such as molybdenum, cobalt, nickel, and copper, has been recognized (Menard, 1964). A great number of analytic work has been undertaken on the different types of marine and lacustrine deposits (e.g., Glasby and Read, Callender and Bowser, 1976). Compared with these investigations, there are only limited data available on the significance of sorption and coprecipitation processes with hydrous Fe/Mn oxides in rivers.

Investigations on the trace metal content in manganese coatings of stream deposits were first performed by Smith (1960), and Canney and Nowlan (1964). Jenne and Wahlberg (1965) studied the enrichment of trace metals in iron oxide compounds of stream sediments. Cutshall (1967) found iron oxides to be the most important single sediment component in the retention of chromium-51 in sediments from the Columbia River.

Chemical leaching of trace elements in carbonate-free sediments from the Amazon and Yukon Rivers by Gibbs (1973) indicate 4%–50% of Cu, Cr, Co, and Ni to be present in the form of hydrous Fe/Mn oxide coatings. Perhac (1974a, b) determined 5%–50% of cobalt, copper, and zinc in the dithionite-reducible fraction (or hydrous iron and manganese oxide fraction) in fluviatile sediments containing principally carbonate and primary silicates.

A practical example from arctic areas of Canada showing the dominant role of *Fe and Mn oxides* in the distribution of trace metals in fluviatile systems is described by Cameron and co-workers (Cameron, 1974; Allan et al., 1974; Cameron and Ballantyne, 1975) and Jonasson (1977). These regions represent somewhat simplified conditions for the study of the interactions between hydrous Fe/Mn oxides and dissolved metal ions because of the scarcity of organic materials.

Two watersheds supplied with metal ions by similar volcanogenic Cu-Zn-sulfide mineralization processes were compared. The first location, Hackett River in the North West Territories, exhibits a pH in excess of 7 due to the presence of carbonate enriched volcanic rocks. The second area, at Agricola Lake, N.W.T., is virtually identical to the Hackett River except that the underlying volcanic rocks are devoid of carbonate minerals; here, pH is commonly between 3 and 4. In the alkaline waters in the area near the source of the Hackett River Fe and Mn were initially present at very low levels, whereas in the Agricola Lake area Fe and Mn concentrations reached 800 μg/l and 90μg/l respectively. At Agricola Lake the levels of dissolved zinc, copper, and arsenic were found to fall off very rapidly, i.e., the content of zinc from about 1000 μg/l at the source to 1 μg/l 5 miles downstream. In the Hackett River area, in contrast, the dispersion of soluble zinc and copper was significantly more extensive. According to Jonasson (1977) this is attributable to the absence of an oxide scavenger coprecipitating trace metals below the source in the Hackett River area, whereas rapid hydrolysis of the high initial content of Fe and Mn in the Agricola Lake waters resulted in the coprecipitation of Cu, Zn, and As as the pH increases with dilution.

2.6.3 Coprecipitation of Trace Elements with Iron Sulfides

From the stability fields and solubility diagrams for the system $Fe-CO_2-S-H_2O$ (see Fig. 21, p. 81) for normal conditions in natural water (10^{-4} mol/l of sulfur $\approx$ 10 mg/l as SO_4^{2-} and 10^{-3} mol/l of carbon = 61 mg/l HCO_3^-), four stable solid phases have to be considered: ferric and ferrous hydroxide, siderite ($FeCO_3$), and pyrite (FeS_2).

Under oxidizing conditions e.g., in the presence of dissolved oxygen, ferric iron is the only species found in the slightly acid to alkaline pH range; ferrous iron is stable in the presence of dissolved oxygen under strongly acid conditions such as would occur in extreme cases of acid mine drainage. In a system where the stable solid is ferric hydroxide, the concentration of dissolved iron should be less than 1 μg/l. Hem (1975) points out that some water bodies contain colloidal ferric hydroxide, other ferric hydroxide or other ferric species that can pass through filters and appear to be part of the dissolved iron. The concentration and chemical behavior of dissolved iron, as well as of other metals, may also be influenced by organic complexing agents. If the reduction potential becomes strongly negative as a result of the lack of oxygen, $Fe(OH)_3$ can, in the presence of CO_2 (from the decomposition of organic substances), be converted to produce $FeCO_3$. Direct precipitation of $FeCO_3$ from solutions containing Fe^{2+} in the pore spaces of river bed deposits, into which river water flows e.g., during bank filtration, is one of the causes of river bed solidification (Förstner and Müller, 1974a).

When an even lower level of oxygen is present, iron sulfide minerals will form, some of which are in metastable phases, e.g., Fe_3S_4 (greigite) and $Fe_{1+x}S$ (mackinawite). Greigite and mackinawite are thermodynamically unstable relative to pyrite. In marine sediments these minerals usually disappear during early diagenesis as a result of transformation to pyrite, as shown by Berner (1970):

$$FeS_{\text{mackinawite}} + S^0 \rightarrow FeS_2 \text{ pyrite}$$

$$Fe_3S_4 \text{ greigite} + 2 S^0 \rightarrow 3 FeS_2 \text{ pyrite}$$

The rate of transformation in the presence of abundant elemental sulfur is relatively rapid, requiring a few years to a few decades for completion. Berner (1970) suggests that the relative long persistence of unstable black Fe-sulfides in Black Sea sediments, in the range of approximately 10,000 years, can be explained by the exhaustion of H_2O from the interstitial waters of the sediments, caused by a high rate of deposition combined with a low SO_4^{2-} concentration in the bottom water. The lowering of SO_4^{2-} contents in turn seem to be caused by freshening of the entire Black Sea during the last Pleistocene glaciation (Manheim and Stoffers, 1969). If a higher sulfur level is present, the stability range of sulfide spreads beyond the carbonates in such a manner that a lowering of the oxygen level directly brings about the transformation of the hydroxide to the sulfide compound. Berner (1964) notes that the major iron-sulfide compound in "hydrotroilite", frequently mentioned in a context pertaining to iron-sulfide minerals in recent anaerobic sediments, is poorly crystallized tetragonal FeS. According to Jenne (1976) hydrotroilite, as the least crystalline form of iron sulfide in natural systems, imposes an upper limit on iron-sulfide solubility – by means of both its amorphous nature and its high solubility. Relative to the crystalline sulfides, "hydrotroilite" is analogous to limonite in the case of iron oxides.

Coprecipitation of Trace Elements. There are only limited data on the significance of coprecipitation effects of trace metals with iron sulfides, in comparison to other processes of trace element sorption. Jenne (1976) cites the work of Bondarenko (1972), who found that the introduction of hydrogen sulfide caused a marked coagulation of copper-fulvic acid complexes, suggesting that this may be one of the significant processes responsible for the inclusion of trace elements in sediments of anaerobic waters. Other possible mechanisms of trace metal incorporation within iron sulfides point to inorganic precipitation. Vertical profiles of dissolved Mn, Cu, Fe, and Zn in waters of the Black Sea show that distribution of these elements are considerably affected by redox reactions at the boundary between oxygenated surface waters and sulfide-containing deep waters. Copper and zinc are probably depleted in deep water by precipitation as insoluble sulfides. Since most of the trace metals generally occur in substantially lower concentrations compared to ferrous iron in interstitial and other anoxic waters, these metals may be expected to coprecipitate with iron sulfides rather than as discrete sulfide crystals (Jenne, 1976).

Metal concentrations have been studied in diagenetic iron sulfides of sedimentary origin, mostly from the Black Sea deposits. From investigations of the distribution of the microelements V, As, Cr, Cu, Pb, Co, Ni and Zn in the pyrite and magnetic sulfides, Butuzova (1969) found that all these elements were—with the exception of vanadium, chromium, and zinc—accumulated in the iron sulfides. The work of Volkov and Fomina (1974) and Pilipchuk and Volkov (1974) reveals characteristic enrichment of molybdenum, copper, nickel, and cobalt in the pyrites from the modern deep-water deposits of the Black Sea, when compared with the enclosing sediments (Table 60). The enrichment of the trace metals in the pyrite of the Neoeuxinian deposits on the continental slope along the Anatolian shore of the Black Sea is quite different from the deep-water sediments, particularly with respect to the concentrations of Cu and Ni. This indicates that the behavior of the elements in these processes of sulfide formation depends on the sediment composition, which is mainly influenced by the concentrations of sulfur and organic carbon.

Unstable iron sulfides of the hydrotroilite type are normally finely dispersed (black mud) and are instantly oxidized under aerobic conditions, rendering the selective analysis of trace components in these phases impossible. Following the description of coarse-grained FeS-concretions from Lake Constance surface sediments by Wagner (1971), the separation of these particles from the fine-grained matrix by sieving (0.5 mm) was attempted. A concentrate was obtained with an iron content approximately six times higher than in the surrounding sediment; cobalt, copper, and chromium were two times greater, nickel 1.5 times, whereas lead, zinc, and manganese did not undergo a significant increase when compared to the composition of the fine-grained sediment matrix. It may therefore be concluded that coprecipitation with iron sulfides is less effective in concentrating trace metals than is the incorporation into hydrous iron oxides.

Table 60. Trace element content in pyrite from Black Sea sediments. F, factor of enrichment relative to the metal concentration in enclosing sediments (from Volkov and Fomina, 1974)

	Molybdenum		Copper		Nickel		Cobalt	
	ppm	F	ppm	F	ppm	F	ppm	F
Recent deep-water sediments	664	35.8	1022	20.5	1190	17.6	134	9.3
Neoeuxinian deposits	14	10	89.5	1.7	164	1.9	67	4.3

2.7 Metal Associations with Organic Substances

The affinity of heavy metals for organic substances and for their decomposition products is of great importance for the behavior of trace substances in aquatic systems. Singer (1977) summarized the influence of dissolved organics on the distribution of metals as follows: Dissolved organic substances are capable of (1) complexing metals and increasing metal solubility; (2) altering the distribution between *oxidized and reduced forms* of metals; (3) alleviating metals *toxicity* and altering metal *availability* to aquatic life; (4) influencing the extent to which metals are *adsorbed on suspended matter*, and (5) affecting the *stability* of metal-containing colloids.

The effect that organic substances have on the *solubilization* of minerals was discovered early in the history of the geologic sciences. The classical studies by Sprengel (1826), Thenard (1870) and Julien (1879) had already then considered the importance of organic ("humic") acids in chemical *weathering processes*. In two papers Bolton (1880, 1882) demonstrated that the power of citric acid to decompose minerals is a little less than that possessed by hydrochloride acid, and that the organic acid can be advantageously employed in field work for assistance in determining mineral species. Bolton's (1882) observations proved to be an indication of the future and "a small contribution to chemical geology." He suggested that an application of these findings to geologic phenomena could possibly lead to the assumption that "the disintegrating effect of the acids of humus differs from that of the other organic acids in degree rather than in kind." The role in *ore genesis* played by the organic substances was also established at a very early date: Gruner (1922), Harrar (1929), and Aschan (1932) found indications for the solution, transport and enrichment of iron by organic substances. Freise (1931) suggested that there is a considerable solubilization of gold by humic acids. These findings were substantiated later by experimental studies (Kee and Bloomfield, 1961, Huang and Keller, 1970, Ong et al., 1970).

A number of studies are concerned with the influence of organic acids during *pedogenesis*. It was shown that chelation is a biochemical weathering factor (Schatz et al., 1954, 1957; Schnitzer and Wright, 1957; Mortensen, 1963), involving the interaction of organic complexing and sorbing agents with trace elements in the soil (Leeper, 1948; Swaine and Mitchell, 1960; Ginzburg et al., 1963; Wright and Schnitzer, 1963; Kononova 1966; Drozdova, 1968; Baker, 1973).

The importance of organic substances for *the transportation of metals in natural aquatic systems* was recognized many years ago. The associations of organic-rich water with iron and manganese were described by Rankama and Sahama (1950) in their textbook on geochemistry. Shapiro (1963, 1964) noted a close correlation between the metal content and the concentration of yellow organic acids in aquatic systems; he found that organic acids in natural water are capable of keeping iron, copper, and other metals in a solution both at high pH and high Eh. Studies performed by Nesterova (1960) on the concentration of copper, nickel, and cobalt in the Ob River indicated that the increase or decrease of dissolved organic matter has a direct bearing on the amount of metals transported. Steelink (1963) postulated that river water containing organic acids should exhibit particular ability to transport gold; experimental studies by Ong et al. (1970) provided further evidence that these mechanisms can serve for transporting colloidal gold under natural conditions.

Corcoran and Alexander (1964), Lee and Hoadley (1967), and Slowey et al. (1967) found that the major portion of copper in tropical seawater is associated with organic matter; other transition metals were suspected to be similarly bonded and transported. The investigations of Bardate and Matson (1967) evidenced a close association of trace metals with organic complexes in arctic and subarctic waters.

Strakhov (1967) observed that more than half the total quantity of iron, manganese, nickel, and copper in the water of the Dnepr River is transported with organic acids. Eremenko (1966) concluded from studies on the Volga and Don Rivers that heavy metals are present in natural waters almost exclusively in the form of charged or uncharged complexes with organic acids, whose concentrations are 20 to 50 times higher than that of the heavy metals.

The percentages of organic complexing of solute form have been found to be 11% for Mn, 23% for Fe, 43% for Cu, 57% for Ni, 75% for Mo, and 78% for V in the Volga River, and 54% for Fe, 77% for Al, and 95% for Cu in the Don River. Similar results were obtained by Goleva et al. (1970) indicating 55%–74% lead present in soluble form and the rest to be inorganic complexes and aqua ions. Hodgson et al. (1966) found 75% of lead and 90% of copper in the interstitial solutions of a soil profile to be in the form of organic complexes.

Several investigations deal with the *stabilization* of iron and other metals in the aqueous phase by organic acids. Lamar's (1968) first result suggested that iron, probably as ferric hydroxide or oxide, forms colloidal sols with the organic matter (polymeric hydroxy carboxyl acid) in natural surface water; a relationship between particle size, pH, and iron concentration is indicated. The investigations of Ghassemi and Christman (1968) on the properties of yellow organic acids in natural aquatic systems show that the degree of association of colored molecules with iron is pH-dependent, being greater at both high and low pH than in the range of 7 to 8. Contrary to these findings, Ong et al. (1970) noted that the stability of metal-organic colloids increases when the organic acid concentration and pH climb. According to these authors, the optimum conditions for stability lie between pH 6 and pH 9.

2.7.1 Organic Substances in Natural Waters

The organic matter of aquatic systems consists of the remains of biologically produced compounds as well as of synthetic organic substances. From a geochemical point of view, organic matter may be subdivided into different groups according to its behavior upon treatment by various solvents (Welte, 1969): (1) Substances extractable with organic solvents such as benzene, ether, and chloroform, are commonly known as *bitumen*. (2) The bitumen-free or insoluble organic matter is frequently referred to as *kerogen*. (3) When the organic matter is treated with weak alkali solution, *humic acids* are extracted. (4) The component of the original mass that remains after the extraction procedures described above may be referred to as *residual organic matter.*

Synthetic organic substances originate from industrial and agricultural applications and have been suggested for use as detergent additives to replace polyphosphates e.g., nitrilotriacetic acid (see Sect. 5.4).

Decomposition of higher-weight organic substances with high molecular mass is mostly due to microbiologic action, thus forming the smaller and more soluble fragments.

Humic Substances. The most important products formed during the composition of organic substances are the *humic acids.* These can be found in soils, in limnic and marine sediments, as well as in the corresponding aqueous solutions. Nissenbaum et al. (1972) show that humic substances are a major component of the organic matter contained in recent marine sediments. Values of 40% on an average and, in some cases, as much as 70% of the matter have been recorded (Nissenbaum and Swaine, 1976). According to Reuter and Perdue (1977) about 60% to 80% of the DOC (dissolved organic carbon) and POC (particulate organic carbon) in fresh waters consist of humic substances.

Humic acids are extremely heterogenous polymers. They are characterized by their widely variable *molecular mass* ranging from less than 700 to more than 2,000,000. It has been suggested by Haworth (1971) that humic acids contain or readily give rise to a complex *aromatic core,* to which (1) polysaccharides, (2) proteins, (3) simple phenols, and (4) metals are attached, chemically or physically. Although the high content of aromatic carbon suggests hydrophobic behavior, humic matter is hydrophilic, probably due to the fact that – centered by a hydrophobic core – the *hydro-*

philic groups turn to the surface like proteins do in their native configurations (Höpner and Orliczek, 1978). On the basis of their molecular mass, humic acids may be subdivided into the following classes (Jonasson, 1977):
a) *Humins,* comprising the highest molecular mass group, are highly polymerized and largely insoluble in aqueous solutions.
b) *Humic acids* belong to the middle molecular size range, are very complex and are soluble under certain conditions.
c) *Fulvic acids,* less condensed humic substances, play an especially important role in the bonding of metals because of their large number of functional groups. Fulvic acids occur in a dissolved state.
d) *Yellow organic acids* are located at the low molecular mass end of the series range. They are frequently found in swamp waters. Yellow organic acids are probably important in pore and interstitial waters of sediments as they represent the last stage of humic matter degradation.

A compilation of the range of concentrations of dissolved organic carbon, particulate organic carbon, and total organic carbon (TOC) in typical aquatic environments has been given by Head (1976; see Table 61). Reuter and Perdue (1977) have noted that man-made contributions of humic substances from effluents of secondary sewage treatment may become significant in rivers of densely populated areas.

Table 61. Concentrations of organic carbon in natural waters (according to Head, 1976). Figures in parentheses are extreme values

Concentrations of C_{org} (mg/l)	River	Estuary	Coastal sea	Open Sea	
				surface	deep
Dissolved	10–20 (50)	1 – 5 (20)	1 –5 (20)	1 –1.5	0.5 –0.8
Particulate	5–10	0.5– 5	0.1–1	0.01–1	0.003–0.01
Total	15–30	1 –10 (25)	1 –6 (21)	1 –2.5	0.5 –0.8

2.7.2 Sorption and Complexation of Metals by Humic Substances

The attractive forces between metal ions with soluble, colloidal or particulate organic material range from weak, which leave the ions easily replaceable (physical adsorption), to strong, thus undistinguishable from chemical bonds as in metal chelation by organic material (Saxby, 1969). According to Krauskopf (1955), the *adsorption* of cations on organic substances is due particularly to the general negative charge of the colloids.

Organic material can sorb between 1% and 10% dry weight of Co, Cu, Fe, Pb, Mn, Mo, Ni, Ag, V, and Zn. It has therefore been concluded that the enrichment of metals in ancient carbonaceous sedimentary rocks may be largely attributable to humates having acted as scavengers or complexing agents (Swanson et al., 1970).

Rashid (1974) has shown that copper is preferentially sorbed (53%), followed by zinc (21%), nickel (14%), cobalt (8%), and manganese (4%). Leaching experiments have demonstrated that copper is more firmly associated with organic material than with other metals. Jonasson (1977) established an order of bonding strength for a number of metal ions onto humic or fulvic acids:

$$UO_2^{2+} > Hg^{2+} > Cu^{2+} > Pb^{2+} > Zn^{2+} > Ni^{2+} > Co^{2+}.$$

Investigations by Hildebrand and Blum (1974c) on the influence of the pH value on the sorption of lead onto organic substances reveal that at alkaline pH values more Pb^{2+} is sorbed than under acidic conditions. This is substantiated by the relatively lower values of Pb released by $MgCl_2$ extraction in an alkaline environment than from an acidic milieu.

It has been estimated that at a total bonding capacity of 200 to 600 meq metal/100 g humic acid, approximately one-third can be designated as the cation exchange capacity; two-thirds of the available positions exist as complexation sites (Rashid, 1971). The same author has demonstrated that fulvic acids with a molecular mass of less than 700 can *complex* two to six times more metal ions than those with a molecular weight exceeding 700; furthermore, the bonding ability of the corresponding humic acids is three to four times larger for divalent metal ions than for trivalent ions.

Fulvic acids play an especially important role in the transport of heavy metals in water. This is due to their lower molecular masses, larger number of functional groups and their much greater solubility of the fulvic than humic fraction (Malcolm, 1964. In: Jenne, 1976). On the other hand, the humic fraction is more important in trace element transport and retention by stream sediments because of the greater quantity present (Jenne, 1976).

In soil science, the sequence of *complex stability* has been established by the Irving-Williams (1948) series:

Pb > Cu > Ni > Co > Zn > Cd > Fe > Mn > Mn

Goldberg (1965) has shown that the ability of marine organisms to concentrate metals generally follows the Irving-Williams succession. A similar sequence was established by Bowen (1966) for stability of metal-humic complexes; Cu, Sn, and Pb are strongly bonded (zinc less strongly), while Ca, Mg, Mn, and alkali metals are only loosely bonded to humic material. This is in general agreement with the HSAB principle of Pearson, discussed in Chap. B.

2.7.3 Coagulation and Flocculation of Metal-Organic Matter

Generally speaking, the trace metal-organic complexes in natural waters represent a mixture of allochthonous and autigenic substances. These originate from solutions seeping through or running over the surface of soils (Shapiro, 1963) or are formed directly in the aquatic system by microbial and chemical processes (Sect. 5.4 and 5.5). It was suggested that the portion of many trace metals present in natural waters as a soluble organic complex is generally greater than that present as the inorganic or the aqua complex (Jenne, 1976). The behavior of humic acids during *coagulation* was summarized by Ong and Bisque (1968) and Ong et al. (1970) in the following schematic description of the reaction types:

Organic Acids + M^+ —Chemical aspect predominates→   Metal-Organic + M   Floss
(Hydrophilic   Complexes   ⇌
colloids)   (Hydrophilic   effect
(negatively   colloids)
charged colloids)

Metal-Organic + M^+ —Physical aspect predominates→   Metal Humates
Complexes
(Hydrophobic   (Precipitates)
colloids)   (neutral colloids)

Mutual repulsion of negative functional groups results in stretched configurations. Upon the addition of salts, the so-called *Fuoss effect* occurs; the cations are attracted to the negative carboxyl- and hydroxyl-groups. This brings about a marked decrease in intramolecular repulsion and thus promotes coiling in the polymer chain. The coiled geometry effects part of the water of hydration surrounding the colloid to be expelled from the structure.

In recent years, the behavior of trace metals and organic material at the *river/seawater interface* finds particular attention. Beck et al. (1974) suggested that a fraction of the river water's organic matter and certain metals co-flocculate in the downstream reaches of the river and in the estuary. Rashid (1974) inferred that several metals, among them Co, Cu, Mo, Ni, Pb, and Zn, may be removed to a considerable extent during or subsequent to *flocculation* of organic substances.

From the experiments of Hahn and Stumm (1970), and Edzwald et al. (1974), four mechanisms can be distinguished leading to destabilization of colloid or fine suspensions in saline waters: (1) neutralization of the negative charge by specific adsorption of positively charged species; (2) compression of the electric double layer as a result of the ionic strength of the solution due to increased charge or increased concentrations of the counter-ions; (3) inter-particle bridging by adsorbed material or (4) enmeshment of clay and hydroxide particles. There is an increasing tendency for *flocculation* to occur when particles collide: the number of these collisions increases due to particle movement and higher concentrations of particulate material (Burton, 1976).

2.7.4 Associations of Metal-Organic Compounds to Sediments

Three major processes leading to the *incorporation* of particular metal-organic species onto a sediment have been suggested by Saxby (1973).
1. Reaction between a metal ion and an organic ligand in solution leading to a species which can either precipitate directly or be adsorbed on sedimentary material.
2. Incorporation in a sedimentary pile of all or part of an organism containing biologic coordination compounds.
3. Adsorption on a sediment of molecules resulting from the solubilization of minerals (sulfides, carbonates, etc.) by natural waters containing organic ligands.

Direct Precipitation or Adsorption of Metal-Organic Species. With respect to the direct association of *metal-organic species* to sedimentary material (process 1), the association with clay minerals is of particular significance. Curtis (1966) has proposed a simplified scheme to explain why certain metals (Cr, Cu, Mn, Mo, Ni, U, V) may show a positive association with organic carbon in a sediment, while other elements exhibit no significant correlation, or may even be negatively correlated with organic carbon. According to that scheme, given in Table 62, positive correlation occurs when metal ions interact in solution with dissolved organic matter that are in turn concentrated by adsorption onto particulates such as clay minerals. Zero correlation may result when there is no interaction between metals and dissolved organic matter, and the latter alone is adsorbed onto the particulates. If there is competition between metal ions and dissolved organics for adsorption or bonding sites, negative carbon-metal correlations result (Jonasson, 1977).

Table 62. Generalized metal-organic-solid reaction scheme as proposed by Curtis (1966)

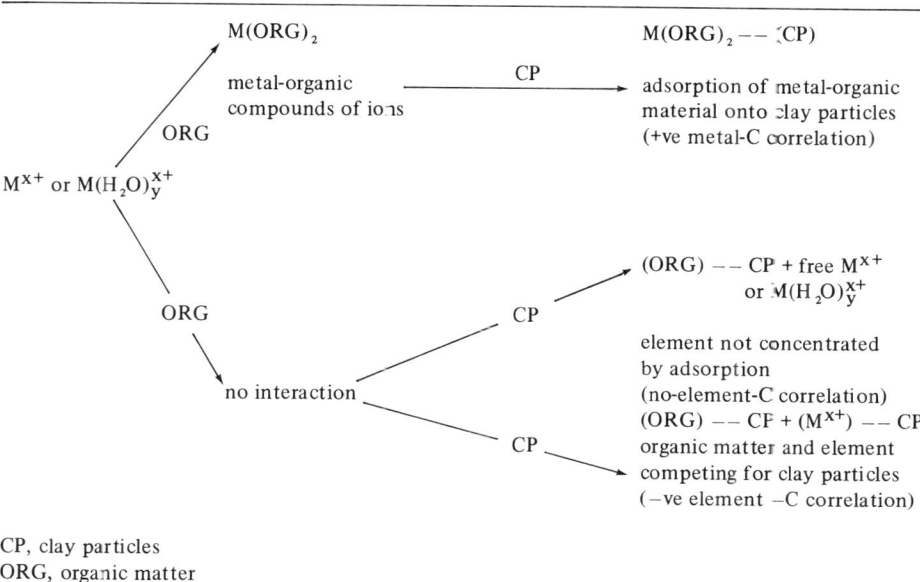

CP, clay particles
ORG, organic matter

Incorporation of Metal-bearing Organisms. Metal enrichment of sediments by incorporation of all or port of dead water organisms having accumulated metal species (process 2) can be expected to result from the high metal enrichment rate of water organisms in relation to their environment (Vinogradov, 1935; Noddack and Noddack, 1940; Stiles, 1946; Berger, 1950 and Black and Mitchell, 1952). In some of the enrichment processes several organic molecules demonstrated a certain selectivity in relation to characteristic metals. An example of extreme metal enrichment is that of vanadium in the blood cells of the tunicate *Pallusia mamillata* which reaches levels one million times that of sea water; the copper content of marine water reaches a level only one one-hundredth that of the blood of the *Octopus vulg.* Bayer (1964) thoroughly discusses the possibilities of an economical enrichment of rare metals in sea water by means of synthetic organic substances with high selectivity.

The enrichment of individual trace metals in fossil-bearing sediments has been known for a long time. There is, for example, a characteristic enrichment of nickel, vanadium, and molybdenum in sapropelic sediments, whereas chromium, lead, zinc, copper, and nickel tend to be enriched in gyttja deposits (Borchert and Krejci-Graf, 1959). The form in which metals actually occur in carbonaceous sediments is, however, not always the same. In some cases a visible amount of sulfides is present: for example, in some black shales zinc is partly present in sulfide form (30% to 50% of the total zinc present), while the remainder may be associated with clay minerals or bound to organic matter. In many cases chemical leaching procedures could be used to distinguish characteristic associations of metals, particularly those bonded to organic material (see Sect. E.4).

A high degree of positive correlation has often been observed between the contents of organic materials and metal concentrations in aquatic sediments. This, however, does not necessarily involve preferential metal bonding by organic substances since a number of mechanisms (for instance, sorption by clay minerals and precipitation of Mn/Fe oxides) produce simultaneous accumulation of organic material as well as typical metals, particularly in the fine-grained sediment fractions. Moreover, in highly polluted areas contamination by characteristic heavy metals coincides with the accumulation of organic substances, derived from both domestic and industrial or rural sludge. Extraction of metals from the organic fraction in sediment samples from Lake Malawi containing 6.8% organic carbon—or approx. 12% organic substances—indicates that only zinc, copper, and vanadium are accumulated in association with organic materials, whereas iron, manganese, chromium, lead, cobalt, and nickel are more or less diluted by these substances (Förstner, 1977c).

On the other hand, Jonasson (1976) determined a typical enrichment of most trace elements in the organic fraction of deep water sediments from Perch Lake in Lanark County/Ontario. It was assumed that gelatinous colloidal substances, which are formed from dissolved organic acids, spores, pollen, and decayed leaves, take up the metal ions from water. Enrichment of heavy metals in organic gels compared to organic sediments of lakes in northern Norway have been described by Tobschall et al. (1978). Timperley and Allan (1974), Jackson (1975), and other researchers have demonstrated the usefulness of organic-rich sediments for the discovery of ore deposits in arctic areas.

Upon selective extraction of various organic phases, Cooper and Harris (1974) determined a clear enrichment of trace metals in humic substances in more heavily contaminated water from River Blyth in Northumberland, United Kingdom. Heavily polluted sediments from Los Angeles Harbor exhibit an obvious affinity of zinc and cadmium for oxidizable fractions (organic substances and sulfides). From samples of Southern Californian waters Chen et al. (1976) analyzed the concentrations of trace metals within the organic fraction that proved to be 2 to 15 times higher than the total sediment on a weight basis. Further examples of the enrichment of trace elements in organic sediment fractions will be shown in Chapter E.4.

An example for the metal/organic carbon interrelation is given by Thomas and Jaquet (1976) in their study on "Mercury in the surficial sediments of Lake Erie". In Figure 61, the trend line "A" shows a high degree of relationship of mercury to organic carbon in samples with mercury concentrations below 0.9 ppm; samples with mercury levels above 0.9 ppm show a wider scattering with a lesser relationship to organic C concentrations (trend line "B"). According to the authors trend line "A" represents the natural situation for the time interval represented by the samples, where the organic matter concentration is proportional to the amount of mercury in the environment. Designation of the samples in zone B, however, show that they occur predominantly in the western basin of Lake Erie and are thus likely to represent more recent influxes of industrial mercury contaminants.

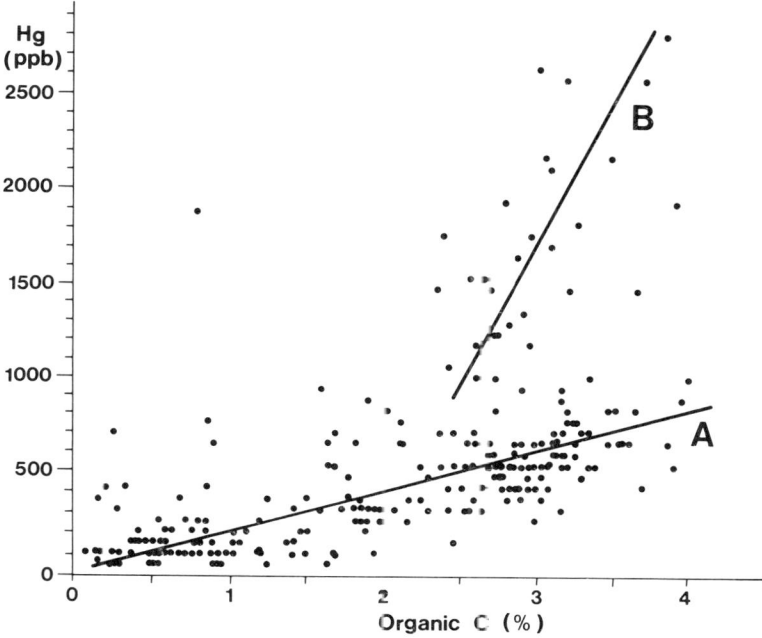

Fig. 61. Interrelation between mercury concentrations and content of organic carbon in sediments from Lake Erie (Thomas and Jaquet, 1976, J. Fish, Res. Board. Canada *33* (3), p. 411, reproduced with permission of Environment Canada, Fisheries and Marine Service, Ottawa). A. B see text

2.8 Sorption of Trace Elements on Carbonates and Phosphates

The extent of heavy metal sorption on *carbonates* has been a source of conflicting opinion for many years. Considering the relatively low contents of most of the heavy metals in carbonates, Krauskopf (1956) suggested that sorption does not play a very significant role. Currently, more has become known about the processes of carbonate sorption; it appears that the surface energy of the substances is generally sufficient to effect adsorption, at least in the order of magnitude observed for clay minerals such as kaolinite (Suess, 1973). Data obtained by Deurer et al. (1978) from lake sediments imply that coprecipitation with carbonate minerals may be an important mechanism for a number of metals e.g., zinc and cadmium (Chap. E.4).

Investigations of Salomons and Mook (1978) on the transition region from the Rhine to the IJssel Sea in Holland show that a decrease of the dissolved loads of zinc, cadmium, and nickel is directly dependent on the precipitation of carbonate minerals, which is chiefly a result of an increase in pH. At Foundry Cove, Cold Spring/New York, where large amounts of heavy metals were discharged from a Cd-Ni-battery factory, the Cd-component is relatively less mobile—when compared to Ni—due to the formation of insoluble $CdCO_3$-phases in the alkaline waste (Bower et al. 1978). The same effects have been proposed by Jenne (1976), when (a) carbonates occur as a major component in soils or the fine-grained fraction of fluvial sediments, or (b) when the trace elements are present in high enough concentrations to saturate other element sinks. Together with sorption processes on Fe-, Mn-, and Al-hydroxides, the mechanisms of coprecipitation with carbonates, whereby the heavy metal cations sorbed onto the surface become part of the crystal lattice, appear to be a very important means of limiting metal concentrations in the marine environment. Seibold (1964) points out that if it were not for these processes, the heavy metal content in seawater would have increased throughout the earth's history to several hundred milligrams per liter as a result of influxes from rivers and diagenetic remobilization from the sea floor.

The conditions for $CaCO_3$ precipitation have been classified by Müller et al. (1972) as follows: (1) CO_2-loss or extraction as a result of changes in the p-T-conditions or by plant assimilation, either by macrophyta (e.g., potamogeton) or by microphyta (especially planktonic algae), (2) evaporation concentration or (3) mixing of different water bodies.

With respect to the coprecipitation of trace elements in carbonate, the third possibility is of interest here. If an alkaline water body comes into contact with and becomes mixed in river water with normal Ca^{2+} and HCO_3^- levels under neutral pH conditions, the pH will increase. Consequently, the solubility product of calcium carbonate is drastically reduced and $CaCO_3$ is precipitated in the mixing zone, carrying heavy metals from solution with it. Coprecipitation experiments on calcium carbonate performed by Popova (1961) show that heavy metal carbonates of low solubility such as $CoCO_3$ and $PbCO_3$ are completely eliminated from solution as a result of $CaCO_3$ precipitation (Table 63). Without a carrier substance, in this case $CaCO_3$, the metal cations would not have been precipitated. However, by adding solid carbonate compounds, the removal of lead from solution was very small.

An example of metal coprecipitation in polluted surface water is described by Patchineelam (1975): calcium carbonate precipitates in the Elbe River estuary as the result of the mixing of normal river water with alkaline effluents from an industrial plant. Compared with the geochemical background values of heavy metals in carbonate rocks (Turekian and Wedepohl, 1961 – see Table 35), enrichment factors of approx. 10 times for cadmium, 6 times for copper and 4.5 times for Zn were found in the calcium carbonate particles.

Sorption Effects. The zero point charge (ZPC) for calcite falls between pH 8 and 9.5 (Somasundaram and Agar, 1967); i.e., below pH 8 the calcite particles have a positive charge and above 9.5 they are negatively charged. This led Leland et al. (1967) to the conclusion that lead in carbonate lacustrine sediments can be adsorbed on calcite, or calcium in calcite can be replaced by lead.

Table 63. Coprecipitation of trace metals with calcium carbonate (from Popova, 1961)

Original concentration	Metal content in the precipitation in % of the original concentration					
(µg/l)	Pb	Co	Cd	Zn	Cu	Ni
25	100	100	100	100	96	66
250	100	100	100	100	–	43
2000	100	96	100	96	75	–
pK_{sp}[a]	13.1	12.8	11.3	10.8	9.6	6.9

[a] pK_{sp}, negative logarithm of the solubility product of simple metal carbonates (see Table 57).

In sorption experiments with organic substances and calcium carbonate Suess (1970) observed that a maximum of 14% of the organic carbon dissolved in seawater could be adsorbed by carbonate minerals. This suggests a similarly strong absorption of heavy metals on carbonates.

Heydemann (1959) investigated the sorption of copper from dilute solutions on pure calcium carbonate: during the first few hours after mixing the copper extraction followed a Freundlich adsorption isotherm. After six hours distinct deviations from the adsorption isotherm could be observed. According to Heydemann (1959) these differences are a function of the ionic concentrations in the original solution. Correns (1969) attributed these deviations to the occurrence of chemical changes taking place: ions situated on the surface tend to enter the adsorbent and finally form *pseudomorphs*.

Little has been investigated as yet of the amounts of heavy metals that are introduced into water bodies by the heavily increased use of phosphorus-containing products such as fertilizers and detergents. In this respect, the enrichment of uranium, cadmium, and arsenic is of particular concern. From a compilation by McKelvey (1956), marine phosphorites contain uranium quantities of 50 to 100 ppm (an average value for "shale" is approx. 4 ppm): residual phosphorites average 50 ppm, fossil bones 50–300 ppm uranium. With these values in mind, real enrichment can be expected in the fertilizers. For detergents similar developments are limited since metal concentrations are confined to a certain level. However, a study from Angino et al. (1970) has shown that *arsenic* occurs at concentrations of 10 to 70 ppm in several common presoaks and household detergents. In areas of repeated usage of surface waters, the concentration of As in water bodies can be expected to rise if the use of detergent products containing arsenic is continued. The data from the Kansas River indicate that at levels of 3 to 8 $\mu g/l$, these concentrations are close to the limit (10 $\mu g/l$) set by the United States Public Health Service as a drinking water standard (Angino et al., 1970).

New data concerning trace metals in phosphate materials is given in Table 64. Measurements of phosphate rocks from three important productive deposits in Florida, Kola, and Morocco (from Langmyhr, 1977) show that there are large differences in trace materials for most of the metals investigated. The values as a rule do not differ greatly from those of average shale data, which are here given for comparison purposes.

Table 64. Trace elements in phosphate materials (ppm)

	Fe	Mn	Zn	Cr	Ni	Cu	Pb	Co	Cd
Phosphate rocks[a]									
Florida, USA	6,915	771	94	69	39	14	15	5	6.8
Kola, USSR	–	266	24	2	2	40	2.4	3	1.0
Morocco	–	10	222	306	69	39	3.2	3	11.0
Superphosphate[b] fertilizer (n = 2)	1,190	225	363	119	34	16	0.7	0.8	4.9
Polyphosphate[b]	15	10	11	5	18	2	0.2	0.1	0.47
Commercial[b] detergents (n = 20)	10	87 (40–175)	27 (8–78)	2	8 (1–18)	1 (0.2–3)	0.1	0.1	0.23 (0.01–0.8)
Shale[c]	43,500	850	95	90	68	45	20	19	0.3

[a] Langmyhr et al., 1977. [b] Unpublished data, Institut für Sedimentforschung, Heidelberg. [c] average shale, for comparison, Turekian and Wedepohl, 1961.

There is, however, at least one exception: the cadmium contents in the phosphate rocks lie about one order of magnitude above these shale values. The cadmium contents are also about as high in phosphate fertilizers, which can lead, as would be expected, to an important influence of water bodies. The polyphosphate additives to detergents contain, on the other hand, very small trace metal contents and, like the detergents themselves, seem unable to affect significantly the metal loads in aquatic systems.

3 Metal Accumulation in Aquatic Sediments—Interactions and Effects of Various Processes and Sinks

It has been stressed by Jenne (1976) in his review on "trace element sorption by sediment and soils—sites and processes", the armoring of sediment particles is of great importance to their sorption characteristics as it slows down *rates of reaction* and prevents portions of the various sinks from *equilibrating* with their associated waters. While the clay minerals themselves are relatively unimportant as trace element sinks, their role as a mechanical substrate for the precipitation and flocculation of organics and of hydrous iron oxides has been pointed out in many studies. In some cases, carbonate coatings on clays and hydrous iron oxides exert a characteristic influence on the sorption properties that sediment particles exercise on trace elements.

3.1 Hydroxidic Coatings on Clay Minerals

In a study on "the role of clay minerals in the transportation of iron", Carroll (1958) dealt extensively with the relationship between the organic phases of *clay minerals* and those of *iron oxides.* Iron oxides, such as goethite, lepidocrocite, hematite or indefinite iron oxides having poor crystallinity—resulting in yellow, brown or red colored soils—are nearly always associated with the clay minerals, although oxide films may also coat mineral grains of sand and silt fractions.

Fripiat and Gastuche (1952) found that iron oxide accumulates in definite quantities on clay minerals; a kaolinite with cation exchange of 5 meq/100 g would, for example, be able to accumulate 50 meq Fe^{3+}/100 g.

The hydrous oxides of iron, manganese, aluminum, and chromium are positively charged and amorphous when formed; with time, they become crystalline and slowly tend to become negatively charged (Rutherford, 1977). Sorption of amorphous colloidal iron oxides would therefore occur preferentially on the plane surfaces (negatively charged) rather than on the edges of kaolinite crystals at pH-values below the isoelectric point of the iron oxide (Follet, 1965; Jenne, 1976). It has been suggested by Follet (1965) that the sorption of positively charged colloidal iron oxide is effected by a coulombic attraction (physical sorption) rather than by specific chemical bonding.

A positive correlation between the quantity of free-iron oxides and clay contents has generally been observed in sediments (Jenne, 1968). This type of relationship has not been found for free-manganese oxide and clay minerals; sometimes even a negative correlation exists between the Mn and clay contents. Hence, free-manganese oxides are presumed to occur largely as discrete particles rather than as coatings, and are frequently found concentrated in the silt-size fractions of soils and sediments (Jenne, 1976). According to the same authors it seems likely that manganese oxide precipita-

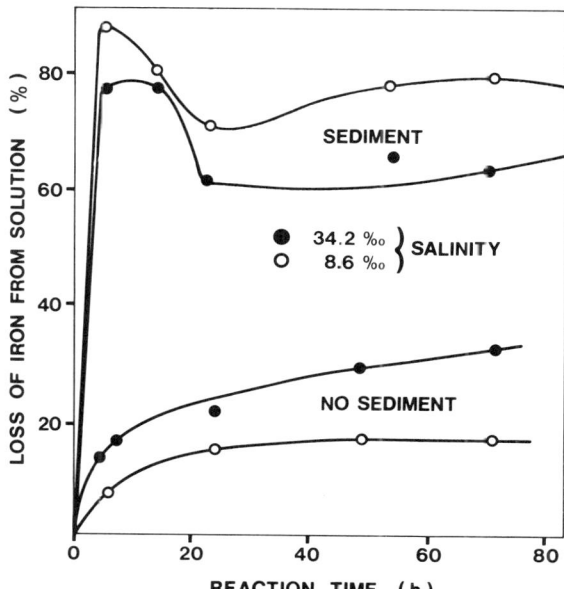

Fig. 62. Influence of suspended particles on the precipitation of iron compounds in the freshwater/seawater mixing zone (after Aston and Chester, 1973)

tion occurs at the interface between oxidizing and reducing conditions induced by micro-environments, preferably adjacent to detrital organic particles.

The influence of suspended particles on the precipitation of iron in natural waters was studied by Aston and Chester (1973) with emphasis on the condition in the river/sea mixing zone (Fig. 62): A preliminary series of experiments was designed to compare the effects of sediment particles on the precipitation of both "aged" and "nonaged" iron in seawater. "Aged" iron has undergone hydrolysis in saline waters before becoming associated with a sediment load and is present in a colloidal form. "Nonaged" iron is simply iron in solution entering a saline environment with a suspended sediment load. It can be seen that the concentration of the "aged" iron remaining in colloidal suspension after sedimentation is proportional to the initial iron concentrations, whereas the concentration of "nonaged" iron remaining in solution is very low and is independent of the initial iron concentration. A second series of experiments was designed to study the precipitation of iron in both the presence and absence of suspended sediment particles during changes in salinity. Figure 62 shows the time reaction curves for dissolved iron ("nonaged"), both at a salinity of 8.6‰, and the seawater salinity at 35‰. It is clear that the presence of sediment particles increases both the rate and the extent of iron precipitation at all degrees of salinity. A rapid initial formation of iron hydrolysis products also takes place, followed by slow redissolution and reprecipitation until equilibrium is achieved. This adjustment may be due to the neutralization of the charged particles by the adhesion of solid hydrolysis products and consequent charge transfer.

The experiments mentioned here have given further evidence that, with regard to the sorption of trace elements in aquatic systems, "the more important sinks generally are comprised of thermodynamically metastable phases" (Jenne, 1976). Such phases are usually formed because they rapidly reduce the excess concentrations of given solute ions.

Recrystallization of amorphous hydrous iron oxides to goethite and hematite is particularly inhibited by coprecipitation of natural organic substances (Schwertmann, 1966), by trace metals (Jenne, 1968), by anions other than oxygen or hydroxyl (Jenne,

1976), and by sorption of silicate (Scheffer et al., 1958) and phosphate (Krause and Borkowsha, 1963). On the other hand, recrystallization of iron and manganese oxides seems to be enhanced in hot arid regions (McLaughlin, 1954).

3.2 Organic Coatings on Clay Minerals

The principles of clay-organic interactions have been discussed in Sect. 2.7.4 and schematically presented by the Curtis diagram (Table 62). It has been shown that the natural silicates provide catalytic surfaces to organic molecules influencing transformations such as polymerization, isomerization, hydrogenation and conversion to cyclic compounds (Degens, 1965). There is an increasing tendency of dissolved organic compounds to adsorb onto solid surfaces as the molecular mass of the organic compound increases (Mortland, 1970); this process can serve to increase the fulvic to humic acid ratio of surface waters (Jenne, 1976).

The behavior of clay-organic compounds under changes in the chemical environment are of particular interest. The experimental data of Rashid et al. (1972) indicate that the Coulombic and van der Waals forces are more important factors influencing the bonding of humic acids to clay minerals than are colloid chemical effects. Adsorption of humic acids on clay minerals is promoted in solutions with low pH and elevated salinities. The latter factor can be clearly seen where river water carrying clays and humic acids enter the marine estuary. Organic coatings on clay minerals may greatly affect size distribution and settling rates of inorganic sediments flocculation at the river/sea mixing zone (Kranck, 1973) and contribute to the ubiquitous increase of organic matter in coastal marine sediments (Johnson, 1974; Sholkovitz, 1976).

3.3 Interactions Between Hydrous Metal Oxides, Organic Substances, Carbonate, and Phosphate

The interaction between *hydrous metal oxides and organic substances* is particularly complex. On the one hand, these oxides are unstable in certain organic-rich sediments, especially under anaerobic conditions (Jonasson, 1977). On the other hand, there is a close correlation between iron, manganese and phosphorus within both river water (dissolved state) and seawater (partly flocculated) organic substances (Sholkovitz, 1976, Sholkovitz et al., 1978).

Theis and Singer (1973, 1974) demonstrated that dissolved organic matter can alter the distribution between oxidized and reduced forms of metals. They have shown that humic acids and model organic compounds formed by the decay of humic material are capable of inhibiting the kinetics of Fe(II) oxidation. During the first stage of their experiments with humic acids, the oxidation of Fe(II) proceeded at the same rate as in simple aqueous media; the reaction then slowed down. This could be an indication of the stabilization of a part of the Fe(II) involved in the reaction (approximately one-fifth of the total Fe concentration), presumably through complexation. Further experiments by Theis and Singer (1973) show that the extent of Fe(II) stabilization is a clear function of the concentration of organic material. The same authors have presented a model to illustrate the manner in which organics influence the redox kinetics and the resultant distribution between Fe(II) and Fe(III) in oxidizing environments. Their diagrammatic description (Table 65) shows that the complexation of Fe^{2+} by organic substances (reaction 3) competes with the oxidation reaction (1) by which the thermodynamically stable phases of Fe(III) should be formed. The existence of reaction (6) indicates that organics are also capable of complexing oxidized forms of metals, thereby increasing its apparent solubility beyond the limitations imposed by the solubility product, and also enhancing its colloidal stability (Singer, 1977).

Table 65. Diagrammatic description of the behavior of iron in the presence of organic matter (after Theis and Singer, 1974; Environ. Sci. Technol. Vol. 8, p. 573)

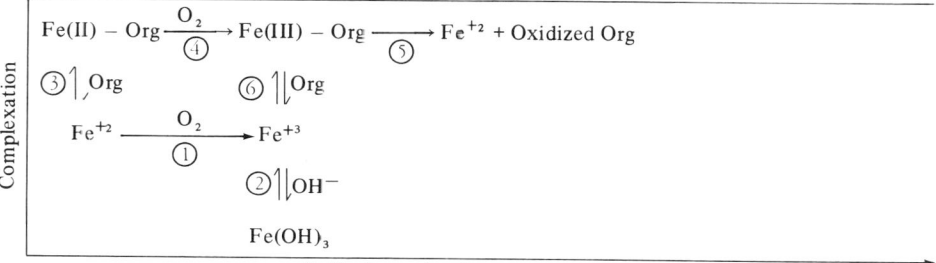

Investigations on *coprecipitation* in hydrous Fe/Mn oxides performed by Groth (1971) also included the competition of organic complexing agents. Precipitation occurred by addition of iron and manganese chloride from distilled water and from natural (filtrated) lake water. In each case, 10 µg/l of the investigated metal cation and, in some cases, organic complexing agents were added to the solution.

Table 66 shows important differences in the behavior of the individual metals: (1) Whereas the added iron component is quantitatively precipitated, the oxidation of Mn(II) is delayed, probably on account of the presence of reducing substances in natural lake water; (2) An almost complete coprecipitation of the copper additive takes place; (3) Zinc and cobalt are precipitated to a much lesser extent in the stable EDTA complexes than in the natural organic metal complexes. The high concentrations of zinc remaining in the distilled water show that Zn^{2+}, as a hydrated ion in solution, is only coprecipitated to a small extent during hydroxide precipitation; the same also applies to cobalt, but to an even lesser degree.

Table 66. Coprecipitation of trace metals with Fe/Mn hydroxides (from Groth, 1971)

Solution	% of metal additive found in the precipitates				
	Fe	Cu	Zn	Co	Mn
distilled water	98	95	14	47	80
inland lake water					
no admixture (a)	99	98	86	67	25
EDTA (b)	99	88	18	25	25
Peat extract (c)	99	97	88	75	25

Metal additive (Cu, Zn, Co) in each case is 10 µg/l. Precipitation with 22 mg/l Fe and 28 mg/l Mn. EDTA additive (b): 30 mg/l; Peat extract (c) corresponds to approximately 2.6 mg/l humic acid.

Carbonate minerals are also represented in this system of sediment interactions, although precipitation of calcium carbonate is blocked in pore waters rich in dissolved organic matter (Berner, 1971). Calcite minerals are important nucleation centers for manganese oxides since there is a microzone of higher pH on the surfaces of the carbonates (Lee, 1975).

Sorption of *phosphates* and polyphosphates onto clay minerals involves chemical bonding of the anions to positively charged edges of the clays as well as substitution of phosphates for silicates in the clay structure (Van Olphen, 1963); in general, high phosphate adsorption by clays is favored by a lower pH (Stumm and Morgan, 1970). Characteristic sorption of phosphate anions occurs on freshly precipitated ferric and aluminum hydroxides (Bache, 1963).

According to Gorbunow (1959) amorphous iron hydroxide absorbs 108 times more phosphate than crystalline iron hydroxide; amorphous aluminum hydroxide absorbs 137 times more phosphate than crystalline aluminum hydroxide. Sorption rate and capacity of crystalline iron and aluminum oxides are at a level between that of the amorphous oxides and the clay minerals (Jenne, 1976). Calcium carbonate deposition is inhibited by adsorption of polyphosphate anions on the $CaCO_3$ crystals (Stumm and Morgan, 1970).

3.4 Significance of the Different Sinks in Natural Systems

The efficiency of the various influencing factors on the bonding and enrichment of heavy metals in aquatic sediment—natural or contaminated—is dependent on the depositional environment, characterized by the chemical composition, particularly on the amount of dissolved iron and carbonate available for reactions, salinity, pH and redox values and by the hydrodynamical conditions. The general effects of these influential factors may be described as follows:

1. Detrital Minerals. The presence of a large proportion of heavy minerals (generally resistant to outside influences) results in characteristic enrichments of trace elements, particularly in the silt and fine-sand fractions of sediments. High proportions of quartz, feldspar, and detrital carbonates tend to have quite the opposite effect on the sedimentary heavy metal budget.

2. Sorption. A generalized sequence of the capacity of solids to sorb heavy metals was established by Guy and Chakrabarti (1975):

MnO_2 > humic acid > iron oxide > clay minerals

The sorption capacity of Fe oxides (crystalline phase goethite) for heavy metals is at least ten times less than that of the Mn oxides (Suarez and Langmuir, 1976). Rashid (1971) estimated that of a total bonding of 200–600 meq metal/100 g humic substance, approximately one-third can be attributed to the cation exchange and two-thirds to chemical sorption and organic complexations.

3. Coprecipitation with Hydrous Fe/Mn Oxides and Carbonates. In aquatic systems where oxidizing occurs, hydrous iron and manganese oxides constitute a highly effective sink for heavy metals (Lee, 1975). Experiments by Groth (1971) show that the metals cobalt, zinc, and copper coprecipitated from natural inland lake water with Fe/Mn hydroxides at a rate of 67%, 86%, and 98% respectively. Coprecipitation with carbonate may be an important elimination mechanism for metals such as zinc and cadmium when carbonate contents occur as a major component i.e., when other substrates, particularly hydrous iron oxides or organic substances, are less abundant.

4. Complexation and Flocculation with Organic Matter. In organic systems the role of Fe and Mn oxides as direct adsorbants of metal ions is either overshadowed by competition from the more reactive humic acids and organo-clays, or is obscured by coatings of organic matter (Jonasson, 1977). Once the metals are complexed by humic acids, the solutions behave as if other ions were not present in the reaction media (Drake, 1967). Furthermore, the metal ions become unavailable to sulfides, hydroxides and carbonates etc., and prevent the formation of insoluble salts (Rashid and Leonard, 1973). Specific chemical and electrostatic processes result in flocculation of Fe, Al, and humates, particularly in marine estuaries (Eckert and Sholkovitz, 1976). Organic flocculant coatings greatly affect the adsorption capacities for trace metals of sediment and suspended matter (Pillai et al., 1971; Sholkovitz, 1976). High enrichments of trace metals have been found by Nissenbaum and Swaine (1976) in the humic substances taken from marine reducing environments.

5. Heavy Metal Precipitates. At present, the amount of sediment formed from the direct precipitation of heavy metal hydroxides, carbonates and sulfides has yet to be observed in a natural water system. In particular, there is usually no means to distinguish between the metals that are precipitated with either iron hydroxide, calcium carbonate or iron sulfide. Nonetheless, it can be assumed that the composition of these precipitates is greatly influenced by the various hydro-chemical conditions of the water body in question, and is subjected to waste metal disposal.

With respect to a characteristic environment of water/sediment interaction, the significance of the various bonding mechanisms and their respective substrates for heavy metals has been estimated for the estuarine mixing zone in Table 67. The processes of *flocculation of metal organic complexes, coprecipitation with hydrous Fe/Mn oxides* and uptake by organisms seem to be of particular importance for the sediment association of heavy metals, especially those introduced by human activities.

Table 67. Factors affecting the accumulation of heavy metals in estuarine sediments

	Detrital minerals, organics	Reactive organics (humic acids, bitumen)	Trace metal hydroxides, carbonates, sulfides	Hydrous iron and manganese oxides	Calcium carbonate
Incorporation in inert positions	××	(×)			
Adsorption = physico-sorption	×	×	(×)	(×)	(×)
Cation exchange = chemi-sorption	×	×	(×)	×	(×)
Precipitation			××		
Coprecipitation				×××	×
Complexation + flocculation		×××			

The term *"flocculation"* summarizes a complex system of physical and chemical interactions (Sect. 2.7.3). It has been emphasized by Turekian (1977) that *particles* play an important role as the sequestering agents for reactive elements in every step of the transport processes from continent to ocean floor. Generally, the adsorption of *humic acids* on clay minerals is enhanced in solution with elevated salinities (Rashid et al., 1972). Experiments performed by Eckert and Sholkovitz (1976) indicate that *calcium ions,* through chemical association, are most effective in destabilizing the more hydrophilic of the humic colloids in river water; NaCl flocculates no more than 20% or 30% of the amount precipitated by $CaCl_2$.

The most important influencing factors in the *precipitation of Fe,* the latter having special significance as a coprecipitation medium for other metals, are increasing ionic strength and *pH-values* (Williams and Chan, 1966). Precipitation takes place during the hydrolysis of hydrated iron(III) ions (Brønstedt, see Stumm and Morgan, 1970, p. 243), with the following sequence:

$$Fe(H_2O)_6^{3+} \xrightarrow{H_2O} Fe(OH)(H_2O)_5^{2+} + H_3O^+ \xrightarrow{H_2O} Fe(OH)_2(H_2O)_4^+ + H_3O^+ + \xrightarrow{H_2O} Fe(OH)_3(H_2O)_3(s) + H_3O^+$$

Such processes can be expected to take place in estuarine zones, where, in addition to an increase of ionic strength, a change in pH conditions frequently occurs, generally ranging from pH values of between 6.5 and 7.5 in river water to approx. 8 in seawater. The absence of a large-scale removal of manganese under the same experimental conditions is explained by the considerably higher pH requirement for the precipitation of manganese than for iron (Morgan, 1967).

It has been established that marine bacteria play an important role in the deposition of heavy metals in estuary sediments (Oppenheimer, 1960). For salt marsh estuaries Windom (1975) observed that higher concentrations of organic matter in the water and sediments and greater *biologic productivity* may lead to even shorter residence times for heavy metals.

Experimental work was performed by Lerran and Holmes (1975) with isolated bacterial cultures to examine their specific role in heavy metal assimilation and precipitation from seawater. This study clearly demonstrated deposition of zinc and cadmium by various bacteria in estuarine sediments. Because of the large amounts of H_2S produced by the bacteria, zinc is probably precipitated as ZnS (Holmes et al., 1974) or as a coprecipitate with FeS.

In this context it should be remembered that the uptake of heavy metals by organisms is greatly influenced by the individual type of bonding of the metals contained in sediments or suspended particles. Data from Luoma (1976) indicate that the biologic availability of zinc in aquatic sediments is controlled by Zn partitioning among manganese oxides, iron oxides, and particulate organics (Chap. F).

3.5 Non-Conservative Effects of Trace Metals in Estuaries

The "interactive" versus "non-interactive" behavior (or "conservative" versus "non-conservative" in relation to dissolved constituents) of trace metals in the river/sea mixing zone has gained much attention during the last few years (Burton, 1976; Liss, 1976). Two indices may be applied to measure conservative mixing: the chlorinity of the water when there is no evidence that halide ions behave non-conservatively in estuarine water, or the salinity gradient.

Coonley et al. (1971) have described an instructive example for the behavior of metals in the estuarine mixing zone from the mouth of the Mullica River in New Jersey.

The Mullica River draws its water from New Jersey pine barrens, which are underlain by Tertiary sand, largely consisting of quartz sand and local gravel and clay beds. Water, percolating through forest debris enters a porous groundwater reservoir, which is remarkably inert chemically. Due to the lack of carbonates, the pH is usually between 4 and 5 in the upper reaches of the river. When the river water enters Great Bay, New Jersey, there is a sudden rapid decrease in the iron concentration with an increasing chloride content, indicating that iron is removed from the mixture of the river water and seawater (Fig. 63). The extensive removal of *iron* was observed in

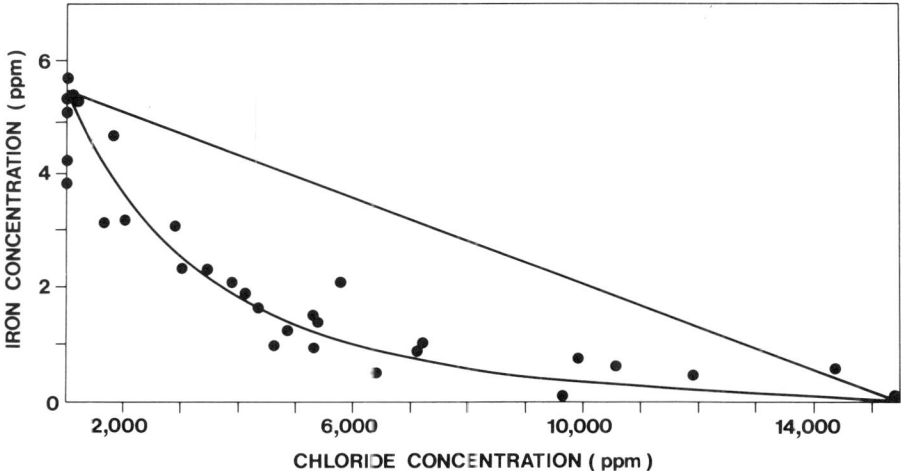

Fig. 63. Nonconservative behavior of dissolved iron in the freshwater/seawater mixing zone of Mullica River/Great Bay, New Jersey (Coonley et al., Geochim. Cosmochim. Acta, 35, 1971. Reproduced with permission of Pergamon Press, Oxford)

a number of locations e.g., in British Columbia (Williams and Chan, 1966), in three estuaries in the southwestern United States (Windom et al., 1971), in the Gulf of St. Lawrence (Brewers et al., 1973), and in the Merrimack River estuary (Boyle et al., 1974). The behavior of iron and manganese at the river/sea boundary was studied by Vandeginste et al. (1976) from the Rhine estuaries, where it was found that the Fe:Mn ratio shows a characteristic development in the transitional zone from river to sea. The occurrence of a maximum in the ratio of iron and manganese concentrations in particulate matter above 2‰ salinity in the Rhine river lends credence to the supposition that particulate *iron hydroxide* species are formed primarily within a restricted area. Obviously, with increasing salinity particulate *manganese* concentrations increase relative to the other elements. This order of precipitation is probably connected with a charge reversal from positively charged sediments in freshwater to negatively charged sediments in seawater, which occurs between 2‰ and 6‰ salinity (Pravdic, 1970)—an electrokinetic variation that could possibly influence the deposition of certain ions and minerals at or near river estuaries.

Recent investigations by Duinker and Nolting (1977) indicate that the removal of metals with respect to conservative estuarine mixing applies to dissolved *copper, zinc, and cadmium* as well. The concentrations of dissolved species of these elements at salinity values within the range of 0.4‰ to 35‰ in the Rhine estuary are lower than they would be if they corresponded with conservative mixing (Fig. 64). The decrease of the concentrations due to removal from the dissolved state is most pronounced in the early stages of mixing: roughly 40% for Cd, 50% for Cu, and 30% for Zn; no evidence of precipitation of *lead,* however, can be seen from these data.

Furthermore, the non-conservative behavior of manganese and zinc is still quite controversial. Conclusive proof of zinc removal was obtained by Mackay and Leatherland (1976) in the Clyde Estuary which is subject to major inputs of industrial and domestic wastes.

Under low-flow conditions the concentrations of zinc found at intermediate salinities lie very much below the theoretical dilution line; under conditions of greater

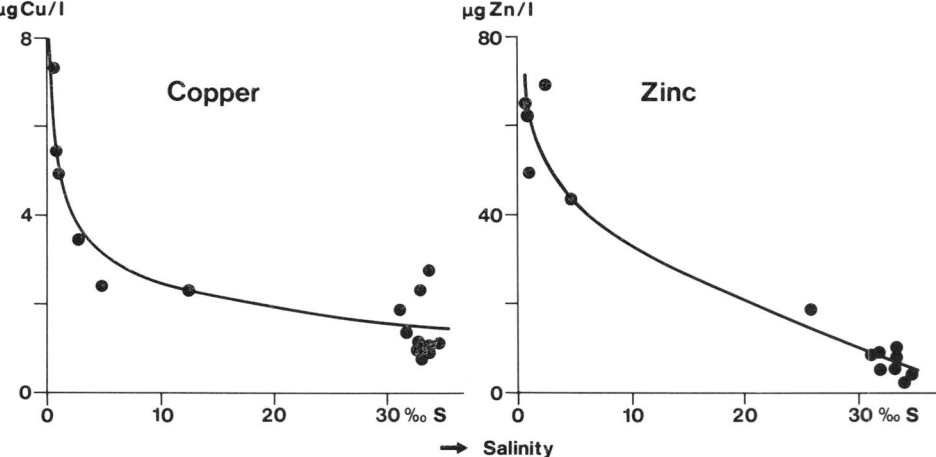

Fig. 64. Nonconservative effects for dissolved copper and zinc (values in µg/l) in the Rhine Estuary (Duinker and Nolting, 1977, Marine Pollution Bulletin *8* (3) p. 70. Reproduced with permission of Pergamon Press, Oxford)

fresh water flow the observed Zn concentrations range closer to the conservative mixing line, indicating less zinc removal, and presumably reflecting a lower proportional rate of sedimentation. The behavior of dissolved iron, manganese, and zinc in the Beaulieu Estuary (Southern England) was investigated by Holliday and Liss (1976). Here, substantial nonlinearity in the iron—salinity relationship indicates large-scale removal of iron from solution during the early mixing of river and sea water. In marked contrast, manganese and zinc appear for the most part to behave conservatively during the estuarine mixing process.

4 Determination of Chemical Phases in Natural and Polluted Sediments

The question of chemical constituents to particulate substances has been under discussion for some time. Problems related to soil science have led to the development of procedures for determining the absorption capacity and leachability of mineral or organic components: both living parameters of great significance for soil fertility. In the sediment-petrographic field, interest has focused mainly on differentiating the metal content in the detrital and nondetrital fraction of carbonate rocks (Hirst and Nicholls, 1958). This method was later used by Chester (1965) in a study of trace element partition in reef and nonreef carbonates, and applied to chemical separation of those trace elements incorporated into pelagic sediments by ferro-manganese concretions (Chester and Hughes, 1967).

In connection with the problems arising from the disposal of contaminated dredge material, further methods of sediment partitioning have been developed. At present, the most advanced techniques include the successive extraction of the metal contents in interstitial waters and also of ion exchangeable, easily reducible, organic and residual sediment fractions (Engler et al., 1974; Burrows and Hulbert, 1975; Gupta and Chen, 1975; Brannon et al., 1976; Gambrell et al., 1976).

A summary of the different methods of extracting heavy metals from aquatic sediments to be used in studies of pollution effects is shown in Table 68. For the determination of residual inorganic substances i.e., detrital silicates, see Chapter D.2.3— *Chemical analyses of sediments*.

4.1 Proportion of the Individual Types of Metal Associations in Natural and Polluted Aquatic Sediments

Chester and Hughes (1967) made the first attempts, by a combination of extraction methods, to establish the main bonding forms of metals in aquatic sediments: chemical techniques employed to study the trace metals incorporated in pelagic sediment by ferro-manganese nodules, carbonate minerals, and surface adsorption onto minerals

Table 68. Extraction of metals associated with different chemical phases in sediments (Patchineelam and Förstner, 1977)

Chemical phase	Extraction methods	Authors
Adsorption and cation exchange	(a) $BaCl_2$, (b) $MgCl_2$, (c) NH_4OAc	Jackson, 1953; Gibbs, 1973
Detrital/autigenic phases	EDTA treatment	Goldberg and Arrhenius, 1958; Gad and Le Riche, 1966
Hydrogenous/lithogenous phases	(a) 0.1 M HCl (c) 0.3 M HCl (b) 0.1 M HNO_3	Piper (1971) Jones (1973) Malo (1977)
Reducible phases	1 M $NH_2OH \cdot HCl$; 25% v/v acetic acid	Chester and Hughes, 1967
Moderately reducible phases (hydrous Fe-oxides)	Reduction with sodium dithionite complexing with sodium citrate	Aguilera and Jackson, 1953; Holmgren, 1967
Easily reducible phases (Mn-oxide and amorph. Fe-oxides)	0.1 M $NH_2OH \cdot HCl$; 0.01 M nitric acid	Chao, 1972
Carbonates	(a) CO_2 treatment (b) Exchange columns	Patchineelam, 1975 Deurer et al., (1978)
Organics, sulphides	30% H_2O_2 at 95 °C, extract with (a) 1 N NH_4OAc or (b) 0.01 M HNO_3, fat solvents: e.g. chloroform, ether, gasoline, benzene, carbon disulfide	Jackson, 1953; Engler et al., 1974; Gupta and Chen, 1975 Bergmann, 1963 Welte, 1969 Cooper and Harris (1974)
Humic and fulvic acids	0.5 N NaOH; 0.1 N NaOH/H_2SO_4	Rashid, 1971; Volkov and Fomina, 1974
Solid organic material	Na hypochlorite, dithionite/citrate	Gibbs, 1973
Detrital silicates	Digestion with $HF/HClO_4$; Lithium metaborate (1000 °C)	see Chap. D.2.3

involving the use of dilute acetic acid and of reducing agents (e.g., 1 M hydroxylamine hydrochloride). The next step involved the use of oxidizing agents as a means of determining the amounts of organic substances and sulfides (see Presley et al., 1972).

Since 1972 new attempts have constantly been made with other extraction procedures; furthermore, interest in the bonding forms within contaminated sediments is growing. This development is shown in five examples in Figure 65. The total metal concentrations are given in Table 69.

The distribution of Fe, Mn, Ni, Zn, Cu, and Sr in a sediment core from the *Sea of Okhotsk* was studied by Nissenbaum (1972). The data indicate that iron and nickel are to a considerable portion incorporated into the residual fraction, whereas manganese, copper and, in particular, zinc are more associated with the hydrogenous phases. With increasing depth, a transfer of metals from the fraction leached by acetic acid to the HCl leachable fraction occurs. This is indicative of changes in the chemical environment and of characteristic chemical reactions taking place after deposition.

Partitioning studies were carried out by Gibbs (1973, 1977) in the suspended load of the *Amazon and Yukon Rivers,* which are less affected by civilisational influences. In the case of iron, manganese, and nickel, the most significant bonding occurs, as expected, in hydroxide "coatings" of particles. This type of bonding is only secondary in the case of copper and chromium, with cobalt situated somewhere in between. As the hydroxide bonding decreases, a strong increase in mineral bonding can be observed, which for chromium constitutes more than 65% and for copper more than 87%. Organic bonding is lowest for copper and highest for nickel (16%). Bonding by cation exchange is relatively insignificant, constituting less than 1% for iron and manganese and varying for the other metals investigated between 2.2% (Cr and Cu) and 4.7% (Co).

In contrast to Gibbs' results, Nissenbaum (1974) detected high levels of carbonates and diluted iron in the sediment of the *Dead Sea,* where pollution effects are of relatively minor importance (Table 69). The sequential leaching with water, hydroxylamine hydrochloride and hydrogen peroxide (according to the analytic procedure of Presley et al., 1972) indicates that manganese and cadmium occur mostly in authigenic carbonates, oxides, and sulfides. The apparent depletion in heavy metals as compared with other examples is according to Nissenbaum (1974) due to the formation of metal complexes in Dead Sea water which enrich the water in many trace components and prevent their removal to the sediment.

Further progress in the determination of sediment-associations of trace metals has been registered since 1974 in studies undertaken by the Environmental Effects Laboratory of the U.S. Army Engineer Waterways Experiment Station in Vicksburg, Miss. (Engler et al., 1974; 1976; Brannon et al., 1976a, b), with the participation of the Environmental Engineering Programs of the University of Southern California, Los Angeles (Chen et al., 1975, 1976).

Table 69. Total metal concentration (in ppm, Fe in %) of the sediment samples to which chemical associations are presented in Figure 65

	Mn	Fe	Ni	Co	Cr	Cu	Pb	Zn	Cd
Sea of Okhotsk (Nissenbaum, 1972)	575	5.5%	79	–	–	44	–	98	–
Yukon River (Gibbs, 1973, 1977)	1270	6.3%	136	40	115	416	–	–	–
Dead Sea (Nissenbaum, 1974)	638	1.9%	46	23	–	43	13	52	(12)
Los Angeles Hr. (Gupta and Chen, 1975)	493	4.5%	47	–	178	568	332	612	2
Lower Rhine (Patchineelam, 1975)	750	3.7%	167	35	397	376	333	1096	28

Proportion of the Individual Types of Metal Associations

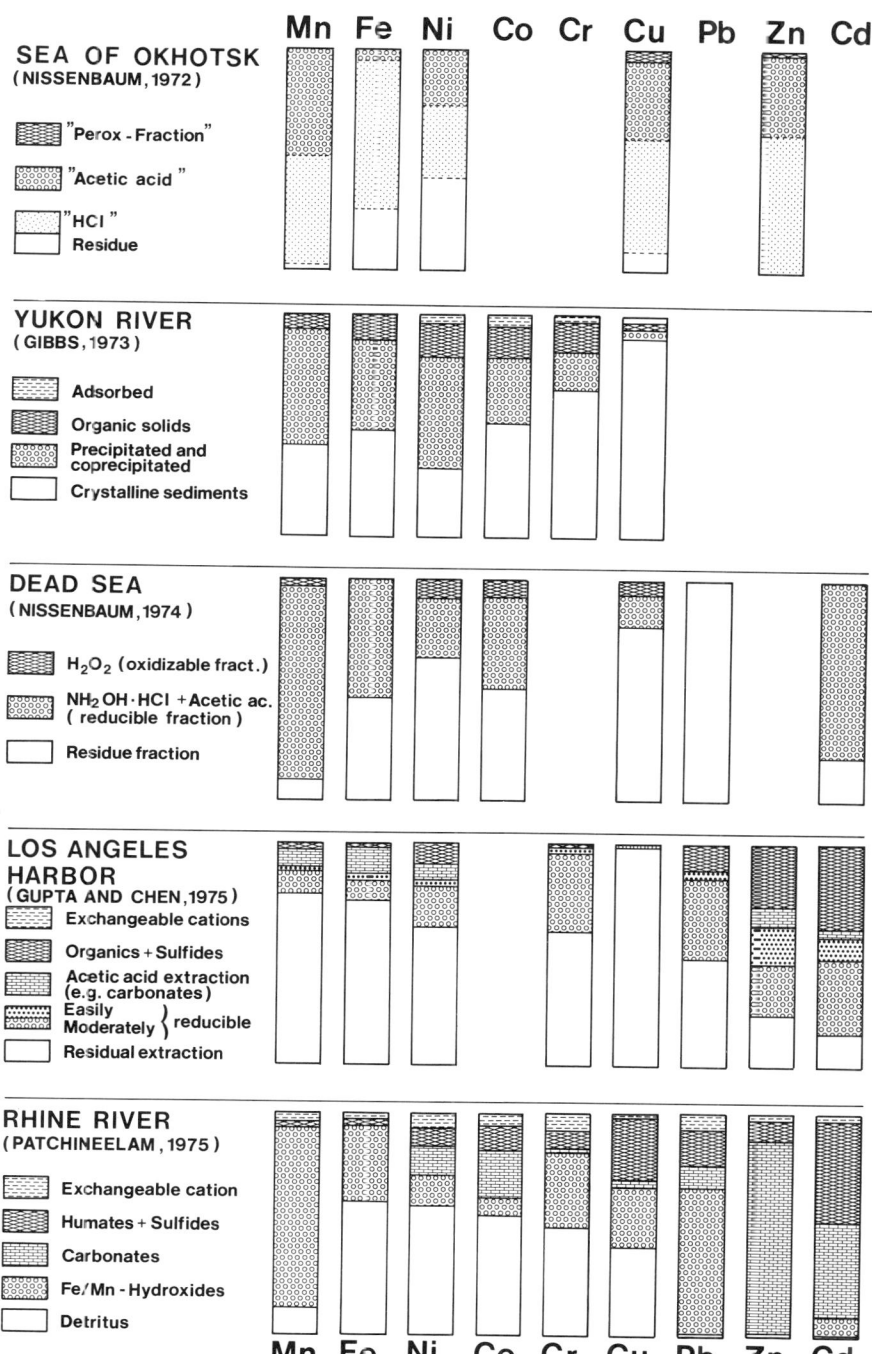

Fig. 65. Chemical associations of heavy metals in sediments from natural and polluted aquatic environments (total metal concentrations are given in Table 69)

Engler et al., (1974) devised a selective extraction procedure to functionally separate interstitial water, exchangeable, easily reducible, moderately reducible, organic and residual fractions from marine and fresh water sediments. The procedure precludes atmospheric oxidation at sensitive steps (during sampling, and during extraction of interstitital water, exchangeable phases and easily reducible phases).

The results of this extraction procedure for nickel, zinc, and cadmium are shown in Table 70 as performed on sediment samples from three harbor areas (Brannon et al., 1976); *Mobile Bay*, Alabama, *Ashtabula* on Lake Erie, and *Bridgeport*, Connecticut. Each of these test locations exceeds the one before in metal pollution.

It is clearly evident that as the metal levels in the sediments rise (usually pointing to a rise in metal pollution), the metal concentrations in the residual fraction decrease. It can therefore be assumed that the additional metals which are carried into the water systems are in relatively instable bonding forms. Further evidence of these effects is given with the last two examples in Figure 65 from *Los Angeles Harbor* and the *lower Rhine River*, both areas being highly affected by civilizational metal inputs.

4.2 Chronological and Grain-Size Variation of Trace Metal Bonding

With regard to the various influencing factors affecting the chemical associations of trace elements in less contaminated as well as in polluted sediments, investigations on sedimentary core profiles and the separate determination of metal contents in single fractions of the grain-size spectrum seem to be particularly useful. Apart from the characteristic developments in the chronological or depositional sequences, there is a major advantage of comparing samples when they have been treated according to the same extraction procedures and analytic methods.

The historical development of the bonding types of a number of heavy metals was studied in six sediment samples taken from different profile depths (1–2 cm, 3–5 cm, 7–8 cm, 14–16 cm, 28–31 cm and 100 cm) in a *core from the German Bight* (southeastern North Sea).

Table 70. Sediment-associations of Ni, Zn, and Cd in harbor sediments (Brannon et al., 1976). Metal concentrations in brackets refer to total particulate content; other values include interstitial water concentrations

	Mobile Bay			Ashtabula			Bridgeport		
Organic carbon	2.0%			2.4%			2.7%		
Inorganic carbon	0.1%			0.6%			2.2%		
Sulfur (ppm)	903			240			2680		
Fraction in percent	Ni	Zn	Cd	Ni	Zn	Cd	Ni	Zn	Cd
Exchangeable cations	1	–	–	1	1	–	1	1	–
Organics and sulfides	7	58	17	10	38	76	26	53	96
Easily reducible	2	4	3	2	5	2	2	13	4
Moderately reducible	–	–	–	23	27	–	3	17	–
Residual fraction	90	38	80	65	29	22	69	17	–
Metal concentration in ppm	(127)	243	3.6	(185)	444	4.1	(182)	1027	(15.2)

Since lead is particularly interesting in the present context, the distribution of its various chemical bonding forms is given in Figure 66, both for the fraction < 2 μm and for the bulk material: these are averaged for two samples from the upper, middle and lower part of the core profile. There are two characteristic findings, which can be considered as contrary to the expected development: (1) the lead concentration of the "residual phases" increases from the bottom to the top of the profiles in both fractionated and bulk samples, and (2) the contents of lead associated with the "detrital minerals" are significantly higher in the < 2 μm samples than in the bulk sediment.

The notable strong increase of lead (and cadmium!) in the detrital fractions, as well as the enrichment of zinc and copper within the *sedimentary profile* taken from the *North Sea*, point toward a potent influence from atmospheric pollutants. The same frequency sequence has been found for the atmospheric metal emissions originating from the burning of fossil fuels (Bertine and Goldberg, 1971; Erlenkeuser et al., 1974). It has been suggested by Suess (1977) that these elements will form a characteristic "coal-residue-assemblage" when they become deposited in the aquatic sediments.

Further insight into the processes of metal accumulation in aquatic sediments could be expected from an analysis of various metal phases within the *grain-size spectrum*:

Two samples from freshwater environments were compared, one from the less contaminated central part of Lake Constance and the other from the lower section of the Rhine River in Germany where a heavy pollution has been registered from sediment studies (see Table 71). In Figure 67 the ratios of copper concentrations are calculated versus the contents of organic carbon (upper part), hydrous iron oxides (middle part), and carbonate minerals (mainly calcite; lower part) within different grain-size classes of both samples.

a) Humate-associated Metal Content Versus Organic Carbon. The contents of copper are strongly enriched in the colloidal and fine-grained solid organic substances, particularly in the sediments from the lower Rhine. This development suggests that the accumulation of copper is partially due to organo-metallic transformations taking place in the aquatic system. Similar grain-size dependencies can be found for lead and iron in the Rhine sample, whereas there is no distinct development of the humate-associated contents of iron with the grain-size spectrum in Lake Constance samples. In the latter materials manganese concentrations increase slightly in the coarser-grained organic substances. It is obvious that the zinc accumulation is confined to detrital organic materials e.g., plant debris.

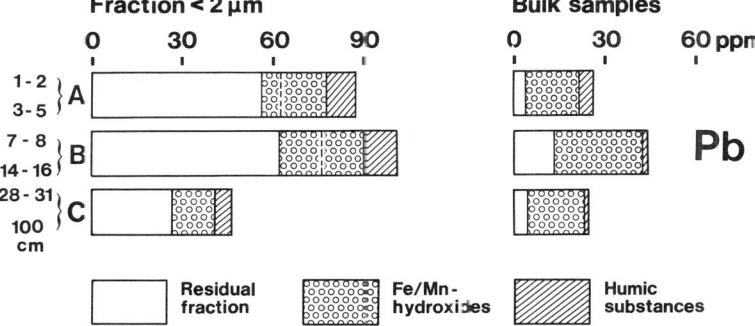

Fig. 66. Chronological variations of sedimentary phases of lead in a core from the German Bight (southeastern North Sea)

Table 71. Phase concentrations factors (PCF) indicating the relative enrichment (or reduction) of metal concentrations in major carrier substances (percent metal phase of total sediment/percent carrier) of clay- and silt-sized sediment particles from the Rhine River

	Organic residues		Inorganic residues		Moderately reducible fraction		Easily reducible fraction		Carbonate fraction		Humic acid fraction	
	Clay	Silt	Clay	Silt	Clay	Silt	Clay	Silt	Clay	Silt	Clay	Silt
Fe	<0.01	0.01	0.5	0.7	–	–	–	–	2	<1	40	30
Mn	0.2	0.2	0.4	0.5	2	2	30	30	10	4	6	4
Cr	3	1	0.5	0.7	16	22	5	7	0	0	3	3
Cu	0.5	0.3	0.1	0.1	8	10	6	4	15	5	50	35
Pb	0.03	0.05	0.2	0.2	8	15	3	5	<1	<1	100	100
Zn	0.1	0.2	<0.1	<0.1	4	7	55	55	25	8	8	17

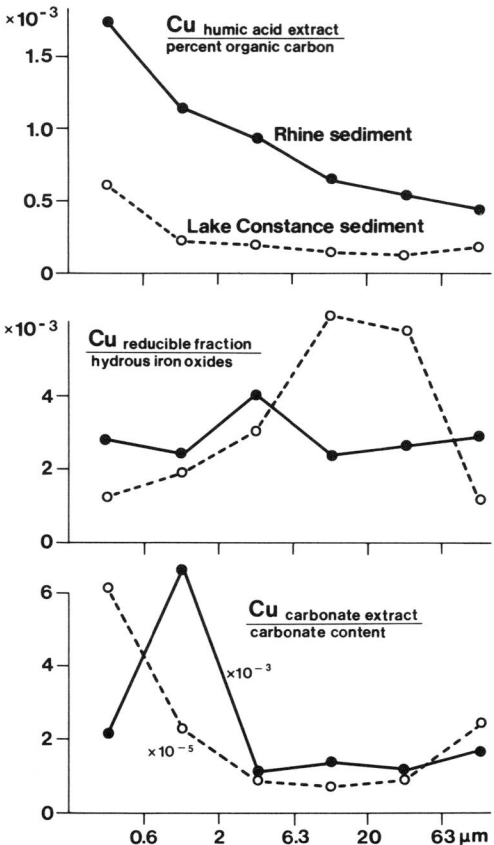

Fig. 67. Grain-size variations of sedimentary phases of copper in samples from the Lake Constance and lower Rhine River

b) Transition Metals in Hydrous Iron Oxides. The comparison between the copper concentrations in hydrous Fe/Mn oxides and the percentages of hydrous Fe-Oxides (given as the reducible fraction of Fe) indicates a distinct increase in the Cu/Fe-oxide ratios in the medium-to-coarse silt fractions of the sample from Lake Constance. A similar development has been observed by Gibbs (1977) for the contents of manganese, nickel, cobalt, chromium, and copper in the suspended solids from the Amazon and Yukon Rivers. It was suggested that these effects are related to the differences in the thickness of the iron-hydroxide layers, as coating on mineral grains. Thick layers on coarse material are considered, among other factors, to be a result of the weathering environment; the coarse material in the soil would have a higher permeability, bringing about a greater supply of the precipitating (and coprecipitated) ions. The contrasting development for the manganese concentrations in Lake Constance, where the Mn/Fe-oxide ratios are highest in the fine-grained fractions, can be explained as a result of diagenetic effects. The Mn contents, which are mobilized under anaerobic conditions and transported by upward diffusion to the sediment/water interface, will there be adsorbed preferentially on fine-grained substances, such as hydrous Fe and Mn oxides.

c) Carbonate Associated Metal Concentrations Versus Carbonate Content. The grain-size dependency curves of the metal concentration within the carbonate phases indicate a characteristic enrichment in the fine-grained fractions for all metals exhibiting carbonate associations—i.e., zinc, manganese, iron, and copper, but not chromium. This effect is particularly pronounced for zinc; it may be concluded that coprecipitation with carbonates is an important mechanism for the distribution of this element in Rhine sediment. Under anoxic conditions precipitation of manganese carbonate seems to be the limiting factor for the Mn concentrations in interstitital solutions in Lake Constance sediments.

In Table 71 (from Förstner and Patchineelam, in prep.) the *phase concentration factors* (PCF) are calculated for the clay- and silt-sized particles of the Rhine sample as the percentage of a certain metal association against the percentual content of the respective carrier material. If the computed values lies above 1, an enrichment has occurred in the studied phase ("phase" is defined here as one of the major sediment fractions, e.g., Fe/Mn-hydroxides, humic acids, carbonates); if the value is less than 1 (as for all metals in the residual phase except for chromium) the metal concentration in the sample has been reduced by the presence of that phase. The PCF is a relative value that is not influenced by the total content and is especially suited for comparison of samples with large differences in the total metal concentrations. PCF values of Cr, Cu, Pb, and Zn in the *moderately reducible* sediment fraction are found to be significantly higher for the silt-sized particles than for the fine-grained substances. Similar effects occur for chromium and lead in the *easily reducible* sediment fractions. Carbonate-associations of heavy metals such as manganese, copper, and zinc are more concentrated in the fine-grained sediment particles than in silt-sized materials. In the *0.1 N NaOH extractable fraction* of the Rhine sediment, PCF-values are particularly high for iron, copper, and lead, and exhibit a slight increase in the fine-grained particles compared with the silt-sized fractions.

This strong enrichment in the sodium hydroxide extract led to the consideration that there might be other extractable metal phases in addition to humic and fulvic acids, for example, phosphates (Förstner and Patchineelam, in prep.). It has been shown by Stumm and Morgan (1970) that for metals such as manganese, zinc, and copper, complex formation reactions of inorganic phosphates might significantly affect distribution of the metal ion, the phosphates, or both; Williams et al. (1971) found that the phosphates associated with iron and aluminum were extracted up to 90% with sodium hydroxide. Data from the grain-size fractionated Rhine sediment indicate a close relationship between the 0.1 N NaOH extractable concentration of Pb, Fe, Cu, and the total P contents; however, Zn and Mn followed the same trend as the humic acid extracts in the different grain-size intervals. It may therefore be inferred that 0.1 N NaOH extraction affects phosphate phases in the sediment, in particular those of iron, lead, and copper.

From the examples presented here, some general conclusions may be drawn, which might be of interest for further discussions of the interactions between solid and aqueous phases in aquatic systems:

1. A high percentage of metals in inert associations with the *crystal lattices* may reflect both *lithogenic influences* from the drainage areas—as in the case of iron and chromium in Lake Constance sediments—and the effects of atmospheric emissions, as suggested for cadmium and lead in the example from the North Sea sedimentary core.

2. Formation of *organo-metal complexes* and their successive adsorption, flocculation, and precipitation on various substrates (organic substances, hydrous Fe/Mn-oxides, clay and carbonate minerals) seem to be important processes controlling the distribution of *copper* and other metals, mostly in the smaller grain-size fractions. The possibility that phases other than humic acids are extracted with sodium hydroxide treatment should, however, be considered.

3. Apart from organic substances, especially those in less condensed forms, hydrous Fe/Mn oxides are the major accumulation phases for heavy metals in aerated waters. Investigations on the chemical associations of heavy metals in lacustrine sediment from different climatic zones indicate that the enrichment factors for Mn in the hydroxide phase is approximately 50–70, for Cu 10 to 80, Ni 20 to 60, Cr 10–40, and Co 15–25; zinc was relatively poorly enriched in the hydrous oxide phase (Schmoll and Förstner, 1979).

4. *Diagenetic redistribution* processes in particular affect the concentrations of *manganese*. These metal contents are predominantly associated with nonlattice phases. The concentrations of manganese in the soluble phase seem to be controlled either by carbonate precipitation (anoxic conditions), coprecipitated with hydrous iron oxides (oxic conditions) or adsorbed onto hydrous Fe/Mn-oxides.

5. The process of *coprecipitation with carbonate minerals* is considered to be the limiting factor for the concentration of zinc in the Rhine and in Lake Constance. A strong enrichment of zinc in the fine-grained carbonate precipitates suggests that this mechanism might be important for the immobilization of metal inputs, particularly of zinc and cadmium, from human activities.

6. The effect of *cation exchange* is relatively variable. In the sediments of Lake Constance, 19%–36% of the copper content and 6%–16% of manganese in the various fractions occur in exchangeable positions, whereas iron, zinc, and chromium indicate practically no cation exchange fractions. The North Sea sediments exhibit highest exchangeable phases for cadmium, 11%–36%, followed by copper, 7%–12%; zinc, manganese, and lead content usually is found to exist with less that 3% in exchangeable form. In a sediment sample from the lower Rhine, between 25% and 35% of the concentrations of Cu and Mn are affected by cation exchange processes with $BaCl_2$-triethanolamine, whereas Cr, Zn, Fe, and Pb are exchangeable only to a minor extent (0.4%).

7. Future research on metal partitioning in sediments should be focussed to anoxic environments; no extraction procedure is available at present to differentiate *organic and sulfidic metal associations*.

5 Mobilization of Heavy Metals from Sediments

A remobilization of heavy metals from suspended material and sediments is potentially hazardous not only for the aquatic eco-system, but also for the drinking water supply. Remobilization is mainly caused by four types of chemical changes in waters:

    1. Elevated salt concentrations, whereby the alkali and alkaline earth cations can compete with the metal ions sorbed onto solid particles.

    2. Changes in the redox conditions, usually in conjunction with a decrease in the oxygen potential due to advanced eutrophication. Iron and manganese hydroxides are partly or completely dissolved, part of the incorporated or sorbed heavy metal load being released.

    3. Lowering of pH which leads to a dissolution of carbonates and hydroxides, as well as to increased desorption of metal cations due to competition with H^+ ions.

    4. Increased use of natural and synthetic complexing agents, which can form soluble metal complexes sometimes of high stability with heavy metals that are otherwise adsorbed to solid particles.

In addition to these four processes, there are other biochemical transformation processes, by means of which the heavy metals are either transferred from the sediment to animal or plant organisms—possibly to be further enriched along the food chain—or are discharged directly or via decomposition products into the water.

5.1 Saltwater/Sediment Interactions

The problem of a desorption of heavy metals from solid material—due to competing cations—has become increasingly more important for inland waters during the last decades. In some rivers and lakes (examples have been noted in Chap. C) the contents of sodium (and of chloride, sulfate, calcium, etc.) have been significantly enriched, mainly as a result of saline effluents. Furthermore, the extensive use of deicing salts has been found to cause considerable changes in the solubility of metal compounds, for example those of mercury (Feick et al., 1972).

One of the most important examples of salt water/sediment interaction can be found in the mixing zone river/sea i.e., the *estuarine environment* (see brief discussion in Sect. D.9). There it is suggested that the distribution of heavy metals in particulate matter and in sediments is dominated by mixing processes between river-induced materials (more or less polluted) and sea-derived (fairly uncontaminated) particulates (Förstner and Müller, 1974a). In addition to this presumption, one must suppose that still other processes are at work which may cause remobilization of metals from particulates by desorption or by dissolution.

The processes chiefly influencing this distribution of dissolved metals are sorption/desorption, precipitation/solubilization, coagulation, flocculation, and complexation. There are, however, processes not only involving elevated salt concentrations but also, and often preferentially, those involving effects of oxidation and reduction, particularly of iron and manganese compounds, and reactions with organic matter, since "estuaries display high levels of biologic activity" (Goldberg, 1975b). Altogether, a very complex system of interreactions exists that is still not completely understood, especially with respect to the behavior of heavy metals.

5.1.1 Desorption Experiments

Although many authors have pointed out that complex chemical environments such as estuaries "cannot be simulated by simple in vitro experiments involving homogenous reaction kinetics" (Turekian, 1977), at distinct phases of the reactions the release of trace metals from river-borne suspended sediments at the seawater interface is possible.

1. Experiments conducted by Kharkar et al. (1968) on the adsorption of cobalt, silver, selenium, chromium, and molybdenum on various mineral phases in distilled water, and on *desorption of adsorbed metals in seawater* show that where a trace element is adsorbed from freshwater solution it is always released – to a greater or lesser degree – upon contact with seawater due to displacement by magnesium and sodium ions. The cobalt released by this mechanism equals about twice the dissolved load, whereas silver and selenium contribute only an additional 10% – and chromium and molybdenum none. However, the authors point out that they cannot draw conclusions with regard to the properties of particles that have been subsequently modified as a result of marine biologic activity or other secondary processes. In this respect, Sholkovitz (1976) suggested that the use of standard clays and distilled water in these experiments restricts the conclusions drawn from artificial conditions and that the absence of dissolved and solid organic matter should be taken into consideration.

2. Patchineelam and Calmano (in prep.) have undertaken several series of experiments on suspended sediment from the heavily polluted Rhine River (250 mg/l and 1000 mg/l) in artificial seawater. Of the investigated elements (zinc, cobalt, chromium, and iron) none released more than 10% of the original metal concentration of the sediment.

In the case of zinc and cobalt, the remobilized metal concentrations were significantly higher than the respective metal fraction, which was found to be in exchangeable position. Therefore, mere cation exchange appears to be unlikely for the remobilization of these two metals. More probably, *redissolution of carbonate minerals* is partially responsible for the increased contents of zinc and cobalt in the aqueous phase after treatment, since a considerable portion of the two metals is associated with the carbonate fraction of the investigated sediment sample.

3. Van Der Weijden et al. (1977) resuspended aliquots of suspended matter collected from the Rhine River in distilled water, in diluted artificial seawater (1:1), in artificial seawater, and in "nitrate seawater" respectively, at pH's of 7.5 and 8.0. The desorption of heavy metals into these solutions was calculated in relation to the concentrations as determined by extraction in 4 M hydrochloric acid. These comparisons demonstrated that *complex formation* (chloride and sulfate complexes in diluted and normal seawater) is important in desorption processes during estuarine mixing. The order of decreasing desorption of metals into 1:1 diluted seawater and normal seawater in the experiments is: Cd > Zn > Mn > Ni > Co > Cu > Cr; for Fe and Pb no desorption was found.

4. An experimental study was performed by Rohatgi and Chen (1975) on the effects of seawater on the concentrations of trace metals in *waste water particulates*. Samples of primary effluents, digested sludge, and mixtures of the primary and secondary effluents were obtained from the Hyperion Treatment Plant of Los Angeles, California. Dry weather flow from the Los Angeles River was also selected because of the past practice of discharging industrial wastes into the river. Table 72 summarizes some of the data obtained by Rohatgi and Chen (1975): in most cases, Cd, Cu, Ni, Pb, and Zn were found to be released to a greater extent than the other metals. Chromium, iron, and manganese were not released except in the case of digested sludge, in which 9% of Cr and 36% of Mn was released from solid phases. However, up to 96% of cadmium was released from suspended particulates. At different dilution rates of waste water effluents and digested sludge with seawater, no significant difference in the release of cadmium was observed. On the other hand, the dilution ratio seems to play an important role in the release of nickel and zinc from suspended particulates.

Generally, two processes seem to be effective in the release of trace metals at the mixing of waste particulates with seawater (Rohatgi and Chen, 1975): (1) *Oxidation* either *of organic particulates* containing trace metals, or *oxidation of metal sulfides* and the

Table 72. Release of trace metals from contaminated particles after contact with seawater (experiments by Rohatgi and Chen, 1975)

	Primary/secondary effluent dilution with seawater 1:5		Los Angeles River, dry weather flow dilution with seawater 1:2	
	initial conc.	% release 4 wks	initial conc.	% release 4 wks
Iron	0.50%	0%	1.22%	0%
Manganese	51 ppm	0%	498 ppm	0%
Chromium	3,296 ppm	0%	452 ppm	0%
Zinc	1,481 ppm	38%	905 ppm	60%
Lead	140 ppm	58%	516 ppm	17%
Nickel	896 ppm	63%	407 ppm	72%
Copper	2,182 ppm	69%	1,063 ppm	66%
Cadmium	87 ppm	93%	–	–

surface desorption of trace metals caused by a high dilution ratio, and (2) *complexation* of trace metals to form soluble complexes of *inorganic* ligands such as Cl^-, and *organic ligands,* possibly resulting from the oxidation of organic particulates.

The dynamic nature of these processes has been described by Ganapathy et al. (1968). In experiments on the uptake of radionuclides by repeated contact of the sediment with fresh quantities of equilibrated seawater, it becomes clear that the sorption/desorption curves for Mn-54, Co-60, and Zn-65 will register a hump followed by a gradual levelling off. This time sequence of early enrichment and later decrease of sediment-associated metal contents is explained by a process of desorption of a part of the initial deposit; the mobilization products are suggested to be deposited by a secondary process in the form of colloidal hydroxides.

5.1.2 Estuary-Sediment Boundary

The fate of heavy metals in estuaries is controlled by processes that take place at two interfaces (Windom, 1975): (a) Those occurring at the *river-estuary boundary* – involving different salt concentrations, suspended material, pH values—control the form and rate of metal input. Examples from experimental investigations in this transition zone are given above. (b) At the second interface, the *estuary/sediment boundary*, metals are lost from the system because of precipitation and accumulation; and conversely, previously accumulated metals are released by chemical reaction and biologic activity.

The above data indicate the particular importance of the processes involved in the removal of nonconservative dissolved metals during estuarine mixing, the most important of which are the precipitation of iron and the coagulation of dissolved organic matter. Should these precipitation products with hydrous Fe/Mn oxides and very fine suspended fractions, consisting of colloidal iron and manganese oxides, enter an environment of reducing conditions, then the prerequisite is given for a remobilization of metals.

The results of isotope studies of Lowman et al. (1966) suggest that the increased coagulation of organic substances may lead to higher oxygen consumption in the bottom sediment which could cause the re-solution of Fe and Mn oxides or the formation of sulfides. Investigations carried out by Carpenter et al. (1975) in the northern Chesapeake Bay indicate that a substantial release of manganese from the sediments takes place during the summer when the concentrations of H^+ and dissolved oxygen in the overlying water are low. The release of zinc from recently deposited sediments was observed in late spring—most of the soluble zinc is complexed during late summer.

In a summary of recent work on trace element distribution in the river-sea mixing zone, Turekian (1977) has noted that in an estuarine system there is continuous movement of some metals in and out of solution, but little is actually lost from the system. It has been shown, according to the data of Evans et al. (1977) for the contents of manganese in Newport Estuary in North Carolina and of Benninger et al. (1975) for the distribution of natural Pb-210 in Long Island Sound, that the dominant cycle of most heavy metals is internal, and whatever will escape from the estuary into the open sea will mainly be (except for elements like uranium and molybdenum) in the form of fine particles.

5.2 Redox Changes and Metal Release

During the past few years, oxygen depletion, usually a result of the process known as "eutrophication", has been observed in a great number of water bodies, particularly in lakes and dams. A rapid spreading of organic substances, mostly of algae, is caused by an increase in nutritious substances, such as nitrogen and phosphorous compounds from fertilizers, detergents, and faeces.

The degradation of these substances requires oxygen. The sediment acts, therefore, as a sink for this gas which is supplied through the sediment surfaces at a rate governed by three factors (Mortimer, 1971): (1) a biologic oxygen demand arising from respiration and metabolic activities in the sediment; (2) a chemical oxygen demand arising from the fact that inorganic elements such as Fe^{2+} released to the sediment from decomposing biologic matter accumulate in reduced form; (3) diffusion, which regulates transport.

The medium of metal transport is the interstitial water; major components which affect the rate of transported metals are organic matter, iron, and manganese, which are available for redox processes. As to the latter substances, it was found that manganese oxides of higher valence precipitate in the oxidized superficial sediment and are reduced and mobilized in the sediment at a higher oxygen potential than are the ferric complexes. A consequence of this difference is that when a lake enters a *reducing phase in its history* or undergoes the *seasonal onset of hypolimnic reduction,* manganese will be mobilized before iron and, conversely, will persist in the water longer than iron after oxygen is reintroduced at the turnover. An example of the processes involved in the *redistribution of Fe and Mn in recent sediments* is given by Tessenow and Baynes (1975) from the Feldsee, a mountain lake in the Black Forest (West Germany). Figure 68 shows the development of the Fe and Mn concentrations in the sediment and in the corresponding pore water in relation to the redox conditions. At a sediment depth of approx. 20 cm the redox values decrease sharply. At the same time, a strong enrichment of iron and manganese in the sediments can be observed between 15 and 10 cm sediment depth. According to Tessenow and Baynes (1975) the gradient of redox potential would be expected to effect an upward diffusion, which together with the reductive solubilization processes in the lower sediment layers and the oxidative precipitation around the root systems of *Isoetes lacustris* L., causes a permanent deposition

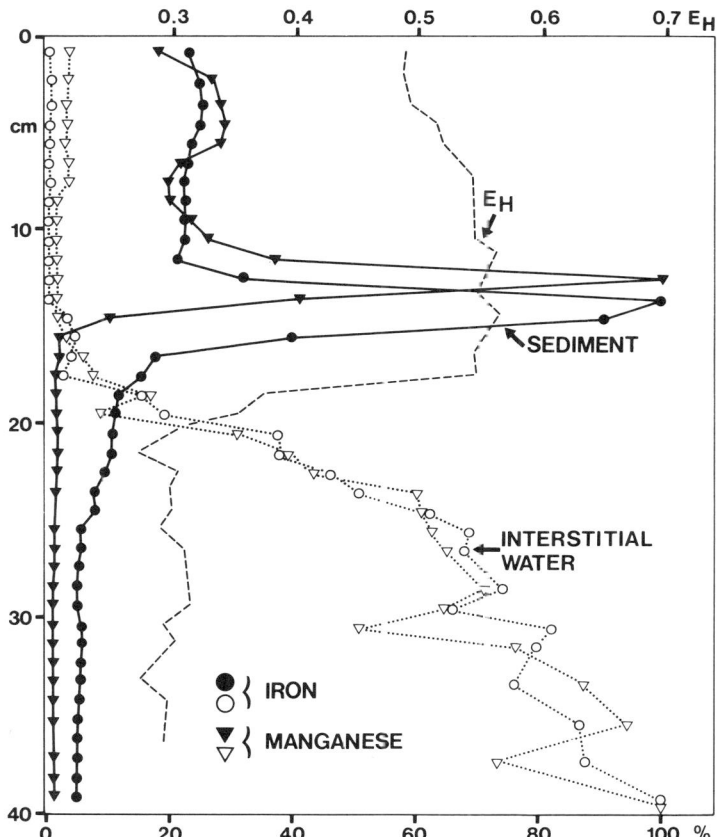

Fig. 68. Redistribution of iron and manganese in recent sediments from the Feldsee (West Germany) as functions of Eh and pH conditions (Tessenow and Baynes, 1975, Naturwissenschaften 62, p. 342)

of the mobile Fe and Mn compounds in the lowermost horizon of the oxidized sediment. With further surface sedimentation the accumulation horizons would become upwardly displaced, thus remaining at the same relative depth in the sediment.

The potential health hazard arising from advanced eutrophication and subsequent dissolution of oxides has been emphasized by Edgington and Callender (1970) in a study of the minor element geochemistry of the Lake Michigan ferro-manganese nodules. These nodules contain unexpectedly large concentrations of arsenic, up to 345 ppm with an average of 180 ppm. Since ferro-manganese concentrations and nodules are stable under aerobic conditions only, further eutrophication would significantly raise the level of arsenic concentrations in the water. According to the calculations of Edgington and Callender (1970) the arsenic release from the total dissolution of the Green Bay nodules in Lake Michigan would increase the content of arsenic, which is highly toxic to mammals and also carcinogenic, up to 90 μg/l—about twice the permissible value in drinking water. Dissolution of nodules has also been observed in the Kingston Basin of Lake Erie, indicating recent changes in lake conditions in the area (Sly and Thomas, 1974).

5.2.1 Chemical Factors Affecting Metal Distribution in Interstitial Water

Metal ions derived from the overlying water or from upward migrating pore solutions are sorbed or coprecipitated onto Mn- and Fe-oxide films or layers near the sediment surface. With continuing sedimentation the surface oxides are buried beneath the oxidized/reduced boundary where they become unstable (Duchart et al., 1973). When a reduction of the hydrous Fe- and Mn-oxides occurs, the heavy metals, which are coprecipitated with or sorbed onto these components should be released (Morgan and Stumm, 1964). It is to be expected, however, that when traces of *sulfide* ions are present in the pore waters, the mobility of most of these metals is significantly lowered. On the basis of the solubility product constants for their sulfide compounds, elements such as Pb, Co, Ni, Hg, Ag, Cu, and Zn remain essentially fixed in reducing sediments of the pore water with a sulfide ion concentration of 10^{-9} mol per liter (Thomson et al., 1975).

A sulfide precipitation of this type and subsequent sorption of heavy metals by sediments, however, appears to contradict all observations made on metal content in pore-water solutions (e.g., Presley et al., 1967, 1972; Bonatti et al., 1971).

Elderfield and Hepworth (1975) studied the pore-water chemistry of estuarine sediments of the river Tees (N.E. England) and the Conway River (N. Wales). They found that the concentrations of most of the metals in the pore fluids were significantly higher than could be predicted from the sulfide solubility. This effect is most striking for copper, lead, and zinc which diverge by factors of 10^{21}, 10^{12} and 4×10^8 respectively from the computed figures (Table 73). A comparison of metal concentrations in the pore water with the overlying water of the Conway Estuary (Elderfield et al., 1971, and the unpublished data of Elderfield) indicates a characteristic increase of manganese and cobalt by factors of 1000 and 10,000 respectively, and a 50- to 100-fold increase of iron, lead, copper, and nickel in the pore fluid of the sediments from that particular estuarine area (Table 73).

Table 73. Comparison of theoretical sulfide-limiting solubility ([a]) for sulfide ion activity of 10^{-10} M) of heavy metals with actual concentrations in pore waters of organic-rich sediments of the Conway Estuary (from Elderfield and Hepworth, 1975; data partly from Elderfield et al., 1971)

| Metal | Concentration (−log M) | | Max. Enrichment Factor | Concentration (µg/l) | | Factor of Enrichment |
	Calc.[a]	Conway Estuary pore water		Conway Estuarine water	Pore water	
Copper	26.1	5.11	10^{21}	3.4	210	62
Lead	17.5	5.52	10^{12}	6	300	50
Zinc	13.8	5.58	4×10^8	56	260	5
Cobalt	10.4	5.31	1.3×10^5	0.05	490	10,000
Nickel	8.5	5.31	1.5×10^3	5	490	100
Iron	7.2	5.28	83	10	520	52
Manganese	2.6	2.33	1.9	2	4,700	2,350

Brooks et al. (1968) determined that none of the metal contents found in pore waters of sediment cores from the coast of Southern California could exist in solution in the amount measured if they were bound as simple sulfides: the authors thus concluded "that it is more probable that complexes are formed which solubilize the metals even in the presence of sulfide." Hallberg (1973, 1974a, b) has expressed the opinion that both precipitation processes and the influences of *organic chelators* are the major counteracting factors controlling metal distribution in interstitital solutions. Presley et al. (1972) ascribe the mobilization of Zn, Ni, and Cu from reducing sediments to the reconstitution of organic complexes and the stabilization of metals initially leached from silicates and oxides. In addition, bacterial action may lead to a further decrease in pH-values, thereby increasing the solubility of various metals.

In laboratory experiments Cline and Upchurch (1973) found that upward migration of heavy metals in pore solutions may occur because of dewatering as a result of compaction and unidirectional ion migration, but is largely attributable to *bacterial activity*. Copper, for example, was complexed to functionally active organic molecules and then, when chemi-specific bacteria metabolized the organics, the metal was released. The decrease of Eh and pH, also due to this bacterial action, is a further factor in the release of the complexed metals. The heavy metal may then be transported upward either on bubble interfaces, in gaseous complexes or as soluble organic complexes. When the metal reaches the biologically active portion of the sediment, it is immobilized by adsorption or as inorganic precipitate.

Experimental data of Lu and Chen (1977) on the migration of trace metals between the interface of seawater and polluted surface sediment suggest that under *reducing conditions* the concentrations of trace metals in the interfacial waters are controlled by sulfide complexes for Cd, Hg, and Pb, by organic complexes for Fe and Ni, chloride complexes for Mn, and by hydroxide complexes for Cr. The high values measured for dissolved copper in comparison with the calculated values may be explained by slow precipitation or nucleation, or may be due to the presence of humic-copper complexes. The association with *humic substances* seems to be the dominant species for soluble zinc and iron under reducing conditions. Under *oxidizing conditions* the controlling solids may change gradually from metallic sulfides to carbonate hydroxides, oxyhydroxides, oxides, or silicates, thus changing the solubility of trace metals. Discrepancies have been found between experimental data and calculated equilibrium concentrations: Cd, Cu, Ni, and Pb are far below, Fe and Mn far above the equilibrium data. Possible explanations are the scavenging effects of Fe/Mn-hydroxides or clay minerals, the low oxidation rates of the reducing solids, and the formation of humic complexes (Lu and Chen, 1977).

Nissenbaum and Swaine (1976) analyzed the metals bound to dissolved organic polymer from interstitial water of a reducing sediment (Saanich Inlet, B.C.). The results show that all of the Zn, almost all of the Cu and a major part of the Fe, Ni, and Co, but only small portions of the Ca and Mn are bound to humic type material. With the exception of Fe, Ni, and Co, which occur mostly as sulfides, the elements concentrated in the interstitial solutions are those which are also enriched in sedimentary humates. One can therefore conclude that the large vertical gradient of dissolved humics is particularly important in controlling the transport of dissolved trace metals in the pore waters (Krom and Sholkovitz, 1977).

5.2.2 Physical Processes Affecting Metal Release from Pore Water

In the high porosity medium of the upper part of the freshly deposited solid particles, diffusion of dissolved materials readily occurs both within the sediment column and across the sediment water interface. Wood (1975) and Petr (1977) suggested that of several mechanisms responsible for the movement of material across the mud-water interface in shallow water deposits, ordinary diffusion is probably the least important, while turbulent exchange, release of gases from microbial activity, and the movement of certain biota are of considerable importance. According to Berner (1970) molecular diffusion gains importance in deep waters, where the other mechanisms are of low intensity. It has been shown by many investigations in both marine and freshwater environments (Duursma and Rosch, 1970; Lerman and Weiler, 1970; Lerman, 1971; Tzur, 1971) that exchange between overlying water and the sediments leads to the establishment of characteristic gradients in dissolved species in the upper part (approx. 30 cm) of the sediment columns (Bricker and Troup, 1975).

Among the physical processes favorable for the transfer of trace elements from the pore solutions into the open water, three mechanisms should be discussed here in more detail: (1) the whirling-up of settled sediments as a result of hydraulic phenomena particularly effective in high energy environments such as rivers and marine tidal areas, but which can also be caused by episodic storm-events in the shallow parts of lakes and seas; (2) the disturbances of surface sediments by burrowing benthic organisms (bioturbation) and by gas bubbles from the decay of organic matter; and (3) the dredging and associated activities in rivers, lakes, and coastal marine areas. Especially the latter activity affecting natural sedimentation relationships has led to intense discussions, not the least of which concerns the question of the liberation of toxic metals.

Hydraulic Phenomena. Information on the physical influences on the redistribution of trace metals in coastal marine enrivonments, e.g., from tidal currents and storm-generated waves, is relatively scarce. From studies on the Chesapeake Bay, Bricker and Troup (1975) suggest that scouring of sediments by tidal or storm-induced current leads to direct mixing and exchange of water across the sediment/water interface, involving chemical reactions between iron, phosphate, and trace metals. Studies by Thomson et al. (1975) from sediments in Long Island Sound indicate that although manganese is remobilized in the reducing portion of the sediment column and is subject to transport to the sediment-water interface, little of it escapes the sediment column because of oxidation and precipitation as an oxide. Results of Lindberg et al. (1975) from experiments on large-scale resuspension of estuarine sediments from Mobile Bay, Ala., indicate a sizeable short-term release of mercury into the surrounding water followed by a decrease to levels close to those predicted by ideal dilution calculations. In the case of organic-rich marsh sediments, both a stronger release and a greater persistence of the mercury concentrations were observed.

The release of toxic substances by whirling-up and aeration of river sediments has been investigated by Müller and Schleichert (1977) in several experimental studies, which had been prompted by problems arising from sludge deposits in navigable waterways and, in particular, by an extensive fishkill occurring in the Middle and Lower Rhine. It has been suggested that unfavorable conditions—rapid increase of flow in industrial agglomeration areas, floods during summer at elevated temperature—may release large quantities of oxygen-consuming substances if the sediments are rich in organic matter.

Time-dependent experiments indicate further details of the behavior of heavy metals during these processes. A characteristic increase in the quantity of the dissolved heavy metals zinc, lead, copper, and chromium and a simultaneous decrease in dissolved organic carbon within the first five days of the experiments confirm the release of organically bound heavy metals (except of mercury and cadmium). After twenty days of aeration, a reduction was observed in dissolved heavy metals, again with the exception of cadmium and mercury. It seems that the second phase—the 5th to the 20th day—is governed by the processes of precipitation, mainly in the form of carbonate precipitates; for some metals, such as chromium, coprecipitation with Fe-oxyhydrate must be taken into account. After an additional twenty days of aeration, the heavy metal contents, with the exception of mercury and chromium, increased again, in some cases greatly. For example, the zinc contents increased from approximately 50 μg/l in phase 2, to 4,660 μg/l. This behavior has been explained by a redissolution of the previously formed carbonate precipitates by organically produced CO_2.

Mobilization of heavy metals from sediments in waters polluted with organic substances has been studied in the Neckar River, a tributary of the Rhine River (Reinhard and Förstner, 1976). Figure 69 shows the zinc levels in a lock reservoir at Besigheim. The highest metal enrichments in each case were found in the center of the approximately 1.50 m long profile and not—as would be expected according to the increasing metal contamination—at the sediment surface. Most of the investigated metal values in the pore water show an increase by factors of 20 to 100 in this section of the river, compared to the corresponding normal levels in the water itself. In the near-surface sediment layers, however, the metal contents of the pore solutions do not differ greatly

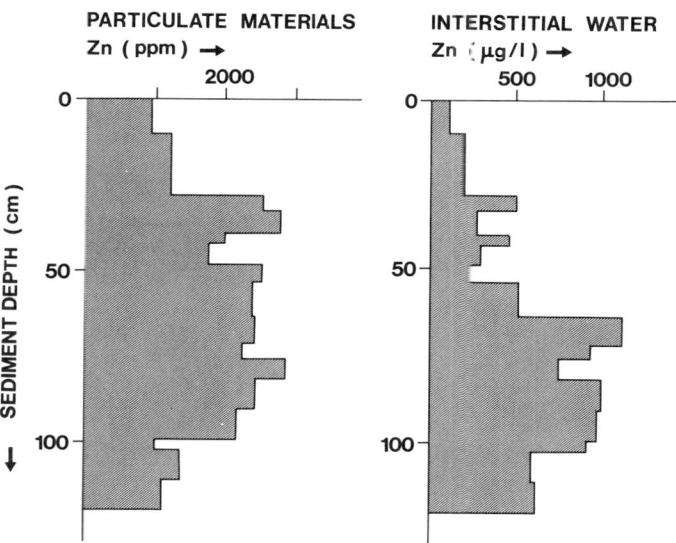

Fig. 69. Concentrations of zinc in sediments and pore waters of a core from the Besigheim lock reservoir of the Neckar River (Federal Republic of Germany) (Reinhard and Förstner, 1976, N. Jb. Geol. Paläontol., p. 307, p. 311)

from the metal concentrations in the surface water. The development of such a concentration gradient can only be explained by an increased metal dissolution in the deeper, reducing sediment parts, probably together with complexing processes due to organic decomposition products and to subsequent exchange at the sediment/water interface. A possible migration mechanism could be an upwards migration in the compaction stream; however, the elevated metal concentrations of the pore solutions can also be released by a redistribution of surface sediments in the flowing water.

Bioturbation. The effects of bioturbation—depending on body size, activity and depth penetration of the animal, as well as on consistency, water content and nutrient concentration of the sediment—are (Petr, 1977): (1) pumping the enriched interstitial water out of the sediment, (2) bringing-in water mostly richer in oxygen, (3) active transport of particulate material to the surface and into deeper layers, and (4) discarding faecal pellets onto the surface. There is a particular important influence of bioturbation on the oxygen conditions in the upper 5–15 cm of the sediments (Schumacher, 1963, Davis, 1974), which in turn—partly via bacterial interactions—affects the concentrations of CO_2, CH_4, NH_4^+, PO_4^{3-}, and H_2S in the pore waters. One major function of bioturbation is therefore the release of the dissolved compounds to the overlying water.

Here too, little is at present known about the quantitative effects and especially the forms and degree of metal enrichment in various organisms. It can be assumed that the processes of bioturbation lead primarily to an acceleration of transformations, the transport of dissolved or complexed metals by means of the gas bubbles that are created possibly playing a role as well (Reeburgh, 1969; Martens and Berner, 1974; Hammond et al., 1975).

Dredging Activities. Bibliographies on environmental aspects of dredging and related activites have been summarized by Cable (1969), Boyd et al., (1972), Windom (1972, 1973). May (1973, 1974), Lee and Plumb (1974), and Oosterbaan (1974). Among the major purposes of dredging activities the following are noted: (1) the maintainance, improvement and extension of navigable waterways and of harbor facilities, (2) the extraction of resource materials for construction and for creation of additional land, and (3) the discharge of undesirable, e.g., contaminated materials from the aqueous environment.

At present the material dredged from waterways and harbors of the United States amounts approx. 250 million m^3 annually (Windom, 1976). The dredging work from the Great Lakes of North America alone, in the period from 1966 to 1972, averaged approximately 9.5 million m^3, 90% of which was undertaken primarily in the interests of commercial navigation. Although the different types of dredging plant are difficult to compare, their practical capacities lie between approx. 120 m^3 per hour for the clamshell dredge and up to 3000 m^3 per hour for the modern cutter-suction dredges [International Working Group on the Abatement and Control of Pollution from Dredging Activities (Anon., 1975c)]. In view of these figures, the following question has been increasingly posed (here formulated by Sly, 1977): Do the activities of dredging, dumping or disposal pose special problems in regard to the availability of nutrient elements and toxic substances and, if so, how? Much work has been performed on this issue in recent years.

Among the authorities particularly dealing with the subject of contaminants in dredged materials the U.S. Army Engineer Waterways Experiment Station at Vicksburg, Mississippi plays a leading role with their Dredged Material Research Program (Keeley and Engler, 1974; Engler et al., 1974, 1976; Brannon et al., 1976a, b; Lee et al., 1976). Cooperating on this program are: The Environmental Engineering Program of the University of Southern California (Chen et al., 1975, 1976; Gupta and Chen, 1975), University of Texas/Dallas (Lee and Plumb, 1974) and Louisiana State University at Baton Rouge (Gambrell et al., 1977; Khalid et al., 1977). Publications from these groups for the present subject are found in the Proceedings of the Special

Conference on Dredging and its Environmental Effects, Mobile, Alabama, January 26–28, 1976 (Brannon et al., 1976; Chen et al., 1976; Gambrell et al., 1976).

Contributions on the effects of dredging operation on metal release, with particular reference to mercury, have been presented at the Proceedings of the Second United States–Japan Experts' Meeting in Tokyo, October 1976 "Management of Bottom Sediments Containing Toxic Substances" (Sameshima; Murakami and Takeishi; Yoshida and Ikegaki; Satoh; Peterson; Baumgartner et al.; Sustar and Wakeman).

Short summaries on the influence of dredging in the Great Lakes and on the release of contaminants from dredged sediments on open water disposal were given during the International Symposium on the Interactions between Sediments and Fresh Water, held at Amsterdam, 1976 (Sly, 1977; Lee, 1977). The subject was further discussed during the Second U.S./European Symposium in the Environmental Impact of Dredging and Dredged Materials, Sept. 12–15, 1978, at Vicksburg. Some of the important results from these investigations are compiled in the following section.

Investigations by Windom (1975) on the exchange of metals between sediment and water in dredge spoil slurries from estuaries and marsh areas of the southeastern United States do not show significant increases of heavy metals such as lead, copper, iron, and mercury above ambient levels. Generally, metal levels have been found to be fairly constant after an initial re-equilibration with the sediments. Even if some metals become more soluble under aerobic conditions, as soon as the sediments are dispersed, reoxidation would lead to their reprecipitation (Windom, 1976). Similar findings have been reported from lake sediments in Japan (Otsuka et al., 1976).

Studies on some influences of dredging in the Canadian portion of the Great Lakes performed by Sly (1977) give similar results with regard to the release of heavy metals. Above all, it was observed that there is a general increase for manganese in the pore waters after dredging, whereas highest Fe concentrations appear during dredging. It was suggested that on a quantitative basis the contribution of particulates to lake systems as results of dredging is insignificant when compared to natural loadings, effect of shipping (local), and lake-wide resuspension by storms. However, since "dilution should not be the solution to pollution", dumped materials should be placed in as great a depth of water as possible with the minimum of mixing and dispersal during the activities of dredging and dumping.

Sly (1977) concluded that the early release of heavy metals in soluble form from pore waters is of little concern since in most cases concentrations are already within the limits set for potable water and, in any case, the scavenging effect of oxides and hydroxides may rapidly coprecipitate phosphorus and other elements in the water mass soon after disposal. On the other hand, biochemical conversions of certain heavy metals (e.g., mercury, arsenic, lead and others), through the process of methylation, provide a mechanism for the long term release of toxic species which may be incorporated into the biota by both respiration and body metabolism. The bigger problems occur when dredge materials are deposited—especially when the depositing takes place on fields that are agriculturally exploited. Although some components, for example phosphorus and nitrogen, which originate in waters heavily polluted by communal waste effluents, can be beneficial for plant growth, other heavy metals, especially cadmium, copper, and zinc, can be very harmful and may even lead to metal enrichments in foodstuffs. This aspect of dredge materials is treated in Chapter G.3.2.

Laboratory experiments by Yeaple et al. (1972) on the effects of dredging of mercury-contaminated sediments indicate that both dissolved and total *mercury* contents in the surface deposits are increased by dredging activities. The release of mercury from the sediments, however, is strongly influenced by the oxidation-reduction conditions and, indirectly, by the type of sediment. In this context it is worthy of note that the concentration of methylmercury in the pore waters has been found to be significantly higher than in the overlying water. Lindberg et al. (1975) have observed up to 15% of the total interstitial mercury in sediments from the Mobile Bay to be present as methylmercury; Frimmel and Winkler (1975) found that 65% of the mercury content in pore water of highly contaminated sediments from the Main River in Germany occurred in the form of methylmercury-chloride.

The general observation that the effects of dredging on both water quality and biota are relatively insignificant due to the slowness of the processes, the largely refractory bonding of metals, or to a rapid reprecipitation of heavy metals (Windom, 1973; May, 1974; Patrick et al., 1977) should be re-examined with respect to the behavior of *cadmium*. Isotope studies performed by Gambrell et al. (1976) with Mississippi River sediment material indicate that exchangeable ^{109}Cd levels were strongly pH-redox potential dependent; a much greater proportion of incubated cadmium isotope was recovered in the readily bio-available forms than for any other potential toxic heavy metal studied. From dredge experiments in the San Francisco Bay (Pacific Northwest Laboratories, 1975) it was found that greater concentrations of cadmium were released to the water column under oxygen-rich conditions than under oxygen-poor conditions. Under oxidized conditions larger releases of cadmium were determined as salinity increased. Investigations performed by Patrick et al. (1977) on the solubility and potential availability of toxic metals in aquatic systems have shown that cadmium is much more active under acid oxidizing conditions than under non-acid reducing conditions. Incubation experiments by Schindler and Albert (1977) with ascorbic acid and Na_2S indicate that cadmium probably associated with the organic fraction of the sediment during reduction treatment. Of particular interest are the studies by Gilbert et al. (1976) on the influence of the sediment/water interface on the aquatic chemistry of chromium, silver and cadmium: with short-term resuspension no significant release of all three metals will occur, but with prolonged resuspension of anoxic sediments some cadmium may redissolve. Thus, due to the high toxicity of cadmium local conditions should be carefully studied. In this context the work of Gambrell et al., (1976, 1977) on the regulatory influence of pH and redox potential on toxic metal and plant nutrient chemistry in sediment-water systems has shown that the study of these two parameters may be particularly useful in predicting the bio-availability of metals from dredged sediments and may contribute in developing effective dredged material disposal criteria.

5.3 Metal Release by Acidic Water

Acidity of surface waters can cause serious problems in all aspects of metal enrichment ranging from the toxification of drinking water to problems concerning the growth and reproduction of aquatic organisms (Gorham and Gordon, 1963; Beamish and Harvey, 1972; Hendrey et al., 1976), the increased leaching of nutrients from the soil and the ensuing reduction in soil fertility (e.g., Whitby et al., 1976; Tamm, 1976), the increased availability and toxicity of metals with regard to essential plants (Lucas and Davis,

1961; Linnman et al., 1973), and finally to the undesirable acceleration of mercury-methylation effects in sediments (Fagerström and Jernelöv, 1972).

5.3.1 Acid Mine Drainage

A very rapid pH decrease is often found in waters seeping from mine refuse. With slow oxidation of sulfide components, partially aided by bacteria, such rock heaps are able to pass increased metal concentrations into waters for decades and more (see Chap. B.4). In waters with little buffer capacity—that is in carbonate-poor areas where dissolved metal pollution can be transported over great distances—the threat is especially great.

The origin of acidic solutions can be seen in the oxidation of sulfidic components, when these reach the surface and are exposed to atmospheric oxygen and moisture. Table 74 (from Hawley, 1972) shows some elements in important sulfides and sulfosalts which are generally formulated $A_m X_n$ and $A_m B_n X_p$ respectively.

Table 74. Elements in sulfides and sulfosalts (from Hawley, 1972)

Sulfides ($A_m X_n$):						Sulfosalts ($A_m B_n X_p$):		
A				B		A	B	X
Ag	Fe	Pb	As	S	As	Cu	As	S
Cu	Co	Hg	Ni	Se	Te	Ag	Sb	
Zn	Cd					Pb	Bi	
						Sn	Sn	

Most important are the levels of FeS_2, pyrite, and marcasite; during oxidation ferrous iron, sulfate and hydrogen ions are formed. The oxidation of FeS_2 and the release of acids into waters draining coal mines had been treated by various authors and is therefore presented here without the exact mechanistic steps (Singer and Stumm, 1970):

Initiator reaction:
$$FeS_2(s) + O_2(aq) \rightarrow Fe^{2+}(aq) + SO_4^{2-}(aq) + H^+(aq) \tag{1}$$
pyrite $\qquad\qquad\qquad\qquad\qquad$ acidity

propagation cycle:
$$\rightarrow Fe^{2+}(aq) + O_2(aq) \xrightarrow{bacteria} Fe^{3+}(aq) \tag{2}$$

$$Fe^{3+}(aq) + FeS_2(aq) \rightarrow Fe^{2+}(aq) + SO_4^{2-}(aq) + H^+(aq) \tag{3}$$
$\qquad\qquad\qquad\qquad\qquad\qquad\qquad$ acidity

The sulfidic component in pyrite is oxidized to sulfate (SO_4^{2-}) whereby H^+ is generated and Fe^{2+} ions are released [Eq. 1]. Once this reaction has been initiated by atmospheric oxidation, a cycle is established whereby Fe^{2+} is oxidized to Fe^{3+} [Eq. 2]; the ferric ion is capable of oxidizing pyrite—thereby taking over the initial role of oxygen —to produce additional Fe^{2+} and acidity [Eq. 3].

The above diagram shows that for each mol of iron sulfide oxidized, four moles of acid are produced.

Of special importance in the rate determination of iron pyrite oxidations and formation of acidity is the oxidation of ferrous iron [Eq. 2]. Singer and Stumm (1970) investigated the catalytic effects of sulfate, iron(III), copper(II), manganese(II), aluminum(III), charcoal, iron pyrite, clay particles, alumina, silica, and micro-organisms. Of these, micro-organisms (thiobacillus ferrooxidans and ferrobacillus ferrooxidans) appeared to exhibit the greatest effect in accelerating the oxidation of Fe^{2+} (see Sect. 5.5). Methods of controlling the problem of acidic mine drainage include thermodynamic measures (elimination of oxygen and the maintenance of reducing conditions), kinetic effects—the latter methods clearly appearing to have the most advantages—and especially the application of bactericides (Singer and Stumm, 1970).

There are many different *pathways* for the release of heavy metals from tailings, the most important of which were summarized by Andrews (1977): (1) structural failures from improper operational techniques and design in regard to possible stimulus of catastrophic events such as earthquakes and floods; (2) direct discharges of mill waste waters or total tailing to surface waters; (3) dust from unstabilized, desiccated, wind-blown surface; (4) biologic concentration in plants and ultimately in animals; (5) erosion of embankment surfaces; (6) leaching to the surface via capillary action due to high groundwater of leaching to subsurface waters by permeation.

A compilation of locations, where acidic mine drainage was observed and studied during the last few decades, has been given in Chapter B.4. In the acidic waters, at pH-values of 2.6 and 2.8 respectively, the contents of manganese, nickel, and cobalt are enriched by factors of 10,000 compared to the natural background values; the concentrations of iron, copper, and lead exceed the normal contents by factors of 2000, 1000, and 500 respectively. Iron and copper values exceed maximum allowed values of drinking water standards by factors of 100 and 1000 respectively.

5.3.2 Acid Precipitation

Acid pH conditions in natural, relatively unpolluted waters occur mainly in conjunction with the oxidation of sulfidic minerals. In addition to the cases mentioned above in mineralized regions, there are known examples of acidification of volcanic crater lakes (e.g., Yoshimura, 1934). Ohle (1936) reported pH values of 3.2 and lower in the water of the "Clay Pond" near Reinbek (West Germany), which are considered to be caused by weathering of pyrite-containing Miocene sediments.

Freshwater bodies in large areas of northern Europe and North America, in and adjacent to areas in which precipitation is strongly acidic, are threatened by further expansion of acid precipitation (Vermeulen, 1978).

Many of these freshwater bodies are poorly buffered and vulnerable to acid inputs (Wright and Gjessing, 1976). Investigations on the major-ion chemistry have been undertaken for acid lakes in southernmost Norway (Wright and Snekvik, 1977), in south-central Norway (Henriksen and Wright, 1976), the west coast of Sweden (Hörnström et al., 1973; Johansson, 1977), west-central Sweden (Grahn et al., 1974), southwest Sweden (Dietrichson, 1977), southeastern Ontario, Canada (Beamish, 1976), Sudbury, Ontario (Scheider et al., 1975), tundra ponds in the Canadian Arctic (Hutchinson and Havas, 1977), and the northeastern United States (Davis et al., 1977; Wright and Galloway, 1977). Soft water, non-acid lakes in regions subject to highly acidic precipitation (pH < 4.4) generally have low pH (< 5) with sulfate as the major anion; otherwise, similar lakes in areas not receiving acid precipitation are less acid (pH > 5.5). Sulfate supplied by acid precipitation has apparently replaced bicarbonate as the predominant anion in the lake waters.

The extent of the different sources of acidic waters has not been completely clarified. In some areas, Sudbury, Ontario for example, there is a direct connection between the emissions of sulfur dioxide from smelters, and the decrease in pH of the adjacent waters. Other regions, however, seem to be only partially affected by this type of atmospheric emission. Rosenqvist (1978) has suggested that a considerable portion of the acidification of river and lake waters in southern Norway is caused by ion exchange reactions in the increased mass of raw humus, which is connected with changes in agriculture, cattle breeding and forestry.

Because of local and regional variations in geology and soils, the buffering capacity of watersheds and lakes varies. Often a distinct bimodality in the distribution of pH values in a series of lakes is observed for many lakes in the well-buffered region above pH 6, for few lakes in the interval of pH 5.5–6, and for still fewer poorly-buffered lakes with pH below 5.5. Figure 70 gives the frequency distribution of pH in a large number of lakes in the Adirondack Mountains in the state of New York (Schofield, 1976). Most of the pH values measured in the 1930s (320) were in the interval of pH 6.0 –7.5; however in June, 1975 the pH levels in 216 lakes displayed characteristic bimodal distributions, with a large group of acid lakes below pH 5.0, a relative large group with pH 6.0–7.5 and only a few examples in the interval pH 5.5 to 6.0.

The effects of these influences on the metal composition of water and sediments in the Adirondack Mountain Lakes were investigated in a study by Galloway et al. (1976). A comparison of two lakes, one with a pH value of 4.7 and the other with 6.7, showed that in the former the content of aluminum, iron, manganese, and zinc were approximately ten times higher than those in the latter example. In the "normal" lake, the levels of Al, Fe, Mn, and Zn in a sediment core were constant with depth. The concentrations of the same elements were greatly depleted in the top 10 cm of the core from the lake with pH 4.7; below this level the sediment metal concentrations in the lake at pH 4.7 were approximately equal to those found—throughout the length of the core—in the lake with pH 6.7.

Characteristic increases of heavy metals resulting from acid redissolution of sediments and from emissions containing both elevated contents of sulfur dioxide and trace metals have been studied in lakes of Scandinavia and North America. Table 75 shows the maximum values of both uncontaminated and acidic areas in these countries from a comparison of average values in Blue Chalk Lake (uncontaminated) and Clear-

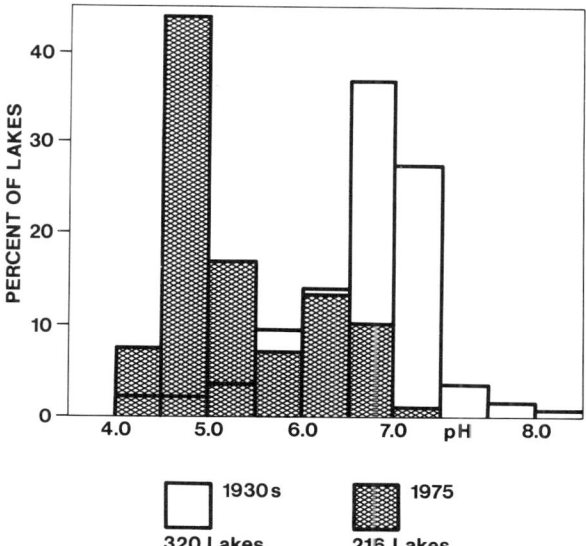

Fig. 70. Comparison of the frequency distribution of pH-values from lakes in the Adirondack Mountains (N.Y.) in the 1930's and 1975 (data after Schofield, 1976, from Wright and Gjessing, 1976, Ambio 5, p. 221)

water Lake (acidic) in Ontario, according to the study of Dillon et al. (1977). It should be noted that a considerable part of the strong enrichment of trace metals in the Sudbury lakes, as compared with the respective contents in the waters from the Experimental Lake area, is due to input from smelter emissions.

5.4 Mobilization of Metals by Organic Complexing Agents

There are several possibilities of solubilization by organic substances: (1) the metal ions may be solubilized by reduction to the more soluble lower valence states. The reduction of iron and manganese by tannic acid (Hem, 1960; Rawson, 1963), and of manganese and vanadium by peat, lignite and humic acids (Andreyev and Chumachenko, 1964; Szalay and Szilagyi, 1967) exemplify this process; (2) the metal ions may form a soluble chemical complex or a colloidal suspension with the organic acids. Humic acids rapidly decompose sulfides such as chalcopyrite, sphalerite, galena or pyrite, and silicates such as clays, micas or chlorite, and also metal oxides such as pyrolusite or goethite. It has been suggested that the solubilization of hydrous Fe/Mn oxides is achieved by reduction processes following complexation by the organic acid (Theis and Singer, 1973; Rashid and Leonard, 1973 and Zajicek and Pojasek, 1976; see Sect. 3.3). Reuter and Perdue (1977) have shown that the stability of the metal-humic complexes in natural waters is higher than that of the corresponding metal complexes with inorganic ligands, and that the amount of organic complexing material is still increasing in many rivers due to secondary sewage treatment effluents.

Synthetic Chelating Substances. Although many investigations have shown that increased pollution of water by natural organic decomposition products can affect the metal remobilization from the sediments, the introduction of synthetic complexing agents can have an even more serious impact on heavy metal remobilization. Among these substances, nitrilotriacetic acid (NTA) as a possible alternative to the polyphosphate in detergents was intensively discussed during the last few years. Despite the great number of studies

Table 75. Metal concentrations (in µg/l) in lake waters affected by acid precipitation and (for comparison) unaffected areas

	Sweden[a]		Norway[b]		Ontario I[c]		Ontario II[d]	
	North Sweden	West coast	Central Norway	South East	Exper. Lake A.	Sudbury	Blue Chalk	Clear Water
Aluminum	–	–	–	–	–	–	13	400
Manganese	–	–	–	–	–	–	40	300
Iron	–	–	–	–	–	–	41	100
Zinc	30	122	17	35	<1	122	9	50
Nickel	–	–	–	–	<3	1850	3	270
Copper	–	–	10	10	2	1120	8	92
Lead	–	5	0.5	10	<1	22	–	–
Cadmium	0.23	0.63	0.5	0.6	<1	5	–	–

[a–c] Maximum values from Wright and Gjessing (1976). [a] Dickson, 1975. [b] Henriksen and Wright, in prep. [c] Beamish, 1976. [d] Average values from Dillon et al. (1977).

and investigations by Thom (1971), Dunlap et al. (1971), Taylor et al. (1971), Epstein (1972), Dept. of National Health and Welfare, Canada (Anon. 1972g), Thayer and Kensler (1973), Mottola (1974) and Prakash (1976), to name only a few, it appears that the influence of NTA on the ecology and drinking water supply has yet to be firmly established. This, in particular, applies to the extent of a complexation of heavy metals in aquatic sediments and the degradation of NTA-metal complexes in natural systems. Depending on the available data concerning re-solution of metals from solid phases (inorganic builders and organic exchanger materials) by means of complexing agents, it is obvious to expect very effective results from the strongly complexing nature of NTA in promoting the release of metals from river and lake sediments (Dietz, 1974, 1977; Koppe, 1976; Dietz and Frank, 1977, Nusch, 1977). In fact, the first investigations of NTA and sediments clearly show the liberation of heavy metals.

By shaking sediments from urban reservoirs with water containing NTA, Gregor (1972) observed in aqueous phases lead concentrations twelve times higher than the maximum permissible limit of 50 μg/l. In a series of experiments Zitko and Carson (1972) treated a less polluted sediment from the northwest Miramichi River, New Brunswick, Canada (extractable metal ions at pH = 2: 23 ppm Zn, 0.6 ppm Cu) and a heavily polluted material from the Tomogonops River, a tributary of the Miramichi (4000 ppm Zn, 2000 ppm Cu) with NTA solutions of different hardness and a pH = 6.8–7.0. At 10 ppm NTA approximately 3.5% of the Zn and 20% of the Cu was remobilized from polluted sediments. Experiments performed by Banat et al. (1974) with polluted river sediments (Cd 75 ppm, Cu 550 ppm, Cr 270 ppm, Pb 150 ppm, Zn 1,700 ppm) indicate a high percentage mobilization of heavy metals with copper, cadmium and—to a lesser extent—nickel showing a positive correlation with both concentration of NTA and time of shaking. The release of zinc and lead is positively correlated with NTA concentration but shows a negative correlation with time of shaking—possibly due to the biodegradation of the (less stable) Pb and Zn chelates. With chromium no mobilization at any concentration or with any time of shaking could be observed.

From their experiments, Zitko and Carson (1972) suggested that increased water hardness suppresses the release of heavy metal ions, and theorized that this is possibly due to competitive formation of Ca-NTA complexes, or may result from an increase in the ionic strength of the solution. Swisher et al. (1973) have shown that protons compete with metals and that pH values below the neutral point indicate distinct decreases in complexing. On the other hand, the mobilization of heavy metals by NTA at higher pH values is also limited, due to their tendency to form hydroxides and oxides. Sanchez and Lee (1973) showed that, when NTA concentrations in natural waters are increased, the Cu concentrations display a decline, suggesting that NTA-Cu chelate adsorbs readily onto particulate matter and probably becomes unavailable in the water column (Prakash, 1976).

From recent investigations Allen and Boonlayangoor (1977) have questioned whether at the NTA levels present in natural river waters—even below sewage outfalls—any mobilization of metals would be measurable. The presence of the soluble NTA concentration of 0.75 mg/l as H_3NTA, a typical effluent concentration from Canadian activated sludge sewage treatment plant, "would probably cause the concentration of no metal in any of the rivers to increase by more than 10%".

Important, however, is the extent of NTA in sewage treatment plants, of which primary input into the system and the effectiveness of biodegradation play the greatest role. Under unfavorable conditions 10 mg/l NTA or more might occur in sewage treatment plants. Epstein (1972) has estimated that if all detergent phosphates now in use were completely substituted, the communal effluents would contain 25 to 75 mg NTA per liter. More recent figures of Prakash (1976) show estimates as high as 20 mg/l NTA for the United States if detergent phosphate would totally be replaced by NTA. At such concentrations a considerable remobilization of trace metals may occur, particularly when elevated contents of labile metal compounds e.g., of fresh precipitates from metallurgical processes are present.

Water hardness, the presence of metal-adsorbing organic matter, and the kinetic characteristics of the systems were factors affecting the fate of the metal chelates (cf. Mottola, 1974): (1) large scale use of NTA will decrease the adsorption of nickel on sewage sludges to the extent that it would not be retained at all in a treatment plant, (2) the corresponding adsorption of copper and cadmium is not affected by NTA, (3) any metal-NTA chelate passing through a treatment plant will be stable for a longer time in soft than in hard receiving water, and (4) humic substances cannot efficiently compete with NTA for the complexation of metals such as copper and nickel.

These findings have been confirmed by other authors. Investigations conducted by Gudernatsch (1970) on activated sewage sludge produced no or only limited degradation of copper or nickel-NTA complexes. The addition of calcium ions causes the rate of degradation of cadmium to increase (Huber and Popp, 1972).

It has been noted by Prakash (1976) that from a strictly chemical point of view most NTA–metal complexes appear to be moderately stable. From an ecological view point, however, chemical stability is not necessarily a measure of biologic stability. In general, NTA complexes of mercury, cadmium, and copper can be expected to be degraded much more slowly than those of nickel, zinc, and iron. Temperature seems to be an important factor. Shannon (1975) found that the NTA complexes of mercury, cadmium, and nickel exhibited poorer degradability than those of lead, zinc, and iron. The results of a test made by Chau and Shiomi (1972) on various NTA metal complexes in natural waters of Lake Ontario show an extremely high stability of copper, nickel and mercury complexes (no degradation even after a period of 100 days), and a very delayed degradation of Cd chelates (degradation after 60 days); Zn– and Pb–NTA complexes required only 16 and 25 days respectively for degradation. In Lake Erie sediment-water experiments (Chau and Shiomi, 1972) NTA complexes of Cu and Ni were more rapidly degraded. These discrepancies might be explained by the difference in organic loading, in particular by the presence of natural chelators e.g., humic substances (Prakash et al., 1975, Prakash, 1976).

Stumm (1972) has called attention to the fact that under constantly spreading anaerobic conditions in the aquatic systems the degradation of NTA is greatly reduced. This is obviously also valid with respect to the NTA–metal complexes.

In this context, the behavior of NTA and its metal complexes in groundwater is of special interest. Studies by Dunlap et al. (1971) showed that NTA infiltration through saturated soils would probably exhibit only very limited degradation, with a major portion entering the groundwater intact. Any NTA which escaped degradation during infiltration through soils could transport such metals as iron, zinc, lead, cadmium, and mercury from soils into the groundwaters (Mottola, 1974). Thus, the greatest danger would seem to lie in the increased amounts of NTA in drinking water obtained from bank filtrations or artificial recharge processes (Chap. G).

Alternatives. Water-soluble ion exchangers, particularly of the Na-Al-silicate type have recently proved to be an interesting alternative to the use of sodium-triphosphate in detergents (Schwuger and Smolka, 1976). Such substances, with the general formula $x\, Na_2O \cdot Al_2O_3 \cdot y SiO_2 \cdot z H_2O$, are to a great extent analogous to naturally existing zeolites, which are primarily found in regions of volcanic activity. Na-Al silicates possess a rather high ion exchange capacity for transition metals (copper, cadmium, zinc, nickel, cobalt, manganese). This ability is considered an advantage because the metals are then able to become incorporated into the sediments (Berth et al., 1975). Schwuger et al. (1976) investigated the ion exchange of heavy metal ions with Ca-charged Na-Al silicates in the range of low ionic concentrations found in the environment, and produced the following sequence of selectivity:

$$Pb^{2+} > Ag^+ > Cu^{2+} \gg Cd^{2+} > Zn^{2+} \gg Co^{2+}, Ni^{2+}, Mn^{2+}$$

It is suggested that the ion exchange will increase with (1) decreasing hardness of water, (2) increasing temperature, (3) increasing pH-value of the solution, (4) increasing concentration of Me-Al silicates, and (5) decreasing concentration of heavy metal ions.

Before the final application of these substances, there are still many questions to be answered, especially considering the effects on various effluent purification processes, and the influences on sedimentation processes in the water bodies.

5.5 Mobilization of Heavy Metals by Microbial Activity

The role of dissolved and particulate organic matter in the transport and accumulation of heavy metals in aquatic systems was described in Sect. 2.7. After deposition by flocculation, adsorption or precipitation, these substances undergo diagenesis, involving the increase of the molecular weight and the loss of some functional groups; they thus form a relatively stable and less reactive *reservoir* for heavy metals in aquatic sediments (Calvert and Morris, 1975). The remobilization and reintroduction of metal components from this reservoir, into the aqueous cycle is largely caused by microbial processes.

5.5.1 Microbial Interactions in Natural Environments

The metabolic activity of organisms is guided by fermentative and enzymatic processes both within and without the living cells. In contrast to the inorganic catalysts, the enzymes are often well adapted to their substrate. This is particularly true for the microorganisms, whose adaptation to the different environments is aided by a rapid generation sequence and mutation frequency. There are close interactions between the chemical environment and the microorganisms (Bertrand, 1972): On the one hand, the mineral elements of the soil influence the metabolism of the organisms, on the other the soil microorganisms can markedly alter the mineral environment.

Microbial processes in the transport and enrichment of metals have long been recognized as an essential mechanism in mineralization, particularly in conjunction with the formation of stratiform sulfide deposits (Bastin, 1926; Baas-Becking, 1959; Baas-Becking et al., 1960; Baas-Becking and Moore, 1961; Temple and Le Roux, 1964). It has been evidenced that microorganisms are capable of carrying out chemical transformations on a large scale of not only organic but also inorganic materials. Evidently, microorganisms can play a more direct role in metal sulfide precipitation than by simply acting as hydrogen sulfide generators (Trudinger and Bubela, 1967). Moreover, it has been suggested that bacterial action is an important factor with regard to the formation of iron-manganese nodules (Ehrlich, 1963) and to the separation of iron and manganese in bog ores and other limnic deposits (Krauskopf, 1957).

There are three major processes leading to the mobilization of metals by microbial activity: (1) *the destruction of organic matter* to lower molecular weight compounds, which are more capable of complexing metal ions; (2) changes in the *physical properties* of the environment by metabolic activities e.g., the oxidation-reduction potential and pH-conditions; (3) *conversion of inorganic* compounds into metal complexation by organic substances, by means of oxidative and reductive processes, catalysed by enzymatic reactions. The first process has been treated in Section 3.3, the second was briefly surveyed in Section 5.3 ("acidic waters") and will be considered in the context of bacterial leaching processes; the third mechanism will be exemplified by the alkylation of the elements mercury, arsenic, and selenium.

5.5.2 Bacterial Leaching of Metals

A group of bacteria, known as thiobacilli, is capable of oxidizing sulfide to sulfate, thus supplying an energy source for autotrophic growth i.e., synthesis of their cells from purely inorganic sources. Certain species require highly acidic conditions for growth; the optimum pH for *Thiobacillus thiooxidans,* for example, is approximately 2.0. Another species under consideration, *Thiobacillus ferrooxidans,* is capable of oxidizing ferrous iron to yield energy as well as an acidic environment. Both species usually thrive in acid mine waters, the acidic conditions being partly due to metabolic activities of these (and possibly other) bacteria (Trudinger, 1971). Although there is no doubt that certain bacterial species are capable of yielding energy from the oxidation of non-ferrous sulfides (Sutton and Corrick, 1964), the presence of iron sulfide stimulates the bacterial oxidation of sulfides such as CdS, ZnS, PbS (Malouf and Prater, 1961). The following reaction of metal sulfides (MS) is described according to the thermodynamic data (Ehrlich and Fox, 1967a, b):

$$MS + 8\,Fe^{3+} + 4\,H_2O \rightarrow M^{2+} + SO_4^{2-} + 8\,H^+ + 8\,Fe^{2+}$$

It is known, however, in acidic ferric media that the dissolution reaction often produces elemental sulfur and very little sulfate; i.e. the actual leaching reaction is more aptly described by the reaction equation (Dutrizac and MacDonald, 1974):

$$MS + 2\,Fe^{3+} \rightarrow M^{2+} + 2\,Fe^{2+} + S^0$$

The divalent iron can be further oxidized under acidic conditions by bacteria to Fe^{3+}, elementary sulfur can thus be biologically transformed to sulfuric acid, depending on various parameters.

Since *thiobacillus ferroxidans* can for example tolerate high concentrations of metals (119 g/l Zn, 28 g/l Cu, 30 g/l Co, 72 g/l Ni, 160 g/l Fe, and pH 5; [Torma and Subramanian, 1974]), these reactions provide the basis for industrial leaching operations, particularly with regard to low-grade ores from copper and uranium spoil heaps. These and other metals may be selectively extracted from ore concentrates (Trudinger, 1971; Dutrizac and MacDonald, 1974).

The fundamental processes involved in the microbial ore leaching are (1): direct bacterial attack on the mineral in the presence of dissolved oxygen, and (2) the indirect leaching action of ferric ion achieved by bacterial oxidation.

Favorable conditions—which should be prevented from taking place uncontrolled—have been established for various processes (Fisher, 1966; Harrison et al. 1966; Napier et al., 1968). These conditions are: (1) large surface area, small particle size, (2) a temperature between 30°–35 °C, (3) sufficient supply of nutrients, e.g. sulfur, phosphate for thiobacillus ferrooxidans, organic carbon, iron sulfate, pyrite, calcium nitrate, ammoniumsulfate, and (4) aeration. The major role played by the bacteria is to generate the acidified sulfate leaching medium from pyrite.

This process of producing acidity from pyrite is considered as the rate-determining step for microbial uranium extraction.

Practical usage of this process—which implies oxidation of U(IV) to acid-soluble U(VI)—is reported from Lake Elliot uranium ore districts. Over a period of 12 months, approximately 50 tons of uranium were recovered from the underground working of the Milliken mine (Fisher, 1966). The overall U-content of water pumped to the surface during this period was 0.135 g/l;

the uranium was extracted from the leachate by means of ion exchangers and successively eluted using a standard nitrate method.

Copper waste dump leaching is used (however not always involving bacterial techniques, Beck, 1967) in many parts of the world, with productions ranging from 2000 to 70,000 tons of cement copper per year for various industrial operations (Woodcock, 1967). Recovery of copper from pregnant solutions is increasingly performed by the application of shredded scrap iron (Environ. Sci. Technol., 1973).

5.5.3 Microbial Action in the Mercury Cycle

When uptake of a toxic substance occurs, microorganisms are frequently able to perform detoxification, thereby yielding a product that can be more toxic to higher organisms (Wood, 1974). This is particularly true for the bacterial production of methylmercury, which may be regarded as a means of resistance against mercurials by conversion to an organic form and subsequent secretion (Hamdy and Noyes, 1975). It thus appears that microorganisms do not require mercury, but rather deal with it when present in their food supply (McIntire and Neufeld, 1975). The ability to transform mercury compounds is not restricted to a small group of microorganisms; aerobic and anaerobic bacteria, as well as fungi, have been noted for instance to possess the capacity to methylate mercury (Vonk and Sijepestein, 1973). Experiments with sediments sterilized by exposure to gamma radiation or autoclaving have demonstrated the essential role of microbial mercury transformation (Sommers and Floyd, 1974).

Given the disproportionation reaction of inorganic mercury

$$Hg_2^{2+} \rightleftharpoons Hg^{2+} + Hg^0$$

several interconversions are possible which are catalyzed or at least promoted by microorganisms. Three characteristic steps are described below; Figure 71 (after Wood, 1974) exemplifies their position in the biologic cycle of mercury.

a) Mercuric Sulfide Transformation. Mercuric sulfide has a very low solubility in water (the theoretical solubility product 10^{-53}) and can be formed by three methods (Fagerström and Jernelöv, 1972): (1) precipitation from Hg^{2+}- and S^{2-}-ions, (2) takeover of sulfide ions from other sulfides like FeS, and (3) interaction with the equilibria of

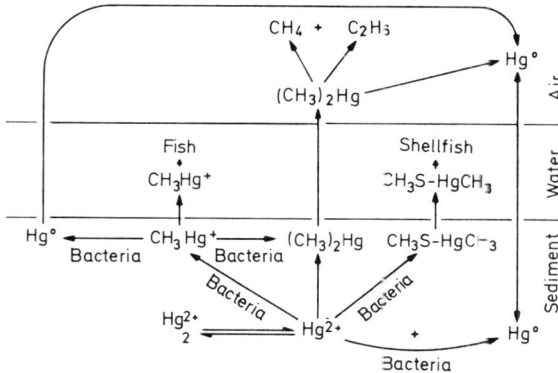

Fig. 71. Biological cycles of mercury in the environment (Wood, 1975)

organomercurials, e.g., $2\ CH_3Hg^+ = Hg^{2+} + (CH_3)_2Hg$. Hg^{2+} from the formulation is removed through formation of HgS, as well as dimethylmercury, which is volatile and slightly soluble in water. It has been shown that microorganisms producing hydrogen sulfide can remove mercury by precipitation as insoluble mercuric sulfide (Muzzarelli and Isolati, 1971/1972). On the other hand, the oxidation of H_2S occurs at a very slow rate as a physicochemical process in aerobic water, dependent on the redox potential; a direct biologic effect on mercuric sulfide as an enzymatic oxidation may lead to a somewhat faster release of divalent mercury (Jensen and Jernelöv, 1969; Fagerström and Jernelöv, 1972) suggested that the oxidation of sulfide is the rate-determining step for the organic-mercurial transformations.

b) Hg^{2+}-Hg^0-Transformations. Dissolved Hg^{2+} can be converted to Hg^0 by enzymatic reactions; Hg^0 may partly be lost from the aqueous environment because of its high vapor pressure. Biologic reduction of mercury as a means of detoxification was first described by Yostida (1967), the volatilization of mercury by bacteria was already reported by Magos et al. (1964). Alberts et al. (1974) demonstrated that elemental mercury is formed in aqueous solution by the chemical reduction of mercuric ion in the presence of humic acid. On the other hand, there is considerable evidence that Hg^0 is readily oxidized to Hg^{2+} in situ under influence of dissolved oxygen and organic matter (Colwell and Nelson, 1974).

c) Organo-Mercurial Transformations. The greatest interest was raised by the processes by which mercury and its compounds are converted to the poisonous neurotin methylmercury. The initial indication for the existence of these processes was given by Westöö (1966) from her findings of methylmercury in fish, which could otherwise not be detected as one of the possible sources of contamination. Jensen and Jernelöv (1967) first suggested from sediment studies that microorganisms could synthesize methylmercury from inorganic mercury. Since then, abundant evidence has been gained for these transformation processes, from both laboratory and natural growth systems (Holm and Cox, 1974; Olson and Cooper, 1974).

The mechanisms of methylation and microbial participation in nature are as yet not fully understood. Wood et al. (1968) investigated the nonenzymatic methylation by the cell-free extract of a methanogenic bacterium with methylcobalamin (vitamin B_{12}-CH_3) as a donor of methyl groups; it was concluded, that all microorganisms capable of vitamin B_{12} synthesis are capable of methylmercury synthesis (Wood, 1972). Similar reactions have been reported from experiments with mercury-resistant bacteria in the presence of cysteine and homocysteine components (Landner, 1971; Yamada and Tonomura, 1972). In a review of the various aspects of mercury transformation Jernelöv et al. (1975) stated that it is difficult to evaluate the ecological significance of these findings since methylcobalamin for example is unstable in a natural environment. To complicate matters, the opposite process, "demethylation", must also be considered. Tonomura and Kanzaki (1969), and Spangler et al. (1973) evidenced that mercury-resistant bacteria metabolize methylmercury to methane and elemental mercury, this process being more extensive under anaerobic than aerobic conditions. Billen et al. (1974) showed by means of the selection of mercury-resistant bacteria that some equilibrium can be attained between the degradation of methylmercury and its addition to, or production in, mineral mercury polluted sediments. These authors suggest that the occurrence of methylmercury-degrading bacteria in a zone where this highly toxic compound can be produced is an example of the efficiency of microbiologic activity in the cleansing of polluted environments.

This form of microbial adaptation can be even more comprehensive; as Walker and Colwell (1974) found from sediment studies in the heavily polluted Baltimore Harbor, where mercury-resistant bacterial populations deserve more importance in evaluating the degradation of oil in high petroleum-laden marine deposits. However, the simultaneous enrichment of heavy metals and chlorinated hydrocarbons in sediment and/or oil can result in a highly toxic environment, i.e., inhibitory to microorganisms capable of degrading each of the components separately but inhibited by the high concentrations in combination (Sayler and Colwell, 1976).

Kinetics of Mercury Transformation. Effective water quality management of mercury-contaminated natural waters necessitates stringent programs based on quantitative models that can predict mercury transformations under various environmental conditions, particularly as a function of concentration and chemical species of inorganic and organic forms of mercury, pH, microbial activity, and redox conditions. Such a model was established by Bisogni and Lawrence (1975) to demonstrate the kinetics of mercury methylation in aerobic and anaerobic aquatic environments: (1) microbial methylation of mercury can occur under both aerobic and anaerobic conditions, (2) methylation conditions are dependent on growth rate or metabolic activity of the methylating organisms, total concentration of mercuric ions, and availability of mercuric ions, (3) the predominant product of microbially methylated mercury at neutral pH is monomethylmercury; dimethylmercury is formed in small amounts, (4) a higher methylation rate results under aerobic conditions for a given inorganic mercury concentration and given microbial growth rate, (5) temperature effects on methylation rates only affect the microbial activity of the methylating organisms. With regard to the engineering significance and application of their model in waste water treatment, these authors state that large amounts of metallic mercury are formed and stripped from the aqueous phase, when a carrier gas such as *air, methane,* or *carbon dioxide* is forced through the following mercuric-mercury-containing aqueous systems: *aerobic microbial reactors, anaerobic microbial reactors,* and *aerobic abiotic systems.*

5.5.4 Bacterial Methylation of Arsenic, Lead, and Selenium

The ability of microorganisms to detoxify their environment through organometallic transformation of mercury is obviously not restricted to this metal, although the processes described here are particularly characteristic and important with regard to the widespread pollution by mercury. Meanwhile, it has been established that the mechanism of biologic methylation is also effective in the formation of volatile compounds of As, Pb, and Se, such as alkyl arsines (Wookon and Kearney, 1973; Cox and Alexander, 1973), tetramethyl lead (Wong et al., 1975; Schmidt and Huber, 1976) and dimethylselenide (Francis et al., 1974). Dimethylarsenic acid and methylarsenous acid were found in natural waters, bird eggshells, seashells, and human urine (Braman and Foreback, 1973). Methylation of tin compounds can be catalyzed by a Sn- and Hg-tolerant strain of Pseudomonas (Huey et al., 1974). Metal-alkyls which are stable in water and which can be synthesized by methylcobalamin reactions further include Te, Pd, Pt, Au, and Tl (Agnes et al., 1971; Wood, 1972, 1974, 1975); the environmental significance of some of these transformations is, however, as yet unclear. In the case of selenium transformations, evidence was obtained that methylation in the aquatic environment only occurs when there is a growth of certain microorganisms (Chau et al., 1976).

Recent investigations have shown that the biologic cycles of the elements are widely interfering, and that this can lead to many unexpected effects. One of the most interesting examples is given by the antagonism of selenium and mercury compounds, both extremely toxic elements and lethal to higher organisms in relatively small doses. It has been found, particularly by the work of Ganther and associates (Ganther et al., 1975), that the toxic effects of high concentrations of methylmercury decreases when a selenium salt is included in the diet. Experiments by Segall and Wood (1975) have shown that selenium salts react directly with both dimethylmercury and methyl-B_{12} to give dimethylselenide as the major product. Wood (1975) concluded from this behavior that methyl-groups can be transferred out of the methylmercury cycle into the selenium cycle, and that the exhalation of the volatile selenium product can be regarded as a mechanism of detoxification.

Chapter F
Heavy Metals in Aquatic Organisms

The role of heavy metals in aquatic organisms has been treated in earlier literature reviews (e.g., Doudoroff and Katz, 1953) with emphasis on the toxicity of the individual metals. Only since mercury and cadmium poisoning in Japan (ref. Chap. B) has the accent been shifted towards investigations dealing with the influence of heavy metals on the metabolism of aquatic organisms and the ability of the latter to accumulate both essential and non-essential metals.

In contrast to the nonessential trace metals, such as lead, cadmium, mercury, arsenic and others, the essential metals such as copper, zinc, iron, and cobalt have important biochemical functions in the organism: they form either an electron donor system or function as ligands in complex enzymatic compounds (cf. Chap. B). Since essential elements are only used by the organisms in trace amounts and, generally speaking, only occur in the environment in small concentrations, their enrichment in the organism does not exceed the level which allows the enzyme system to function without interference. This means that the concentrations of essential trace elements are generally higher in the organism than in water. If there is too great an abundance of essential heavy metals, the metal content in the organism can be regulated by homeostatic control mechanisms (Bryan and Hummerstone, 1973b). However, if the heavy metal concentration at the source of supply (e.g., water, food) is too high, the homeostatic mechanisms cease to function and the essential heavy metals (cf. also Fig. 3) act in an either acutely or chronically toxic manner. Thus in the event of a resulting extended bioaccumulation of heavy metals the organism may be damaged.

Investigations on heavy metal pollution of the environment have shown that, depending on the compartment (e.g., water, suspended matter, sediment), different interpretations regarding the toxicity of the metal concentrations measured in respect to biocenosis are possible.

Meaningful biotope analysis in respect to toxic effects of heavy metals on organisms cannot be carried out merely by determining the enrichment in water and sediment, for the following reasons:

Water Analyses

a) Water analyses only produce useful data for the biotic environment if multiple sampling is carried out to determine the degree of fluctuation of metal pollution in water. Very often only a single sample is taken to serve as a basis for subsequent conclusions.
b) Heavy metal concentrations in water very often barely reach the levels at which they

can be readily determined with the result that it is often impossible to produce any accurate data without costly extraction processes.
c) Since metal enrichment in aquatic organisms is heavily dependent upon the metal content in the water, a slight increase in the concentration in the water could lead to a significant metal increase in the organism.
d) Moreover, a basic analysis of heavy metals in water gives little indication of the toxicity or availability of the investigated metal in the organism since these factors are mainly governed by the chemical characteristics of the metal, in particular the predominant species existing under prevailing physico-chemical conditions in the water.

Analysis by techniques such as atomic absorption spectrometry provide a useful guide for determining overall metal concentrations of standardized water samples (filtered using a 0.45 μm membrane filter), but fail to produce evidence of chemical speciation.

Although a certain degree of standardization in the water samples can be achieved by means of filtration over a 0.45 μm membrane filter, the above-mentioned factors cannot be determined by routine analysis using atomic absorption, for example.

Sediment Analysis

In many investigations evidence of a water system, which is highly polluted by heavy metals, is provided by sediment analyses. Due to physico-chemical processes, the major part of heavy metals introduced into a water system are deposited in the sediment (cf. Chap. E). Here they can become dangerous to benthic organisms. The toxicity of the metals contained in the sediment is to a large extent dependent—as was the case with heavy metal toxicity in water—on the availability and ion activity of the metals, with the result that high accumulation in sediment does not necessarily pertain to in organisms living in the sediment. Particularly in polluted rivers, therefore, a strong fluctuation in the heavy metal content in the sediment (as a result of flood water drainage, varying sedimentation rates, and inhomogenous distribution of heavy metals in the sediment) can result in wrong conclusions being drawn as to the overall heavy metal pollution of a river (cf. Chap. D). Furthermore, sediment analysis alone provides little information regarding the amount of metal which enters the biomass in water.

Consequently, if heavy metals in certain environments occur in increased concentrations, this does not mean that the metal concentrations are likewise elevated in organisms. Many investigations have demonstrated that in order to determine the overall heavy metal pollution of a biotope, it is necessary to determine the heavy metal concentrations in as many trophic levels as possible in the aquatic system.

Few investigations have been conducted on the biologic aspect (mercury has, however, been well-documented) of metal availability for the organism in the aquatic environment. However, especially laboratory experiments have revealed that many physico-chemical parameters have a considerable influence on the toxicity and accumulation of metals in organisms (Lloyd, 1965; Tabata, 1969). In general toxicity of heavy metals in water is accordingly determined by:
a) The toxicity of the metal itself (dependent on its electronegativity; cf. also Chap. B);
b) The synergic or antagonistic aspects of heavy metals;

c) The influence of physico-chemical parameters which determine a metal's availability by activation or deactivation.

Investigations on synergism and antagonism of heavy metals have long been conducted (Jones, 1938); experiments on the toxicity of heavy metals with regard to water organisms have also been carried out for many years (Carpenter, 1924; Steinmann, 1928; Weber, 1934; Dawson, 1935, Jones, 1946; Schweiger, 1956).

In most of the investigations aimed at establishing the toxicity limits of heavy metals for lower as well as higher water organisms, the assumption was often made that a complete dissociation of the metal into cations and anions occurred and that the metal salt added was thereby available in full to the organism in the solution.

Detailed investigations on the behavior of heavy metal solutions on organisms with particular respect to the other dissolved substances present in the water have not been carried out until more recently. These showed that the toxicity limits for organisms can be considerable lowered or raised depending on which parameters are altered. Since the experimental conditions in the earlier toxicity tests were only seldom standardized, the toxicity levels determined vary according to the quality of the water used, thus making a comparison almost impossible (cf. review by Doudoroff and Katz, 1953).

Considering that the parameters of oxygen content, temperature, pH-value, hardness, salinity, organic solution components, heavy metal complexation, and the chemical characteristics of individual metal species in the aquatic environment are determining factors of the heavy metal effect on organisms, these parameters deserve closer examination.

1 Physico-Chemical Influences on the Toxicity and the Uptake of Heavy Metals with Respect to Organisms

1.1 Temperature and Oxygen Content

Water temperature and dissolved oxygen content are, due to metabolic processes, vital factors in water for stenobiontic organisms. The mechanism whereby heavy metal toxicity increases with higher water temperatures (Lloyd, 1965; Fig. 72) can be explained by elevated respiratory activity. Thus, rainbow trout in a zinc sulfate solution survived 2.35 times longer when the temperature was lowered from 22° to 12 °C.

Moreover, the metal solution itself causes increased respiratory activity. This was tested for sticklebacks subjected to copper sulfate and lead nitrate solutions by Jones, 1946 (Fig. 73). The absorption and release of metals can also depend on temperature. This was established for mercury, methylmercury, and phenylmercury acetate in experiments using rainbow trout (*Salmo gairdneri*) (Macleod and Pessah, 1973; Reinert et al., 1974; Ruohtula and Miettinen, 1975). Lloyd (1965) investigated the effects of oxygen content in water on rainbow trout. The dependence on oxygen saturation in water was the same for copper, lead, and zinc (Fig. 74).

Apart from the effects of water temperature and oxygen content on the toxicity of heavy metals as a result of physiologic changes in the organism, these two parameters can, due to chemical processes in water and sediment (e.g., oxidizing-reducing

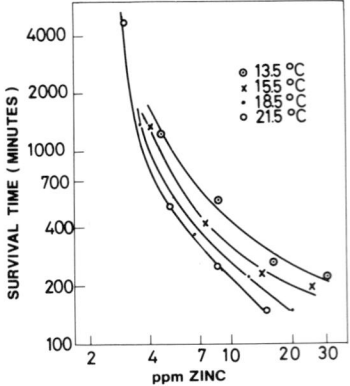

Fig. 72. Influence of the water temperature on the toxicity of zinc to rainbow trout (*Salmo gairdneri* Rich.) (Lloyd, 1965)

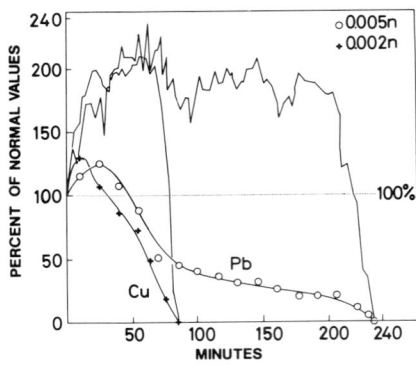

Fig. 73. Respiratory frequence (*upper curves*) and oxygen uptake (*lower curves*) of sticklebacks (*Gasterosteus aculeatus*) as a function of toxicity and time in 0.002 N $CuSO_4$ and 0.005 N $Pb(NO_3)_2$ solutions (Jones, 1946)

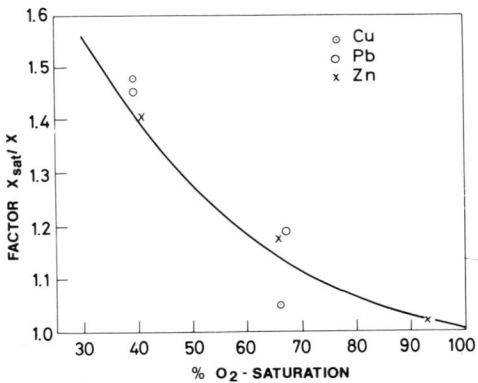

Fig. 74. Influence of oxygen saturation on the toxicity of copper, lead, and zinc to rainbow trout (*Salmo gairdneri* Rich.). Factor x_{sat}/x: x_{sat} = lethal toxicity limit of the metal concentration when water is saturated in oxygen; x = lower oxygen contents in the water (Lloyd, 1965)

environment, see Chap. E). decisively influence heavy metal availability. Thus, the concentration of heavy metals in interstititial waters associated with anaerobic sediments can be up to 10 times higher than in supernatant water (Reinhard and Förstner, 1976).

1.2 Water Hardness

Early experiments were conducted to establish the influence of water hardness on the toxicity of metals (e.g., Schweiger, 1956). The degree of water hardness influences the toxicity—and consequently the activity of heavy metals—by forming insoluble carbonates or by adsorption on calcium carbonate. This is especially true for the mixing zone of water from different sources, e.g., waste water flow into a natural water body (Patchineelam, 1975; cf. Chap. C), where the acute toxicity of metals decrease when their bioavailability is less. Thus in hard water, the concentrations of heavy metals necessary to reach the level of lethal dosage must be greater, and lower when the water hardness is less. This often described characteristic of water has, however, rarely been investigated in any great detail. Apart from this interaction of heavy metals with inorganic compounds, Zitko and Carson (1976) demonstrated that magnesium and calcium ions compete with heavy metal ions for active sites in fish tissues, so that this may be a mechanism effecting the lethality of heavy metals by water hardness.

Lloyd (1965) and Tabata (1969) conducted experiments on rainbow trout (*S. gairdneri*), *Daphnia* sp., and Japanese killifish to determine to what extent the toxicity limits depend on the degree of water hardness. Tabata's (1969) experiments on *Daphnia* clearly demonstrate that in spite of the reduced toxicity resulting from increased water hardness, the order of toxicity for the metals investigated remains the same (cf. Fig. 75) and occurs more or less in the electronegativity sequence—even when there is a high degree of water hardness.

The various slopes of individual graph curves provide evidence of the different reactions of the individual metals with increasing water hardness; mercury, for example,

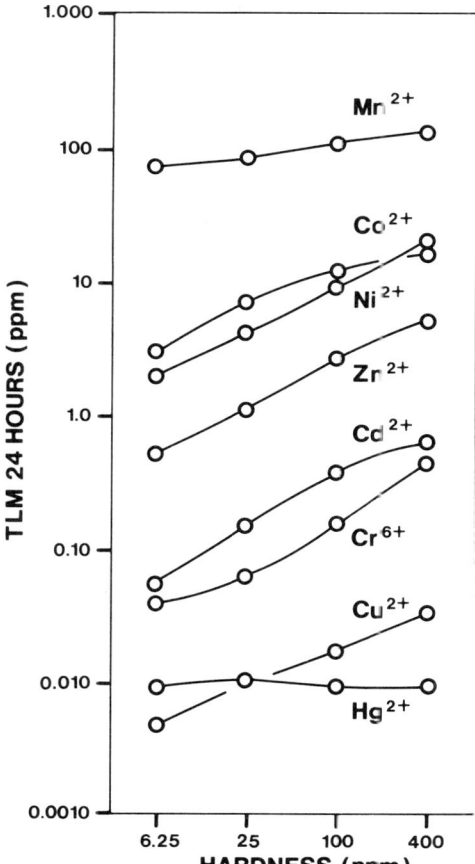

Fig. 75. Effect of hardness (ppm $CaCO_3$) on the amount of metals needed to affect TLM 24 h on *Daphnia* sp. (TLM 24 h, median toxicity limit at which 50% of organisms are dead within 24 hours; Tabata, 1969)

is hardly affected by water hardness. By differentiating between calcium and magnesium hardness, Tabata (1969) established that for Daphnia the influence of calcium hardness was dominant (Fig. 76).

The mechanism whereby heavy metal toxicity is reduced in the face of increasing water hardness can be explained as chemical reactions between metal ions and carbonates (cf. Chap. C).

1.3 Organic Compounds

Apart from natural changes occurring in the dissolved heavy metal content in water as a result of bioaccumulation (Morris, 1971; Knauer and Martin, 1973), metal concentrations in water can be altered by organic substances and synthetic organic compounds (such as NTA or EDTA), some of which have been released by biodegradation.

The influence of organic compounds on the toxicity and enrichment of heavy metals with regard to organisms should be particularly considered for highly eutrophied and anthropogenically polluted waters. Normally, the amount of organic substances in the natural aquatic environment is determined by the balance between biomass production and biodegradation leading finally to mineralization. The concentration of released and particulate organic compounds is therefore generally low in water (an exception to

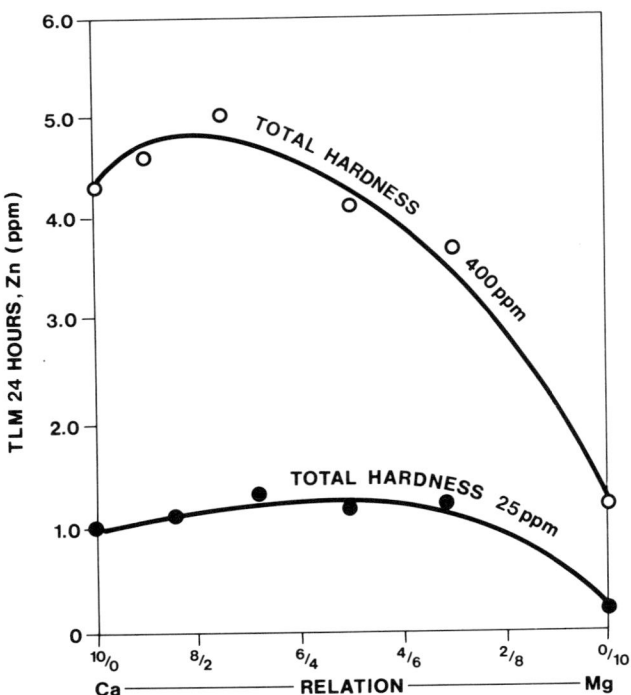

Fig. 76. Quantitative influence of calcium and magnesium hardness on the toxicity of zinc to the TLM 24 h on *Daphnia* sp. (for TLM see Fig. 75; Tabata, 1969)

this is found in moor lakes, for example). The potential of natural complexing organic compounds is therefore also low.

Field investigations on the complexing capacity of organic compounds in respect to the uptake of heavy metals by organisms prove to be controllable only with some difficulty. Bearing this in mind, several important laboratory experiments were carried out to investigate this aspect. Using copper as an example, Gächter et al., (1973) showed, in investigations to determine the complexing capacity of a nutrient medium and natural waters, that changes occurred in the photosynthesis rate of algae (*Chlorella* sp.). Even before the complexing equilibrium had been attained, photosynthesis was reduced by 70%–95%. Since it can be assumed that the copper concentrations are masked by organic substances and thereby have only a small degree of toxicity, this reduction must be ascribed to very low concentrations of available Cu^{2+}.

On the other hand, these investigations have also shown the high complexing capacity of natural water.

The complexing capacity in the water of Lake Ontario (Hamilton Harbor) reached 1.23 μmole Cu^{2+}/l. This complexing capacity was therefore very nearly the same as the algal nutrient solution used in the experiment (1.67 μmole Cu^{2+}/l with added EDTA of 1 μmole/l).

If additional organic components are present in the solution, e.g., in different bacterial nutrient solutions, then metal activity is reduced even further. The activity of free cations can—depending on the metal involved—be almost completely suppressed (Ramamoorthy and Kushner, 1975). In experiments conducted by the above authors the metal affinity toward the different microbial growth media largely follows the stability constants of metals to organic ligands (Irving and Williams, 1953). The metals investigated show the following availability of free cations (see Fig. 77):

$$Cd^{2+} \gg Cu^{2+} \gg Pb^{2+} > Hg^{2+}$$

It can therefore be generally concluded that the "nominal concentrations" used in the toxicity experiments and the resulting toxicity limits of heavy metals for aquatic organisms are much too high in many cases since it is only the toxic active metal contents of the total metal added which produce any effect (cf. Reddy and Patrick, 1977a, Harrison, 1978). Differentiated investigations on the detoxification of heavy metals using organic complexing agents were conducted by Sprague (1968), Shaw and Brown (1974), and Tabata (1969) in laboratory experiments on organisms.

Sprague's investigations on brook trout and Shaw and Brown's experiments on rainbow trout using NTA show a multiple reduced toxicity of the metals copper and zinc (Fig. 78).

In natural waters the presence of NTA as well as EDTA as an additive in detergents has the same effect (cf. Chap. E). Sprague (1968) has therefore advocated that complexing agents should be introduced into metal-polluted industrial effluents as a safety precaution.

However, negative effects can also result from such substances. On the one hand, the complexes themselves are poisonous when in excess of certain concentrations, and on the other hand, they can remobilize heavy metals from sediments as a result of their

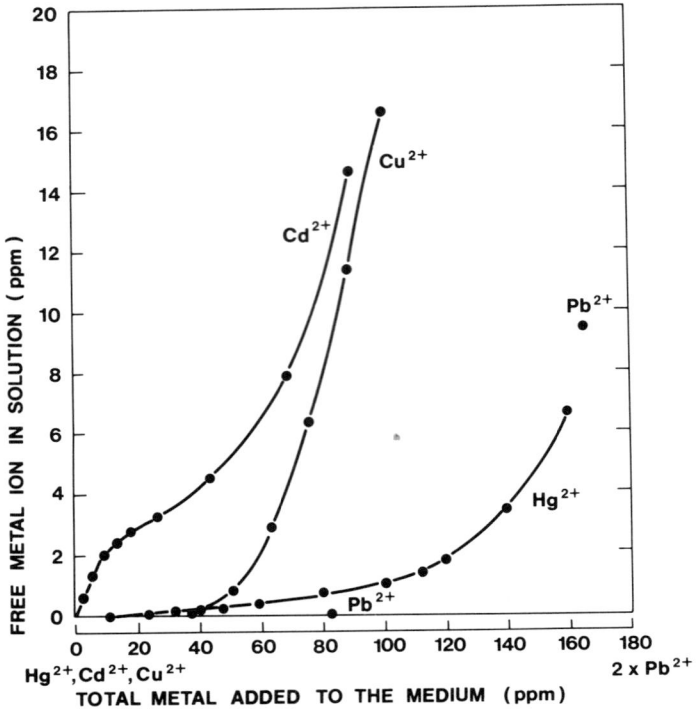

Fig. 77. Free metal ion concentration in solution after added as metal ion to bacteria growth medium (Ramamoorthy and Kushner, 1975)

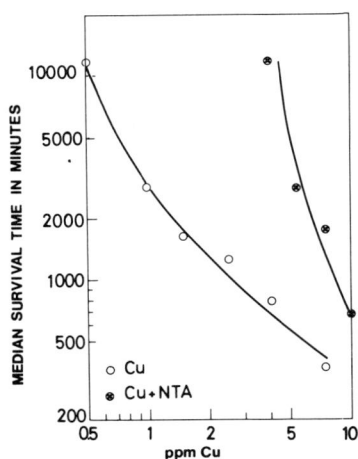

Fig. 78. Reduced copper toxicity for rainbow trout (*Salmo gairdneri* Rich.) by added equivalents of NTA (Shaw and Brown, 1974, Water Research, *8*, p. 380. Reproduced with permission of Pergamon Press, Oxford)

complexing effect. Metal ions can thereby be released as a result of the rapid degradability of NTA (Thompson and Duthie, 1968; cf. Chap. E).

Tabata (1969) likewise determined a considerable reduction of metal toxicity for other organic substances (EDTA, sodium citrate). Experiments on oysters were also conducted to test the organism's accumulation of copper when different metal-complexing substances were added. The reduction of copper incorporation was greatest when EDTA was added, so that the bioavailability of Cu^{2+} in this solution will necessarily be lower than that of the control experiments (Fig. 79).

Tabata's results clearly demonstrate that reduced toxicity also signifies reduced heavy metal incorporation in the organism. This means, however, that not only the acute toxicity but also the otherwise chronic toxic effects of heavy metals (enrichment and subsequent damage on individual organs) can be suppressed.

In addition to the dissolved components of organic compounds, an equally important part is played by colloidal and suspended organic matter. In heavily polluted water bodies such as the Rhine River, considerable amounts of metals are sorbed onto suspended particles (Schleichert, 1975). Here the metal concentrations are found in quantities similar to those found in sediments. This explains why, in the case of many organisms, the major part of the metals introduced into the water are not readily available. Only a few groups of organisms depend directly on the highly polluted suspended matter or the sediments as their basic source of nutriment.

1.4 pH Values

pH values play an important role in the interactions between heavy metals and parameters such as carbonate hardness and organic compounds. In experiments by Shaw

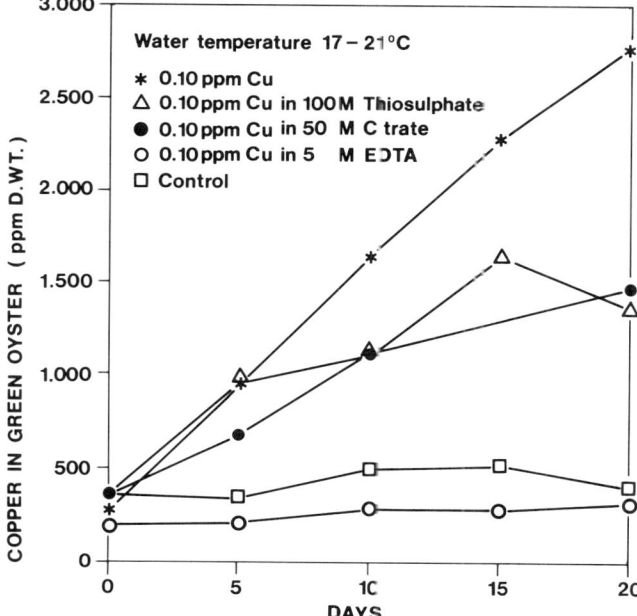

Fig. 79. Different suppression of copper uptake in green oyster caused by various compounds (Tabata, 1969)

and Brown (1974), in which a copper solution in hard water (at pH = 8.0) was not renewed, the concentration of dissolved copper dropped to 0.5 mg/l from the original concentration of 1.1 mg/l within two hours, thereby reducing the toxicity of the "nominal copper concentration". Consequently it is of utmost importance to consider the metal species in solution when establishing toxicity limits. That the toxicity of heavy metals can increase with basic pH values—apparently the exact opposite effect— was recorded by Whitley (1968) in experiments with lead conducted on tubificid worms; lead was found to be more toxic at a pH value of 8.5 than at pH 6.5. In these experiments, lead that was precipitated at a pH value of 8.5 was most probably incorporated orally; this confirms the fact that lead has a low distribution coefficient in the aquatic environment (Koppe, 1973). The lead ions, therefore, may precipitate when the pH is higher (Stumm and Morgan, 1970). In respect to organic substances, changes in pH values in the water can strongly influence the adsorption or desorption of cations. For example, amino acids, which occur in substantial amounts in both free water and in sediments in eutrophied or polluted waters, can adsorb or desorb cations due to the pH-dependent amphoteric character of the acids.

Further influences effected by the pH value are described in detail in Chap. E.

1.5 Salinity

Generally speaking, salinity in the marine environment is relatively constant and has little influence on the heavy metal concentrations. In estuaries, where fresh- and saltwater intermix, salinity, however, plays a dominant role in influencing metal concentrations in free water. In seawater the contents of dissolved heavy metals are generally much lower than in fresh water. Moreover, the high salt content alters the pH-value of the milieu and consequently the metal solubility (freshwater environment pH 7–7.5; marine environment approx. 8.0). In the case of organisms in brackish water, the negative potential difference of the inner body wall increases with lower salinity (proved by Fletcher, 1970 on *Nereis diversicolor*); ion transport into organisms consequently increases. In addition, there is no competition in the freshwater environment between calcium and magnesium as cation exchangers in respect to heavy metals.

For zinc, Bryan and Hummerstone (1973b) demonstrated in laboratory experiments on *Nereis diversicolor* that the absorption rate per mass unit and time period decreases with rising salinity; on the other hand, in field investigations no rag-worms were found in the saline environment that had less Zn than those in areas with low salinity. Here, the high uptake of Zn is possibly equalized by an increased zinc excretion.

In the case of copper, dependencies were found to exist between salinity and copper concentration in *Nereis diversicolor* (Fig. 80; Bryan and Hummerstone, 1971). However, no direct correlation could be established in this case. Salinity apparently plays only a secondary role (e.g., through pore water; Chap. E).

The combined effect of the parameters described here—temperature, oxygen content, water hardness, organic compounds, pH-value, and salinity—shows the many influences which vary the toxicity and the incorporation of heavy metals in aquatic organisms.

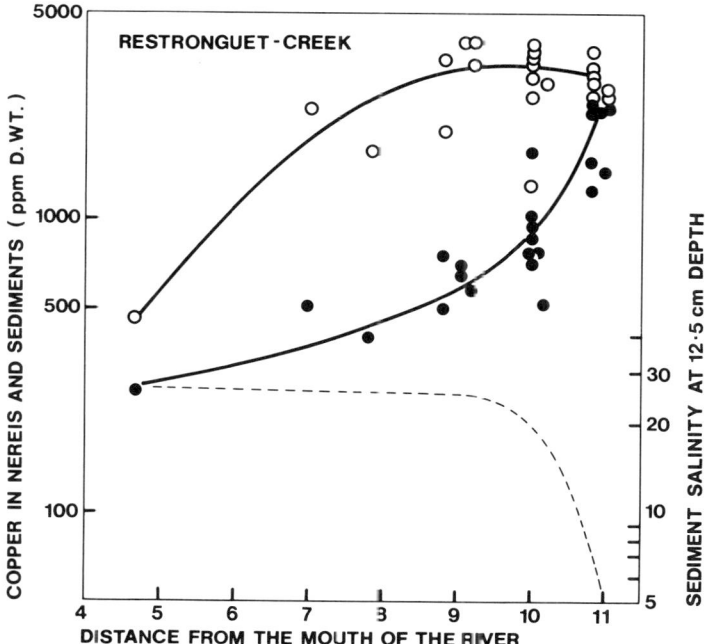

Fig. 80. Correlation of sediment salinity (in per mille) and copper concentrations in sediments to copper in *Nereis diversicolor* at different points of Restronguet Creek, Southwest England. *Solid line, Nereis diversicolor; light line,* sediment, *broken line,* salinity (Bryan and Hummerstone, 1971, J. Mar. Biol. Assoc. U. K. *51,* p. 850. Reproduced with permission of Cambridge University Press, London)

The heavy metal content determined in an organism should by no means be regarded as a constant value, but rather as a factor subject to the influence of varying biotic (endogenic and exogenic) and abiotic environmental conditions.

The abundance of organisms sensitive to heavy metals will therefore correspond to the dynamic balance of the influence of these factors; the heavy metal concentration in the organism itself will also adjust accordingly.

The above factors deserve special consideration when dealing with heavy metal enrichment in aquatic organisms, particularly in respect to food chains.

2 Biologic Factors Affecting Heavy Metal Concentrations in Aquatic Organisms

In addition to the abiotic factors already mentioned, metal enrichment can also be affected to an important degree by biotic factors which arise in the organism itself and through other organisms. Which of these factors are to be considered depends largely on the aspect under which the investigation is being conducted.

When considering the heavy metal content in organisms suitable for human consumption (e.g., snails, bivalves, crustaceans, and fish as well as marine algae), the most important aspect of the heavy metal is its toxicity toward humans. In this context the maximum tolerable levels for foodstuffs determined from heavy metal data are set according to human toxicologic as well as economic considerations, and are in no way identical with the natural metal background of the individual aquatic species. For that reason the criterion for indicating heavy metal contamination is the increased metal contents of the organism in respect to its natural background and not to the maximum tolerable level for human consumption. Thus, the question of natural background arises; for example, how it can be determined for certain sediments and water in the abiotic environment (Turekian and Wedepohl, 1961).

Environmental pollution by heavy metals is very often determined simply by analyzing such heavy metals present. The metal concentrations measured in organisms may lie within the normal range or above it, according to the degree of pollution in the biotope. The normal range is determined by comparison with investigations conducted in areas with relatively little or no pollution.

In investigations on heavy metal enrichment in the food chain, the concentrations in organisms of various trophic levels are usually directly compared to each other without taking the background data into consideration. The basic point of reference here is generally the water or the phytoplankton. Even in unpolluted ecosystems heavy metal enrichment can take place at each of the next-highest trophic level. If this is the case in the absolute data, the natural metal background of the higher predatory organisms is even higher in the unpolluted biotope, an increase which is due to natural enrichment.

Since the normal metal content of an organism is dependent on many variables, a range of concentration is usually given, the scope of which is generally dependent upon how well the organism in question has been investigated.

In addition to the abiotic factors mentioned above, both the organism's natural metal background and its metal content from contaminated areas can essentially be determined by its biologic behavior.

2.1 General Physiologic Behavior

Due to osmotic regulation of organisms in freshwater and marine waters the flux of ion incorporation and excretion in these hyper- and hypo-osmotic media is principally different (Florey, 1970). Therefore, heavy metal concentration in marine and limnic organisms must be regarded separately.

2.2 Life Cycle and Life History of the Organism

The individual and species-related age of an organism can increase or reduce the possibility of heavy metals being enriched on a long-term basis. Matsunaga (1975) found that the overall mercury content in fishes was clearly dependent upon age (crucian carp, *Carassius carassius langsdorfii*). Hasselrot and Göthberg (1974) arrived at similar results in the axillary muscle of northern pike (*Esox lucius*) and roaches (*Rutilus rutilus*). Bache et al.

(1971) established a correlation between age and the concentrations of mercury in lake trout (*Salvelinus namaycush*). This dependence could be determined for the whole fish and not merely from the muscle tissue as in the case of Matsunaga (1975), and Hasselrot and Göthberg (1974; see Fig. 81). Mackay et al. (1975) report that the total mercury and selenium content in marine black marlin (*Makaira indica* Cuvier) from unpolluted waters to the northeast of Australia are dependent upon weight.

Apart from mercury enrichment, Müller and Prosi (1978) also established a dependence between cadmium enrichment and the age of certain fish (roaches *Rutilus rutilus* L.). However, in agreement with Lovett et al. (1972), this age-dependence could not be determined in muscular tissue, but only for the kidney and the liver. This means that in roaches and probably for all fish species the latter organs provide a better indication of the extent of cadmium contamination than the muscle tissue.

In all investigations dealing with the dependence of metal concentrations on the age of aquatic organisms, the species were taken from the same biotope. This must be basically done in order to exclude local factors that might influence the metal content in the individual organism.

In Pacific oysters (*Crassostrea gigas* Thunberg; Ayling 1974) show that the dependence of the cadmium and zinc concentrations on age and weight is lost if the results from various sampling sites in the same river system (the Tamar River in this study) are combined for evaluation. However, when the populations are separated, there are distinct correlations between metal incorporation and the overall weight. Further investigations and weight dependant metal concentrations of marine and estuarine mollusks have been carried out by Boyden (1977).

In addition to the fact that age-dependence of metal concentrations becomes indistinct if data from different localities are compiled, the extent of the organism's mobility—its sessility or its wandering—is of importance for the uniform influence of the contaminating metal sources. The importance of mobility for the metal content was clearly demonstrated by Nisimura (1974) for stationary and nonstationary fish using mercury in fluviatile sediments as contaminants (Fig. 82). In this case a clear correla-

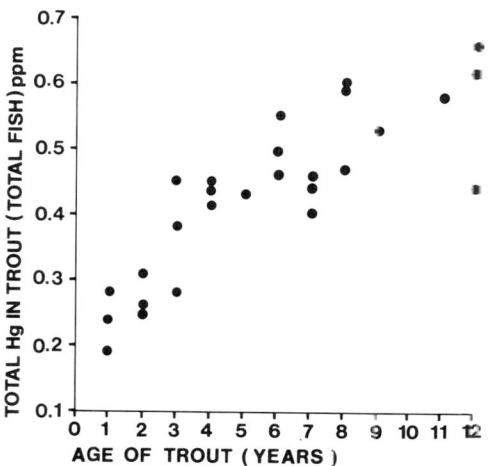

Fig. 81. Total mercury residues in chopped lake trout (*Salvelinus namaycush*) as a function of age (Bache et al., 1971, Science *172*, p. 952. Reproduced by permission of the authors; copyright 1971 by the American Association for the Advancement of Science)

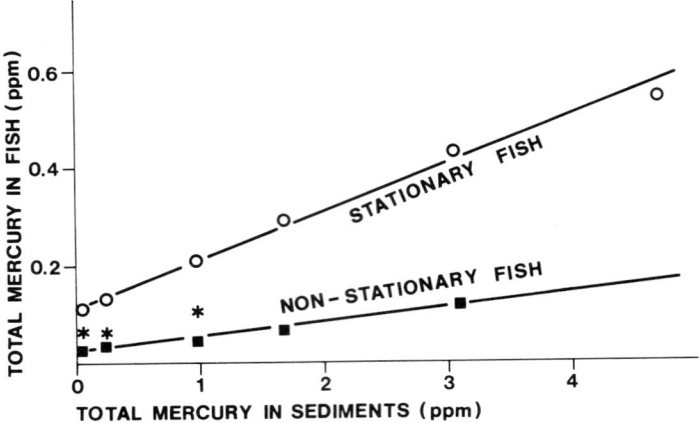

Fig. 82. Difference of total mercury concentration between stationary and nonstationary fish in relation to the mercury content in the sediments at the different sampling station (after Nisimura, 1974)

tion between the mercury content of stationary fish and sediment was found. The correlation was not distinct for nonstationary fish.

The species' individual mode of development, e.g., molting (crustaceans, larva stages of water insects) and the diet which is different in young and adult stages for many species of fish can also influence the degree of metal incorporation.

Renfro et al. (1975) report on experiments with crabs (*Carcinus maenas* Rath.) in which an average of 61% of the total ^{65}Zn present was lost with every discarded exoskeleton.

Metal enrichments dependent upon the sex of animal species have been observed in fish (Rehwoldt and Karimian-Teherani, 1975). Adult zebrafish were fed on food contaminated with Cd (10 ppm Cd) for an exposure time of six weeks. After this period the Cd concentrations determined in the female fish were more than double those found in the male (12.95 ppm/5.15 ppm dry weight).

In uncontaminated fish (control), significant differences in the natural background ($P = 99\%$) of a similar magnitude (male ~40 ppb, female ~80 ppb) were observed. The sex-related difference for cadmium enrichment increases in the same ratio (male:female = 1:2) as in the case when the fish are fed with Cd-contaminated food. The dissimilar deposit of Cd is therefore probably attributable to different physiologic conditions of male and female fish, for example during reproduction.

2.3 Seasonal Variations of Metal Content in Organisms

The natural seasonal variation in heavy metal content can depend on several factors which are not easily differentiated. The biologic activity or the metabolism rates of an organism often drops when the surrounding environment becomes cooler, whereby the rates of incorporation and release of heavy metal may change. In laboratory experiments on oysters (*Crassostrea virginica* Gmelin) Cunningham and Tripp (1975a) simulated the temperature-sensitive mercury enrichment in the summer months and mercury loss in the winter months. The seasonal difference in fat accumulation could therefore have a considerable influence on the storage capacity for organo-metal compounds, which are known to have a great affinity for lipids.

The mass of algae present can influence the availability of metals, since as primary consumers they can strongly affect the dissolved metal content in open water (Morris, 1971; Murphy et al., 1976).

Morris (1971) showed that the concentration of dissolved manganese in the water decreases by about half when marine flagellate *Phaeocystis* are in bloom conditions. The availability of heavy metals can therefore be influenced by the activity of organisms and the seasonal variation of their biomass.

2.4 Species-Specific and Individual Variability

Metal concentrations in the different types of organisms can vary greatly. The various trophic levels of several species must therefore be simultaneously investigated, since the metal concentrations within a certain trophic level can fluctuate considerably according to different dietary habits (Potter et al., 1975).

In addition to its position within the food chain, the morpho-physiologic structure of a particular species can also help account for the differences in metal content encountered in different species, since the structure and function of the organs in contact with the contaminated medium can be instrumental in governing the metal content. Thus, an important aspect of metal enrichment is whether respiration occurs via the gills, the skin or atmospherically. In the case of *Annelida,* for example, the body covering is formed of thin, relatively porous protein and chitin substances, whereas in the case of higher crustaceans it consists of thick chitin skeletons with calcium deposits. For many water organisms body excretions are responsible for the spontaneous adsorption of heavy metals.

2.5 Contamination by Food and Intestine Content

Organisms which feed on strongly contaminated material should be examined in order to determine how such material affects the overall metal concentration in the organism. This is particularly important for organisms which live in or feed on sediments (Flegal and Martin, 1977).

All those abiotic and biotic factors affecting heavy metal concentration in aquatic organisms: temperature and oxygen content, hardness, organic compounds, pH values, general physiologic behavior, life cycle and life history, seasonal variations, species-specific and individual variability, contamination by food and intestine content—Bryan (1971) listed similar points on the toxicity of metals—must be taken into consideration when evaluating the metal contamination of biologic systems. This is particularly true for determining the degree of metal enrichment within the food chain. The omission of any single factor can lead to serious misinterpretations.

In the following sections, therefore, organisms of various trophic levels will be investigated in respect to their metal uptake and storage properties in both the limnic and marine environment.

3 Heavy Metal Enrichment in Limnic and Marine Organisms at Different Trophic Levels

Contamination of the aquatic ecosystem by heavy metals can be confirmed in the water, organisms and sediment. In water, which is the primary contaminated medium, introduced heavy metals are exposed to many different chemical changes whereby a high degree of variation in metal concentrations arises. Thus, analyses of this medium supply only a very transitory view of the metal load (cf. Chap. E).

Metals that do not remain soluble in water eventually reach the sediment, where they become bonded to various components of the sediment such as clay minerals and organic matter (cf. Chap. E). In the two spearate media, water and sediment, the organism is a constantly active component within the aquatic milieu that transforms the chemical-static equilibrium of the abiotic phase into a biologic steady state.

3.1 Autotrophic Organisms

In the case of photo- and chemo-litho-autotrophic organisms, heavy metals are incorporated directly from the water. Several special species of the latter group (*Thiobacillus ferrooxidans* and *Ferrobacillus ferrooxidans*) can incorporate metals directly from inorganic compounds (such as metal sulfides). Some use the metal as an oxidation-reduction system (Fe^{2+}/Fe^{3+}; *Ferrobacillus ferrooxidans*) for their basic energy source. An incorporation of heavy metals does not occur from solid food particles so that the metal concentrations in autotrophic organisms are considered solely in the light of their uptake from water.

The biomass in a biotope steadily decreases from the primary producer level (autotrophic organisms) through the primary and secondary consumer levels to the end consumers. In the aquatic environment, this biomass concentration at the lowest trophic level can bond a large part of the heavy metals in the water to the organic substance. With regard to heavy metal enrichment in the food chain, the primary producers, since they are the basic organisms, represent the first stage of enrichment.

3.1.1 Phytoplankton

The relatively short lifetime of phytoplankton (few days to a few months) makes it very susceptible to changes in the abiotic parameters. The metal concentrations, therefore, can vary considerably within a short period of time.

Knauer and Martin (1973) established in the case of cadmium, copper, manganese, zinc, and lead that the respective metal concentrations underwent comparable changes in phytoplankton (diatoms) in Monterey Bay, California (Fig. 83). The cadmium concentrations in the water were seen to undergo a distinct decrease while at the same time the cadmium contents in phytoplankton increased. A similar observation was made in the case of manganese during a blooming period of the phytoflagellate *Phaeocystis* (Morris, 1971). When plankton algae are used as bio-indicators for metal pollution in water, it is possible that the metal concentrations in the water may still be low even if the algae are highly contaminated. That algae can store heavy metals over long periods even when the metal content in the water is low, has been demon-

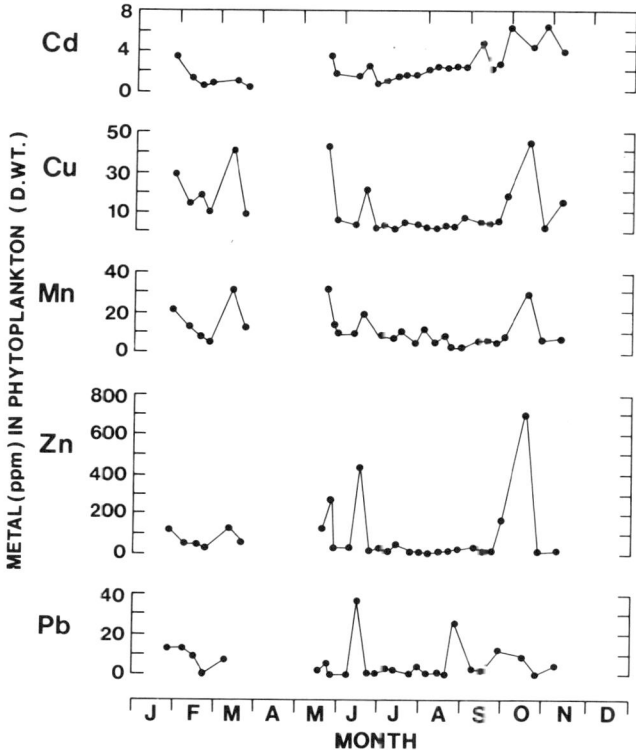

Fig. 83. Annual variations of heavy metal concentrations in diatom phytoplankton algae (Knauer and Martin, 1973, Limnology and Oceanography *18*, p. 601. Reproduced with permission of the American Society of Limnology and Oceanography, Ann Arbor, Mich.)

strated for lead in *Phaeodactylum tricornutum* (diatom) and *Platymonas subcordiformis* (flagellate) (Schulz-Baldes and Lewin, 1976).

Even after treatment with 10^{-2} M EDTA, a reduction in the lead content was only maintained during the first 20 min., but then settled at a level still many times higher than the normal degree of contamination. It must therefore be assumed that lead becomes strongly bonded to the cell, whereby, at least for lead, a relatively long half-life[1] must be assumed. Plankton algae thus seem to be well suited as medium-term indicators of metal contamination in water.

Although algae can eliminate heavy metals from water, phytoplankton do not generally have any great influence on the heavy metal content as the concentrations rarely occur as minimum factors for biologic growth. When comparing the metal content in the phytoplankton biomass with the metal concentration in the corresponding water, the increase in metal concentration in the phytoplankton is not indicative of the actual metal content in the water. But depending on their individual capacity to

---

[1] Biologic half-life: the species- and pollutant-specific time after which 50% of the total original metal content incorporated by a species still remains.

accumulate metals from the water, plankton algae as bio-indicators always reflect the average metal content of the water.

The interpretation of metal content in phytoplankton and in zooplankton is quite often difficult due to complications generally encountered in separating them, and to the fact that each can be contaminated to a different degree. Moreover, it is seldom possible to obtain the plankton separated in quantities adequate for metal analysis. Therefore, only those studies will be discussed in which a separation of phyto- and zooplankton has been successful, in order to avoid confusion with regard to both trophic levels and their respective metal concentrations.

It was established from diatom plankton that cadmium concentrations in Monterey Bay (Calif.) submarine canyon increased due to the influence of nutrient rich upwelling water. In contrast to this natural enrichment, the cadmium content in algae in near-shore waters is about twice as high as in algae in off-shore waters as a result of anthropogenic influences. In San Diego (Calif.) coastal waters, maximum values for cadmium were measured of up to 19.5 ppm (dry weight) (Knauer and Martin, 1973; Martin and Broenkow, 1975).

In the Monterey submarine canyon (Appendix B), the concentrations of mercury are, on the average, twice as high as those of cadmium. Both metal concentrations varied considerably during the period of investigation (0.104–0.590 ppm Hg). Moreover, the maximum value observed (0.590 ppm) can be attributed to changes in the species composition of the "phytoplankton", which are caused by an increased occurrence of radiolarians.

In *freshwater*, phytoplankton analysis, which may appear to be of value only in stagnant waters, is complicated due to the suspended matter content resulting from anthropogenic contamination of the water, with the effect that is is usually only possible to carry out an analysis of the seston. Many laboratory tests on freshwater plankton therefore concentrate exclusively on the incorporation kinetics of heavy metals (Chipman et al., 1958; Filip and Lynn, 1972; Fujita and Hashizume, 1975; Gilbert et al., 1976; Lorch et al., 1977). Geographic differences in metal pollution originating from lead (air) and cadmium (fertilizers) were determined using colonies of *Scenedesmus obliquus* as a monitoring system in culture basins in Thailand and the Federal Republic of Germany (Payer et al., 1976; see data in Appendix B).

3.1.2 Macroalgae

In addition to planktonic algae, macroalgae can also be used as indicators of metal pollution. Their metal contents are directly related to the metal concentrations in the water in the same way as those of phytoplankton (Bryan, 1971). The sessility of most marcoalgae clearly enables them to be associated with local concentrations of pollution and thereby facilitates an explicit interpretation of the increased metal concentrations in the plant. Among the primary producers, therefore, macrophytes are better suited as indicator organisms than the phytoplankton. Macroalgae have a longer live span than phytoplankton and thus represent a much longer period of metal contamination in the water. For *Fucus vesiculosus* (brown algae) the older parts of the plant show considerably higher heavy metal concentrations than the area around the growing point (Bryan, 1971).

In the marine environment, investigations have been carried out on macroalgae particularly in near-shore waters. A comparison of different geographic regions is possible since many species occur worldwide (Appendix B). Higher species of water plants have only rarely been the subject of investigation in the marine environment.

A detailed analysis of bladder wrack (*Fucus vesiculosus*) populations in coastal waters around Great Britain revealed significantly higher contamination in the eastern Irish Sea and North Sea by zinc, iron, manganese, copper, nickel, lead, copper silver, and cadmium than in other coastal waters surrounding Great Britain (Preston et al., 1972). The maximum values in the entire investigation area were zinc 962 ppm, iron 1517 ppm, manganese 190 ppm, copper 28.4 ppm, nickel 18.0 ppm, lead 9.0 ppm, silver 0.79 ppm and cadmium 20.8 ppm (dry weight). A comparison of these *Fucus* sp. analyses with earlier analyses in 1961 on the same type of algae in the same area of investigation did not reveal any significant changes during this ten-year period—except that cadmium concentrations had decreased considerably.

Other authors have found still higher concentrations of copper, lead, and silver in *Fucus vesiculosus* (Stenner and Nickless, 1975; Foster, 1976; Lande, 1977). In each case, however, the maximum values only occur in regionally limited pollution zones. The high metal concentrations in *Fucus vesiculosus* in the Anglesey Island/Menai straits region for example (Foster, 1976), can be traced to the Afon Goch River which is polluted by acid mine effluents (pH 2.62–6.80) (Foster et al., 1978).

Brown algae of the *Fucus* species are therefore excellently suited for the analysis of metal pollution of water. In addition to bladder wrack, other species from the brown algae genus (Phaeophyta) have also been used as biologic indicators of heavy metals. The different species are presented in Appendix B. Although relatively closely related, the different species reveal considerable differences in metal concentrations, even when taken from the same sampling site (Lande, 1977; Foster, 1976; Preston et al., 1972).

Similar differences can be observed in the case of green and red algae (Appendix B); they likewise must be regarded as species-dependent variations with respect to heavy metal incorporation. In one particular case, Stenner and Nickless (1975) found a high degree of heavy metal contamination in higher marine water plants which, on account of their distribution, are less suitable for comparative metal analyses. In an examination of sea grass (*Zostera* sp., *Potamogetonaceae*) in the estuary of Rio Tinto in southwestern Spain, the following concentrations were determined: 1430 (100) ppm for zinc, 1350 (9) ppm for copper, and 1800 (16) ppm for lead. These metal concentrations are clearly dependent on mining effluents. For comparison, other values (in brackets) are given of concentrations in polluted Cadiz Bay. Particularly conspicuous in the Rio Tinto estuary is the high lead content, which originates from the lead mining district of the Rio Tinto, as do the metals cadmium, copper, and zinc. These high values have possibly resulted from additional metal incorporation through the roots of the plant, since all bottom sediments are also contaminated to a high degree by these metals.

In the compiled data of marine algae investigations (Appendix B), the *zinc concentrations* do not appear to lie below 40 ppm or peak above 700 ppm. Higher zinc contamination, nevertheless, seems to be quite possible in marine plants, as demonstrated by the investigations on *Zostera* species.

In the case of *cadmium* the values are generally between 1 and 2 ppm. Minimum values occur in *Porphyra* sp. (0.05 ppm) and maximum values in *Fucus* sp. (20.8 ppm).

The *copper* concentrations vary widely. Minimum contents are near 6 ppm, with one exception of 1.5 ppm. Only in one case was 100 ppm exceeded (*Ascophyllum nodosum* 123 ppm).

In the case of *lead* the lower concentrations are between 2 and 3 ppm, with a minimum of 0.5 ppm. Maximum values are found in green algae (*Ulva lactuca*) at 18 ppm.

Chromium appears in an average concentration of 2–3 ppm with a fluctuation range of 1–13 ppm.

Iron has a very wide fluctuation range in algae and appears to undergo maximum enrichment of almost 4 g/kg with the *Fucus* and *Porphyra* species. Average concentrations lie between 100 and 200 ppm, the lower concentrations range between 15 and 50 ppm.

In the case of *manganese* the minimum concentrations lie between 6 and 10 ppm, with the maximum values reaching almost 200 ppm.

Nickel demonstrates a concentration range similar to that of lead. Lowest concentrations have a range of approx. 2 ppm. with one exception where a minimum of 0.2 ppm was reached. The maximum concentration is 22 ppm.

All values cited here refer to the dry mass of the plants investigated.

Although no global norms have been established for heavy metal concentrations in marine algae, it can be assumed that despite the individual, species-related, as well as seasonal variations in metal concentrations in plants, the lower concentration range corresponds to normal contents in algae. The higher concentration ranges—and this is the case with most of the data from the investigations decribed—should be attributed to general sources of pollution. The fact that the investigations on algae were carried out preferably in near-shore waters evidences the particular danger to the marine environment by heavy metals in the dissolved phase.

3.1.3 Freshwater Algae

Just as flora can be used as an indicator of heavy metals in marine environments, freshwater flora, ranging from algae to mosses and higher water plants, have been proven to be good indicators of heavy metals in the freshwater environment (Appendix B).

However, as a result of the general pollution of freshwater in industrialized countries, problems are encountered in the determination of metal concentrations in phytoplankton communities and microbenthic flora.

When analyzing algae for heavy metal content, a number of difficulties may arise during sample preparation that render the interpretation of the measured heavy metal concentrations difficult, e.g.:

1. Contamination of the algal aufwuchs by particulates and fine sediments which in many cases cannot be removed even by intensive washing.
2. The various species within the algal aufwuchs and phytoplankton associations can only be separated with difficulty. Furthermore, the metal concentration varies from one species to another and the combination of species present within the various samples of a series is rarely the same.

Among the freshwater flora plankton algae have a particularly short life span and are thus only able to take up those amounts of heavy metals as are available in the water during this period. In the epilimnion of southern Lake Michigan, Leland (1975) found that freshwater phytoplankton, like marine phytoplankton (Martin and Broenkow, 1975), only influence the metal content of water in certain near-shore areas of high biomass production. In all other cases, the trace elements (Cr, Cu, Fe, Pb, Mn, Zn) in the lake water were not significantly related to chlorophyll a.

Of the multicellular sedentary algae, the filamentous algae have been studied most thoroughly (*Cladophora* species). Because of size, abundance and wide distribution, this algae genus can be readily separated from others and thus lends itself well to comparative local studies as an isolated species (Leland and McNurney, 1974; Abo-Rady, 1977; Bibo, 1977).

Leland and McNurney (1974) found that the lead concentrations in the *Cladophora* species in the Vermillion River (Ill.) drainage net varied between 14.9 ppm in rural areas and 347 ppm in urban areas.

Keeney et al. (1976) likewise established that the metal contamination in the *Cladophora glomerata* algae was dependent on the heavy metal concentrations in their environment. The *Cladophora* population extracted from the heavily polluted Deadman's Bay on Lake Ontario had much higher zinc and cadmium concentrations that those on a remote island of the lake (Appendix B).

Cladophora algae were also studied by Abo-Rady (1977) in the upper Leine River downstream from Göttingen (Federal Republic of Germany). Here he found a significantly higher contamination by Cd, Cu, Hg, Ni, Pb, and Zn than further upstream. Compared to other aquatic plants (see below), *Cladophora* had the highest metal enrichments, in particular of lead with which this algae seems to become enriched preferentially (max. 49.2 ppm; other water plants in this investigation, max. 8.0 ppm).

Upon studying the amounts of zinc and copper enriched in *Cladophora rivularis*, Bibo (1977) found that the heavy metal contamination in two separate sections of the Elsenz River (a tributary of the Neckar River) differed significantly. The minimum and maximum lead concentrations in the algae compared favorably with the results of Abo-Rady on the Leine River. Bibo also established that the zinc concentrations in the *Cladophora* of the Elsenz River (max. 375 ppm) were up to ten times those of the Leine River (see Appendix B). Even higher metal concentrations have been found in freshwater algae (Bartelt and Förstner, 1977; Vogt and Kittelberger, 1977, see Appendix B), although in these studies partial contamination by suspended matter and sediment particles cannot be excluded as they were undertaken on algal aufwuchs. Algal aufwuchs (other than *Cladophora*) from the Elsenz River (Bibo, 1977), which was freed of particulates, displayed concentrations comparable to those of *Cladophora* algae. Unrinsed samples of the same algal aufwuchs with amounts of organic and inorganic particulates from sewage effluents had seven to ten times higher concentrations depending on the metals. The differences in the cadmium concentrations in algal aufwuchs, the total aufwuchs including bacteria and fungi, and particulates was lower than all other metals investigated (Pb, Zn, Cu). Leland and McNurney (1974) found high Pb concentrations (max. 265 ppm) in the periphyton of the Vermillion River, a water body heavily polluted by urban waste water.

Interesting results were obtained by Reay (1972) in his study of the arsenic enrichments in algae and aquatic flora. Algae (*Nitella hookeri*) from geothermically heated areas of New Zealand contained ten times more arsenic than algae from non-heated areas (at Hastings: 13 ppm). Fish (1963) performed similar studies on mosses (*Lagarosiphon major*) from Lake Rotura (New Zealand) and established maximum values of up to 120 ppm. Here again the metal concentrations of the various water plant species fluctuated despite their shared locality.

3.1.4 Mosses

It has generally been observed that mosses have a particular storage capacity for lead. In that respect, Heydt (1977) found levels of up to 277.4 ppm lead in *Fontinalis antipyretica* in the Elsenz River (Federal Republic of Germany); Dietz (1972) established absolute peak values of 2180 ppm in the same moss species taken from the heavily polluted Ruhr River (Federal Republic of Germany) as well as excessive lead values in *Hygroamblystegium* sp. (800 ppm). Cadmium concentrations between 0.17 and 13.2 ppm were determined in mosses (Appendix B), and the record amount of 13.2 ppm was established in *Lagarosiphon major* from the Lago Maggiore in Italy (Ravera et al., 1973). On the average, the mercury concentrations recorded by Dietz (1972) in the two moss species *Fontinalis antipyretica* and *Hygroamblystegium* sp. did not differ greatly from each other. It would thus seem that mosses are particularly valuable as indicators of lead enrichments. Nonetheless, it is important to note that in some localities such use is often severely limited because the ubiquitous distribution of these plants can be negatively influenced in polluted waters.

3.1.5 Higher Plants

Higher water plants are enabled by their roots to incorporate additional metals from the sediment (interstitial water). Here too, the heavy metal enrichments differ from species to species: hence the importance of using only one plant species when determining these enrichments.

Reay (1972) found high arsenic enrichments in the different families of higher water plants from the Waikito River, New Zealand, which were similar to enrichments determined in the algae *Nitella hookeri*. Maximum amounts of up to 0.5 g arsenic per kg of dry mass were recorded. The arsenic concentrations of the individual species studied varied by as much as 100%, even though they were measured at the same sampling station. Thus, these variations in enrichment can only be interpreted as being dependent on the physiologic and biochemical attributes particular to each species. Similar results were obtained for other metals by Dietz (1972), Abo-Rady (1977), and Heydt (1977) (see Appendix B).

Heydt (1977) also found different amounts of Cd, Zn, Pb, and Cu in the roots, stems and leaves of the different species. These differences decisively influence the total content in the plant shoot (see Table 76); the highest concentrations were found in the leaves. Although Abo-Rady (1977) found even higher lead concentrations in *Potamogeton pectinatus* from a section of the Leine River downstream from Göttingen (Federal Republic of Germany), Heydt (1977) found no difference in the concentration in either *Potamogeton pectinatus, Potamogeton crispus* or *Callitriche palustris*

Table 76. Heavy metal concentrations in different organs of higher water plants (from Heydt, 1977). n.d. = not determined, all data are dry mass related

Species	Roots				Stems				Leaves				Total (without roots)			
	Cd	Zn	Pb	Cu	Cd	Zn	Pb	Cu	Cd	Zn	Pb	Cu	Cd	Zn	Pb	Cu
Potamogeton pectinatus	0.33	54.9	2.20	16.09	n.d.				n.d.				1.18	241.5	5.80	49.87
Potamogeton crispus	n.d.				1.60	222.7	6.43	21.34	3.59	500.2	8.58	99.97	2.24	380.3	7.53	45.46
Callitriche palustris	n.d.				0.96	348.1	15.86	22.92	1.58	690.5	37.46	49.72	1.28	555.9	18.27	31.49

in two different sections of the Elsenz River. A possible homogenous pollution of Pb in an entire section of the river was thereby indicated. However, the amounts of cadmium, zinc, and copper varied significantly for all three plant species in the various river sections.

Considering the fluctuation range of metal enrichments in freshwater as a whole, no general pattern has been determined for the investigated metals and the various plant species. As in marine plants, the concentration of Zn and Pb in freshwater plants is higher than those of less common metals. All the species studied showed above-average heavy metal enrichments, whereby the point source of heavy metals are readily identified, particularly in flowing water. Depending on the plant species and family, and the physico-chemical conditions in the sediment (Gambrell et al., 1976; Patrick et al., 1977; Reddy and Patrick, 1977b), varying amounts of enrichment can be found within the same sampling locality.

Sedentary algae have a particularly high adsorption capacity for lead, and are thus well suited as indicator organisms. The use of terrestrial plant families as indicators of heavy metals, a process which was observed at a very early date, can be seen in the genetic adaption of plant communities in soils with natural heavy metal enrichments (cf. also Ernst, 1974). In recent years the same phenomenon was observed in aquatic plants (e.g., Russel and Morris, 1970; Stokes et al., 1973; Stokes, 1975). It is for this reason that such plants are becoming increasingly important for heavy metal research.

3.2 Heterotrophic Organisms

The metal uptake of the above mentioned plants occurs only via water, or for higher plants additionally via the roots. Corresponding to the nutrient uptake of the plants, heavy metals reach the plants only by mechanisms of passive and active transport, or by adsorption on the plant, e.g., enrichment in the mucus of *Cyanophytes*. The heavy metals entering these plants must necessarily be in a soluble phase.

In aquatic fauna, however, there is a greater variety of ways in which heavy metals can be introduced into the organism. Modes of incorporation can vary widely according to the species; three basic processes occur:
a) via respiration (gills, skin surface),
b) by adsorption onto the body surface,
c) from foods (solids and dissolved matter).

In respect to metal enrichment of heterotrophic organisms, the proportion of metal concentrations in water to those in nutriments is of decisive importance. The breathing mechanism in all aquatic organisms permits the uptake of heavy metals from water in which the metal supply is constant for all species. For this reason, the uptake from food assumes a greater importance because its heavy metal concentrations are prone to greater variations.

The feeding habits and rate of resorption are important factors in the appraisal of metal concentrations in organisms. In polluted water systems of freshwater or marine environments the heavy metal concentrations in sediments, particulates and detritus are many times higher than those in living organisms. It is thus necessary to distinguish between organisms which obtain their nutriments from such heavily

polluted sources and those which prefer other dietary sources. In order to establish the source of heavy metal contamination ensuing through dietary practices, the following feeding habits should be distinguished:
a) phytophagous (e.g., gastropods, crustaceans),
b) filter feeding (e.g., zooplankton, barnacles, bivalves),
c) sediment feeding (e.g., poly- and oligochaetes),
d) detritus feeding (e.g., gastropods, isopods and amphipods, chironomid larvae),
e) carnivorous (e.g., zooplankton, polychaetes, gastropods, cephalopods, crustaceans, freshwater insect larvae, fish).

Even from this short over-simplified listing the great variety of feeding habits of the different animal species is evident, so that a systematic classification into groups according to heavy metal uptake may not be very meaningful. Therefore, only those organisms will be mentioned which are either unequivocally related to one of the above mentioned groups or have a clear position within an investigated food chain or food web.

3.2.1 Zooplankton

Generally speaking, zooplankton are normally considered to be primary cons
This is, however, only true for the phytoplankton filter-feeding species (e.g., Daphnia). Very often there is a mixture of phytoplankton-feeding and carnivorous zooplankton in the sample, which increases the difficulty in the determination of the corresponding heavy metal contents in both trophic levels. Knauer and Martin (1973) found a high variation of Hg concentrations which they determined to be due to season- and species-related changes in the individual composition of the zooplankton (euphausiids, copepods, ctenophores, coelenterates, and radiolarians; see Appendix B). Furthermore, overlapping influences arising from the species composition depending on the season and locality, and which affect zinc and iron composition, were found by Van As et al. (1975) in plankton along the south west coast of South Africa. Variations of at least a factor of 100 in certain metals were determined here (Fe: 2.7–2900 ppm, Zn: 1.9– 710 ppm; wet weight related). With the latter zinc and cobalt data, the difficulty encountered in the interpretation of plankton data becomes clear (Table 77). An unequivocal classification of Zn and Co variations according to seasonal and zoocoenosis changes cannot be made. The variation coefficients from both monitoring sites differ widely indicating that the pollution affect at station III is markedly dissimilar to that of station IV.

In a field study (Skei et al., 1976) the heavy mercury contamination of the Sörfjørden in Norway could be detected by the use of zoo- and phytoplankton as indicator organisms, whereby maximum concentrations of 25.21 ppm Hg dry weight were found. The plankton analyses at the different sampling points strongly indicated the effect of point source contamination. The mercury concentrations in plankton samples gradually diminished with distance from the source, although the effects of increasing salinity on heavy metal elimination was not investigated. The lowest mercury concentrations occurred in the seawards situated Hardangerfjørd, which borders on the Sörfjørden, with a minimum of 0.52 ppm Hg dry weight.

Table 77. Variations of zinc and cobalt concentrations in zooplankton communities from two monitoring stations (III, IV) in South African coastal waters (data from Van As et al., 1975)

| Date | | Species | | ppm wet weight | | | |
				Zinc		Cobalt	
		III	IV	III	IV	III	IV
1971	January	Copepods	Copepods Euphausiids	13	1.9	–	0.004
	April	Copepods	Copepods Sagitta Euphausiids	37	10	0.025	0.015
	July	Copepods	Copepods	15	14	0.007	0.005
	October	Copepods	Sagitta	132	12	0.26	0.020
1972	January	Copepods Sagitta	–	4.6	–	0.007	–
	October	Classification incomplete		0.62	0.78	0	0.012
1973	January	Zooplankton: classification incomplete		2.1	5.1	0.003	0.023
		Variation coefficient		161.0	75.3	204.8	66.7

A clear differentiation on zooplankton samples was possible from only two sites, where no mercury concentrations higher than in phytoplankton samples could be observed.

The contamination of the low trophic levels in the same part of Sörfjørden by mercury has been reported by Haug et al. (1974), where maximum levels in brown seaweed (*Ascophyllum nodosum*) reached 20 ppm (dry weight), thereby indicating a high amount of dissolved mercury in the fjord.

Zooplankton samples taken off the eastern coast of the United States between Cape Hateras and Cape Cod by Windom et al. (1973) had an average mercury content of 0.5 ppm (dry weight related, range 0.06–5.3 ppm). Whereas the zooplankton taken near shore were usually higher and sometimes even exceeded 1 ppm Hg, levels in zooplankton from further off-shore averaged 0.2 ppm. There again, the results of Knauer and Martin (1973) for the influence of coastal pollution could be reproduced by plankton analysis. Cocoros et al. (1973) undertook similar studies in three bays along the eastern coast of the United States. Significant levels of mercury concentrations (0.11–0.19 Hg, dry weight) in 153 μm zooplankton in Core Sound (North Carolina), in Oyster Bay (New York), and in phytoplankton from Chesapeake Bay were found. However, it was not possible to take account of the species-specific differences in concentration.

Although zooplankton appears to be well-suited as an indicator of a metal uptake from water and from organisms in lower trophic levels that serve as nutriment sources (for zooplankton), it finds little use in comparative field studies as indicator substances. However, the importance of zooplankton cannot be overlooked in the evaluation of heavy metal enrichment in food chains.

3.2.2 Bivalves

Bivalves are exclusively made up of filtering organisms. Unlike the filter-feeding zooplankton organism, these are only slightly mobile, and thereby reflect the heavy metal composition of a narrowly limited space. Bivalves can even serve as replacements of the filtrating zooplankton to a certain extent, since they exhibit a similar diet pattern. On the basis of their biology, the bivalves are excellently suited for use as heavy metals indicator organisms. It is, however, not only for this reason that the effects of heavy metals on bivalves have been emphasized in investigations, but also because they represent an important nutrition source for man, and are in that respect an economic factor.

Many underlying principles important for the evaluation of heavy metal pollution in organisms have been studied in investigations on bivalves. Some of these studies have already been introduced in Section 2.

Other investigations have been carried out in the field and in the laboratory on the kinetics involved in the uptake of heavy metals; the kinetics depend on: the metal concentration in the water with respect to time, the seasonal variations, the distribution of the metals in the various organs, how one metal influences the content of another, the toxicity, the heavy metal uptake rates of the various bivalve species, photo sensitivity, the enrichment period in the organism for the given metal compound and the enrichment from the diet.

Investigations on the variations of metal content in bivalves show how the organism reacts to changes in its environment (Kečkeš et al., 1968; Bryan, 1973; Cunningham and Tripp, 1973, 1975a, b). Over a period of three years Bryan (1973) determined that zinc contents in *Chlamys opercularis* and *Pecten maximus* (queen scallop) had more than doubled. Minimum concentrations of zinc in kidneys of *Chlamys opercularis* occur seasonally in the spring, whereas the maximum rates are reached towards the end of the year (cf. upper curve of Fig. 84a). In the lower curve of Fig. 84a the zinc concentration of kidneys of *Pecten maximus* is presented. No significant seasonal changes are indicated. However the digestive gland of this species is of greater importance for zinc accumulation and therefore shows similar seasonal variation as did the kidney of *Chlamys*.

Influence of temperature and changes in the food supply are important factors influencing zinc content in *Mytilus*; the temporary zinc minimum, which coincides with the spawning period early in the year, is surely not accidental. Dare and Edwards (1975) investigating seasonal physiologic changes in respect to body weight/spawning period, and to the biochemical composition in *Mytilus edulis*, were able to present a further indication of the origins of the metal content variations in bivalves, as previously determined by Bryan (1973). They found that the gonads as well as the kidneys and the digestive glands have higher zinc concentrations than other organs. The influence of the spawning period on mercury content has been demonstrated by Cunningham and Tripp (1973) on *Crassostrea virginica*. Another factor for variations in heavy metal content in bivalves is the seasonally varying freshwater input into coastal waters. Bryan (1973) determined in investigations on the Tamar River estuary in England that high water occurring regularly in the autumn and winter can give rise to a higher metal pollution in scallop kidneys even after the water has entered the estuary (cf. Fig. 84a, b). The significant correlation between water flow and zinc content of *Chlamys opercularis*

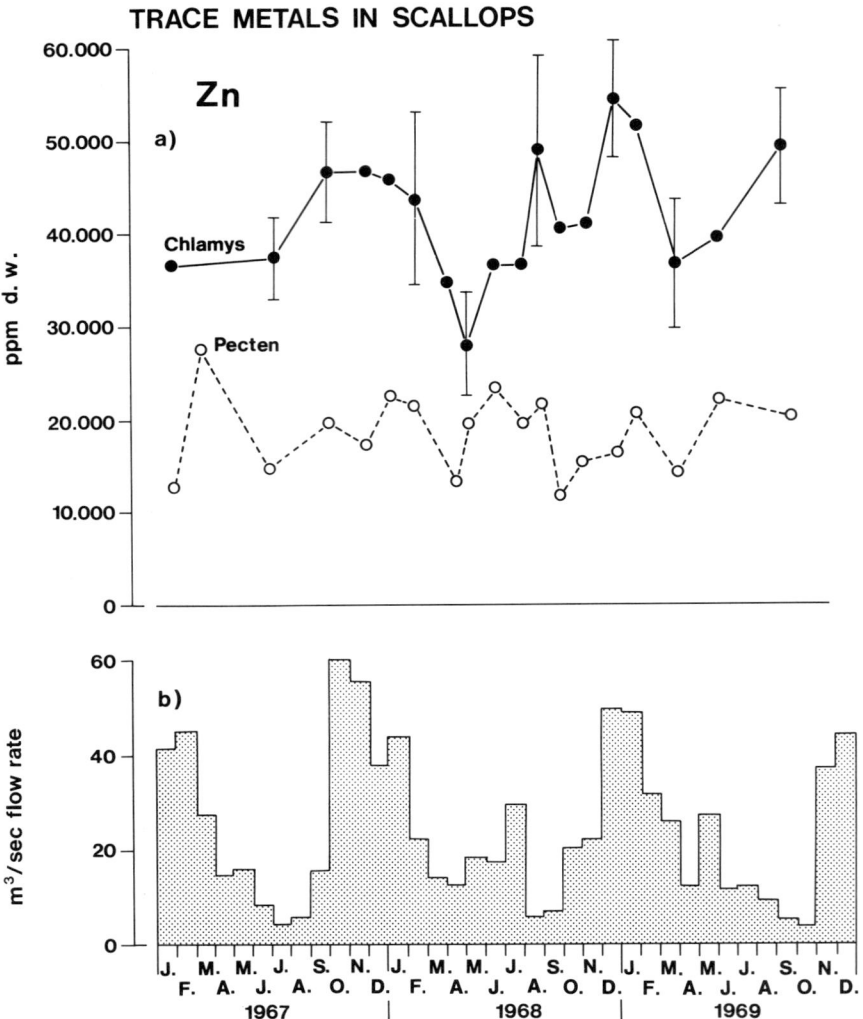

Fig. 84. Seasonal variations of zinc content in kidneys of scallops from the English Channel near Plymouth (a) and alterations of the flow rate of the Tamar river entering the sampling area (b) (after Bryan, 1973, J. Mar. Biol. Assoc. U. K. 53, p. 151, 159. Reproduced with permission of Cambridge University Press, London)

(upper curve of Fig. 84a) can be possibly explained as being due to an increased zinc load on particulate substances and the resulting high availability of that metal. For other metals Bryan (1973) established only a partial correlation. High metal concentrations in coastal waters on particulate substances originating from inland waters could thereby play an important role in the contamination of filter-feeding scallops (cf. Elderfield et al., 1971).

Phillips (1976a) established a similar seasonal variation in the metal content of the bivalve *Mytilus* in Port Phillip Bay, Australia, which is clearly influenced by the inflow of a fresh water tributary. Zinc and cadmium concentrations (Table 78) here proved to be about four times higher in winter than in summer, an effect that was especially pronounced in the upper water layers. Whereas during winter a concentration gradient is formed with increasing water depth, no definite variations could be determined for the summer. Phillips (1976a) attributes these differences to conditions of the Yarra River tributary system, where in winter—when water flow is greatest—effects similar to those in the investigations of Bryan (1973) were documented.

The investigation of time-dependent metal uptake by bivalves provides important knowledge as to the suitability of bivalves as indicator organisms. This is because of the rapid heavy metal uptake of these organisms even over brief periods of exposure to heavily contaminated water. The net uptake—the relation of heavy metal uptake to heavy metal loss—is decisive for these indicators.

Many authors have reported a rapid enrichment of ^{65}Zn in bivalves (Chipman et al., 1958; Romeril, 1971; Pentreath, 1973; Renfro et al., 1975). Renfro et al. (1975) reported a rapid uptake of ^{65}Zn during the first 20 days of an experiment on *Mytilus galloprovincialis;* after 50 days no net accumulation was observed, which is probably due to the steady state being reached between the exchange of ^{65}Zn and water. The incorporation rate of heavy metals is also dependent on their individual physico-chemical behavior in the aquatic environment. In a graphic representation various slopes of heavy metal uptake curves can be expected for each metal investigated. Friedrich and Filice (1976) report that nickel uptake and accumulation in *Mytilus edulis* did not differ significantly at concentrations of 18 and 30 ppb Ni/l over a 4-week period. Only at higher concentrations (more than 50 ppb) did a significant uptake of nickel occur; the slope remained much lower than for other metals (e.g., cobalt and zinc) when the available concentrations remain the same.

The species of chemical compound of the heavy metal also influences its incorporation in organisms and can affect concentrations of other metals. The mercury uptake rate in the American oyster (*Crassostrea virginica*) is different for mercury(II) chloride, methylmercuric chloride, and phenylmercuric acetate. The copper and zinc content in the oyster is slightly decreased by the influence of the former compound and heavily decreased by the latter two metal compounds (Kopfler, 1974). However, this decrease

Table 78. Variation of zinc concentrations in *Mytilus edulis* due to season and depth variations (data from Phillips, 1976a)

	Zinc in *Mytilus edulis* in whole soft parts (ppm, w.wt)	
Depth, m	Winter ($\bar{x}$, s)	Summer ($\bar{x}$, s)
0.5	97.1 ± 67.0	18.3 ± 6.5
1.5	44.2 ± 12.5	13.9 ± 8.1
3.5	29.9 ± 6.4	18.0 ± 5.7
6.5	31.7 ± 9.6	17.8 ± 7.2
9.5	20.6 ± 4.1	15.5 ± 6.1

was not observed in another experiment using only 1 ppb mercury in the water (50 ppb were used in the first). Here, the uptake of organic mercury was much higher and that of mercury (II) chloride lower than in the first experiment.

Although the kinetics involved in the uptake of heavy metals in the various organs may appear to be similar, the contents in the organs differ by several orders of magnitude when maximum enrichment has been reached.

For cobalt[58] Shimizu et al. (1971) determined after approx. ten days an enrichment in *Mytilus*, which only changed negligibly during further experimentation. At the same time, however, a change in the excretion rate of ^{58}Co—occurring through the faeces—was registered, which increased approaching the tenth day, whereas the ^{58}Co content in the organs remained constant (Fig. 85a, b). It can therefore be assumed that the steady state between the organism and the water, which is reached after about ten days, is due to an increased excretion of cobalt.

No difference in the enrichment of ^{58}Co could be established in a comparison of the shells of live and dead freshly separated mussels. This indicates that the metal uptake in the shells occurs passively and is not influenced by the mussel itself. It was subsequently shown that in the shells of the dead mussels the metal was mainly adsorbed from the exterior wall. A passive uptake of ^{58}Co was also determined for the *byssus* of the investigated mussels.

Romeril (1971) found that the uptake of ^{65}Zn in the shell of *Ostrea edulis* increased with the addition of Fe and Co; a linear uptake of ^{65}Zn in soft organs was however suppressed. This effect supports the results of Shimizu et al. (1971) concerning the passive uptake of metals in bivalve shells. Therefore, mussel shells—and thereby probably all mollusc shells—are not suitable as bio-indicators. In investigations on the enrichment and desorption of ^{65}Zn, Kečkeš et al. (1968) found that the uptake of ^{65}Zn by the shell of *Mytilus edilus* is not greatly affected by biologic mechanisms. These authors postulated that a passive uptake occurs rather by mechanisms of physicochemical adsorption. The shells were therefore exposed to a ^{65}Zn solution only for a short period of time; the rapid removal (about 70% of the heavy metal ions—much quicker than from the soft tissues of the mussel— correspond to the cation exchange between the water and the shell.

Apart from the varying distributions of ^{65}Zn in bivalve organs, a species-related accumulation was observed in specimen from the same collection point (Chipman et al., 1958). Here, the ^{65}Zn accumulation was greatest in oysters, less in hard shell clams and least in bay scallops. A local variation of ^{65}Zn behavior within a single species is also reported by Romeril (1971): The uptake slopes of Portuguese oysters (*Crassostrea angulata*) established in laboratory experiments were different in Sado and Tagus river waters. The different ^{65}Zn uptake must be due to the higher zinc content of native Tagus oysters, which was found to be about three times greater than those of the Sado River. With these factors in mind, it is evident that the soft parts of bivalves are excellently suited as metal indicators in water.

The uptake of heavy metals from food substances—in this case lead—is shown by Schulz-Baldes (1974) on artificially contaminated algae and mussels (*Mytilus edulis*). The rate of metal uptake at the beginning of the experiment is directly proportional to the lead contamination of the food.

Bivalves

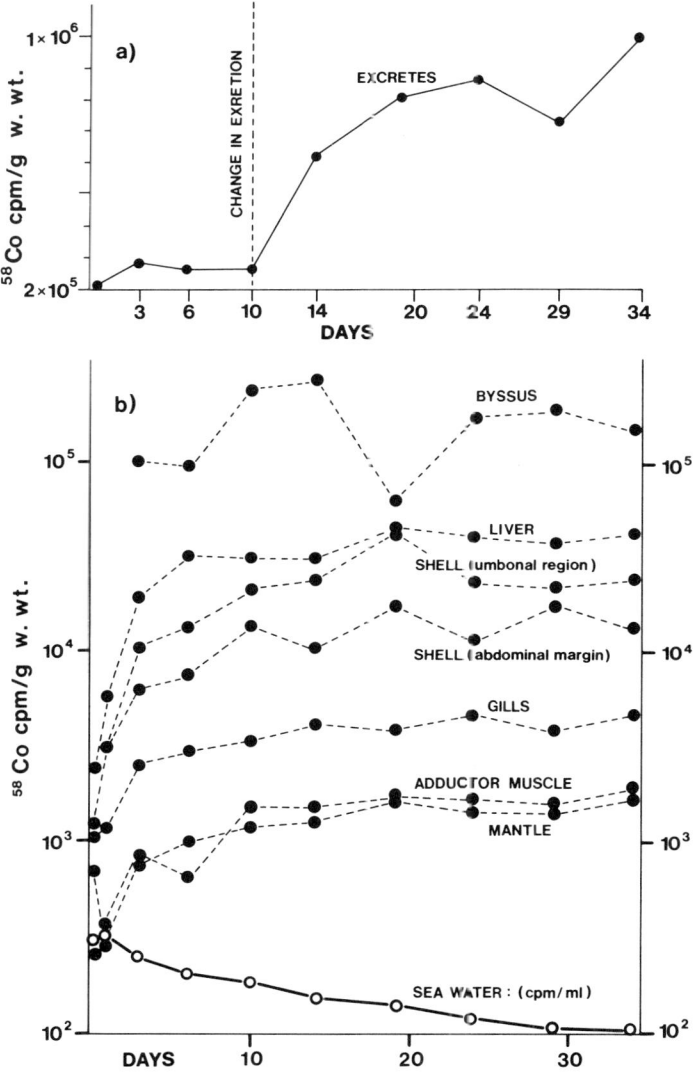

Fig. 85. Uptake and distribution of ^{58}Co in different organs of the mussel *Mytilus edulis* (b) and changes of cobalt excretion in feces (a) (after Shimizu et al., 1971)

However, the conditions of this experiment indicating food chain enrichment were unrealistic since the lead (600 ppm, dry weight rel.) contamination of the algae was much higher than that occurring normally. Furthermore, in polluted waters the representation of a low trophic level by the filter-feeding organisms does not correspond to the actual conditions. As the organism ingests suspended matter originating from waste water and thereby containing high metal concentrations, the filter-feeding bivalves then feed on the highest level, which is caused by man and his activities.

The acute toxicity of heavy metals for embryos of marine bivalves is reported e.g., by Calabrese and Nelson (1974). The range of the heavy metal concentrations in the water for the LC_o and LC_{50} limits (lethal concentration limit where 50% of the organisms die) are in agreement with the metal concentration of polluted waters (e.g., Hg: $LC_o - LC_{50} = 1-5.6$ ppb, Zn: 75–310 ppb for oysters). Therefore a chronic toxicity of heavy metals by a long-term exposure in subacute metal concentrations can be assumed. In this respect, the uptake of heavy metals in bivalves through water and food and the resulting metal concentrations of bivalves in the natural environment is an important aspect of the composition of bivalve species in an aquatic zoocenosis. Concentrations of heavy metals in marine bivalves, along with examples in freshwater are compiled in Appendix B.

The variations in heavy metal content in bivalves and in snails reveal the ability of these organisms to reflect the heavy metal contamination of their environment, whereby a differentiation of local pollution effects is possible. Regarding the usefulness of a species as an indicator of heavy metal contamination, variations occurring at a certain locality are decisive.

Due to their wide distribution in the marine environment *Mytilus* species (especially *Mytilus edulis*) and oyster species (*Ostrea edulis* and *Crassostrea* species) have proved to be especially useful indicator organisms. In his "mussel watch", Goldberg (1975c) has even suggested that as long-term indicators bivalves can make certain water and sediment sampling procedures unnecessary. However, other indicators are necessary for copper, as this metal can be metabolized and forms hemocyanin in molluscs (Phillips, 1976a). This is seen in the high copper concentrations in the soft parts of freshwater gastropods from unpolluted water (*Radix peregra:* 116–160 ppm Cu, dry weight; Heydt, 1977). The contamination of marine coastal regions and especially the origin of the contamination—be it through point source waste water effluents or through estuaries where the rivers are highly polluted by heavy metals—has been shown by various authors (e.g., Boyden and Romeril, 1974; Stenner and Nickless, 1975; Thornton et al., 1975; Fowler and Oregioni, 1976; Bryan and Hummerstone, 1977).

The usefulness of gastropods as indicators is substantiated for several species (*Patella, Nucella, Littorina;* cf. also Appendix B). However, due to their behavior differences in respect to the food supply (phytophagous, deposit-feeding, and carnivorous), they are not as suitable as the filter-feeding bivalves. Their importance lies rather in the uptake of heavy metals within a food chain in which the species can be at any of the various trophic levels. Mussels taken from the Derwent Estuary, by Bloom and Ayling (1977) were found to be unfit for human consumption because of their extremely high metal content (Hg, Zn, Cd, Pb; cf. also Fig. 86 and Table 79). Equally high mercury rates (1000 ppb, wet weight) were registered in *Mytilus edulis* by De Wolf (1975) on the English coast of the North Sea and neighboring countries.

In freshwater, monitoring of heavy metals using bivalves is difficult since heavily polluted rivers or lakes automatically inhibit the growth of these organisms (Nancy and George, 1977; Merlini et al., 1978).

Investigations of different bivalve species that were transferred from an unpolluted lake into the Elbe River near Hamburg revealed—depending on the section of river— various contaminations of Hg, Cr, Fe, and Co (Karbe et al., 1975). Similar results for

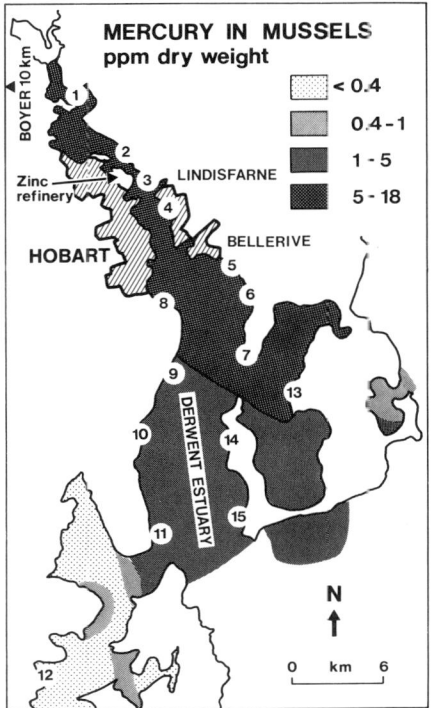

Fig. 86. Mercury contamination of mussels in the Derwent estuary (Tasmania) (after Bloom and Ayling, 1977)

Table 79. Heavy metal concentrations in mussels, sediments and filtered water of the Derwent estuary (Tasmania). Site numbers are identical with locations in Figure 86 (after data from Bloom and Ayling, 1977)

Site No.	Conc. in dried sediments, ppm				Conc. in mussels (d.w.), ppm				Conc. in filtered water, ppb			
	Cd	Pb	Zn	Hg	Cd	Pb	Zn	Hg	Cd	Pb	Zn	Hg
1	40	830	4090	7.5	23	58	228	1.6	1.7	5	46	–
2	50	1000	2500	5.8	25	64	171	12	1.0	10	36	16
3	50	1820	9120	13	38	121	180	13	<0.5	5	78	0.5
4	15	1000	3900	18	23	378	1350	3.0	–	16	230	–
5	2.5	266	851	3.6	19	291	854	7.7	0.8	11	128	0.1
6	13	384	1100	12	22	226	556	4.8	<0.5	10	64	0.1
7	8.0	600	2000	10.5	20	452	725	5.0	0.6	16	112	0.2
8	15	950	3180	15	29	532	744	6.7	2.1	10	230	–
9	–	13	50	0.1	13	127	322	2.1	2.1	9	210	–
10	–	13	50	0.1	9.3	103	401	1.8	0.8	6	120	–
11	–	–	250	1.3	6.8	59	354	1.6	<0.5	12	70	–
12	–	36	172	0.5	6.2	16	296	0.4	<0.5	10	6	–
13	–	59	169	1.6	26	239	572	5.1	–	11	232	0.6
14	9.3	678	1810	12	27	255	702	3.6	–	11	232	0.6
15	–	5	100	0.6	23	376	699	3.0	0.5	17	144	–

mercury were found for snails (*Radix balthica*) taken from their natural environment and released in the same river section (Zauke, 1977). Long-term metal contamination control could then be easily performed by systematically using uncontaminated bivalves in bioassays in water bodies in which no bivalves are otherwise found. In Appendix B, data on the metal contents in freshwater bivalves and gastropods are compiled.

3.2.3 Higher Marine Crustaceans

The metal content in higher crustaceans is of bearing in respect to toxicity and human dietary practices on the one hand, and to the marine food chain on the other, because such crustaceans are on a trophic level higher than bivalves. However, their feeding habits happen to be of many different forms; they are phytophagous, carnivorous, or both. An additional complication is their ability to change certain basic aspects of their feeding habits during the life cycle.

Relatively speaking, from laboratory experiments much is already known about the uptake kinetics of metals and their storage and distribution in the body organs of crustaceans (cf. Bryan, 1967; Bryan, 1971; O'Hara, 1973; Renfro et al., 1975; Jones, 1975, and Luoma, 1976). According to observations (of these and other researchers), biologic factors—as similarly ascertained in bivalves in respect to heavy metal enrichment—such as organ distribution, and seasonal and individual variations must be considered. Especially important is the high percentage of the total zinc lost each time *Carcinus maenas* moults its exoskeleton (cf. Renfro et al., 1975). The degree to which this applies to other metals has yet to be established. The organ-specific metal distribution here plays an important role. In *Carcinus maenas* the highest cadmium concentration was in the exoskeleton, and the lowest in the muscles, whereas for zinc the highest concentration was found in the muscle, and the lowest in the gills. At 63.8 ppm (wet weight) the highest amount of copper was in the hemolymph (hemocyanin!), the concentration in muscle tissue being only one-tenth of that (Wright, 1976).

In the field, the musculature of crustaceans was investigated most frequently as different contamination rates occur in that body part, i.e., mercury between chela and body muscle (Luoma, 1976).

The compiled data for crustaceans in Appendix B are not readily comparable because too little information on the individual species has been collected. In contrast to other metals the copper concentration is on the average relatively high, perhaps because the copper is complexed in hemocyanin. Although all metals have a high rate of variation according to locality and crustacean species, extremely high contaminations, such as were found in bivalves, rarely occur in the higher crustaceans.

In the sedentary *Cirripedia*, a lower sub-class of crustaceans, very high metal concentrations appear, however, which are possibly due to the sessile and filtrating characteristics of their life cycle. In specimens from southwest Spain and southern Portugal Stenner and Nickless (1975) compared metal concentrations of *Cirripedia* and higher crustaceans. It was found that *Cirripedia* appear to be good zinc indicators—with a maximum enrichment of > 3,000 ppm (dry weight related). Additionally, the high zinc enrichment for *Balanus* was confirmed by Ireland (1973) with maximum values of 23,000 ppm zinc (dry weight related) and by Walker et al. (1975; cf. Appendix B for both results).

3.2.4 Freshwater Crustaceans

Data on freshwater crustaceans are unfortunately quite limited. Zauke (1977) investigated the mercury content of Gammarids from the lower Elbe River. As the limnic environment gradually becomes brackish, a resulting decrease of mercury is observed. In the Gammarids and *Asellus aquaticus* samples the Hg levels of those from the brackish region were comparable to the ratios of the concentrations in the sediment (cf. Table 80).

Table 80. Relative Hg enrichment in crustaceans and sediments in the Elbe delta (from Zauke, 1977; sediment data of Förstner and Müller, 1975)

Crustaceans	12	:	7	:	4	:	1
Sediment ($< 2 \mu m$)	14	:	5	:	3	:	1
Station	a		b		c		d
Salinity	limnic				brackish		

The distribution of Hg in the crustaceans from the four sample sites is identical with that in the sediment fraction $< 2 \mu m$. However, it cannot be concluded that the mercury in the crustaceans originates from the sediment; rather an ecologically active presence of Hg can be assumed on the strength of the same distribution pattern in crustacean and sediment. Additionally, the osmoregulatory effects of animals that inhabit brackish water play an important role in their heavy metal regulation.

Among the freshwater isopods, *Asellus aquaticus* has been investigated for mercury, especially in Sweden (Johnels et al., 1967; Hasselrot and Göthberg, 1974), for which various rates of contamination in lakes and rivers were established. Local mercury contamination rates—for example downstream of paper mills, which use organic mercury compounds as wood preservatives, where they were 30 times greater than upstream— were registered as well (cf. Appendix B).

Other metals (Pb, Zn, Cd, Cu) were determined in a tributary of the Neckar River, the Elsenz River (Federal Republic of Germany; Prosi, 1977). At two sections of the river with different metal pollution levels, a significantly higher metal enrichment in *Asellus* was registered for copper only, although significant increases were recorded for the other investigated metals (except for Pb) in sediment from the same sampling site. In conformity with general metal frequency zinc showed the highest absolute concentrations in that species followed by copper, lead, and cadmium. However, corresponding to the minimum local values of *Asellus* the maximum enrichment factors (max/min) were: for Pb 6.1, Cd 6.3, Zn 1.7, and Cu 2.3. It is evident that in relation to minimum values, the nonessential metals Pb and Cd are enriched to a greater extent than are the essential metals Zn and Cu. Homeostatic control may be at work for the latter.

In urban-influenced drainage Leland and MnNurney (1974) established in decapods significantly higher contamination with lead (11 ppm, dry weight) than those subjected to rural drainage (4.7 ppm, dry weight).

These few data do not give a complete picture of metal enrichments in freshwater crustaceans, but such values as are determined do show that, in freshwater, crustaceans are also suitable indicators of heavy metal pollution.

3.2.5 Marine and Freshwater Fish

Compared to bivalve mollusks, a great deal of data is available on the heavy metal content of fish in both the marine and freshwater environments (cf. Appendix B). Three major aspects must be considered in the investigation of fish:

Firstly, *the use of fish as indicator organisms for the heavy metal pollution of their environment* and their possible unfitness for human consumption from a toxicological point of view (Johnels et al., 1976; Uthe and Bligh, 1971; Ui and Kitama, 1971; Lovett et al., 1972; Dworsky et al., 1973; Krüger et al., 1975). It is mainly the muscle tissue which is investigated in this context.

Secondly, *the use of fish to study the physiological behavior of heavy metals.* Here the most important factors are: distribution of heavy metals in body organs and the respective affinity of these organs for metals, uptake kinetics, regulatory mechanisms (especially for essential metals), effects on the metabolism by heavy metals (e.g. Hg), the synergism of metals and their uptake (Goldberg, 1962; Jefferies and Hewett, 1971; Merlini et al., 1971; Pentreath, 1973; Suzuki et al., 1973; Koli et al., 1977, Müller and Prosi, 1978). Thirdly, *fish as the end consumer in the aquatic food chain* and thus their *use as an indicator of heavy metal enrichment* (cf. Section 3.4). Knowledge of biologic factors such as age and size (Fig. 81), life cycle and life history, seasonal and local variations of heavy metal content in the animal, and the trophic level of the species, as well as of the biologic half-life of the metal is essential (Bache et al., 1971; Uthe et al., 1973; Brook and Rumsey, 1974; Hardisty et al., 1974a, b; Hasselrot and Göthberg, 1974; Nisimura, 1974; Scott, 1974; Mackay et al., 1975; Matsunaga, 1975; Brown and Chow, 1977).

The study of fish muscle tissue is one of the means for investigating the amount of heavy metals entering the human by food chain enrichment and has therefore been investigated more than other organs.

On the basis of the varying affinities of the metals for the individual organs, the musculature proves not to be the most suitable body part for determining the extent of the heavy metal contamination of the entire organism. The absolute increase of heavy metals in muscle tissue of contaminated fish is often much lower than in other organs. A generally higher species-specific variation in metal content can, therefore, not always be determined by analysis of the muscle system (Fig. 87). Furthermore, it appears that the musculature only then becomes enriched by metals when the contamination is extremely high (cf. also Jernelöv and Lann, 1971). In the musculature of the roach (*Rutilus rutilus*) from a section of the Neckar River a ~25-fold relative increase between minimum and maximum cadmium contamination was registered. In the liver there was a 156-fold increase and in the kidney, a 196-fold increase (Müller and Prosi, 1978). In one individual, the maximum enrichment in the kidney was 35.3 ppm (w.wt.), whereas in the musculature of the same fish, only 18.0 ppb (w.wt) were measured. Organs with the greatest affinity to heavy metals would therefore appear to be more suited for the evaluation of metal contamination in fish. Moreover, with severe contamination, it is primarily the gills which show elevated cadmium concentrations, whereas the other organs are less readily contaminated (Mount and Stephan, 1967).

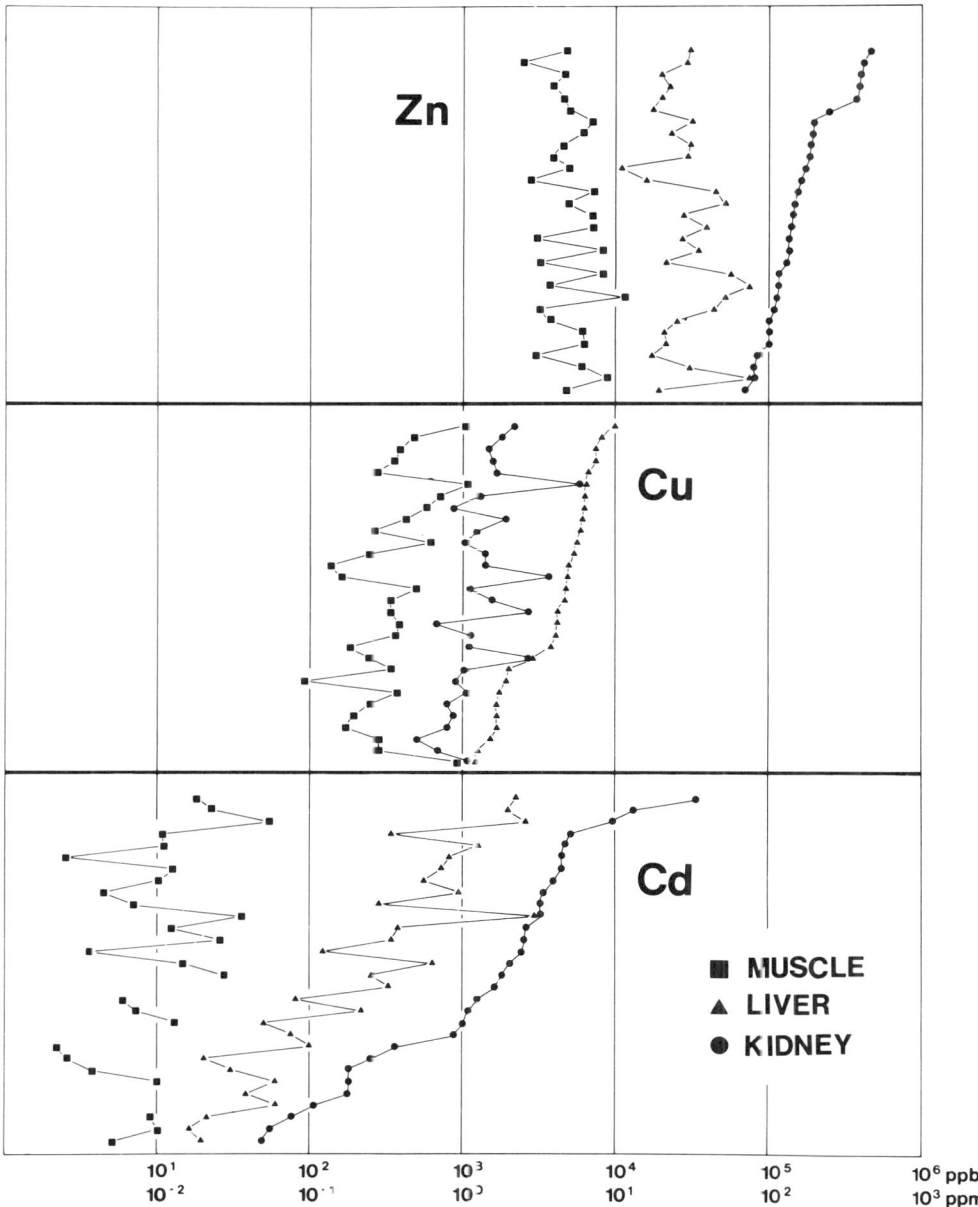

Fig. 87. Distribution and variation of Zn, Cu, and Cd in organs of roaches (*Rutilus rutilus*) of the Neckar and Elsenz rivers (Federal Republic of Germany). Each point represents one fish; ppb Cd, ppm Zn and Cu; wet weight related (Müller and Prosi, 1978)

The scales of Atlantic salmon (*Salmon salar*) and brown trout (*Salmo trutta*) were analyzed by Abdullah et al. (1976) in order to determine heavy metal contamination. In part of the sample a distinct correlation was established between concentrations in the scales and in the environment. The analysis of zinc in the otoliths of mackerel (*Scomber japonicus colias*) was also discussed by Papadopoulou et al. (1978) as a possible method; a negative linear function was found between the age and the zinc content of the fish. Hardisty et al. (1974a, b) found similar results in all parts of flounder (*Platichthyes flesus*) from the Severn Estuary as did Cross et al. (1973) in the muscle of morids (*Antimora rostrata*). Many of the biotic and abiotic factors that influence the uptake and release of metals have been investigated for mercury, which, especially due to bacterial methylization, can be taken up by organisms as organic compounds. For this reason, it is most frequently found in the organism in the form of organic compounds (e.g., Olson et al., 1975; Ruohtula and Miettinen, 1975; Lindahl and Hell, 1970; Fagerström et al., 1975; Freitas et al., 1974; Bishop and Neary, 1974).

Taking characteristic examples from the numerous investigations on the marine and freshwater environment, the suitability of fish as heavy metal indicators will be discussed in the next section.

Marine Fish. The dependency of heavy metal enrichment on the age and size of the fish has often been discussed. One of the most striking investigations is that of Miller et al. (1972) who detected similar mercury concentrations in museum specimens of tuna and swordfish species (less than 0.5 ppm), and freshly caught fish. In the muscle tissue of black marlin *(Makaira indica)* from apparently unpolluted waters off northeast Australia, Mackay et al., (1975a) found very high mean mercury and selenium levels of 7.3 and 2.2 ppm respectively (wet weight). Such values exceed those allowed for human consumption. A highly significant correlation between Hg and Se on the one hand and the weight of the fish on the other was reported. The high metal values in muscle tissue can only be explained by the longevity of the blue marlin which enables a weight-age related bio-magnification. The highest values of all metals were found in the liver, which is expressed by a greater variation coefficient in respect to the musculature (cf. Table 81).

Table 81. Variation coefficient in muscle and liver tissues of 42 black marlin taken from northeast Australian waters (data from Mackay et al., 1975a)

Metal	Muscle %	Liver %
Hg	8.2	18.8
Se	7.0	15.7
Cd	1.0	22.8
Zn	3.3	20.7
Cu	6.3	16.7
Pb	3.5	4.4
As	3.0	8.3

In the pelagic northern anchovy the same distribution in muscle and liver tissues was found for mercury, although the concentrations were much lower (0.04 ppm in muscle and 0.09 ppm in the liver—wet weight; Knauer and Martin, 1972).

Similar tendencies for mercury (as shown by Mackay et al., 1975a) have been found in Pacific blue marlin (*Makaira nigricans*) taken in Hawaiian waters (Schultz et al., 1976), mean values of 2.06 ppm (w.wt.) in muscle tissue and 6.3 ppm in liver tissue were registered. A significant correlation between weight and mercury content of the fish muscle was determined. A natural pollution of the fish by active volcanoes and geothermal activity in their habitat has been discussed, but has yet to be confirmed.

The analysis of 12 specimens of blue shark (*Prionace glauca*) caught in the northwest Atlantic revealed only low concentrations of Cu, Zn, Cd, and Pb in the muscle tissue (cf. Appendix B). Variations in different muscle sections of each individual fish were detected for copper and zinc. The heavy metal levels in muscle tissue of the shark did not vary from their fish prey (Stevens and Brown, 1974). However, high Hg concentrations have been determined in other *Chondrichthyes* from the North Atlantic and the North Sea (Krüger et al., 1975). In the basking shark (*Cetorhinus maximus*) a mean value of 2.1 ppm was measured (values range from 0.9 to 4.0 ppm), and in 3 porbeagle samples (*Lamna nasus*) 1.5, 2.2., and 3 ppm. All these fish, therefore, have metal contents above the allowable limits for human consumption in Federal Republic of Germany, which is set at 1 ppm Hg. The Hg concentrations in the muscle tissue of picked dogfish (*Squalus accanthias*) are considerably lower—a mean value of 0.24 ppm was measured.

All of the above fish species are generally found in many parts of the Atlantic Ocean and therefore cannot be associated with any specific limited region. Because of their longevity and possibly also their position as end consumers, such high metal concentrations (at least for mercury) can be determined. Local variations of mercury contamination were recorded by Krüger et al. (1975) in Norway haddock (*Sebastes marinus*). Off the Norwegian coast concentrations were clearly higher than in the fishing areas near the Faeroe Islands and Iceland. Species from the gadidae family that were studied all revealed Hg concentrations under 0.5 ppm.

In halibut (*Hippoglossus hippoglossus*) a clear-cut interrelation was established between Hg concentration in the muscle tissue (< 0.05 to 1.0 ppm Hg) and weight (< 2 to > 100 kg). Fish with a weight of > 20 kg ≙ a Hg content > 1 ppm are therefore unsuitable for human consumption as well. The origin of this high Hg-contamination could not be determined, as another closely related fish (*Reinhardtius hippoglossoides*) had much lower Hg quantities ($\bar{x}$ = 0.15 ppm).

Similarly contaminated are fish species from the Elbe Estuary. The mean values of *Anguilla anguilla, Osmerus eperlanus, Alosa fallax,* and *Acerina cernua* are all higher than 0.5 ppm in the musculature; furthermore, in the latter species a mean value of 1.5 ppm was registered. An acute danger for humans is thus present, especially since the coastal fishing industry is concentrated within the Elbe Estuary.

Other localized pollution was found by Peden et al. (1973) in the Bristol Channel. Dogfish (*Scyliorhinus* sp.) had high Cd and Zn concentrations (cf. Appendix B). Cd concentrations were generally higher than 100 ppb in muscle tissue (wet weight).

The local pollution effect of river inflows in the vicinity of industrialized inland regions was confirmed by Stenner and Nickless (1975) in coastal areas of southwestern Spain and southern Portugal. High cadmium concentrations were found in the muscle of sole (*Solea solea*) of up to 2.1 ppm (dry wt.), and with regard to the high cadmium concentration in the gills (up to 4.3 ppm, dry wt.) a metal intoxication of the fish has apparently occurred. However, too few specimens were investigated and no clear explanation was given as to the pollution origin.

In a study by Windom et al. (1973), in an area of the North Atlantic east of the Gulf Stream in the vicinity of the Sargasso Sea, similar levels of heavy metals in both coastal and off-shore species of *Chondrichthyes* and *Osteichthyes* were found. In muscle tissue only the metal arsenic showed a marked difference; for the former species 8.5 ± 6.2 ppm (dry wt.) was recorded and only < 1.0–2.5 ppm (dry wt.) for the latter. Although a difference in heavy metal contamination between coastal and off-shore fish as well as between different classes (*Chondrichthyes* and *Osteichthyes*) had been expected, again no clear pattern evolved, probably because too few specimens—in some cases only one—were analyzed. In respect to the Hg content of *Chondrichthyes*, similar results are indicated but were not as evident as for As. This is substantiated by the investigation of Krüger et al. (1975), whose findings reveal higher Hg concentrations in *Chondrichthyes* than in *Osteichthyes*. Local arsenic contamination of fish was found by Bohn (1975) at the inlet area of Marmorilik in western Greenland where lead and zinc mining waste is dumped. The overall range in fish muscle was 11.1–307 ppm As (dry wt. related), which is much higher than the concentration measured in *Osteichthyes* by Windom et al. (1973) and in black marlin muscle tissue ($\bar{x}$ = 0.6 ppm, dry wt.) (Mackay et al., 1975a). Investigations of flatfish, such as the American plaice (*Hippoglossoides platessoides*), revealed very high concentrations in muscle and liver tissue. In the muscle tissue size-dependent concentrations were found.

In the marine environment the heavy metal content in fish is usually low with the exception of coastal areas with high local pollution. Such regions are generally not very extensive, but the pollution present can impair the local fishing industry as shown by Krüger et al. (1975) for the German Bight, which is heavily polluted with mercury originating from the Elbe River.

In contrast, high metal concentrations in edible finfish with a life span similar to that of marlin, tuna, and shark can be attributed to the biology of the species. The result is a naturally higher background level of heavy metals such as Hg and As. Due to their longevity, however, these species have much more time to adapt to slight increases of heavy metals in their habitat. In these cases it is therefore especially difficult to establish whether an enrichment is due to natural causes or arises through anthropogenic influences.

Freshwater Fish. The question of heavy metals in freshwater fish first became an issue only in the sixties when serious mercury contamination was detected in Sweden and later in Canada (cf. review of Ackefors et al., 1970). The great amounts of data available on mercury in fish has led to a good understanding of the concentration- and time-dependent kinetics of mercury uptake from the various milieu as well as the biochemical metabolism of mercury in the fish. Freitas et al. (1974) reported a different biologic half-life of mercury for various fish species and for various Hg compounds (e.g., CH_3HgCl and $HgCl_2$). The biological half-life of methylmercury in perch (*Perca flavescens*) taken up from water or food was 69 days.

For mercury (II) chloride the biological half-life of the metal was 34 days when uptake occurred from water and only 14 days by incorporation through food. In field studies (Lake St. Clair, Canada) Bishop and Neary (1974) found that on the average

88.9% of the total mercury in fish musculature was in the form of methylmercury irrespective of the weight, length, and species of the fish. The assumption can be made that most of the mercury in the ecosystem was available as organic mercury compounds to fish. According to Freitas et al. (1974) the high retention rate of methylmercury (85% or more, compared to only 10% to 15% for $HgCl_2$) was thus responsible for this high amount of organic mercury in the organism. Bache et al. (1971) reported an increase of methylmercury in whole fish with age in lake trout (*Salvelinus namaycush*). Age-dependent Hg enrichments were also found by Johnels et al., (1967) in northern pike and by Hasselrot and Göthberg (1974) in roach. Due to this age-dependent bioaccumulation and in order to be able to more accurately compare data, concentrations of the metal must be corrected according to fish size (cf. Fagerström et al., 1975). Johnels et al. (1967) further determined that slopes of age-dependent accumulation vary according to the extent of the pollution of the geographical location.

In other areas not affected by man, Abernathy and Cumbie (1977) reported an age-dependent Hg enrichment in largemouth bass (*Micropterus salmoides*) taken from recently impounded water. But here the elevated mercury concentrations in the fish samples have been attributed to the mobilization of mercury from the suspended load which had not yet been in balance within the newly impounded reservoir. In another area of no known mercury inputs, Smith et al., (1975) also found relatively high concentrations; those in pike (*Esox lucius*) exceeded those of lake whitefish (*Coregonus clupeaformis*). It can be derived from these investigation results in "clean" areas that a significant mercury enrichment with age is generally dependent on the basic level present in the environment of the fish. This basic level is determined by the equilibrium between mercury input (e.g. from weathering material) and the transport out of the ecosystem.

The highest levels of mercury in freshwater fish were reported by Scandinavian authors who assumed the source to be in the mercury-contaminated biota downstream from wood pulp factories which had used organo-mercurials as wood preservatives. In southern Norway Underdal and Hastein (1971) found an average concentration of 0.21 ppm mercury in brown trout (*Salmo trutta*) caught upstream from a wood pulp factory. Downstream from the factory the average concentration was 4.35 ppm (a 20-fold increase). Similarly high concentrations were found in perch (*Perca fluviatilis*) (cf. Appendix B). More than 8 ppm Hg were measured in pike (*Esox lucius*) taken from comparable areas in Sweden (Johnels et al., 1967). These extreme Hg levels are mainly associated with local point-source pollution by mercury-containing waste effluents.

In investigations accounting for background values in the Great Lakes area the mercury concentrations in fish muscle tissue were less than 0.5 ppm. Only some locally elevated metal concentrations could be detected (e.g., Lucas et al., 1970; Uthe and Bligh, 1971). In Austrian lakes Krocza et al., (1974) determined that mercury concentrations in predatory and nonpredatory fish were generally low. Of 62 Cyprinids investigated for their mercury content, none had concentrations greater than 0.25 ppm; of 68 pike and perch specimens only four pikes had a mercury concentration exceeding 0.5 ppm. In a study on the Danube River, low Hg concentrations were found in the muscle tissue of nonpredatory fish ($\bar{x} = 0.105$ ppm, wet weight). In fish from the Rhine River, however, Vogt (1974) detected significantly higher mercury levels. Upstream from Ludwigshafen and Mannheim, average Hg concentrations of 0.2 ppm in roach

(*Rutilus rutilus*) and 0.3 ppm in pike (*Esox lucius*) were found, whereas the average concentrations downstream revealed an increase to 0.9 and 1.7 ppm respectively. Maximum values of up to 6 ppm Pb in pike muscle tissue substantiate the fact that this section of the river can be generally considered as heavily contaminated. Other metals are reported in fish species from the Great Lakes by Lucas et al. (1970). In fish samples of trout perch (*Percopis omiscomaycus*), significantly higher Cd values were found (140 ± 60 ppb) in Lake Superior than in Lake Michigan (76 ± 8 ppb). However, copper and arsenic concentrations in the same fish species were higher for specimens from Lake Michigan than from Lake Superior (cf. Appendix B). An extensive study on several hundred specimens of fish from 49 New York water bodies for Cd was carried out by Lovett et al. (1972). In samples of decapitated and eviscerated lake trouts (*Salvelinus namaycush*) of known age from Cayuga Lake, no relation was apparent between the size of the fish and Cd content, which is in contrast to the results for mercury for the same fish species from the same sampling site (Bache et al., 1971; cf. Fig. 81). Of the total fish investigated 68.5% had Cd concentrations of less than 20 ppb, from an overall range of 10–170 ppb. Natural higher background values of up to 100 ppb and more have been detected in the Adirondacks, where cadmium is typically associated with metallic ore deposits. Similar results for tin have been reported in the nonindustrialized Moose Lake in Canada by Uthe and Bligh (1971) where tin-bearing minerals are common. For northern pike significantly higher Hg levels were found in industrialized areas near Lake St. Pierre and Lake Erie, whereas for Pb, Ni, Cd, As, Cu, Zn, and Mn no differences occurred (cf. Appendix B).

Lovett et al. (1972) and Havre et al. (1973) established no correlation between fish weight and cadmium content in the muscle tissue. Nonetheless, there is reasonable doubt as to whether muscle tissue analysis as used in determining Hg content is also suitable for the determination of cadmium content. The reasons for this are: (a) Hg has a much higher affinity to muscle tissue as compared to cadmium, and (b) this type of analysis does not take the organ-specific distribution of cadmium into consideration. In a section of the Neckar River (Federal Republic of Germany) polluted by cadmium, Müller and Prosi (1977, 1978) found a highly significant weight/cadmium content correlation (98% and 99% probability level respectively) in the livers and kidneys of roaches (*Rutilus rutilus*). As in the case of Lovett et al. and Havre et al. (see above), this correlation was not found in muscle tissue. Furthermore, in these studies of the internal organs differing levels of Zn, Cd, and Cu contamination could be determined in the various populations of roach in the Neckar River and one of its tributaries. Further data on other heavy metals in freshwater fish are compiled in Appendix B.

Despite the great number of variables affecting the heavy metal contents in fish, the above data show that marine and freshwater fish are to a certain extent suitable as indicators of heavy metals. Local sources of pollution can be detected by comparing the verified metal content to the background level. The background can differ widely within species of fish and from region to region. Any fish species is therefore suitable as an indicator for ecological studies such as food chain enrichment, but should not be used as a monitoring system for heavy metal pollution. In addition, even though the heavy metal concentrations in the fish musculature are usually low, the amounts found in the viscera and waste products of fish often exceed those permitted for

3.3 The Mobilization of Heavy Metals from Sediment by Aquatic Biota

human consumption. This is of relevance when one considers that both the viscera and waste products play an important role in the food industry (e.g., cod liver oil, animal feed) and still present a danger to man.

3.3 The Mobilization of Heavy Metals from Sediment by Aquatic Biota

The distribution of heavy metals in the aquatic milieu shows that the highest metal concentrations are generally found in the bottom sediment. The amounts of heavy metals concentrated here present, however, a particular danger to organisms, especially to those living in the sediment and which enter the food chain (Luoma and Bryan, 1978). The mobilization of mercury in sediment has been thoroughly studied thanks to research performed on the methylzation of this metal by bacteria (Wood, 1973). Kushner (1974) found bacteria in the Ottawa River of which 50% and more were resistant to 1 ppm Hg^{2+} and some were even resistant to 10 ppm Hg^{2+}. These bacteria could transform Hg^{2+} into a volatile mercury compound, probably metallic Hg. This ability of bacteria to produce metallic Hg is genetically determined by plasmids in Hg-resistant *Escherichia coli* and *Pseudomonas aeruginosa* (Schottel et al., 1974). The occurrence of bacteria with a developed resistance to other heavy metals was reported by Mills and Colwell (1977). The mobilization of ^{65}Zn and ^{109}Cd from water into the biomass of the sediment by marine bacteria was proven by experiments carried out by McLerran and Holmes (1974).

In two Swiss lakes (Lake Lucerne and Alpnach Lake) Baccini (1976) found that the sedimentation of metals is dominated by allochthonous particles and the biogenic particularization is partially reversed by the internal decomposition processes in the lake. Therefore, the trace metals reaching the sediment are practically redissolved.

As for sediment-feeding macrobenthos, Kranz (1976) reported only a slight uptake of ^{57}Co by sediment-associated *Nereis diversicolor*. Similarly, low concentrations of mercury were found in *Nereis diversicolor* and *Arenicola marina* taken from the Elbe Estuary by Zauke (1977). In estuaries highly polluted by copper Bryan and Hummerstone (1971) detected a close correlation between copper in *Nereis* and in sediment adhering to worms (Figs. 88, 89). The Cd content in these worms is also reported to be roughly proportional to that in the sediment, whereas Zn is not. It has been suggested that the essential element zinc is regulated by *Nereis* populations. Considering such populations from sediments highly polluted with zinc and those from less polluted sediments, it has been found that the former populations can regulate this metal better than the latter when exposed to excessive Zn concentrations in sediment. The accumulation of radioactive cadmium (^{115}Cd) from sediments and seawater by polychaetes was observed by Ueda et al. (1976). Worms in contact with sediments accumulated six times more cadmium than worms not in contact with sediment. Renfro (1973) even determined a loss of ^{65}Zn in sediments by the activity of marine polychaetes. This loss was 3–7 times higher in sediments with worms than in sediments without worms. It is therefore not surprising that in sediment-associated bivalves, heavy metal concentrations is dependent on the metal content of the bottom sediment (Bryan and Hummerstone, 1978; Bryan and Uysal, 1978). In that respect straight-line relationships have been found for cadmium and copper in pacific oysters

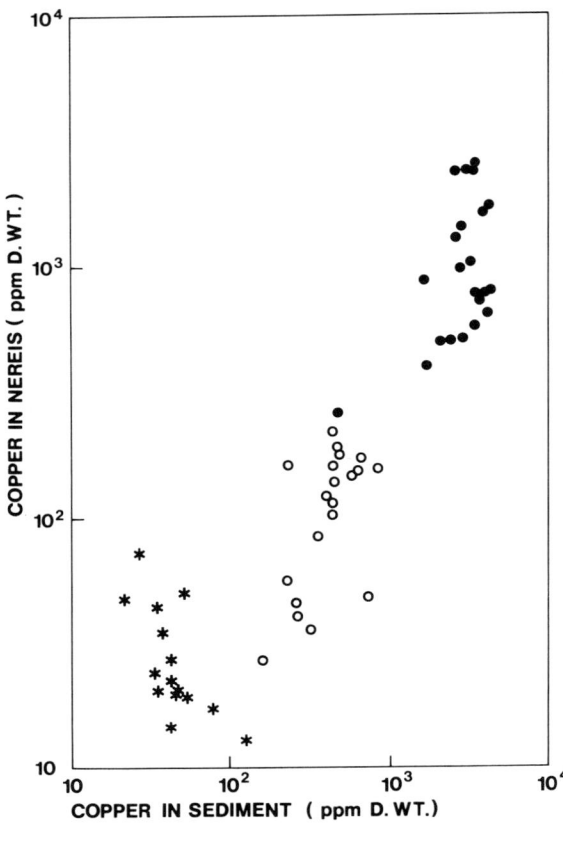

Fig. 88. Dependence of copper in *Nereis diversicolor* from different estuaries and copper concentrations in corresponding sediments in southwest England. *Asterisk*, Plym, Dart, Avon, Camel estuaries; *open circle*, Tamar/Tiddy estuary; *closed circle*, Restronguet creek/Fal estuary (Bryan and Hummerstone, 1971, J. Mr. Biol. Assoc. U.K. *51*, p. 853. Reproduced with permission of Cambridge University Press, London)

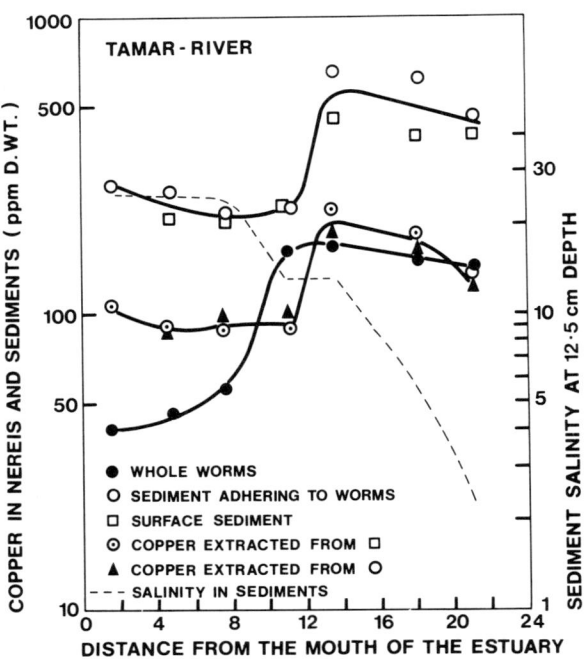

Fig. 89. Changes of copper concentrations in *Nereis diversicolor* and different sediment fractions with distance from the mouth of the Tamar river estuary in southwest England (Bryan and Hummerstone, 1971, J. Mar. Biol. Assoc. U.K. *51*, p. 850. Reproduced with permission of Cambridge University Press, London)

of the Tamar Estuary in Tasmania (Fig. 90). Similar dependencies have also been found in the Derwent Estuary of Tasmania by Bloom and Ayling (1977; cf. Table 79).

In respect to the high mobilization rate of Hg from sediment, Sayler et al. (1975) tested the uptake and metabolization of ^{203}HgCl$_2$ on oyster tissues with mercury metabolizing bacteria (Hg^{2+} → volatile Hg^0, *Pseudomonas*) extracted from Chesapeake Bay. With the presence of these bacteria, mercury concentrations in tissue fractions were 200 times higher than in control experiments. Several authors have reported that Hg-methylization from sediment to water—due to bacterial activities—results in higher mercury levels in fish as well (e.g., Hasselrot and Göthberg, 1974; Walter et al., 1974; Abernathy and Cumbie, 1977). In experiments where Hg compounds were added to sediment, Gillespie and Scott (1971) and Gillespie (1972) recorded a rapid uptake to Hg in guppies (*Poecilia reticulata;* cf. Fig. 91). The uptake of artificially added mercury (II) chloride or -sulphide was similar to that in control sediment; however, the amount of methyl-mercury in fish was higher in experiments with artificially contaminated sediments. An even more rapid accumulation of mercury was indicated upon addition of metallic mercury. Once again methylmercury concentrations in the guppies were similar as in the mercury(II) chloride and -sulphide experiments. In other experiments on goldfish and catfish an uptake from heavily contaminated sediments was found

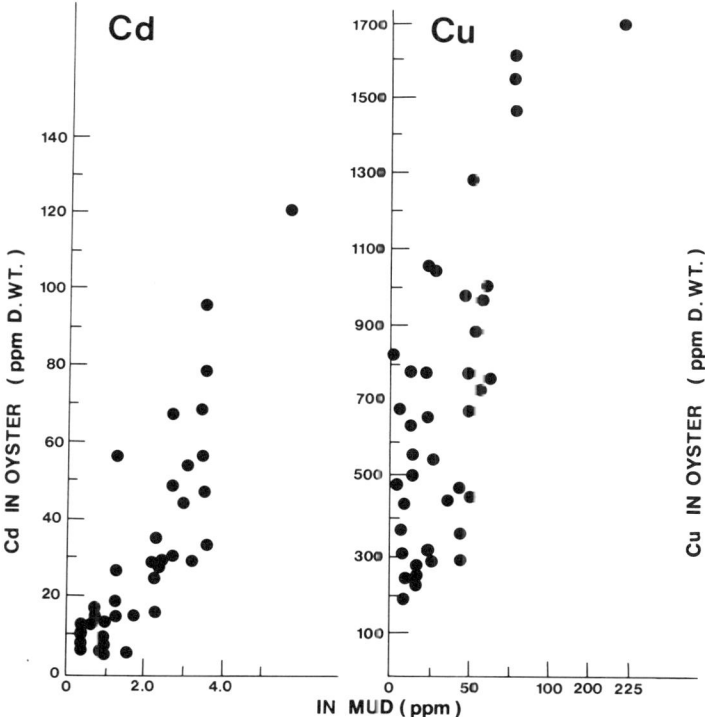

Fig. 90. Cadmium and copper concentrations in the Pacific oyster (*Crassostrea gigas*) and in dried mud of the same habitat, the Tamar estuary, Tasmania (Ayling, 1974, Water Research *8*, p. 734. Reproduced with permission of Pergamon Press, Oxford)

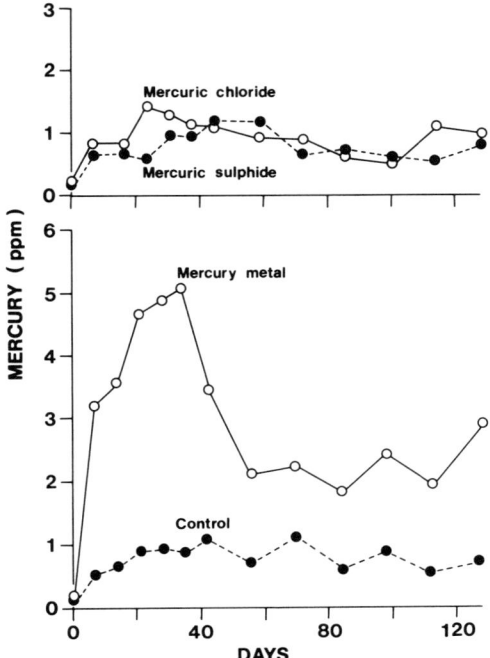

Fig. 91. Uptake of mercury in guppies (*Poecilia reticulata*) exposed to sediments in an aquarium ecosystem. Mercury concentration of artificially contaminated sediments: 50 ppm, dry weight related, control 0.1 ppm. Each *point* represents an average of 5 whole fishes, wet weight related (after Gillespie, 1972)

(Delisle et al., 1975; Fig. 92). For the investigated metals Zn, Cu, Pb, and Cd metal values in goldfish were much higher than in catfish, although goldfish generally have lower metal background concentrations (This discrepancy can probably be explained as due to the different feeding habits of the species; the goldfish burrow when searching for food whereas the catfish is a predatory fish species). Maximum levels in both fish species had been attained after two months and remained on this niveau. If these maximum levels are compared to the background values, the relative bioaccumulation in both fish species is higher for the nonessential metals Pb and Cd than for Zn and Cu.

In respect to the investigations by Bryan and Hummerstone (1971, 1973a, b) in the marine environment mentioned above, a similar influence of heavy metals in sediment on the corresponding sediment-associated macrobenthos in freshwater, such as tubificid worms, can be expected.

In waters with low dissolved oxygen concentration and in anaerobic sediments, the respiratory end of the worm generally reaches into the overlying water column for oxygen uptake. In this environment the worms will take up heavy metals both from the sediment and the water. Therefore, tubificids are of the organisms most affected by heavy metals in sediments (including interstitital water) and by available metals at the sediment/water interface.

In laboratory experiments Patrick and Loutit (1976) reported that metal accumulation of tubificid worms can occur via contaminated bacteria cells, which are a basic source of nutriment. Field studies on the detection of heavy metals in tubificid worms were carried out by Leland and McNurney (1974), Mathis and Cummings (1973), Hasselrot and Göthberg (1974), Hölzinger (1977), and Prosi (1977). All these investi-

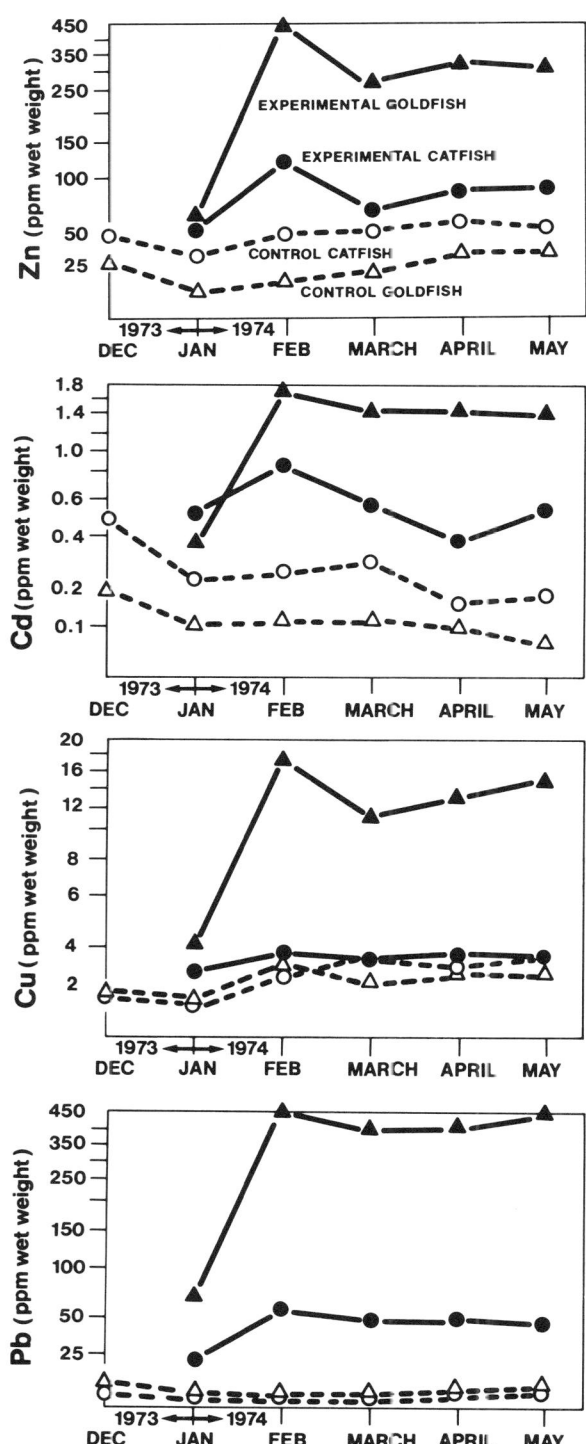

Fig. 92. Uptake of Zn, Cd, Cu, and Pb from sediments taken from a tailing area, in two fish species, goldfish (*Carassius auratus*) and catfish (*Corydoras aeneus*) (after Delisle et al, 1975)

gations revealed high trace element concentrations in tubificid worms of the biota, and in even some examples the sludge worms had higher metal concentrations than the sediment. None of the authors found a simple correlation between the heavy metals in the sediment and in the worms but it is assumed that a real relationship between heavy metal content and worms does exist. This relationship may be dependent on parameters such as the heavy metal content in interstitial water, grain size distribution, the amount of organic matter in sediment (including microorganisms), chemical bonding forms of the metals in sediment, etc. (Prosi, in prep.). Furthermore, the abundant tubificids as occur in highly polluted sediment (cf. Aston, 1973), can probably contribute to a release of heavy metals from sediments by activities such as the release of interstitial water with excretion of faeces and transport of anaerobic sediment to aerobic surface layers (Woods, 1975).

As the above examples show, large amounts of heavy metals can be taken up by sediment-associated organisms, of which tubificid worms or marine polychaetes are only two of many. This is of even greater importance as several sediment-feeding macrobenthic organisms especially in polluted waters develop a large biomass and become thereby a major nutriment source for many organisms in higher trophic levels.

3.4 Food Chain Enrichment in Aquatic Life

Many investigations have been centered on determining the extent of heavy metal enrichment within a food chain, a food web or just within various trophic levels. An essential requirement for the study of heavy metal enrichment is that the various metal contents of the entire organism must be brought into relation with one another or, at the least, the individual organs containing the greatest percentage of the total metals. Organs for which the respective metals show the highest affinity can be good indicators for metal contamination of the organism, but still are not suited for studying food chain transfers of heavy metals. For example, Cumbie (1975) detected very high Hg levels in hair of otter and mink which reached the values in human hair reported to indicate an acute mercury intoxication.

An enrichment of mercury from fish to otter and mink via the food chain apparently occurs due to the investigated organ's different mercury affinity. However, it should not be assumed that mink and otter have higher Hg levels than fish. Thus, as suggested by Thomann et al. (1974), estimates of heavy metal transfer rates and biomass of the trophic levels must be available. In this respect a model analysis by Fagerström et al. (1975) reveals that direct uptake of methylmercury from water is a most important aspect, but that the amount of mercury taken up from food has only a small additional effect (cf. also Jernelöv and Lann, 1971).

In experiments on marine shrimps, crabs, and fish Renfro et al. (1975) revealed a predominating uptake of ^{65}Zn from water (Fig. 93). Additional uptake of ^{65}Zn via the feed of labelled *Artemia* gives rise to only a small increase of the total content of this metal in the animal. However, from highly contaminated natural nutriments this additional metal uptake can have a decisive influence on the metal content of the organism, especially when heavy metal concentrations in the aquatic ecosystems are low (cf. Schulz-Baldes, 1974; Schulz-Baldes and Lewin, 1976).

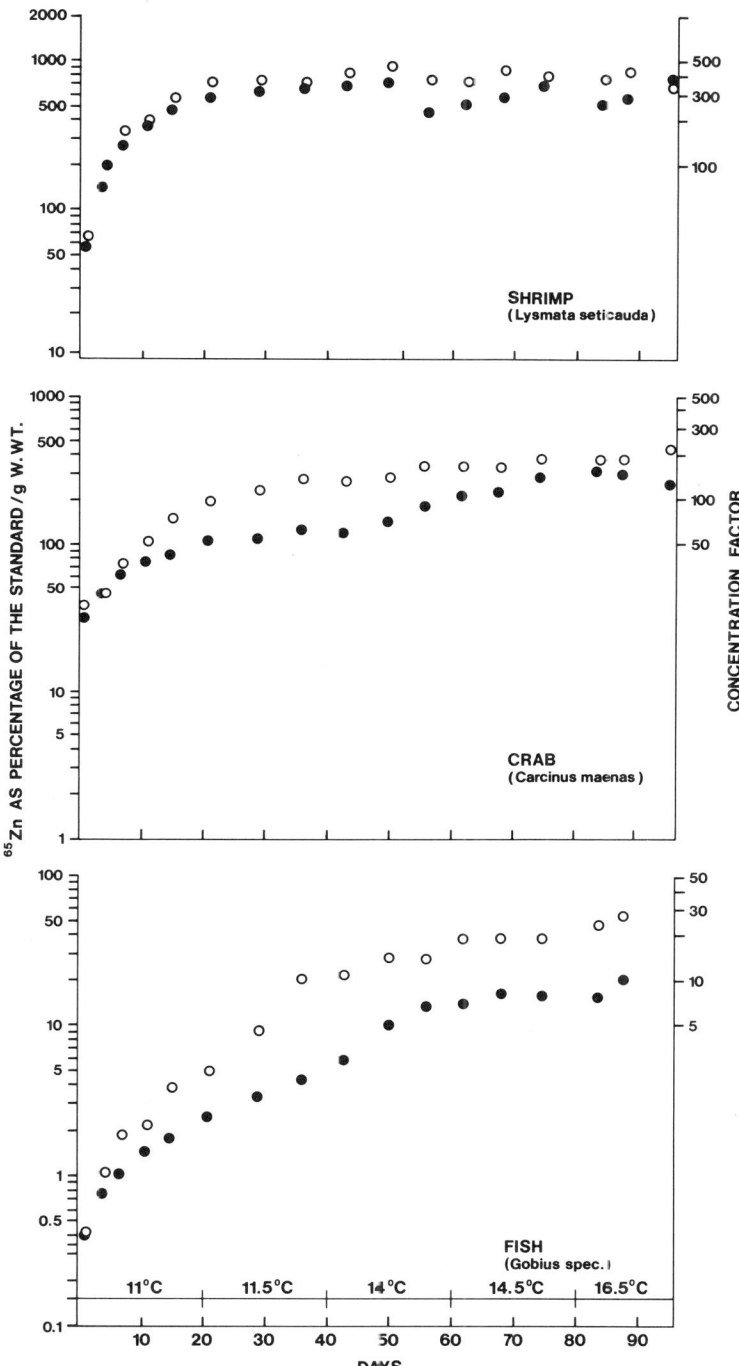

Fig. 93. Accumulation of ^{65}Zn in shrimp, crab, and fish from food (^{65}Zn labeled *Artemia*) and water (*open circle*) and from water only (*closed circle*). (Renfro et al., 1975, J. Fish, Res. Board Canada *32*, p. 1343. Reproduced with permission of Environment Canada, Fisheries and Marine Service, Ottawa)

Many field studies have indicated that mercury can be enriched in aquatic food chains. In the highly polluted Derwent Estuary of Tasmania Ratkowsky et al. (1975) established mercury concentrations in various fish species ranging from minimal amounts to 2.0 ppm (wet weight); the highest values were recorded in Ralphs Bay independent of individual species. These results agree with those of Bloom and Ayling (1977) done on mussels in the same area. The heavy metal concentrations in mussels are similarly high, although they are on a lower trophic level in comparison to the fish species. Nevertheless, a relationship of heavy metal enrichment to the food chain can be readily seen because approx. 51% of the individual fish, whose nutriment consists mainly of other fish, reveal mercury concentrations of more than 0.5 ppm. Individual Hg concentrations of more than 0.5 ppm are shown by 24% of the invertebrate predators and only 7% of the herbivorous fish. Similar dependencies of Zn contents from the dietary habits of fish species feeding on contaminated invertebrate fauna were reported by Hardisty et al. (1974a).

In a more detailed investigation on the Severn Estuary, England, Hardisty et al. (1974b) determined that the quantity of Cd and Pb contamination in seven different fish species was directly related to the content of crustaceans in the diet. The zinc content was not significantly affected by the extent of their crustacean diet, but in every case heavy metal concentration of organisms in the diet was higher than in the consuming fish species. Another example of an apparent Hg enrichment via detritic nutrient sources, primary production, herbivorous and small predatory fish, and up to larger predatory fish in these different trophic levels of a freshwater hydroelectric reservoir on the upper Colorado River was reported by Potter et al. (1975). On the other hand the high Hg content in plant debris (148 ppb), which is similar to that in predatory fish indicates that a high mercury concentration takes place even on the low trophic levels. Therefore a close relationship between biomagnification and the different levels of the investigated food chain can not be ensured. Johnels et al. (1967) were among the first to report high Hg concentrations on low trophic levels. In a comparison of leeches (*Helobdella* sp.) and pike (*Esox lucius*) caught above and below a paper mill, it was found that the Hg content in the leeches below the mill was 124-fold higher than above the mill; for pike as end consumers the increase was only 4.4-fold. No food chain enrichment of Hg in the marine environment between plankton and plankton-feeding northern anchovy (*Engraulis mordax*) was found by Knauer and Martin (1972). In Southhampton waters and the Solent region of England, Leatherland and Burton (1974) found no higher arsenic and mercury concentrations on upper trophic levels. On the contrary, the lowest investigated level (red algae) had even higher concentrations than the other classes of organisms (porifora, gastropods, and bivalves). For Cd, Cu, Pb, and Zn, unusually high concentrations were detected in the marine biota of the Hardangerfjorden in Norway (Stenner and Nickless, 1974a), but uptake of these metals may have occurred via water pathways and not by food because all fish had very high heavy metal levels in the gills indicating an increased amount of available metal in the water.

In another example a food chain enrichment as described for mercury between benthic invertebrates and benthic fish from the Elbe Estuary was not confirmed by Zauke (1977). Stenner and Nickless (1975) reported extremely high concentrations of zinc in barnacles from Southwest Spain and Southern Portugal (cf. also Appendix B) although the amounts in fish as endconsumers were low.

A significant enrichment of As along the food chain within a defined polluted area was documented by Bohn (1975): Zooplankton 6.0 ppm, brown seaweed 35.5 ppm, mussels 14.1–16.7 ppm, prawn 62.9–80.2 ppm, different fish species 38.4 ppm (range 43.4–188.0 ppm); all data are dry weight related (cf. Appendix B). The highest arsenic content was found in American plaice (188 ppm), which can point to the benthic habits of this species; As, similar to Hg, can be released from the sediment by bacterial mobilization.

In an urban-influenced river section Prosi (1977) determined a significant increase of Cd and Pb in the foodweb of benthic invertebrates compared to fish, whereas the essential metals Zn and Cu had an intermediate behavior. Fish (roaches and sticklebacks) had the lowest concentrations for all four investigated metals (Fig. 94). It was generally found that according to feeding habits sediment-dependent organisms (Tubificidae) had greater metal concentrations than other biota. Metal contents of the benthic food web, sludgeworms (*Tubifex tubifex* and *Limnodrilus hoffmeisteri*), isopods (*Asellus aquaticus*) and leeches (*Herpobdella octoculata*) constantly decrease, so that the lowest concentrations appear in the fish (cf. Fig. 94). Similar observations have been made by Hölzinger (1977) in an ecosystem of the Danube River for the metals Cd, Hg, Pb, Zn, Co, Cr, Cu, and Ni. Mathis and Cummings (1973) found even higher metal contents in omnivorous than in carnivorous fish; benthic organisms revealed higher concentrations than fish. In a recent investigation of Enk and Mathis (1977) the following increase sequence for cadmium and lead was determined in a stream ecosystem:

water < fish < sediments < aquatic invertebrates.

From the investigations mentioned above the conclusion should not be drawn that there is no food chain enrichment in waters polluted with heavy metals, since most organisms showed metal concentrations that were higher than background levels. There is, however, no food chain enrichment in the classic sense of the term, as found for organic pesticides (e.g., DDT) where the highest trophic levels coincided with the highest concentrations of the pollutant. Furthermore, in polluted environments the sediment has generally the highest heavy metal concentrations (cf. Chap. C), and consequently if a significant heavy metal contamination through food ingestion does occur, the sediment-feeding organisms will have higher metal concentrations than other organisms. As to species on higher trophic levels it must be additionally considered that they very often have a longer lifespan as those on lower trophic levels. Thus metals can be enriched over a longer period of time. This typical heavy metal distribution in aquatic systems was clearly pointed out by Leland and McNurney (1974) in the case of lead, a metal which has a very low distribution coefficient in aquatic systems (Fig. 95). Therefore with the exception of As and Hg, with their high affinity to organic substances, the heavy metal enrichment in higher trophic levels of polluted water systems is generally low, but still higher than the background. Detritus-, sediment- and filter-feeding organisms could be affected by heavy metals occurring in their environment. Nonetheless, these organisms apparently only transfer a small amount of their metal content to higher levels of the food chain so that if these higher levels are highly contaminated, they must receive the major part of their metal content from the water. In this context the speciation of heavy metals in the water

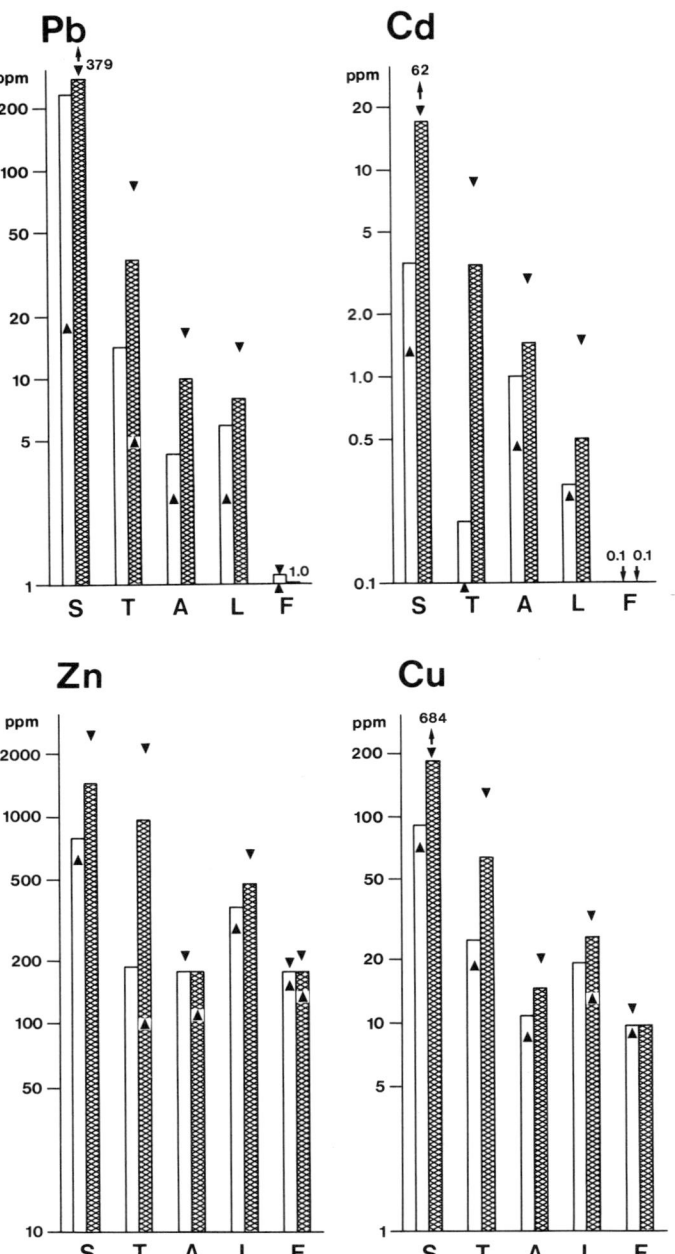

Fig. 94. Heavy metal distribution in two sections of the Elsenz river at different trophic levels *S,* sediments < 2 μm; *T,* tubificid worms (*Tubifex* sp. and *Limnodrilus hoffmeisteri*); *A, Asellus aquaticus; L,* leeches (*Herpobdella octoculata*); *F,* fish (roaches: *Rutilus rutilus,* sticklebacks: *Gasterosteus aculeatus*). Mean concentrations of whole animals (dry weight related): *arrows,* minimum and maximum values; *light columns,* rural; *shaded columns,* urban-influenced river section (after Prosi, 1977)

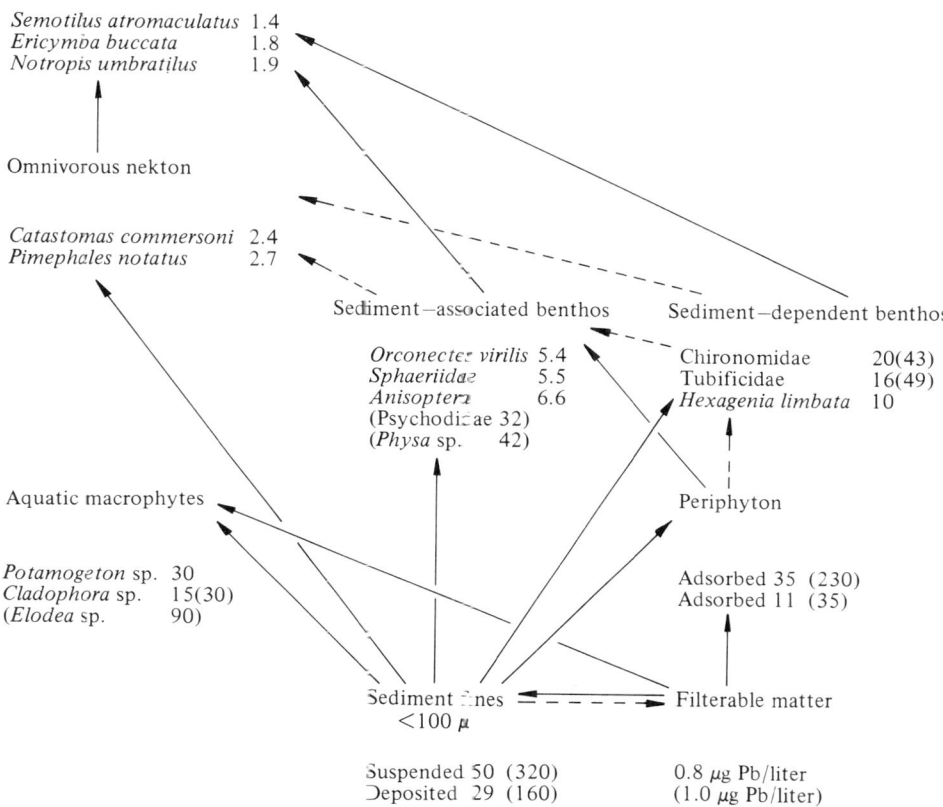

Fig. 95. Lead distribution in a river system of a rural area. ppm related to dry weight of whole animals; *brackets,* mean values for urban area (after Leland and McNurney, 1974)

plays a role which is decisive not only for the acute toxicity but also for the bioavailability of the metals. The reason that the thesis of bioamplification of heavy metals along the "food chain" isn't generally valid lies in the discrepancy between overly simplified investigations of the past and more recent better differentiated studies of food web enrichment processes.

Chapter G
Trace Metals in Water Purification Processes

The heavy pollution of many inland waterways, with respect to the amount as well as to the variety of the contaminants, has put the water supply utilities in an increasingly difficult situation. It has basically been shown that "every type of fresh water recovery from surface waters is accompanied by problems that increase overproportionally to the actual pollutant load of the used water" (Haberer, 1975). Special problems arise when water is scarce even though the pollution intensity may be relatively low; such a situation is found mainly in arid areas.

1 Heavy Metal Removal for the Production of Drinking Water

In this section bank filtration (1.1), artificial recharge by land spreading and injection (1.2), and several physico-chemical methods will be discussed in relation to the production of potable water from surface waters. A general scheme is given in Figure 96 (from Haberer, 1975).

1.1 Obtaining Water by Bank Filtration

Under favorable conditions surface waters attain the quality of groundwater when river and lake water is filtered through gravel or sand layers. The organic substances are degraded by microbial processes or by reaction with oxygen, suspended particles are deposited and inorganic substances can be sorbed to mineral surfaces or removed

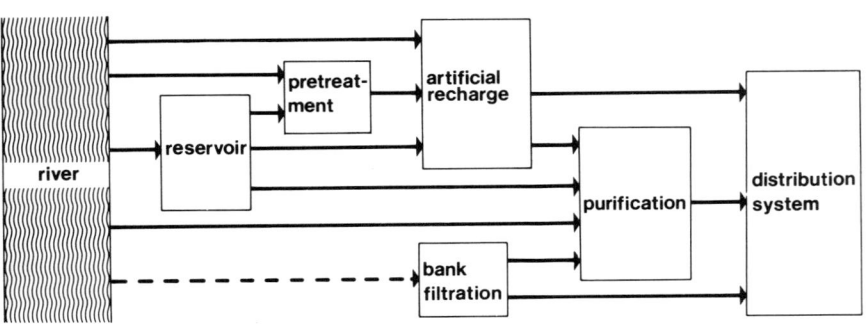

Fig. 96. Schematic presentation of procedures for production of potable water from surface waters (after Haberer, 1975)

by precipitation. Processing of surface waters by such methods of infiltration has been proven to be effective for more than 80 years (Thiem, 1898; Scheelhase, 1912). Due to the sharp rise in demand for water in industrial areas, and which could only be met by increased use of surface waters, the inexpensive method of groundwater enrichment by bank filtration has become very important.

The examples presented here are from an area where the processing methods of bank filtration and artificial recharge play an important role (in the Federal Republic of Germany, from a total drinking water production from surface waters of 1300 million m^3, 500 million m^3 was gained through bank filtration and another 400 million m^3 by artificial recharge—1976 values).

Investigations carried out by Kludig (1968) on "obtaining bank-filtrated groundwater and the influence of the Rhine pollution" and by Holluta et al. (1968) "on the effect of increasing river water pollution on water quality and the capacity of bank filtrates" contained pessimistic predictions regarding the future application of this method, particularly in the Middle and Lower Rhine sections, which are heavily polluted with effluent contaminants. Apart from the biologic aspects of the problem, the migrational behavior of inorganic pollutants in the bank filter is of primary importance. Since the "elimination of inorganic, nonvolatile elements and compounds by bank filtration always constitutes a temporary process" (Koppe, 1970), the retention effect on such substances is reduced with increased usage and intensity. Although Kölle et al. (1971) were able to prove that on the whole a decontaminating effect occurs within the filter channels, they also point out the danger of "breakthroughs", whereby elevated concentrations of pollutants from the surface waters enter the drinking water supply.

The migration behavior of heavy metals in the bank filtration is influenced by immobilization and remobilization processes. Generally speaking, there are three different mechanisms involved in immobilization: (a) sorption on marginal surfaces due to intermolecular forces, (b) cation exchange, whereby negative charges are neutralized by positively charged cations which in turn are exchanged for others with predominantly higher charges, (c) precipitation products, mainly carbonates, hydroxides and sulfides, which form less soluble compounds with heavy metals. Coprecipitates of trace metals also occur in iron and manganese compounds.

These three forms of "temporary elimination capacity" have a ratio of 1:1,000:1,000,000 in respect to the effective duration (Koppe, 1973).

Mobilization of heavy metals is particularly likely to occur when iron and manganese compounds are dissolved under reducing conditions and the incorporated trace metals are released or when as Hausen (1972) points out, complexation takes place in the river water or in the reduction zone of the bank filtration. In a study on the chemical reactions between surface water and groundwater in the Mosel Valley between Trier and Koblenz, Schwille (1973) arrived at the conclusion that the intrusion of groundwater in the sub-soil can lead to very pronounced reducing conditions.

An investigation conducted by Förstner and Müller (1975) was undertaken to establish the relationship between river water and bank filtrate on the behavior of heavy metals under reducing conditions. The site chosen for these bank filtrate investigations was at a steam power plant on the Neckar River. The following factors proved particularly advantageous in the choice of this site: (1) the water level in the Neckar was constant and the influence of the groundwater was minimal, (2) chemical marking of the river water by saline effluents originated a few hundred meters upstream. The experiment strip encompassed three boreholes dug in a line perpendicular to the course of the river. The distance to the borehole No. 1 was measured at 40 m. There was a distance of approx. 40 m as well between the boreholes No. 1 and No. 2, and No. 2 and No. 3; from a site near borehole No. 3 bank filtrate water had been extracted for many years. For comparison pure groundwater was recovered from a borehole situated on the opposite side of the river about 10 m from the river bank.

Twenty-two water samples were collected from four sampling stations in the Neckar River and vicinity, at monthly intervals between January 1972 and October 1973. One main feature of the experiments was that the relatively high oxygen content in the Neckar River is drastically reduced in the bank area i.e., in the groundwater immediately adjacent to the river.

Table 82 shows the average values of the individual heavy metals for the period 1972/73 (for cadmium: January to October 1973) in the above mentioned areas.

Table 82. Average values of heavy metals (μg/l). (Values in brackets based on Neckar standard = 100)

Metal	Groundwater	Neckar River	Distance from river					
			1 (40 m)		2 (80 m)		3 (120 m)	
Iron	33	(54)	61	1066	(1740)	1233	(2030)	127	(207)
Manganese	17	(32)	53	588	(1110)	761	(1440)	798	(1510)
Nickel	4.4	(31)	14.0	5.8	(41)	5.7	(40)	6.5	(47)
Chromium	2.5	(72)	3.5	1.8	(52)	1.3	(37)	1.4	(40)
Zinc	17.8	(42)	42.8	41.1	(33)	18.3	(43)	24.9	(48)
Copper	1.1	(20)	5.6	1.3	(23)	1.6	(29)	2.8	(50)
Lead	1.5	(35)	4.3	1.7	(40)	2.0	(46)	2.2	(51)
Cadmium	0.3	(33)	0.9	0.3	(33)	0.3	(33)	0.5	(56)

The fluctuations within the filtration stretch are evident if one equates the individual data on the basis of the Neckar River value = 100. Infiltration into the Neckar bank causes the iron and manganese concentrations to rise by a factor of 10 to 20. The remaining trace metals in the filtered stretch are all lower than the corresponding concentrations in the Neckar River water.

Reducing conditions in groundwater and bank filtrate is in many ways a disadvantage to this process of water reclamation because of increased corrosion in the pipe network, dissolution of iron and manganese compounds, and consequent precipitation as well as a massive increase of anaerobic bacteria. Such reducing conditions, however, appear ambivalent for the migration behavior of heavy metals. On the one hand, an undesirable release of trace metals occurs when manganese and iron hydroxides and oxides are dissolved. Their mobility is further increased in this environment due to increased complexing. On the other hand, the low degree of solubility of heavy metal sulfides constitutes an effective retention mechanism, which in the present example led to the almost complete elimination of dangerous cadmium concentrations prior to filtration.

Under *oxidizing conditions* an elimination of heavy metals would probably occur primarily as a result of coprecipitation with hydrous Fe- and Mn-oxides. In this case study, no reliable data from pure bank filtrates are available. Investigations by Kölle et al. (1971) on the water-processing plants along the Rhine River show that a particularly strong reduction of the nickel, zinc, and copper concentrations may occur in the river water.

In a state-of-affairs report, *Water Supply and Bank Filtration of the Federal Ministry of the Interior* (Anon, 1975e), summarizing the different aspects of bank filtration in West Germany, a group of experts concluded that "in normal operation, it appears that it has not as yet come to dangerous situations, at least in respect to heavy metal pollution. However, it cannot be excluded that a further contamination of the surface water will lead to an increasing pollution of the processed water."

1.2 Artificial Recharge of Groundwater by Land Spreading and Injection

Water extracted from the river is infiltrated into either natural or artificially prepared sand beds and later extracted once more by means of boreholes or springs. Consequently, the infiltration area is usually situated close to a river.

As inventor of the process for the artificial enrichment of groundwater, Richert (1900) brought in 1898 two filtration ponds of 500 m^2 each into operation in Göteborg, Sweden, that functioned as slow-sand filters. Before that, in 1875, the water works of Chemnitz (Karl-Marx-Stadt) allowed water, at a rate of 10,000 m^3/d, to seep through sand-filter canals in order to improve water quality (see Schmidt, 1975, Trüeb, 1975 for further information on the methods of artificial enrichment of groundwater).

Since there are many deposits of iron- and manganese-oxides in the filter material as well as in the underground water passages, it is essential to maintain aerobic subsoil conditions. Examples of artificial groundwater enrichment are, therefore, well suited to demonstrate the behavior of heavy metals in an oxidizing environment.

In several studies on the behavior of heavy metals during slow-sand filtration, Schöttler (1975, 1977) investigated the different aspects of metal immobilization and remobilization during artificial groundwater enrichment. The results may be summarized as follows: (1) Under normal conditions most of the heavy metal load, bound to organic and inorganic suspended material, is eliminated in the gravel bed and other preliminary stages of purification. By regularly changing the filter material the degree of efficiency can be maintained. (2) In a long-term experiment lasting 109 days, the initial phase (I) consisted in adding river water for 37 days. It is demonstrated that almost constant concentrations are released by the filter (6 μg/l Cu and 0.1 μg/l Cd), and that together approx. 60% of the copper input and 70% of the cadmium are retained by the filter. Even in the subsequent treatment (II), in which metals were added for 104 days in concentrations of 10–100 μg/l, the copper concentrations were only slightly changed. However, cadmium exhibits a constant increase in concentration. Nonetheless, copper and cadmium exhibited elimination rates of 98% and 99% respectively during this study. In the following normalization phase (III) a partial remobilization of the previously retained heavy metals takes place. Compared with similar experiments, without the addition of doses of metals, the concentrations of copper and cadmium in the filtrate were found to have increased by factors of 5 and 50 respectively. The remobilization of heavy metals is preferentially influenced by the addition of 0.1–1 ppm complexing agent such as EDTA. (3) The investigation of impact (= slug) contamination by heavy metals on slow-sand filters provides a different picture. In the case of copper a high metal dosage of 5 mg/l kills algae, which are renowned for their high capacity to assimilate Cu. Hydroxide and carbonate precipitation occurs, but in spite of this high Cu concentrations soon make their way into the filter body. This development is even more pronounced in the case of Cd where an almost complete breakthrough of Cd occurs in filter run-off. (4) Apart from this, cadmium also affects the ecosystem as a considerable bacteria attack on the algae occurs, without immediately causing them to die (the contrary is true for copper). Algae die off approximately 30 days later and cause an additional Cd contamination in the filter medium. The high Cd-concentrations in the filtrate only later decrease slowly. Schöttler (1975) concluded that a slow-sand filter coupled to a subsoil passage is well-suited to eliminate most of the Cu contamination up to 100 μg/l over a long period of time (approx. 100 days), whereas a long-term contamination by the same concentration of cadmium would only be retained for approx. 50 days.

An example of migration behavior of trace metals in sand filters is presented in Fig. 97 (from unpublished data of Förstner and Schöttler). Water from the Ruhr River was allowed to flow through three laboratory filter columns, (1) without additives in the first column, (2) with 100 mg/l humic acid added to the second column and 100 mg/l humic acid + 20 mg/l Cu added to the third column. After the experiment, sand samples were extracted from intervals of 2 to 10 cm and the sediment fractions < 2 μm were separated in settling tubes. Various chemical forms of Cu in the solid substances were extracted according to the methods given in Chap. E.4. There is a clear enrichment of copper contents in exchangeable forms, humic acids and hydroxide bonding in the upper 30 cm of the profile, but not in the residual fraction. These enrichments can be traced to the presently considerable metal percentages in water of the Ruhr, which was used for the experiments (column I). The addition of humic acids (column II) does not change these distribution patterns significantly. However, the addition of dissolved copper (column III) effects a characteristic increase of all three non-detrital copper associations. Fig. 97b compares the concentration of hydrous iron oxides and the distribution of acid-reducible copper within the three columns. It is shown that the original copper contents in the Ruhr River water (both dissolved and in solid particles) are effectively eliminated in the upper 10 cm of the filter sands in the form of acid-reducible precipitates or coprecipitates (column I); addition of humic acids obviously does not interfere with these mechanisms (column II). The effect of acid-reducible bonding of dissolved copper, which is added together with humic acids to the original river water, is however more limited (column III) since even in the lower section of the sand profile a characteristic enrichment of copper can be observed. In Fig. 97c the concentration of humic acids and the distribution of humate-associated copper in the pelitic fractions of the filter material is compared; it is suggested that the humic substances may play a considerable role in the transport of copper into the deeper parts of the filter profile. In column III a distinct correlation is evident between elevated concentrations of humic acids and an increase of the 0.1 N NaOH extractable fraction of Cu, e.g., at 10–20 cm and 40–60 cm filter depth.

Anodic stripping voltammetry determinations on the speciation of copper and zinc in the aqueous phase show that zinc in the inflow to the sand filter is basically labile-bound, whereas copper is for the most part in a relative stable chemical association at the same stage. Upon passing through the filter column, a relative increase of the labile bonding form of copper occurs. The varying rates of elimination of both metal examples can be explained as a result of a more active interaction of zinc with the hydroxides of the filter material, which has a high uptake capacity for free metal ions (and labile metal complexes), during flow through the filter. The copper component, which is in a more stable association with complexing substances, remains partially unaffected by the physico-chemical interactions with the filter material. We therefore conclude that the cleansing capacity of a sand filter is determined both by its material composition (particle specific surface, amount of hydrous Fe/Mn-oxides, clays, organic particles) and by the chemical form of the metals in the dissolved phase (Förstner et al., 1979).

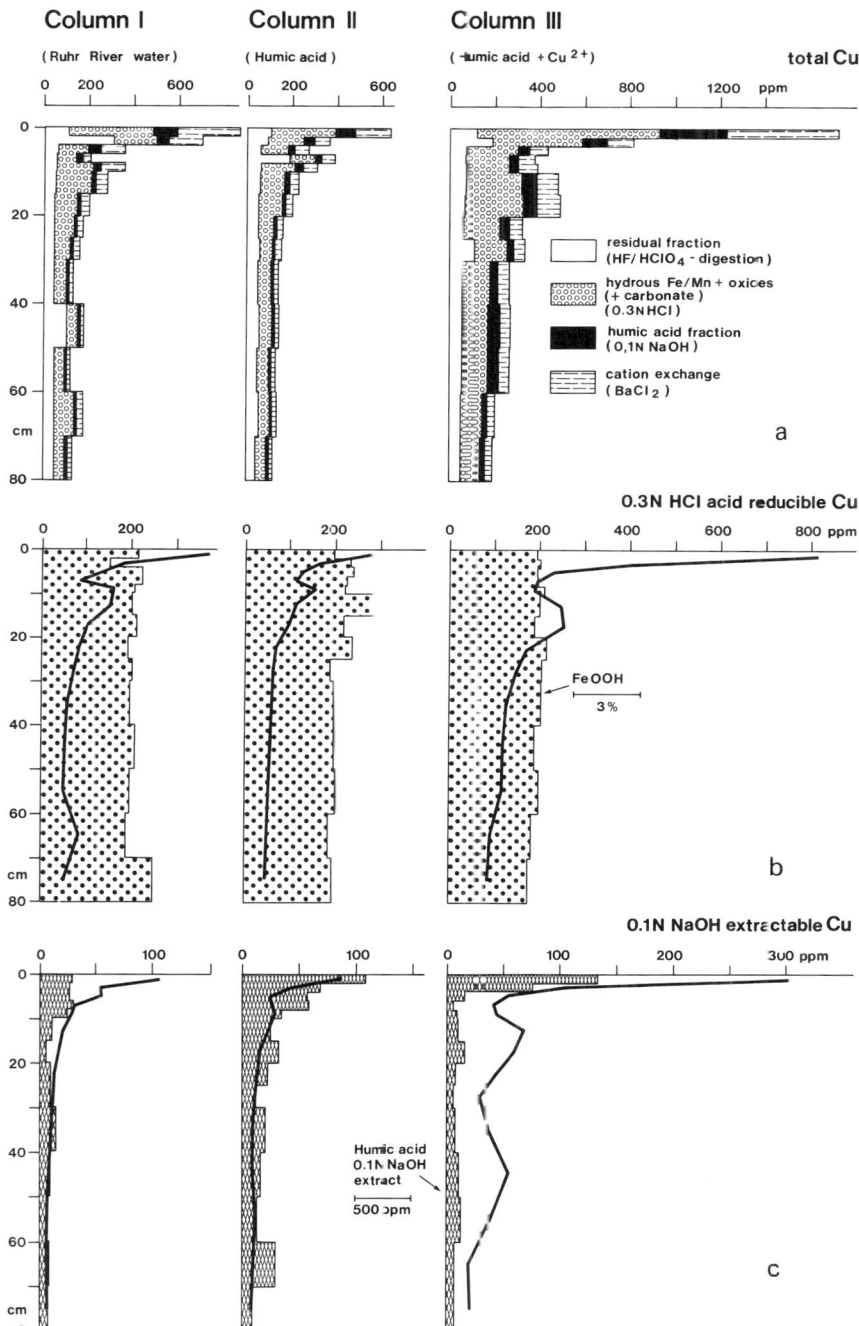

Fig. 97a–c. Sand filter column experiments with Ruhr river water (column I) and humic acid (II) and copper (III) additives. (a) Copper phases in clay sized particles, (b) comparison of the content of FeOOH (*dotted area*) and the distribution of acid-reducible Cu concentrations, (c) comparison of humic acid concentrations and 0.1 N NaOH extractable copper in the pelitic fraction of the filter material (Förstner et al., unpublished data, Research Program "Water Supply by Artificial Recharge" Ministry of the Interior, Federal Republic of Germany)

1.3 Direct Water Purification by Traditional Physico-Chemical Treatment (PCT) and Related Advanced Methods

Conventional PCT and some advanced methods are discussed here with regard to the removal of heavy metals from domestic effluents, while the specific treatment of sewage effluents and sludges from mixed and pure industrial origin will be dealt with in the following sections.

Although very important in practice, cost aspects will generally be omitted in this discussion.

Experimental data should always be considered in relation to the scale on which they were obtained (bench, pilot plant or full scale).

Because of their general interest, O'Connors's classification of inorganic aqueous constituents in natural waters, trace metal drinking water standards, and corresponding toxicity levels are here again referred to (Robeck, 1974; refer to Chap. B.3).

The large-scale planned purification of surface water is gaining considerable importance, especially in areas where groundwater is becoming scarce and where relatively natural methods of purification such as direct seepage and bank filtration can no longer guarantee the quality of drinking water to meet set standards. In such areas a secondary treatment of available water is often necessary. The water works situated in these areas can avail themselves of various methods, depending on the characteristics of the water. Examples will be given from the water works at Wiesbaden-Schierstein on the Rhine, where due to unfavorable conditions, physico-chemical methods had to be employed at a very early stage. In the past few years these methods were continually adapted and accompanied by a large number of research investigations, including the behavior of heavy metals during traditional and recently developed or improved treatment stages. However, it should be kept in mind that originally none of the traditional purification methods was aimed at the specific removal of toxic trace elements. From the technological point of view the main purpose of drinking water purification processes was very simple: To soften water and to oxidize and remove certain substances. This method differs from industrial-technological purification methods due to the larger quantities of water containing pollutants even at low concentrations. Some advantages for PCT over biologic processes are (Cooper and Thomas, 1974):

1. Greatly reduced capital costs (operating costs would be higher) and land requirements resulting from the usage of smaller units made possible by the enhanced settling rates, and lower strength of the chemically treated wastes.
2. Easier control of the plant if the system can be made to respond more rapidly to fluctuations in the quality and nature of the sewage treated.
3. Reduced susceptibility of the chemical treatment stage to materials which might be toxic to a biologic plant. The chemical clarification stage can remove an appreciable proportion of many toxic metals by precipitation.
4. Almost complete removal of phosphorus and small quantities of nitrogen may be achieved at sites where eutrophication is a problem.

Conventional water treatment may consist of the following processes:

| Aeration | Precipitation/ Sedimentation (CaO, Na_2CO_3) biologic (algae) pre-purification | Oxidation $(Cl_2, KMnO_4)$ (O_2, O_3) | Sorption and flocculation (Al, Fe) (polyelectrolytes) | Sand (multimedia) filtration | Activated carbon adsorption |

Aeration is effected by means of cascades, whereby an oxygen saturation of 70–80% is achieved (Rincke and Haberer, 1974). Following aeration, suspended solids are removed by settling in a pre-basin. Here, increased algal activity also occurs, the biologic effect of which is still not fully understood. On the one hand it causes the biomass to increase, and on the other accelerated $CaCO_3$-precipitation takes place due to elevated pH-values. Both effects are a strain on the subsequent filtration stages. The reduction of ammonium concentrations, however, and the partial elimination of iron and phosphate as well as the temperature adjustment in the pre-basin must be regarded as positive influences.

The next steps are of a chemical and physical nature. Suggestions by Patterson and Minear (1975) on the form of precipitation will be covered, with emphasis on the various compositions of waste water. Chlorination at sufficient levels (occasionally after *ammonia stripping,* in order to reduce chlorine costs) causes oxidation of residual ammonium compounds. Chlorine concentrations in river water fluctuate just as much as the nitrate contents due to the varying biologic nitrification at different temperatures. Moreover, the use of nitrate fertilizers during spring must also be taken into account (Haberer, 1974). Direct addition of fertilizers to irrigation water should account for the N, P and K contents of the water body itself. Varying quantities of chlorine are therefore added seasonally (*breakpoint chlorination,* also effective in virus destruction) to ensure that a chlorine content of 0.3–1 mg/l reaches the *activated carbon filter,* which is the final stage in most conventional plants. This chlorine concentration protects the filter from excessive fouling. In addition to alleviating organic interference, strong oxidants ensure the conversion of the relatively soluble Fe(II) and Mn(II) to the highly insoluble $Fe(OH)_3$ and MnO_2, both of which have high cation exchange capabilities and therefore serve as excellent metal adsorbents (Morgan and Stumm, 1964b). The destabilization and removal of the largely organic suspended material and their sorbed metal load takes place by means of *aluminum* or iron *flocculation.* In the Rhine water works at Wiesbaden, the ratio $FeCl_2$:CCD (chemical oxygen demand, a typical pollution indicator) = 1.2 was used as a guide for the addition of extra iron(II) chloride. The degree of flocculation effectively depends upon the type of pollutant e.g., algae, especially diatamaceous algae which are not included in the hydrous iron oxide flocculate, and which can only be removed by the subsequent *sand/multimedia filtration.*

However, the lignosulfonic acids, which originate from effluents of the cellulose industry (and cause very heavy pollution of the Rhine water), are almost as completely flocculated as the other organic acids. Since iron flocculation takes place ideally at pH 7.6, *lime neutralization* (with hydrous CaO) must be carried out after chlorination and the addition of iron flocculant ($FeCl_3$, $FeSO_4$). During the past few years this treatment has undergone continual technological improvement, as an insufficiency of lime results in unstable water with low pH-values, whereas an oversupply causes a strain on the subsequent sand filtration and activated carbon adsorption due to excessive $CaCO_3$-precipitation (Rassbach, 1974).

1.3.1 Traditional Removal of Trace Metals by Pre-clarification, Chlorination, Flocculation, and Filtration

The efficiency of trace metal elimination during these two stages of purification e.g., pre-clarification in the open basin (1), and chlorination, flocculation and filtration (2), was studied in the Rhine water works at Wiesbaden (Federal Republic of Germany) by Reichert et al. (1972). Table 83 summarizes the data of several important trace metals.

During the pre-clarification in the open basin, sedimentation of suspended material occurs eliminating part of the potentially soluble trace metal content in the untreated raw water. This must be taken into account when evaluating the retention effect on dissolved metal components at this stage, which is generally regarded as rather ineffective. Only beryllium is significantly eliminated, mercury by 30%, zinc by less than 20%, while the concentrations of most of the elements investigated, arsenic, cadmium, co-

Table 83. Trace metal elimination during pre-clarification (aeration, sedimentation basin) and chemical-physical water purification (chlorination, flocculation, sand-filtration, activated carbon filtration) (after Reichert et al., 1972)

Element	Rhine Water Mean µg/l (n = 17)			Sedimentation basin effluent Mean (n = 10) µg/l		reduction	Clean water effluent Mean (n = 10) µg/l		reduction
As	1.2	(0.1–	5.5)	1.1	(0.1– 5.5)	–	1.0	(0.1– 3.0)	–
Be	0.3	(0.2–	0.5)	0.2		90%	0.2		–
Cd	2.3	(1.5–	4.1)	3.3	(1.4– 3.9)	–	3.8	(1.9– 5.0)	–
Co	8	(4 –	12)	10	(9 –15)	–	10	(7 –14)	–
Cr	10.1	(4.5–	26)	12	(3.5–30)	–	6.0	(2.9–20)	50%
Cu	13.4	(3.6–	21)	17	(8 –33)	–	15	(4 –27)	–
Hg	0.5	(0.2–	0.6)	0.3	(0.12–17)	30%	0.3	(0.1– 0.6)	–
Ni	14.4	(8.9–	24)	17	(10 –17)	–	23	(9 –33)	–
Pb	27	(13.8–	43)	29	(13 –55)	–	35	(21 –62)	–
Se	5.9	(0.1–	10)	5.4	(1.3 – 9.0)	–	1.8	(0.1– 3.7)	70%
Zn	62.9	(24 –	176)	50	(20 –77)	20%	31	(16 –41)	40%

balt, copper, nickel, and lead are not reduced at all. On the contrary, in some instances elements such as nickel, cadmium, and lead enter the drinking water system in even higher concentrations than one would expect from their contents in the heavily polluted Rhine. It is possible, as Reichert et al. (1972) point out that in the case of lead and where the activated carbon has been in use for some time, part of the metal adsorbed as organic complexes is desorbed again, for instance after chemical reduction. Further experiments showed that subsequent treatment of the water by land spreading and slow sand filtration can lead to a substantial decrease of the mercury concentration as well as to an additional reduction of the selenium and zinc levels. Generally, investigations at various water works, using traditional methods of purification. show deviating results (Haberer and Normann, 1972). At three waterworks near Denver, Colorado, which apply aluminum dosing (Barnett et al., 1969), a substantial reduction in the concentrations of copper, iron, and manganese was observed, whereas lead and zinc were only eliminated to a small degree. With aluminum flocculation, the water from three German drinking water reservoirs was treated in the same way and a reduction in the zinc content of 50%–80% could be measured. Over a large range of initial concentrations lead exhibited a reduction of 30%, iron over 75%, and manganese between 30% and 90%. Rhine water treated by means of iron flocculation showed a reduction in the arsenic, copper and lead levels. Chromium, nickel, and zinc on the other hand were scarcely eliminated at all (Kölle et al., 1971).

Logsdon et al., (1974) reported on the conventional removal of MeHgCl, inorganic Hg, Ba(II), Se(IV), Se(VI), As(III), and As(V) from drinking water by Fe coagulation, Al coagulation, lime softening, high lime treatment, and activated carbon adsorption, using water from one river and three ground well sources. Three of the processes were effective in reducing one or more of the trace inorganics, activated carbon for organic and inorganic Hg, $Fe_2(SO_4)_3$ coagulation for Se(IV), As(III) and As(V), high lime for inorganic Hg, Ba(II), As(III) and As(V), but no single process was the most effective for all of the contaminants.

As a rule, improved water quality is observed when polyelectrolytes (= polyacrylamide-type-flocculants) are added to the coagulants mentioned above. At the same time the amounts of chemicals needed to enhance settling rates are reduced.

Shen (1973) reported the practical removal of As from drinking water by simple oxidation with chlorine (20 ppm) or permanganate, coagulation with iron(III) chloride (60 ppm), and sand filtration. Grigor et al. (1975) studied the conditions of sorption of As(III) to synthesized iron(II) sulfide and showed that, given a ratio of iron(II) sulfide to arsenic of 170, the residual As(III) does not exceed the maximum permissible level. Laguitton (1976) recommended lime addition for the removal of As in gold mine waste waters as the most economical treatment, provided careful control of the oxidation of As(III) to As(V), pH > 12 and effective filtration of the precipitate is exercised. If As levels below 0.5 mg/dm^3 are required, a modification of the method by phosphate addition must be considered.

Data taken from Dutch waterworks (Molt, 1967; Coutris and Goumella, 1969) show that using traditional methods, the elements As, Cr, Cu, Pb, and Se can be eliminated to such an extent that their values correspond to those of the groundwater. The elimination of mercury, however, is not effective enough. (Further details regarding mercury removal will be discussed in Sect. 2.2). In recent years several attempts have been made in Germany to introduce new technology into the traditional methods of water purification with the object of reducing water hardness. The results of these efforts showed that a simultaneous reduction of most heavy metal concentrations had been achieved. According to Haberer and Normann (1971–76) the following are economically interesting possibilities for softening processes in the preparation of drinking water:

(1) the addition of lime to the existing flocculation tank immediately before precipitation of hydrous iron oxide takes place; (2) refiltration flocculation combined with decarbonation and coprecipitation with magnesium hydroxide, which displays a particularly good flocculation quality; (3) softening by direct dosing of lime, soda ash (Na_2CO_3) or sodium hydroxide in the sedimentation basin; (4) application of the pellet reactor process, which has been in use for a long time in the boiler feed treatment, and (5) continuously operating ion exchange, especially for removing residual hardness.

Concerning the simultaneous reduction of trace metals during PCT, the processes of ion exchange, precipitation and coprecipitation and some other recent developments can be further elaborated.

1.3.2 Heavy Metal Removal by Chemical Precipitation

Precipitation is traditionally used for the removal of water hardness [Ca(II), Mg(II)]. Sulfides, carbonates, phosphates, and especially hydrous metaloxides are the most common inorganic precipitates.

Numeric estimations on this means of metal removal should always be treated carefully, as it is evident that oversimplification of theoretical solubility data can lead to errors of several orders of magnitude, due to the ionic strength of some waterbodies, the presence of complexing agents, temperature deviations or settling kinetics causing nonequilibrium systems (Patterson and Minear, 1975). Obviously, optimum pH-values for the removal of several metals by, for instance, lime neutralization differ widely.

In order to achieve pH 11,5 high lime doses are required. Besides the removal of heavy metals, color and turbidity, this high lime process is particularly advantageous for sterilization and is necessary for the removal of high ammonia concentrations by air stripping. Furthermore, sludges resulting from lime treatment possess superior thickening and dewatering characteristics and are suitable for filter pressing at lower cost than coagulates from Fe(III) chloride or alum. Lime is also a more popular base causing coagulation and sedimentation than sodium carbonate or hydroxide, due to its general availability at low cost.

Pretreatment and/or separation prior to precipitation is often essential for effective metal removal. For instance, Cr(VI) should be reduced to Cr(III) in order to form the poorly soluble chromium(III) hydroxide.

Some complexing agents (NH_3, ^-CN, organics) can prevent precipitation thus severely reducing the efficiency of the treatment.

Nilsson (1971) reports on aluminum sulfate and lime treatment of waste water to reduce Pb(II), Cu(II), Cr(III), Hg(II), Cd(II), and As(V) to low levels by both precipitants, although Zn(II), Ni(II), and Co(II) are precipitated only at pH < 9.5, while Cu(II) and Pb(II) precipitation is greatly inhibited by the presence of NTA at pH < 9.

This indicates that a well planned sequence of several purification steps. adapted to specific raw water compositions is essential for obtaining optimal results, Alternatively, post-treatment such as recarbonation with CO_2 gas might also be desirable for the stability of the product water, while optimal metal removals will only be achieved after improved separation of the solids. As Weeks (1975) states it: "At high volume throughput, a highly retentive filter to replace sand filtration is necessary because the more insoluble compounds apparently escape as extremely small crystals and afterwards become available for solution, thus causing low efficiency."

The classic method of decarbonation is based on the following reaction:

$$Ca^{2+} + 2\,HCO_3^- + \frac{Ca^{2+}+2\,OH^-}{\text{lime dosing}} \rightarrow 2\,CaCO_{3(s)} + 2\,H_2O$$

In suitable reactors the precipitation products can be removed after growing on contact grains which are easier to separate than fine lime precipitates. In their experiments on the reducibility of trace metals simultaneously with lime precipitation, Haberer et al. (1976) chose experimental conditions in the pellet reactor for maximum decarbonation. Saturated lime (hydrated CaO) was added to the raw water. Due to the very high pH-values many trace metals are converted to insoluble hydroxides or coprecipitated with $CaCO_{3(s)}$ and $Mg(OH)_{2(s)}$ causing a dual effect of metal elimination. The latter effect is also considered to reduce soluble anionic species as arsenate and selenite.

Table 84 shows the removal of trace elements in the treatment of raw water from primary-clarified Rhine water with and without metal dosing.

Very good reduction takes place in the case of zinc, cobalt, and copper in both series of experiments. The elimination capacity is also effective with regard to the concentrations of cadmium, iron, manganese, nickel, lead, vanadium, and—for the first time, considered in the water purification methods—mercury. Nevertheless, arsenic and lithium pass through this stage of treatment with almost unchanged concentrations.

1.3.3 Activated Carbon Filtration in Drinking Water Purification

Activated carbon (powdered or granular) filtration was formerly used for dechlorination purposes and removal of organic color, taste, and odor. It plays now also a large role in the elimination of part of the dissolved organic substances which remain after chemical clarification. The process involved also acts as an additional filter for residual iron concentrations. In some cases possible manganese tailings can also be retained on this filtering adsorbant (Haberer and Normann, 1974a).

Robeck (1974) and Suffet et al. (1978) elaborated on the use of activated carbon for reducing the level of trace organics in drinking water. Conventional U.S. plants, adding only 1 or 2 ppm powdered activated carbon (PAC) for odor control, obtain

Table 84. Reduction of metal contamination (maximum values) during decarbonation in the pellet reactor (after Haberer et al., 1976)

	Maximal values in µg/l					
	Before dosing			After dosing		
	Raw water	Reactor water	% Removal	Raw water	Reactor water	% Removal
As	0.1	–		0.6	1.6	+
Cd	5.9	3.1	50	50	4.4	90
Co	11	2.8	75	656	21	95
Cr	4	2.5	40	43	33	20
Cu	32	9.6	70	100	24	75
Fe	29	15	50	100	23	75
Hg	0.10	0.03	70	4.6	2.5	45
Li	17	19	0	56	54	0
Mn	3.4	–		18	0.7	95
Ni	15	7.6	50	56	28	50
Pb	42	17	60	35	19	50
V	5	1	80	26	14	50
Zn	97	7.7	90	156	14	90

minimal inherent protection against pesticides. Other plants possessing granular activated carbon (GAC) beds provide better occasional protection, also against organic and inorganic mercurial pesticides.

In fact, carbon can be used in either powdered or granular form, the former requiring a batch-type contacting system and the latter requiring a flow-through or bed-type operation, which offers an additional filtering action thereby removing metal containing suspended solids. As Singer (1974) points out, the mechanism of adsorption can be one of simple physical adsorption due to Van der Waals forces of attraction, exchange adsorption due to electrostatic forces induced by charged functional groups on the carbon surface, or chemisorption in which the adsorbate reacts with the carbon surface and is chemically bound to the carbon. The first two mechanisms are reversible in nature and represent rather weak energies of interaction, while chemisorption is an irreversible process and is characterized by high energies of interaction. Several mechanisms can effect trace metal removal. The carbon generally contains noncarbon impurities, particularly oxygen and sulfur. The oxygen content can range between 2% and 25% by weight, depending upon the temperature and method of activation (Snoeyink and Weber, 1967; Hassler, 1963). The sulfur content of the carbon ranges from trace amounts up to 2% by weight, depending upon impurities in the base material (wood, lignite, coal, etc.) and upon the method of activation.

The presence of oxygen results in the formation of surface oxides on the carbon which undergoes hydrolysis when the carbon is in contact with water. Hydrolysis of these surface oxides impart weak acid cation exchange properties to the carbon. As such, the carbon functions partly as a cation exchanger and can be expected to attract trace metals from the solution with which it is in contact. This shows at the same time why observed exchange properties may be greatly pH-dependent.

The presence of sulfur leads to the formation of sulfide groups at the carbon surface. Many heavy metals tend to interact quite strongly with the sulfur (e.g., mercury, cadmium, copper) and, consequently, the sulfido groups cause the carbon to exhibit chemisorptive behavior towards metals.

The carbon surface also provides nucleation sites for the precipitation of metals from solution. Localized high metal concentrations in the vicinity of the carbon surface due to adsorption can bring about oversaturated conditions leading to the precipitation of metalhydroxides from otherwise undersaturated solutions.

Activated carbon can also affect trace metal removal by chemical reduction. Carbon behaves as a reducing agent in the presence of oxidized metals such as dichromate and permanganate, forming Cr(III) and Mn(IV) which are readily insoluble at neutral pH-values and precipitate as $Cr(OH)_3$ and MnO_2, also having good adsorbing properties.

Huang (1975) described Cr(VI) removal, mainly as $HCrO_4^-$, by carbon adsorption using calcinated coke at laboratory level. Decreasing chromium concentrations and low pH values improved the removal efficiency. Duncan (1974) reported the successful use of activated carbon in the loading of gold in a countercurrent expanded-bed-system, after cyanide leaching of low-grade ore. This procedure proved most advantageous in the recovery of gold after amalgamation was banned at Lead, South Dakota.

Logsdon and Symons (1973) found that inorganic and methylmercury could both be removed by contact with PAC at doses of 10 mg carbon/μg of mercury. Granular carbon columns were shown to be effective in treating large volumes of water containing 20 to 29 $\mu g/dm^3$ with more than 90% removal of methylmercury even after processing 20,000 bed volumes of water. The greater retention of methylmercury over the inorganic mercury can be attributed to the affinity of the organic methyl-group for the activated carbon in addition to retention of the mercury by the sulfido and oxo groups. Logsdon et al. (1974) concluded that for the control of mercury in surface water where the pollution rates are greater, high-dose PAC treatment should be effective.

Thiem et al. (1976) deduced from test results that the presence of chelating agents will enhance the removal of mercury from drinking water using PAC as adsorbent, and that neutral pH and small additions of tannic acid increased the effectiveness.

Linstedt et al. (1971) established removals exceeding 95% for silver(I), cadmium(II), and chromium(VI) after GAC treatment. The cations Ag(I) and Cd(II) were probably retained by a combination of cation exchange and chemisorption by the sulfido groups as well as adsorption of organic cation complexes. Removal of Cr(VI) anions can be attributed to the chemical reduction to insoluble Cr(III) hydroxide. Anionic SeO_3^{2-} however was only removed for 35%, which is twice as much as after lime coagulation but only about one third of the almost complete removal by cation and anion exchange treatment.

1.3.4 Heavy Metal Removal by Ion Exchange

During treatment ion exchange involves the reversible exchange of ions between a solution and a solid phase that are in direct contact.

The solid phase can be a natural zeolite or a synthetic resin consisting of a crosslinked polymeric network with charged functional groups which attract oppositely charged ionic species and retain them by electrostatic forces (Singer, 1974).

Resins exhibit a relative affinity to all ionic species of opposite charge, depending upon the specific ionic charge, the hydrated ionic radius, the concentration in solution, the degree of resin cross-linking, and the nature of the functional group on the resin (e.g., sulfonic, phosphonic or carbonic acid groups).

In fixed-bed type units, this process continues until the solution being treated exhausts the resin exchange capacity. At that stage, the exhausted resin must be regenerated by an acid/base or other chemical replacing the ions and converting the resin back to its original composition and yielding up to a 500 times concentrated regenerant brine. In some cases valuable metals can be recovered from the brine, for instance by electrolysis. This method is widely used in the potable water supply treatment for final softening and demineralization. The cation exchange resins remove divalent Ca(II) and Mg(II) ions and are also expected to remove Cd(II), Zn(II), Ag(I), etc. effectively. Similarly, anion exchange resins exhibit an affinity not only for the common hydrogencarbonate, sulfate and chloride, but also for trace metal anions such as hydrogenarsenate, selenite, hydrogenchromate, etc. It is of the greatest importance to always keep in mind that the effectiveness of ion exchange for trace metal removal depends significantly on the chemical form of the given trace metals (e.g., Cr(III) vs. $HCrO_4^-$, Zn(II) vs. $ZnOH^+$, Hg(II) vs. CH_3Hg^+ ...).

Anion exchange was reported by Logsdon and Symons (1973) to be the only process of those studied to provide satisfactory removal of Se(VI)/SeO_4^{2-}. Shen (1973) and Calmon (1973) noted almost complete removal of As(V). Hexavalent chromate and dichromate are successfully removed from chromium waste waters (Richardson et al., 1968; Anon, 1972 f; Patterson and Minear, 1975). Chromic acid may be recovered through substitution by sodium by subsequent cation exchange.

Linstedt et al. (1971) observed 85.8% removal of Ag(I) and 99% of Cd(II) following the passage of a high quality waste water through a cation exchange bed. Subsequent anion exchange increased the percentage removal of dichromate and selenate from 5.4% to 96% and from 0.9% to 99% respectively.

Copper, lead, mercury, and nickel have also been successfully treated by cation exchange (Schore, 1972; Cheremisinoff and Habib, 1972; Gardiner and Munoz, 1971; Patterson and Minear, 1975). Linstedt et al. (1971) used lime coagulation, sedimentation, sand filtration, and activated carbon adsorption prior to ion exchange for conditioning the waste water in order to reduce interference of organics by complexing cations or competing directly with anionic trace metals, thus fouling the exchange column and interfering with regeneration.

Continuous ion exchange was originally introduced (Higgins, 1973) for isotope separations and the removal of trace radioactive elements from waste water. The major advantages of continuous countercurrent operation applied to ion exchange over fixed-bed units are: high throughput rate per unit size, higher chemical utilization, lower dilution of process streams, lower process water requirements and more uniform product quality over the total cycle.

Other processes, not yet widely used but which also have potential in some aspects of water purification, are reverse osmosis, electrodeposition, electrodialysis, cementation, solvent extraction, γ/UV irradiation, evaporation (distillation), ion flotation, and freeze concentration.

1.3.5 Potential Metal Enrichments in the Water Distribution System

During successive stages of treatment waters repeatedly come into contact with metal-bearing containers, pipes, and armatures; this often leads to an increase in the heavy metal concentrations of drinking water. These effects are even stronger during the further transport of the water, particularly in the water piping of individual households, where long periods of stagnation can result in intensive corrosion. Striking examples of such corrosion are known especially from lead pipe networks e.g., the "lead epidemics" of Dessau (1886) and Leipzig-Naunhof (1930).

In order to obtain a general view, Benger and Kempf (1972) took two samples in each case from cold water taps in flats in different parts of the city of West Berlin: The first one after at least 12 hours stagnation and the second after 10 min flowing time (i.e., equivalent to approx. 50 l water). Since significant differences could be expected according to the type of plumbing, the analysis data were tabulated according to different construction periods and evaluated accordingly. In this manner the following periods were obtained: 1900–1914, 1920–1939, 1945–1960, and 1961–1970. Mention must be made in this context that after 1935 lead piping was no longer used in the drinking and domestic water supply and effluent systems in Berlin due to a metal saving campaign. It is of interest to note that the use of lead piping had been previously banned in other countries.

In Table 85 (after Benger and Kempf, 1972) the concentrations of the trace metals lead, copper, and iron in domestic plumbing from the building periods 1900–1914 and 1961–1970 are compared.

The following conclusions can be made for the individual heavy metals: The highest lead concentrations were found in the group of houses built before 1914 which still had lead pipe plumbing. After the water had been allowed to stagnate, mean values of 100 µg/l were determined; after the water had been allowed to flow for 10 min, the values dropped to 20 µg/l. The maximum values in each case were 280 µg/l and 50 µg/l Pb respectively. However, in the houses built between 1961 and 1972, the water sampled after a 12 h period of stagnation contained lead concentrations below the limit of 100 µg/l. Whereas the lead values are seen to decrease to about one-fifth of the initial concentrations in old houses (1900–1914) after draining the water pipes for 10 min, the decrease in the newer houses is much less, approximately half of the initial lead concentrations. The decrease in iron contents in the piping system after withdrawal of approximately 50 l water, was between 20% and 80%, regardless of the building period. The highest Fe-concentrations were determined in houses built between the two World Wars (1920–1939). These buildings were generally equipped with galvanized iron pipes, in which the zinc layer had either corroded or was covered by protective deposits. Although the iron concentrations present no problem, they do reflect that even after several decades stable conditions are not created, so that the water continues to oxidize the pipe material. During the period 1961–1970 the installation of copper pipes was preferred because of easier handling. For this reason the highest copper contents are found largely in newer houses. A marked decrease can be observed in the copper contents after the draining of the water, with the mean values decreasing to approximately one-tenth of the original

Table 85. Trace metal contents in the water supply system of West Berlin (after Benger and Kempf, 1972)

		1900–1914 0 min 10 µg/l		1961–1970 0 min 10 µg/l		Limit[a] µg/l	Remarks[a]
Lead	Mean value	100	20	25	14		Not more than 300 µg/l when
	Maximum					100	there is long contact with Pb
	value	280	50	57	26		piping
Iron	Mean value	71	52	125	61		Up to 300 µg/l in smaller water
	Maximum					100	supply plants if in stable form
	value	170	100	650	160		
Copper	Mean value	130	19	340	28		Up to maximum 3.0 mg/l after
	Maximum					50	16 hours standing in the pipes
	value	1030	70	2450	70		

[a] Suggestions of a study group set up by the European office of the World Health Organization, Copenhagen.

values. The decrease in the maximum concentrations after draining the water for ten minutes is even higher. Similar behavior can be witnessed in the case of zinc contents.

Since commercial zinc used for galvanizing can contain up to 1.4% lead, small quantities of lead may be dissolved. The same holds true for cadmium, which is associated with commercial zinc in quantities of approx. 0.2 mass percentage. Kempf (1974) reported 1–2 µg/l cadmium in the Berlin pipe water after nightly stagnation. After the water had been drained for some time, all values decreased to less than 0.5 µg/l. Schroeder et al. (1967) described an example of elevated cadmium concentrations in the pipe water of Brattleboro (Vermont) of 15 to 77 µg/l Cd, while municipal water contained as much as 14–21 µg Cd/l.

The corrosive behavior of the individual metals is controlled by different factors, such as pH and the hardness of the water. Stumm (1960) found that in the case of iron the corrosion by water increases with rising pH values, although in practice pH elevation is used as a protection against corrosion. According to Grohmann (1973) the protective layer in the iron pipes results from several effects: Chloride ions, low flow rate, inhomogeneous surface deposits and certain bacteria inhibit its formation. On the other hand, high flow rates, the presence of calcium ions, buffer capacity, and preventive chemical additives create favorable conditions. With zinc and lead pipes it is evident that slightly acidic soft water can solubilize these metals during stagnation. In order to limit effectively the potential uptake of zinc into the water during stagnation in galvanized piping, it is sufficient to raise the pH without exceeding the pH of lime saturation in order to prevent precipitation of calcium as scale. If the piping consists of synthetic material it should not be forgotten that such pipes contain metals as stabilizers and antioxidants. In this manner lead stearate, lead phosphate, and several basic lead sulfates may be incorporated in plastic pipes in concentrations of up to 2%. These substances, which may also include metal-containing pigment substances, are, to a limited extent, soluble in water. Investigations on PVC hardened synthetic pipes have demonstrated, however, that under normal pH-conditions metal dissolution is limited (Herzel, 1968).

In a large-scale investigation programme McCabe et al. (1970) analyzed a total of 2595 drinking water samples taken from 969 public water supply systems for their trace metal contents. Table 86 contains some of the results of this study.

The maximum limits for drinking water, as recommended by the U.S. Public Health Service were exceeded by Fe and Mn in more than 8% of the samples analyzed, by Cu and Pb in approximately 1.5%, and by As, B, Cr, Se and Zn in less than 1% of the samples.

In special cases, there must be an absolute guarantee that limits are not exceeded, for example in hemodialysis. Since the heavy metal concentrations in the pipe water often exceed the contents of these metals in the blood plasma, special precautionary measures should be taken against the possible corrosion of pipes, containers and armatures (Carlson and Hässelbarth, 1971). This is especially important when softened, oxygen-rich water is used; fatal cases of copper intoxication have occurred due to corroded pipes. These pipes and containers must therefore be composed of corrosion-proof material e.g., stainless steel.

With regard to metal elimination in the different processes of drinking water purification, it is possible to establish that ion exchange and carbonate precipitation are effective processes, particularly when dealing with the elimination of heavy metals from

Table 86. Drinking water investigations on 2595 samples from public water supply systems in the USA (McCabe et al., 1970)

Metal	Maximum concentrations µg/l	Exceeded USPHS guidelines in %	Exceeded USPHS threshold values in %
Ag	30		0
As	100	0.4	0.2
B	3280	0.8	0
Cd	3940		0.2
Cr(VI)	79		0.2
Cu	8350	1.6	
Fe	2600	8.6	
Mn	1320	8.6	
Pb	640		1.4
Se	70		0.4
Zn	1300	0.3	

strongly polluted waters. There is clear evidence, on the other hand, that some methods of conventional drinking water purification e.g., sedimentation in clarifying basins, sand filtration and bank filtration are not well suited to safeguard against the passages of trace metals. This fact, together with the high costs involved in implementing the more effective combination of ion exchange after precipitation and activated carbon adsorption, necessarily led to the conclusion that "pollutants should not be present in surface water to be used for drinking water purification" (Hässelbarth, 1972). In practice this means "a removal of anthropogenic trace elements *before* they are introduced into the surface waters, at the point where they are most concentrated and easiest to remove" (Reichert et al., 1972).

2 Heavy Metals in Industrial and Domestic Effluents

2.1 Effluents from the Electroplating Industry

One of the most important sources of heavy metal pollution in surface waters is the direct or indirect discharge of effluents from the electroplating industry. In the United States there are 20,000 plants in operation, and 800 in the Federal Republic of Germany. Attempts have been made in the past to reduce the worst dangers by introducing a standard discharge rate not to be exceeded. The two major pollution effects not only pose a direct threat to the aquatic habitats but also interfere with biologic purification in sewage treatment works. Different limits were therefore set for both needs (examples in Table 87 from the Federal Republic of Germany):

The values represented under (1) are only recommendations enabling water control authorities to enforce stricter or more generous measures depending on the individual

Table 87. Guidelines for effluent discharge metals in the Federal Republic of Germany. 1 direct discharges from electroplating industries into surface water–LAWA (1970). 2 discharges to urban sewage treatment plants–ATV (1970)

	Effluent guidelines	
	1	2
Chromium (total)	2 mg/l	4.0 mg/l
(hexavalent)	0.5 mg/l	0.5 mg/l
Copper (Cu)	1 mg/l	3.0 mg/l
Nickel (Ni)	3 mg/l	5.0 mg/l
Zinc (Zn)	3 mg/l	5.0 mg/l
Iron (Fe)	2 mg/l	
Cadmium (Cd)	3 mg/l	

conditions. It is of the utmost importance that when permission is granted by the authorities to run a sewage treatment plant, some directives must be given regarding the maximum quantities of effluent which can be dealt with. In the case of factories, which discharge their effluent into smaller surface waters, difficulties arise mainly when the volume of discharge and the water level of the receiving waterbody are subjected to considerable fluctuations.

The metal concentrations introduced into urban sewage systems are subjected to ever-increasing biologic treatment stages, which can also be used as a basis for recommendations. Several large cities have developed their own tabulations giving the maximum permissible quantities of pollutants. These are based on the specific capacities of their treatment plants and sometimes contain additional information e.g., maximum values of 20 mg/l for cadmium, copper, chromium, nickel, zinc, and iron; 5 mg/l for lead, and the total of the former six elements combined not exceeding 50 mg/l (City of Johannesburg).

Funke (1975) states that no distinction is made in these standards between dissolved and undissolved metals. Generally, however, it is possible that dilution with other effluents takes place, as at this stage control is not carried out at the exit of a factory's own treatment plant but is done at the point where the effluent joins the urban sewage system (Hartinger, 1976).

In cases where the standard values are exceeded, this can usually no longer be attributed to technical shortcomings. Such are more likely to be the result of insufficient knowledge and inadequate consideration of the physical and chemical conditions (Bucksteeg, 1967). Several basic principles will be listed below for reactions which can occur during the treatment of effluents from the electroplaing industry:

Electroplating is the process, whereby an object is coated with a metallic layer in order to improve its appearance and/or resistance to corrosion. The plating process itself is preceded by suitable *surface preparation* such as mechanical finishing (polishing, buffering), degreasing, and the removal of oxides, rust or scale by chemical means e.g., by pickling (see Chap. 3.4). (After Funke and Coombs, 1973). Alkaline cleaner may contain sodium cyanide in concentrations of 4–8 g/l as CN^-. Acid dips for neutralizing after electrolytic cleaning produce elevated metal contents in the rinse waters e.g., between 2 and 4 g/l of zinc. Acidic electrolytes are mainly employed for the plating of copper, nickel, zinc, and chromium. Alkaline electrolytes, containing the metal as cyanide complexes, are used for the deposition of copper, zinc, cadmium, gold, silver, and brass.

An exchange of the solutions occurs with the cleaners at intervals of a few days to several weeks. With the plating solutions the interval can be one to two years. The life of chromic acid solutions is especially influenced by contamination with reducing agents, resulting in the formation of trivalent chromium. Only approximately 13% of the chromium content in the solution is deposited as plating, the rest is dispersed as drag-out loss, or loss in the first running rinse (67%) and loss in condensation (20%). The drag-out losses for nickel were estimated by Harris (1960) at approximately 60% of the expenditure for nickel salts.

Funke and Coombs (1973) found that from 32 electroplating works in the Johannesburg area approximately 270,000 liters of nickel plating solution were discharged each year into the sewage works of Klipspruit and Palmietfontein. This was equivalent to a loss of some 300,000 dollars. Funke (1975), reports examples from certain Witwatersrand sewage works, where sludge discharges of zinc or chromium resulted in a drastic reduction in gas production of the anaerobic digesters. The zinc concentration was reported to have reached 2%–3% of the sludge dry solids when the drop in performance was observed at sewage works, east of Johannesburg. A peak concentration of 700 mg/l for chromium was observed at the inflow into the Krugersdorp sewage works where the digesters subsequently failed a few days later.

In-plant measures are applied to reduce metal concentrations of the effluents: Large quantities of metal salts are lost during the normal rinsing operations. Even larger amounts are lost by accidental wastage through pipeline breakages, cracks in tanks, negligent operations or the wilful dumping of contaminated processing solutions (Funke and Coombs, 1973). This is why control measures have to be introduced to prevent a possible threat to treatment plants and receiving water bodies. Of particular importance, however, are the cleansing measures within the plating plant itself where a large quantity of the metals is retained by means of PCT and can be reused for production. The costs of energy supply and additional water requirements should be taken into account. Where recovery is not practical, destruction can be resorted to. Of primary importance here is the treatment of cyanide and compounds of hexavalent chromium, which in this context are the most dangerous toxins that can be discharged into the sewage effluents (Hartinger, 1968a). Evaporation, counterflow, reverse osmosis, and ion exchange methods may all be employed for recovery of valuable metals, and eventually make pollution control an asset rather than a liability.

Open and closed loop evaporation systems have been designed for full and partial recovery of the processing chemicals otherwise lost in the dragout (Ciancia, 1973). Application of such principles as enacted in Sweden are described by Göransson and Moberg (1975) for chromium and zinc electroplating. In Canada toxic ions are reclaimed by selective precipitation and vacuum filtration. The *integrated* effluent treatment (LANCY-process of controlled recirculation), in contrast to conventional detoxification, takes place within the finishing stage. Mixing of the metal solutions is avoided; frequently an economically viable recovery is possible, especially in the case of nickel, cadmium and copper.

The June 1973 edition of "Environmental Science and Technology" (Anon. 1973b) presented a *zero effluent discharge* system in the metal finishing industry, similar to the one which had been put into operation at the U.S. Army's Rock Island Arsenal in Illinois. Figure 98 indicates the most important runoffs of this fully automated system, engineered by DMP Corp. (Charlotte, N.C.). Water is recycled and all wastes are reclaimed; it is estimated to save the Arsenal more than 200 million liters of water per

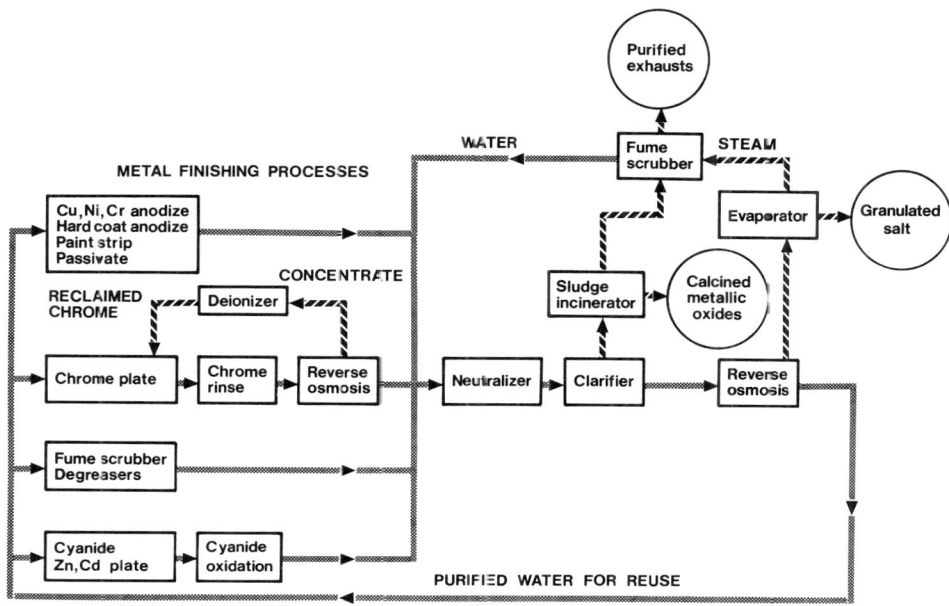

Fig. 98. Zero effluent discharge system for electroplating wastes (reprinted with permission of Environmental Science and Technology, 7, (6), June 1973, Copyright by the American Chemical Society)

annum, 9% of its total requirement. The system utilizes the largest reverse osmosis unit ever constructed for industrial purposes.

2.2 Mercury Removal from Chlor-Alkali Plant Effluents

In recent years great efforts were made to reduce the dangerous mercury emissions from chlor-alkali plant effluents. These include improvements in the methods of chlor-alkali electrolysis, where the mercury loss has been reduced from 100 g per metric ton of manufactured chlorine to approx 2 g/ton.

New methods of purifying Hg-effluents have been developed (See for example Cheremisinoff and Habib, 1972): (1) Originally mercury effluents were treated with inorganic sulfides, which were allowed to settle in order to bring the reaction to completion. (2) Later ferrous chloride was introduced for reduction and precipitation as metallic mercury, cleaning an effluent to residual concentrations of 0.1 to 0.3 mg/l. (3) Mercury recovery from brine has been accomplished by a strong base anion exchange resin (using tertiary alkyl groups); mercury is reduced to 0.1 mg/l after passing through these ion exchangers, and to a few parts per billion after passing through an adsorption tower. (4) In the Ventron process, $NaBH_4$ has been used as a reducing agent. The effluent concentration is claimed to reach levels well below 10 ppb, while the caustic solution is maintained at between pH 9 and 11. Almost all mercury is recovered by cyclone and polishing filter after reduction of the mercury compounds. (5) The Osaka Soda Process in Japan results in effluent concentrations of 5 ppb; ion exchange stripping recovers mercury which is reduced to elemental form by addition of sodium amalgam. (6) The FMC Corp. process, protected by United States patent, requires only standard mercury cell chlorine plant engineering facilities and yields, on an average, liquid effluents with 3 ppb mercury (Anon. 1973c). (7) Heavy metals have also been successfully removed from chlor-alkali brines by one-step exchange treatment using silicon alloys; ferrosilicon is fairly selective in removing mercury without significant iron contamination of the water (Mc Kaveney et al., 1972). (8) One of the latest developments is mercury removal from waste water with starch xanthate-cationic polymer complexes (Swanson et al., 1973); residual mercury concentrations were as low as 3.8 ppb after single treatment of solutions that had initial mercury concentrations of 100 mg/l.(9) The Akzo process (de Jong and Rekers, 1974) uses a cation exchanger with thiol as active groups. In order to obtain residual mercury concentrations < 0.005 mg/l, the following pretreatment stages were introduced: Oxidation of metallic mercury with chlorine, and pH adjustment to approximately 3. This stage was followed by filtration to prevent clogging, and dechlorination with a special activated carbon column in order to protect the thiol groups of the resin against oxidation. (10) Finally, ASAHI Chemicals of Japan introduced a new ion exchange technology for the production of caustic soda with a similar NaCl content as from mercury cells, and capital and operation costs apparently lower than for the diaphragm cells.

For the most recent developments the interested reader is referred to Sect. 60 (Sewage, Water) of the respective editions of Chemical Abstracts.

2.3 Prevention and Control of Acidic Mine Drainage

The protection measures against the formation of acidic mine effluents begin with the choice of the deposition site. Several procedures for minimizing effects have been collected by Andrews (1975), who discusses in detail the different criteria (geotechnic, hydrologic, climatological, economic, aesthetic, of safety, social, and ecological) which must be considered in addition to the actual engineering problems. Of prime importance is the protection of air and water. High sulfide-bearing material should be exposed for the shortest time possible; as part of the mining operation, this material should get adequate cover containing little or no sulfides (Hill, 1973). Since the primary obstacle to growing grass is the presence of acidic conditions, the cover material should be vegetated, as vegetation serves to stabilize the cover material, and upon dying decays and acts as an oxygen absorber (Smith and Bradshaw, 1970). Recent experiments have indicated that metal-tolerant plant populations can grow very successfully on heavily contaminated materials (Bradshaw, 1975). In this context one should also consider the introduction of strongly oxygen-consuming sewage sludge as tailing cover. Singer and Stumm (1970) discuss the possibilities of mine-sealing e.g., by flooding the mines and introducing organic material in order to reduce iron(III) and sulfate chemically. They determined that a kinetic control which requires suppression of the catalytic agent can be achieved by bactericides; strip (surface) mines and coal refuse piles are especially amenable to such treatment.

Where mine drainage cannot be prevented, the effluents thus formed must be subsequently treated. According to Hill (1973) this is possible by means of the following procedures (also refer to Sect. 1.3 on PCT):

1. Neutralization: A typical system would include adding an alkaline reagent, mixing, aerating, and removing the precipitate. Alkaline reagents are ammonia, sodium carbonate, sodium hydroxide, limestone, and lime. Bergmann (1971) described a two-step precipitation by chlorine and lime (reagents) to recover metal from mine water of Meggen (Federal Republic of Germany), the costs of which tend to zero after 20 years. Two-stage treatment with limestone followed by lime appears to be advantageous on discharges containing iron, zinc, cadmium, and manganese; two- or more-stage treatments with the proper selection of pH can be used to separate heavy metals with subsequent recovery (Dean et al., 1972; Hill, 1973; Van der Walt et al., 1975).

2. Ion Exchange: This method has until now been used predominantly in the treatment of coal acid mine drainage. Due to the high costs of operation, the application of this method was not widespread and further developments will therefore be necessary. Most promising is the hydrometallurgical treatment of low grade ores and mineral conversion. Rosenbaum (1976) reported on in situ uranium leaching in Texas, a technique low in capital and operation costs with minimum land-surface disruption and the avoidance of disposal or storage of mill tailings, using a dilute alkaline leach solution (instead of MnO_2/H_2SO_4 acid) with a low uranium loading but a high circulation rate.
3. Reverse osmosis: Since this is a concentration system, the disposal of the waste stream is a major problem. Hill et al. (1971) have developed a system with the Environmental Protection Agency (United States), whereby the waste stream is neutralized, the sludge removed and the neutralized water returned to the influent of the reverse osmosis unit, a procedure referred to as neutrolysis.
4. Cementation: Metals are removed from mine drainage by the electromotive force of other metals, e.g., by passing the copper-bearing water through shredded iron (see Chap. H, on recycling of metals).
5. Electrolysis: This method is again only applicable in waters with very high concentrations of the metal. Kendrick (1977) reported promising investigations including the use of microorganisms inhibiting the oxidation of sulfide, hydraulically sealing abandoned mines by flooding and diverting water to prevent it entering mining areas. Such action is already successfully studied and implemented in Australia (Anon, 1974).

2.4 Heavy Metals in Urban Drainage Systems—Biologic Treatment (BT)

Although the heavy metal contents in urban drainage systems generally do not reach the proportions of industrial effluents—certainly not those of metal processing industries—the problems caused by their presence, particularly in large cities, should not be underestimated.

The enormous quantities of ever-increasing material passing through the urban drainage system must eventually be deposited in a manner which does not lead to a deterioration of the environment. Furthermore, the unfavorable effects which some metal-containing effluents have on the urban sewage systems should not be overlooked. These consist largely of an inhibition of nitrification and oxidation processes in the biologic stages of the systems.

A two-stage purification plant consists of methods of a mechanical as well as a biological nature. After eliminating the coarser components by means of bar screens and a detritus removal system—as well as primary clarification in a settling tank—the predominantly organic effluent substances reach the biologic stage. This type of effluent treatment is regarded as an ideal method in purification, since it resembles and even accelerates natural processes of pollutant degradation. At present two methods are in use: Firstly, *trickling filter* (now often called biological filter) treatment, whereby the effluent is sprayed from nozzles on a rotating distributor onto filter media converged with a film of microorganisms. Secondly the *activated sludge* process in which the mechanically pretreated water is aerated and agitated for several hours in a suitable basin. Similar organism cultures, as in the films on the media of the trickling filters, live in flocs of bacterially active sludge which have the ability to adsorp and degrade the dissolved and colloidal components in the presence of oxygen (Rüb, 1974). From the aeration basin the mixed liquor is transferred into a secondary clarifier where the sludge is separated by sedimentation. Most of the settled sludge is returned (pumped back) to the aeration basin as innoculation sludge. The smaller fraction of surplus sludge is removed with the primary sludge for further treatment.

2.4.1 Metal Extraction in the Mechanical (Primary) Sedimentation Unit

The data in Table 88 were measured in a two-stage sedimentation basin of the purification plant in North Johannesburg; they show the mean values of seven samples (Funke, 1975). In the mechanical treatment phase the relatively high extraction rate of most metal examples from waste water outflows is supposedly accounted for by the presence of only a part of the heavy metals in solution. The major part is either bound to fine suspended substances or is transported into the plants in colloidal form, where it at first is deposited on solid matter.

2.4.2 Reduction of Metal Loads in the Biologic Stage

In the biologic (secondary) treatment stage metal concentrations in most cases clearly decrease. The anaerobic treatment in anaerobic digester is evidently quite effective, since metal sulfides form here, which are soluble only with difficulty (Wood and Tchobanoglous, 1975). At the same time this treatment stage also tolerates higher metal contents in waste water effluents, while under aerobic conditions the delayed nitrification in the biologic stage points to the presence of toxic metal concentrations (Wood and Tchobanoglous, 1975). High extraction rates (50%–60%) were measured for mercury in activated sludge treatment (Ghosh and Zugger, 1973).

Investigations on purification plants in New York, New Jersey, and Connecticut by Mytelka et al. (1973) generally show a minor elimination rate for the metals zinc, chromium, and cadmium in both the mechanical sedimentation and biologic stages (both were carried out under aerobic conditions). Data on activated sludge from British plants show for copper an extraction rate of between 54% and 93%, for chromium between 10% and 100%, and for zinc between 60% and 100% (Jackson and Brown, 1970).

Table 88. Removal of heavy metals in waste water treatment in urban purification plants in Johannesburg, South Africa (Funke, 1975)

	Cd ($\mu g/l$)	Cr ($\mu g/l$)	Cu ($\mu g/l$)	Ni ($\mu g/l$)	Pb ($\mu g/l$)	Zn ($\mu g/l$)
Primary Sedimentation						
Screened sewage after primary settling						
Influent	–	120	100	160	60	–
Effluent	–	50	40	90	30	–
Removal	–	58%	60%	44%	50%	–
Activated sludge unit						
Influent	63	1570	890	2370	180	3900
Effluent	26	610	210	2190	70	690
Removal	59%	61%	76%	8%	61%	83%
Digested sludge treatment						
Influent	10	170	80	130	70	1200
Effluent	2	20	10	50	10	70
Removal	80%	88%	87%	61%	86%	94%

The kinetics of activated sludge processes have been recently investigated in detail (for example by Parsons and Dugan, 1971; Cheng et al., 1975; Neufeld and Hermann, 1975). A rapid uptake of metals in the biomass occurs here (the sequence Pb > Cu > Cd > Ni corresponds to increasing solubility), where a longer phase with delayed metal assimilation follows. Organic polymers in the activated floc form the functional groups to which the metals tend to be absorbed. Moreover, metal cations can be adsorbed or incorporated in mainly anionic cell material and on mineral particulates. With the use of suitable organism species and respective correct control of the plant, much higher heavy metal concentrations seem to be tolerated than it has been assumed to be possible (Neufeld and Hermann, 1975). However, with an eye to the possible application of such sludges for agricultural, forest, and garden purposes (cf. Sect. 5.2), it appears doubtful that this is a solution to the problem of high heavy metal concentrations in certain urban waste waters.

Digested sludge processes are detrimentally influenced by increased metal concentrations, which lead to a diminished production of methane. Hayes and Theis (1978) found a sequence of decreasing toxicity: Ni > Cr > Pb > Cr > Zn > Cd. Since the heavy metal concentrations in the biologic treatment stage are not only controlled by assimilation into the biomass but also through precipitation reactions, it has been suggested that suitable ligands, such as sulfide, should be added in order to obtain precipitation, from which heavy metals are readily reprocessed, for instance by weak acid treatment (Hayes and Theis, 1978).

2.5 Tertiary Physico-Chemical Treatment of Wastewater

The relatively low rate of elimination for several metals has led to the introduction of a tertiary stage based on physico-chemical treatment (PCT) in certain plants. These processes have already been discussed in the first section of this chapter with regard to the processing of drinking water; i.e., precipitation, sand filtration, activated carbon, and ion exchange processes. Several detailed studies have been carried out on the efficacy of simultaneous *trace element removal in advanced PCT waste water treatment.* (See Table 89).

Research by Maruyama et al. (1975) and Hannah et al. (1977) is here of special interest. The studies were carried out on raw wastewater from a residential suburb of Cincinnati, Ohio, where pilot plants were in use. Each plant consisted of a flash mix (either ferric sulfate, ferric chloride, lime or alum would be added—for particulars refer to the original studies), flocculator, settler with sludge collection tank, dual-media filter with backwash collection tank, and two carbon columns. A carbon column, called "old carbon", had been in operation for about a year; a parallel column of virgin carbon ("new carbon") was used to determine the effect of carbon exhaustion on metals removal. Several of the results, with some 20 trace metals, are compiled in Table 90 (from data in the text and various tables of the mentioned studies). The results of the concentration measurements of metals in the raw wastewater influents to the pilot plant (second column) should be considered as normal; however, special industry inflows could bring specific pollution loads.

Table 89. Examples of studies on tertiary physical-chemical treatment of wastewater

Authors	Lime treatment	Ferric chloride	Ferric sulfate	Alum treatment	Activated carbon
Linstedt et al. (1971)	X				X
Nilsson (1971)	X				
Argo and Culp (1972)	X				X
Cheremisinoff and Habib (1972)					X
Dean et al. (1972)					X
Graeser (1972)					X
Sigworth and Smith (1972)					X
Gulledge and O'Connor (1973)			As	As	
Linstedt and Bennett (1973)	X			X	
Logsdon and Symons (1973a, b)	Hg		Hg	Hg	
	X		X	X	X
Shen (1973)	As	As	As	As	
Humenick and Schnoor (1974)					Hg
Linstedt et al. (1974)				X	
Rohrer (1975)	Pb	Pb			
Kunz et al. (1976)			V		V
Maruyama et al. (1975)	X	X			X
Hannah et al. (1977)	X		X	X	X

The data show that the given standards were exceeded in three cases with the ferric system, in six cases with the lime system, and in four with alum. For cadmium none of the applied methods reduced the concentration below the standard value of 10 μg Cd/l. Recently proposed effluent standards, however, permit 40 μg/l cadmium when the receiving streams' low flow equals or exceeds ten times the waste flow. Similar regulations have been proposed for mercury, where for potable water 2 μg/l is the limit, whereas the effluent standards permit 20 μg/l mercury when the above mentioned prerequisites are fulfilled (U.S. Federal Register, Dec. 27, 1973, Anon. 1973e). Also, the selenium content will not be reduced to a level under the upper limit of the standard; cation exchange has practically no reducing effect on the anions of selenium-(IV), which can, as chromium (VI), be readily eliminated by anion exchange (Linstedt et al., 1970). To sum up, with the ferric system the rate of elimination is almost negligible for Mn, Co, Tl, Mo, Sb, and Se. On the other hand, the rate is high for Ag, Be, Bi, Cr, Pb, Cu, Hg, Sn, and V. In a lime system the elimination rate is insignificant with Mo, Sb, Se, As, and Zn, but good with Ag, Be, Bi, Co, Cr, Pb, Ni, and Cd. In alum systems the rate is under 50% for, among others, No, Tl, Zn, Mn, and Ni, while a 95% –100% elimination takes place for Ag, Be, Bi, Hg, Ti, V, Cr, Cd, and Pb under certain conditions. Further details on the behavior of trace metals and their removal in the various systems is given in Table 90.

The effect of activated carbon in a column 5.5 m long was investigated by Maruyama et al. (1975). The column had been in operation a total of three months during metals addition in the plant (Fig. 99). Samples representing 0.9 m or 1.2 m segments were taken for metal analysis; samples were extracted by refluxing with hydrochloric acid (for 24 h), air dried and then weighed. For comparison, a sample of the virgin

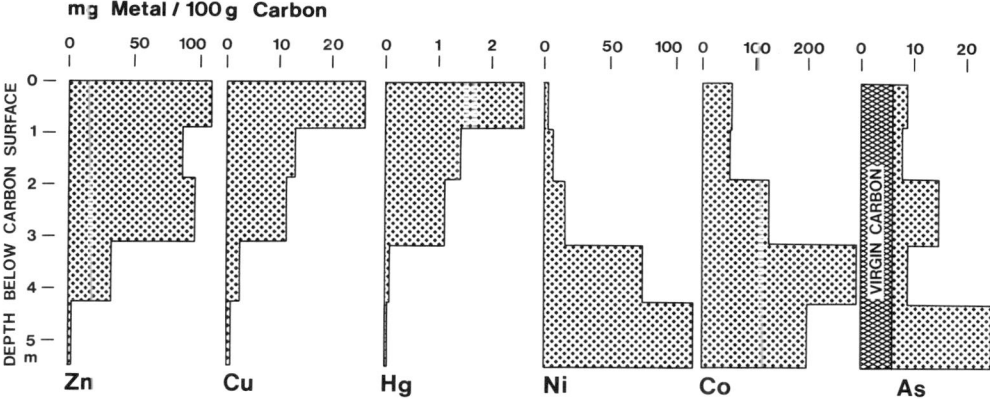

Fig. 99. Vertical distribution of chromium, zinc, copper, nickel, cobalt, and arsenic in a column of activated carbon (after data from Maruyama et al., 1975 and Hannah et al., 1977)

activated carbon was also extracted and analyzed for metals. Chromium, zinc, copper, mercury, and the trace metals antimony and vanadium (Hannah et al., 1977) accumulated in the upper part of the column, indicating an adsorption or filtration mechanism of removal. To a lesser degree lead, manganese, cadmium, molybdenum, beryllium, and bismuth accumulated in the upper part of the column. In contrast, nickel, cobalt, and arsenic tended to accumulate in the lower section of the column (see Fig. 99). Precipitation as sulfides is the probable mechanism of removal of the latter three metals, although this condition—resulting from the action of sulfate-reducing bacteria under anaerobic conditions—is not desirable from a general operating standpoint. Sulfide should actually be beneficial for precipitating heavy metals (Maruyama et al., 1975; see also findings from bank filtration, Chap. G.1.1). Hannah et al. (1977) state that "age" of the carbon defined as "old" or "new" carbon had no discernible effect on metals removed.

There are several other proposals for removal systems; one is the combination of lime and discarded automobile tires for elimination of trace metals from waste water (Netzer et al., 1974). Continuous bench-scale studies showed that removals in excess of 99.5% for most of the metals can be achieved.

In general, these investigations indicate that the introduction of tertiary treatment, especially ion exchange and chemical precipitation, can lead to a very effective simultaneous elimination of trace metals from municipal effluents. Coupled with a reduction of other major constituents which are mainly organic pollutants, drinking water can be obtained directly from such effluents. This is often more economical in the long run than introducing primary-purified effluents into surface water and subsequently withdrawing it again for renewed and more intensive treatment, frequently with the same methods used in advanced waste water treatment.

Table 90. Removal of trace metals by physical and chemical treatment processes (from data of Maruyama et al., 1975 and Hannah et al., 1977)

Metal	Water Standard ($\mu g/l$)	Concentration in raw wastewater		Initial concen. ($\mu g/l$)	Ferric system 45 (40) mg Fe/l pH = 6.0 (6.2)		Lime system 600 (415) mg/l pH = 11.5		Alum system 220 mg/l pH = 8.4		Comments: General chemistry and removal effects
		Median ($\mu g/l$)	Maximum ($\mu g/l$)		Old C ($\mu g/l$)	New C ($\mu g/l$)	Old C ($\mu g/l$)	New C ($\mu g/l$)	Old C ($\mu g/l$)	New C ($\mu g/l$)	
Ag	50	5	23	500 600	5	5	14	10	3	5	Unstable Ag-hydroxide decomposes to Ag-oxide. All three coagulant-systems may be selected
As	100			5000	(70)	58	(900)[a]	770[a]	n.a.	n.a.	Fe-compounds slightly soluble; Fe systems preferred
Be	–	<1	<3	100	1	1	1	1	1	1	Slightly soluble hydroxide + carbonate. Lime preferred
Bi	–	25	160	500 600	24	9	19	24	17	19	Basic salts, containing BiO^+ group, are only slightly soluble. All three systems were equivalent
Cd	10	43	215	5000 700	(60)[a]	50[a]	14[a]	16[a]	13[a]	312[a]	Insoluble hydroxides at high pH; precipitation by sulfide over a broad pH-range. Lime system preferred
Co	–	89	750	500 800	330	350	25	25	208	352	Slightly soluble carbonates, phosphates, particularly hydroxides and sulfides. Lime only effective precipitant
Cr^{3+}	50			5000 700	6	(8)	(20)	18	6	5	Hydrous Cr^{3+}-oxide amphoteric, dissolves in an excess of a strong base. All three coagulant systems equal
Cr^{6+}		128	250	5000 700	21	(24)	(85)[a]	76[a]	18	18	No complete precipitation; good effect of activated carbon; reduction to Cr^{3+} before clarification recommended
Cu	1000	174	368	5000 700	(200)	155	(500)	352	22	12	Cupric salts somewhat amphoteric forming cuprate ions. Precipitation of cupric sulfide or alum + activated carbon
Hg	2	<0.5	1.6	500 50	1	1	40[a]	45[a]	1	1	Sparingly soluble precipitates with phosphate + sulfide. Ferric chloride preferred. Activated carbon recommended

Tertiary Physico-Chemical Treatment of Wastewater

Metal											Remarks
Mn	50	223	—	5000/700	3820[a]	(4500)[a]	(280)[a]	235[a]	469[a]	469[a]	Under alkaline conditions oxidation to insoluble manganese dioxide. Lime treatment preferred
Mo	—	<6	30	500/600	186	120	500	500	600	540	Exists in all oxidation states between +2 and +6. Forms soluble sulfides. Ferric systems of some benefit
Ni	200	363	750	5000/900	37	3150[a]	12	(25)	495[a]	567[a]	Slightly soluble hydroxide and sulfide precipitates. Preferred lime treatment followed by activated carbon
Pb	50	101	150	5000/600	(50)	30	(25)	19	21	20	Insoluble salts with sulfate, carbonate, and sulfide. All three systems (without activated carbon) are equal
Sb	—	60	340	500/600	145	140	216	288	150	174	Slightly soluble basic salts; sulfides dissolve in excess alkali; ferric chloride or alum preferred
Se	10	<10	60	500/100 and 50	22[a] / 12[a]	20[a] / 13[a]	20[a]	25[a]	90[a]	220[a]	Properties similar to those of sulfur. Form stable selenite (SeO_3^{2-}) or selenate (SeO_4^{2-}) anions. Lime + activated carbon or iron systems preferred
Sn	—	83	200	500/600	8	8	40	40	21	36	Slightly soluble hydrous oxides and sulfides dissolve in excess alkali. Iron systems are most effective
Ti	—	<25	160	500/600	50	50	22	24	25	25	Ti^{4+} hydrolyzes to form insoluble Ti-dioxide. Lime and alum more effective than iron systems
Tl	—	<13	30	500/600	318	330	80	140	300	366	Most of the thallous compounds are relatively soluble. Best removal by lime treatment
V	—	<13	<25	500	13	11	45	45	23	23	Iron and alum effective without activated carbon
Zn	5000	2023	3380	5000/2500	(250)	192	561	(700)	1600	1800	Hydroxide in excess alkali form soluble zincate ions. Combination of lime and iron at pH 10 recommended

Data in parenthesis according to values from figures; 45 mg/l in ferric system amd 600 mg/l in lime system refer to initial metal concentrations of 5000 μg/l (Maruyama et al., 1975).

Water standards refer to U.S. Water Quality Criteria 1972 (Anon. 1973d); U.S.National Interim Primary Drinking Water Regulations (Dec. 1975 Anon. 1975f), and USEPA (United States Environmental Protection Agency) Quality Criteria for Water (July, 1976 Anon 1976c).

Ni, irrigation water.

[a] Exceeding water standard.

3 Heavy Metals in Sewage Sludges

The disposal of sludge is at present a delicate and costly problem. Traditionally, sludge is dewatered, dried and dumped at safely selected areas or disposed at sea through pipelines or barging. Incineration has only a limited future due to the cost of the needed fuel and expensive plant alterations which become necessary when air pollution control measures are to be met. Sewage sludges are chiefly civilizational waste products, and as such are dependent on the quantity and quality of applied technology in a particular area, agricultural influences, the type of water treatment and on several other factors.

With regard to domestic wastes a rather constant pattern can be observed with respect to the heavy metal load, with a few exceptions (e.g., Hg in certain cosmetics, Ag in amateur photography).

If the treatment plants have to cope with industrial effluents, certain distinct influences can be observed which are dependent upon the type of industry and the variety of its heavy metal emissions (Berrow and Webber, 1972). The strong fluctuations in the heavy metal concentrations as shown in Table 91 (Page, 1974) are therefore not surprising.

A comparison with the corresponding soil data shows that even when a relatively unpolluted sewage sludge is deposited on soils, a marked increase in the cadmium, zinc, and copper contents can be expected.

In respect to the environmental behavior of heavy metals in sludges it can be determined that the possibility of a remobilization is greater from solid substances than from sediments (Chap. E.5), and is likewise greater with the highest enriched metals, e.g., cadmium, zinc, and lead. Present problems involved in the land application of sewage sludge contaminated by heavy metals include a potential impact on groundwater quality and on sedimentation in coastal waters (Förstner and Stiefel, 1978).

3.1 Land Application of Sewage Sludges

Land application of wet sludge is both economical and beneficial as it supplies all major essential nutrients (N, P, K, S, Ca, and Mg) to crop land. Two of the main advantages in applying wet sludge is that it increases the humus content of soils and

Table 91. Metal content in sewage sludges (Michigan: Berrow and Webber, 1972; England/Wales: Blakeslee, 1973; Sweden: Bergren and Oden, 1972). All values in ppm

Element	Sweden Range	Median	England and Wales Range	Median	Michigan Range	Median	Soil
Cd	2.3– 172	6.7	60– 1,500	–	2– 1,100	12	0.06
Cr	20 –40,615	86	40– 8,800	250	22–300,000	380	100
Cu	52 – 3,300	560	200– 8,000	800	84– 10,400	700	20
Mn	73 – 3,861	384	150– 2,500	400	–	–	850
Ni	16 – 2,120	51	20– 5,300	80	52– 2,977	52	40
Pb	52 – 2,917	180	120– 3,000	700	80– 2,600	480	10
Zn	705 –14,700	1,567	700–49,000	3,000	72– 16,400	2,200	50

recycles phosphates, otherwise wasted in sewage. However, sludges, especially from industrialized or mixed areas, contain elements in concentrations that cause direct or indirect symptoms of toxicity to vegetation. In addition to Cu, Zn, Cd, B, and Ni (particularly in acidic soils), which are typical pollutants of residential origin, certain industries also release large amounts of V, Cr, Mn, Co, As, Ba, and Pb which can affect crop yields. Furthermore, some plants accumulate high concentrations of Se, Mo, Cd, and possibly Pb (a cumulative toxin in the food chain), which are toxic to man and animals but apparently do not affect normal plant growth (Page, 1974), The *availability* of these elements to plants is frequently more important than their total concentrations in sludge treated soils

According to a recent summary on land disposal of toxic substances and water related problems given by Epstein and Chaney (1978) the availability of heavy metals to plants, their uptake, and their accumulation depend on a number of soil, plant, and other factors, as follows (detailed information is given after these general observations):

Soil Factors:
1. Soil pH—toxic metals are more available to plants below pH 6.5
2. Soil phosphorus—phosphorus interacts with certain metal cations to decrease their availability to plants
3. Organic matter—organic matter can chelate and complex heavy metals so that they are less available to plants
4. Cation exchange capacity (CEC)—this factor is important in the binding of metal cations. Soils with a high CEC are safer for the disposal of sludges
5. Moisture, temperature, and aeration—these can affect plant growth and uptake of metals.

Plant Factors:
6. Plant species and varieties—vegetable crops are more sensitive to heavy metals than are grasses
7. Organs of the plant—grain and fruit accumulate lower amounts of heavy metals than leafy tissues
8. Plant age and seasonal effects—the older leaves of plant will contain higher amounts of metals.

Other Factors:
9. Reversion—with time, metals may revert to unavailable forms in soils
10. Metals—Zn, Cu, Ni, and other metals differ in their relative toxicities to plants and in their reactivity in soils.

Bingham et al. (1976) correlated the decrease in plant growth with the contents and soil availability of Cd. Keeney and Walsh (1977) found that DTPA (diethylene-triamine-pentaacetic-acid) extractable metals generally were highly related to plant tissue metal concentrations. Similar results were obtained by Webber and Corneau (1977) from studies on sludge treated by lime, by ferric chloride or left untreated, which were subsequently applied to soils with wide ranges of exchange capacities and clay and organic matter content; also HOAc-extraction was proposed as a valuable procedure to determine the availability of metals in sludge-amended soils. John (1977) found that concentrations of Cd in lettuce were significantly related to the NH_4OAc-extractable Cd in the contaminated soil. Bloomfield and Pruden (1975) reported that incubating digested sludge causes wide variations in the solubilities of Cu, Ni, Zn, Pb, Cd, and Cr in water, as well as in the reagents commonly used to assess availabilities of trace metals to plant i.e., acetic acid and EDTA. Depending on the element and the extractant, and whether the conditions are aerobic or anaerobic, the extractability may increase or decrease.

A clear dependency of dilute acetic acid and 0.05 M diammonium EDTA extractable Pb, Zn, Cu, and Cd concentrations in soils with those in plants was determined by Davies and co-workers (Alloway and Davies, 1971; Davies, 1975, 1978) from Welsh soils, which were contaminated by air- and water-borne heavy metal compounds. Adding soil and/or lime to the sludge influences the extractability of metals during aerobic incubation, although no consistent trends may be evident. Rhode (1972) demonstrated that the uptake of copper, zinc, cobalt, nickel, and manganese is heavily dependent on the acidity of the soil. By adding lime, the pH value is raised and the heavy metals generally become more immobilized. Wagner (1977), however, stated that Pb^{2+} ions reach the surface of the plant's root, are fixed there and then absorbed. From experiments performed by Hodenberg (1974) in the flood plain region of the Oker and Innerste Rivers in northern Germany (see Chap. D.8.2), it is clear that with an increase of soil contamination by copper, the copper content in the plants is also magnified. The poorest growth level is found in the soils enriched by metals. On the other hand, Giordano et al. (1975) reported that the application of rather high rates of garbage compost and sewage sludge resulted in increased yields of corn although tissue concentrations of several heavy metals were higher. In contrast, sludge treated plots had lower yields of beans, possibly due to their greater sensitivity to high rates of available zinc. Although soil extraction with 0.5 molar HCl did not indicate a difference in zinc extractability between compost and sludge, plant concentrations of zinc were always higher with sludge. Furthermore, these toxic effects may be manifested in decreased yields and elevated levels of heavy metals in the edible portions of plants, while the vegetative parts remain relatively unaffected (Van Loon, 1974).

The role of organic matter was studied by Webber and Beauchamp (1977); it seems that Cd is weakly chelated by soil organic matter as low solubility complexes and is also adsorbed on cation exchange sites of organic matter. There is an inverse correlation of Cd concentrations in lettuce with soil organic matter content and cation exchange capacity, indicating that Cd is more strongly adsorbed by organic and heavy clay soils than by coarse-textured soils (John, 1977).

Data from Iskandar (1977) indicate that mobility of heavy metals is due to a redistribution of organic matter (hydraulic effect), a decrease in soil pH, or both.

Spray irrigation of heavy metal-spiked wastewater resulted in much higher concentrations in the plant tissue than in the case of *flood irrigation treatment*. This could be due to absorption of heavy metals by the leaves of the sprayed forages.

In summarizing the various effects of metals Van Loon (1974) determined that although repeated application of the municipal waste products did not result in proportionally higher concentrations of heavy metals in plants, the potential for toxicity still exists with high rates of application over a longer period of time.

A direct approach to the specification of maximum concentrations for certain trace elements in sludges was used by Chaney (1974). Table 92 gives the maximum permissible concentrations of toxic elements destined for land application; sludges in

Table 92. Maximum permissible concentrations of toxic elements in sludge destined for land application (Chaney, 1974)

Toxic element	Max. permissible concentration (ppm)
Cadmium	1% of Zinc
Chromium	1000 ppm
Copper	1000 ppm
Mercury	10 ppm
Nickel	200 ppm
Lead	1000 ppm
Zinc	2000 ppm

which even one parameter exceeds recommended maxima are considered unacceptable for agricultural purposes.

Application rates for unamended soils are usually based primarily on the concentrations of phytotoxic elements, such as Zn, Cu, and Ni. The "zinc equivalent" (ZE) is a single figure approximation used as a proposed standard for the heavy metal contents (Zn–Cu–Ni) of some sludges, and is equal to the ppm (on a dry weight basis) of Zn + 2 x ppm of Cu + 8 x ppm of Ni (coefficients are based on relative toxicity considerations to plants, while chromium, mercury, lead, and cadmium are not accounted for. The term "ZE" was introduced by Chumbley (1971) who proposed that available data indicates that copper is two times and nickel eight times more phytotoxic than zinc.

The transfer of Cd to the food chain warrants particular concern since pot experiments indicated that plants readily absorb and translocate Cd to above-ground tissues (John, 1977). Application rates based on cadmium have been proposed, which are more conservative than the abovementioned guidelines.

3.2 Impact of Heavy Metals on Groundwater Quality

A second problem area in connection with sewage sludge disposal and landfills is the presence of enriched metal contents in percolating-, and more recently, groundwater. With the heavy increase of waste materials and their widespread "unregulated disposal", the groundwater can be affected by various factors that are often only on a global level meaningful.

Table 93 (adopted by Matthess, 1974) gives the contents of trace elements in unpolluted groundwaters, in mineral water, thermal water, and in groundwater in mineralized zones; the last column shows some of the data from the compilation of Brinkmann (1974) on groundwater from polluted areas in the Netherlands. The various geogenic influence factors in the groundwater samples (elevated temperature, ore deposits) indicate that extreme metal enrichments can occur, which are especially widespread in the proximity of sulfide deposits and their respective acidic effluents (see Chap. E). Among the six mechanisms controlling the development of chemical concentrations in the groundwater in areas that are polluted from the domestic and industrial sectors, the last factor has the strongest effect on enriched trace metal contents (Neumair and Matthess):

(1) contacts of the geologic sediments with the interstital water, (2) interface- and mixed-water along freshwater/saltwater boundaries, (3) permanent periodic and episodic admixture-processes caused by leakage factors (–2– and –3– mainly in coastal marine areas), (4) admixture of deep groundwater along tectonic structures, (5) solution-processes in confined groundwater with a different solubility behavior, (6) anthropogenic influences (fertilizing, sewage, industrial effects).

The de-icing salts NaCl and $CaCl_2$, due to their recent more widespread use endangering water supplies, vegetation, and wildlife (Young, 1974), are further anthropogenic factors which have to be considered. A particular hazard is the mobilization of the constituents in contaminated snow, such as heavy metals, oils, phenols and BOD (biochemical oxygen demand) from decaying organic matter (Van Loon, 1972).

Studies performed by Quasim and Burchinal (1970), Walker (1973), Meyer (1973), and Hughes (1975) on groundwater pollution from sanitary landfill leachate, simulated

Table 93. Preliminary list of heavy metal concentrations in natural groundwaters. (Matthess, 1974, supplemented by data of Brinkmann, 1974)

Element	Fresh ground water	Mineral water oilfield brines	Thermal water	Groundwater in mineralized zones	Polluted groundwater (Netherlands)
Cd	< 7 µg/l	5–71 µg/l (California)[a]	–	up to 207 µg/l[b]	0.2–1.0 µg/l[c]
Cr	< 1 µg/l	up to 21 µg/l (California)[a]	–	–	0.5–20 µg/l[c]
Co	< 1 µg/l	–	–	20 µg/l (average S-Ural)[d]	1–11 µg/l[c]
Cu	< 10 µg/l	up to 1 mg/l[e,f] (USA)	up to 5 mg/l[g] (USA)	200 µg/l (average S-Ural)[d]	8–470 µg/l[c]
Fe	< 0.01–10 mg/l	up to 1000 mg/l[d]	–	–	–
Hg	~ 0.03 µg/l	–	–	–	–
Mn	1 mg/l	up to 30 mg/l	up to 42 mg/l[h] (Japan)	–	–
Mo	up to 3 µg/l	–	up to 10 mg/l[i,j] (Japan, USSR)	–	–
Ni	< 4 mg/l	up to 40 mg/l[k]	–	40 µg/l (average S-Ural)[d]	8–22 µg/l[c]
Pb	< 10 µg/l	up to 380 µg/l[k,l]	up to 80 mg/l[g] (USA)	up to 1300 mg/l[m]	5–124 µg/l[c]
Ra	< 1 pCi/l	up to 720 pCi/l[d]	up to 70000 pCi/l[o] (Japan, Austria, Germany)	up to 100 pCi/l[n]	–
Sb	up to 0.2 µg/l	–	up to 0.93 µg/l (USA, U.S.S.R., Japan, New Zealand)[p]	–	–
Sn	< 1 µg/l	up to 670 µg/l	up to 1 µg/l (Japan)[q]	–	–
U	< 0.01–10 µg/l	–	–	up to 460 µg/l[r]	–
W	< 1 µg/l	–	–	up to 64 mg/l (U.S.S.R., USA)[r,s]	–
Zn	< 10 µg/l	up to 27.5 mg/l (Canada, USA)[t]	0.5–5000 µg/l; average: 192 g/l (Japan)[t]	up to 177 mg/l[u]	50–740 µg/l[c]

[a] Silvey, 1967; [b] Udodov and Parilov, 1961; [c] Brinkmann, 1974; [d] Hem, 1970; [e] Rittenhouse et al., 1969; [f] Fricke and Werner, 1957; [g] Skinner et al., 1967, Helgeson, 1968; [h] White et al., 1963; [i] Sugawara et al., 1961; [j] Vinogradov, 1957; [k] Krejci-Graf, 1963; [l] Hermann, 1961; [m] Nowak and Preul, 1971; [n] Scott and Baker, 1961, 1962; [o] Haberer, 1969; [p] Onishi, 1970; [q] Hamaguchi and Kuroda, 1970; [r] Rogers and Adams, 1970; [s] Krauskopf, 1970; [t] Wedepohl, 1972; [u] van Everdingen, 1970.

landfills, and areas treated with sewage sludges indicate that deeper fills pose less of a pollution problem than the shallower fills, which may leach the bulk of pollution in a shorter period of time, thereby exceeding the dilutional capacity of the moving groundwater.

From the Federal Republic of Germany, a series of investigations have clearly shown that the decrease in the quality of groundwater brought about by increased heavy metal concentrations is a widespread phenomenon.

The contamination of such river flood plains as the Oker and Innerste in the Harz, which arises through heavy metals traditionally occurring in waste water from the upper Harz silver mines, is well known (Knickmann, 1959). This enrichment of the soil is not limited to the river plain itself, where the heavy metals are carried and deposited by the river, but also spreads through the neighboring land with the groundwater.

Schöttler (1972) investigated over a period of 2 years and on 600 samples, ten various depositional areas, chosen for investigation according to geologic, hydrogeologic, and material criteria. The deposits occured in an industrial and mining region near Aachen. In the proximity of domestic refuse bins usually high concentrations of cadmium and copper ions were registered. Lead and zinc also increase near mine tailing heaps. Heitfeld and Schöttler (1973) arrived at the conclusion "that due to trace metals there is a potential danger to water in the proximity of tailing heaps", and suggested that in the future it would probably be necessary to consider these elements when drawing up drinking water standards.

Golwer (1973) investigated the effects of roads on groundwater quality in the Frankfurt airport area, in the Taunus Mountains, and in the city of Hanau. It was found that the extent of contamination depends mainly on the traffic volume, the solubility of the involved materials, the purifying effect of the strata above the groundwater table, and on the ground dilution. The lead polluted groundwater does not occur only in small strips alongside traffic routes (as in the case of chloride contents), but rather in widespread regions in the entire area of groundwater formation. Balke et al. (1973) report that for several decades industrial waste waters had infiltrated the substratum under a zinc processing plant at Nievenheim, in the lower reaches of the Rhine River. This led to a precipitation and accumulation of chemical substances. Materials dissolved into the groundwater far exceed the limits set for potable water. In the groundwater the concentrations of arsenic surpassed 50 mg/l; maximum concentrations have been measured for cadmium of 600 μg/l, for thallium 800 μg/l, mercury 50 μg/l, and zinc 40 mg/l. It was shown that the concentration of the salts and trace elements in the contaminated groundwater had only very insignificantly decreased, even 18 months after the percolation was stopped.

Investigations performed by Golwer et al. (1976) in the Frankfurt, Federal Republic of Germany, area, both for municipal and industrial waste deposits, supply criteria for the choice of sites and feasible methods for controlled dumping: the most suitable form of groundwater protection is central sanitary landfills, where waste material of different origin and composition are deposited principally *above the groundwater table*. (Excluded are radioactive substances and other materials, the removal of which is fixed in special laws or governmental decrees). Suitable sites are areas with relatively little precipitation, large distances between the ground surface and the groundwater table, aquifers of good self-purification capacity, poor groundwater quality and low groundwater velocity.

Investigations on the impact of leachate from landfill on groundwater quality, which was reported among others by McLellon et al. (1974), Atwell (1970), Kelly (1976), and Shuster (1976), indicate that the trace element contents of surface drainage water is mainly influenced by the *adsorption capacities* of the specific soil. This is in accordance with Wentink and Etzel (1972) who stressed that the ion exchange capacity increases as the type of soil becomes finer grained (sand–loams–clays); similar results have been presented from lysimeter studies performed by Schöttler (1972). Investigation on the attenuation of pollutants in municipal landfill leachate by clay minerals, with particular reference to heavy metal adsorption, clearly indicate that the potential usefulness of clay materials as liners for waste disposal sites depends to a large extent on the *pH of the leachate solutions;* precipitation of the heavy-metal cations in leachate was an important mechanism at pH values of 5 and above (Griffin et al., 1977).

Hartinger (1976) shows in a comprehensive presentation on the behavior of metallic hydroxide containing muds in disposal areas, that even with these substances no danger arises if the proposed disposal site is chosen with care and expertise in a geologic favorable area. However, it should also be taken into consideration that with the process of *aging,* as well as with the largely aerobic processes of *weathering,* sparingly soluble phases may form. A resulting *leaching* will probably be hindered by permanently high pH-values, the addition of water containing CO_2 changing the solubility only slightly. For exactly this type of metally enriched muds there is a strong possibility for recycling, so that the deposits should only be seen as a temporary phase before the metals can be extracted once more (Chap. H).

3.3 Sewage Sludge Disposal to the Sea

Officials in coastal cities have found discharge of sludge at sea to be a convenient and economical means of sludge disposal. Sometimes sludge is anaerobically digested to reduce the amount of decomposable components prior to discharge to the sea; when deep water can be reached within a reasonable distance of shore, sewer outfalls have been used, as for example at Los Angeles, where a sludge outfall extends 11 km into Santa Monica Bay. More often, tankering or barging have been used to transport sludge to sea (Dick, 1975).

Several examples of ocean sludge disposal and the distribution of trace metals in such areas have been treated in Sect. D.9. It was shown that heavy metals such as mercury, lead, zinc, and silver are typical indicators of trace sewage sludge dispersal in coastal waters.

In many countries, regulations have recently been made to control marine sludge disposal. Examples of such regulations in the United States are the Marine Protection, Research and Sanctuaries Act, the 1972 Amendments to the Water Pollution Control Act, and the 1973 "Ocean Dumping Criteria" (Anon. 1973f, g). Such criteria now include the maximum permissible concentration of heavy metals in the discharges.

Based on food quality considerations, Inoue (1974) proposed an evaluation method for estimating the maximum permissible concentrations of As, Cd, Cr, Hg, and Pb in sludges to be dumped at sea. This is necessary since a portion of trace metals associated with suspended solids may go through various chemical changes in the presence of high concentrations of sodium chloride (Chen et al., 1974). From studies on the trace metal accumulation around the outfalls of major discharges in Southern California Galloway (1972) found that on the average less than 15% of the transported metals could be accounted for in the adjacent sediments. The remaining fraction may be widely dispersed because of high dilution and mixing, or the trace metals may be mobilized in solution, which results in the long-distance transport of trace metals in the ocean (see Chap. E.5.1).

These effects do indeed seem to belong to the especially detrimental capacity of heavy metals in the aquatic system, while it is generally not possible to differentiate between the input from direct sewage disposal in coastal areas and the indirect influence from rivers. In Chapter C examples have been presented for increased heavy metal concentration in effluent plumes of marine coastal regions. There have also been various indications of generally increased metal contents in estuaries, not only in suspended substances but also in water, as shown clearly in the example of metal enrichment in the region of the southwestern North Sea. Krüger et al. (1975) have furnished clear-cut evidence of mercury enrichment in young fish from the Elbe Estuary fishing grounds which is more than likely due to excessive pollution of the region. Should the metal input continue to increase, the time is not far when the coastal fishery industry will have to be closed down in certain areas. Immediate measures have been suggested to combat this heavy metal enrichment, the most promising being toxic waste storage in anoxic, sulfide-rich environments. For example, according to an investigation carried out by Brown and Caldwell Consulting Engineers (Faisst, 1978), the anoxic basins of San Pedro and Santa Monica are particularly well suited for future sludge disposal from California coastal areas since these deep geologic structures are both close to the shore and nearly devoid of life near the bottom. From chemical-equilibrium computer modeling it was predicted that the chemistry of all trace metals except Cr and Mn will be controlled by the precipitation of metal sulfide solids. For freshwater systems Jackson (1978) has similar results from lakes in Canada. The results of this investigation suggest that deliberate stimulation of algal blooms, with attendant H_2S production, in small lakes or settling ponds serving as disposal sites for toxic metal wastes may be an effective, economic method for preventing heavy-metal pollution of natural waters. Sewage from the nearest community would be a cheap, convenient, and virtually limitless source of P and N for the algal bloom, although orthophosphate and other inorganic fertilizers could also be used. In addition, sulfate could be used to stimulate SO_4-reducing bacteria (Jackson, 1978).

As an alternative to dumping wastes into rivers and coastal areas, these methods should be only regarded as provisional emergency measures. A much more important goal is to recover the lost metals, beginning with those in waste of certain industry works.

Chapter H
Concluding Remarks

In the present study we have concerned ourselves with heavy metal pollution in aquatic systems, and found that a series of metals, many of which are toxic, have become accumulated in sediments associated with inland water bodies, estuaries and coastal zones. A comparison of the metal values found in the sediments, collected from various areas, reveals that in industrialized and urbanized areas, the metal enrichment must be ascribed to anthropogenic influences rather than to natural enrichment of sediments by geological weathering (as evidenced by the analyses of the lower sections of sediment profiles).

An example of these effects is given in Figure 100 from the lower Rhine River in the Federal Republic of Germany. The greatest increases in concentration levels are found in the particularly dangerous heavy metals: more than 90% of the concentrations of copper, zinc, lead, mercury, and cadmium in recent sediments of the lower Rhine originate from man-made sources. Similar results are found from a comparison of the consumption rates of heavy metals with the natural concentrations of the respective elements in rocks, soils, and sediments. The "index of the relative pollution potential" and the "technophility index" (see Chap. B) show that particularly large man-made enrichments can be expected from emissions of lead, mercury, copper, cadmium, and zinc in surfaced sediments and soils. In several instances the metal enrichment surpasses that of the natural environment many times over: metal levels of organisms are 100 times greater in polluted inland and coastal waters than in less contaminated areas (Chap. F);

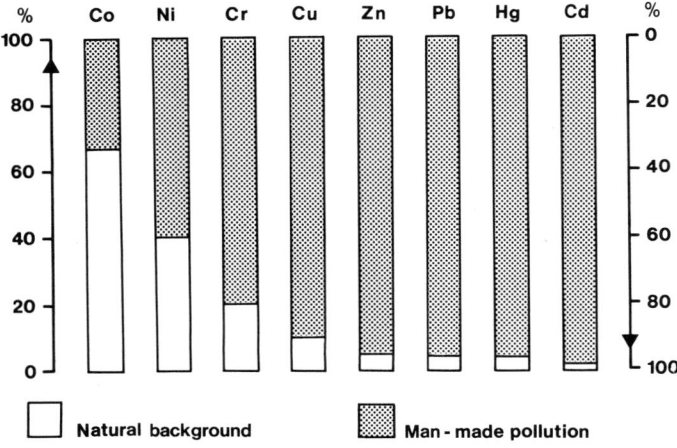

Fig. 100. Sources of heavy metals in pelitic sediment from the lower Rhine River

Concluding Remarks

up to 100 times "normal" levels of cadmium were established in water samples (Chap. C), and sediment samples revealed mercury levels over 10,000 times those of uncontaminated areas (Chap. D). In some areas, this type of pollution is still on the increase. By analyzing sediments collected since 1922, Salomons and de Groot (1978) have been able to trace the development of metal pollution in the lower Rhine section of the Netherlands over a period of more than 50 years (Figure 101). The samples taken from the river flood plain at the beginning of this century were apparently already anthropogenically influenced with respect to the concentrations of copper, zinc, lead, cadmium, and mercury, as can be shown by comparison with sediment data from polders reclaimed in the 15th and 18th centuries. Between 1920 and 1958 all trace element concentrations studied have increased in the sediment from the Rhine River; increase continued for cadmium, copper, and chromium between 1958 and 1975. The concentrations of lead, zinc, mercury, and arsenic decreased however (the strong decline of arsenic is probably caused by a ban on the use of arsenic containing pesticides). For some metals no real improvement in aquatic systems can be expected in the near future. On the contrary, a further increase of metal pollution is more likely to occur. This is due to a number of toxic metal reservoirs built up in the past and which are gradually being emptied into the aquatic systems: problems arising from the release of metals from sediments have already been thoroughly discussed ("sediment is our greatest pollutant"). The restoration of contaminated lakes pose special difficulties as they are, in regard to metal pollution, particularly conservative systems. Furthermore, large amounts of potentially

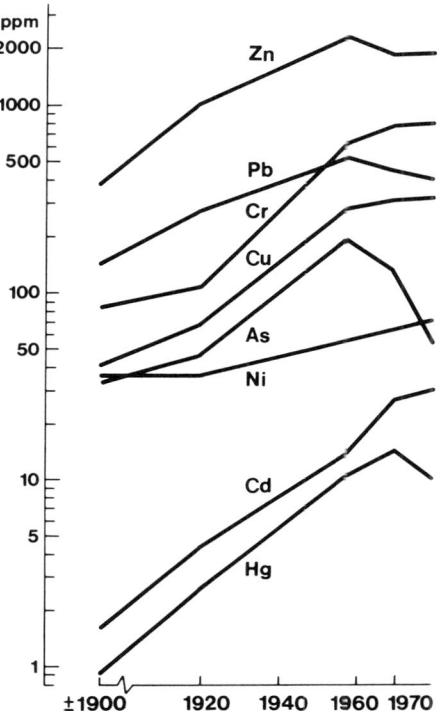

Fig. 101. History of trace metals in Rhine sediments (after Salomons and De Groot, 1978)

dangerous substances are being stored in public refuse dumps. Many poisonous metals are still in daily use e.g., in machinery construction, pesticides, and in the public water system. Surface erosion of contaminated soils—for a long time to come—will continue to contribute great amounts of metals to water.

Although a large proportion of these sparingly soluble metal components are in most cases immobilized in particulates and, for the time being, remain relatively harmless, one cannot overlook the fact that these metals can be released by the numerous processes due to anthropogenic influences.

Remobilized metals can subsequently be readily absorbed by plant and animal life in the aquatic system, the effects of which are numerous (described in Chapter E), i.e. rise in salinity, oxygen depletion and lowered pH, high inputs of natural and synthetic complexing agents and the influence of microbial processes. Furthermore, certain environmental factors can also promote metal enrichment in organisms (Chap. F). In this respect, more important than the actual concentrations of these metals, is their availability to organisms, which depends on the species of the metal in question. As yet, the total scope of latent toxic effects of heavy metals is unknown. Thus, until all doubt about the effects of potentially toxic metals on the biologic food chain and drinking water supply is removed, heavy metals should be kept to a minimum in all aquatic systems.

Apart from the direct intake of heavy metals such as through fish, a further danger arises from the deposition of waste materials onto agricultural land. Here the concentrations of heavy metals enriched in river and estuary sediments, have direct access to plant life and in turn are introduced to the food chain. Further research into the uptake mechanism of heavy metals by organic living material appears to be of primary importance before the role of the individual can be assessed.

The various water purification processes for drinking water were commented upon in Chapter G. Although the standards set for drinking water quality are usually set very low to avoid any health hazard, it is impossible to eliminate toxic metals by means of conventional water purification methods to achieve some limits. Nonetheless, progress is constantly being made in the processing of waste water and thus decisive improvements in the quality of the receiving water can be expected by the end of the century in most industrialized regions, on condition that a tertiary physico-chemical purification step is introduced. When setting norms for waste water, the size of the receiving waters must be taken into consideration. This is already being practiced in certain parts of the United States. The proposed effluent standards for cadmium and mercury permit higher concentrations in the waste water, when the receiving low flow equals or exceeds 10 times the waste flow. Furthermore, greater control of the effluents from small and middle-sized production units (such as in the electroplating industry) into lakes and rivers is necessary in order to ensure their ecological balance. *Water* is generally being regarded as an unlimited source of fundamental importance for the sustaining of human life and the efficient functioning of industry and agriculture. However, the ever-increasing demand on the quality and quantity of water supplies to be met with is likely to cause grave concern within the next decades, especially in semi-arid industrialized countries. It has been prophesized, that water of a *high quality standard,* is bound to become a cardinal problem and a critical factor for most human activities by the end of this century (Gabor et al., 1976).

1 Disposal Versus Reuse

The past and usually prevailing practice in dealing with municipal wastes and other waste materials from specialized industrial undertakings—containing diversified materials such as oil sludges, paint residues, spent catalysts, obsolete or surplus pharmaceutical supplies or agricultural pesticides, materials from factory catastrophes, etc.—has been to dispose of such wastes in the proximity of producers in a safe, unobtrusive, and aesthetically acceptable site. Whether the principal aim, to pose no threat to the water supplies and the environment in general, has always been achieved in this manner, is highly improbable. Numerous examples may be cited from literature where ignorance has undermined good intentions. For example, the stockpiling of slimes, which result from gold mining activities, in the form of self-contained earth dams has resulted in insiduous leaching of embodied minerals by acid mine drainage and the destruction of the environment.

Apart from critical environmental arguments, economic aspects speak for the reduction of metal wastes. The period from 1972 and 1974 has already become recognized as forming the transition from an era of abundance to an age of scarcity. Although the energy crisis (fuel shortage) has featured prominently since 1973, it would appear that there is far less awareness of shortages which already exist in virtually all raw materials and many secondary materials. Some shortages may be ascribed to temporary production set-backs, but for the greater part they are of a long-term nature and threaten to become increasingly severe.

The processing of raw materials forms the backbone of innumerable branches of industry. It therefore stands to reason that depletion of the source(s) of a particular raw material—through natural cessation of mining deposits, or in consequence of political incentives—underlines the vulnerability of the raw material industry.

Principally, energy reserves in the form of fossil fuels may be compared to mineral deposits. However, once the conserved energy has been dissipated, it cannot be regained. On the other hand, recovered minerals or raw materials which have been processed into materials, do not necessarily have to go to waste. In many instances, the processed products can be recycled, i.e. be put to use again by physical and/or chemical reprocessing. Unfortunately, such conversion processes require technology and the availability of energy, and cannot be performed without emissions of waste products. In consequence, they may not be economically viable and environmentally acceptable.

Figure 102 (from Rose et al., 1972) depicts relative energy requirements for the various processes involved in recycling. The figure relates the enrichment of metals from different sources to the relevant energy requirements. It is obvious that not only the costs involved to obtain enrichment increase rapidly with lower metal contents, but that the reversed process—scrapping and waste disposal—also require energy.

The effectiveness of recycling depends on the material concerned and its usage. For example, copper, mainly used in household pipelines, becomes waste to an amount of 45%, whereas, recycling would reduce the annual loss to 5% (Gabor et al., 1976).

The actual data for the recycling of some metals are given in Table 94 (Mann, 1977), which differentiates between metal uses with good recovery and uses with poor recovery.

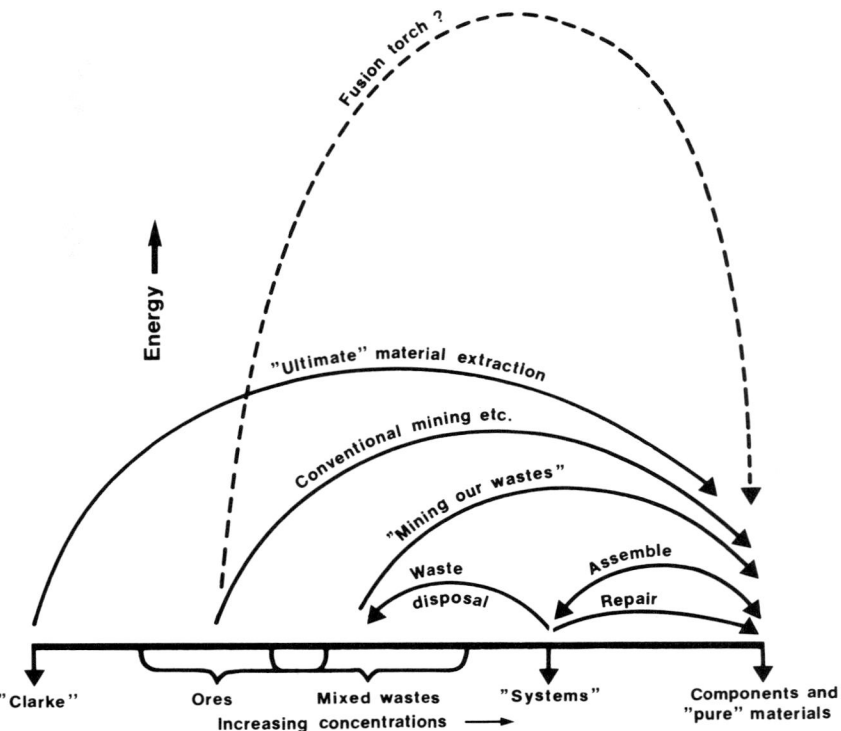

Fig. 102. Recovery of metals from different source material (with corresponding input of energy; after Rose et al., 1972)

Due to the decrease in metal reserves the enrichment methods have continuously increased in expense. However, deposits which have until now been considered uneconomical can be exploited; eventually metal may even be obtained from normal rock.

By means of more advanced technological developments and future innovation, it would appear that problems associated with recovery and recycling can be overcome (even in the advanced field of uranium extraction, where recycling necessitates the reworking of previously discarded slime dams; on the East Rand, South Africa, the Ergo Company is presently extracting gold and uranium from disused slimes dams on a profitable economical basis). Even the exploitation of certain lake and river sediments, particularly in areas where metal pollution has reached high levels, should prove within the near future to be more economical than traditional mining.

2 Alternative Materials

The main goal in the development of new techniques should be to prevent metal pollution in the environment. This can be achieved in two ways: firstly, by means of appropriate cleaning methods in production together with thorough recovery of scrap

Table 94. Present recycling percentages, usage patterns and alternative materials of selected metals (from Mann, 1977)

Metal	Percent recycled	Uses with good recovery	Uses with poor recovery	Uses	Alternatives	RUI status (metal/alternative)
Copper	40.9%	Brass alloys Coinage Electrical	Chemical Fungicides Fertilizers	Electrical Construction Plumbing	Al, Na alloys Stainless steel Plastics	1.0 (Al) 1.5 (Fe) −1.2 (Petrol.)
Gold	15.9%	Alloys	Jewelry Electronics Bullion	Jewelry Electronics Dental	Pt, Pd, Ag, Al, Plastics	– 7.9 (Ag) 9.8 (Petrol.)
Lead	40.0%	Storage batteries Pb- and Cu- based alloys	Gasolines additives, Solder, Pigments	Storage cells Gasoline additives Pigments Cable cover Plumbing	Cd, Hg, Ni, Ag, Zn Ni (catalytic reforming) Ti, Zn Plastics Plastics	−1.0 (Zn) 2.8 (Ni) 3.0 (Ti) 0.4 (Petrol.) 0.4 (Petrol.)
Mercury	20.6%	Hg cells in Cl_2 plants Electrical Amalgams	Fungicides Germicides Paints	Hg cells Medical Germicides Protective paint	Other processes S, Organics Organics Plastics, Cu paint	– 4.6 (Petrol.) 4.6 (Petrol.) 5.8 (Cu)
Silver	47.2%	Stampings Bimetal scrap Batteries	Photographic solutions	Photographic Reflectors Coinage	Se Al, Rh Cu-Ni alloys	– 4.1 (Au) 3.1 (Cu)
Zinc	27.0%	News scrap (75%) Brass, bronze alloys, Batteries	Galvanized products Pigments	Diecasting Anti-corrosion agent Reducing agent	Al, Mg, Plastics Al sheet Sn ceramic Al, Mg	3.8 (Al) −0.8 (Sn) 4.5 (Mg)

metals; secondly, by means of substituting metals and metal compounds in manufactured products. Substitution of the highly toxic metals such as mercury, cadmium, and lead is of particular importance. For example, mercury losses from chlor-alkali products could be eliminated by the use of the diaphragm process instead of the conventional mercury electrode procedure. This would account for an estimated one-third of the present mercury pollution. A further 12% reduction could be achieved by employing substitutes for mercury-containing dyes. Similarly, the elimination of cadmium-containing pigments would reduce environmental cadmium pollution by more than 20%; a reduction of atmospheric emissions from smelting operations by means of improved exhaust purification would be even more rewarding. A lowering of the lead pollutant level would of course mean the discontinuance of the use of additives in gasoline, which account for 20% of the total lead consumption.

In Table 94 further examples of the use of alternative materials for selected metals have been summarized. The reserve usage status of the alternative and the original material is quoted as a qualitative indication as to whether the use of an alternative is advantageous as far as the overall use of reserves is concerned [the reserve usage index (RUI) indicates the percentage of annual exploitation of presently known reserves of a particular resource, e.g., RUI = 4.5 for zinc implies that without recycling and without discovery of further reserves, the metal would be exhausted in 22 years—100/4.5]. Thus, in the case where copper is replaced by aluminum in electrical equipment, we find that the original metal, copper, has a reserve usage index 1.0% higher than that of aluminum, and—as far as the overall replacement is concerned—this development would be desirable. On the other hand, the replacement of zinc by tin as an anti-corrosion agent is— from a purely economic view—not as favorable since the reserve usage index for zinc is 0.8% less than that for tin at the present time (Mann, 1977).

In such calculations not only the pure RUI data in the long run will have to be considered, but also such difficultly describable parameters as "environmental compatibility" of a certain element. Here, too, are undeniable economic factors, especially when it is realized that the extraction of nutrients from the sea and the production of drinking water can be very negatively influenced by such substances as discussed in this book.

Appendix

A. Heavy metals in freshwater (to Chap. C)

Mean value		Range (μg/l)			Example	Reference
Aluminum						
15	μg/l	(<7	–	30)	California rivers	Silvey (1967)
		(0.5	–	3)	California (6 R)	Jones et al. (1974)
Antimony						
		(<0.1	–	31.2)	Siberia (3490 R/L)	Ududov and Parilov (1961)
		(0.05	–	0.08)	Sweden (3 R)	Landström and Wenner (1965)
1.1	μg/l				U.S. rivers, Rhone, Amazon	Kharkar et al. (1968)
Arsenic						
		(0.4	–	5)	Waikato River (N.Z.)	Ritchie (1961)
1.6	μg/l	(0.45	–	4.87)	Columbia River (U.S.A.)	Silker (1964)
1.7	μg/l	(0.25	–	7.7)	Japan	Kanamori and Sugawara (1965)
0.3	μg/l	(0.2	–	0.4)	Sweden (3 R)	Landström and Wenner (1965)
		(1.9	–	54.5)	Greece (11 L)	Grimanis et al. (1965)
		(0.38	–	1.9)	Japan (5 L)	Kanamori (1965)
		(1	–	1.10)	U.S.A. (720 R & L)	Durum et al. (1971)
3.1	μg/l				Rhine (R & L)	Kölle et al. (1971)
20	μg/l	(10	–	1,100)	U.S.A.	Ferguson and Gavis (1972)
3.2	μg/l	(1	–	8)	Danube (Ulm/F.R.G.)	Quentin and Winkler (1974)
3.7	μg/l	(2	–	5)	Lake Constance (Lindau)	Quentin and Winkler (1974)
5.3	μg/l				Ruhr (F.R.G.)	Ruhrverband (Anon. 1975g)

Appendix A (continued)

Mean value		Range (μg/l)			Example	Reference
Barium						
45	μg/l	(9	–	152)	North America (15 R)	Durum and Haffty (1961)
		(10	–	50)	Orange River (R.S.A.)	De Villiers (1962)
		(40	–	70)	Mecong (Cambodia)	Durum and Haffty (1963)
		(12	–	50)	Patuxent (Md., U.S.A.)	Heidel and Frenier (1965)
10	μg/l	(2	–	>80)	Connecticut Streams	Turekian (1966)
		(5.7	–	22)	Neuse River (N.C.)	Turekian et al. (1967)
Beryllium						
0.2	μg/l	(0.05	–	0.67)	Delaware and Hudson	Merill et al. (1960)
		(0.1	–	1.1)	Siberia (3490 R&L)	Ududov and Parilov (1961)
		(0.0	–	0.03)	U.S.A. (R&L)	Meenan and Smythe (1967)
0.01	μg/l	(<0.005	–	0.02)	Rhine/Main R. (F.R.G.)	Reichert (1973)
0.01	μg/l	(<0.005	–	0.04)	Ruhr River (F.R.G.)	Ruhrverband (Anon. 1975g)
Boron						
100	μg/l				Rivers in volcanic areas	Gmelin (1954)
9,000	μg/l				Borax Lakes	Gmelin (1954)
11.6	μg/l				North America (15 R)	Durum and Haffty (1961)
3,000	μg/l				U.S.A., Norway (R)	Durum and Haffty (1963)
13	μg/l	(15	–	58)	Lakes in volcanic areas	Livingstone (1963)
					U.S.S.R. (R)	Konovalov (1969)
201	μg/l	(<10	–	400)	Ruhr R. (F.G.R.)	Dietz (1975), Ruhrverband (1975)
Cadmium						
3	μg/l				Siberia (3490 R&L)	Ududov and Parilov (1961)
0.7	μg/l				California (R)	Silvey (1967)
0.5	μg/l	(0	–	1.2)	Conway River (Wales)	Elderfield et al. (1971)
0.2	μg/l				Cayuga Basin (N.Y.) (R)	Mills and Oglesby (1971)
		(0.4	–	3.7)	Wales (R)	Abdullah and Royle (1972)
2	μg/l	(0.6	–	14)	Delaware (R)	Biggs et al. (1972)
		(0.02	–	0.12)	Australia (3 R)	Doolan and Smythe (1973)
		(1.2	–	16.4)	Lower Rhine (F.R.G.)	Kempf (1973)

Appendix

Conc.	Units	Range	Location	Reference
1.0	µg/l		Austria (R)	Korkisch and Dimitriadis (1973)
0.8	µg/l		Missouri River	Proctor et al. (1973)
<0.5	µg/l		Australia (2 R)	Phillip et al. (1975)
2.7	µg/l	(0.3 – 1.9)	Ruhr (F.R.G.)	Ruhrverband (Anon. 1975g)
0.2	µg/l	(0 – 5)	Pamir (Afghanistan) (R)	Buchroithner and Förstner (unpubl.)

Chromium

Conc.	Units	Range	Location	Reference
0.2	µg/l	(2 – 4)	Maine (U.S.A.) (439 R&L)	Turekian and Kleinkopf (1956)
<0.7	µg/l		California (R)	Silvey (1967)
1.4	µg/l		U.S.A. (12 R), Rhone, Amazon	Kharkar et al. (1968)
1.2	µg/l	(0.1 – 4.1)	L. Superior, Huron, Erie, Ontario	Weiler and Chawla (1969)
<1	µg/l	(0.7 – 19)	U.S.A. (720 R&L)	Durum et al. (1971)
3	µg/l	(0.3 – 69)	Streams to L. Michigan	Robbins et al. (1972)
7.2	µg/l	(3 – 18)	Danube (Ulm)	Quentin and Winkler (1974)
5.7	µg/l	(2 – 10)	L. Constance (Lindau)	Quentin and Winkler (1974)
25	µg/l	(10 – 115)	Ruhr (F.R.G.)	Ruhrverband (Anon. 1975g)

Cobalt

Conc.	Units	Range	Location	Reference
0.2	µg/l	(0.06 – 6.1)	Connecticut Streams	Turekian (1966)
<0.5	µg/l (median)		California (R)	Silvey (1967)
0.05	µg/l	(0.03 – 0.85)	Neuse River (U.S.A.)	Turekian et al. (1967)
0.19	µg/l	(0.037 – 0.36)	U.S. (12 R), Rhone, Amazon	Kharkar et al. (1968)
0.04	µg/l		Cayuga Lake	Mills and Oglesby (1971)
		(1 – 4,500)	U.S.A. (720 R&L)	Durum et al. (1971)
6.3	µg/l	(<5 – 11)	Ruhr (F.R.G.)	Ruhrverband (Anon. 1975g)

Copper

Conc.	Units	Range	Location	Reference
15	µg/l	(8 – 29)	Saale (G.D.R.)	Heide and Singer (1954)
1.4	µg/l	(0.4 – 3.2)	Japan (R)	Morita (1955)
1.9	µg/l	(0.5 – 7)	Japan (L)	Morita (1955)
12	µg/l		Maine (439 R&L)	Turekian and Kleinkopf (1956)
10.3	µg/l	(0.1 – 20,710)	Siberia (3490 R&L)	Ududov and Parilov (1961)
5.3	µg/l	(0.83 – 105)	North America (15 R)	Durum and Haffty (1963)
3.8	µg/l	(0.7 – 27.5)	Columbia River	Silker (1964)
1.2	µg/l	(0.4 – 5)	Sierra Nevada (1)	Bradford et al. (1968)
1.5	µg/l		Lake Erie	Chawla and Chau (1969)

Appendix A (continued)

Mean value	Range (μg/l)	Example	Reference
Copper (cont.)			
10 μg/l		U.S.S.R. (R)	Konovalov (1969)
8.3 μg/l		Rhine (F.R.G.)	Kölle et al. (1971)
0.7 μg/l	(0.7 – 5.5)	Stream to Lake Cayuga	Mills and Oglesby (1971)
2 μg/l	(<1 – 10)	Wales (R&L)	Abdullah and Royle (1972)
3 μg/l	(0.3 – 52)	Mackenzie (Canada)	Reeder et al. (1972)
10 μg/l	(5 – 26)	Streams to L. Michigan	Robbins et al. (1972)
3 μg/l	(0 – 31)	Danube (Austria)	Ebner and Gams (1973)
40 μg/l	(<5 – 120)	Missouri (R)	Proctor et al. (1973)
8.2 μg/l	(4 – 19)	Kansas (12 R)	Angino et al. (1974)
30 μg/l		L. Constance (Lindau)	Quentin and Winkler (1974)
3 μg/l	(22 – 70)	Ruhr (F.R.G.)	Ruhrverband (Anon. 1975g)
		Pamir (Afghanistan, R)	Buchroithner and Förstner (unpubl.)
Iron			
670 μg/l	(10 – 1,400)	World average	Livingstone (1963)
300 μg/l	(31 – 1,670)	North America (15 R)	Durum and Haffty (1963)
33 μg/l	(us. – <100)	California Rivers	Silvey (1967)
39 μg/l	(6 – 309)	Stream to L. Michigan	Robbins et al. (1972)
74 μg/l	(12 – 130)	Danube (Ulm)	Quentin and Winkler (1974)
32 μg/l	(5 – 53)	L. Constance (Lindau)	Quentin and Winkler (1974)
<10 μg/l	(0.7 – 8.8)	California Coast Range Streams (1972)	Jones et al. (1974)
Lead			
2.6 μg/l		Maine (439 R&L)	Turekian and Kleinkopf (1956)
3.9 μg/l	(1.4 – 10.2)	Saale, Elbe (G.D.R.)	Heide et al. (1957)
1.5 μg/l	(0.1 – 2.071)	Siberia (3400 R&L)	Ududov and Parilov (1961)
4 μg/l	(<1 – 55)	North America (15 R)	Durum and Haffty (1963)
0.5 μg/l		Sierra Nevada (L)	Bradford et al. (1968)
2.7 μg/l	(2.2 – 3.3)	Lake Superior, Huron, Erie, Ontario	Weiler and Chawla (1969)
2 μg/l (median)	(<1 – 890)	U.S.A. (720 R&L)	Durum et al. (1971)
			Durum and Hem (1972)

Appendix

Value	Unit	(Range)	Location	Reference
3.2	µg/l		Rhine (F.R.G.)	Kölle et al. (1971)
0.9	µg/l		Streams to Cayuga Lake	Mills and Oglesby (1971)
3	µg/l	(0.7 – 17.6)	Wales (R & L)	Abdullah and Royle (1972)
6	µg/l	(0.6 – 14)	Delaware River	Biggs et al. (1972)
9	µg/l	(5 – 50)	Danube, Austria (11 R)	Ebner and Gams (1972)
3	µg/l	(0 – 70)	Missouri Rivers	Proctor et al. (1973)
41	µg/l	(<10 – 78)	Kansas (12 R)	Angino et al. (1974)
15	µg/l	(5 – 48)	Ruhr (F.R.G.)	Ruhrverband (Anon. 1975g)
Lithium				
1.1	µg/l	(0.075 – 37)	North America (15 R)	Durum and Haffty (1963)
17,600	µg/l		Dead Sea	Rathner and Ludmer (1964)
220,000	µg/l	(0 – 40)	Kansas Basin (R)	Galle and Angino (1970)
10	µg/l	(4.1 – 11)	Tüz Gölü (Turkey)	Irion (1972)
10	µg/l	(1 – 25)	Rhine River	Reichert et al. (1972)
			Illinois River	Mathis and Cummings (1973)
Manganese				
4	µg/l		Maine (439 R & L)	Turekian and Kleinkopf (1956)
5	µg/l		Columbia River	Silker (1964)
7	µg/l (mean)		California Rivers	Silvey (1967)
<0.6	µg/l (median)			
26	µg/l		Lake Erie	Chawla and Chau (1969)
12	µg/l		U.S.S.R. (R)	Konovalov (1969)
12	µg/l	(0.8 – 28)	Wales	Abdullah and Royle (1972)
34	µg/l	(0.3 – 130)	Streams to L. Michigan	Robbins et al. (1972)
29	µg/l	(<4 – 170)	Kansas (12 R)	Angino et al. (1974)
5.4	µg/l	(4 – 64)	Danube (Ulm/Regensburg)	Quentin and Winkler (1974)
117	µg/l	(3 – 10)	L. Constance (Lindau)	Quentin and Winkler (1974)
		(58 – 240)	Ruhr (F.R.G.)	Ruhrverband (Anon. 1975g)
Mercury				
0.03	µg/l	(0.035 – 0.145)	Saale River (D.R.G.)	Heide et al. (1957)
		(0.01 – 0.05)	Toscana and Latium (300 R & L)	Dall'Aglio (1968)
		(0.006 – 0.515)	Thames	Smith et al. (1971)

Appendix A (continued)

Mean value	Range (µg/l)	Example	Reference
Mercury (cont.)			
	(0.030 – 0.075)	Greenland ice (before 1946)	Weiss et al. (1971)
	(0.087 – 0.230)	Greenland ice (after 1946)	Weiss et al. (1971)
0.055 µg/l	(0.02 – 2.8)	Northeastern U.S.A. (67 R&L)	Klein (1972)
	(0.02 – 0.13)	Clay Lake (Ontario)	Armstrong and Hamilton (1973)
0.18 µg/l	(0.07 – 0.75)	Danube, Austria (11 R)	Dworky et al. (1973)
0.08 µg/l	(0.02 – 0.38)	Danube (Ulm/Regensburg)	Quentin and Winkler (1974)
0.18 µg/l	(0.03 – 0.47)	L. Constance (Lindau)	Quentin and Winkler (1974)
0.3 µg/l	(<0.03 – 2.3)	Ruhr (F.R.G.)	Ruhrverband (Anon. 1975g)
Molybdenum			
0.6 µg/l		Japan	Sugawara and Okabe (1960)
4 µg/l	(1 – 5)	California Rivers	Silvey (1967)
1.8 µg/l		U.S.A. (R)	Kharkar et al. (1968)
<0.1 µg/l		Rhine (F.R.G.)	Reichert et al. (1972)
7 µg/l		Stream to L. Michigan	Robbins et al. (1972)
Nickel			
0.4 µg/l	(0 – 71)	Maine (U.S.A., 439 R&L)	Turkian and Kleinkopf (1956)
10 µg/l	(<0.3 – 5)	North America (15 R)	Durum and Haffty (1963)
3 µg/l	(1 – 130)	California Rivers	Silvey (1967)
19 µg/l	(2 – 5.6)	U.S.A.	Kopp and Kroner (1968)
3.5 µg/l	(0.5 – 3.5)	L. Superior, Huron, Erie, Ontario	Weiler and Chawla (1969)
		Wales	Abdullah and Royle (1972)
27 µg/l	(<10 – 89)	Kansas (92 R)	Angino et al. (1974)
10 µg/l	(2 – 17)	Danube (Ulm/Regensburg)	Quentin and Winkler (1974)
5 µg/l	(3 – 7)	L. Constance (Lindau)	Quentin and Winkler (1974)
62 µg/l	(35 – 100)	Ruhr (F.R.G.)	Ruhrverband (Anon. 1975g)
	(3.2 – 6.9)	St. Lawrence River	Subramanian and d'Anglejan (1976)
Selenium			
0.20 µg/l	(0.11 – 0.33)	U.S.A. (R), Rhone, Amazon	Kharkar et al. (1968)
2.4 µg/l	(<1 – 8)	Danube (Ulm/Regensburg)	Quentin and Winkler (1974)

Appendix

1.3	μg/l	(<1	–	4)	L. Constance (Lindau)	Quentin and Winkler (1974)
0.7	μg/l	(0.6	–	1.5)	Ruhr (F.R.G.)	Ruhrverband (Anon. 1975g)
Silver						
0.5	μg/l	(0.2	–	3.5)	Connecticut Streams	Turekian (1966)
3.3	μg/l				Neuse River (U.S.A.)	Turekian et al. (1967)
		(0.0	–	0.94)	North America (15 R)	Durum and Haffty (1961)
0.39	μg/l				U.S.A. (R), Rhone, Amazon	Kharkar et al. (1968)
Strontium						
		(35	–	8,000)	U.S.A. (R)	Skougstad and Horr (1960)
		(6	–	802)	North America (R)	Durum and Haffty (1961)
46	μg/l				Easter U.S.A. (R)	Turekian et al. (1967)
Tin						
0.03	μg/l				Maine (R & L)	Kleinkopf (1960)
0.09	μg/l				Siberia (R & L)	Ududov and Parilov (1961)
Zinc						
2.5	μg/l	(0.25	–	34)	Maine (438 R & L)	Kleinkopf (1960)
13	μg/l	(0.1	–	5,770)	Siberia (4374 R & L)	Ududov and Parilov (1961)
29	μg/l	(<14	–	?)	California Rivers	Silvey (1967)
1.5	μg/l	(0.3	–	100)	Sierra Nevada (170 L)	Bradford et al. (1968)
64	μg/l	(2	–	1,183)	U.S.A.	Kopp and Kroner (1968)
39	μg/l				U.S.S.R. (R)	Konovalov (1969)
35	μg/l	(11	–	71)	L. Superior, Huron, Erie, Ontario	Weiler and Chawla (1969)
		(121	–	566)	Rhine (F.R.G.)	Haberer (1969)
20	μg/l	(<12	–	42,000)	U.S.A. (720 R & L)	Durum et al. (1971)
180	μg/l				Rhine (F.R.G.)	Kölle et al. (1971)
3	μg/l				Streams to L. Cayuga	Mills and Oglesby (1972)
		(11	–	600)	Wales	Abdullah and Royle (1972)
60	μg/l	(3.6	–	176)	Rhine (F.R.G.)	Reichert et al. (1972)
4	μg/l	(0.9	–	32)	Streams to L. Michigan	Robbins et al. (1972)
72	μg/l	(21	–	296)	Danube, Austria (11 R)	Ebner and Gams (1973)
31	μg/l	(3	–	140)	Missouri River	Proctor et al. (1973)
70	μg/l	(11	–	120)	Danube (Ulm/Regensburg)	Quentin and Winkler (1974)
37	μg/l	(19	–	95)	L. Constance (Lindau)	Quentin and Winkler (1974)
		(3.5	–	40)	Marbyrnong, Yarra (Australia)	Phillip et al. (1975)
209	μg/l	(100	–	300)	Ruhr (F.R.G.)	Ruhrverband (Anon. 1975g)
		(29.2	–	32.5)	St. Lawrence River	Subramanian and d'Anglejan (1976)

B. Heavy metal concentrations in organisms (to Chap. F)

Organism	ppm As	Hg	Zn	Cd	Pb	Cr	Fe	Mn	Sb	Co	Locality/Reference
I. Phyto- and Zooplankton											
Marine Phytoplankton											
Phytoplankton 2			35–44			0.54	73–170	0.95–1.1		0.094	Cape West Coast, S. Africa, Van As et al., 1975
Phytoplankton 1 (pure diatoms)				4.8–16.5							Monterey Bay, Baja Calif./Pacific, Martin and Broenkow, 1975
Phytoplankton 1 (≤ 64 μm)				13.2 (8.9–19.5)							San Diego/Calif./Pacific, Martin and Broenkow, 1975
Phytoplankton 1 (76 μm) with pteropods, radiolarians, small copepods		0.410 (0.115–0.713)									Hawaii-Monterey transect, Knauer and Martin, 1972
Phytoplankton 1		0.207 (0.104–0.509)									Monterey submarine canyon, Knauer and Martin, 1972
Phytoplankton 1		0.370–0.713									Oregon coast, Knauer and Martin, 1972
Marine Zooplankton											
Zooplankton 2 copepods, sagitta, euphausids			1.9–7.0			0.03–18	2.7–2900	0.14–7.8	0.022–6.1	0.004–0.52	Cape West Coast, S. Africa, Van As et al, 1975
Zooplankton pure copepods				6.0–15.2							Monterey Bay/Baja Calif./Pacific, Martin and Broenkow, 1975
Zooplankton 1 (≤ 366 μm)				<5							2 transects San Diego-Hawaii, Monterey-Hawaii, Martin and Broenkow, 1975
Zooplankton 1		0.130 (0.039–0.448)									Hawaii-Monterey transect, Knauer and Martin, 1972

Appendix

Zooplankton euphausids ctenophores et al.	1			0.119 (0.051–0.290)	Monterey submarine canyon, Knauer and Martin, 1972
Zooplankton euphausids, ctenophores et al.	1			0.122–0.166	Oregon coast, Knauer and Martin, 1972
Zooplankton	1			0.11–0.19	Core Sound (N.C.), Chesapeake Bay (Md.), Oyster Bay (N.Y.), U.S.A., Cocoros et al., 1973
Zooplankton	1			0.5 (0.06–5.3)	Cape Hatteras–Cape Cod, N.C. U.S.A Windom et al., 1973
Zooplankton predominantly copepods	1		6.0		West Greenland, Bohn, 1975
Freshwater phytoplankton					
Scenedesmus obliquus	1	0.06	0.35	0.07 (air test)	6.03 (air test)	0.43 (air test)	Thailand without fertilizer, Payer et al., 1976
Scenedesmus obliquus	1	2.36	1.67			Thailand with fertilizer, Payer et al., 1976
Scenedesmus obliquus	1	2.36	2.46	0.09 (air test)	34.8 (air test)	0.83 (air test)	F.R.G. with fertilizer, Payer et al., 1976

1, dry weight; 2, wet weight

Appendix B (continued)

Organism	ppm As	Hg	Zn	Cd	Cu	Pb	Cr	Fe	Mn	Ni	Sb	Co	Ag	Locality/Reference
I. Marine Algae														
Green algae														
Chlorophyta ash			330	5	11			700	43	35		20		Puerto Rico, Lowman et al., 1966
Ulva lactuca	1		59–160	0.5–2.0	5.5–26	10–18								Portugal a. Spain (Atlantic) Stenner and Nickless, 1975
Ulva spp.	2		5.6±4.1				0.45±0.38	39±13	4.1±2.8		0.16±0.09	0.038±0.013		Cape West Coast (S. Africa) Van As et al, 1975
Brown algae														
Fucus vesiculosus	1		116 (98–138)	2.1	9 (7.4–10)	3.2	4.5	218 (146–360)	103 (89–130)	8.1 (7.1–8.9)				Menai Straits/Anglesey/ G.B., Foster, 1976
Fucus vesiculosus	1		306 (228–398)		71 (49–97)			75 (41–168)	71 (52–97)	6.0 (7.6–7.0)				Afon Goch/Anglesey, G.B. Foster, 1976
Fucus vesiculosus	1		88.4–262	3.82–19.5	3.82–19.5				38.7–89.2					Bristol Channel, G.B., Fuge and James, 1974
Fucus vesiculosus	1	35.8												West Greenland, Bohn, 1975
Fucus sp.	1		110–345	1.7–3.2	9–31	5–13								Portugal a. Spain (Atlantic) Stenner and Nickless, 1975
Fucus spp.	1		42–450	1.4 (0.4–20.8)	1.7–28.4	0.5–9.0		56–1517	64 (33–190)	1.8–18.0			0.30 (0.07–0.79)	Irish Sea, North Sea, Channel, Preston et al. 1972
Fucus vesiculosus	1	0.5 f.w.	55–666	1	35–85		1–4	140–1168		2–7			2–20	Trondheimsfjorden/Norway, Lande, 1977
Ascophyllum nodosum	1		149 (82–236)	1.8	12 (6–18)	2.6	2.8	86 (54–120)	21 (10–35)	5.5 (4.5–6.3)				Menai Straits/Anglesey G.B., Foster, 1976

Appendix

Species	Type												Location/Reference
Ascophyllum nodosum	1		199 (130–278)	1.5	68 (46–96)	2.2		30 (15–40)	16 (9–25)	4.6 (3.9–5.2)			Mouth of Dulas Bay, Anglesey/G.B., Foster, 1976
Ascophyllum nodosum	1	0.1 f.w.	198 (59–446)	<1 (<1–1)	32 (6–123)		3 (1–13)	166 (51–467)		7 (1–22)		1.5 (<1--2)	Trondheimsfjorden/Norway, Lande, 1977
Phaeophyta	ash		500	6	31			1100	120	72	31		Puerto Rico, Lowman et al., 1966
Chorda filum	1		85	0.7	5.5	3							Portugal a. Spain (Atlantic) Stenner and Nickless, 1975
Red algae													
Porphyra capensis	2		11±7				0.47±0.30	37±22	6.1±2.0		0.21±0.15		Cape West Coast, S. Africa, Van As et al., 1975
Porphyra spp.	1		66 (35–177)	0.25 (0.05–0.87)	11.5 (6.6–19.5)	3.1 (0.8–10.5)		309 (104–3800)	29 (14–93)	2.2 (0.2–9.6)		0.13 (0.01–0.30)	Irish Sea, Preston et al., 1972
Rhodophyta	ash		450	6	27			820	99	48	26		Puerto Rico, Lowman et al., 1966
Corallina officinalis	1		53–55	4.4–6.2	7.5–8	8							Portugal a. Spain (Atlantic) Stenner and Nickless, 1975
Chondrus crispus	1		145	4.9	8								Portugal a. Spain (Atlantic) Stenner and Nickless, 1975
Delesseria sanguina	1		72	4.1	9	10							Portugal a. Spain (Atlantic) Stenner and Nickless, 1975
Higher plants													
Zostera spc. (Potamogeton.)	1		100–1480	2.0–5.3	9–1350	6–1800							Portugal a. Spain (Atlantic) Stenner and Nickless, 1975

1, dry weight; 2, wet weight

Appendix B (continued)

Organism	ppm	As	Hg	Zn	Cd	Cu	Pb	Ni	Locality/Reference
III. Freshwater Algae and Higher Plants									
Algae									
Spirogyra sp.	2				1.8				Danube F.R.G./Hölzinger, 1977
Cladophora glomerata	1			23.7	3.9	7.2	9.5		Deadman Bay Kingston/Ontario, Keeney et al., 1976
Cladophora glomerata	1			8.2	1.4	6.4	12.2		Main Duck (Lake Ontario), Keeney et al., 1976
Cladophora sp.	1						14.9–347		Vermillion River/Illinois, Leland and McNurney, 1974
Cladophora glomerata	1		0.53–0.68	62–190	0.29–0.94	9.1–23.0	5.2–49.2	11.9–23.8	Leine River/F.R.G Abo Rady, 1977
Cladophora glomerata	1				0.05–22.93				Bavarian rivers (Danube excluded) Bayerische L.f.W., 1977
Cladophora rivularis	1			139 (24–375)	0.45 (0.08–1.37)	4.9 (0.8–31.7)	28.9 (5.8–43.2)		Waikito River/Hastings/N.Zealand Reay, 1972
Enteromorpha nana	1	20–40							
Lyngbya sp.	1		0.6 (0.5–0.8)	564 (400–1320)	3.5 (1.7–9.5)	195 (120–335)	98 (55–190)		Neckar River/F.R.G. Bartelt and Förstner, 1977
Aufwuchs	1			151–242	1.4–2.0	22.7–45.5	10.5–20.4		Elsenz River/F.R.G., Bibo, 1977
Algal Aufwuchs	1								
Periphyton	2		0.032						Lake Powell, Colorado River, Potter et al., 1975
Periphyton	1							46–265	Vermillion River/Illinois Leland and McNurney, 1974
Algal Aufwuchs	1		0.5–25		10.3–16.8		0.43–0.87		Vogt and Kittelberger, 1977
Moss									Rhine River
Fontinalis antipyr.	1			169–270	0.17–0.21	13.2–13.5	56.3–217.5		Elsenz River/F.R.G. Heydt, 1977
Fontinalis antipyr.	1		0.25 (0.12–0.36)				450–2180		Ruhr River/F.R.G. Dietz, 1972
Hygroamblystegium sp.	1		0.25 (0.03–0.64)				440–800		Ruhr River/F.R.G. Dietz, 1972

Appendix

Species	Type							Source
Lagarosiphon major	1		3.02 (0.44–13.2)					Lago Maggiore/Italy Ravera et al., 1973
Lagarosiphon major	1	20–120						L. Rotura/New Zealand Fish, 1963
Higher plants								
Potamogeton sp.	1					30.5±5.4		Vermillion, R. (rural)/Illinois Leland and McNurney, 1974
Elodea sp.	1		11.00					Bavarian rivers (Danube excluded) Bayerische L.f.W., 1977
Elodea sp.	1					90±93		Vermillion R. (mainstream)/Illinois, Leland and McNurney, 1974
Potamogeton pectinatus	1			0.37–0.46	137–213	0.49–0.85	14.6–19.7	2.4–6.8	9.2–15.8	Leine River/F.R.G. Abo Rady, 1977
Potamogeton pectinatus	1				165–517	0.13–1.29	5.1–34.3	8.0–9.7	Elsenz River/F.R.G. Heydt, 19//
Potamogeton crispus	1				280–331	0.21–1.25	8.2–41.3	7.4–14.8	Elsenz River/F.R.G. Heydt, 1977
Callitriche palustris	1				255–511	0.25–1.34	7.6–25.4	33.5–21.3	Elsenz River/F.R.G. Heydt, 1977
Ceratophyllum demersum	1	1.4–650						Waikito River/Hastings, N.Z. Reay, 1972
Elodea canadensis	1	3.0–307						Waikito River/Hastings, N.Z. Reay, 1972
Potamogeton sp.	1	<6–178						Waikito River/Hastings, N.Z. Reay, 1972
Ranunculus fluitans	1		0.46					Bavarian rivers (Danube excluded) Bayerische L.f.W., 1977
Ranunculus fluitans	1			0.18 (0.11–0.28)			130–2500	Ruhr River/F.R.G. Dietz, 1972
Nuphar luteum	1			0.11 (0.03–0.43)			20	Ruhr River/F.R.G. Dietz, 1972
Sagittaria sagittifolia	1			0.15 (0.08–0.33)			130	Ruhr River/F.R.G. Dietz, 1972
Myriophyllum spicatum	1			0.20 (0.09–0.33)			30	Ruhr River/F.R.G. Dietz, 1972
Myriophyllum spicatum	1		1.06 (0.48–3.04)					Lago Maggiore/Italy, Ravera et al., 1973

1, dry weight; 2, wet weight

Appendix B (continued)

IVa. Mollusks: Bivalves, Gastropods (Marine)

Bivalves

Organisms	ppm As	Hg	Zn	Cd	Cu	Pb	Cr	Fe	Mn	Ni	Bi	Sb	Co	Ag	Locality/Reference
Mytilus edulis	1	–	–	31 (50–180)	10	9 (5–11)	12 (3–25)	16 (9–24)	1960 (960–2640)	27 (12–38)	7 (1–17)	–	–	–	0.1 (0.1–0.3)	New Zealand/Brooks and Rumsby, 1965
Mytilus edulis	2	–	–	14 (3.8–26.0)	0.63 (0.26–1.6)	8.3 (1.7–18)	–	–	–	–	–	–	–	–	–	New Zealand/Nielsen and Nathan, 1975
Mytilus edulis	1	–	–	–	24.6 ± 21.87	–	–	–	–	–	–	–	–	–	–	Port Phillip/Talbot et al., 1976b
Mytilus edulis	2	–	–	42.7 (16.4–37.1)	2.09 (0.2–18.16)	0.73 (0.47–1.45)	–	–	–	–	–	–	–	–	–	Port Phillips Bay/Western Port Bay, Australia/Phillips, 1976b
Mytilus edulis	1	–	4.2 (0.4–13)	516 (171–1350)	18.6 (4.3–38)	–	199 (3–352)	–	–	–	–	–	–	–	–	Derwent Estuary/Tasmania Bloom and Ayling, 1977
Mytilus edulis	1	–	–	91	5.1	9.6	9.1	1.5	1700	3.5	3.7	–	–	1.6	0.03	Irish Sea/GB/Segar et al., 1971
Mytilus edulis	1	–	–	147 (62–250)	18 (4–60)	–	11 (1–30)	–	–	–	–	–	–	–	–	Bristol Channel/GB/Nickless et al., 1972
Mytilus edulis	1	–	–	253–779	–	–	–	–	–	–	–	–	–	–	–	Cardigan Bay/Wales/GB/Ireland, 1973
Mytilus edulis	2	–	0.11 (0.06–0.2)	–	–	–	–	–	–	–	–	–	–	–	–	Jutland/DeWolf, 1975
Mytilus edulis	2	–	0.18 (0.06–0.83)	–	–	–	–	–	–	–	–	–	–	–	–	North Sea, Germany/France/DeWolf, 1975
Mytilus edulis	2	–	0.29 (0.11–1.65)	–	–	–	–	–	–	–	–	–	–	–	–	North Sea/GB DeWolf, 1975

Appendix

Species											Location/Reference
Mytilus edulis	2	–	0.29 (0.07–0.8)	–	–	–	–	–	–	–	English Channel/GB DeWolf, 1975
Mytilus edulis	2	–	0.09 (0.05–0.16)	–	–	–	–	–	–	–	English Channel/France DeWolf, 1975
Mytilus edulis	2	–	0.16 (0.05–0.29)	–	–	–	–	–	–	–	Atlantic/France DeWolf, 1975
Mytilus edulis	1	9.45–14.8	0.43–1.86	–	–	–	–	–	0.04	–	Southampton Waters/GB Raymont, 1972
Mytilus edulis	1	–	–	–	3.1–3.5	–	–	–	–	–	North Sea/Weser Estuary Schulz-Baldes, 1974
Mytilus edulis	1	–	–	169 (85–359)	2 (1–5)	24 (5–88)	20 (4–49)	363 (112–1623)	–	15 (6–43)	–	3 (1–6)	Trondheimsfjorden/Norway Lande, 1977
Mytilus edulis	1	14.1–16.7	–	–	–	–	–	–	–	–	–	West Greenland/ Bohn, 1975
Mytilus edulis	1	5.3±5.4	–	–	–	–	–	–	–	–	–	Morton's Harbor/Newfoundl. Penrose et al., 1975
Mytilus edulis	1	–	–	130–170	1.7–3.6	6.5 14	2 15	–	–	–	–	S.W. Spain/Portugal Mediterr. Stenner and Nickless, 1975
Mytilus galloprovincialis	1	–	–	209 (97–644)	1.9 (0.4–5.9)	18.0 (2.4–154)	21.5 (2.7–117)	7.5 (0.5–288)	443 (149–2220)	21.1 (3.3–69.8)	4.3 (0.9–14.1)	–	2.8	Mediterranean Sea/NW coast (France/Italy) Fowler and Oregioni, 1976
Mytilus edulis	1	–	0.16–0.629	–	–	–	–	–	–	–	–	St. Lawrence estuary, Canada Bourget and Cossa, 1976
Mytilus edulis	1	–	–	–	–	–	–	–	–	5–19	–	Berkeley Marine, Berkeley Calif. Friedrich and Filice, 1976
Mytilus edulis	1	–	–	–	–	0.414–9.358	–	–	–	–	–	La Jolla, Calif./ Chow et al., 1976
Choromytilus meridionalis	2	–	–	16±1	–	–	–	–	0.20	–	0.038±0.006	Cape, West Coast South Africa Van As et al., 1975
Choromytilus meridionalis	2	–	–	12.65±2.39	0.26±0.13	1.12±0.34	0.15±0.14	1.05±0.29	14.5±8.16	1.83±0.46	0.26±0.13	2.3±0.69	–	Saldanha Bay, Langebaan Bay R.S.A./Fourie, 1976

1, dry weight; 2, wet weight

Appendix B (continued)

Organisms	ppm As	Hg	Zn	Cd	Cu	Pb	Cr	Fe	Mn	Ni	Bi	Sb	Co	Ag	Locality/Reference
Crassostrea margaritacea	2	–	120± 41	0.88± 0.27	1.9± 0.9	0.42± 0.12	–	4.7± 2.5	0.55± 3.6	0.20± 0.05	0.68± 0.13	1.8± 0.13	–	–	Saldanha Bay, Langebaan Bay R.S.A./Fourie, 1976
Crassostrea gigas	2	–	21±5	0.87± 0.20	3.7± 1.0	0.34± 0.09	0.54	12±5	1.3± 0.5	0.25± 0.06	0.49± 0.20	1.4± 0.4	–	–	Saldanha Bay, Langebaan Bay R.S.A./Fourie, 1976
Crassostrea gigas	1	–	396	3.7	32	–	–	128	16	1.6	–	–	–	1.9	Knysna Estuary, R.S.A./ Watling and Watling, 1976
Crassostrea margaritacea	1	–	886	2.5	17	–	–	57	2	1.6	–	–	–	2.6	Knysna Estuary, R.S.A./ Watling and Watling, 1976
Ostrea edulis	1	–	660	3.1	38	–	–	167	6	1.7	–	–	–	6.4	Knysna Estuary, R.S.A./ Watling and Watling, 1976
Crassostrea gigas	1	–	1350– 2670	2.2– 26.7	207– 643	–	–	–	–	–	–	–	–	–	Helford, Colne, Poole Estuaries/GB, Thorntonetal.,1975
Crassostrea gigas	1	–	9860– 35120	17–40	1760– 6480	15–17	–	214– 422	16–46	3.8– 6.5	–	–	2.2– 3.5	–	Hinkley Pt, Lower Station Severn Est./Boyden and Romeril, 1974
Ostrea sinuata	1	–	1103 (850– 1500)	35 (10– 43)	41 (21– 53)	10 (6–14)	3 (2– 6)	682 (630– 750)	8 (1– 11)	2 (1– 3)	–	–	–	5.6 (4.5– 7.3)	Tasman Bay, N. Z. Brooks and Rumsby, 1965
Crassostrea virginica	2	–	1428	3.1	91.5	0.47	0.40	67	4.3	0.19	–	–	0.1	–	Atlantic coast U.S.A. Pringle et al., 1968
Crassostrea glomerata	2	–	337 (97– 900)	1.3 (0.12– 5.0)	40 (4– 380)	0.9 (0.2– 21.0)	–	–	–	–	–	–	–	–	New Zealand Nielsen and Nathan, 1975
Crassostrea gigas	1	–	7227± 4010	33.18± 7.64	691.6± 420.7	0.82± 1.47	12.45± 7.49	–	–	–	–	–	–	–	Tamar R. Est./Tasmania Ayling, 1974
Crassostrea commercialis	2	1.2 (0.3– 3.4)	227 (80– 665)	0.2 (0.1– 1.0)	20 (3–48)	0.8 (0.3– 1.3)	–	–	–	–	–	–	–	–	N.S.W. estuaries Australia Mackay et al., 1975b
Ostreidea angasi	1	–	–	91.6± 73.14	–	–	–	–	–	–	–	–	–	–	Port Phillip Bay Talbot et al., 1976b

Appendix

Species															Reference
Crassostrea virginica	2	—	0.47±0.125	—	—	—	—	—	—	—	—	—	—	—	Mispillion R., Delaware, USA Cunningham and Tipp, 1973
Crassostrea virginica	1	1.3	0.05	322	3.2	161	0.8	—	—	—	—	—	—	—	San Antonio Bay/Texas USA Sims and Presley, 1976
Cardium edule	1	—	—	100–200	0.3–0.6	6–26	0.7–2	—	—	—	—	—	—	—	S.W. Spain a. Portugal, Mediterr. Sea/Stenner and Nickless, 1975
Cardium edule	1	—	—	130	1.5	11	0.76	—	590	6.3	7.9	—	7.1	0.04	Irish Sea, GB/Segar et al., 1971
Cardium edule	2	—	0.03–0.04	—	—	—	—	—	—	—	—	—	—	—	Elbe River/Zauke, 1977
Cardium edule	1	4.5–6.3	0.16–0.80	—	—	—	—	—	—	—	—	0.01–0.04	—	—	Southampton waters, GB/Raymont, 1972
Pecten maximus	1	—	—	230	13	3.3	8.3	—	170	140	49	—	8.5	—	Irish Sea, GB/Segar et al., 1971
Pecten novae-zelandiae	1	—	—	283 (195–368)	249 (210–299)	9 (2–14)	16 (10–23)	10 (3–23)	2915 (1140–6000)	111 (12–306)	6 (2–17)	—	—	0.7 (0.2–2.3)	Tasman Bay, N.Z. Brooks and Rumsby, 1965
Mya arenaria	2	—	—	17	0.27	5.8	0.7	0.52	405	6.7	0.27	—	0.1	—	Atlantic coast waters Pringle et al., 1968
Mercenaria mercenaria	2	—	—	20.6	0.19	2.6	0.52	0.31	30	5.8	0.24	—	0.2	—	Atlantic coast waters Pringle et al., 1968
Rangia cuneata	1	—	—	51	0.5	25	1.1	—	—	—	—	—	—	—	San Antonio Bay, Texas, USA Sims and Presley, 1976
Tellina tenuis	1	—	—	—	—	≤50	—	—	—	—	—	—	—	—	Loch Ewe, Scottland, Saward et al., 1975
Mactra clabrata	2	—	—	7.9±2.7	0.19±0.09	0.22±0.08	—	0.54±0.17	—	41±14	—	n.d.	—	—	Saldanha Bay, Langebaan Bay R.S.A./Fourie, 1976
Macoma baltica	2	—	0.03–0.07	—	—	—	—	—	—	0.34±0.13	—	n.d.	—	—	Elbe Estuary, F.R.G./Zauke, 1977
Gastropods															
Patella vulgata	1	—	—	158 (103–274)	13.1 (3.8–23)	14.4 (3.5–22.0)	7.9 (5.4–12.5)	—	2060 (1380–3270)	42 (23–88)	7.0 (4.5–9.9)	—	—	2.7 (1.3–3.6)	Irish Sea/Preston et al., 1972
Patella vulgata	1	—	—	84	31	7.7	32	—	150	13	2.5	—	0.4	—	Irish Sea, GB/Segar et al., 1971

1, dry weight; 2, wet weight

384   Appendix

Appendix B (continued)

Organisms	ppm As	Hg	Zn	Cd	Cu	Pb	Cr	Fe	Mn	Ni	Bi	Sb	Co	Ag	Locality/Reference
Patella vulgata 1	–	–	340 (290–410)	220 (67–440)	30 (n.d.–50)	–	–	–	–	–	–	–	–	–	Bristol Channel/N. Somerset, GB/ Stenner and Nickless, 1974b
Patella vulgata 1	–	–	91 (55–130)	11 (3.5–28)	7 (3.5–12)	–	–	–	–	–	–	–	–	–	Bristol Channel/Dorset, GB/ Stenner and Nickless, 1974b
Patella vulgata 1	–	–	256 (100–580)	158 (30–550)	–	6 (3–9.5)	–	–	–	–	–	–	–	–	Severn Estuary, Bristol Channel, GB/ Butterworth et al., 1972
Patella vulgata 1	–	–	193 (65–375)	127 (9–500)	–	9 (2–27)	–	–	–	–	–	–	–	–	Bristol Channel, GB/ Nickless et al., 1972
Patella vulgata 2	2.7 (1.0–3.9)	0.07 (0.02–0.31)	74.6 (43.2–129)	66.1 (10.3–118.5)	7.9 (5.4–12)	0.51 (0.17–0.75)	–	–	–	–	–	–	–	–	Bristol Channel, GB/Somerset Peden et al., 1973
Patella vulgata 1	–	–	181 (127–238)	10 (2–22)	18 (12–30)	–	10 (7–17)	1817 (1285–2505)	–	7 (4–11)	–	–	2 (1–4)	–	Morton's Harbor, Newfoundl. Penrose et al., 1975
Nucella lapillus 1	–	–	1101 (110–4200)	168 (31–725)	–	9 (1–38)	–	–	–	–	–	–	–	–	Bristol Channel, GB/ Nickless et al., 1972
Nucella lapillus 1	–	–	345 (110–480)	36 (11–62)	70 (20–210)	–	–	–	–	–	–	–	–	–	Bristol Channel/Dorset, GB/ Stenner and Nickless, 1974
Nucella lapillus 1	–	–	2900 (1460–4030)	780 (500–1120)	950 (385–1750)	–	–	–	–	–	–	–	–	–	Bristol Channel, N. Somerset GB/ Stenner and Nickless, 1974
Nucella lapillus 1	–	–	860	73	150	4.9	–	65	12	2.4	–	–	0.3	–	Irish Sea, GB/Segar et al., 1971
Littorina littorea 1	–	–	219 (100–520)	68 (15–210)	–	0.6 (0.1–3.0)	–	–	–	–	–	–	–	–	Severn Estuary, Bristol Channel, GB/ Butterworth et al., 1972

Appendix

Species															Location/Reference
Littorina littorea	1	–	–	117 (60–210)	27 (8–75)	–	7 (1–19)	–	–	–	–	–	–	–	Bristol Channel, GB/ Nickless et al., 1972
Littorina littorea	2	0.06– 0.156	–	–	–	–	–	–	–	–	–	–	–	–	Elbe Estuary, F.R.G. Zauke, 1977
Nerita picea	2	0.03± 0.01	–	–	–	–	–	–	–	–	–	–	–	–	Coastal waters of Hawaii Klemmer et al., 1976
Cellana exerata	2	0.02± 0.03	–	–	–	–	–	–	–	–	–	–	–	–	Coastal waters of Hawaii Klemmer et al., 1976

IVb. Mollusks: Bivalves, Gastropods (Freshwater)

Bivalves

Species															Location/Reference
Anodonta sp.	1	–	–	120	1.2	3.0	1.2	0.84	5500	2100	0.35	–	0.42	0.28	L. Aberffraw/Segar et al., 1971
Anodonta anatina	2	0.022	–	–	–	–	–	–	–	–	–	–	–	–	Schaal Lake/Elbe Est. (transplanted) Karbe et al., 1975
Anodonta anatina	2	0.034– 0.075	–	–	–	–	–	–	–	–	–	–	–	–	Schaal Lake/Elbe Est. (transplanted) Karbe et al., 1975
Dreissena polymorpha	2	0.020	–	–	–	–	–	–	–	–	–	–	–	–	Schaal Lake/Karbe et al., 1975
Dreissena polymorpha	2	0.04– 0.135	–	–	–	–	–	–	–	–	–	–	–	–	Elbe Estuary (transplanted) Karbe et al., 1975
Dreissena polymorpha	1	0.08	–	113	–	–	–	6.6	–	250	–	0.7	–	–	Schaal Lake/Karbe et al., 1975
Dreissena polymorpha	1	0.16– 0.67	–	140– 208	–	–	–	9.4– 60	–	531– 916	–	1.2– 2.8	–	–	Elbe Estuary (transplanted) Karbe et al., 1975
Unio tumidus	2	0.035	–	–	–	–	–	–	–	–	–	–	–	–	Schaal Lake Karbe et al., 1975
Unio tumidus	2	0.040– 0.135	–	–	–	–	–	–	–	–	–	–	–	–	Elbe Estuary (transplanted) Karbe et al., 1975
Fusconaria flava	2	–	–	66 (25– 120)	0.69 (0.36– 1.17)	1.7 (0.9– 2.0)	3.7 (1.8– 5.1)	7.7 (1.1– 11.6)	–	–	2.1 (0.7– 3.0)	1.2 (0.6– 1.6)	–	–	Illinois River, USA. Mathis and Cummings, 1973
Amblema plicata	2	–	–	95 (40– 178)	0.38 (0.15– 1.41)	1.2 (0.3– 3.2)	2.7 (1.1– 7.6)	4.4 (0.6– 9.9)	–	–	1.1 (0.4– 2.3)	0.7 (0.4– 1.2)	–	–	Illinois River, USA. Mathis and Cummings, 1973
Quadrula quadrula	2	–	–	48 (28– 64)	0.56 (0.31– 1.37)	1.7 (1.1– 3.6)	2.2 (0.9– 3.8)	4.7 (1.8– 8.3)	–	–	0.9 (0.4– 1.6)	0.8 (0.5– 1.3)	–	–	Illinois River, USA. Mathis and Cummings, 1973

1, dry weight; 2, wet weight

Appendix B (continued)

Organisms	ppm As	Hg	Zn	Cd	Cu	Pb	Cr	Fe	Mn	Ni	Bi	Sb	Co	Ag	Locality/Reference
Gastropods															
Physa sp.	1	–	–	–	–	42±25	–	–	–	–	–	–	–	–	Illinois River, USA. Mathis and Cummings, 1973
Radix peregra	1	–	131.1–167.2	0.094–0.770	116.2–160.9	1.8–4.04	–	–	–	–	–	–	–	–	Elsenz River, F.R.G., Heydt, 1977
Radix balthica	2	0.25–0.40	–	–	–	–	–	–	–	–	–	–	–	–	Elbe/Hamburg (upstream), Zauke, 1977
Radix balthica	2	0.08–0.09	–	–	–	–	–	–	–	–	–	–	–	–	Elbe/Hamburg, Zauke 1977

1, dry weight; 2, wet weight

Appendix   387

Appendix B (continued)

Organisms	ppm	As	Hg	Zn	Cd	Cu	Pb	Cr	Fe	Mn	Ni	Bi	Sb	Co	Locality/Reference
Va. Crustaceans (Marine)															
Decapoda															
Crangon crangon	w.a. 2		0.03–0.12												Elbe estuary, F.R.G./Zauke, 1977
Crangon crangon	w.a. 1	0.6	<0.02	14	<0.4	34	<0.2								San Antonio Bay/Texas/Sims and Presley, 1976
Crangon vulgaris	m 2					25.9					8.8				Northumberland Coast/GB/Wright, 1976
Crangon vulgaris	w.a. 2			23.73± 9.98	3.5± 2.1										Northumberland Coast/GB/Wright, 1976
Carcinus maenas	w.a. 2			43.24± 9.94	0.98± 0.32	10.90± 5.50					6.50				Northumberland Coast/GB/Wright, 1976
Carcinus maenas	m 2				0.51± 0.18	6.00± 3.00					6.20				Northumberland Coast/GB/Wright, 1976
Carcinus macnas	m 1			121	<1	175		8	666		11				Orkdalsfjorden/Hardangerfjorden/Norway/Lande, 1977
Cancer irroratus	w.a. 2			30.5± 10.2		19.9± 11.8			60.0± 28.0	10.6± 4.2					Terrence Bay/Nova Scotia, Canada/Martin, 1974
Cancer pagurus	w.a. 2			28.91± 11.69	0.36± 0.42	6.80					9.80				Northumberland Coast/GB/Wright, 1976
Maia squinada	m 1		0.03	240	0.46–0.74	19–95	1.2–5.1								Quarteira, S. Portugal/Stenner and Nickless, 1975
Thalamita crenata	m 2		0.061± 0.020												Hawaiian Est., Ala Wai Canal/Luoma, 1976
Thalamita crenata	w.a. 2		0.09± 0.10												Hawaii/Kilauea, Kauai Oahu Klemmer et al., 1976
Pandalus borealis	m 1	61.6 (52.5–70.6)													West Grønland/mining area Bohn, 1975
Homarus americanus	m 1	3.8–7.6													Moreton's Harbour, Newfoundl./Penrose et al., 1975

1, dry weight; 2, wet weight; w.a., whole animal; m, muscle

Appendix B (continued)

Organisms	ppm	As	Hg	Zn	Cd	Cu	Pb	Cr	Fe	Mn	Ni	Bi	Sb	Co	Locality/Reference
Jasus lalandii	m 2			14–17	0.03–0.15	3.3–8.2	0.4–0.59	0.03	0.62–1.6	0.24–0.42	0.15–0.26	0.53–1.1	2.2–3.3		Saldanha Bay/S.A./Fourie, 1976
Jasus lalandii	m 2			17±2				0.08±0.08	2.7±1.3	0.27±0.06			0.12±0.07	0.004±0.002	Melkbosch Strand/West Coast/S.A./Van As et al., 1975
Nephrops norvegicus	m 1			79	0.7	48	3.5								Faro/Portugal Stenner and Nickless, 1975
Nephrops norvegicus	m 1			90	3.3	41	–								Ayamonte/Guadiana Estuary, S.W. Spain, Stenner a. Nickless, 1975
Cirripedia															
Chthamalus stellatus	s.p.1			237	6.3	6.3									S. Portugal/Stenner and Nickless, 1975
Chthamalus stellatus	s.p. 1			100–158	5.1	10–									S.W. Spain/Atlantic/Stenner and Nickless, 1975
Balanus perforatus	s.p.1			40–60	4.5–5.8	8.0–									S. Portugal/Stenner and Nickless, 1975
Balanus amphitrites	s.p. 1			1780–3300	10.8–12.1	550–									Rio Tinto Estuary, S.W. Spain/Stenner and Nickless, 1975
Balanus balanoides	s.p. 2			1028											Menai Strait/GB/Walker et al., 1975
Balanus balanoides	s.p. 2			1770–3438											Cardigan Bay/GB/Walker et al., 1975
Balanus balanoides	s.p. 1			4500–23100											Cardigan Bay/GB/Ireland, 1973

Vb. Crustanceans (Freshwater)

Gammarids	2	0.02–0.20					Elbe Estuary, limnic and brackish F.R.G./Zauke, 1977
Asellus aquaticus	2	0.35 (0.22–0.56)					Elbe Estuary, limnic, F.R.G./Zauke, 1977
Asellus aquaticus	2	0.06; 1.9					Above and below a paper mill, Sweden/Johnels et al., 1967
Asellus aquaticus	2	0.05					Lake Åsjön/Sweden/Hasselrot and Göthberg, 1974
Asellus aquaticus	2	0.1–0.2					Lake Kyrksjön/Sweden/Hasselrot and Göthberg, 1974
Asellus aquaticus	w.a. 1	181.9 (118.6–203.1)	1.21 (0.47–2.97)	12.99 (8.64–19.91)	6.80 (2.69–16.50)		Elsenz River, F.R.G./Prosi, 1977
Freshwater Decapoda	1					4.7/11	Vermillion River, rural/urban drainage/Leland and McNurney, 1974

1, dry weight; 2, wet weight; w.a., whole animal; m, muscle, s.p., soft parts

Appendix B (continued)

Organisms	ppm	As	Hg	Zn	Cd	Cu	Pb	Cr	Fe	Mn	Ni	Co	Locality/Reference
VIa. Fish (Marine)													
Chondrichthyes													
Raja clavata	m. 1		0.53	52	2.45	1.6	7						Sagres, S. Portugal/ Stenner and Nickless, 1975
Scyliorhinus spp.	m. 2			11.5 (8.9–13.7)	0.32 (0.07–0.59)								Bristol Channel, GB/ Peden et al., 1973
Cetorrhinus maximus	m. 2		2.1 (0.9–4.0)										N. Atlantic/ Krüger et al., 1975
Lamna nasus	m. 2		2.2 (1.5–3.0)										N. Atlantic/ Krüger et al., 1975
Squalus acanthias	m. 2		0.24										N. Atlantic/Krüger et al., 1975
Prionace glauca	m. 1			35	<0.05	4.4	<0.2						N.E. Atlantic, Cornwall/GB Stevens and Brown, 1974
Osteichthyes													
Pleuronectes flesus	m. 2		0.05–0.86										Swedish Sound/ Ackefors et al., 1970
Pleuronectes platessa	m. 2		0.07–3.10										Swedish Sound/ Ackefors et al., 1970
Pleuronectes platessa	m. 2			30.3	1.44	0.5					2.8		Northumberland, Power station filter at Lynemouth/ Wright, 1976
Platichthyes flesus	w.a. 1			76.3–175.6	3.4–5.4		13.8–29.2						Oldbury/Severn Estuary/GB Power station Hardisty et al., 1974a
Platichthyes flesus	w.a. 1			195.2–224.5	1.1–1.7		14.1–19.1						Barnstable Bay/Severn Est./GB Hardisty et al., 1974a
Hippoglossoides platessiodes	m. 1	188 (17.0–290)											West Greenland (mining area)/ Bohn, 1975

Appendix

Species	Type										Location/Reference
Solea solea	m. 1	0.17	57	2.1	4.4	2					Sagres, S. Portugal/Stenner and Nickless, 1975
Brevoortia tyrannus	visc. 2	0.095/0.166									Core Sound, W. Atlantic/Chesapeake Bay, Md, USA/Cocoros et al., 1973
Fundulus heteroclitus	w.a. 2			0.33							Rhode Island, USA Eisler et al., 1972
Sardina pilchardus	m. 1		24	<0.05	6.6	<0.2					N.E. Atlantic/Cornwall/GB/Stevens and Brown, 1974
Belone belone	m. 1		42–54	<0.05–0.9	3.0–5.1	<0.2					N.E. Atlantic/Cornwall/GB/Stevens and Brown, 1974
Pomatomus saltatrix	m. 2	0.37	6.4	0.136	0.61		7.0	0.22			Cape Hatteras, N.C. USA/Cross et al., 1973
Scomber scombrus	m. 2	0.15									diff. areas of N. Atlantic and N. Sea/Krüger et al., 1975
Clupea harengus	m. 2	0.09									diff. areas of N. Atlantic and N. Sea/Krüger et al., 1975
Clupea sprattus	m. 2		119	0.24±0.136	1.61±4.07				7.2±4.8		Northumberland, Power station filter at Lynemouth/Wright, 1976
Engraulis mordax	m. 2	0.04 (0.01–0.08)	54 ±38.69								Monterey Bay, Calif., USA/Knauer and Martin, 1972
Scomber japonicus	m. 2		7.2±2.2				0.65±0.85	17±2	0.28±0.14	0.02±0.017	Cape West Coast, S.A. Van As et al., 1975
Merluccius capensis	m. 2		3.7±0.4				0.26±0.44	4.3±3.3	0.22±0.11	0.004±0.001	Cap West Coast, S.A./Van As et al., 1975
Anarrhichas lupus	m. 1(2)	0.2( )	20	0.02	2		1	29	4		Trondheimsfjorden/Lande, 1977
Anarrhichas minor	m. 1	78.3 171–19.5									West Greenland mining area Bohn, 1975
Seriola grandis	m. 2		9.5 2.8–56.0	0.006 0.002–0.014	0.58 0.20–3.36	0.40 0.20–0.97	0.02 0.01 0.03	6.8 2.0 42.0	0.34 0.12 1.10	0.03 0.02 0.08	New Zealand (North Island)/Brooks and Rumsey, 1974
Seriola pappei	m. 2		6.7±1.7				1.0±1.8	14±3	0.29±0.08	0.008±0.005	Cape West Coast, S.A./Van As et al., 1975

m, muscle; visc., viscera; w. a., whole animal; w.a.[+], decapitated and eviscerated; 1, dry weight; 2 wet weight

Appendix B (continued)

Organisms	ppm	As	Hg	Zn	Cd	Cu	Pb	Cr	Fe	Mn	Ni	Co	Locality/Reference
Mugil cephalus	m. 1	<1.0	0.1	17	<0.1	1.9							N. Atlantic, east of Gulf Stream/ Windom et al., 1973
Trigla kumu	m. 2			5.5 2.5– 56.0	0.015 0.008– 0.024	0.34 0.15– 0.75	0.16 0.13– 0.40	0.01 0.01– 0.02	4.2 2.1– 13.0	1.15 0.10– 3.70	0.02 0.01– 0.03		N. Zealand (North Island)/ Brooks and Rumsey, 1974
Sebastes mentella	m. 2		0.24 0.07– 0.54										Iceland/Faroer/Shetland/ Krüger et al., 1975
Sebastes marinus	m. 2		0.14 0.07– 0.22										Iceland/Faroer/Shetland/ Krüger et al., 1975
Gadus virens	m. 2		0.11 0.06– 0.15										Iceland/Faroer/Norway/ Krüger et al., 1975
Gadus morrhua	m. 2		0.15 0.04– 0.45										Iceland/Faroer/Bornholm/Elbe/ Krüger et al., 1975
Gadus morrhua	m. 2		1.06 0.12– 3.18					1–2	28–30	4–5			Kilsfjord, S. Norway/ Underdal and Håstein, 1971
Gadus morrhua	m. 1(2)		0.2( )	38–48	0.01– 0.02	3–4							Trondsheimfjorden/Norway/ Lande, 1977
Melanogrammus aeglefinus	m. 2		0.08 0.05– 0.12										Iceland/Faroer/Bornholm/Elbe/ Krüger et al., 1975
Molva byrkelange	m. 2		0.34 0.27– 0.52										Iceland/Faroer/Norway/ Krüger et al., 1975
Tuna fish (museum sp.)	m. 1		0.95 0.53– 1.51										Museum specimen, USA/ Miller et al., 1972

Appendix 393

Species	Tissue							Location / Reference
Tuna fish (recent spp.)	m. 1		0.91 (0.44–1.53)					Pacific/Calif. Miller et al., 1972
Sword fish (recent spp.)	m. 1		3.1± 1.5 (0.94–5.08)					Pacific/Calif. Miller et al., 1972
Makaira indica	m. 2	0.6 (0.1–1.65)	8.6 (5.8–14.6)	0.09 (0.05–0.40)	0.4 (0.9–1.2)	0.6 (0.1–0.9)		Northeastern Australia (apparently unpolluted waters) Mackay et al., 1975
Makaira nigricans	m. 2		2.06 (0.30–8.35)					Hawaiian waters Schultz et al., 1976
Anguilla anguilla	m. 2		0.93 max. 2.02					Elbe estuary Krüger et al., 1975
Osmerus eperlanus	m. 2		0.54					Elbe estuary/Krüger et al., 1975
Alosa fallax	m. 2		0.76					Elbe estuary/Krüger et al., 1975
Acerina cernua	m. 2		1.50					Elbe estuary/Krüger et al., 1975
VIb. Fish (Freshwater)								
Phytophagous-omnivorous species								
Cyprinus carpio	m. 2	0.25 (0.14–0.26)						Lake Oahe, S. Dakota, USA / Walter et al., 1974
"carp"	m. 2	0.366	0.129		0.014		0.211	0.524	Lower Mississippi river/ Hartung, 1974
Cyprinus carpio	m. 2		10.2 (1.1–16.1)	0.035 (0.011–0.069)	0.24 (0.12–0.41)	0.56 (0.15–2.13)	0.16 (0.02–0.46)	0.19 (0.04–0.28)	Illinois river, USA/ Mathis and Cummings, 1973
Cyprinus carpio	m. 2	0.25 (0.18–0.34)						Lake Powell, N. M., USA/ Potter et al., 1975
Carpiodes cyprinus	m. 2		3.4 (2.1–5.5)	0.024 (0.004– −0.046)0.30)	0.17 (0.10–0.30)	0.64 (0.09–1.30)	0.21 (0.02–0.60)	0.18 (0.15–0.45)	Illinois river, USA/ Mathis and Cummings, 1973

m, muscle; visc., viscera; w.a., whole animal; w.a.[+], decapitated and eviscerated; 1, dry weight; 2 wet weight

Appendix B (continued)

Organisms	ppm	As	Hg	Zn	Cd	Cu	Pb	Cr	Fe	Mn	Ni	Co	Locality/Reference
Dorosoma cepedianum	m. 2			4.0 (2.7–6.3)	0.033 (0.005–0.068)	0.26 (0.18–0.39)	0.84 (0.19–1.78)	0.45 (0.10–1.06)			0.28 (0.06–0.52)		Illinois river, USA/ Mathis and Cummings, 1973
Ictiobus cyprinellus	m. 2			3.5 (2.6–5.1)	0.032 (0.001–0.055)	0.18 (0.07–0.26)	0.57 (0.35–0.95)	0.13 (0.02–0.53)			0.10 (0.02–0.18)		Illinois river, USA/ Mathis and Cummings, 1973
Rutilus rutilus	m. 2		0.22–0.65										Lake Åsjön, Sweden/ Hasselrot and Göthberg, 1974
Rutilus rutilus	m. 2			4.7 (2.5–7.1)	0.015 (0.002–0.055)	0.48 (0.14–1.07)							Neckar river, F.R.G./ Müller and Prosi, 1978
Rutilus rutilus	m. 2			8.4 (6.1–11.9)	<0.006 (<0.005–<0.01)	0.27 (0.17–0.38)							Elsenz river, F.R.G./ Müller and Prosi, 1978
Scardinius erythrophthalmus phytophygous fish	m. 2 m. 2		0.03–0.21 0.105±0.053										Wörther Lake, Austria/ Krocza et al., 1974 Danube river/Krems/Wien/ Austria/Dworsky et al., 1973
Carnivorous species													
Esox lucius	w.a.+ 2	<0.05/ <0.05	0.11/ 0.49	19/11	<0.05/ <0.05	0.7/0.7	<0.5/ <0.5	<0.035/ <0.031		2.98/ 0.93	<0.2/ <0.2		Moose Lake/Lake Erie, USA/ Uthe and Bligh, 1971
Esox lucius	w.a. 2		0.38–0.75		0.001–0.008								Skotfoss/S. Norway/ Havre et al., 1973
Esox lucius	m. 2		2.3										Stockholm archipelago/ Johnels et al., 1967
Esox lucius	m. 2		0.3										Tidö-Lindö, Mälaren/ Johnels et al., 1967
Esox lucius	m. 2		2.3										Bråviken, Östergötland/ Johnels et al., 1967
Esox lucius	m. 2		1.8–9.9										Lake St. Clair, Canada/ Bishop and Neary, 1974

Appendix

Species	Tissue									Location / Reference
Esox lucius	m. 2	1.14								Lake Åsjön, Sweden / Hasselrot and Göthberg, 1974
Esox lucius	m. 2	0.76 (0.45–1.21)								La Grande river, Canada / Smith et al., 1974
Esox lucius	m. 2	0.37 (0.20–0.52)								Lake Oahe, S. Dakota, USA / Walter et al., 1974
Esox lucius	m. 2	2.6 (2.3–2.8)	0.022 (0.013–0.031)	0.07 (0.05–0.08)	0.34 (0.17–0.61)	0.13 (0.02–0.22)	0.15 (0.08–0.19)	0.07 (0.03–0.11)		Illinois river, USA / Mathis and Cummings, 1973
Micropterus salmoides	m. 2	3.4 (0.8–5.4)	0.022 (0.004–0.060)	0.10 (0.08–0.13)	0.59 (0.36–1.13)	0.11 (0.04–0.24)	0.11 (0.05–0.23)	0.09 (0.06–0.18)		Illinois river, USA / Mathis and Cummings, 1973
Micropterus dolomieu	m. 2	3.8 (3.5–4.1)	0.005	0.15 (0.14–0.16)	0.98 (0.68–1.28)	0.16 (0.04–0.27)	0.13 (0.08–0.19)	0.15 (0.14–0.16)		Illinois river, USA / Mathis and Cummings, 1973
Micropterus salmoides	m. 2	0.31 (0.19–0.69)								Lake Powell, N. M., USA / Potter et al., 1975
Micropterus dolomieu	w.a.+ 2		0.016							Lake Erie, USA / Lovett et al., 1972
Micropterus dolomieu	w.a.+ 2		0.014							Lake Ontario, USA / Lovett et al., 1972
Micropterus dolomieu	w.a.+ 2		0.025							St. Lawrence River, USA / Lovett et al., 1972
Micropterus salmoides	m. 2	1.87–4.49								Lake Jocassee, USA / Abernathy and Cumbie, 1977
Lucioperca sandra	m. 2	0.051–0.080								Swedish Sound / Ackefors et al., 1970
Perca fluviatilis	m. 2	1.39–4.16								Swedish Sound / Ackefors et al., 1970
Perca fluviatilis (juven.)	m. 2	0.11 (0.06–0.24)								Millstätter Lake/Austria / Krocza et al., 1974

m, muscle; visc, viscera; w.a., whole animal; w.a.+, decapitated and eviscerated; 1, dry weight; 2 wet weight

Appendix B (continued)

Organisms	ppm	As	Hg	Zn	Cd	Cu	Pb	Cr	Fe	Mn	Ni	Co	Locality/Reference
Perca flavescens	m. 2		0.19 (0.04–0.32)										Lake Oahe, S. Dakota, USA/ Walter et al., 1974
Perca fluviatilis	m. 2		0.30 (0.13–0.59)										above a paper mill, Kammerfoss river/Underdal and Håstein, 1971
Perca fluviatilis	m. 2		4.93 (3.86–6.08)										below a paper mill, Kammerfoss river/Underdal and Håstein, 1971
Salmo trutta	m. 2		0.21 (0.09–0.47)										above a paper mill, Kilsfjord, S. Norway/ Underdal and Håstein, 1971
Salmo trutta	m. 2		4.35 (2.66–7.38)										below a paper mill, Kilsfjord, S. Norway/ Underdal and Håstein, 1971
Salmo gairdneri	m. 2		0.08 (0.07–										Lake Powell, N.M., USA/ Potter et al., 1975
Coregonus clupeaformis	w.a.+ 2	0.09/ 0.70	0.07/ 0.17	14/12	<0.05/ <0.05	0.50/ 0.94	<0.5/ <0.5	0.033/ <0.017		0.69/ 0.66	<0.2/ <0.2		Moose Lake/Lake Erie/USA/ Uthe and Bligh, 1971
Coregonus clupeaformis	m. 2		0.09 (0.06–0.14)										La Grande river/Canada/ Smith et al., 1975
Salvelinus namaycush	w.a. 1		0.43 (0.19–0.66)										Lake Cayuga, N.Y., USA/ Bache et al., 1971
Oncorhnchys kisutch	w.a.+ 2				0.024								Cattarangus Creek, N.Y., USA/ Lovett et al., 1972
Oncorhnchys kisutch	w.a.+ 2				0.013								Lake Ontario/USA/ Lovett et al., 1972

Appendix   397

Species						Location/Reference
Stizostedion v. vitreum	m. 2	0.30 (0.12–0.39)				Lake Oahe, S. Dakota, USA/ Walter et al., 1974
Stizostedion v. vitreum	m. 2	0.43 (0.21–)				Lake Powell, N. M./USA/ Potter et al., 1975
Lepomis macrochirus	m. 2	0.09 (0.06–0.13)				Lake Powell, N. M./USA/ Potter et al., 1975
Percopsis omiscomaycus	w.a. 2	0.043± 0.012	0.076± 0.008	2.7±0.6	1.6±0.2	Lake Michigan/ Lake Superior/USA
		0.007± 0.002	0.140± 0.060	0.8±0.3	≤3	Lucas et al., 1970
Alosa pseudoharengus	w.a. 2	0.023	0.062± 0.015	0.85± 0.05	1.1±0.5	Lake Michigan, USA/ Lucas et al., 1970
Notropis hudsonius	w.a. 2	0.003/ 0.0035 ±0.003	0.100± 0.030	0.80± 0.15 1.2± 0.2	0.9±0.5 10	Lake Michigan/Lake Erie, USA Lucas et al., 1970
Pseudogobio brevirostris	w.a. 2		70.7	<0.03	<0.1	0.3	Ta-Tu river/Taiwan/ Chung and Jeng, 1974
Rhinogobio similis	w.a. 2		36.7	<0.02	0.5	<0.1	0.3	Ta-Tu river/Taiwan/ Chung and Jeng, 1974
"catfish"	m. 2	0.281		0.043		0.479	0.424	Lower Mississippi river/ Hartung, 1974
Anguilla anguilla	m. 2	0.81–2.03					Swedish Sound/ Ackefors et al., 1970

Additional values column: 0.024±0.002, 0.022±0.005, 0.029±0.014, 0.042±0.15, 0.025±0.004

m, muscle; visc, viscera; w.a., whole animal; w.a.⁺, decapitated and eviscerated; 1, dry weight; 2 wet weight

References

Abdullah, M.I., Royle, L.G.: Heavy metal content of some rivers and lakes in Wales. Nature (London) *238*, 329–330 (1972)

Abdullah, M.I., Royle, L.G.: Cadmium in some British coastal and fresh water environments. Proc. Int. Symp. "Problems of the Contamination of Man and his Environment by Mercury and Cadmium". Luxembourg, 1973, 69–81 (1974a)

Abdullah, M.I., Royle, L.G.: A study of the dissolved and particulate trace elements in the Bristol Channel. J. Mar. Biol. Assoc. U.K. *54*, 581–597 (1974b)

Abdullah, M.I., Royle, L.G., Morris, A.W.: Heavy metal concentration in coastal waters, Nature (London) *235*, 158–160 (1972)

Abdullah, M.I., Banks, J.W., Miles, D.L., O'Grady, K.T.: Environmental dependence of manganese and zinc in the scales of Atlantic salmon, Salmo salar (L.) and Brown trout, Salmo trutta (L.). Freshwater Biol. *6*, 161–166 (1976)

Abdul-Razzak, A.K.: Geochemisch-sedimentpetrographischer Vergleich lakustrischer Sedimente aus verschiedenen Klimabereichen. Chem. Erde *33*, 154–184 (1974)

Abernathy, A.R., Cumbie, P.M.: Mercury accumulation by largemouth bass (Micropterus salmoides) in recently impounded reservoirs. Bull. Environ. Contam. Toxicol. *17*, 595–602 (1977)

Abo-Rady, M.D.K.: Die Belastung der oberen Leine mit Schwermetallen durch kommunale und industrielle Abwässer, ermittelt anhand von Wasser-, Sediment-, Fisch- und Pflanzenuntersuchungen. Diss. Univ. Göttingen (FRG), 120 p. (1977)

Ackefors, H.: Effects of particulate pollutants – mercury pollution in Sweden with special reference to conditions in the water habitat. Proc. R. Soc. London B *177*, 365–387 (1971)

Ackefors, H., Löfroth, G., Rosen, C.-G.: A survey of the mercury pollution problem in Sweden with special reference to fish. Oceanogr. Mar. Biol. *8*, 203–224 (1970)

Adams, C.E., Eckenfelder, W.W., Goodman, B.L.: The effects and removal of heavy metals in biological treatment. In: Heavy Metals in the Aquatic Environment. Krenkel, P.A. (ed.). Oxford: Pergamon Press 1975, pp. 277–292

Adams, W.J., Johnson, H.E.: Survey of the selenium content in the aquatic biota of the western Lake Erie. J. Great Lakes Res. *3*, 10–14 (1977)

Addis, G., Moore, M.R.: Lead levels in the water of suburban Glasgow. Nature (London) *252*, 120–121 (1974)

Agnes, G., Hill, H.A.D., Pratt, J.M., Ridsdale, S.C., Kennedy, F.S., Williams, R.J.P.: Methyl transfer from methyl vitamin B_{12}. Biochim. Biophys. Acta *252*, 207–211 (1971)

Aguilar-Ravello, A.N.: The regional distribution of arsenic in southwest England. Unpubl. Ph.D. Thesis, Univ. London (1974). Lit. cit. Colbourne et al. (1975)

Aguilera, N.H., Jackson, M.L.: Iron oxide removal from soils and clays. Soil. Sci. Soc. Amer. Proc. *17*, 359–364 (1953)

Ahl, T.: River discharges of Fe, Mn, Cu, Zn, and Pb into the Baltic Sea from Sweden. In: 3rd Soviet-Swedish Symp. on the Pollution of the Baltic. Åkerblom, A. (ed.). Ambio Spec. Rept. *5*, 219–228 (1977)

Ahrens, L.H.: The lognormal distribution of the elements – a fundamental law of geochemistry. Geochim. Cosmochim. Acta *11*, 205–212 (1957)

Ahrens, H.L.: Ionization potentials and metal-amino acid complex formation in the sedimentary cycle. Geochim. Cosmochim. Acta *30*, 1111–1119 (1966)

Ahrland, S.: Thermodynamics of complex formation between hard and soft acceptors and donors. Nature and scope of the classification of acceptors and donors as hard and soft. Struct. Bonding (Berlin) *5*, 118–123 (1968)

Ahrland, S.: Metal complexes present in seawater. In: The Nature of Seawater. Goldberg, E.D. (ed.). Berlin: Dahlem Konferenzen, 1975, pp. 219–244

Alberts, J.J., Schindler, J.E., Miller, R.W., Nutter, D.E.: Elemental mercury evolution mediated by humic acid. Science *184*, 895–896 (1974)

Ali, S.A., Gross, M.G., Kishpaugh, J.R.L.: Cluster analysis of marine sediments and waste deposits in New York Bight. Environ. Geol. *1*, 143–148 (1976)

Allan, R.J.: Lake sediment, a medium for regional geochemical exploration of the Canadian Shield. Can. Inst. Min. Met. Bull. *64*, 43–59 (1971)

Allan, R.J.: Metal contents of lake sediment cores from established mining areas: An interface of exploration and environmental geochemistry. Geol. Surv. Can. *74-1/B*, 43–49 (1974)

Allan, R.J.: Heavy metal concentrations in lake sediments in Canada: A review. Abstr. Int. Conf. Heavy Met. Environ., Toronto, C-5 (1975)

Allan, R.J.: Natural versus unnatural heavy metal concentrations in lake sediments in Canada. Proc. Int. Conf. Heavy Met. Environ. Toronto 1975, *II/2*, 785–808 (1977)

Allan, R.J., Brunskill, G.J.: Relative atomic variation (RAV) of elements in lake sediments: Lake Winnipeg and other Canadian lakes. In: Interactions Between Sediments and Fresh Water. Golterman, H.L. (ed.). The Hague: Junk Publ. 1977, pp. 108–118

Allan, R.J., Cameron, E.M., Durham, C.C.: Lake geochemistry – a low density technique for reconnaissance geochemical exploration and mapping of the Canadian Shield. In: Exploration Geochemistry, Proc. 4th Int. Geochem. Explor. Symp., 131–160 (1972)

Allan, R.J., Cameron, E.M., Durham, C.C.: Reconnaissance geochemistry using lake sediments of a 36,000 square-mile area of the northwestern Canadian Shield. Geol. Surv. Can. Pap. *72/50*, 70 p. (1973)

Allan, R.J., Cameron, E.M., Jonasson, I.R.: Mercury and arsenic levels in lake sediments from the Canadian Shield. In: Primero Congr. Int. Mercurio, Barcelona, 93–119 (1974)

Allen, H.E., Boonlayangoor, C.: Mobilization of metals from sediment by NTA. Abstr. SIL-Congr., Copenhague, 7 (1977)

Alloway, B.J., Davies, B.E.: Heavy metal content of plants growing on soils contaminated by lead mining. J. Agric. Sci. Camb. *76*, 321–323 (1971a)

Alloway, B.J., Davies, B.E.: Trace element content of soils affected by base metal mining in Wales. Geoderma *5*, 197–208 (1971b)

Alten, G.R.: Geochemical analysis of stream sediments as a tool for environmental monitoring: A pigyard case study. Geol. Soc. Am. Bull. *86*, 174–176 (1975)

Amiel, A.J., Navrot, J.: Nearshore sediment pollution in Israel by trace metals derived from sewage effluents. Mar. Pollut. Bull. *9*, 10–14 (1978)

Anderson, B.J., Jenne, E.A., Chao, T.T.: The sorption of silver by poorly crystallized manganese oxides. Geochim. Cosmochim. Acta *37*, 611–622 (1973)

Anderson, J.: A study of the digestion of sediment by the HNO_3-H_2SO_4, and the HNO_3-HCl procedures. At. Absorpt. Newsl. *13*, 31 (1974)

Anderson, J.B., Wheeler, R., Dunning, C.P.: Geologic assessment of environmental impact in Lake Macatawa. Environ. Geol. *2*, 67–78 (1978)

Anderson, T.W.: Historical evidence of land use in a pollen profile from Osoyoos Lake, British Columbia. Geol.Surv. Can. Rept. *73-1/A*, 178–180 (1973)

Anderson, W.L., Smith, K.E.: Dynamics of mercury at coal-fired power plant and adjacent cooling lake. Environ. Sci. Technol. *11*, 75–80 (1977)

Anderson, A., Nilsson, K.O.: Enrichment of trace elements from sewage sludge fertilizer in soils and plants. Ambio *1*, 176–179 (1972)

Andersson, A.C., Abedelghani, A.A., Smith. P.M., Mason, J.W., Englande, A.J.: The acute toxicity of MSMA to Black bass (Micropterus dolomieu), Crayfish (Procambarua sp.) and Channel catfish (Ictalurus lacustris). Bull. Environ. Contam. Toxicol. *14*, 330–333 (1975)

Andren, A.W.: The geochemistry of mercury in three estuaries from the Gulf von Mexico. Ph. D. Thesis, Florida State Univ. (1973)

Andren, A.W., Harriss, R.C.: Methylmercury in estuarine sediments. Nature (London) *245*, 256–257 (1973)

Andren, A.W., Harriss, R.C.: Observations on the association between mercury and organic matter dissolved in natural waters. Geochim. Cosmochim. Acta *39*, 1253–1257 (1975)

Andrew, R.W.: Toxicity relationships to copper forms in natural waters. In: Toxicity to Biota of Metal Forms in Natural Water. Andrew, R.W., Hodson, P.V., Konasewich, D.E. (eds.). Int. Joint Comm., Windsor, Ontario, 1976, pp. 127–143

Andrews, R.D.: Tailings: Environmental consequences and controls. Proc. Int. Conf. Heavy Met. Environ., Toronto 1975, *II/2*, 645–675 (1977)

Andreyev, P.F., Chumachenko, L.M.: Reduction of uranium by natural organic substances. Geochem. Int. *1*, 3–7 (1964)
Angino, E.E., Magnuson, L.M., Waugh, T.C.: Mineralogy of suspended sediment and concentration of Fe, Mn, Ni, Zn, Cu, and Pb in water and Fe, Mn, and Pb in suspended load of selected Kansas streams. Water Resour. Res. *10*, 1187–1191 (1974)
Angino, E.E., Magnuson, L.M., Waugh, T.C., Galle, O.K., Bredfeldt, J.: Arsenic in detergents – possible danger and pollution hazard. Science *168*, 389–390 (1970)
Anon.: Rivers Pollution Commission. Fifth report of the commisioners appointed in 1868 to inquire into the best means of preventing the pollution of rivers (1874). Lit. cit. Lewin, J. et al. (1977)
Anon.: U.S. National Bureau of Standards: Maximum permissible amounts of radioisotopes in the human body and maximum permissible concentrations in air and water. Handbook *52*, 45 (1955)
Anon.: U.S. Atomic Energy Commission: Status report on handling and disposal of radioactive wastes in the AEC program. Rep. WASH-*742*, 41 p. (1957)
Anon.: Mercury in water, a bibliography. U.S. Office of Water Resources Research. U.S. Natl. Tech. Inf. Serv., Springfield, Rept. 201267 (1970a)
Anon.: Normalwerte für Abwassereinleitungsverfahren. Länderarbeitsgemeinschaft Wasser – LAWA – (FRG), 2nd ed. Hamburg: Verlag Wasser und Boden 1970b
Anon.: Hinweise für das Einleiten von Abwasser aus gewerblichen und industriellen Betrieben in eine öffentliche Abwasseranlage. BDI-Printed matter No. 90 – ATV-Arbeitsblatt A 111 Köln: BDI-Abt. für Umweltfragen (1970c). Cit. L. Hartinger (1976)
Anon.: Metals focus shift to cadmium. Environ. Sci. Technol. *5*, 754–757 (1971a)
Anon.: Strontium in water, a bibliography. U.S. Office of Water Resources Research. U.S. Natl. Tech. Inf. Serv., Springfield, Rep. PB 201268 (1971b)
Anon.: Manganese in water, a bibliography. U.S. Office of Water Resources Research. U.S. Natl. Tech. Inf. Serv., Springfield, Rep. PB 201270 (1971c)
Anon.: Arsenic and lead in water, a bibliography. U.S. Office of Water Resources Research. U.S. Natl. Tech. Inf. Serv., Springfield, Rep. PB 202578 (1971d)
Anon.: Trace elements in water, a bibliography. U.S. Office of Water Resources Research. U.S. Natl. Tech. Inf. Serv., Springfield, Rep. PB 201266 (1971e)
Anon.: Copper in water, a bibliography U.S. Office of Water Resources Research. U.S. Natl. Tech. Inf. Serv., Springfield, Rep. PB 201269 (1971f)
Anon.: Zinc in water, a bibliography. U.S. Office of Water Resources Research. U.S. Natl. Tech. Inf. Serv., Springfield, Rep. PB 201272 (1971g)
Anon.: Water quality criteria data book. Vol. 2: Inorganic chemical pollution of freshwater. Arthur D. Little Inc. U.S. Natl. Tech. Inf. Serv., Springfield, Rep. 208988 (1971h)
Anon.: Blei und Umwelt: Kommission für Umweltgefahren, Arbeitsgruppe Blei, des Bundesgesundheitsamtes, Berlin (1972a)
Anon.: Baseline studies of pollutants in the marine environment and research recommendations. Deliberations of the International Decade of Ocean Exploration (IDOE) Baseline Conference. New York (1972b)
Anon.: Chromium in water, a bibliography. U.S. Office of Water Resources Research. U.S. Natl. Tech. Inf. Serv., Springfield, Rep. PB 210921 (1972c)
Anon.: Review of toxicity and potential biological effects of NTA. 2nd Rept. Dept. of National Health and Welfare, Canada. Environmental Health Directorate (1972d)
Anon.: Northeastern New Brunswick mine water quality program. Montreal Engineering Co., Fredericton, N.B. *4* (1972e). Cit. Harvey (1976)
Anon.: Ion exchange solves chrome waste water problem. Environ. Sci. Technol. *10*, 865 (1972f)
Anon.: Natl. Monit. Program Progress Rept. No. 2. Water Qual. Branch, Inland Waters Directorate. Dept. Environ., Ottawa, Canada (1973a)
Anon.: Cleaning up an industrial discharge. Environ. Sci. Technol. *7*, 678–679 (1973b)
Anon.: Removal of mercury from chlorine plant effluents. Environ. Sci. Technol. *7*, 185 (1973c)
Anon.: Water Quality Criteria 1972. Report of the Committee on Water Quality Criteria, Natl. Acad. Sci. EPA-R3-033, Washington D.C. (1973d). Cit. Maruyama et al. (1975)
Anon.: Proposed Toxic Pollutant Effluent Standards. U.S. Federal Register *38*, 247 – 35388 (1973e). Cit. Maruyama et al. (1975)
Anon.: Environmental Protection Agency (U.S.) "Ocean Dumping Criteria". Federal Register *38* (94), 12872–12877 (1973f)
Anon.: Environmental Protection Agency (U.S.) "Ocean Dumping Final Criteria". Federal Register *38* (198), 28610–28621 (1973g)

Anon.: Technical Committee on Mine Waste Pollution of the Molonglo River. Final Report on Remedial Measures. Aust. Govt. Publ. Serv. (1974)

Anon.: Controlling the Radiation Hazard from Uranium Mill Tailings. U.S. Energy Research and Development Administration, U.S. Nuclear Regulatory Commission. Rept. to Congress by the Comptroller General, Washington D.C. (1975a)

Anon.: Arbeitsgruppe „Metalle" im Schwerpunktprogramm „Schadstoffe im Wasser" der Deutschen Forschungsgemeinschaft. Internal Report of the German Research Society, DFG, Bad Godesberg (1975b)

Anon.: International Working Group to the Abatement and Control of Pollution from Dredging Activities. Ottawa and Washington D.C. (1975c)

Anon.: Characterization of pollutant availability for San Francisco Bay dredge sediments. Pacific Northwest Laboratories (1975d). Cit. Patrick et al. (1977)

Anon.: Uferfiltration − Bericht des Fachausschusses „Wasserversorgung und Uferfiltrat". Edited by the German Federal Ministry of the Interior, 192 (1975e)

Anon.: National Interim Primary Water Regulations. Federal Register 40, 248, 59566 (1975f). Cit. Hannah et al. (1977)

Anon.: Ruhrwassergüte-Berichte. Publications of the Ruhrverband, D-43 Essen, F.R.G., 1973−1976 (1975g)

Anon.: Umweltfragen des Rheins. 3. Sondergutachten des Rates von Sachverständigen für Umweltfragen. Stuttgart−Mainz: Kohlhammer 1976a

Anon.: Mercury Levels in the Rivers of Western Canada 1970−1976. Can. Inland Waters Directorate, Water Qual. Branch. Ottawa. Soc. Sci. Ser. 16, 46 p. (1976b)

Anon.: Quality Criteria for Water. U.S. Environ. Protect. Agency, Prepubl. Copy Washington, D.C. (1976c). Cit. Hannah et al. (1977)

Anon.: Bayerische Landesanstalt für Wasserforschung: Untersuchungen über die Belastung bayerischer Gewässer mit Cadmium (1972−1977), 78 p. (1978)

Applequist, M.D., Katz, A., Turekian, K.K.: Distribution of mercury in the sediments of New Haven (Conn.). Harbor. Environ. Sci. Technol. 6, 1123−1124 (1972)

Argo, D.G., Culp. G.L.: Heavy metals in wastewater treatment processes. Water Sewage Works 119, 62−65, 128−132 (1972)

Armstrong, F.A.J., Hamilton, A.L.: Pathways of mercury in a polluted northwestern Ontario lake. In: Trace Metals and Metal-Organic Interactions in Natural Waters. Singer, P.C. (ed.). Ann. Arbor Sci. Publ. 1973, pp. 131−156

Armstrong, F.A.J., Alton, F.M., Royer, L.: Mercury in sediments and waters of Clay Lake, northwestern Ontario. In: Mercury in the Aquatic Environment. Uthe, J.F. (ed.). Fish. Res. Board Can. Misc. Spec. Publ. 1167, 46−67 (1972)

Armstrong, P.B., Hanson, G.M., Gaudette, H.E.: Minor elements in sediments from Great Bay estuary. New Hampshire. Environ. Geol. 1, 207−214 (1976)

Arnold, R.G.: The concentrations of metals in lake waters and sediments of some Precambrian lakes in the Flin Flon and La Ronge areas. Saskatchewan Res. Counc. Div. Circ. 4, 30 p. (1970)

Artmann, N.R.: Safety considerations for detergents. In: Progress in Water Technology. Water Quality Management and Pollution Control Problems. Jenkins, S.H. (ed.). Oxford: Pergamon Press 1973, Vol. 3, pp. 277−288

Asami, T.: Environmental pollution by cadmium and zinc discharged from a braun tube factory. Ibaraki Daigaku Nogakubu Gakujutsu Hokaku 22, 19−23 (1974)

Aschan, O.: Water humus and its role in the formation of marine iron ore. Ark. Kemi Mineral. Geol. $10A$, 1−143 (1932)

Aston, R.J.: Tubificids and water quality: a review. Environ. Pollut. 5, 1−34 (1973)

Aston, S.R., Chester, R.: The influence of suspended particles on the precipitation of iron in natural water. Estuarine Coastal Mar. Sci. 1, 225−231 (1973)

Aston, S.R., Thornton, I.: The application of regional geochemical reconnaissance surveys in the assessment of water quality and estuarine pollution. Water Res. 9, 189−195 (1975)

Aston, S.R., Thornton, I.: Regional geochemical data in relation to seasonal variations in water quality. Sci. Total Environ. 7, 247−260 (1977)

Aston, S.R., Bruty, D., Chester, R., Padgham, R.: Mercury in lake sediments. A possible indicator of technological growth. Nature (London) 241, 450−451 (1973)

Aston, S.R., Thornton, I., Webb, J.S., Purves, J.B., Milford, B.L.: Stream sediment composition, an aid to water quality assessment. Water Air Soil Pollut. 3, 321−325 (1974)

Aston, S.R., Thornton, I., Webb, J.S., Milford, B.L., Purves, J.B.: Arsenic in stream sediments and waters of south-west England. Sci. Total Environ. 4, 347−358 (1975)

References

Atwell, J.S.: Identifying and correcting groundwater contamination at a land disposal site. Proc. 4th Natl. Congr. Waste Management.Technol. Res. Energy Rec. U.S. EPA SW-8p, 278–298 (1976)

Aubert, M.: Le problème du mercure en Méditerranée. Rev. Int. Oceanogr. Méd. 37/38, 215–231 (1975)

Auer, M.T., Canale, R.P., Freedman, P L.: The limnology of Grand Traverse Bay, Lake Michigan. U.S. Natl. Tech. Inf. Serv. PB 25614 (1976)

Ault, W.U., Senechal, G., Erlebach, W.E.: Composition as a natural tracer of lead in the environment. Environ. Sci. Technol. 4, 305–317 (1970)

Aurand, K., Behrens, H.: Ein einfaches Gerät zur Sammlung und Entnahme von Sedimentproben aus Oberflächengewässern. Arch. Hydrobiol. 62, 104–110 (1966)

Axelsson, V., Håkanson, L.: Sambendet mellan kvicksilverförekomst och sedimentologisk miljö i Ekoln. Del 1: målsättning och analysmethodik. Univ. Uppsala, Ungi Rapport 11 (1971)

Axelsson, V., Håkanson, L.: Sambandet mellan kvicksilverförekomst och sedimentologisk miljö i Ekoln. Del 2: Sedimentens egenskaper och kvicksilverinnehåll. Univ. Uppsala, Ungi Rapport 14 (1972)

Axelsson, V., Håkanson, L.: Kvicksilver i Södra Vätterns, Ekolns och Björkens sediment. Univ. Uppsala, Ungi Rapport 25 (1973)

Axelsson, V., Håkanson, L.: Sambandet mellan kvicksilverförekomst och sedimentologisk miljö i Ekoln. Del 3: Transport och deposition av kvicksilver. Univ. Uppsala, Ungi Rapport 35 (1975)

Axtmann, R.C.: Environmental impact of a geothermal power plant. Science 187, 795–802 (1975)

Ayling, G.M.: Uptake of cadmium, zinc, copper, lead, and chromium in the Pacific oyster, Crassostrea gigas, grown in the Tamar River,Tasmania. Water Res. 8, 729–738 (1974)

Azumi, T., Yoneda, A.: Environmental analysis. I. Content of heavy metals in the sediment of rivers near Himeji City. Himeji Kogyo Daigaku Kenkyu 28A, 110–114 (1975)

Baas-Becking, L.G.M.: Geology and microbiology. New Zealand Dept. Sci. Ind. Res. Inf. Ser. 22, 48–64 (1959)

Baas-Becking, L.G.M., Moore, D.: Biogenic sulfides. Econ. Geol. 56, 259–272 (1961)

Baas-Becking, L.G.M., Kaplan, I.R., Moore,D.: Limits of the natural environment in terms of pH and oxidation-reduction potentials. J. Geol. 3, 243–286 (1960)

Baccini, P.: Untersuchungen über den Schwermetallhaushalt der Seen. Schweiz. Z. Hydrol. 38, 121–158 (1976)

Baccini, P., Roberts, P.V.: Die Belastung der Gewässer durch Metalle. Beil. Forsch. Tech. Neue Zürcher Z. 18, 57–58 (1976)

Bache, B.W.: Aluminium and iron phosphate studies relating to soils. J. Soil. Sci. 14, 113–123 (1963)

Bache, C.A., Gutemann, W.H., Lisk, D.J : Residues of total mercury and methylmercuric salts in lake trout as a function of age. Science 172, 951–952 (1971)

Baier, R.W., Healy, M.L.: Partitioning and transport of lead in Lake Washington. J Environ. Qual. 6, 291–296 (1977)

Bailey, S.M., Helz, G.R., Harris, R.L.: Investigation of the transport of metals and orthophosphate away from a sewage treatment plant outfall. Environ. Lett. 10, 159–169 (1975)

Bails, J.D.: Mercury in fish in the Great Lakes. In: Environmental Mercury Contamination. Hartung/Dinman (eds). Ann. Arbor Sci. Fubl. 1972 pp. 31–37

Baker, C.W.: Mercury in surface waters of seas around the United Kingdom. Nature (London) 270, 230–232 (1977)

Baker, W.E.: The role of humic acids from Tasmanian podzolic soils in mineral degradation and metal mobilization. Geochim. Cosmochim. Acta 37, 269–281 (1973)

Baker-Blocker, A., Callender, E., Josephson, P.D.: Trace element and organic carbon content of surface sediment from Grand Traverse Bay, Lake Michigan. Bull. Geol. Soc. Am. 86, 1358–1362 (1975)

Bakir, F., Damluji, S.F., Amin-Zaki, L., Murtadha, M., Khalidi, A., Al-Rawi, N.Y., Tikriti, S., Dhahil, H.I., Clarkson, T.W., Smith, J.C., Doherty, R.A.: Methylmercury poisoning in Iraq. Science 181, 230–241 (1973)

Balke,K.D., Kussmaul, H., Siebert, G.: Chemische und thermische Kontamination des Grundwassers durch Industriewässer. Z. Dtsch. Geol. Ges. 124, 447–460 (1973)

Banat, K., Förstner, U., Müller, G.: Schwermetalle in den Sedimenten des Rheins. Umsch. Wiss. Tech. 72, 192–193 (1972a)

Banat, K., Förstner, U., Müller, G.: Schwermetalle in Sedimenten von Donau, Rhein, Ems, Weser und Elbe im Bereich der Bundesrepublik Deutschland. Naturwissenschaften 12, 525–528 (1972b)

Banat, K., Förstner, U., Müller, G.: Schwermetall-Anreicherungen in den Sedimenten wichtiger Flüsse im Bereich der Bundesrepublik Deutschland, eine Bestandsaufnahme. Internal Rep. Lab. Sedimentforsch., Heidelberg (1972c)

Banat, K., Förstner, U., Müller, G.: Experimental mobilization of metals from aquatic sediments by nitrilotriacetic acid. Chem. Geol. *14*, 199–207 (1974)

Banus, M.D., Valiela, I., Teal, J.M.: Export of lead from salt marshes. Mar. Pollut. Bull. *5*, 6–9 (1974)

Banus, M.D., Valiela, I., Teal, J.M.: Lead, zinc and cadmium budgets in experimentally enriched salt marsh ecosystems. Estuarine Coastal. Mar. Sci. *3*, 421–430 (1975)

Barber, R., Ryther, J.: Organic chelators: Factors affecting primary production in the Cromwell Current upwelling. J. Exp. Mar. Biol. Ecol. *33*, 191–199 (1969)

Barendrecht, E.: Stripping voltammetry. In: Electroanalytical Chemistry. Bard, A.J. (ed.), Vol. 2. New York: Marcel Dekker 1956, pp. 53–109

Bärlocher, F., Kendrick, B.: Assimilation efficiency of Gammarus pseudolimnaeus (Amphipoda) feeding on fungal mycelium or autumn-shed leaves. Oikos *26*, 55–59 (1974)

Barnes, R.S., Schell, W.R.: Physical transport of trace metals in the Lake Washington watershed. In: Cycling and Control of Metals, Proc. of an Environmental Resources Conference (compiled by M.G. Curry and G.M. Gigliotti). National Environmental Research Center U.S. EPA, Cincinnati, Ohio 1973, pp. 45–53

Barnett, P.R., Skougstad, M.W., Miller, K.J.: Chemical characterization of a public water supply. J. Am. Water Works Assoc. *61*, 61–67 (1969)

Barnhart, B.J.: The disposal of hazardous wastes. Environ. Sci. Technol. *12*, 1132–1136 (1978)

Barsdate, R.J., Matson, W.R.: Trace metals in arctic and sub-arctic lakes with the reference to the organic complexes of metals. In: Radioecological Concentration Processes. Aberg, B., Hungate, F.P. (eds.). Oxford: Pergamon 1967, pp. 711–734

Bartelt, R.D., Förstner, U.: Schwermetalle im staugeregelten Neckar. Untersuchungen an Sedimenten, Algen und Wasserproben. Jahresber. Mitt. Oberrheinischen Geol. Ver. *59*, 247–263 (1977)

Barth, T.W.F.: Theoretical Petrology. New York: Wiley 1952

Bastin, E.S.: A hypothesis of bacterial influence in the genesis of certain sulfide ores. J. Geol. *34*, 773–792 (1926)

Batti, R., Magnaval, R., Lanzola, E.: Methylmercury in river sediments. Chemosphere *1*, 13–14 (1975)

Baturin, G.N., Kochenov, A.V., Serin, Y.M.: Uranium concentrations in recent ocean sediments in zones of rising current. Geochem. Int. *8*, 281–286 (1971)

Baumann, A., Best, G., Kaufmann, R.: Hohe Schwermetall-Gehalte in Hochflut-Sedimenten der Oker (Niedersachsen). Dtsch. Gewässerkd. Mitt. *21*, 113–117 (1977)

Baumgartner, D.J., Schults, D.W., Ingle, S.E., Specht, D.T.: Interchange of nutrients and metals between sediments and water during dredged material disposal in coastal waters. In: Proc. 2nd U.S.-Japan Experts Meeting "Management of Bottom Sediments Containing Toxic Substances". Corvallis, Or.: U.S. EPA 1977, pp. 229–245

Bayer, E.: Struktur und Spezifität organischer Komplexbildner. Angew. Chem. *76* (2), 76–83 (1964)

Beals, H.L.: Manganese-iron concentrations in Nova Scotia lakes. Marit. Sediments *2*, 70–72 (1966)

Beamish, R.J.: Long-term acidification of a lake and resulting effects on fishes. Ambio *4*, 98–102 (1975)

Beamish, R.J.: Acidification of lakes in Canada by acid precipitation and the resulting effects on fishes. Water Air Soil Pollut. *6*, 501–514 (1976)

Beamish, R.J., Harvey, H.H.: Acidification of the La Cloche Mountain lakes, Ontario and resulting fish mortalities. J. Fish. Res. Board. Can. *29*, 1131–1143 (1972)

Beavington, F.: Pollution by heavy metals of rivulet and harbour water in Wollongong. Search *6*, 390–391 (1975)

Beck, J.V.: The role of bacteria in copper mining operations. Biotech. Bioeng. *9*, 487–497 (1967)

Beck, K.C.: Sediment-water interactions in some Georgia rivers and estuaries. Compl. Rept. OWRR-Project B-033-GA. Environ. Resour. Cent., Georgia Inst. Technol. (1972)

Beck, K.C., Reuter, J.H., Perdue, E.M.: Organic and inorganic geochemistry of some coastal plain rivers of the southeastern United States. Geochim. Cosmochim. Acta *38*, 341–364 (1974)

Becker, R.: Geochemische Untersuchung der Sedimente des Flusses Blies, Saarland. Naturwissenschaften *63*, 144 (1976)

Beckert, W.F., Moghissi, A.A., Au, F.H.F., Bretthauer, E.W., McFarlane, J.C.: Formation of methylmercury in a terrestrial environment. Nature (London) *249*, 674–675 (1974)

Bellinger, E.G., Benham, B.R.: The levels of metals in dock-yard sediments with particular reference to the contributions from ship-bottom paints. Environ. Pollut. *15*, 71–81 (1978)

Bender, J.A.: Trace metal levels in beach dipterans and amphipods. Bull. Environ. Contam. Toxicol. *14*, 187–192 (1975)

Bender, M.L., Gagner, C.L.: Dissolved copper, nickel and cadmium in the Sargasso Sea. J. Mar. Res. *34*, 327–339 (1976)

Bender, M., Broecker, W., Gornitz, V., Middle, U., Kay, R., Sun, S.S., Biscaye, P.: Geochemistry of three cores from the east Pacific rise. Earth Planet. Sci. Lett. *12*, 425–433 (1971)

Benedek, P., Literáthy, P., Puskas, M.: Wasserqualitätsprobleme des ungarischen Donauabschnitts. GWF Wasser Abwasser *113*, 310 (1972)

Beneš, P., Steinnes, E.: In situ dialysis for the trace determination of the state of trace elements in natural waters. Water Res. *8*, 947–953 (1974)

Benger, H., Kempf, Th.: Vorkommen von Schwermetallionen im Trinkwasser. Bundesgesundheitsblatt *15*, 17–20 (1972)

Bengtson, L., Fleischer, S.: Sediment investigations in the Lakes Trummen and Hinnasjön 1968–1970. Vatten *22*, 73–94 (1971)

Bennett, H., Hawley, W.G.: Methods of Silicate Analysis. London, New York: Academic Press 1965

Benninger, L.K., Lewis, D.M., Turekian, K.K.: The use of natural Pb-210 as a heavy metal tracer in the river-estuarine system. In: Marine Chemistry in the Coastal Environment. Church, T.M. (ed.). Amer. Chem. Soc. Symp. Ser. *18*, 202–210 (1975)

Benon, P., Blanc, F., Bourgade, B., David, P., Kantin, R., Leveau, M., Romano, J.-C., Sautriot, D.: Distribution of some heavy metals in the Gulf of Fos. Mar. Pollut. Bull. *9*, 71–75 (1978)

Ben-Yaakov, S.: pH buffering of pore water of recent anoxic marine sediments. Limnol. Oceanogr. *18*, 86–94 (1973)

Berger, W.: The geochemical role of organisms. Tschermaks Mineral. Petrogr. Mitt. *2*, 136–140 (1950)

Berggren, B., Odén, S.: Analyseresultat rorande Fungmetaller och Klorerade Kolväten i Rötslam fra Svenska Renningsverk 1968–1971. Inst. f. Markventeskap Lantbrukshögskolan, Uppsala/Schweden, 1972 (ct. A.L. Page 1974)

Bergmann, A.: How Meggen purifies mine water and recovers marketable Zn precipitate. World Min. 48–51 (1971)

Bergmann, W.: Geochemistry of lipids. In: Organic Geochemistry. Breger, I.A. (ed.). New York: Pergamon Press 1963

Berner, R.A.: Stability fields of iron minerals in anaerobic marine sediments. Geochim. Cosmochim. Acta *28*, 1497–1503 (1964)

Berner, R.: Iron – abundance in natural waters. In: Handbook of Geochemistry. Wedepohl, K.H. (ed.). Berlin, Heidelberg, New York: Springer 1970, pp. 26–I/1–2

Berner, R.A.: Sedimentary pyrite formation. Am. J. Sci. *268*, 1–23 (1970a)

Berner, R.A.: Pleistocene sea levels possibly indicated by buried black sediments in the Black Sea. Nature (London) *227*, 700 (1970b)

Berner, R.A.: Principles of Chemical Sedimentology. New York: McGraw Hill 1971

Bernhardt, H., Wilhelms, A.: Einfluß chelatbildender Substanzen auf die Flockung mit Aluminiumsalzen in der Trinkwasseraufbereitung. Vom Wasser *38*, 217–231 (1971)

Bernhard, M., Goldberg, E.D., Piro, A.: Zinc in seawater – an overview 1975. In: The Nature of Seawater. Goldberg, E.D. (ed.). Berlin: Dahlem Konferenzen 1975, pp. 43–68

Berrow, M.L., Webber, J.: Trace elements in sewage sludges. J. Sci. Food. Agric. *23*, 93–100 (1972)

Berth, P., Jakobi, G., Schmadel, E., Schwuger, M.J., Krauch, C.H.: The replacement of phosphates in detergents – possibilities and limits. Angew. Chem. Int. Ed. (Engl.) *14*, 94–102 (1975)

Bertine, K.K., Goldberg, E.D.: Fossil fuel combustion and the major sedimentary cycle. Science *178*, 233–235 (1971)

Bertine, K.K., Goldberg, E.D.: History of heavy metal pollution in southern California coastal zone – reprise. Environ. Sci. Technol. *11*, 297–299 (1977)

Bertine, K.K., Mendeck, M.F.: Industrialization of New Haven, Conn., as recorded in reservoir sediments. Environ. Sci. Technol. *12*, 201–207 (1978)

Bertine, K.K., Chan, L.H., Turekian, K.K.: Uranium determinations in deep sea sediments and natural waters using fission tracks. Geochim. Cosmochim. Acta *34*, 641–648 (1970)

Bertrand, D.: Interactions entre éléments minéraux et microorganismes du sol. Rec. Ecol. Biol. Soc. *3*, 349–396 (1972)

Bevan, C.D., Harbison, S.A., Nelson, L.A., Lakey, J.R.A.: A trace element study in the Thames estuary. Proc. Int. Symp. Impacts Nucl. Releases Aquat. Environ. IAEA, Vienna, 93–106 (1975)

Beveridge, T.J.: The interaction of metals in aqueous solution with bacterial cell walls from Bacillus subtilis. In: Environmental Biogeochemistry and Geomicrobiology. Krumbein, W.E. (ed.). Ann. Arbor Sci. Publ. 1978, Vol. 3, pp. 975–985

Bewers, J.M., MacAulay, I.D., Sundby, B.: Trace metals in the waters of the Gulf of St. Lawrence. Can. J. Earth Sci. *11*, 939–950 (1974)

Bibo, J.: Schwermetalluntersuchungen an Wasser, Schwebstoffen, Aufwuchs und Cladophora rivularis der Elsenz. Diplom-Arbeit, Univ. Heidelberg, 111 p. (1977)

Biggs, R.B., Flemer, D.A.: The flux of particulate carbon in an estuary. Mar. Biol. *12*, 11–17 (1972)

Biggs, R.B., Miller, J.C., Otley, M.J.: Trace metals in several Delaware watersheds – a progress report. Newark: Water Resources Center, Univ. Delaware 1972

Bilinski, H., Schindler, P., Stumm, W., Zobrist, J.: Kupfer und Blei in natürlichen Gewässern. Vom Wasser *43*, 107–116 (1974)

Bilinski, H., Huston, R., Stumm, W.: Determination of the stability constants of some hydroxo and carbonato-complexes of Pb (II), Cu (II), Cd (II) and Zn (II) in dilute solutions by anodic stripping voltammetry and differential pulse polarography. Anal. Chim. Acta *84*, 157–164 (1976)

Billen, G., Joiris, C., Wollast, R.: A bacterial methylmercury mineralizing activity in river sediments. Water Res. *8*, 219–225 (1974)

Billings, C.E., Matson, W.R.: Mercury emissions from coal combustion. Science *176*, 1232–1233 (1972)

Bingham, F.T., Page, A.L.: Cadmium accumulation by economic crops. Proc. Int. Conf. Heavy Met. Environ. Toronto 1975, *II*/1, 433–442 (1977)

Bingham, F.T., Page, A.L., Mahler, R.J., Ganje, T.J.: Growth and cadmium accumulation of plants grown on a soil treated with a cadmium-enriched sewage sludge. J. Environ. Qual. *4*, 207–211 (1975)

Bingham, F.T., Page, A.L., Mahler, J., Ganje, T.J.: Cadmium availability to rice in sludge-amended soil under "flood" und "non-flood" culture. Soil Sci. Soc. Am. Proc. *40*, 715–719 (1976)

Birch, H.F.: Nitrification in soils after different periods of dryness. Plant Soil *12*, 81–96 (1960)

Bishop, J.N., Neary, B.P.: The form of mercury in freshwater fish. Proc. Int. Conf. Transp. Persist. Chem. Aquatic Ecosyst., Ottawa, III-25–III-29 (1974)

Bisogni, J.J., Jr., Lawrence, A.W.: Kinetics of mercury methylation in aerobic and anaerobic aquatic environments. J.W.P.C.F. *47*, 135–153 (1975)

Bittell, J.E., Miller, R.J.: Lead, Cadmium, and calcium selectivity on a montmorillonite, illite, and koalinite. J. Environ. Qual. *3*, 250–253 (1974)

Black, W.A.P., Mitchell, R.L.: Trace elements in the common brown algae and in sea water. J. Mar. Biol. Assoc. U.K. *30*, 575–584 (1952)

Blackburn, T.R.: Mercury in sediments of the Horwer Bucht, Lake Lucerne, and tributary streams, Switzerland. Schweiz. Z. Hydrol. *35*, 201–205 (1973)

Blackburn, T.R., Cornwell, J.C., Fogg, T.R.: Mercury and zinc in the sediments of Seneca Lake and the Seneca River. Abstr. 20th Conf. Great Lakes Research. Int. Assoc. Great Lakes Res. (1977)

Blakeslee, P.A.: Monitoring considerations for municipal wastewater effluent and sludge application to the land. U.S. Environ. Protection Agency U.S. Dept. Agric., Univ. Workshop, Champaign, Urbana 1973. Cit. A.L. Page (1974)

Block, W., Schneider, H.: Zur Frage der Belastbarkeit des Rheins mit radioaktiven Nukliden. I. Mitt.: Sorption von Radionukliden durch den organischen Anteil der Schwebstoffe. GWF-Wasser/Abwasser *108*, 1249–1257 (1967)

Bloom, H., Ayling, G.M.: Heavy metals in the Derwent Estuary. Environ. Geol. *2*, 3–22 (1977)

Bloomfield, C., Pruden, G.: The effects of aerobic and anaerobic incubation on the extractibilities of heavy metals in digested sewage sludge. Environ. Pollut. *8*, 217–232 (1975)

Bloxam, T.W., Aurora, S.N., Leach, L., Rees, T.R.: Heavy metals in some river and bay sediments near Swansea. Nature (Phys. Sci.) *239*, 158–159 (1972)

Bohn, A.: Arsenic in marine organisms from West Greenland. Mar. Pollut. Bull. *6*, 87–89 (1975)

Bolter, E., Jennett, J.C., Wixson, B.G.: Geochemical impact of lead mining waste waters on streams in southeastern Missouri. Trans. 27th Purdue Ind. Waste Conf. 679–694 (1972)

Bolter, E., Wixson, B.G., Butherus, D.L., Jennett, J.C.: Distribution of heavy metals in soils near an active lead smelter. Issues Confronting Min. Ind., Ann. Meet. Sect. AIME 47th. Fedkenheuer, P.J. (ed.). Dept. Conf. Cont. Educ. Ext. Univ. Minn.. Minneapolis, Minn. 1974, pp. 73–76

Bolter, E., Butz, T., Arseneau, J.F.: Mobilization of heavy metals by organic acids in the soils of a lead mining and smelting district. In: Trace Substances in Environmental Health. Hemphill, D.D. (ed.), Vol. IX. Univ. of Missouri, Columbia 1975, pp. 107–112

Bolton, H.C.: Action of organic acids on minerals. Mineral. Mag. *19*, 1–8 (1880)

Bolton, H.C.: Application of organic acids to the examination of minerals. Proc. Am. Assoc. Adv. Sci. *31*, 3–7 (1882)

Bombace, M.A., Cigna-Rossi, L., Clemente, G.F., Zucca-Labellarte, G., Allerini, M., Lanzola, E.: Recherche écologique sur le mercure dans la région du Monte Amiata (Toscane, Italie). Proc. Int. Symp. "Problems of the Contamination of Man and his Environment by Mercury and Cadmium". Luxembourg 1974, pp. 47–67

Bonatti, E., Fisher, D.E., Joensuu, O., Rydell, H.S.: Postdepositional mobility of some transition elements, phosphorous, uranium and thorium in deep sea sediments. Geochim. Cosmochim. Acta 35, 189–201 (1971)

Bondam, J., Asmund, G., Schrøder, S.: Miljøkontrol ved Marmorilik in Nordvest Grønland. In: Cadmium Forskning i Danmark. Rapport Danmarks Tekniske Højskole, 1976, pp. 21–32

Bondarenko, G.P.: Stability of soluble coordination compounds of copper with humic and fulvic acids. Geochem. Int. 702–711 (1972)

Bondietti, E.A., Sweeton, F.H., Tamura, T., Perhac, R.M., Hulett, L.D., Kneip, T.J.: In: Proc. 1st Annu. NSF Trace Contaminants Conf., Oak Ridge Nat. Lab. Oak Ridge, Tenn. 1973

Bongers, L.H., Khattak, M.N.: Sand and gravel overlay for control of mercury in sediments. Water Pollution Control Res. Ser. U.S. Env. Protection Agency, Washington 1972. Cit. Wolery and Walters (1974)

Bopp, F. III, Biggs, R.B.: Trace metal environments near shell banks in Delaware Bay. Delaware Bay Report Series 3, Rep. No. 2. Trace Met. Geochem. Estuarine Sediment, 23–69 (1973)

Bopp, F. III, Lepple, F.K., Biggs, R.B.: Trace metal baseline studies on the Murderkill and St. Jones Rivers, Delaware Coastal Plain. Delaware Bay Report Series 3, Rep. No. 3. Trace Met. Geochem. Estuarine Sediment, 71–96 (1973)

Borchert, H., Krejci-Graf, K.: Spurenmetalle in Sedimenten und ihren Derivaten. Bergbauwissenschaft 6, 205–215 (1959)

Boström, K., Fisher, D.E.: Distribution of mercury in East Pacific sediments. Geochim. Cosmochim. Acta 33, 743–745 (1969)

Bothner, M.H., Carpenter, R.: Sorption-desorption reactions of mercury with suspended matter in the Columbia River. IAEA-SM-158/5, 73–87 (1974)

Bothner, M.H., Robertson, D.E.: Mercury contamination of sea water samples stored in polyethylene containers. Anal. Chem. 47, 592–595 (1975)

Bourg, A.C.M., Filby, R.H.: Adsorption isotherms for the uptake of Zn^{2+} by clay minerals in a freshwater medium. Proc. Int. Conf. Transp. Persist. Chem. Aquatic Ecosyst., Ottawa, Canada, II–1–18 (1974)

Bourget, E., Cossa, D.: Mercury content of mussels from the St. Lawrence Estuary and northwestern Gulf of St. Lawrence, Canada. Mar. Pollut. Bull. 7, 237–240 (1976)

Bowden, K.F.: Circulation and diffusion. In: Estuaries. Lauff, G.H. (ed.). AAAS-Publ. Washington D.C. 83, 15–36 (1967)

Bowen, H.J.M.: Trace elements in biochemistry, the biochemistry of the elements Chapter 12, Bowen, H.J.M.: Trace Elements in Biochemistry. London, New York: Academic Press 1966, pp. 173–210

Bowen, H.J.M.: Residence times of heavy metals in the environment. Proc. Symp. Heavy Met. Environ., Toronto 1975, I/1–9 (1977)

Bowen, W.S., Steele, K.F.: Relation of lead mineralization and bottom sediment composition of streams, Ponca-Boxley District, Arkansas. Proc. Arkansas Acad. Sci. 29, 24–26 (1975)

Bower, P.M., Simpson, H.J., Williams, S.C., Li, Y.H.: Heavy metals in the sediments of Foundry Cove, Cold Spring, New York. Environ. Sci. Technol. 12, 683–687 (1978)

Boyd, B., Saucier, R.T., Keeley, J.W., Montgomery, R.L., Brown, R.D., Mathis, D.B., Guice, C.J.: Disposal of dredge spoil, problem identification and assessment and research program development. Techn. Rept. H-72-8, U.S. Army Engineer Waterways Experiment Station, Vicksburg 1972

Boyden, C.R.: Distribution of some trace metals in Poole Harbour, Dorset. Mar. Pollut. Bull. 6, 180–187 (1975)

Boyden, C.R.: Effect of size upon metal content of shellfish. J. Mar. Biol. Assoc. U.K. 57, 675–714 (1977)

Boyden, C.R., Romeril, M.G.: A trace metal problem in pond oyster culture. Mar. Pollut. Bull. 5, 74–78 (1974)

Boyle, E.A.: Trace element geochemistry of the Amazon and its tributaries. Abstr. 1978 Spring Meeting AGU. EOS 59, 276 (1978)

Boyle, E.A., Edmond, J.M.: Copper in surface waters south of New Zealand. Nature (London) 253, 107–109 (1975)

Boyle, E.A., Collier, R., Dengler, A.T., Edmond, J.M., Ng, A.C., Stallard, R.F.: On the chemical mass-balance in estuaries. Geochim. Cosmochim. Acta 38, 1719–1738 (1974)

Boyle, E.A., Sclater, F.R., Edmond, J.M.: Copper profiles from Pacific Geosecs stations. Abstr. Ann. Fall Meet. Am. Geophys. Union 0–77, EOS *57*, 938 (1976a)

Boyle, E.A., Sclater, F.R., Edmond, J.M.: On the marine chemistry of cadmium. Nature (London) *263*, 42–44 (1976b)

Boyle, E.A., Sclater, F.R., Edmond, J.M.: The distribution of dissolved copper in the Pacific. Earth Planet. Sci. Lett. *37*, 38–54 (1977)

Boyle, R.W., Illsley, C.I., Green, R.N.: Geochemical investigation of the heavy metal content of stream and spring waters in the Keno-Hill-Galena-Hill area, Yukon Territory. Bull. Geol. Surv. Can. *32* (1955)

Boyle, R.W., Jonasson, I.R.: The geochemistry of arsenic and its use as an indicator element in geochemical prospecting. J. Geochem. Explor. *2*, 251–296 (1973)

Bradford, G.R., Bair, F.L., Hunsker, V.: Trace and major element content of 170 High Sierra Lakes in California. Limnol. Oceanogr. *13*, 526–529 (1968)

Bradford, W.L.: Distribution and movement of zinc and other heavy metals in south San Francisco Bay, California. U.S. Natl. Tech. Inf. Serv. Rep. PB-251111, 64 (1976)

Bradford, W.L.: Urban stormwater pollutant loadings: a statistical summary through 1972. J.W.P.C.F. *49*, 613–622 (1977)

Bradshaw, A.D.: Evolution of metal tolerance and its significance. Abstr. Int. Conf. Heavy Met., Toronto, C-311 (1975)

Brainina, K.Z.: Stripping Voltammetry in Chemical Analysis. Toronto: John Wiley and Sons 1974

Braman, R.S., Foreback, C.C.: Methylated forms of arsenic in the environment. Science *182*, 1247–1249 (1973)

Brannon, J.M., Engler, R.M., Rose, J.R., Hunt, P.G., Smith, I.: Distribution of toxic heavy metals in marine and freshwater sediments. In: Proc. Conf. Dredging and its Environmental Effects, Mobile/Ala. Krenkel, P.A., Harrison, J., Burdick, J.C. (eds.). New York: Amer. Soc. Civil Eng. 1976a, pp. 455–495

Brannon, J.M., Rose, J.R., Engler, R.M., Smith, I.: The distribution of heavy metals in sediment fractions from Mobile Bay, Alabama. In: Chemistry of Marine Sediments. Yen, T.F. (ed.). Ann Arbor Sci. Publ. 1976b, pp. 125–149

Brannon, J.M., Plumb, R.H., Jr., Smith, I.: Long-term release of contaminants from dredged material. Dredged Material Research Program U.S. Army Engineer Waterways Experiment Station, Vicksburg, Miss., Final Report D-78-49 (1978)

Brewer, P.G.: Minor elements in seawater. Chapter 7. In: Chemical Oceanography. Riley, J.P., Skirrow, G. (eds.). London: Academic Press 1975, pp. 415–496

Brewer, P.G., Spencer, D.W.: Minor element models in coastal waters. In: Marine Chemistry in the Coastal Environment. Church, T.M. (ed.). Am. Chem. Soc. Symp. Ser. *18*, 80–96 (1975)

Brewer, P.G., Spencer, D.W., Robertson, D.E.: Trace element profiles from the Geosecs II test station in the Sargasso Sea. Earth Planet. Sci. Lett. *16*, 111–116 (1972)

Bricker, O.: Some stability relationship in the system MnO_2-H_2O at $25°$ and one atmosphere total pressure. Am. Mineral. *50*, 1296–1354 (1965)

Bricker, O.P., Troup, B.N.: Sediment-water exchange in Chesapeake Bay. In: Estuarine Research. Cronin, L.E. (ed.). New York: Academic Press 1975, Vol. 1, pp. 3–27

Brinckman, F.E., Iverson, W.P.: Chemical and bacterial cycling of heavy metals in the estuarine system. In: Marine Chemistry in the Coastal Environment. Church, T.M. (ed.). Am. Chem. Soc. Symp. Ser. *18*, 319–337 (1975)

Brinckman, F.E., Jewett, K.L., Blair, W.R., Iverson, W.P., Huey, C.: Mercury distribution in the Chesapeake Bay. In: Heavy Metals in the Aquatic Environment. Krenkel, P.A. (ed.). Oxford: Pergamon Press 1975, pp. 251–252

Brinkman, F.J.: Inventory of trace elements in groundwater of the Netherlands. Geol. Mijnbouw *53*, 157–161 (1974)

Brkovic-Popovic, M.: Effects of heavy metals on survival and respiration rate of tubificid worms: II. Effects on respiration rate. Environ. Pollut. *13*, 93–101 (1977)

Broadbent, F.E., Bradford, G.R.: Cation-exchange groupings in the soil organic fraction. Soil Sci. *74*, 447–452 (1952)

Brongersma-Saunders, M.: Metals of Kupferschiefer supplied by normal sea water. Geol. Rundsch. *55*, 365–375 (1965)

Brooks, R.R., Quin, B.F.: Heavy metals in stream sediments of the Port Pegasus area of Stewart Island. N. Z. J. Sci. *14*, 25–30 (1971)

Brooks, R.R., Rumsby, M.G.: The biogeochemistry of trace element uptake by some New Zealand bivalves. Limnol. Oceanogr. *10*, 521–527 (1965)

Brooks, R.R., Rumsey, D.: Heavy metals in some New Zealand commercial sea fishes. N. Z. J. Mar. Freshwater Res. *8*, 155–166 (1974)

Brooks, R.R., Presley, B.J., Kaplan, I.R.: Trace elements in the interstitial waters of marine sediments. Geochim. Cosmochim. Acta *32*, 397–414 (1968)

Browman, M.G., Chesters, G.: Transfer of organic pollutants across the solid-water interface. In: Fate or Pollutants in the Air and Water Environments. Part. 1. Suffet, I.H. (ed.). New York: Wiley 1977, pp. 49–105

Brown, B.E.: Effects of mine drainage on the River Hayle, Cornwall. A) Factors affecting concentrations of copper, zinc and iron in water, sediments and dominant invertebrate fauna. Hydrobiologia *52*, 221–233 (1977)

Brown, G.: The X-ray Identification and Crystal Structures of Clay Minerals. Mineral. Soc. (Clay Mineral Group) London, 1961

Brown, J.R., Chow, L.Y.: Heavy metal concentrations in Ontario fish. Bull. Environ. Cont. Toxicol. *17*, 190–195 (1977)

Brown, M.J.F.: A development consequence – disposal of mining waste on Bougainville, Papua, New Guinea. Geoforum *18*, 19–27 (1974)

Brown, R.P., Smith, D.D.: Marine disposal of solid wastes. An interim summary. Mar. Pollut. Bull. *1*, 12–16 (1969)

Brown, V.M.: Aspects of heavy metal toxicity in freshwater. In: Toxicity to Biota of Metal Forms in Natural Waters. Andrew, R.W., Hodson, P.V., Konasewich, D.E. (eds.). Int Joint Comm., Windsor, Ontario, 1976, pp. 59–75

Brown, V.M., Shaw, T.L., Shurben, D.G.: Aspects of water quality and toxicity of copper to rainbow trout. Water Res. *8*, 797–803 (1974)

Brügmann, L.: Zur Verteilung einiger Schwermetalle in der Ostsee – eine Übersicht. Acta Hydrochim. Hydrobiol. *5*, 3–21 (1977)

Bruland, K.W., Bertine, K., Koide, M., Goldberg, E.D.: History of metal pollution in Southern California coastal zone. Environ. Sci. Technol. *8*, 425–432 (1974)

Bruland, K.W., Knauer, G.A., Martin, J.H.: Zinc in north-east Pacific water. Nature (London) *271*, 741–743 (1978)

Brumsack, H.J.: Potential metal pollution in grass and soil samples around brickworks. Environ. Geol. *2*, 33–41 (1977)

Brunskill, G.J., Povoledo, D., Graham, B.W., Stainton, M.P.: Chemistry of surface sediments of sixteen lakes in the experimental lake area, northwestern Ontario. J. Fish Res. Board. Can. *28*, 277–294 (1971)

Bryan, G.W.: Zinc regulation in the freshwater crayfish (including some comparative copper analysis). J. Exp. Biol. *46*, 281–296 (1967)

Bryan, G.W.: The absorption of zinc and other metals by the brown seaweed Laminaria digitata. J. Mar. Biol. Ass. U.K. *49*, 225–243 (1969)

Bryan, G.W.: The effects of heavy metals (other than mercury) on marine and estuarine organisms. Proc. R. Soc. Lond. *B177*, 389–410 (1971)

Bryan, G.W.: The occurrence and seasonal variation of trace metals in the scallops Pecten maximus (L.) and Chlamys opercularis (L.). J. Mar. Biol. Assoc. U.K. *53*, 145–166 (1973)

Bryan, G.W.: Heavy metal contamination in the sea. In: Marine Pollution. Johnston, R. (ed.). London: Academic Press 1976, pp. 185–302

Bryan, G.W., Hummerstone, L.G.: Adaptation of the polychaete Nereis diversicolor to estuarine sediments containing high concentrations of heavy metals. I. General observations and adaptation to copper. J. Mar. Biol. Assoc. U.K. *52*, 845–863 (1971)

Bryan, G.W., Hummerstone, L.G.: Brown seaweed as an indicator of heavy metals in estuaries in south-west England. J. Mar. Biol. Assoc. U.K. *53*, 705–720 (1973a)

Bryan, G.W., Hummerstone, L.G.: Adaptation of the polychaete Nereis diversicolor to estuarine sediments containing high concentrations of zinc and cadmium. J. Mar. Biol. Assoc. U.K. *53*, 839–857 (1973b)

Bryan, G.W., Hummerstone, L.G.: Indicators of heavy-metal contamination in the Looe Estuary (Cornwall) with particular regard to silver and lead. J. Mar. Biol. Assoc. U.K. *57*, 75–91 (1977)

Bryan, G.W., Hummerstone, L.G.: Heavy metals in the burrowing bivalve Scrobicularia plana from contaminated and uncontaminated estuaries. J. Mar. Biol. Assoc. U.K. *58*, 401–419 (1978)

Bryan, G.W., Uysal, H.: Heavy metals in the burrowing bivalve Scrobicularia plana from the Tamar estuary in relation to environmental levels. J. Mar. Biol. Assoc. U.K. *58*, 89–108 (1978)

Bucksteeg, W.: Beseitigung anorganischer Schmutzstoffe – Forderungen und Erfüllung. GWF-Wasser/Abwasser *108*, 962–965 (1967)

Buffa, L.: Review of Environmental Control of Mercury in Japan. Canada Environmental Protection Service. Econ. Technol. Rev. Rep. EPS 3-WP-76-7, 81 (1976)

Bukenberger, U., Lodemann, C.K.W., Loeschke, J.: Die Verteilung der Schwermetalle in ober- und unterirdischen Wässern sowie in den Böden des Neckartales oberhalb Tübingen. Oberrheinische Geol. Abh. *21*, 43–62 (1972)
Bull, R.J.: Toxicological research, its application to the setting of drinking water standards. Proc. 16th Water Qual. Conf., Trace Metals in Water Supplies: Occurrence, Significance and Control. Univ. Ill., Coll. Engin., 49–64 (1974)
Burns, N.M., Ross, C.: Oxygen-nutrient relationship within the central basin of Lake Erie. In: "Project Hypo". Burns, N.M., Ross, C.(eds.). Can. Cent. Inland Waters Pap. *6*, 85–119 (1972)
Burrows, K.C., Hulbert, M.H.: Release of heavy metals from sediments, preliminary comparison of laboratory and field studies. In: Marine Chemistry in the Coastal Environment. Church, T.M. (ed.). Am. Chem. Soc. Symp. Ser. *18*, 382–393 (1975)
Burton, J.D.: Basic properties and processes in estuarine chemistry. In: Estuarine Chemistry. Burton, J.D., Liss, P.S. (eds). London: Academic Press 1976, pp. 1–36
Burton, J.D., Leatherland, T.M.: Mercury in a coastal marine environment. Nature (London) *231*, 440–441 (1971)
Burton, J.D., Liss, P.S.: Estuarine Chemistry. London, New York, San Francisco: Academic Press 1976
Buser, W., Graf, F.: Differenzierung von Mangan (II)-Manganit und δ-MnO_2 durch Oberflächenmessung nach Brunauer-Emmet-Teller. Helv. Chim. Acta *38*, 830–842 (1955)
Butterworth, J., Lester, P., Nickless, G.: Distribution of heavy metals in the Severn Estuary. Mar. Pollut. Bull. *3*, 72–74 (1972)
Butuzova, G.Y.: K mineralogii i geokhimii sul'fidov zhelza v osadkakh Chernogo morya. (Mineralogy and geochemistry of iron sulfides in Black Sea sediments). Litol. Polezn. Iskop. *4*, 3–16 (1969)
Cable, C.C.: Optimum dredging and disposal practices in estuaries. ASCE J. Hydraul. Div. *95*, 103–115 (1969)
Cahill, R.A., Kuhn, J.K., Dreher, G.B.: Distribution of major, minor and trace elements in sediments of northern Lake Michigan. Abstr. 20th Conf. Great Lakes Res. Int. Assoc. Great Lakes Res. (1977)
Caines, L.A., Holden, A.V.: Stream pollution by an organomercury compound. Bull. Environ. Contam. Toxicol. *16*, 383–391 (1976)
Calabrese, A., Nelson, D.A.: Inhibition of embryonic development of the hard clam, Mercenaria mercenaria, by heavy metals. Bull. Environ. Contam. Toxicol. *11*, 92–97 (1974)
Callender, E., Bowser, C.J.: Freshwater ferromanganese deposits. In: Handbook of Strata-Bound and Stratiform Ore Deposits. Wolf, K.H. (ed.), Vol. 7. New York: Elsevier 1976, pp. 343–394
Calmon, C.: Notes and comments. J. A.W.W.A. *65*, 568 (1973)
Calvert, S.E., Morris, R.J.: Geochemical studies of organic-rich sediments from the Namibian Shelf. II. Metal-organic associations. Deep-Sea Res.: Deacon Volume (1978)
Calvert, S.E., Price, N.B.: Recent sediments of the south-west African Shelf. ICSU/SCOR Working Party, Cambridge, 1971. The Geology of the East Atlantic Continental Margin. Delany, F.M. (ed.). Inst. Geol. Sci. Rep. *70/16*, 171–185 (1971)
Calvert, S.E., Price, N.B.: Geochemical variation in ferromanganese nodules and associated sediments from the Pacific Ocean. Mar. Chem. *5*, 43–74 (1977)
Cambray, R.S., Jefferies, D.F., Topping, G.: An estimate of the input of atmospheric trace elements into the North Sea and the Clyde Sea (1972–1973). U.K. At. Energy Auth. Harwell Rep. *30* (1975)
Cameron, E.M.: Geochemical methods of exploration for massive sulphide mineralization in the Canadian Shield. In: Geochemical Exploration-Proc. 5th Int. Geoch. Expl. Symp. Vancouver. Elliott, I.L., Fletcher, W.K. (eds.). 1974, pp. 21–49
Cameron, E.M., Ballantyne, S.B.: Experimental hydrogeochemical surveys of the High Lake and Hackett River areas, Northwest Territories.Geol. Surv. Can. Pap. *75/29*, 19 p. (1975)
Canney, F.C., Nowlan, G.A.: Determination of ammonium citrate-soluble cobalt in soils and sediments. Econ. Geol. *59*, 1361–1367 (1964)
Capuzzo, J.M., Anderson, F.E.: The use of modern chromium accumulations to determine estuarine sedimentation rates. Mar. Geol. *14*, 225–235 (1973)
Carlson, S., Hässelbarth, U.: Wasseraufbereitung für die Hämodialyse. Bundesgesundheitsblatt *14*, 256–262 (1971)
Carmody, D.J., Pearce, J.B., Yasso, W.E.: Trace metals in sediments of New York Bight. Mar. Pollut. Bull. *4*, 132–135 (1973)
Carpenter, J.H., Bradford, W.L., Grant, V.: Processes affecting the composition of estuarine waters (H_2CO_3, Fe, Mn, Zn, Cu, Ni, Cr, Co and Cd). In: Estuarine Research. Cronin, L.E. (ed.), Vol. I. New York: Academic Press 1975, pp. 137–152

Carpenter, K.E.: A study of the fauna of rivers polluted by lead mining in the Aberystwyth district of Cardiganshire. Ann. Appl. Biol. *11*, 1–23 (1924)
Carr, R.A., Jones, M.M., Warner, T.B., Cheek, C.H., Russ, E.R.: Variation in time of mercury anomalies at the Mid-Atlantic-Ridge. Nature (London) *258*, 588–589 (1975)
Carroll, D.: Role of clay minerals in the transportation of iron. Geochim. Cosmochim. Acta *14*, 1–27 (1958)
Caspers, H.: Estuaries: Analysis of definitions and biological considerations. In: Estuaries. Lauff, G.H. (ed.), AAAS Publ. *83*, 6–8 (1967)
Caspers, H.: Pollution in Coastal Waters: An Interim Report on the Results of the Priority Programme from 1966–1974. A Report of the German Research Society. Boppard: Boldt–Verlag 1975
Cato, I.: Recent sedimentological and geochemical conditions and pollution problems in two marine areas in southwestern Sweden. Striae Uppsaliensis pro Geologia Quaternaria *6*, 158 p. (1977)
Chaney, R.L.: Recommendations for management of potentially toxic elements in agricultural and municipal wastes. In: Factors Involved in Land Application of Agricultural and Municipal Wastes. U.S. Dep. Agric., Beltsville, Maryland, 1974, pp. 97–120
Chao, L.L.: Selective dissolution of manganese oxides from soils and sediments with acidified hydroxylamine hydrochloride. Soil Sci. Soc. Am. Proc. *36*, 764–768 (1972)
Chau, Y.K., Lum-Shue-Chan, K.: Determination of labile and strongly bound metals in lake water. Wat. Res. *8*, 383–388 (1974)
Chau, Y.K., Saitoh, H.: Mercury in the international Great Lakes. Proc. 16th Conf. Great Lakes Res., Int. Assoc. Great Lakes Res., 221–232 (1973)
Chau, Y.K., Shiomi, M.T.: Complexing properties of nitrilotriacetic acid in the lake environment. Water Air Soil Pollut. *1*, 149–164 (1972)
Chau, Y.K., Wong, P.T.S., Silverberg, B.A., Luxon, P.L., Bengert, G.A.: Methylation of selenium in the aquatic environment. Science *192*, 1130–1131 (1976)
Chawla, V.K., Chau, Y.K.: Trace elements in Lake Erie. Proc. 12th Conf. Great Lakes Res. Int. Assoc. Great Lakes Res. 760–769 (1969)
Chen, K.Y., Lockwood, R.A.: Evaluation strategies of metal pollution in oceans. J. Environ. Eng. Div. ASCE *102*, EE2, 347–359 (1976)
Chen, K.Y., Young, T.K.J., Rohatgi, N.: Trace metals in waste-water effluents. J.W.P.C.F. *46*, 2663–2675 (1974)
Chen, K.Y., Gupta, S.K., Sycip, A.U., Lu, J.C.S., Knezevic, M., Choi, W.-W.: Research study on the effect of dispersion, settling, and resedimentation on migration of chemical constituents during openwater disposal of dredged materials. U.S. Army Corps of Engineers, Dredged Material Research Program, Vicksburg Miss., Rept. D-76-1, 221 (1976a)
Chen, K.Y., Lu, J.C.S., Sycip, A.Z.: Mobility of trace metals during open water disposal of dredged material and following resedimentation. Proc. Spec. Conf. Dredging and its Environ. Effects. Mobile, Ala. New York: ASCE 1976, pp. 435–454
Cheng, M.H., Patterson, J.W., Minear, R.A.: Heavy metals uptake by activated sludge, J.W.P.C.F. *47*, 363–376 (1975)
Cheremisinoff, P.N., Habib, Y.H.: Cadmium, chromium, lead, mercury: A plenary account for water pollution. Water Sewage Works *119*, 46–53 (1972)
Chester, R.: Geochemical criteria for the differentiating reef from non-reef facies in carbonate rocks. Bull. Am. Assoc. Petrol. Geol. *49*, 258–276 (1965)
Chester, R., Aston, S.R.: The geochemistry of deep-sea sediments. In: Chemical Oceanography. Riley, J.P., Chester, R. (eds.). London: Academic Press 1976, pp. 281–390
Chester, R., Hughes, M.J.: A chemical technique for the separation of ferro-manganese minerals, carbonate minerals and adsorbed trace elements from pelagic sediments. Chem Geol. *2*, 249–262 (1967)
Chester, R., Stoner, J.H.: Average trace element composition of low level marine atmospheric particulates. Nature (London) *246*, 138–139 (1973a)
Chester, R., Stoner, J.H.: Pb in particulates from the lower atmosphere of the eastern Atlantic. Nature (London) *245*, 27–28 (1973b)
Chester, R., Stoner, J.H.: The distribution of zinc, nickel, manganese, cadmium, copper and iron in some surface waters from the world ocean. Mar. Chem. *2*, 17–32 (1974)
Chester, R., Stoner, J.H.: Trace elements in sediments from the lower Severn and Bristol Channel. Mar. Pollut. Bull. *6*, 92–96 (1975a)
Chester, R., Gardner, D., Riley, J.P., Stoner, J.H.: Mercury in some surface waters of the world ocean. Mar. Pollut. Bull. *4*, 28–29 (1973)

Chesterikoff, A., Carru, A.M., Garban, B., Ollivon, D., Chesterikoff, C.: La pollution de la Basse-Seine par le mercure (du Pecq à Tancarville). Cent. Belge Etude Doc. Eaux *356*, 269–276 (1973)

Chipman, W.A., Rice, T.R., Price, T.J.: Uptake and accumulation of radioactive zinc by marine plankton, fish and shellfish. Fish. Bull. Fish. Wildlife Serv. U.S. *58*, 279–292 (1958)

Chisolm, J.J.: Lead poisoning. Sci. Am. *224*, 15–23 (1971)

Chow, T.J.: Lead isotopes in sea water and marine sediments. J. Mar. Res. *17*, 120–127 (1958)

Chow, T.J.: Proc. 2nd Int. Clear Air Congr. 348–352 (1971). Cit. E.D. Goldberg (1975)

Chow, T.J., Goldberg, E.D.: On the marine geochemistry of barium. Geochim. Cosmochim. Acta *20*, 192–198 (1960)

Chow, T.J., Goldberg, E.D.: Mass spectrometric determination of lithium in seawater. J. Mar. Res. *20*, 163–167 (1962)

Chow, T.J., Patterson, C.C.: The occurrence and significance of lead isotopes in pelagic sediments. Geochim. Cosmochim. Acta *26*, 263–308 (1962)

Chow, T.J., Patterson, C.C.: Concentration profiles of barium and lead in Atlantic waters of Bermuda. Earth Planet. Sci. Lett. *1*, 397–400 (1966)

Chow, T.J., Thompson, T.G.: Flame photometric determination of strontium in seawater. Anal. Chem. *27*, 18–21 (1955)

Chow, T.J., Bruland, K.W., Bertine, K., Soutar, A., Koide, M., Goldberg, E.D.: Lead pollution, records in southern California coastal sediments. Science *181*, 551–552 (1973)

Chow, T.J., Snyder, H.G., Snyder, C.B.: Mussels (Mytilus sp.) as an indicator of lead pollution. Sci. Total Environ. *6*, 55–63 (1976a)

Chow, T.J., Snyder, C.B., Snyder, H.G., Earl, J.L.: Lead content of some marine organisms. J. Environ. Sci. Health *A11*, 33–44 (1976b)

Christensen, E.R., Scherfig, J., Koide, M.: Metals from urban runoff in dated sediments of a very shallow estuary. Environ. Sci. Technol. *12*, 1168–1173 (1978)

Chumbley, C.G.: Permissible levels of toxic metals in sewage used in agricultural land. ADAS Adv. Paper No. 10. Min. Agric. Fish Food (1971)

Chung, I.H., Jeng, S.S.: Heavy metal pollution of Ta-Tu River. Bull. Inst. Zool., Acad. Sin. *13*, 69–73 (1974)

Chynoweth, D.P., Black, J.A., Mancy, K.H.: Effects of organic pollutant on copper toxicity to fish. In: Toxicity to Biota of Metal Forms in Natural Water. Andrew, R.W., Hodson, P.C., Konasewich, D.E. (eds.). Int. Joint Comm., Windsor, Ontario 1976, pp. 145–157

Ciancia, J.: Pollution abatement in the metal finishing industry. In: Cycling and Control of Metals. Curry, M.G., Gigliotti, G.M. (eds.). Cincinnati: Natl. Environ. Res. Cent. U.S. EPA 1973, pp. 83–90

Clark, A., Condon, W.A., Hoare, J.M., Sorg, D.H.: Analysis of rocks and stream sediment samples from the Taylor Mountains C-8 quadrangle, Alaska. U.S. Geol. Surv. Open-File Rep. 110 (1970)

Clifton, A.P., Vivian, C.M.G.: Retention of mercury from an industrial source in Swansea Bay sediments. Nature (London) *253*, 621–622 (1975)

Cline, J.T., Chambers, R.L.: Numerical investigation of heavy metals in sediments at Sleeping Bear Point, Lake Michigan. Abstr. 37th Annual Meeting Am. Soc. Limnol. Oceanogr., Seattle, Wash. (1974)

Cline, J.T., Upchurch, S.B.: Mode of heavy metal migration in the upper strata of lake sediment. Proc. 16th Conf. Great Lakes Res.. Int. Assoc. Great Lakes Res. 349–356 (1973)

Cline, J.T., Hillson, J.B., Upchurch, S.B.: Mercury mobilization as an organic complex. Proc. 16th Conf. Great Lakes Res. 233–242 (1973)

Cocoros, G., Cahn, P.H., Siler, W.: Mercury concentrations in fish, plankton and water from three western Atlantic estuaries. J. Fish. Biol. *5*, 641–647 (1973)

Coenen, R., Fehrenbach, R., Fritsch, W., Goetzmann, S., Piotrowski, H.D., Schladitz, R.: Alternativen zur Umweltmisere – Raubbau oder Partnerschaft? München: Carl Hanser 1972a

Coenen, R., Fritsch, W., Goetzman, S., Kesberger, H., Langheim, J., Piotrowski, H.D., Schladitz, R.: Chemisch-toxikologische Probleme des Umweltschutzes. Naturwissenschaften *59*, 106–111 (1972b)

Coker, W.B., Nichol, I.: The relation of lake sediment geochemistry to mineralization in the northwest Ontario region of the Canadian Shield. Econ. Geol. *70*, 202–218 (1975)

Colbourne, P., Alloway, B.J., Thornton, I.: Arsenic and heavy metals in soils associated with regional geochemical anaomalies in southwest England. Sci. Total Environ. *4*, 359–363 (1975)

Colby, B.R.: Fluvial sediments, a summary of source, transportation, deposition, and measurement of sediment discharge. Geol. Surv. Bull. *1181-A*, 47 (1963)

Collinson, C., Shimp, N.F.: Trace elements in bottom sediments from upper Peoria Lake, middle Illinois River – a pilot project. Environ. Geol. Notes *56*, 1–21 (1972)

Colwell, R.R., Nelson, J.D.: Bacterial mobilization of mercury in Chesapeake Bay. Proc. Int. Conf. Transp. Persist. Chem. Aquatic Ecosyst., Ottawa III–1–10 (1974)

Coonley, L.S., Baker, E.B., Holland, H D.: Iron in the Mullica River and in Great Bay, New Jersey. Chem. Geol. 7, 51–63 (1971)

Cooper, B.S., Harris, R.C.: Heavy metals in organic phases of river and estuarine sediment. Mar. Pollut. Bull. 5, 24–26 (1974)

Cooper, P.F., Thomas, E.V.: Recent developments in sewage treatment based on physico-chemical methods. Water Pollut. Control 73, 505–520 (1974)

Copeland, R.A.: Selenium, the unknown pollutant. Limnos 3, 7–9 (1971)

Copeland, R.A.: Mercury in the Lake Michigan environment. In: Environmental Mercury Contamination. Hartung, R., Dinman, B.D. (eds.). Ann Arbor Sci. Publ. 1972, pp. 71–76

Copeland, R.A., Ayers, J.C.: Trace element distributions in water, sediment, phytoplankton, zooplankton, and benthos of Lake Michigan. Environ. Res. Group, Inc., Ann Arbor, Mich. (1972). Cit. D.H. Klein (1975)

Corcoran, E.F., Alexander, J.E.: The distribution of certain trace elements in tropical sea water and their biological significance. Bull. Mar. Sci. Gulf Carib. 14, 594–602 (1964)

Correns, C.W.: Adsorptionsversuche mit sehr verdünnten Kupfer- und Bleilösungen und ihre Bedeutung für die Erzlagerstättenkunde. Kolloidzeitschrift 34, 341–349 (1924)

Correns, C.W.: Über die Herkunft der Elemente in Sedimentgesteinen. Geol. Rundsch. 58, 365–379 (1969)

Costescu, L.M., Hutchinson, T.C.: The ecological consequences of soil pollution by metallic dust from the Sudbury smelters. Inst. Environ. Sci. Proc. 18, 540–545 (1972)

Coutris, R., Gomella, C.: Micropollution and treatment of water with special reference to the prevention of undesirable tastes and odours. Gen. Rep. No. 2, IWSA-Congr., Vienna 1969 (cit. Haberer, K., Normann, S., 1972)

Covill, R.W.: The quality of the Forth Estuary. Proc. R. Soc. Edinburgh 71 (B), 143–170 (1971/1972)

Cowgill, U.M.: Mercury contamination in a 54-m core from Lake Huleh. Nature (London) 256, 475–478 (1975)

Cowgill, U.M., Hutchinson, G.E.: A general account of the basin and the chemistry and mineralogy of the sediment cores. In: History of the Laguna de Petenxil. Mem. Connecticut Acad. Arts Sci. 17, 2–62 (1966)

Cox, D.P., Alexander, M.: Production of trimethylarsine gas from various arsenic compounds by three sewage fungi. Bull. Environ. Contam. Toxicol. 9, 84–91 (1973)

Craig, P.J., Morton, S.F.: Mercury in Mersey estuary sediments, and the analytical procedure for total mercury. Nature (London) 261, 125–126 (1976)

Cranston, R.E.: Geochemical interaction in the recently industrialized Strait of Canso. Proc. Int. Conf. Transport of Persistent Chemicals in Aquatic Ecosystems. Ottawa I–59–68 (1974)

Cranston, R.E.: Accumulation and distribution of total mercury in estuarine sediments. Estuar. Coast. Mar. Sci. 4, 695–700 (1976)

Cranston, R.E., Buckley, D.E.: Mercury pathways in a river and estuary. Environ. Sci. Technol. 6, 274–278 (1972)

Cranston, R.E., Murray, J.W.: Dissolved chromium species in seawater. Abstr. 1978 Spring Meeting AGU, EOS 59, 306 (1978)

Crecelius, E.A.: The geochemical cycle of arsenic in Lake Washington and its relation to other elements. Limnol. Oceanogr. 20, 441–451 (1975)

Crecelius, E.A., Carpenter, R.: Arsenic distribution in waters and sediments of the Puget Sound region. Proc. 1st Annu. NSF Trace Contam. Conf. 1973, 615–625 (1974)

Crecelius, E.A., Piper, D.Z.: Particulate lead contamination recorded in sedimentary cores from Lake Washington, Seattle. Environ. Sci. Technol. 7, 1053–1055 (1973)

Crecelius, E.A., Bothner, M.H., Carpenter, R.: Geochemistries of arsenic, antimony, mercury and related elements in sediments of Puget Sound. Environ. Sci. Technol. 9, 325–333 (1975)

Crocket, J.H.: Gold – abundance in natural waters and in the atmosphere. In: Handbook of Geochemistry. Wedepohl, K.H. (ed.) Vol. 79-I, Berlin, Heidelberg, New York: Springer 1974, 6 p.

Crocket, J.H., Teruta, Y.: Pt, Pd, Au and Ir content of Kelley Lake bottom sediments. Can. Mineral. 14, 58–61 (1976)

Crocket, J.H., Winchester, J.W.: Coprecipitation of zinc with calcium carbonate. Geochim. Cosmochim. Acta 30, 1093–1109 (1966)

Cronan, D.S.: Manganese nodules and other ferro-manganese oxide deposits. In: Chemical Oceanography. Riley, J.P., Chester, R. (eds.), Vol. 5, London: Academic Press 1976, pp. 217–263

Cross, F.A., Duke, T.W., Willis, J.N.: Biochemistry of trace elements in a coastal plain estuary: Distribution of manganese, iron, and zinc in sediments, water and polychaetous worms. Chesapeake Sci. *11*, 221–234 (1970)

Cross, F.A., Hardy, L.H., Jones, N.Y., Barber, R.T.: Relation between total body weight and concentration of manganese, iron, copper, zinc and mercury in white muscle of blue fish and bathy-dermersal fish. J. Fish. Res. Board Can. *30*, 1287–1291 (1973)

Culkin, F.: The major constituents of sea water. In: Chemical Oceanography. Riley, J.P., Skirrow, G. (eds.), Vol. 1. London: Academic Press 1965, pp. 121–161

Cumbie, P.M.: Mercury levels in Georgia otter, mink and freshwater fish. Bull. Environ. Contam. Toxicol. *14*, 193–196 (1975)

Cumont, G., Montiel, A.: Etude de l'accumulation du mercure dans un ecosystème aquatique. Cent. Belge Etude Doc. Eaux *352*, 124–126 (1973)

Cunningham, P.A., Tripp, M.R.: Accumulation and depuration of mercury in the American oyster Crassostrea virginica. Mar. Biol. *20*, 14–19 (1973)

Cunningham, P.A., Tripp, M.R.: Factors effecting the accumulation and removal of mercury from tissues of the American oyster (Crassostrea virginica). Mar. Biol. *31*, 311–320 (1975a)

Cunningham, P.A., Tripp, M.R.: Accumulation, tissue distribution, and elimination of $^{203}HgCl_2$ in the tissues of the American oyster Crassostrea virginica. Mar. Biol. *31*, 321–334 (1975b)

Curtis, C.D.: The incorporation of soluble organic matter into sediments and its effect on trace element assemblages. In: Advances in Organic Geochemistry. Hobson, G.D., Louis, M.C. (eds.). Oxford: Pergamon 1966, pp. 1–13

Cutshall, N.H.: Chromium-51 in the Columbia River and adjacent Pacific Ocean. Ph. D. Thesis, Oregon State Univ., Corvallis, 1967. Cit. Jenne, E.A. (1976)

Cutshall, N.H., Renfro, W.C., Evans, D.W., Johnson, V.: Zinc-65 in Oregon-Washington continental shelf sediments. In: Radionuclides in Ecosystems. Nelson, D.J. (ed.). Proc. 3rd Nat. Symp. Radioecol. Washington, D.C. U.S. AEC 1973, pp. 694–702

Dahlberg, E.C.: Application of a selective simulation and sampling technique to the interpretation of stream sediment copper anomalies near South Mountain, Pa. Econ. Geol. *63*, 409–417 (1968)

Dahmen, F.W., Dahmen, G., Heiss, W.: Neue Wege der graphischen und kartographischen Veranschaulichung von Vielfaktorenkomplexen. Decheniana (Bonn) *129*, 145–178 (1976)

Dall'Aglio, M.: The abundance of mercury in 300 natural water samples from Tuscany and Latium. In: Origin and Distribution of the Elements. Ahrens, L.H. (ed.). Oxford: Pergamon 1968

Damiani, V., Thomas, R.L.: Mercury in sediments of the Pallanza Basin. Nature (London) *251*, 696–697 (1974)

Damiani, V., Morton, T.W., Thomas, R.L.: Freshwater ferro-manganese nodules from the Big Bay section of the Bay of Quinte, northern Lake Ontario. Proc. 16th Conf. Great Lakes Res. 397–403 (1973)

Dare, P.J., Edwards, D.B.: Seasonal changes in flesh weight and biochemical composition of mussels (Mytilus edulis) in the Conway Estuary, North Wales. J. Exp. Biol. Ecol. *18*, 88–97 (1975)

Darracott, A., Watling, H.: The use of molluscs to monitor cadmium levels in estuaries and coastal marine environments. Trans. R. Soc. S. Afr. *41*, 325–338 (1975)

Datsko, V.G., Klimov, I.T., Krasnov, V.N.: Content of certain heavy metals in the waters and silts of the Tsimlyansky Reservoir. Gidrokhim. Mater. *36*, 50–55 (1964)

Datsko, V.G., Krasnov, V.N.: Content of certain trace elements (of heavy metals) in the waters and silts of the Staro-Beshevo Reservoir. Gidrokhim. Mater. *40*, 99–108 (1965)

Davaud, E.: Contribution à l'étude géochemique et sédimentologique de dépots lacustres récents (Lac de Morat, Suisse). Genève, Ecole de Physique, Thèse No. 1745 (1976)

Davaud, E.: Chemical evolution of muds in a eutrophic lake. In: Proc. Int. Symp. Interactions Between Sediments and Fresh Water, Amsterdam, Sept. 6–10, 1976. Golterman, H.L. (ed.). Wageningen: Junk B.V. 1977, pp. 378–381

Davaud, E., Viel, M., Vernet, J.P.: Contamination des sédiments côtiers par les métaux lourds. Etude de la Pollution des Sédiments du Léman et du Bassin du Rhône. Vernet, J.-P., Scolari, G. (eds.). Rapp. Comm. Int. Protect. Eaux Léman (in press, 1978)

Davey, E.W., Morgan, M.J., Erickson, S.J.: A biological measurement of the copper complexation capacity of seawater. Limnol. Oceanogr. *19*, 993–997 (1974)

Davey, H.A., Moort, J.C. van: Current mercury deposition at Ngawha Springs, New Zealand. Search *5*, 154–156 (1974)

Davies, B.E.: Occurrence and distribution of lead and other metals in two areas of unusual disease incidence in Britain. Proc. Intern. Symp. Environmental Health Aspects of Lead. Amsterdam (Nederland), October 2–6, 1972, pp. 125–134

Davies, B.E.: The role of organic materials in heavy metal problems of soil and water. Welsh Soils Discussion Group Report "Soil Organic Matter" 16, 120–123 (1975)

Davies, B.E.: Mercury content of soils in western Britain with special reference to contamination from base metal mining. Geoderma 16, 183–192 (1976)

Davies, B.E.: Heavy metal pollution of British agricultural soils with special reference to the role of lead and copper mining. Proc. Int. Seminar Soil Environment and Fertility Management in Intensive Agriculture. Tokyo/Japan. 1977, pp. 394–401

Davies, B.E.: Plant-available lead and other metals in British garden soils. Sci. Total Environ. 9, 243–262 (1978)

Davies, B.E., Lewin, J.: Chronosequences in alluvial soils with special reference to historic lead pollution in Cardiganshire, Wales. Environ. Pollut. 6, 49–57 (1974)

Davies, B.E., Roberts, L.J.: The distribution of heavy metal contaminated soils in northwest Clwyd, Wales. Water Air Soil Pollut. 9, 507–518 (1978)

Davies, P.H., Goettl, J.P., Jr., Sinley, J.R., Smith, N.F.: Acute and chronic toxicity of lead to rainbow trout Salmo gairdneri, in hard and soft water. Water Res. 10, 199–206 (1976)

Davis, R.B.: Tubificids alter profiles of redox potential and pH in profundal lake sediment. Limnol. Oceanogr. 19, 342–346 (1974)

Davis, R.B., Bailey, J.H., Norton, S.A.: Decrease in the pH of lakes in Maine, probably a result of acid precipitation. Abstr. SIL-Congr., Copenhague 1977, p. 57

Dawson, A.B.: The hemopoietic response in the catfish, Ameiurus nebulosus to chronic lead poisoning. Biol. Bull. 68, 335–346 (1935)

Day, J.H.: The ecology of South African estuaries I. A review of estuarine conditions in general. Trans R. Soc. S. Afr. 33, 53–91 (1951)

Day, J.P., Hart, M., Robinson, M.S.: Lead in urban street dust. Nature (London) 253, 343–345 (1975)

Dean, J.G., Bosqui, F.L., Lanouette, V.H.: Removing heavy metals from waste water. Environ. Sci. Technol. 6, 518–522 (1972)

Degens, E.T.: Geochemistry of Sediments. New Jersey: Prentice Hall 1965

Degens, E.T., Valeton, I.: Lebensraum Alster, eine Geodokumentation. Uni-Forschung, Wissenschaftsber. Univ. Hamburg VII, 1975

Delfino, J.J.: Effects of river discharge and suspended sediment on water quality in the Mississippi River. J. Environ. Sci. Health A12, 79–94 (1977)

Delisle, C.E., Hummel, B., Wheeland, K.G.: Uptake of heavy metals from sediment by fish. Proc. Int. Conf. Heavy Met. Environ., Toronto 1975, Vol. II/2, 821–827 (1977)

Derban, L.K.A.: Outbreak of food poisoning due to alkyl-mercury-fungicide. Arch. Environ. Health 28, 49–52 (1974)

Derryberry, O.M.: Investigation of mercury contamination in the Tennessee Valley region. In: Environmental Mercury Contamination. Hartung, R., Dinman, B.D. (eds.). Ann Arbor Sci, Publ. Inc. 1972, pp. 76–79

Deurer, R., Förstner, U., Schmoll, G.: Selective chemical extraction of carbonate-associated trace metals in recent lacustrine sediments. Geochim. Cosmochim. Acta 42, 425–427 (1978)

Dewling, R.T., Anderson, P.W.: New York Bight. I. Ocean dumping policies. Oceanus 19, 2–10 (1976)

Dick, R.I.: Alternatives to marine disposal of sewage sludge. In: Marine Chemistry in the Coastal Environment. Church, T.M. (ed.). Am. Chem. Soc. Symp. Ser. 18, 453–466 (1975)

Dick, R.I.: Sludge treatment, utilization and disposal. J.W.P.C.F. 49, 1040–1067 (1977)

Dietrichson, W.: The acidity and the anionic composition of Swedish lake waters. Abstr. SIL-Congress, Kopenhagen 1977, p. 63

Dietz, F.: Die Anreicherung von Schwermetallen in submersen Wasserpflanzen. GWF-Wasser/Abwasser 113, 269–273 (1972)

Dietz, F.: Bestimmung einzelner Komplexbildner in Wässern, speziell der synthetischen Komplexbildner Nitrilotriessigsäure (NTA) und Äthylendiamintetraessigsäure (AeDTA) Z. Wasser Abwasserforsch. 7, 74–80 (1974)

Dietz, F.: Die Borkonzentration in Wässern als ein Indikator der Gewässerbelastung. GWF-Wasser/Abwasser 116, 301–308 (1975)

Dietz, F.: Zusammenfassende Bewertung der Ergebnisse der methodischen Untersuchungen zur Bestimmung des Remobilisierungsvermögens von Komplexbildnern. Z. Wasser Abwasserforsch. 10, 20–24 (1977)

Dietz, F., Frank, H.-D.: Anwendung anorganischer und organischer Austauschermassen als Bodenkörper für Remobilisierungsversuche. Z. Wasser Abwasserforsch. 10, 109–115 (1977)

Dietz, F., Frank, H.D., Koppe, P.: Bestimmung des Remobilisierungspotentials von Wässern gegenüber Schwermetallverbindungen. Z. Wasser Abwasserforsch. 4, 104–113 (1975)

Dillon, P.J., Yan, N.D., Scheider, W.A., Conroy, N.: Acidic lakes in Ontario, Canada, their extent and responses to base and nutrient additions. Paper presented at Jubilee Symp. on Lake Metabolism and Lake Management, Uppsala Univ. 1977, p. 25.
Dominik, J., Förstner, U., Mangini, A., Reineck, H.E.: Pb-210 and Cs-137 chronology of heavy metal pollution in a sediment core from the German Bight (North Sea). Senckenberg. Mar. 10, 213–227 (1978)
Doolan, K.J., Smythe, L.E.: Cadmium content of some New South Wales waters. Search 4, 162–163 (1973)
Dörjes, J., Little-Gadow, S., Schäfer, A.: Zur Schwermetallverteilung in litoralen Sedimenten der ostfriesischen Küste. Senckenbergiana Marit. 8, 103–109 (1976)
Doudoroff, P., Katz, M.: Critical review of literature on the toxicity of industrial wastes and their components to fish. II. The metals as salts. Sewage Ind. Wastes 25, 802–839 (1953)
Downing, A.L., Edwards, R.W.: Effluent standard and the assessment of the effects of pollution on rivers. Water Pollut. Control 68, 283–299 (1969)
Drake, M.: Soil chemistry and plant nutrition. In: Chemistry of the Soil. Bear, F.E. (ed.), 2nd ed. New York: Reinhold 1967, pp. 395–444
Driel, W. van, de Groot, A.J.: Heavy metals in river sediments. Geol. Mijnbouw 53, 201–203 (1974)
Drifmeyer, M.S., Odum, W.E.: Lead, zinc and manganese in dredge-spoil pond ecosystems. Environ. Conserv. 2, 39–45 (1975)
Drozdova, T.V.: The role of humic acids in concentrating rare elements in soils. Sov. Soil Sci. 10, 1393–1396 (1968)
Duce, R.A., Hoffmann, G.L., Fasching, J.L., Moyers, J.L.: The collection and analysis of trace metals in atmospheric particulate matter over the North Atlantic. Proc. WMO/WHO Techn. Conf. on the Observation and Measurement of Atmospheric Pollution. Helsinki, Finland, 30 July–4 August 1973, WMO 368, 370–379 (1974a)
Duce, R.A., Hoffmann, G.L., Zoller, W.H.: Atmospheric trace metals at remote northern and southern hemisphere sites – pollution or natural? Science 187, 59–61 (1975)
Duce, R.A., Parker, P.L., Giam, C.S.: Pollutant transfer to the marine environment. Deliberations and Recommendations of the NSF/IDOE Pollutant Transfer Workshop held in Port Arcansas, Texas, Jan. 11–12 1974, 55 p. (1974b)
Duce, R.A., Quinn, J., Olney, C.M., Piotrowitz, S., Ray, B., Wade, T.: Enrichment of heavy metals and organic compounds in the surface microlayer of Narragansett Bay, Rhode Island. Science 176, 161–163 (1972)
Duchart, P., Calvert, S.E., Price, N.B.: Distribution of trace metals in the pore waters of shallow water marine sediments. Limnol. Oceanogr. 18, 605–611 (1973)
Duinker, J.C., Nolting, R.F.: Distribution model for particulate trace metals in the Rhine Estuary, the Southern Bight and Dutch Wadden Sea. Neth. J. Sea Res. 10, 78–102 (1976)
Duinker, J.C., Nolting, R.F.: Dissolved and particulate trace metals in the Rhine Estuary and the Southern Bight. Mar. Pollut. Bull. 8, 56–71 (1977)
Duinker, J.C., Van Eck, G.T.M., Nolting, R.F.: On the behaviour of copper, zinc, iron and manganese in the Dutch Wadden Sea. Neth. J. Sea Res. 8, 214–239 (1974)
Duncan, D.M.: In: Mineral Yearbook, Denver: Colorado Mining Association 1974, p. 92. Cit. Rosenbaum, J.B.
Dunlap, W.J., Cosby, R.L., McNabb, J.F., Bledsoe, B.E., Scalf, M.R.: Investigation concerning probable impact of nitrilotriacetic acid on ground waters. Environmental Protection Agency. Washington D.C.: U.S. Gov. Printing Office 1971
Dunning, C., Shepley, S., Wheeler, R., Anderson, J.B.: Sedimentary, geochemical and fossil diatom analysis of sediment cores from Lake Macatawa. An assessment of man's impact. Can. Mineral. 13, 307–322 (1975)
Durum, W.H., Haffty, J.: Occurrence of minor elements in water. U.S. Geol. Surv. Circ. 445, 11 (1961)
Durum, W.H., Haffty, J.: Implication of minor element content of some major streams in the world. Geochim. Cosmochim. Acta 27, 1–11 (1963)
Durum, W.H., Hem, J.D.: An overview of trace element distribution patterns in water. Ann. N.Y. Acad. Sci. 199, 26–36 (1972)
Durum, W.H., Hem, J.D., Heidel, S.G.: Reconnaissance of selected minor elements in surface waters of the United States. U.S. Geol. Surv. Circ. 643, 49 (1971)
Duthie, H.C., Sreenivasa, M.R.: Evidence for the eutrophication of Lake Ontario from the sedimentary diatom succession. Proc. 14th Conf. Great Lakes Res. 1–13 (1971)
Dutrizac, J.S., MacDonald, R.J.C.: Ferric iron as a leaching medium. Min. Sci. Eng. 6, 59–100 (1974)

Dutton, J.W.R., Jefferies, D.F., Folkard, A.R., Jones, P.G.W.: Trace metals in the North Sea. Mar. Pollut. Bull. *4*, 135–138 (1973)

Duursma, E.K.: Radioactive tracers in estuarine chemical studies. In: Estuarine Chemistry. Burton, J.D., Liss, P.S. (eds.). London: Academic Press 1976, pp. 159–183

Duursma, E.K., Gross, M.G.: Marine sediments and radioactivity. In: Radioactivity in the Marine Environment. Washington, D.C.: Natl. Acad. Sci. 1971, pp. 147–160

Duursma, E.K., Rosch, C.J.: Theoretical, experimental and field studies concerning diffusion of radioisotopes in sediments and suspended particles of the sea. Neth. J. Sea Res. *4*, 395–469 (1970)

Dworsky, R., Ebner, F., Gams, H., Ottendorfer, L.J.: Untersuchungen über den Quecksilbergehalt in österreichischen Oberflächengewässern. Österr. Abwasserrundsch. *18*, 22–27 (1973)

Dyck, W.: Lake sampling re. stream sampling for regional geochemical surveys. Geol. Surv. Can. Pap. *71-1* (B), 70–71 (1971)

Dyer, K.R.: Sedimentation in estuaries. In: The Estuarine Environment. Barnes, R.S.K., Green, J. (eds.). London: Applied Science Publ. 1972, pp. 10–32

Dyrrsen, D., Wedborg, M.: Equilibrium calculations of the speciation of elements in seawater. In: The Sea. Goldberg, E.D. (ed.). New York: Wiley 1974, pp. 181–195

Ebner, F., Gams, H.: Schwermetalle in der Salzach. Österr. Abwasser Rundsch. *17*, 53–60 (1972)

Ebner, F., Gams, H.: Schwermetalle in der österreichischen Donau. Österr. Abwasser Rundsch. *18*, 47–48 (1973)

Ebner, F., Gams, H.: Schwermetalle in der Salzach. Österr. Abwasser Rundsch. *20*, 30–32 (1975a)

Ebner, F., Gams, H.: Schwermetalle in den Flüssen Glan und Gurk unter besonderer Berücksichtigung der Quecksilbergehalte. Österr. Abwasser Rundsch. *20*, 51–53 (1975b)

Ebner, F., Dworsky, R., Gams, H.: Die Bestimmung von Schwermetallen in österreichischen Oberflächengewässern. Österr. Abwasser Rundsch. *4*, 53–59 (1972)

Eckert, J.M., Sholkovitz, E.R.: The flocculation of iron, aluminium and humates from river waters by electrolytes. Geochim. Cosmochim. Acta *40*, 847–848 (1976)

Edgington, D.H., Callender, E.: Minor element geochemistry of Lake Michigan ferromanganese nodules. Earth Planet. Sci. Lett. *8*, 97–100 (1970)

Edgington, D.H., Robbins, J.A.: Records of lead deposition in Lake Michigan sediments since 1800. Environ. Sci. Technol. *10*, 266–273 (1976)

Edgington, D.H., Robbins, J.A.: Lead and lead-210 in the sediments of southern Lake Michigan. Abstr. Meet. Am. Soc. Limnol. Oceanogr. (1974)

Edgren, M.: Sediment as indicator of pollution. In: 3rd Soviet-Swedish Symp. on the Pollution of the Baltic. Åkerblom, A. (ed.). Ambio Special Report *5*, 133–139 (1977)

Edwards, A.C., Davis, D.E.: Effects of an organic arsenical herbicide on a salt marsh ecosystem. J. Environ. Qual. *4*, 215–219 (1975)

Edzwald, J.K.: Phosphorus in aquatic systems, the role of the sediment. In: Fate of Pollutants in the Air and Water Environment. Part. 1. Suffet, I.H. (ed.). New York: Wiley 1977, pp. 183–214

Edzwald, J.K., Upchurch, J.B., O'Melia, C.R.: Coagulation in estuaries. Environ. Sci. Technol. *8*, 58–63 (1974)

Eginhouse, R.P., Jr.: Mercury in sediments. Southern Calif. Coastal Wat. Res. Project, El Segundo, Cal. 90245. Annu. Rep. 1975, 83–89 (1976)

Eginhouse, R.P., Young, D.R., Johnson, J.N.: Geochemistry of mercury in Palos Verdes sediments. Environ. Sci. Technol. *12*, 1151–1157 (1978)

Ehrlich, H.L.: Bacteriology of manganese nodules. I. Bacterial action on manganese in nodule enrichments. Appl. Microbiol. *11*, 15–19 (1963)

Ehrlich, H.L.: Reactions with manganese by bacteria from marine ferromanganese nodules. Dev. Ind. Microbiol. *7*, 279–286 (1966)

Ehrlich, H.L., Fox, S.I.: Copper sulfide precipitation by yeasts from acid minewaters. Appl. Microbiol. *15*, 135–139 (1967a)

Ehrlich, H.L., Fox, S.I.: Environmental effects on bacterial copper extraction from low-grade copper sulfide ores. Biotechnol. Bioeng. *9*, 471–485 (1967b)

Eisler, R., Lapan, R.L., Telek, G., Davey, E.W., Soper, A.E., Barry, M.: Survey of metals in sediments near Quonset Point, Rhode Island. Mar. Pollut. Bull. *8*, 260–264 (1977)

Eisler, R., Zaroogian, G.E., Hennekey, R.J.: Cadmium uptake by marine organisms. J. Fish. Res. Board Can. *29*, 1367–1369 (1972)

Elderfield, H., Hem, J.D.: The development of crystalline structure in aluminium hydroxide polymorphs on aging. Min. Mag. *39*, 89–96 (1973)

Elderfield, H., Hepworth, A.: Diagenesis, metals and pollution in estuaries. Mar. Pollut. Bull. *6*, 85–87 (1975)

Elderfield, H., Thornton, I., Webb, J.S.: Heavy metals and oyster culture in Wales. Mar. Pollut. Bull. 2, 44–47 (1971)

Ellis, D.V.: Pollution controls on mine discharges to the sea. Proc. Int. Conf. Heavy Met. Environ., Toronto, *II/2*, 677–685 (1977)

Ellis, M.M.: Pollution of the Coeur d'Alène River and adjacent waters by mine wastes. Rept. to U.S. Bureau of Fisheries, 55 p. (1932)

Emery, K.O., Rittenberg, S.C.: Early diagenesis of California Basin sediments in relation to origin of soil. Bull. Am. Ass. Petrol. Geol. *36*, 735–806 (1952)

Engler, R.M., Brannon, J.M., Rose, J.: A practical selective extraction procedure for sediment characterization. 168th Meeting Am. Chem. Soc. Atlantic City, N.Y. 17 p. (1974)

Engler, R.M., Brannon, J.M., Rose, J., Bigham, G.: A practical selective extraction procedure for sediment characterization. In: Chemistry of Marine Sediments. Yen, T.F. (ed.). Ann Arbor: Science Publ. Inc. 1976, pp. 163–171

Engler, R.M., Patrick, W.H., Jr.: Stability of sulfides of manganese, iron, zinc, copper and mercury in flooded and non-flooded soil. Soil. Sci. *119*, 217–221 (1975)

Enk, M.D., Mathis, B.J.: Distribution of cadmium and lead in a stream ecosystem. Hydrobiologia *52*, 153–158 (1977)

Epstein, E., Chaney, R.L.: Land disposal of toxic substances and water-related problems. J.W.P.C.F. *50*, 2037–2042 (1978)

Epstein, S.S.: Toxicological and environmental implications on the use of nitrilotriacetic acid as a detergent builder. Int. J. Environ. Stud. *2*, 291–300 (1972)

Erlenkeuser, H., Suess, E., Willkomm, H.: Industrialization affects heavy metal and carbon isotope concentration in recent Baltic Sea sediments. Geochim. Cosmochim. Acta *38*, 823–842 (1974)

Ermenko, V.Y.: Gidrokhim. Mater. *41*, 152–157. Cit. Jenne, E.A. (1976)

Ernst, W.: Schwermetallvegetation der Erde. Stuttgart: Fischer (FRG), 1974

Eshleman, A., Siegel, S.M., Siegel, B.Z.: Is mercury from Hawaiian volcanoes a natural source of pollution? Nature (London) *233*, 471–472 (1971)

Eustace, I.J.: Zinc, cadmium, copper, and manganese in species of finfish and shellfish caught in the Derwent Estuary, Tasmania. Aust. J. Mar. Freshwater Res. *25*, 209–222 (1974)

Evans, D.W., Cutshall, N.H.: Effects of ocean water on the soluble suspended distribution of Columbia radionuclides. In: Radioactive Contamination of the Marine Environment. Vienna: Int. Atomic Energy Agency 1973, pp. 125–140

Evans, D.W., Cutshall, N.H., Cross, F.A., Wolfe, D.A.: Manganese cycling in the Newport River Estuary, N.C. Estuarine Coastal Mar. Sci. *5*, 71–80 (1977)

Evans, R.J., Bails, J.D., D'Itri, F.: Mercury levels in muscle tissues of preserved museum fish. Environ. Sci. Technol. *6*, 901–905 (1972)

Evans, R.L., Sullivan, W.T., Li, S.: Mercury in public sewer systems. Water Sewage Works *120*, 74–76 (1973)

Evans, W.D.: The organic solubilization of minerals in sediments. In: Advances in Organic Geochemistry. Colombo, U., Hobson, G.D. (eds.). Earth Sci. Ser. Monograph *15*, 1964, pp. 263–270

Everdingen, R.O. van: The Paint Pots, Kootenay National Park, British Columbia – acid spring water with extreme heavy metal content. Can. J. Earth Sci. *7*, 831–852 (1970)

Exler, H.J.: Ausbreitung und Reichweite von Grundwasserverunreinigungen im Unterstrom einer Mülldeponie. GWF-Wasser Abwasser *113*, 101–148 (1972)

Fabricand, B.P., Sawyer, R.R., Ungar, S.G., Adler, S.: Trace metals in the ocean by atomic absorption spectroscopy. Geochim. Cosmochim. Acta *26*, 1023–1027 (1962)

Fagerström, T., Åsell, B.: Methyl mercury accumulation in an aquatic food chain. A model and some implications for research planning. Ambio *2*, 164–171 (1973)

Fagerström, T., Jernelöv, A.: Aspects of the quantitative ecology of mercury. Water Res. *6*, 1193–1202 (1972)

Fagerström, T., Kurte, N., Åsell, B.: Statistical parameters as criteria in model evaluation: kinetics of mercury accumulation in pike Esox lucius. Oikos *26*, 109–116 (1975)

Faisst, W.K.: Digested sludge: characterization and modelling for ocean disposal. Preprints of papers presented at the 175th national meeting, ACS, Anaheim/Calif., Div. Environm. Chem. *18/1*, 194–197 (1978)

Fanning, K.A., Pilson, M.E.Q.: Interstitial silica and pH in marine sediments, some effects of sampling procedures. Science *173*, 1228–1231 (1971)

Farmer, J.G.: Lead concentration profiles in lead-210 dated Lake Ontario sediment cores. Sci. Total Environ. *10*, 117–127 (1978)

Fauth, M.I., Houser, M.E.: Determination of the metal concentration in sediments of the Potomac River Estuary. Proc. Inst. Environ. Sci., Wash. D.C. (1973)

Feather, C.E., Koen, G.M.: The significance of the mineralogical and surface characteristics of gold grains in the recovery process. Afr Inst. Min. Metall. 73, 223–234 (1973)

Feick, G., Horne, R.A., Yeaple, D.: Release of mercury from contaminated freshwater sediments by the runoff of road deicing salt. Science 175, 1142–1143 (1972)

Feitknecht, W., Schindler, P.: Solubility constants of metal oxides, metal hydroxides and metal hydroxide salts in aqueous solution. London: Butterworths 1963

Ferguson, J.F., Gavis, J.: A review of the arsenic cycle in natural waters. Water Res. 6, 2159–2174 (1972)

Field, R.A., Lager, T.A.: Urban runoff pollution control – State-of-the-art. J. Environ. Eng. Div. ASCE 101, EE-1, 107–125 (1975)

Filby, R.H., Shah, K.R., Funk, W.H.: Role of neutron activation analysis in the study of heavy metal pollution of a lake-river system. Proc. 2nd Int. Conf. Nuclear Methods Environ. Res. Vogt, J.R., Meyer, W. (eds.). 10–23 (1974)

Filip, D.S., Lynn, R.I.: Mercury accumulation by the fresh water alga Selenastrum capricornutum. Chemosphere 6, 251–254 (1972)

Fimreite, N.N.: Mercury used in Canada and their possible hazards as sources of mercury contamination. Environ. Pollut. 1, 119–131 (1970)

Fish, G.R.: Observations on excessive weed growth in two lakes in New Zealand. N.Z.J. Bot. 1, 410–418 (1963)

Fisher, J.R.: Bacterial leaching of Elliot Lake uranium ore. Bull. Trans. Can. Inst. Min. Met. 69, 169–171 (1966)

Fitchko, J., Hutchinson, T.C.: A comparative study of heavy metal concentrations in river mouth sediments around the Great Lakes J. Great Lakes Res. 1, 46–78 (1975)

Fitchko, J., Hutchinson, T.C.: Fluvial input of copper and nickel to Georgian Bay, Lake Huron. Abstr. 20th Conf. on Great Lakes Res. Int. Assoc. Great Lakes Res. (1977)

Fitzgerald, B.W., Rankin, J.S., Skauer, D.M.: Zinc-65 levels in oysters in the Thames River, Connecticut. Science 135, 926 (1962)

Fjerdingstad, E.: Accumulated concentrations of heavy metals in red snow algae in Greenland. Schweiz. Z. Hydrol. 35, 247–251 (1973)

Flegal, A.R.: Mercury in the seston of the San Francisco Bay Estuary. Bull. Environ. Cont. Toxicol. 17, 733–738 (1977)

Flegal, A.R., Martin, J.H.: Contamination of biological samples by ingested sediment. Mar. Pollut. Bull. 8, 90–92 (1977)

Fleischer, M.: Recent estimates of the abundance of elements in the earth's crust. U.S. Geol. Surv. Circ. 285, 7 pp. (1953)

Fleischer, M., Richmond, W.E.: The manganese oxide minerals, a preliminary report. Econ. Geol. 38, 269–286 (1943)

Fleischer, M., Sarofim, A.F., Fassett, D.W., Hammond, P., Shacklette, H.T., Nisbet, I.C.T., Epstein, S.: Environmental impact of cadmium. A review by the panel of hazardous trace substances. Environ. Health 253–323 (1974)

Fletcher, C.R.: The regulation of calcium and magnesium in the brackish water polychaete Nereis diversicolor. J. Exp. Biol. 53, 425–443 (1970)

Flögl, H.: Grundwasserverunreinigung durch Halden. Österr. Wasserwirtsch. 11, 148–152 (1958)

Florence, T.M., Batley, G.E.: Determination of chemical forms of trace metals in natural waters with special reference to copper, lead, cadmium and zinc. Talanta 24, 151–158 (1977)

Florey, E.: Lehrbuch der Tierphysiologie. Stuttgart: Thieme 1970

Folk, R.L., Ward, W.C.: Brazos River bar, a study in the significance of grain-size parameters. J. Sediment. Petrol. 27, 3–26 (1957)

Follet, E.A.C.: The retention of amorphous colloidal 'ferric hydroxide' by kaolinites. J. Soil. Sci. 16, 334–341 (1965)

Folsom, T.R.: Plutonium chemistry of the ocean INR Workshop on Chemical Oceanography. Monterey, Calif., 8 pp (1972)

Förstner, U.: Hydrochemische Entwicklungen in Flüssen und Seen Afghanistans. Chem. Erde 32, 216–238 (1973)

Förstner, U.: Lake sediments as indicators of heavy-metal pollution. Naturwissenschaften 63, 465–470 (1976)

Förstner, U.: Mineralogy and geochemistry of sediments in arid lakes of Australia. Geol. Rundsch. 66, 146–156 (1977a)

Förstner, U.: Metal concentration in recent lacustrine sediments. Arch. Hydrobiol 80, 172–191 (1977b)

Förstner, U.: Metal concentrations in freshwater sediments – natural background and cultural effects. In: Interactions between Sediments and Fresh Water. Golterman, H.L. (ed.). Wageningen/The Hague: Pudoc/Junk B.V. Publ. 1977c, pp. 94–103

Förstner, U.: Geochemische Untersuchungen an den Sedimenten des Ries-Sees (Forschungsbohrung Nördlingen 1973). Geol. Bavarica 75, 37–48 (1977d)

Förstner, U.: Metallanreicherungen in rezenten See-Sedimenten — geochemischer background und zivilisatorische Einflüsse. Mitt. Nationalkomm. B.R. Deutschland IHP 2, 66 pp. (1978a)

Förstner, U.: Ursachen und Folgen der Verschmutzung des Neckars. Polizei Technik Verkehr 20, 101–108 (1978b)

Förstner, U.: Sources and sediment associations of heavy metals in polluted coastal regions. In: Origin and Distribution of the Elements. Ahrens, L.H. (ed.). Oxford: Pergamon Press 1978c, pp. 849–866

Förstner, U.: Cadmium in polluted sediments. In: Biogeochemistry of Cadmium. Nriagu, J.O. (ed.). New York: Wiley 1979 (in press)

Förstner, U., Müller, G.: Heavy metal accumulation in river sediments, a response to environmental pollution. Geoforum 14, 53–62 (1973a)

Förstner, U., Müller, G.: Anorganische Schadstoffe im Neckar. Ruperto Carola 51, 67–71 (1973b)

Förstner, U., Müller, G.: Schwermetalle in Flüssen und Seen. Berlin, Heidelberg, New York: Springer 1974a

Förstner, U., Müller, G.: Schwermetallanreicherungen in datierten Sedimentkernen aus dem Bodensee und aus dem Tegernsee. Tschermaks Mineral. Petrogr. Mitt. 21, 145–163 (1974b)

Förstner, U., Müller, G.: Hydrochemische Beziehungen zwischen Flußwasser und Uferfiltrat. GWF-Wasser Abwasser 116, 74–80 (1975)

Förstner, U., Müller, G.: Heavy metal pollution monitoring by river sediments. Fortschr. Mineral. 53, 271–288 (1976)

Förstner, U., Patchineelam, S.R.: Bindung und Mobilisation von Schwermetallen in fluviatilen Sedimenten. Chem. Z. 100, 49–57 (1976)

Förstner, U., Patchineelam, S.R.: Chemical associations of heavy metals in polluted sediment from the lower Rhine River. Paper submitted for publication in "Advances in Chemistry", American Chemical Society, 1979

Förstner, U., Reineck, H.E.: Die Anreicherung von Spurenelementen in den rezenten Sedimenten eines Profilkernes aus der Deutschen Bucht. Senckenbergiana Marit. 6, 175–184 (1974)

Förstner, U., Stiefel, R.: Umweltprobleme durch Metallanreicherungen in kommunalen Abwässern. Chem. Z. 102, 161–168 (1978)

Förstner, U., Wittmann, G.T.W.: Metal accumulations in acidic waters from gold mines in South Africa. Geoforum 7, 41–49 (1976)

Förstner, U., Müller, G., Wagner, G.: Schwermetalle in Sedimenten des Bodensees, natürliche und zivilisatorische Anteile. Naturwissenschaften 61, 270 (1974)

Förstner, U., Nähle, C., Schöttler, U.: Sorption of heavy metals in sand filters in the presence of humic acids. Summaries of Papers of the International Symposium on Artificial Groundwater Recharge. Dortmund (F.G.R.), May 14–18, 1979. Institute of Water Research, Dortmund, 61/1–3

Foster, P.: Concentration and concentration-factors of heavy metals in brown algae. Environ. Pollut. 10, 45–53 (1976)

Foster, P., Hunt, T.E., Morris, A.W.: Metals in acid mine stream and estuary. Sci. Total Environ. 9, 75–86 (1978)

Foster, R.F.: Evaluation in the vicinity of Hanford for 1963. U.S. Atomic Energy Comm. HW-90991, 198 pp (1964)

Fourie, H.O.: Metals in organisms from Saldanha Bay and Langebaan Lagoon prior to industrialization. S. Afr. J. Sci. 72, 110–113 (1976)

Fowler, S.W., Oregioni, B.: Trace metals in mussels from the N.W. Mediterranean. Mar. Pollut. Bull. 7, 26–29 (1976)

Føyn, E.: Disposal of waste in the marine environment and the pollution of the sea. Oceanogr. Mar. Biol. Ann. Rev. 3, 95–114 (1965)

Francis, A.J., Duxbury, J.M., Alexander, M.: Evolution of dimethylselenide from soils. Appl. Microbiol. 28, 248–250 (1974)

Frank, W.H.: Zur Geschichte der künstlichen Grundwasseranreicherung. Veröffentl. Hydrolog. Forschungsabt. Dortmunder Stadtwerke 9 (1966)

Fredriksson, I., Ovarfort, U.: The mercury content of sediments from two lakes in Dalarna, Sweden. Geol. Fören. Stockholm Förh. 95, 237–242 (1973)

Freeland, G.N., Hoskinson, R.M., Mayfield, R.J.: Adsorption of mercury from aqueous solutions by polyethylnimime-modified wool fibers. Environ. Sci. Technol. 8, 943–944 (1974)

Freise, F.W.: The transportation of gold by organic underground solution. Econ. Geol. 26, 421–431 (1931)

Freitas, A.S.W. de, Qadri, S.U., Case, B.E.: Origins and fate of mercury in fish. Proc. Int. Conf. Transp. Persist. Chem. Aquatic Exosyst., Ottawa, III, 31–36 (1974)

Freundlich, H.: Colloid and Capillary Chemistry. London: Methuen, 1926
Frey, D.G.: Evidence for eutrophication from remains of organisms in sediments. In: Eutrophication: Causes, Consequences, Correctives. Washington D.C.: Nat. Acad. Sci. 1964, pp. 594–613
Friberg, L., Piscator, M., Nordberg, G.F., Kjellström, T.: Cadmium in the Environment. Cleveland: CRC 1974
Fricke, K., Werner, H.: Geochemische Untersuchungen von Mineralwasser auf Kupfer, Blei und Zink in Nordrhein-Westfalen und angrenzenden Gebieten. Heilbad Kurort 45–46 (1957)
Friedländer, S.K.: Chemical element balances and identification of air pollution sources. Eviron. Sci. Technol. 7, 235–240 (1973)
Friedman, G.M.: Dynamic processes and statistical parameters compared for size frequency distribution of beach and river sands. J. Sediment. Petrol. 37, 327–354 (1967)
Friedrich, A.R., Filice, F.P.: Uptake and accumulation of the nickel ion by Mytilus edulis. Bull. Environ. Contam. Toxicol. 16, 750–755 (1976)
Friedrich, G., Kulms, M.: Eine Methode zur Bestimmung geringer Konzentrationen von Quecksilber in Gesteinen und Böden und ihre Anwendung in der geochemischen Exploration von Erzlagerstätten. Erzmetall 22, 74–80 (1969)
Frimmel, F., Winkler, H.A.: Differenzierte Bestimmung verschiedener Quecksilberverbindungen in Wasser und Sediment. Vom Wasser 45, 285–298 (1975)
Frink, C.R.: Nutrient budgets and rational analysis of eutrophication in a Connecticut lake. Environ. Sci. Technol. 1, 425–428 (1967)
Fripiat, J.J., Gastuche, M.C.: Etude physio-chimique des surfaces des argiles. Les combinaisons de la kaolonite avec des oxides de fer trivalent. Publ. Inst. Natl. Etudes Agron. Congo Belge 54, 7–35 (1952)
Fripiat, J.J., Jelli, A., Poncelet, G., André, J.: Thermodynamic properties of adsorbed water molecules and electrical conduction in montmorillonites and silicas. J. Phys. Chem. 69, 2185–2197 (1965)
Frye, J.C., Shimp, N.F.: Major, minor and trace elements in sediments of late pleistocene Lake Saline compared with those in Lake Michigan sediments. Environ. Geol. Note 60, 14 pp. (1973)
Füchtbauer, H., Müller, G.: Sedimente und Sedimentgesteine. Sedimentpetrologie, Teil II. Stuttgart: Schweizerbart 1970
Fuge, R., James, K.H.: Trace metal concentrations in Fucus from the Bristol Channel. Mar. Pollut. Bull. 5, 9–12 (1974)
Fujiki, M.: The transitional condition of Minamata Bay and the neighbouring sea polluted by factory waste water containing mercury. 6th Int. Conf. Water Pollut. Res. Paper No. 12 (1972)
Fujino, S.: Using sand fill to cover dredge spoil containing mercury. In: Management of Bottom Sediments Containing Toxic Substances. Proc. 2nd U.S.-Japan Experts Meeting 1976, Tokyo. Corvallis, Or.: U.S. EPA 1977, pp. 144–154
Fujita, M., Hashizume, K.: Status of uptake of mercury by the fresh water diatom, Synedra ulna. Wat. Res. 9, 889–894 (1975)
Fukai, R.: A contribution to the stability aspects of metal organic complexes in sea water. Proc. Symp. Hydrogeochem. Biogeochem., Tokyo 1970. Washington: The Clarke Co. 1973, pp. 562–570
Fukai, R., Huynh-Ngoc, L.: Studies on the chemical behaviour of radionuclides in sea-water. I. General considerations and study of precipitation of trace amounts of Cr, Mn, Fe, Co, Zn and Ce. In: Radioactivity in the Sea. Vienna Int. Atomic Energy Agency, No. 22, 1968
Fukai, R., Murray, C.N., Huynh-Ngoc, L.: Variations of soluble zinc in the Var River and its estuary. Estuarine Coastal Mar. Sc. 3, 177–188 (1975)
Funke, J.W.: Pollution abatement in the electroplating industry. Technol. Guide K 29a, CSIR, 11 (1973)
Funke, J.W.: Metals in urban drainage systems and their effect on the potential reuse of purified sewage. Water S.A. 1, 36–44 (1975)
Funke, J.W., Coombs, P.: Water and effluent mangement in the electroplating and anodizing industry. CSIR (Natl. Inst. for Water Res.). Technol. Guide K 29 (1973)
Gabor, D., Colombo, U., King, A., Galli, R.: Das Ende der Verschwendung (Beyond the Age of Waste. Science, Technology and the Management of Natural Resources, Energy, Materials, Food. A Report to the Club of Rome). Stuttgart: Deutsche Verlags-Anstalt 1976
Gächter, R., Chan, K.L.-S., Chau, Y.K.: Complexing capacity of the nutrient medium and its relation to inhibition of algal photosynthesis by copper. Schweiz. Z. Hydrol. 35, 252–261 (1973)
Gad, M.A., Le Riche, H.H.: A method for separating the detrital and non-detrital fractions of trace elements in reduced sediments. Geochim. Cosmochim. Acta 30, 841–846 (1966)
Gadde, R.R., Laitinen, H.A.: Study of the sorption characteristics of synthetic hydrous ferric oxide. Environ. Lett. 223–228 (1973)

Gadow, S., Schäfer, A.: Die Sedimente der Deutschen Bucht: Korngrößen, Tonmineralien und Schwermetalle. Senckenbergiana Mar. 5, 165–178 (1973)

Gaggino, G.F., Gerletti, M., Marchetti, R., Pennacchioni, A.: Inquinamento da mercurio nelle aque interne Italiane: Il fiume Toce. Instituto di Recerca sulle Aque (CNR), Sezione di Idrobiologia Applicata, Milan Italy, 1975, 10 pp.

Galbraith, J.H., Williams, R.E., Siems, P.L.: Migration and leaching of metals from old mine tailings deposits. Ground Water 10, 33–44 (1972)

Gale, N.L., Bolter, E., Wixson, B.G.: Investigation of Clearwater Lake as a potential sink for heavy metals from lead mining in southeast Missouri. In: Trace Substances in Environmental Health. Hemphill, D.D. (ed.), Vol. X. Columbia: Univ. Missouri 1976, pp. 187–196

Gale, N.L., Hardie, M.G., Jennett, J.C., Aleti, A.: Transport of trace pollutants in lead mining wastewaters. In: Trace Substances in Environmental Health. Hemphill, D.D. (ed.), Vol. VI. Columbia: Univ. Missouri 1972, pp. 95–106

Gale, N.L., Wixson, B.G., Hardie, M.G., Jennett, J.C.: Aquatic organisms and heavy metals in Missouri's New Lead Belt. Am. Water Res. Bull. 9, 673–688 (1973)

Galloway, J.N.: Man's alteration of the natural geochemical cycle of selected trace metals. Ph. D. Thesis Univ. of California, San Diego (1972)

Galloway, J.N., Schofield, C.L., Yost, E., Likens, G.E.: Influence of acid precipitation on the water and sediment geochemistry of Adirondack Mountain lakes. Abstr. 39th Ann. Meeting ASLO, Savannah 1976

Gambrell, R.P., Khalid, R.A., Patrick, W.H., Jr.: Physicochemical parameters that regulate mobilization and immobilization of toxic heavy metals, Proc. Spec. Conf. on Dredging and its Environ. Effects, Mobile, Ala. 1976. New York: ASCE 1976, pp. 418–434

Gambrell, R.P., Khalid, R.A., Collard, V.R., Reddy, C.N., Patrick, W.H., Jr.: The effect of pH and redox potential on heavy metal chemistry in sediment-water systems affecting toxic metal bioavailability. In: Dredging: Environmental Effects and Technology. Proc. World Dredging Conference, San Pedro/Calif., 1976, 24 pp.

Gambrell, R.P., Khalid, R.A., Verloo, M.G., Patrick, W.H.: Transformations of heavy metals and plant nutrients in dredged sediments as affected by oxidation reduction potential and pH. II. Materials and methods / results and discussion. U.S. Army Corps of Engineers, Dredged Material Research Program. Vicksburg Miss., Rept. D-77-4, 309 pp. (1977)

Ganther, H.E., Goudie, C., Sunde, M.L., Kopecky, M.J., Wagner, Q., Sang-Hwan, O., Hoekstra, W.G.: Selenium, relation to decreased toxicity of methylmercury added to diets containing tuna. Science 175, 1122–1124 (1972)

Gapon, J.N.: Specific surface area of soil humus. Kolloid Z. 9, 329–334 (1947)

Garcia, W.J., Blessin, C.W., Inglett, G.E., Carlson, R.O.: Physical-chemical characteristics and heavy metal content of corn grown on sludge-treated strip-mine soil. J. Agric. Food Chem. 22, 810–815 (1974)

Gardiner, J., Stiff, M.J.: The determination of cadmium, lead, copper and zinc in ground water, estuarine water, sewage, and sewage effluent by anodic stripping voltammetry. Water Res. 9, 517–523 (1975)

Gardiner, W.G., Munoz, F.: Mercury removed from waste effluent via ion exchange. Chem. Eng. 78, 57–61 (1971)

Gardner, D.: Obervations on the distribution of dissolved mercury in the ocean. Mar. Pollut. Bull. 6, 43–46 (1975)

Gardner, D., Riley, J.P.: The distribution of dissolved mercury in the Bristol Channel and Severn Estuary. Estuarine Coastal Mar. Sci. 1, 191–192 (1973a)

Gardner, D., Riley, J.P.: Distribution of dissolved mercury in the Irish Sea. Nature (London) 241, 526–527 (1973b)

Gardner, L.R.: Organic versus inorganic trace metal complexes in sulfidic marine waters – some speculative calculations based on available stability constants. Geochim. Cosmochim. Acta 38, 1297–1302 (1974)

Gardner, L.R.: Runoff from the intertidal marsh during tidal exposure–recession curves and chemical characteristics. Limnol. Oceanogr. 20, 81–89 (1975)

Gardner, L.R.: Exchange of nutrient and trace metals between marsh sediments and estuarine waters – a field study. Final Rept. No. 63 OWRT-Proj., Water Resources Res. Inst. Clemson Univ., South Carolina, 1976

Garrels, R.M., Mackenzie, F.T., Hunt, C.: Chemical Cycles and the Global Environment. Los Altos/ Calif.: William Kaufmann, Inc. 1975, 206 pp.

Garrett, W.D.: The organic chemical composition of the ocean surface. Deep-Sea Res. 14, 221–227 (1967)

Garrigan, G.A.: Land application guidelines for sludges contaminated with toxic elements. J.W.P.C.F. 49, 2380–2389 (1977)

Gast, P.W.: Isotopic composition as a natural tracer of lead in the environment. Environ. Sci. Technol. 4, 313–314 (1970)

Gavis, J., Ferguson, J.: The cycling of mercury through the environment. Water Res. 6, 989–1008 (1972)

Ghassemi, M., Christman, R.F.: Properties of the yellow organic acids of natural waters. Limnol. Oceanogr. 13, 583–597 (1968)

Ghosh, M.M., Zugger, P.D.: Toxic effects of mercury on the activated sludge process. J.W.P.C.F. 45, 424–429 (1973)

Gibbs, R.J.: Amazon River: Environmental factors that control its dissolved and suspended load. Science 156, 1734–1736 (1970a)

Gibbs, R.J.: Mechanisms controlling world water chemistry. Science 170, 1088–1090 (1970b)

Gibbs, R.: Mechanisms controlling world water chemistry. Evaporation crystallization process. Science 172, 870–872 (1971)

Gibbs, R.J.: Water chemistry of the Amazon River. Geochim. Cosmochim. Acta 36, 1061–1066 (1972)

Gibbs, R.: Mechanisms of trace metal transport in rivers. Science 180, 71–73 (1973)

Gibbs, R.J.: Transport phases of transition metals in the Amazon and Yukon Rivers. Geol. Soc. Am. Bull. 88, 829–943 (1977)

Gibbs, R., Jarosewich, E., Windom, H.L.: Heavy metal concentrations in museum fish species: Effects of preservatives and time. Science 184, 475–477 (1974)

Giddings, J.C.: Chemistry, man, and environmental change. An integrated approach. San Francisco: Canfield Press 1973

Gilbert, T.R., Clay, A.M., Leighty, D.A.: Influence of the sediment/water interface on the aquatic chemistry of heavy metals. Environmental Chemistry Research Div., Tyndall Air Force Base, Fla., Rep. AFCEC-TR-6-22, 89 pp. (1976)

Gillespie, D.C.: Mobilization of mercury from sediments into guppies (Poecilia reticulata). J. Fish. Res. Board. Canada 29, 1035–1041 (1972)

Gillespie, D.C., Scott, D.P.: Mobilization of mercuric sulfide from sediment into fish under aerobic conditions. J. Fish. Res. Board Can. 28, 1807–1808 (1971)

Ginzburg, I.L., Yashina, R.S., Matveeva, L.A., Belyatskii, V.V., Nuzhdelovskaya, T.S.: Decomposition of certain minerals by organic acids. In: Chemistry of the Earth Crust. Vinogradov, A.P. (ed.). Jerusalem: Israel Program for Scientific Translation 1963, pp. 304–320

Giordano, P.M., Mortvedt, J.J., Mays, D.A.: Effect of municipal wastes on crop yields and uptake of heavy metals. J. Environ. Qual. 4, 394–399 (1975)

Giovanoli, R.: A simplified scheme for polymorphism in the manganese dioxides. Chimia 23, 420–470 (1969)

Gish, C.D., Christensen, R.E.: Cadmium, nickel, lead, and zinc in earthworms from roadside soil. Environ. Sci. Technol. 7, 1060–1062 (1973)

Gissel-Nielsen, G., Gissel-Nielsen, M.: Ecological effects of selenium application to field crops. Ambio 2, 114–117 (1973)

Gissy, M.: Neueste Analysen an Rheinfischen bestätigen die hochgradige Quecksilberverseuchung des Stromes. Unsere Umwelt 11, 5 (1973)

Gjessing, E.T.: Ultrafiltration of aquatic humus. Environ. Sci. Technol. 4, 437–438 (1970)

Glasby, G.P., Read, A.J.: Deep-sea manganese nodules. In: Handbook of Strata-Bound and Stratiform Ore Deposits. Wolf, K.H. (ed), Vol. VII. Amsterdam: Elsevier 1976

Gluskoter, H.J., Lindahl, P.: Cadmium: Mode of occurance in Illinois coals. Science 181, 264–266 (1973)

Gmelin: Handbuch der Anorganischen Chemie, Bd. 13. Weinheim: Verlag Chemie 1954

Goebgen, H.G., Brockmann, J.: Bindungsvermögen von anaerobem Faulschlamm für Schwermetallionen. Wasser, Luft Betr. 13, 409–412 (1969)

Goldberg, A.: Drinking water as a source of lead pollution. Environ. Health 7, 103–105 (1974)

Goldberg, E.D.: Marine geochemistry. Chemical scavengers of the sea. J. Geol. 62, 249–266 (1954)

Goldberg, E.D.: Biogeochemistry of trace metals. Geol. Soc. Am. Mem. 1, 345–358 (1957)

Goldberg, E.D.: Elemental composition of some pelagic fishes. Limnol. Oceanogr. 7, 22–25 (1962)

Goldberg, E.D.: Minor elements in seawater. In: Chemical Oceanography. Riley, J.P., Skirrow, G. (eds.), Vol. I. London: Academic Press 1965

Goldberg, E.D.: River-ocean interactions: In: Fertility of the Sea. Costlow, J.D., Jr. (ed.), Vol. I. New York: Gordon and Breach 1971, pp. 143–156

Goldberg, E.D.: Man's role in the major sedimentary cycle. In: The Changing Chemistry of the Oceans. Nobel Symposium. Dyrssen, D., Jagner, D. (eds.). Uppsala: Almquist and Wiksell 1972, pp. 267–288

Goldberg, E.D.: North Sea Science. Cambridge, Mass., London: MIT Press 1974
Goldberg, E.D.: Marine Pollution. In: Chemical Oceanography, 2nd ed. Riley, J.P., Skirrow, G. (eds.), Vol. 3. London, New York, San Francisco: Academic Press 1975a, pp. 39–89
Goldberg, E.D.: Pollution history of estuarine sediments. Oceanus 18–26 (1975b)
Goldberg, E.D.: The mussel watch. A first step in global marine monitoring. Mar. Pollut. Bull. 6, 111 (1975c)
Goldberg, E.D.: The Health of the Oceans. Paris: UNESCO Press 1976
Goldberg, E.D., Arrhenius, G.O.: Chemistry of Pacific pelagic sediments. Geochim. Cosmochim. Acta 13, 153–212 (1958)
Goldberg, E.D., Broecker, W.S., Gross, M.G., Turekian, K.K.: Radioactivity in the Marine Environment. Washington D.C.: Natl. Acad. Sci. 1971
Goldberg, E.D., Hodge, V., Koide, M., Griffin, J.J.: Metal pollution in Tokyo as recorded in sediments of the Palace Moat. Geochem. J. 10, 165–174 (1976)
Goldberg, E.D., Gamble, E., Griffin, J.J., Koide, M.: Pollution history of Narragansett Bay as recorded in its sediments. Estuarine Coastal. Mar. Sci. 5, 549–561 (1977)
Goldwater, L.: Mercury in the environment. Sci. Am. 224, 15–21 (1971)
Goleva, G.A., Polyakov, V.A., Nechayeva, T.P.: Distribution and migration of lead in ground waters. Geochem. Int. 256–268 (1970)
Golwer, A.: Beeinflussung des Grundwassers durch Straßen. Z. Dtsch. Geol. Ges. 124, 39–50 (1973)
Golwer, A., Knoll, K.H., Matthess, G., Schneider, W., Wallhäuser, K.H.: Belastung und Verunreinigung des Grundwassers durch feste Abfallstoffe. Abh. Hess. Landesamtes Bodenforsch. 73, 131 pp. (1976)
Göransson, B., Moberg, P.O.: Metal-finishing waste treatment in Sweden. J.W.P.C.F. 47, 764–772 (1975)
Gorbunow, N.L.: The significance of minerals for soil fertility. Soviet Soil Sci. 7, 757–767 (1959)
Gorham, E.: Factors influencing supply of major ions to inland waters with special reference to the atmosphere. Geol. Soc. Am. Bull. 72, 795–840 (1961)
Gorham, E., Gordon, A.G.: Some effects of smelter pollution upon aquatic vegetation near Sudbury, Ontario. Can. J. Bot. 41, 371–378 (1963)
Gorham, E., Swaine, D.J.: The influence of oxidizing and reducing conditions upon the distribution of some elements in lake sediments. Limnol. Oceanogr. 10, 268–279 (1965)
Goto, M.: Irorganic chemicals in the environment – with special reference to the pollution problems in Japan. Environ. Qual. Saf. 2, 72–77 (1973)
Gott, G.B., Botbol, J.M., Billings, T.M., Pierce, A.P.: Geochemical abundance and distribution of nine metals in rocks and soils of the Coeur d'Alene district, Shoshone County, Idaho. U.S. Geol. Surv. Open-File Rept., 3 pp. (1969)
Goyer, R.A.: Perspective on low level lead toxicity. Environ. Health 7, 102–105 (1974)
Graeser, H.J.: Research activities by water utilities. J.A.W.W.A. 64, 638–645 (1972)
Grahn, O., Hultberg, H., Landner, L.: Oligotrophication – a self-accelerating process in lakes subjected to excessive supply of acid substances. Ambio 3, 93–94 (1974)
Grancini, G., Stievano, M.B., Girardi, F., Guzzi, G., Pietra, R.: The capability of neutron activation for trace element analysis in sea water and sediment samples of the northern Adriatic Sea. J. Radioanal. Chem. 34, 65–72 (1976)
Gray, J.S., Ventilla, R.J.: Growth rates of sediment living marine protozoan as a toxicity indicator for heavy metals. Ambio 2, 118–121 (1973)
Greenland, D.J.: Interactions between humic and fulvic acids and clays. Soil. Sci. 111, 34–41 (1971)
Gregor, C.D.: Solubilization of lead in lake and reservoir sediments by NTA. Environ. Sci. Technol. 6, 278–279 (1972)
Greig, R.A., McGrath, R.A.: Trace metals in sediment of Raritan Bay. Mar. Pollut. Bull. 8, 188–192 (1977)
Greig, R.A., Reid, R.N., Wenzloff, D.R.: Trace metal concentrations in sediments from Long Island Sound. Mar. Pollut. Bull. 8, 183–188 (1977)
Grice, G.D., Reeve, M.R., Koeller, P., Menzel, D.W.: The use of large volume, transparent, enclosed sea-surface water columns in the study of stress on plankton ecosystems. Helgoländer Wiss. Meeresunters. 30, 118–133 (1977)
Griel, J.V., Robinson, R.J.: Titanium in sea water. J. Mar. Res. 11, 173–179 (1952)
Grieve, D., Fletcher, K.: Trace metals in Fraser Delta sediments. Geol. Surv. Can. Pap. 75-1B, 161–163 (1975)
Grieve, D.A. Fletcher, W.K.: Heavy metals in deltaic sediments of the Fraser River, British Columbia. Can. J. Earth Sci. 12, 1683–1693 (1976)

Grieve, D., Fletcher, K.: Interactions between zinc and suspended sediments in the Fraser River Estuary, British Columbia. Estuarine Coastal. Mar. Sci. *5*, 415–419 (1977)

Griffatong, A., Hellmann, H.: Neue Untersuchungen zur Bestimmung von gelösten und ungelösten Schwermetallen in Gewässern durch Röntgenfluoreszenz. Vom Wasser *40*, 59–87 (1973)

Griffin, R.A., Cartwright, K., Shimp, N.F., Steele, J.D., Ruch, R.R., White, W.A., Hughes, G.M., Gilkeson, R.H.: Attenuation of pollutants in municipal landfill leachate by clay minerals. I. Column leaching and field verification. Environ. Geol. Notes *78*, Ill. State Geol. Surv., 34 pp. (1976)

Griffin, R.A., Frost, R.R., Au, A.K., Robinson. G.D., Shimp, N.F.: Attenuation of pollutants in municipal landfill leachate by clay minerals. II. Heavy metal adsorption. Environ. Geol. Notes *79*, Ill. State Geol. Surv., 47 pp. (1977)

Griffith, J.J.: Influence of mines upon land and livestock in Cardiganshire. J. Agric. Sci. *9*, 241–271 (1918)

Griffith, R.G., Austin, J.: The Schuylkill metals and oil sludge in a river. Abstr. Int. Conf. Heavy Met. Environ. *C-139* (1975)

Griffith, S.M., Schnitzer, M.: The isolation and characterization of stable metal-organic complexes from tropical volcanic soils. Soil Sci. *120*, 126–131 (1975)

Griggs, G.B., Johnson, S.: Bottom sediment contamination in the Bay of Naples, Italy. Mar. Pollut. Bull. *9*, 208–214 (1978)

Griggs, G.B., Grimanis, A.P., Griman., M.V.: Bottom sediments in a polluted marine environment. Upper Saronikos Gulf, Greece. Environ. Geol. *2*, 97–106 (1978)

Grim, R.E.: Clay Mineralogy. 2nd ed. New York: McGraw-Hill 1968

Grimanis, A.P., Pantazis, G., Papadopoulus, C., Tsanos, N.: Determination of trace elements in Greek lakes by neutron activation analysis. Proc. U.N. Conf. Peaceful Uses Atom. Energy, Geneva, 1964, *15*, 412–423 (1965)

Grimshaw, D.L., Lewin, J., Fuge, R.: Seasonal and short-term variations in the concentration and supply of dissolved zinc to polluted aquatic environments. Environ. Pollut. *11*, 1–7 (1976)

Grimstone, G.: Mercury in British fish. Chem. Br. *8*, 244–247 (1972)

Grohmann, A.: Bewertung der Korrosionsneigung von Trinkwasser aus der Sicht der öffentlichen Gesundheitspflege. Bundesgesundheitsblatt *5*, 66–70 (1973)

Groot, A.J. de: Origin and transport of mud in coastal waters from the western Scheldt to the Danish frontier. In: Dev. Sedim., 1964, pp. 91–103

Groot, A.J. de: Mobility of trace elements in deltas. Trans. Comm. Comm. II and IV, Int. Soc. Soil Sci. Aberdeen, 1966, pp. 267–297

Groot, A.J. de, Allersma, E.: Field observations on the transport of heavy metals in sediments. In: Heavy Metals in the Aquatic Environment. Krenkel, P.A. (ed.). Oxford: Pergamon Press 1975, pp. 85–101

Groot, A.J. de, Salomons, W.: Influence of civil engineering projects on water quality in deltaic regions. Proc. Symp. on Effects of Urbanization and Industrialization on the Hydrological Regime and on Water Quality. Amsterdam: IAHS Publ. *123* (1977), pp. 351–357

Groot, A.J. de, Zschuppe, K.H., Bruin, M. de, Houtman, J.P.W., Singgih, P.A.: Activation analysis applied to sediments from various river deltas. Proc. Int. Conf. on Modern Trends in Activation Analysis. Gaithersberg (USA) 62–71 (1968)

Groot, A.J. de, Goeij, J.J.M., Zegers, C.: Contents and behaviour of mercury as compared with other heavy metals in sediments from the rivers Rhine and Ems. Geol. Mijnbouw *50*, 393–398 (1971)

Groot, A.J. de, Allersma, E., Driel, W. van: Zware metalen in fluviatile en marine ecosystemen. Symp. Water in Dienst van Industrie en Milieu. Publikatie No. 110 N, Sekt. 5, 24 (1973)

Groot, A.J. de, Salomons, W., Allersma, E.: Processes affecting heavy metals in estuarine sediments. In: Estuarine Chemistry. Burton, J.D., Liss, P.S. (eds.). London, New York, San Francisco: Academic Press 1976, pp. 131–157

Gross, M.G.: Concentrations of minor elements in diatomaceous sediments of a stagnant fjord. In: Estuaries. Lauff, G.H. (ed.). Am. Assoc. Adv. Sci. Publ. *83*, 273–282 (1967)

Gross, M.G.: Analysis of dredged wastes, fly ash and waste chemicals. Mar. Sci. Res. Cent. Tech. Rep. *7*, 42 (1970)

Gross, M.G.: Geologic aspects of waste solids and marine waste deposits, New York metropolitan region. Geol. Soc. Am. Bull. *83*, 3163–3176 (1972)

Gross, M.G., Black, J.A., Kalin, R.J., Schramel, J.R., Smith, R.N.: Survey of marine waste deposits, New York metropolitan region. Mar. Sci. Res. Cent., State Univ. N.Y. Tech. Rep. *8*, 72 pp. (1971)

Groth, P.: Untersuchungen über einige Spurenelemente in Seen. Arch. Hydrobiol. *68*, 305–375 (1971)

Gruner, J.W.: The origin of sedimentary iron formations, the Biwabik formation of the Mesabi Range. Econ. Geol. *17*, 407–460 (1922)

Gudernatsch, H.: Verhalten von Nitrilotriessigsäure im Klärprozess und im Abwasser. GWF Wasser Abwasser *111*, 511–516 (1970)

Gudernatsch, H.: Biologischer Abbau von Schwermetallkomplexen der Nitrilotriessigsäure in Labor-Belebtschlammanlagen. GWF Wasser Abwasser *116*, 512–517 (1975)

Gulledge, J.H., O'Connor, J.T.: Removal of As(V) from water by adsorption in aluminum and ferric hydroxides. J.A.W.W.A. *65*, 548–553 (1973)

Gunton, J.E., Nichol, I.: Delineation and interpretation of metal dispersion patterns related to mineralization in the Whipsaw Creek area. CIM Bull. 66–75 (1974)

Gupta, R.S.: On some trace metals in the Baltic. Ambio 1, 226–230 (1973)

Gupta, S.K., Chen, K.Y.: Partitioning of trace metals in selective chemical fractions on nearshore sediments. Environ. Lett. *10*, 129–158 (1975)

Gutknecht, J.: Uptake and retention of cesium-137 and zinc-65 by seaweeds. Limnol. Oceanogr. *10*, 58–66 (1965)

Guy, R.D., Chakrabarti, C.L.: Distribution of metal ions between soluble and particulate forms. Abstr. Int. Conf. Heavy Met. Environ. Toronto, Ont., Can. 1975, pp. D-29-30

Guy, R.D., Chakrabarti, C.L.: Analytical techniques for speciation of trace metals. Proc. Int. Conf. Heavy Met. Environ. Toronto 1975, pp. 275–294 (1977)

Haberer, K.: Erfahrungen mit der künstlichen Grundwasseranreicherung in den Wassergewinnungsanlagen Wiesbaden-Schierstein. GWF Wasser Abwasser *109*, 636–640 (1968)

Haberer, K.: Physikalische und chemische Eigenschaften des Wassers. In: Wasser und Luft – Handbuch Lebensmittelchemie. Souci, S.W., Quentin, K.E. (eds.). Berlin, Heidelberg, New York: Springer 1969, pp. 1–50

Haberer, K.: Trinkwasser im Umweltschutz. GWF Wasser Abwasser *113*, 555–561 (1972)

Haberer, K.: Ergebnisse langjähriger Untersuchungen zur Rheinwasserqualität. In: Künstliche Grundwasseranreicherung am Rhein. Wiss. Ber. Stadtwerke Wiesbaden 2, 17–42 (1974)

Haberer, K.: Probleme der Wassergewinnung durch Grundwasseranreicherung aus der Sicht der Stadtwerke Wiesbaden AG. GWF Wasser Abwasser *55*, 828–836 (1975)

Haberer, K., Normann, S.: Metallspuren im Wasser. Vom Wasser *38*, 157–182 (1971)

Haberer, K., Normann, S.: Die Bedeutung der Metallspuren in den Gewässern für die Trinkwasserversorgung. GWF Wasser Abwasser *113*, 382–386 (1972)

Haberer, K., Normann, S.: Untersuchungen zur Trinkwasserenthärtung an einer kontinuierlich arbeitenden Ionenaustauscheranlage. Vom Wasser *41*, 277–307 (1973)

Haberer, K., Normann, S.: Untersuchungen zur zentralen Trinkwasserenthärtung durch Schnellentkarbonisierung. Vom Wasser *42*, 399–430 (1974a)

Haberer, K., Normann, S.: Zur Wirksamkeit der physikalisch-chemischen Rheinwasseraufbereitung. Künstliche Grundwasseranreicherung am Rhein, Wiss. Ber. Stadtwerke Wiesbaden AG 2, 93–110 (1974b)

Haberer, K., Normann, S.: Untersuchungen zum Einsatz der zentralen Enthärtung in Wiesbaden-Schierstein. Forschung und Entwicklung in der Wasserwerkspraxis, Wiss. Ber. Stadtwerke Wiesbaden AG 3, 29–46 (1976)

Haberer, K., Normann, S., Reichert, J.K.: Untersuchungen zur Entnahme von Metallspuren bei der Entkarbonisierung von Trinkwasser. Forschung und Entwicklung in der Wasserwerkspraxis, Wiss. Ber. Stadtwerke Wiesbaden AG 3, 129–140 (1976)

Hagino, N., Yoshioka, K.: A study on the cause of itai-itai disease. J. Jpn. Orthop. Assoc. *35*, 812 (1961), cit. Friberg et al. 1974

Hahn, H.H., Stumm, W.: The role of coagulation in natural waters. Am. J. Sci. *268*, 368–374 (1970)

Hahne, H.C.H., Kroontje, W.: Significance of pH and chloride concentration on behaviour of heavy metal pollutants: Mercury (II), cadmium (II), zinc (II), and lead (II). J. Environ. Qual. 2, 444–450 (1973)

Håkanson, L.: Sambandet mellan kvicksilverförekomst och sedimentologisk miljö i Ekoln. Del 3. Transport och deposition av kvicksilver. Uppsala Naturgeogr. Inst. Rep. *15* (1972)

Håkanson, L.: Mercury in some Swedish lake sediments. Ambio *3*, 37–43 (1974)

Håkanson, L.: Kvicksilver i Vänern – nuläge och prognos. Naturvårdsverkets Limnol. Undersökning *80*, 121 (1975)

Håkanson, L.: A bottom sediment trap for recent sedimentary deposits. Limnol. Oceanogr. *21*, 170–174 (1976)

Håkanson, L.: Sediments as indicators of contamination – investigations in the four largest Swedish lakes. Naturvårdsverkets Limnol. Undersökning *92*, 159 (1977)

Håkanson, L., Ahl, T.: Vättern – recenta sediment och sediment-kemi. Naturvårdsverkets Limnol. Undersökning *88* (1976)
Håkanson, L., Uhrberg, R.: Determination of mercury from lake sediments. Vatten *29*, 4–13 (1973)
Håkanson, L., Uhrberg, R.: Undersökning av bly, zink och kvicksilver i recenta sediment i norra Vättern.NLU Inf. *12* (1976)
Halbach, P.: Vergleich stofflicher Eigenschaften limnischer und mariner Manganknollen. Erzmetall *27*, 161–168 (1974)
Halcrow, W., MacKay, D.W., Thornton, I.: The distribution of trace metals and fauna in the Firth of Clyde in relation to the disposal of sewage sludge. J. Mar. Biol. Ass. U.K. *53*, 721–739 (1973)
Hall, A., Valente, I.: Effects of port and town activity on Lourenzo Marques Bay. Mar. Pollut. Bull. *4*, 165–171 (1974)
Hall, K., Fletcher, K.: Trace metal pollution from a metropolitan area: sources and accumulation in the lower Fraser River and estuary. Proc. Int. Conf. Persist. Chem. Aquatic Ecosyst., I-69-74 (1974)
Hall, K., Lee, G.F.: Molecular size and spectral characterization of organic matter in a meromictic lake. Water Res. *8*, 239–251 (1974)
Hallberg, R.O.: Sedimentary sulfide mineral formation – an energy circuit system approach. Miner. Deposita *7*, 189–201 (1972)
Hallberg, R.O.: The microbiological C-N-S cycles in sediments and their effect on the ecology of the sediment-water interface. Oikos *15*, 51–62 (1973)
Hallberg, R.O.: Paleoredox conditions in the eastern Gotland Basin during the recent centuries. Merentutkimuslait. Julk./Havsforskninginst. Skr. *238*, 3–16 (1974a)
Hallberg, R.O.: Metal distribution along a profile of an intertidal area. Estuar. Coast. Mar. Sci. *2*, 153–170 (1974b)
Hamaguchi, H., Kuroda, R.: Tin – abundance in natural waters and in the atmosphere. In: Handbook of Geochemistry, 50-I/1–2. Wedepohl, K.H. (ed.). Berlin, Heidelberg, New York: Springer 1970
Hamdy, M.K., Noyes, O.R.: Formation of methyl mercury by bacteria. Appl. Microbiol. *30*, 424–432 (1975)
Hamilton, A.L., Gillespie, D.C., Barica, J., McRae, G.P.: The effect of chelating agents on the mobilization of mercury in an aquarium ecosystem. Fish. Res. Board. Can. Manuscript Rep. Ser. *1167*, 107–114 (1972)
Hamm. A.: Limnologische Untersuchungen am Tegernsee und Schliersee nach Abwasserfernhaltung (Stand 1970). Wasser Abwasserforsch. *5*, 131–150 (1971)
Hammond, A.L.: Mercury in the environment: natural and human factors. Science *171*, 788–789 (1971)
Hammond, D.E., Simpson, H.J., Matheu, G.: Methane and radon-222 as tracers for mechanisms of exchange across the sediment-water interface in the Hudson River Estuary. In: Marine Chemistry in the Coastal Environment. Church, T.M. (ed.). Am. Chem. Soc. Symp. Ser. *18*, 119–132 (1975)
Handa, B.K.: Chemistry of manganese in natural waters. Chem. Geol. *5*, 161–165 (1969/1970)
Hannah, S.A., Jelus, M., Cohen, J.M.: Removal of uncommon trace metals by physical and chemical treatment processes. J.W.P.C.F. *49* 2297–2309 (1977)
Hantge, E.: Einfluß der BASF-Kläranlage auf die Beschaffenheit des Rheinwassers. BASF *2*, 33–40 (1976)
Harbridge, W., Pilkey, O., Whaling, P., Swetland, P.: Sedimentation in the lake of Tunis: a lagoon strongly influenced by man. Environ. Geol. *1* 215–225 (1976)
Harding, S.C., Brown, H.S.: Distribution of selected trace elements in sediments of Pamlico River Estuary, South Carolina. Environ. Geol. *1*, 181–191 (1976)
Hardisty, M.W., Huggins, R.J., Kartar, S., Sainsbury, M.: Ecological implications of heavy metal in fish from the Severn Estuary. Mar. Pollut. Bull. *5*, 12–15 (1974a)
Hardisty, M.W., Kartar, S., Sainsbury, M.: Dietary habits and heavy metal concentrations in fish from the Severn Estuary and Bristol Channel. Mar. Pollut. Bull. *5*, 61–63 (1974b)
Hardy, E.P., Kreym, P.W., Volchok, H.L.: Global inventory and distribution of fallout plutonium. Nature (London) *241*, 444–445 (1973)
Hargrave, B.T., Phillips, G.A.: Adsorption of ^{14}C-DDT to particle surfaces. Proc. Intern. Conf. Transp. of Persistent Chem. Aquatic Ecosyst. *II*, 13–18 (1974)
Harrar, N.J.: Solvent effects of certain organic acids upon oxides of iron. Econ. Geol. *24*, 50–61 (1929)
Harris, E.P.: A survey of the electroplating industry to determine the sources and extent of nickel losses and to consider the technical and financial aspects of recovery. London: The British Non-Ferrous Metals Res. Assoc. (1960)

Harrison, F.L.: Effect of the physicochemical form of trace metals on their accumulation by bivalve molluscs. Reprints of papers pres. at the 176th Nat. Meet. ACS, Miama Beach, Fla., Sept. 10–15, 1978, Div. Environ. Chem., pp. 415–416

Harrison, V.F., Gow, W.A., Hughson, M.R.: Factors influencing the application of bacterial leaching to a Canadian uranium ore. J. Metals (N.Y.) *18*, 1189–1194 (1966)

Harriss, R.C.: Mercury content of deep-sea manganeous nodules. Nature (London) *219*, 54–55 (1968)

Harriss, R.C.: Ecological implications of mercury pollution in aquatic systems. Biol. Conserv. *3*, 279–283 (1971)

Harriss, R.C., Troup, A.G.: Freshwater ferromanganese concretions, chemistry, and internal structure. Science *166*, 604–606 (1969)

Harriss, R.C., White, D.B., MacFarlane, R.B.: Mercury compounds reduce photosynthesis by plankton. Science *170*, 736–737 (1970)

Hart, B.T., Davies, S.H.R.: Physico-chemical forms of trace metals and the sediment-water interface. In: Interactions Between Sediments and Fresh Water. Goltermann, H.L. (ed.). The Hague: Junk Publ. 1977, pp. 398–402

Hart, B.T., Davies, S.H.R.: A study of the physico-chemical forms of trace metals in natural waters and wastewaters. Caulfield Inst. Technol., Water Studies Centre, Tech. Rep. *6*, 188 pp. (1978)

Hartinger, L.: Die Metalle im Abwasser – ihre Toxikologie und die Chemie ihrer Ausfällung. IWL Forum *66*, 1-43 (1968a)

Hartinger, L.: Die Beseitigung von Schwermetallen aus Abwasser. Z. Wasser Abwasserforsch. *1*, 30–40 (1968b)

Hartinger, L.: Taschenbuch der Abwasserbehandlung für die metallverarbeitende Industrie, Bd. 1. Chemie. München, Wien: Hanser 1976

Hartung, R.: The role of food chains in environmental mercury contamination. In: Environmental Mercury Contamination. Hartung, R., Dinman, B.D. (eds.). Ann Arbor: Science Publ. Inc. 1972, pp. 172–174

Hartung, R.: Heavy metals in the lower Mississippi. Proc. Int. Conf. Persist. Chem. Aquatic Ecosyst., Ottawa, I-93-98 (1974)

Harvey, H.H.: Aquatic environmental quality: problems and proposals. J. Fish. Res. Board. Can. *33*, 2634–2670 (1976)

Harvey, H.W.: Recent Advances in the Chemistry and Biology of Sea Water. Cambridge Univ. Press 1945

Hässelbarth, U.: Wassergefährdende Stoffe in Oberflächengewässern aus der Sicht der Trinkwasserversorgung. GWF Wasser Abwasser *113*, 509–512 (1972)

Hasselrot, T.B.: Report on current field investigations concerning the mercury content in fish, bottom sediments, and water. Dep. Int. Fresh Res. *48*, 103–111 (1968)

Hasselrot, T.B.: Mercury in fish, water, and bottom sediment. Investigations made by the Research Laboratory of the National Swedish Environment Protection Board. Naturvårdsverkets Limnol. Undersökning SNV PM-239, 922 pp. (1972)

Hasselrot, T.B., Björklund, I.: Undersökning i samband med uddevallavarvets muddringsarbeten i Byfjorden. Statens Naturvårdsverk SNV PM-512, 7 pp. (1974)

Hasselrot, T.B., Göthberg, A.: The ways of transport of mercury to fish. Proc. Int. Conf. Persist. Chem. Aquatic Ecosyst., Ottawa *III*, 37–43 (1974)

Hassler, J.W.: Activated Carbon. New York: Chemical Publ. Co. 1963

Hattingh, W.H.J.: Reclaimed water: a health hazard? Water S.A. *3*, 104–112 (1977)

Haug, A., Melsom, S., Omang, S.: Estimation of heavy metal pollution in two Norwegian fjord areas by analysis of the brown alga ascophyllum nodosum. Environ. Pollut. *7*, 179–192 (1974)

Hausen, B.M.: Eliminierung von Schwermetallverunreinigungen in Oberflächengewässern durch die Filterwirkung von Flußufern. Mitt. Dtsch. Bodenkundl. Ges. *16*, 225–263 (1972)

Havre, G.N., Underdal. B., Christiansen, C.: Cadmium concentration in some fish species from a coastal area in southern Norway. Oikos *24*, 155–157 (1973)

Hawkes, H.E.: Downstream dilution of stream sediment anomalies. J. Geochem. Explor. *6*, 345–358 (1976)

Hawkes, H.E., Webb, J.S.: Geochemistry in Mineral Exploration. New York: Harper and Row 1962

Hawkes, H.E., Williston, S.H.: Mercury vapour as a guide to lead-zinc-silver deposits. Min. Congr. J. *48*, 30–32 (1962)

Hawkes, H.E., Bloom, H., Riddell, J.E.: Stream sediment analysis discovers two mineral deposits. 6th Commonw. Min. Metall. Congr., 259–268 (1957)

Hawley, J.R.: The problem of acid mine drainage in the province of Ontario. Ont. Water Resour. Commiss., Toronto (1972)

Haworth, R.D.: The chemical nature of humic acid. Soil Science *111*, 71–79 (1971)

Hayes, T.D., Theis, T.L.: The distribution of heavy metals in anaerobic digestion. J.W.P.C.F. *50*, 61–72 (1978)

Head, P.C.: Observations on the concentration of iron in sea water, with particular reference to Southampton water. J. Mar. Biol Assoc. U.K. *51*, 981–1003 (1971)

Head, P.C.: Organic processes in estuaries. In: Estuarine Chemistry. Burton, J.D., Liss, P.S. (eds.). London, New York, San Francisco: Academic Press 1976, pp. 53–91

Heath, G.R.: Barriers to radioactive waste migration. Oceanus *20*, 26–30 (1977)

Heichel, G.H., Hankin, L.: Particles containing lead, chlorine and bromine detected on trees with an electron microprobe. Environ. Sci. Technol. *6*, 1121 (1972)

Heide, F., Singer, E.: Der Gehalt des Saalewassers an Kupfer und Zink. Naturwissenschaften *41*, 498 (1954)

Heide, F., Herz, H., Böhm, G.: Gehalt des Saalewassers an Blei und Quecksilber. Naturwissenschaften *44*, 441–442 (1957)

Heidel, S.G., Frenier, W.W.: Chemical quality of water and trace elements in the Patuxent River Basin. Md. Geol. Surv. Rep. Invest. (1965)

Heidman, J.A., Brunner, D.R.: Solid wastes and water quality. J.W.P.C.F. *49*, 1188–1192 (1977)

Heier, K.S., Billings, G.K.: Lithium – abundance in natural waters. In: Handbook of Geochemistry 3-I. Wedepohl, K.H. (ed.). Berlin, Heidelberg, New York: Springer 1970

Heinrichs, H.: Die Untersuchung von Gesteiner und Gewässern auf Cd, Sb, Hg, Tl, Pb und Bi mit der flammenlosen Atom-Absorptions-Spektralphotometrie. Diss. Univ. Göttingen (1975)

Heinrichs, H.: Emissions of 22 elements from brown-coal combustion. Naturwissenschaften *64*, 479–481 (1977)

Heitfeld, K.H., Schöttler, U.: Versickert wohin? Kontamination des Wassers im Bereich von Abfallhalden durch Spurenmetalle. Umwelt *1*, 57–58 (1973)

Helgeson, H.C.: Geologic and thermodynamic characteristics of the Salton Sea geothermal system. Am. J. Sci. *266*, 129–137 (1968)

Hellmann, H.: Die Charakterisierung von Sedimenten auf Grund ihres Gehalts an Spurenmetallen. Dtsch. Gewässerkundl. Mitt. *14*, 160–164 (1970a)

Hellmann, H.: Die Absorption von Schwermetallen an den Schwebstoffen des Rheins – eine Untersuchung zur Entgiftung des Rheinwassers (ein Nachtrag). Dtsch. Gewässerkundl. Mitt. *14*, 42–47 (1970b)

Hellmann, H.: Die Ermittlung der Herkunft der Schlammablagerungen im Ginsheimer Altrhein. Forschungsber. Bundesanst. Gewässerkd., 16 (1970c)

Hellmann, H.: Untersuchungen zum Beitrag von Abwässern an der Schlammbildung in Bundeswasserstraßen. Z. Binnenschiffahrt Wasserstraßen *11*, 427–431 (1971)

Hellmann, H.: Herkunft der Sinkstoffablagerungen in Gewässern. Dtsch. Gewässerkundl. Mitt. *16*, 131–141 (1972a)

Hellmann, H.: Definition und Bedeutung des Backgrounds für umweltschutzbezogene gewässerkundliche Untersuchungen. Dtsch. Gewässerkundl. Mitt. *16*, 170–174 (1972b)

Hellmann, H.: Matrix-Effekt und Korngrößenverteilung bei der Röntgenfluoreszenzanalyse von Feststoffen der Gewässer. Z. Anal. Chem. *263*, 14–19 (1973)

Hellmann, H.: Mineralölprodukte in den Westdeutschen Schiffahrtskanälen. Z. Binnenschiffahrt Wasserstraßen *2*, 48–52 (1975)

Hellmann, H., Griffatong, A.: Herkunft der Sinkstoffablagerungen in Gewässern (1. Mitt.: Chemische Untersuchungen der Schwermetalle). Dtsch. Gewässerkundl. Mitt. *16*, 4–18 (1972)

Helmer, R.: Pollutants from land-based sources in the Mediterranean. Ambio *6*, 312–316 (1977)

Helmke, P.A., Koons, R.D., Iskandar, I.K.: The determination of heavy metal pollution of sediments by analysis of the $<2\mu$ fraction. Abstr. Int. Conf. Heavy Met. Environ., Toronto, D-31 (1975)

Helmke, P.A., Koons, R.D., Schomberg, P.J., Iskandar, I.K.: Determination of trace element contamination of sediments by multielement analysis of clay size fraction. Environ. Sci. Technol. *11*, 984–989 (1977)

Helsinger, M.H., Friedman, G.M.: The effects of industrialization and urbanization on the upper Hudson River Basin, New York. Abstr. 10th Annu. Meet. North. Sect. Geol. Soc. Am. *7*, 73 (1975)

Helz, G.R.: Trace element inventory for the northern Chesapeake Bay with emphasis on the influence of man. Geochim. Cosmochim. Acta *40*, 573–580 (1976)

Helz, G.R., Huggett, R.J., Hill, J.M.: Behavior of manganese, iron, copper, zinc, cadmium, and lead discharged from a wastewater treatment plant into an estuarine environment. Water Res. *9*, 631–636 (1975)

Hem, J.D.: Complexes of ferric iron with tannic acid. U.S. Geol. Surv. Water Supply Paper *1459-D*, 75–94 (1960)

Hem, J.D.: Deposition and solution of manganese oxides. U.S. Geol. Surv. Water Supply Pap. *1667-B*, 42 (1964)
Hem, J.D.: Study and interpretation of the chemical characteristics of natural water. U.S. Geol. Surv. Water Supply Pap. (2nd ed.) *1473*, 363 (1970)
Hem, J.D.: Chemistry and occurrence of cadmium and zinc in surface water and groundwater. Water Resour. Res. *8*, 661–679 (1972)
Hem, J.D.: Role of hydrous metal oxides in the transport of heavy metals in the environment (discussion of the paper of G.F. Lee). In: Heavy Metals in the Aquatic Enrivonment. Krenkel, P.A. (ed.). Oxford: Pergamon Press 1975, pp. 149–153
Hem, J.D.: Geochemical controls on lead concentrations in stream water and sediments. Geochim. Cosmochim. Acta *40*, 599–609 (1976)
Hem, J.D., Durum, W.H.: Solubility and occurrence of lead in surface water. J. Am. Water Works Assoc. *65*, 562–567 (1973)
Hem, J.D., Roberson, C.E., Lind, C.J., Polzer, W.L.: Interactions of aluminium with aqueous silica at 25°C. U.S. Geol. Surv. Water Supply Pap. *1827-E*, 33 (1973)
Hendrey, G.R., Borgstrøm, R., Raddum, G.: Acid precipitation in Norway: effects on benthic faunal communities. Abstr. 39th Annu. Meet. ASLO, Savannah (1976)
Henriksen, A., Wright, R.F.: Effects of acid precipitation on a small acid lake. Paper presented at Nordic Hydrological Conf., Reykjavik (1976)
Hermann, A.G.: Über die Einwirkung Cu-, Sn-, Pb- und Mn-haltiger Erdölwässer auf die Stassfurt-Serie des Südharzbezirkes. Neues Jahrb. Mineral. Monatsh. 60–67 (1961)
Hermann, A.G.: Praktikum der Gesteinsanalyse – Chemisch-instrumentelle Methoden zur Bestimmung der Hauptkomponenten. Berlin, Heidelberg, New York: Springer 1975
Herrig, H.: Untersuchungen an Flußwasser-Inhaltsstoffen. GWF Wasser Abwasser *110*, 1385–1391 (1969)
Herzel, F.: Hygienische Probleme beim Wassertransport in Kunststoffrohren. Schriftenr. Ver. Wasser Boden Lufthyg. *27*, 13–19 (1968)
Hesse, J.L., Evans, E.D.: Heavy metals in surface water, sediments, and fish in Michigan. Mich. Water Resour. Comm. Dep. Nat. Resour., 58 pp. (1972)
Hesse, P.R.: A Textbook of Soil Chemical Analysis. London: John Murray Publ. 1971
Hesslein, R.: An in situ sampler for close interval pore water studies. Limnol. Oceanogr. *21*, 912–914 (1976)
Heydemann, A.: Adsorption aus sehr verdünnten Kupferlösungen an reinen Tonmineralien. Geochim. Cosmochim. Acta *15*, 305–329 (1959)
Heydt, G.: Schwermetallgehalte von Wasser, Wasserpflanzen, Chironomidae und Mollusca der Elsenz. Dipl. Arbeit Univ. Heidelberg, 143 pp. (1977)
Heyn, A.: Die Bestimmung von Schwermetallen in der Donau im Raum Wien unter Anwendung der Flammenatomabsorption. Österr. Abwasserrundsch. *3*, 41–43 (1974)
Hibbard, P.L.: The chemical status of zinc in the soil with method of analysis. Hilgardia *13*, 1–29 (1940)
Higgins, I.R.: Ion exchange. Environ. Sci. Technol. *7*, 1110–1114 (1973)
Higgins, T.E.: The fate of copper in a eutrophic lake following treatment by copper sulfate. M.S. Thesis, Univ. of Notre Dame, Notre Dame, Indiana (1972). Cit. P.C. Singer (1977)
Hildebrand, E.E., Blum, W.E.: Lead fixation by clay minerals. Naturwissenschaften *61*, 169 (1974a)
Hildebrand, E.E., Blum, W.E.: Lead fixation by iron oxides. Naturwissenschaften *61*, 169–170 (1974b)
Hildebrand, E.E., Blum, W.E.: Lead fixation by soil humic acids. Naturwissenschaften *61*, 128–129 (1974c)
Hildebrand, S.G., Andren, A.W., Huckabee, J.W.: Distribution and bioaccumulation of mercury in biotic and abiotic compartments of a contaminated river-reservoir system. Oak Ridge Natl. Lab. Rep., 28 pp. (1975)
Hill, R.D.: Control and prevention of mine drainage. In: Cycling and Controls of Metals. Cincinnati, Ohio: Natl. Environ. Res. Center, U.S. Environ. Protect. Agency 1973, pp. 91–94
Hill, R.D., Wilmoth, R.C., Scott, R.B.: Neutrolysis treatment of acid mine drainage. 26th Ann. Purdue Ind. Waste Conf. Lafayette, Indiana, 1971
Hilmer, E.: Geochemische Untersuchungen im Bereich der Lagerstätte Meggen, Rheinisches Schiefergebirge. Fak. Bergbau Hüttenwes., Tech. HS Aachen, Diss. (1972)
Hinneri, S.: Enrichment of elements, especially heavy metals in recent sediments of the freshwater reservoir of Uusikaupunki, SW coastland of Finland. Turun Yliopiston Julkaisuja Annales Universitatis Turkuensis – Ser. A, II Biol. Geogr. Geol. *56*, 30 (1974)

Hinrich, H.: Rhein bei Karlsruhe und Koblenz; Gegenüberstellung von Abfluß, Schwebstoffgehalt und Salzgehalt (einschl. der Frachten). Wasser Luft Betr. *16*, 421–424 (1972)

Hirst, D.M.: The geochemistry of modern sediment from the Gulf of Paria – II: The location and distribution of trace elements. Geochim. Cosmochim. Acta *26*, 1147–1187 (1962)

Hirst, D.M., Nicholls, G.D.: Techniques in sedimentary geochemistry: 1. Separation of the detrital and non-detrital fractions of limestone. J. Sediment. Petrol. *28*, 461–468 (1958)

Hjulström, F.: Studies of the morphological activity of rivers as illustrated by the River Fyris. Bull. Geol. Inst. Univ. Uppsala *25*, 221–452 (1934)

Hodenberg, A.v.: Ermittlung von Toxizitätsgrenzwerten von Kupfer, Zink und Blei in Getreide, Rotklee und Rüben sowie Aufklärung der Toxizitätsschäden an Feldpflanzen im Harzvorland. Diss. Kiel 1974. Cit. Wagner, K.-H. (1977)

Hodgson, J.F., Lindsay, W.L., Trierweiler, J.F.: Micronutrient cation complexing in soil solution: II. Complexing of zinc and copper in displaced solution from calcareous soils. Soil Sci. Soc. Am. Proc. *30*, 723–726 (1966)

Hoffmann, G.L., Duce, R.A., Walsh, P.R., Hoffman, E.J., Fasching, J.L., Ray, B.J.: Residence time of some particulate trace metals in the oceanic surface microlayer: significance of atmospheric deposition. J. Rech. Atmos. *8*, 745–759 (1974)

Hogson, G.F.: Cobalt reactions with montmorillonite. Soil. Sci. Soc. Am. Proc. *24*, 165–168 (1960)

Hohl, H., Stumm, W.: Interaction of Pb^{2+} with hydrous γ-Al_2O_3. J. Colloid Interface Sci. *55*, 281–288 (1976)

Holdren, G.R., Bricker, O.P., Matisoff, G.: A model for the control of dissolved manganese in the interstitial waters of the Chesapeake Bay. In: Marine Chemistry in the Coastal Environment. Church, T.M. (ed.). Am. Chem. Soc. Symp. Ser. *18*, 364–381 (1975)

Holliday, L.M., Liss, P.S.: The behaviour of dissolved iron, manganese and zinc in the Beaulieu Estuary, S. England. Estuar. Coast. Mar. Sci. *4*, 349–353 (1976)

Hollister, C.D.: The seabed option. Oceanus *20*, 18–25 (1977)

Holluta, J., Bauer, L., Kölle, W.: Über die Einwirkung steigender Flußwasserverschmutzung auf die Wasserqualität und die Kapazität der Uferfiltrate. GWF Wasser Abwasser *109*, 1406–1409 (1968)

Holm, H., Cox, M.F.: Mercury in aquatic systems. Methylation, Oxidation, Reduction and Bioaccumulation. Natl. Environ. Res. Cent. U.S. EPA, Corvalis, Oregon, EPA 660/3-74-021, 1974, 38 pp.

Holmes, C.W., Slade, E.A., McLerran, C.J.: Migration and redistribution of zinc and cadmium in marine estuarine system. Environ. Sci. Technol. *8*, 255–259 (1974)

Holmgren, G.S.: A rapid citrate-dithiorite extractable iron procedure. Soil. Sci. Soc. Am. Proc. *31*, 210–211 (1967)

Holtzman, R.B.: Isotopic composition as a natural tracer of lead in the environment. Environ. Sci. Technol. *4*, 314–317 (1970)

Hölzinger, J.: Der Einfluß von Sulfit-Zellstoff-Abwässern und Schwermetallen auf das Ökosystem des Öpfinger Donaustausees. J. Ornit. *11*, 329–415 (1977)

Hoos, R.A.W., Holman, W.N.: A preliminary assessment of the effects of Anvil Mine on the environmental quality of Rose Creek, Yukon. Environ. Can. Surv. Rep. EPS 5-PR-73-8 (1973)

Höpner, T., Orliczek, C.: Humic matter as sediment component in estuaries. Paper presented at the Unesco Workshop on the Biochemistry of Estuarine Sediments. Melreux/Belgium (1976)

Horn, M.K., Adams, J.A.S.: Computer-derived geochemical balances and element abundances. Geochim. Cosmochim. Acta *30*, 279–297 (1966)

Hörnström, E., Ekström, C., Miller, U., Dickson, W.: Försurningens inverkan pa vastkustsjöar. Inform. Sötvattens-Laboratoriet Drottningsholm *4*, 81 (1973)

Horvath, G.J., Harriss, R.C., Mattraw, H.C.: Lead development and heavy metal distribution in the Florida Everglades. Mar. Pollut. Bull. *3*, 182–184 (1972)

Hosohara, K.: Mercury content of deep-sea water. Nippon Kagaku Zasshi *82*, 1107–1108 (1961)

Hosokawa, I., Ohshima, F., Kondo, N.: On the concentration of the dissolved chemical elements in the estuary of the Chikugogawa River. J. Oceanogr. Soc. Jpn. *26*, 1–5 (1970)

Houtman, J.P.W.: Trace elements behaviour in soil of some Indonesial sawahs and in sludge of an Indonesian river. Chem. Phys. Eng., Delft, Progr. Rep. *1*, 5–16 (1973)

Huang, C.P.: The removal of chromium by carbon adsorption. J.W.P.C.F. *47*, 2437–2439 (1975)

Huang, C.P., Stumm, W.: Specific adsorption of cations on hydrous γ-Al_2O_3. J. Colloid Interface Sci. *43*, 409–420 (1973)

Huang, W.H., Keller, W.D.: Dissolution of rock-forming silicate minerals in organic acids. Am. Mineral. *55*, 2076–2094 (1970)

Huang, W.H., Keller, W.D.: Dissolution of clay minerals in dilute organic acids at room temperature. Am. Mineral. *56*, 1082–1095 (1971)

Huang, W.H., Keller, W.D.: Geochemical mechanics for dissolution, transport, and deposition of aluminum in the zone of weathering. Clays Clay Miner. *20*, 69–74 (1972)

Huber, W., Popp, K.H.: Der biologische Abbau der Nitrilotriessigsäure in Gegenwart von Cadmium-Ionen. Fette Seifen Anstrichm. *71*, 166–168 (1972)

Huey, C., Brinkman, F.E., Grim, S., Iverson, W.P.: The role of bacterial methylation. Proc. Int. Conf. Transp. Persist. Chem. Aquatic Ecosyst. Ottawa, Canada, II, 1974, pp. 73–78

Huggett, R.J., Bender, M., Slone, D.H.: Mercury in sediments from three Virginia estuaries. Chesapeake Sci. *12*, 280–282 (1972)

Hughes, J.L.: Evaluation of ground-water degradation resulting from waste disposal to alluvium near Barstow, California. U.S. Geol. Survey Prof. Paper *878*, 33 pp. (1975)

Huhn, F.J.: Lake sediment records of industrialization in the Sudbury area of Ontario. M. Sc. Thesis, Univ. Toronto 1974. Cit. R.G. Semkin, J.R. Kramer (1976)

Hullinger, D.L.: A study of heavy metals in Illinois impoundments. J. A.W.W.A. 572–576 (1975)

Humenick, M.J., Schnoor, J.L.: Improving mercury (II) removal by activated carbon. J. Environ. Eng. Div. ASCE *100*, 1249–1253 (1974)

Hung, T.C., Li, T.H., Wu, D.C.: The pollution of heavy metals in the Kaohsiung Harbor, Taiwan. Proc. Int. Conf. Heavy Met. Enrivon., Toronto 1975, *II/2*, 809–820 (1977)

Hunt, B.S., Henson, E.B.: Recent sedimentation and water properties. N.Y. State Geol. Assoc. 41st Annu. Meet., 21–35 (1969)

Huntzicker, J.J., Friedländer, S.K., Davidson, C.I.: Material balance for automobile-emitted lead in Los Angeles Basin. Environ. Sci. Technol. *9*, 448–457 (1975)

Hutchinson, G.E.: A Treatise on Limnology. New York: Wiley 1957

Hutchinson, G.E., Wollack, A.: Studies on Connecticut lake sediments. II. Chemical analysis of a core from Linsley Pond, N. Branford. Am. J. Sci. *238*, 493–497 (1940)

Hutchinson, G.E., Wollack, A., Setlow, J.K.: The chemistry of lake sediments from Indian Tibet. Am. J. Sci. *241*, 533–542 (1943)

Hutchinson, T.C.: The effects of acid rainfall and heavy metal particulates on a boreal forest ecosystem near the Sudbury smelting region of Canada. Proc. 1st Int. Symp. Acid Precipitation and the Forest Ecosystem, USDA Report NE-23, 1976

Hutchinson, T.C., Czyrska, H.: Cadmium and zinc toxicity and synergism to floating aquatic plants. Water Pollut. Res. Can. Inst. Environ. Sci. Eng., WI-3, 59–65 (1972)

Hutchinson, T.C., Fitchko, J.: Heavy metal concentrations in river mouth sediments around the Great Lakes. Proc. Int. Conf. Persist. Chem. Aquatic Ecosyst. *1*, 69–77 (1974)

Hutchinson, T.C., Havas, M.: The effect of long-term acidification of tundra ponds in the Canadian Arctic at Smoking Hills, N.W.T. Abstr. SIL-Congr., Copenhagen, 124 (1977)

Hutchinson, T.C., Stokes, P.M.: Heavy metal toxicity and algal bioassay. ASTM *573*, 320–343 (1975)

Hutchinson, T.C., Whitby, L.M.: A study of airborne contamination and soils by heavy metals from the Sudbury copper-nickel smelter, Canada. Inst. Environ. Sci. Eng. Univ. Toronto, Publ. EL-3 (1973)

Hutchinson, T.C., Whitby, L.M.: Heavy-metal pollution in the Sudbury mining and smelting region of Canada: I. Soil and vegetation contamination by nickel, copper, and other metals. Environ. Conserv. *1*, 132–132 (1974)

Hutchinson, T.C., Fedorenko, A., Fitchko, J., Kuja, A., Van Loon, J.: Nickel and copper movement in an aquatic ecosystem. 2nd Int. Symp. Environ. Biogeochem. (1975)

Hyde, H.C.: Utilization of wastewater sludge for agricultural soil enrichment. J.W.P.C.F. *48*, 77–84 (1976)

Ikeya, M.: Trace elements in total particulate material from surface sea water. Nature (London) *255*, 50–51 (1975)

Imura, N., Sukegawa, E., Pan, S.K., Nagao, K., Kim, J.Y., Kwan, T., Ukita, T.: Chemical methylation of inorganic mercury with methylcobalamin, a vitamin B12 analog. Science *172*, 1248–1249 (1971)

Inoue, Y.: Safety evaluation for marine disposal of sludges contaminated with heavy metals, Mizu Shori Gijutsu *15*, 753–766 (1974). Cit. CA *84*, 8733 (1976)

Ireland, M.P.: Result of fluvial zinc pollution on the zinc content of littoral organisms in Cardigan Bay, Wales, Environ. Pollut. *4*, 27–35 (1973)

Ireland, M.P.: Variations of the zinc, copper, manganese, and lead content of Balanus balanoides in Cardigan Bay, Wales, Environ. Pollut. *7*, 65–75 (1974)

Irion, G.: Lithium als Anreicherungsprodukt in zwei türkischen Salzseen. Naturwissenschaften *59*, 467 (1972)

Irion, G., Förstner, U.: Chemismus und Mineralbestand amazonischer See-Tone. Naturwissenschaften 62, 179 (1975)

Irving, H., Williams, R.J.P.: Order of stability of metal complexes. Nature (London) 162, 746–747 (1948)

Irving, H., Williams, R.J.P.: The stability of transition metal complexes. J.Chem. Soc. 3182–3210 (1953)

Isensee, A.R., Kearney, P.C., Woolson, E.A., Komes, G.E., Williams, V.P.: Distribution of alkyl arsenicals in model ecosystem. Environ. Sci. Technol. 7, 841–845 (1973)

Ishibashi, M., Ueda, S., Yamamoto, Y.: The chemical composition and the cadmium, chromium, and vanadium contents of shallow-water deposits in Tokyo Bay. J. Oceanogr. Soc. Jpn. 26, 189–194 (1970)

Iskandar, I.K.: Urban waste as a source of heavy metals in land treatment. Proc. Int. Conf. on Heavy Metals in the Environ., Toronto, 1975. II/1, 417–432 (1977)

Iskandar, I.K., Keeney, D.R.: Concentration of heavy metals in sediment cores from selected Wisconsin lakes. Environ. Sci. Technol. 8, 165–170 (1974)

Iskandar, I.K., Syers, J.K., Jacobs, L.W., Keeney, D.R., Gilmore, J.T.: Determination of total mercury in sediments and soils. Analyst 97, 388–399 (1972)

Ito, K.: Heavy metal pollution in the sediments of Nagoya Harbor. Kogai To Taisaku 11, 650–659 (1975). Cit. CA 83, 208520r

D'Itri, F.M.: Sources of mercury in the environment. In: Environmental Mercury Contamination. Hartung, R., Dinman, B.D. (eds.). Ann Arbor Science Publishers Inc. 1972, pp. 5–25

D'Itri, F.M., Annett, C.S., Fast, A.W.: Comparison of mercury levels in an oligotrophic and an eutrophic lake. Mar. Technol. Soc. J. 5, 10–14 (1971)

Jackson, G.A., Morgan, J.J.: Theory of chelator-metal ion interaction and phytoplankton growth. Unpubl. manuscript, W.M. Keck Lab. Environ. Eng. Sci. Calif. Inst. Technol., Cit. Sibley, T.H., Morgan, J.J. (1975)

Jackson, K.S.: Geochemical dispersion of elements via organic complexing. Unpubl. Thesis, Carleton Univ., Ottawa, Canada, 344 (1975)

Jackson, M.L.: Soil Chemical Analysis. Englewood Cliffs: Prentice Hall Inc. 1958

Jackson, M.L., Whittig, L.D., Pennington, R.P.: Segregation procedure for mineralogic analysis of soils. Proc. Soil Sci. Soc. Am. 14, 77–81 (1950)

Jackson, S., Brown, V.: Effect of toxic wastes on treatment processes and water-courses. Water Pollut. Control (London) 1970, 292–296

Jackson, T.A.: The biogeochemistry of heavy metals in polluted lakes and streams at Flin Flon, Canada, and a proposed method for limiting heavy-metal pollution of natural waters. Environ. Geol. 2, 173–189 (1978)

Jacobs, L.W., Keeney, D.R.: Methylmercury formation in mercury-treated river sediments during in situ equilibration. J. Environ. Qual. 3, 121–126 (1974)

Jaffé, D., Walters, J.K.: Trace metals in sediments from the Humber Estuary. Abstr. Int. Conf. Heavy Met. Enrivon. C-270 (1975)

Jaffé, D., Walters, J.K.: Intertidal trace metal concentrations in some sediments from the Humber Estuary. Sci. Total Environ. 7, 1–15 (1977)

James, R.O., Healy, T.W.: Adsorption of hydrolyzable metal ions at the oxide-water interface. J. Colloid Interface Sci. 40, 42–52, 65–81 (1972)

James, R.O., McNaughton, M.G.: The adsorption of aqueous heavy metal on inorganic minerals. Geochim. Cosmochim. Acta 41, 1549–1555 (1977)

Jefferies, D.F.: Fission-product radionuclides in sediments from the north-east Irish Sea. Helgol. Wiss. Meeresunters. 17, 280–290 (1968)

Jefferies, D.F., Hewett, C.J.: The accumulation and excretion of radioactive caesium by the Plaice (Pleuronectes platessa) and the Thronback Ray (Raia clavata). J. Mar. Biol. Assoc. U.K. 51, 411–422 (1971)

Jeffries, D.S., Stumm, W.: The metal-adsorption chemistry of buserite. Can. Min. 14, 16–22 (1976)

Jenkins, R., Vries, J.L. de: An Introduction to X-ray Powder Diffractometry. Eindhoven, Holland: N.V. Phillips Gloeilampenfabrieken 1970

Jenne, E.A.: Controls on Mn, Fe, Co, Ni, Cu, and Zn concentrations in soils and water: the significant role of hydrous Mn- and Fe-oxides. Am. Chem. Soc. Adv. Chem. Ser. 73, 337–387 (1968)

Jenne, E.A.: Trace element sorption by sediments and soils – sites and processes. In: Symposium on Molybdenum. Chappell, W., Petersen, K. (eds.). Vol. 2. New York: Marcel Dekker 1976, pp. 425–553

Jenne, E.A., Luoma, S.N.: Forms of trace elements in soils, sediments, and associated waters: an overview of their determination and availability. Paper presented at 15th Life Sciences Symp. Biological Implications of Metals in the Environment, Hanford 1975

Jenne, E.A., Wahlberg, J.S.: Manganese and iron oxide scavenging of cobalt-60 in White Oak Creek sediment (Oak Ridge, Tenn.). Am. Geophys. Union Trans. *46*, 170 (1965)

Jennett, J.C., Wixson, B.G.: Treatment and control of lead mining wastes in S.E. Missouri. Trans. 26th Purdue Ind. Waste Conf., 476–491 (1971)

Jennett, J.C., Wixson, B.G.: The New Lead Belt: aquatic metal pathways control. Proc. Int. Conf. Heavy Met. Environ., Toronto 1975, *II/1*, 247–255 (1977)

Jennett, J.C., Wixson, B.G., Bolter, E., Gale, N.: Transport mechanisms of lead industry wastes. Trans. 28th Purdue Ind. Waste Conf., 496–512 (1973)

Jenny, H., Elgabaly, M.M.: Cation and anion interchange with zinc montmorillonite clays. J. Phys. Chem. *47*, 399–410 (1943)

Jensen, S., Jernelöv, A.: Biosynthesis of methyl mercury. Nordforsk Biocidinform. (Swed.) *10*, 4–12 (1967)

Jensen, S., Jernelöv, A.: Biological methylation of mercury in aquatic organisms. Nature (London) *233*, 753–754 (1969)

Jernelöv, A.: Release of methyl mercury from sediments with layers containing inorganic mercury at different depths. Limnol. Oceanogr. *15*, 958–959 (1970)

Jernelöv, A.: Factors in the transformation of mercury to methyl-mercury. In: Environmental Mercury Contamination. Hartung, R., Dinman, B.D. (eds.). Ann Arbor Sci. Publ. 1972, pp. 167–172

Jernelöv, A.: Microbial alkylation of metals: Proc. Int. Conf. Heavy Met. Environ., Toronto 1975, *II/2*, 845–859 (1977)

Jernelöv, A., Åsell, B.: The feasibility of restoring mercury-contaminated waters. Proc. Int. Conf. Heavy Metals in the Aquatic Environment. Vanderbilt Univ. Nashville, Tenn. 1974

Jernelöv, A., Landner, L., Larsson, T.: Swedish perspectives on mercury pollution. J.W.P.C.F. *47*, 810–822 (1975)

Jernelöv, A., Lann, H.: Mercury accumulation in food chains. Oikos *22*, 403–406 (1971)

Jernelöv, A., Lann, H.: Studies in Sweden on feasibility of some methods for restoration of mercury-contaminated bodies of water. Environ. Sci. Technol. *7*, 712–718 (1973)

Joensuu, O.I.: Fossil fuels as a source of mercury pollution. Science *172*, 1027–1928 (1971)

Johansson, K.: The fundamental chemical and physical characteristics of Swedish lakes; heavy metal content in lake sediment from some lakes on the Swedish west coast and its connection with the atmospheric supply. Abstr. SIL-Congress, Copenhagen, 133 (1977)

John, M.K.: Lead contamination of some agricultural soils in western Canada. Environ. Sci. Technol. *5*, 1199 (1971)

John, M.K.: Transfer of heavy metals from soils to plants. Proc. Int. Conf. Heavy Met. Environ., Toronto 1975, *II/1*, 365–378 (1977)

Johnels, A.G., Westermark, T., Berg, W., Persson, P.I., Sjöstrand, B.: Pike (Esox lucius L.) and some other aquatic organisms in Sweden as indicators of mercury contamination in the environment. Oikos *18*, 323–333 (1967)

Johnson, D.L., Pilson, M.E.Q.: Arsenate in the western North Atlantic and adjacent regions. J. Mar. Res. *30*, 140–149 (1972)

Johnson, H.: Determination of selenium in solid wastes. Environ. Sci. Technol. *4*, 850–853 (1970)

Johnson, P.G., Villa, O., Jr.: Distribution of metals in Elizabeth River sediments. U.S. Natl. Tech. Inf. Serv. Rep. PB-260501 (1974)

Johnson, R.D., Miller, R.J., Williams, R.E., Wai, C.M., Wiese, A.C., Mitchell, J.E.: The heavy metal problem of Silver Valley, Northern Idaho. Proc. Int. Conf. Heavy Met. Environ., Toronto 1975, *II/2*, 465–486 (1977)

Johnston, L.H.: Geochemistry of selected sediment cores from the Kingston Basin – upper St. Lawrence River area. Abstr. Conf. Great Lakes Res. (1977)

Johnston, R.: Sea water, the natural medium of phytoplankton. II. Trace metals and chelation, and general discussion. J. Mar. Biol. Assoc. U.K. *44*, 87–109 (1964)

Jonasson, I.R.: Mercury in the natural environment: a review of recent work. In: Geological Survey of Canada – Department of Energy, Mines and Resources. Ottawa: Queen's Printer for Canada 1970

Jonasson, I.R.: Migration of trace metals in snow. Nature (London) *241*, 447–448 (1973)

Jonasson, I.R.: Detailed hydrogeochemistry of two small lakes in the Greenville Geological Province. Geol. Surv. Can. Pap. *76-13*, 37 (1976)

Jonasson, I.R.: Geochemistry of sediment/water interactions of metals, including observations on availability. In: The Fluvial Transport of Sediment-Associated Nutrients and Contaminants, IJC/PLUARG. Shear, H., Watson, A.E.P. (eds.). Windsor/Ont.: 1977, pp. 255–271

Jonasson, I.R., Boyle, R.W.: Geochemistry of mercury and origins of natural contamination of the environment. CIM Trans. *75*, 8–15 (1972)

Jones, A.S.G.: The concentration of copper, lead, zinc, and cadmium in shallow marine sediments, Cardigan Bay, Wales. Mar. Geol. *14*, M1–M9 (1973)
Jones, B.F., Kennedy, V.C., Zellweger, G.W.: Comparison of observed and calculated concentrations of dissolved Al and Fe in stream water. Water Resour. Res. *10*, 791–793 (1974)
Jones, J.R.E.: Antagonism between two heavy metals in their toxic action on freshwater animals. Proc. Zool. Soc. Ser. A., 481–499 (1938)
Jones, J.R.E.: The oxygen consumption of Gasterostens aculeatus L. in toxic solutions. J. Exp. Biol. *23*, 298–311 (1946)
Jones, M.B.: Synergistic effects of salinity, temperature, and heavy metals on mortality and osmoregulation in marine and estuarine isopods (crustacea). Mar. Biol. *30*, 13–20 (1975)
De Jong, G.J., Rekers, C.J.N.: The Akzo Process for the removal of mercury from waste water. J. Chromatogr. *102*, 443–450 (1974)
Julien, A.A.: On the geological action of the humus acids. Proc. Am. Assoc. Adv. Sci. *28*, 311–410 (1879)
Jung, K.D.: Wirkungskonzentration (gesundheits-)schädigender Stoffe im Wasser für niedere Wasserorganismen sowie kalt- und warmblütige Wirbeltiere einschliesslich des Menschen bei oraler Aufnahme des Wassers oder Kontakt mit dem Wasser. Gelsenkirchen: Hygiene-Institut des Ruhrgebiets 1973
Kaback, D.S.: Transport of molybdenum in mountainous streams, Colorado. Geochim. Cosmochim. Acta *40*, 581–582 (1976)
Kamps, L., Carr, R., Miller, H.: Total mercury – monomethylmercury content of several species of fish. Bull. Environ. Contam. Toxicol. *8*, 273–279 (1972)
Kanamori, S.: Geochemical study of arsenic in natural water. III. The significance of ferric hydroxide precipitate in stratification and sedimentation of arsenic in lake waters. J. Earth Sci. *13*, 46–51 (1965)
Kanamori, S., Sugawara, K.: Geochemical study of arsenic in natural waters, Part 2: Arsenic in river water. Res. Lab. Rep., Fac. Sci., Nagoya Univ. *13*, 36–45 (1965)
Karbe, L., Antonacopoulos, N., Schnier, C.: The influence of water quality on accumulation of heavy metals in aquatic organisms. Verh. Int. Ver. Limnol. *19*, 2094–2101 (1975)
Kato, K., Takahashi, H., Morishita, Y., Mori, H., Umemura, M., Watanabe, N., Hayakawa, T., Yamada, F.: Influence of the Togane arsenic mine on the Wada River. Gifu-ken Eisei Kenkyusho Ho *18*, 31–38 (1973). Cit. CA *85*, 25117m
Kazantzis, G.: The poison chain for mercury in the environment. Int. J. Environ. Stud. *1*, 301–306 (1971)
Kečkeš, S., Ozretić, B., Krasnović, M.: Loss of Zn-65 in the mussel Mytilus galloprovincialis. Malacologia *7*, 1–6 (1968)
Kee, N.S., Bloomfield, G.: The solution of some minor element oxides by decomposing plant material. Geochim. Cosmochim. Acta *24*, 206–225 (1961)
Keeley, J.W., Engler, R.M.: Discussion of regulatory criteria for ocean disposal of dredged materials: elutriate test rationale and implementation guidelines. U.S. Army Corps of Engineers, Dredged Material Research Program, Vicksburg, Miss., Rep. D-74-14, 13 (1974)
Keeney, D.R., Walsh, L.M.: Heavy metal availability in sewage-sludge-amended soils. Proc. Int. Conf. Heavy Met. Environ., Toronto 1975, *II/1*, 379–402 (1977)
Keeney, W.L., Breck, W.G., Van Loon, G.W., Page, J.A.: The determination of trace metals in Cladophora glomerata as a potential biological monitor. Water Res. *10*, 981–984 (1976)
Kelley, W.P.: Cation Exchange in Soils. A.C.S. Monograph No. 109. New York: Reinhold 1948
Kelly, W.E.: Ground-water pollution near a landfill. J. Environ. Eng. Div. Am. Soc. Civ. Eng. *102*, 1189–1193 (1976)
Kemp, A.L.W., Thomas, R.L.: Impact of man's activities on the chemical composition in the sediments of Lakes Ontario, Erie and Huron. Water Air Soil Pollut. *5*, 469–490 (1976a)
Kemp, A.L.W., Thomas, R.L.: Cultural impact on the geochemistry of the sediments of Lakes Ontario, Erie and Huron. Geosci. Can. *3*, 191–207 (1976b)
Kemp, A.L.W., Anderson, T.W., Thomas, R.L., Mudrochova, A.: Sedimentation rates and recent sediment history of Lakes Ontario, Erie and Huron. J. Sediment. Petrol. *44*, 207–218 (1974)
Kemp, A.L.W., Thomas, R.L., Dell, C.I., Jaquet, J.M.: Cultural impact on the geochemistry of sediments in Lake Erie. J. Fish. Res. Board Can. *33*, 440–462 (1976)
Kempf, T.: Hygienisch-toxikologische Bewertung von Trinkwasserinhaltsstoffen. Schriftenr. Ver. Wasser Boden Lufthyg. *40*, 149–153 (1973)
Kempf, T.: Die gesundheitlichen Aspekte der Umweltverschmutzung durch Blei. Bundesgesundheitsblatt *17*, 84–86 (1974)
Kempf, T., Lüdemann, D.: Gefährdung des Biotops Wasser durch Quecksilber und seine Verbindungen. Bundesgesundheitsblatt *15/16*, 225–223 (1971)

Kendrick, P.J.: Acid mine drainage – an old problem with a new dimension. J.W.P.C.F. *49*, 1576–1577 (1977)

Kennedy, E.J., Ruch, R.R., Shimp, N.F.: Distribution of mercury in unconsolidated sediments from southern Lake Michigan. Environ. Geol. Notes *44* (1971)

Kennedy, V.C., Zellweger, G.W., Jones, B.F.: Filter pore-size effects on the analysis of Al, Fe, Mn, and Ti in water. Water Resour. Res. *10*, 785–790 (1974)

Kester, D.R. (rapp.) et al.: Chemical speciation in seawater – group report. In: The Nature of Seawater. Goldberg, E.D. (ed.). Berlin: Dahlem Konferenzen 1975, pp. 17–41

Khalid, R.A., Gambrell, R.P., Verloo, M.G., Patrick, W.H., Jr.: Transformations of heavy metals and plant nutrients in dredged sediments as affected by oxidation reduction potential and pH. I. Literature review. U.S. Army Corps of Engineers, Dredged Material Research Program. Vicksburg, Miss., Rep. D-77-4, 221 (1977)

Khalid, R.A., Patrick, W.H., Jr., Gambrell, R.P.: Effect of dissolved oxygen on chemical transformations of heavy metals, phosphorus, and nitrogen in an estuarine sediment. Estuar. Coast. Mar. Sci. *6*, 21–35 (1978)

Khalid, R.A., Gambrell, R.P., Patrick, W.H., Jr.: Chemical transformations of cadmium and zinc in Mississippi River sediments as influenced by pH and redox potential. In: Proc. 2nd Mineral Cycling Symp., Environmental Chemistry and Cycling Processes. Adriano, D.C., Brisbin, I.L. (eds.). Oak Ridge/Tenn.: U.S. ERDA 1978 b (in press)

Kharkar, D.P., Turekian, K.K., Bertine, K.K.: Stream supply of dissolved silver, molybdenum, antimony, selenium, chromium, cobalt, rubidium, and cesium to the oceans. Geochim. Cosmochim. Acta *32*, 285–298 (1968)

Kiba, T., Terada, K., Honjo, T., Matsumoto, R., Ameno, K.: Generic relationships among samples of river sediments with the aid of concentration correlation matrix. Bunseki Kagaku *24*, 18–25 (1975)

Kim, K.C., Chu, R.C., Barron, G.P.: Mercury in tissues and lice of northern fur seals. Bull. Environ. Contam. Toxicol. *11*, 281–284 (1974)

Kirby, R., Parker, W.R.: Fluid mud in the Severn Estuary and its relevance to pollution studies. Inst. Chem. Eng. Ann. Symp. Estuarine Coastal Pollut. (1973)

Kitamura, S.: Determination of mercury content in bodies of inhabitants, cats, fishes, and shells in Minamata District and in the mud of Minamata Bay. Minamata Dis. 257–266 (1968)

Kitano, Y., Tokuyama, A., Kanamori, N.: Measurement of the distribution coefficient of zinc and copper between carbonate precipitate and solution. J. Earth Sci. Nagoya Univ. *16*, 1–12 (1968)

Klein, D.H.: Some estimates of natural levels of mercury in the environment. In: Environmental Mercury Contamination. Hartung, R., Dinman, B.D. (eds.). Ann Arbor: Ann Arbor Sci. Publ. 1972, pp. 25–29

Klein, D.H.: Fluxes, residence times and sources of some elements to Lake Michigan. Water Air Soil Pollut. *4*, 3–8 (1975)

Klein, D.H., Goldberg, E.D.: Mercury in the marine environment. Environ. Sci. Technol. *4*, 765–768 (1970)

Klein, D.H., Russel, P.: Heavy metals – fallout around a power plant. Environ. Sci. Technol. *7*, 354–357 (1973)

Klein, L.A., Lang, M., Nash, N., Kirschner, S.L.: Sources of metals in New York City waste-water. J.W.P.C.F. *46*, 2653–2662 (1974)

Kleinkopf, M.D.: Spectrographic determination of trace elements in lake water of northern Maine. Bull. Geol. Soc, Am. *71*, 1231–1234 (1960)

Klemmer, H.W., Unninayer, C.S., Okubo, W.I.: Mercury content of biota in coastal waters in Hawaii. Bull. Environ. Contam. Toxicol. *15*, 454–457 (1976)

Kludig, K.H.: Die Gewinnung von uferfiltriertem Grundwasser und der Einfluß der Rhein-Verschmutzung. GWF Wasser Abwasser *109*, 1401–1405 (1968)

Klusman, R.W., Edwards, K.W.: Toxic metals in ground water of the Front Range, Colorado. Ground Water *15*, 160–169 (1977)

Knauer, G.A.: Immediate industrial effects on sediment mercury concentrations in a clean coastal environment. Mar. Pollut. Bull. *7*, 112–115 (1976)

Knauer, G.A.: Immediate industrial effects on sediment metals in a clean coastal environment. Mar. Pollut. Bull. *8*, 249–254 (1977)

Knauer, G.A., Martin, J.H.: Mercury in a marine pelagic food chain. Limnol. Oceanogr. *17*, 868–876 (1972)

Knauer, G.A., Martin, J.H.: Seasonal variations of cadmium, copper manganese, lead and zinc in water and phytoplankton in Monterey Bay, California. Limnol. Oceanogr. *18*, 597–604 (1973)

Kneip, T.J., Re, G., Hernandez, T.: Cadmium in an aquatic ecosystem: distribution and effects. In: Trace Substances in Environmental Health. Hemphill, D.D. (ed.), Vol. 8. Columbia: Univ. Missouri 1974, pp. 172–177

Knickmann, E.: Zur Nutzung unfruchtbarer Böden mit hohen Gehalten an Blei und Zink. Z. Pflanzenernähr. Düng. Bodenkde. *84*, 255–258 (1959)

Knipling, J.J.: Adsorption from Solution of Non-Electrolytes. New York: Academic Press 1965

Kobayashi, J.: Relation between the ' Itai-Itai" disease and the pollution of river water by cadmium from a mine. Adv. Water Pollut. Res., Proc. 5th Int. Conf. San Francisco, Hawaii *I-25*, 1–7 (1971)

Kobayashi, S., Lee, G.F.: Accumulation of arsenic in sediments of lakes treated with sodium arsenite. Environ. Sci. Technol. *12*, 1195–1200 (1978)

Koch, O.G., Koch-Dedic, G.A.: Handbuch der Spurenanalyse. 2. Aufl., Teil 1 und 2. Berlin, Heidelberg, New York: Springer 1974

Koch, P.: Discharge of wastes into the sea in European coastal areas. In: Proceedings of the First International Conference on Waste Disposal in the Marine Environment. Pearson, E.A. (ed.). New York: Pergamon Press 1960, pp. 122–152

Koczy, F.F., Tomic, E., Hecht, F.: Zur Geochemie des Urans im Ostseebecken. Geochim. Cosmochim. Acta *11*, 86–102 (1967)

Koeman, J.H., Peeters, W.H.M., Koudstaal-Hol, C H.M., Tjide, P.S., De Goeij, J.J.M.: Mercury-selenium correlations in marine mammals. Nature (London) *245*, 385–386 (1975)

Koirtyohann, S.R., Wixson, B.G., Edwards, H.W.: Environmental lead distribution in relation to automobile and mine and smelter sources. In: Heavy Metals in the Aquatic Environment. Krenkel, P.A. (ed.). Oxford: Pergamon Press 1975, pp. 243–245

Koli, A.K., Williams, W.R., McClary, E B., Wright, E.L., Burrell, T.M.: Mercury levels in freshwater fish of the State of South Carolina. Bull. Environ. Contamin. Toxicol. *17*, 82–89 (1977)

Kölle, W., Dorth, K., Smiricz, G., Sontheimer, H.: Aspekte der Belastung des Rheins mit Schwermetallen. Vom Wasser *38*, 183–196 (1971)

Kononova, M.M.: Soil Organic Matter, 2nd ed. London: Pergamon Press 1966

Konovalov, G.S.: Transport of microelements by the most important rivers of the U.S.S.R. (engl. translation by M. Fleischer). Dokl. Akad. Nauk. USSR *129*, 912–915 (1969)

Konovalov, G.S., Nazarova, L.N.: Mapping trace elements in river waters (russ.). Gidrokhim. Mater. *62*, 37–42 (1975)

Konrad, J.G.: Mercury contents of bottom sediments from Wisconsin rivers and lakes. In: Environmental Mercury Contamination. Hartung, R., Dinman. B.D. (eds.). Ann Arbor Sci. Publ. 1972, p. 52

Kopfler, F.C.: The accumulation of organic and inorganic mercury compounds by the eastern oyster (Crassostrea virginica). Bull. Environ. Contam. Toxicol. *11*, 275–280 (1974)

Kopp, J.F.: Current status of analytical methodology for trace metals. In: Proc. Int. Conf. Heavy Metals Environ., Toronto 1975, *1*, 251–274 (1977)

Kopp, J.F., Kroner, R.C.: Trace metals in waters of the United States. Fed. Water Pollut. Control admin. Div. Pollut. Surveillance (1968)

Koppe, P.: Grundlegende Überlegungen und Untersuchungen über die hydrochemischen Beziehungen zwischen Flußwasser und dem Wasser ufernaher Brunnen. Schriftenr. Wasser Boden Lufthyg. *33*, 129–142 (1970)

Koppe, P.: Untersuchungen über das Verhalten von Inhaltsstoffen der Abwässer der metallverarbeitenden Industrie im Wasserkreislauf und ihren Einfluß auf die Wasserversorgung. GWF Wasser Abwasser *114*, 170–175 (1973)

Koppe, P.: Theoretische und experimentelle Grundlagen zur standardisierten Untersuchung der Wechselwirkung von Komplexbildnern in wässrigen Lösungen mit ungelösten Metallverbindungen als Bodenkörper. Z. Wasser Abwasserforsch. *9*, 153–160 (1976)

Korkisch, J., Dimitriadis, D.: Anwendung von Ionenaustauschverfahren zur Bestimmung von Spurenelementen in natürlichen Wässern. II. Cadmium. Talanta *20*, 1295–1301 (1973)

Kovacik, T.L., Walters, L.J., Jr.: Mercury distribution in sediment cores from western Lake Erie. Proc. 16th Conf. Great Lakes Res. 252–259 (1973)

Koziorowski, B., Kucharski, J.: Industrial Waste Disposal. New York: Pergamon Press 1972, pp. 196–202

Kranck, K.: Flocculation of suspended sediment in the sea. Nature (London) *246*, 348–350 (1973)

Kranck, K.: The role of flocculation in the transport of particulate pollutants in the marine environment. Proc. Int. Transp. Persist. Chem. Aquatic Ecosyst. Ottawa, May 1–3 (1974)

Kranck, K.: Sediment deposition from flocculated suspensions. Sedimentology *22*, 111–123 (1975)

Kranz, H.: Tracerversuche über die Kontamination durch Radionuclide in Sedimenten und ausgewählten Benthosorganismen in mesohalinen Wattenzonen des Elbeästuars. Diplomarbeit, Univ. Hamburg 1976, 93 pp.

Krasnicki, K., Szczepanski, A.: Occurrence of mercury in sediments of Lake Beldany at the mouth of the Krutynia River (north-east Poland). Bull. Acad. Pol. Sci. *24*, 463–467 (1976)

Krasnov, V.N., Kuz'menko, A.L.: Content of certain trace elements (of heavy metals) in the waters and silts of the Volgograd Reservoir. Gidrokhim. Mater. *43*, 182–190 (1967)

Krauskopf, K.B.: Sedimentary deposits of rare metals. Econ. Geol. *50*, 411–463 (1955)

Krauskopf, K.B.: Factors controlling the concentration of thirteen rare metals in sea-water. Geochim. Cosmochim. Acta *9*, 1–32 (1956)

Krauskopf, K.B.: Separation of manganese from iron in sedimentary processes. Geochim. Cosmochim. Acta *12*, 61–84 (1957)

Krauskopf, K.B.: Introduction to Geochemistry. New York, St. Louis, San Francisco: McGraw-Hill 1967

Krauskopf, K.B.: Tungsten, abundance in natural waters. In: Handbook of Geochemistry. Wedepohl, K.H. (ed.). Berlin, Heidelberg, New York: Springer 1960

Krejci-Graf, K.: Diagnostik der Salinitätsfazies der Ölwässer. Fortschr. Geol. Rheinl. Westfalen *10*, 367–448 (1963)

Krishnaswamy, S., Martin, J.M., Meybeck, M.: Geochronology of lake sediments. Earth Planet. Sci. Lett. *11*, 407–414 (1971)

Krocza, W., Glantschig, P., Stöckl, W.: Über den Quecksilbergehalt von Fischen aus Seen des Landes Kärnten. Wien. Tierärztl. Monatsschr. *61*, 169–178 (1974)

Krom, M.D.: Chemical speciation and diagenetic reactions at the sediment-water interface in a Scottish fjord. Edinburgh: Abstr. Joint Oceanogr. Assembly 1976, p. 89

Krom, M.D., Sholkovitz, E.R.: Nature and reactions of dissolved organic matter in the interstitial waters of marine sediments. Geochim. Cosmochim. Acta *41*, 1565–1573 (1977)

Kronfeld, J., Navrot, J.: Transition metal contamination in the Qishon River system, Israel. Environ. Pollut. *6*, 281–287 (1974)

Kronfeld, J., Navrot, J.: Aspects of trace metal contamination in the coastal rivers of Israel. Water Air Soil Pollut. *4*, 127–134 (1975)

Kropf, R., Geldmacher, M., v. Mallinckrodt: Der Cadmiumgehalt von Nahrungsmitteln und die tägliche Cadmiumaufnahme. Arch. Hyg. *152*, 218–224 (1968)

Krüger, K.E., Nieper, L., Auslitz, H.-J.: Bestimmung des Quecksilbergehaltes der Seefische auf den Fangplätzen der deutschen Hochsee- und Küstenfischerei. 1. Mitt. Arch. Lebensmittelhyg. *26*, 201–207 (1975)

Kubota, J., Mills, E.L., Oglesby, R.T.: Lead, cadmium, zinc, copper, and cobalt in streams and lake waters of Cayuga Basin, New York. Environ. Sci. Technol. *8*, 243–248 (1974)

Kudo, A., Hart, J.S.: Uptake of inorganic mercury by bed sediments. J. Environ. Qual. *3*, 273–278 (1974)

Kudo, A., Mortimer, D.C., Hart, J.: Factors influencing desorption of mercury from bed sediments. Can. J. Earth Sci. *12*, 1036–1040 (1975)

Kullenberg, B.: The piston core sampler. Sven. Hydrogr.-Biol. Komm. Skr. Ser. *3*, Hydrogr. *1*, 2 (1947)

Kullenberg, B.: On the salinity of water contained in marine sediments. Medd. Oceanogr. Inst. Göteborg *21*, 38 pp. (1952)

Kunz, R.G., Gianelli, J.F., Stensel, H.D.: Vanadium removal from industrial wastewaters. J.W.P.C.F. *48*, 762–770 (1976)

Kuroda, R.: Vanadium, chromium and molybdenum contents of deep-sea deposits. J. Chem. Soc. Jpn. *61*, 1060 (1940)

Kuroda, R.: Vanadium, chromium and molybdenum contents in deep-sea deposits. J. Chem. Soc. Jpn. *63*, 496 (1942)

Kushner, D.J.: Microbial dealing with heavy metals. Proc. Int. Conf. Transp. Persist. Chem. in Aquatic Ecosyst., Ottawa (1974)

LaBarre, N.: Lead contamination of snow. Water Res. *7*, 133–137 (1973)

Lackey, J.B.: The microbiota of estuaries and their roles. In: Estuaries. Lauff, G.H. (ed.). AAAS Publ. *83*, 291–302 (1967)

Lagerwerff, J.V., Biersdorf, G.T.: Interaction of zinc with uptake and translocation of cadmium in radish. In: Trace Substances in Environmental Health, Vol. V. Hemphill, D.D. (ed.). Columbia Univ. Missouri 1972, pp. 515–519

Lagerwerff, J.V., Brower, D.L.: Exchange adsorption of trace quantities of cadmium in soils treated with chlorides of aluminum, calcium and sodium. Soil Sci. Soc. Am. Proc. *36*, 734–737 (1972)

Lagerwerff, J.V., Brower, D.L.: Exchange adsorption or precipitation of lead in soils treated with chlorides of aluminum, calcium and soidum. Soil Sci. Soc. Am. Proc. *37*, 11–13 (1973)

Laguitton, D.: Arsenic removal from gold-mine waste waters: basic chemistry of the lime addition method. CIM Bull. *69*, 105–109 (1976)

Lamar, W.L.: Evolution of organic color and iron in natural surface water. U.S. Geol. Surv. Prof. Pap. *600-D,* 24 pp. (1968)

Lande, E.: Heavy metal pollution in Trondheimsfjorden, Norway, and the recorded effect on the fauna and flora. Environ. Pollut. *12,* 187–198 (1977)

Landner, L.: Biochemical model for the biological methylation of mercury suggested from methylation studies in vivo with Neurospora crassa. Nature (London) *230,* 452–453 (1971)

Landström, O., Wenner, C.G.: Neutron-activation analysis of natural waters applied to hydrogeology. Aktiebolaget Atomenergi *AE-204,* (1965). Cit. Onishi, H. (1970)

Landström, O., Samsahl, K., Wenner, C.G.: An investigation of trace elements in marine and lacustrine deposits by means of a neutron activation method. Mod. Trends Activ. Anal. NBS Spec. Publ. *312,* 353–367 (1969)

Lange-de la Camp, M., Steinmann, W.: Speicherung von Schwermetallen in niederen Organismen. Arch. Mikrobiol. *19,* 87–106 (1953)

Langmuir, I.: The adsorption of gases on plane surfaces of glass, mica and platinum. J. Am. Chem. Soc. *40,* 1361–1403 (1918)

Langmyhr, F.J., Solberg, R., Thomassen, Y.: Atom absorption spectrometric determination of thirteen minor and trace metals in phosphate rock concentrates. Anal. Chim. Acta *92,* 105–109 (1977)

Larsen, H.P.: Chemical treatment of metal-bearing mine drainage. J.W.P.C.F. *45,* 1582–1695 (1975)

Laskowski, N., Pommerenke, D., Schäfer, A., Tobschall, H.J.: Hohe Quecksilberkonzentration in Sedimenten des Ginsheimer Altrhein. Naturwissenschaften *61,* 681 (1974)

Laskowski, N., Kost, T., Pommerenke, D., Schäfer, A., Tobschall, H.J.: Heavy-metal and organic-carbon content of recent sediments near Mainz. Naturwissenschaften *62,* 136 (1975)

Laszlo, F., Literathy, P., Benedek, P.: Heavy metal pollution in the Sajo River, Hungary. Proc. Int. Conf. Heavy Met. Environ., Toronto, 1975, *II/2,* 923–932 (1977)

Leatherland, T.M., Burton, J.D.: The occurance of trace metals in coastal organisms with particular reference to the Solent region. J. Mar. Biol. Assoc. UK *54,* 457–468 (1974)

Leatherland, T.M., Burton, J.D., McCartney, M.J., Culkin, F.: Mercury in north-eastern Atlantic Ocean water. Nature (London) *232,* 112 (1971)

Lee, C.R., Engler, R.M., Mahloch, J.L.: Land application of waste materials from dredging, construction and demolition processes. Misc. Pap. D-76-5. Prepared for Office, Chief of Engineers, U.S. Army Washington (1976)

Lee, C.R., Smart, R.M., Sturgis, T.C., Gordon, R.N, Landin. M.C.: Prediction of heavy metal uptake by marsh plants based on chemical extraction of heavy metals from dredged material. U.S. Army Engineer Waterway Experiment Station, Vicksburg, Miss. Dredged Material Res. Program, Tech. Rep. D-78–6 (1978)

Lee, G.F.: Role of hydroxous metal oxides in the transport of heavy metals in the environment. In: Heavy Metals in the Aquatic Environment. Krenkel, P.A. (ed.). Oxford: Pergamon Press 1975, pp. 137–147

Lee, G.F.: Contamination in the New York Bight. Discussion of the paper of J.A. Mueller, A.R. Anderson, J.S. Jeris (1976). J.W.P.C.F. *49,* 1920–1921 (1977a)

Lee, G.F.: Summary of studies on the release of contaminants from dredged sediments on open-water disposal. In: Interactions Between Sediments and Fresh Water. Golterman, H.L. (ed.). The Hague: Junk Publ. 1977b, pp. 444–446

Lee, G.F., Hoadley, A.W.: Biological activity in relation to the chemical equilibrium composition of natural waters. In: Equilibrium Concepts in Natural Systems. Am. Chem. Soc. Adv. Chem. Ser. *67,* 319–338 (1967)

Lee, G.F., Plumb, R.H.: Literature review on research study for the development of dredged material disposal criteria. U.S. Army Corps of Engineers, Dredged Material Research Program. Vicksburg, Miss. Rep. D-74–1, 145 (1974)

Leeper, G.W.: Factors affecting availability of inorganic nutrients in soils with special reference to micronutrient metals. Nature (London) *162,* 1–15 (1948)

Leland, H.V.: Distribution of solute and particulate trace elements in southern Lake Michigan. Proc. Int. Conf. Heavy Met. Environ. *II/2,* 709–730 (1977)

Leland, H.V., McNurney, J.M.: Lead transport in a river ecosystem. Proc. Int. Conf. Transp. Persist. Chem. Aquatic Ecosyst., Ottawa III, 17–23 (1974)

Leland, H.V., Shukla, S.S., Shimp, N.F.: Factors affecting distribution of lead and other trace elements in sediments of southern Lake Michigan. In: Trace Metals and Metal-Organic Interactions in Natural Waters. Singer, P.C. (ed.). Ann Arbor: Science Publ. 1973, pp. 89–129

Lemberg, R., Legge, J.W.: Hematin Compounds and Bile Pigments, their Constitution, Metabolism, and Function. New York: Interscience 1949

Lengweiler, H., Buser, W., Feitknecht, W.: Die Ermittlung der Löslichkeit von Eisen (III)-hydroxiden mit ^{59}Fe. II. Der Zustand kleinster Mengen Eisen (III)-hydroxid in wässriger Lösung. Helv. Chim. Acta *44*, 805–812 (1961)

Lerman, A.: Time to chemical steady-states in lakes and oceans. In: Nonequilibrium Systems in Natural Water Chemistry. Hem, J.D. (ed.). Adv. Chem. Ser. *106*. Washington: Am. Chem. Soc. 1971, pp. 30–76

Lerman, A., Childs, C.W.: Metal-organic complexes in natural waters: Control of distribution by thermodynamic, kinetic and physical factors. In: Trace Metals and Metal-Organic Interactions in Natural Waters. Singer, P.C. (ed.). Ann Arbor Sci. Publ. 1973, pp. 201–235

Lerman, A., Weiler, R.: Diffusion and accumulation of chloride and sodium in Lake Ontario sediment. Earth Planet. Sci. Lett. *10*, 150–156 (1970)

Lesaca, R.M.: Monitoring of heavy metals in Philippine rivers, bay waters and lakes. Proc. Int. Conf. Heavy Met. Environ., Toronto 1975, *II/1*, 285–307 (1977)

Leutwein, F., Weise, L.: Hydrogeochemische Untersuchungen an erzgebirgischen Gruben- und Oberflächengewässern. Geochim. Cosmochim. Acta *26*, 1333–1348 (1962)

Lewin, J., Davies, B.E., Wolfenden, P.J.: Interactions between channel change and historic mining sediments. In: River Channel Changes. Gregory, K.J. (ed.). New York: Wiley and Sons 1977, pp. 353–367

Leyden, D.E.: X-ray emission spectrometry and environmental water analysis. Am. Lab. *6*, 24–26 (1974)

Lichtfuß, R., Brümmer, G.: Schwermetallbelastung von Elbe-Sedimenten. Naturwissenschaften *64*, 122–125 (1977)

Liebenberg, W.R.: Mineralogical features of gold ores in South Africa. In: Gold Metallurgy in South Africa. Adamson, R.J. (ed.). Johannesburg: Chamber of Mines of South Africa 1972, pp. 352–426

Liebmann, H.: Handbuch der Frischwasser- und Abwasserbiologie, Band II. München: R. Oldenbourg 1958

Lindahl, P.E., Hell, C.E.B.: Effects of short-term exposure of Leuciscus rutilus L. (pisces) to phenylmercuric hydroxide. Oikos *21*, 267–275 (1970)

Lindberg, S.E., Andren, A.W., Harriss, R.C.: Geochemistry of mercury in the estuarine environment. In: Estuarine Research, Vol. I. Cronin, L.E. (ed.). New York: Academic Press 1975, pp. 64–107

Linnman, L., Andersson, A., Nilsson, K.O., Lind, B., Kjellström, T., Friberg, L.: Cadmium uptake by wheat from sewage sludge used as a plant nutrient source. Arch. Environ. Health *27*, 45–57 (1973)

Linstedt, K.D., Bennett, E.R.: Evaluation of treatment for urban wastewater reuse. Environ. Prot. Technic. Ser. EPA-R2-73-122, Washington, D.C. (1973)

Linstedt, K.D., Bennett, E.R., Fox, R.L., Heaton, R.D.: Alum clarification for improving wastewater effluent quality. Water Res. *8*, 753–760 (1974)

Linstedt, K.D., Houck, C.P., O'Connor, J.T.: Trace element removals in advanced wastewater treatment processes. J.W.P.C.F. *43*, 1507–1513 (1971)

Liperovskaya, E.S., Drozhbina, T.M.: Pollution of Moscow River muds and their effect on the distribution of oligochaetes. Protsessy Zagryaz. Samochishcheniya Reki Moskvy *1972*, 130–139

Liss, P.S.: Conservative and non-conservative behaviour of dissolved constituents during estuarine mixing. In: Estuarine Chemistry. Burton, J.D., Liss, P.S. (eds.). London: Academic Press 1976, pp. 93–174

Literáthy, P.: Study of river pollution caused by micropollutants. Water Res. *9*, 1001–1003 (1974)

Literáthy, R., László, F.: New method for characterizing the chemistry of bottom sediments. Int. Symp. Geochem. Nat. Waters. Burlington, Ont., Canada, 1975

Literáthy, P., László, F.: Uptake and release of heavy metals in the bottom silt of recipients. In: Interactions Between Sediments and Fresh Water. Golterman, H.L. (ed.). The Hague: Junk Publ. 1977, pp. 403–409

Little, P., Martin, M.H.: A survey of zinc, lead and cadmium in soil and natural vegetation around a smelting complex. Environ. Pollut. *3*, 241–245 (1972)

Little-Gadow, S., Schäfer, A.: Schwermetalle in den Sedimenten der Jade. Senckenbergiana Mar. *6*, 161–174 (1974)

Littlepage, J.: Heavy metals in a northern inlet. Abstr. Int. Conf. Heavy Metals in the Environment. Toronto, Ontario, Canada (1975)

Livingstone, D.A.: Chemical composition of rivers and lakes. In: Data of Geochemistry, 6th ed. Fleischer, M. (ed.). U.S.Geol. Surv. Prof. Paper *440-G*, 64 pp. (1963)

Livingstone, D.A., Boykin, J.C.: Vertical distribution of phosphorous in Linsley pond mud. Limnol. Oceanogr. *7*, 57–63 (1962)

Ljungreen, P.: Some data concerning the formation of manganiferous and ferriferous bog ores. Geol. Fören. Stockholm Förh. *75*, 277–297 (1953)

Lloyd, N.A., et al.: Mercury concentrations in sediment samples from the Tennessee, Mobile, Warrior and Tombigbu Rivers, Alabama. Geol. Surv. Alabama Circ. *79*, (1972)

Lloyd, R.M.: A technique for separating clay minerals from limestones. J. Sediment. Petrol. *24*, 218–220 (1954)

Lloyd, R.M.: Factors that affect the tolerance of fish to heavy metal poisoning. Biol. Problems in Water Pollution, 3rd Seminar 1962. U.S. Dept Health, Education and Welfare, 181 pp. (1965)

Lockwood, R.A., Chen, K.Y.: Adsorption of Hg (II) by hydrous manganese oxides. Environ. Sci. Technol. *7*, 1028–1034 (1973)

Lockwood, R.A., Chen, K.Y.: Adsorption of Hg (II) by ferric hydroxide. Environ. Lett. *6*, 151–166 (1974)

Lodemann, C.K.W., Bukenberger, U.: Schwermetallspuren im Bereich des oberen Neckars. GWF Wasser Abwasser *114*, 478–487 (1973)

Logsdon, G.S., Symons, J.M.: Mercury removal by conventional water treatment techniques. J.A.W.W.A. *65*, 554–558 (1973a)

Lodgsdon, G.S., Symons, J.M.: Removal of trace inorganics by drinking water treatment unit processes. Paper pres. at 75th Ann. Meet. Am. Inst. Chem. Eng. (1973b)

Lodgson, G.S., Sorg, T.J., Symons, J.M.: Removal of heavy metals by conventional treatment. Proc. 16th Water Qual. Conf., Trace Metals in Water Supplies: Occurrence, Significance and Control. Univ. of Illinois, College of Engineering, 139 pp. (1974)

Loon, J.C. van: The snow removal controversy. Water Pollut. Control *110*, 16–22 (1972)

Loon, J.C. van: Mercury contamination of vegetation due to the application of sewage sludge. Environ. Lett. *6*, 211–218 (1974)

Lorch, D.W., Melkonian, M., Weber, A. Wettern, M.: Reasons for different lead accumulations by green algae, some experiments, some suggestions. SIL-Congr., Copenhagen 1977

Loring, D.H.: Mercury in the sediments of the Gulf of St. Lawrence. Can. J. Earth Sci. *12*, 1219–1237 (1975)

Loring, D.H.: Distribution and partition of cobalt, nickel, chromium, and vanadium in the sediments of the Saguenay Fjord. Can. J. Earth Sci *13*, 1706–1718 (1976)

Lovett, R.J., Gutemann, W.H., Pakkala, I.S., Youngs, W.D., Lisk, D.J., Burdick, G.E., Harris, E.J.: A survey of the total cadmium content of 406 fish from 49 New York State freshwaters. J. Fish Res. Board Can. *29*, 1283–1290 (1972)

Lowe, W.: The origin and characteristics of toxic wastes, with particular reference to the metal industries. Water Pollut. Control *69*, 270–280 (1970)

Lowman, F.G., Phelps, D.K., McClin, R., De Vega, V.R., De Padovani, I.O., Garcia, R.J.: Interactions of the environmental and biological factors on the distribution of trace elements in the marine environment. In: Disposal of Radioactive Wastes into Seas, Oceans and Surface Waters. Vienna: International Atom. Energy Assoc. 1966, pp. 248–266

Lu, C.S.J., Chen, K.Y.: Migration of trace metals in interfaces of seawater and polluted surficial sediments. Environ. Sci. Technol. *11*, 174–182 (1977)

Lucas, H.F., Jr., Edgington, D.N., Colby, P.J.: Concentrations of trace elements in Great Lakes fishes. J. Fish. Res. Board Canada *27*, 677–684 (1970)

Lucas, R.E., Davis, J.F.: Relationships between pH values of organic soils and availabilities of 12 plant nutrients. Soil Sci. *92*, 177–182 (1961)

Luh, M.D., Baker, R.A., Henley, D.E.: Arsenic analysis and toxicity – a review. Sci. Total Environ. *2*, 1–12 (1973)

Lund, L.J., Page, A.L., Nelson, C.O.: Movement of heavy metals below sewage disposal ponds. J. Environ. Qual. *5*, 330–334 (1976)

Luoma, S.N.: The uptake and interorgan distribution of mercury in a carnivorous crab. Bull. Environ. Contam. Toxicol. *16*, 719–723 (1976)

Luoma, S.N., Bryan, G.W.: Trace metal bioavailability: modelling chemical and biological interactions of sediment-bound zinc. Reprints of papers pres. at the 176th Nat. Meet. ACS, Div. Environ. Chem., Miami Beach, Fla., 1978, pp. 413–414

Lynn, D.C., Bonatti, E.: Mobility of manganese in diagenesis of deep-sea sediments. Mar. Geol. *3*, 457–474 (1965)

MacKay, D.W., Leatherland, T.M.: Chemical processes in an estuary receiving major inputs of industrial and domestic waters. In: Estuarine Chemistry. Burton, J.D., Liss, P.S. (eds.). London, New York, San Francisco: Academic Press 1976, pp. 185–218

MacKay, D.W., Halcrow, W., Thornton, I.: Sludge dumping in the Firth of Clyde. Mar. Pollut. Bull. *3*, 7–11 (1972)

MacKay, N.J., Kazacos, M.N., Williams, R.J., Leedow, M.I.: Selenium and heavy metals in Black Marlin. Mar. Pollut. Bull. *6*, 57–61 (1975a)

MacKay, N.J., Williams, R.J., Kaprzak, J.L., Kazacos, M.N., Collins, A.J., Auty, E.H.: Heavy metals in cultivated oysters (Crassostrea commercialis – Saccostrea cucullata) from the estuaries of New South Wales. Aust. J. Mar. Freshwater Res. *26*, 31–46 (1975b)

Mackenzie, F.T., Garrels, R.M.: Chemical mass balance between rivers and oceans. Am. J. Sci. *264*, 507–525 (1966)

MacLeod, J.C., Pessah, E.: Temperature effects on mercury accumulation, toxicity, and metabolic rate in rainbow trout (Salmo gairdneri). J. Fish. Res. Bourd Can. *30*, 485–492 (1973)

Magos, L., Tuffery, A.A., Clarkson, T.W.: Volatilization of mercury by bacteria. Brit. J. Industr. Med. *21*, 294–298 (1964)

Maienthal, E.J., Becker, D.A.: A survey on current literature on sampling, sample handling for environmental materials, and long-term storage. Interface *5*, 49–62 (1976)

Maksimovic, Z., Dangic, A.: Mercury mine at Mount Avala, a source of environmental pollution by mercury and arsenic. Geol. An. Balk. Poluostrva (Serbian) *38*, 349–358 (1973)

Malcolm, R.L.: Mobile soil organic matter and its interactions with clay minerals and sesquioxides. Ph. D. Thesis, North Carolina State Univ., Raleigh, 1964

Malmquist, P.-A.: Heavy metals in urban storm water. Abstr. Int. Conf. Heavy Met. Environ., Toronto, C-46/48 (1975)

Malo, B.A.: Partial extraction of metals from aquatic sediments. Environ. Sci. Technol. *11*, 277–282 (1977)

Malouf, E.E., Prater, J.D.: Role of bacteria in the alteration of sulphide minerals. J. Met. (N.Y.) *13*, 353–356 (1961)

Manahan, S.E., Smith, M.J.: Copper micronutrient requirement for algae. Environ. Sci. Technol. *7*, 829–833 (1973)

Mancy, K.H., Allen, H.E.: A controlled bioassay system for measuring toxicity of heavy metals. U.S. EPA, Ecol. Res. Serv. EPA-600/3-77-037, Duluth, Minnesota (1977)

Manheim, F.T.: A geochemical profile in the Baltic Sea. Geochim. Cosmochim. Acta *25*, 52–70 (1961)

Manheim, F.T.: Symposium on marine geochemistry. Rhode Island Univ. Narrangansett Marine Lab. Occasional Publ. *3*, 217–276 (1965)

Manheim, F.T.: A hydraulic squeezer for obtaining interstitial water from consolidated and unconsolidated sediments. U.S. Geol. Surv. Prof. Pap. *550-C*, 256–261 (1966)

Manheim, F.T., Stoffers, P.: Composition of interstitial waters in sediments of the Black Sea. Abstr. Annu. Meet. Geol. Soc. Am., Atlantic City 1969

Manly, R., George, W.O.: The occurrence of some heavy metals in populations of the freshwater mussel Anodonta anatina from the Thames. Environ. Pollut. *14*, 139–151 (1977)

Mann, A.W.: Resources. In: Environmental Chemistry. Bockris, J.O.M. (ed.). New York, London: Plenum Press 1977, pp. 121–178

Mantell, C.L.: Solid Wastes: Origin, Collection, Processing and Disposal. New York: John Wiley and Sons 1975

Marshall, C.E.: The Colloid Chemistry of the Silicate Minerals. New York: Academic Press 1949

Marshall, C.E.: The physical chemistry and mineralogy of soils. In: Soil Materials, Vol. 1. New York, London, Sydney: John Wiley and Sons 1964

Martens, C.S., Berner, R.A.: Methane production in the interstitial waters of sulfate-depleted marine sediments. Science *185*, 1167–1169 (1974)

Martin, A.E.: Chemical studies of podsolic illuvial horizons. V: Flocculation of humus by ferric and ferrous iron and nickel. J. Soil Sci. *11*, 382–393 (1960)

Martin, J.B.M.: Metals in Cancer irroratus (Crustacea Decapoda): Concentrations, concentration factors, discrimination factors, correlations. Mar. Biol. *28*, 245–251 (1974)

Martin, J.H.: Trace metal transport via copipod moults. Limnol. Oceanogr. *15*, 756–761 (1970)

Martin, J.H., Broenkow, W.W.: Cadmium in plankton: Elevated concentrations of Baja California. Science *190*, 884–885 (1975)

Martin, J.H., Flegal, A.R.: High copper concentrations in squid livers in association with elevated levels of silver, cadmium and zinc. Mar. Biol. *30*, 51–55 (1975)

Martin, J.H., Bruland, K.W., Broenkow, W.W.: Cadmium transport in the California current. In: Marine Pollutant Transfer. Windom, H.L., Duce, R.A. (eds.). Lexington: D.C. Heath 1976, pp. 159–184

Martin, J.M., Meybeck, M., Heuzel, M.: A study of the dynamics of suspended matter by means of natural radioactive tracers: an application to the Gironde Estuary. Sedimentology *14*, 27–37 (1970)

Martin, J.M., Jednačak, J., Pravdić, V.: The physico-chemical aspects of trace element behaviour in estuarine environments. Thalassia Jugosl. *7,* 619–637 (1971)

Martin, J.M., Kulbicki, G., De Groot, A.J.: Terrigenous supply of radioactive and stable elements to the ocean. In: Proc. Symp. Hydrogeochemistry and Biogeochemistry, Tokyo, 1970, Vol. 1. Washington: The Clarks Co., pp. 463–483

Martin, J.M., Høgdahl, O., Phillpott, J.C.: Rare earth element supply to the ocean. J. Geophys. Res. *81,* 3119–3121 (1976)

Maruyama, T., Hannah, S.A., Cohen, J.M.: Metal removal by physical and chemical treatment processes. J.W.P.C.F. *47,* 962–975 (1975)

Mason, B.: Principles of Geochemistry. New York: Wiley 1958, 310 pp.

Mathis, B.J., Cummings, T.F.: Selected metals in sediments, water and biota in the Illinois River. J.W.P.C.F. *45,* 1573–1583 (1973)

Mathis, B.J., Kevern, N.R.: Distribution of mercury, cadmium, lead and thallium in a eutrophic lake. Inst. Water Res. Michigan State Univ., Technol. Dep. *34,* 23 pp. (1973)

Matida, Y., Kumada, H.: Distribution of mercury in water, bottom sediments and aquatic organisms of Minamata Bay, the River Agano and other water bodies in Japan. Bull. Freshwater Fish Res. Lab. *19,* 73 (1969)

Matson, W.R.: Trace metals, equilibrium and kinetics of trace metal complexes in natural media. Ph. D. Thesis, Mass. Inst. Technol., 1968. Cit. R.D. Guy, D.L. Chakrabarti (1977)

Matsumoto, E., Yokota, S.: History of heavy metal pollution in Tokyo Bay and Osaka Bay. Kaguku *46,* 182–184 (1976)

Matsunaga, K.: Concentration of mercury by three species of fish from Japanese rivers. Nature (London) *257,* 49–50 (1975)

Matsunaga, K., Nishimura, M., Konishi, S.: Mercury in the Kuroshio and Oyashic regions and the Japan Sea. Nature (London) *258,* 224 (1975)

Matthess, G.: Heavy metals as trace constituents in natural and polluted groundwaters. Geol. Mijnbouw *53,* 149–155 (1974)

Maxfield, D., Rodriguez, J.M., Buettner, M., Davis, J., Forbes, L., Kovacs, R., Russel, W., Schultz, L., Smith, R., Stanton, J., Wai, C.M.: Heavy metal pollution in the sediments of the Coeur d'Alène River Delta. Environ. Pollut. *7,* 1–6 (1974)

Maxwell, J.A.: Rock and Mineral Analysis. New York, London, Sydney, Toronto: Interscience Publ. 1968

May, E.B.: Environmental effects of hydraulic dredging in estuaries. Ala. Mar. Resour. Bull. *9,* 1–85 (1973)

May, E.B.: Effects on water quality when dredging a polluted harbour using confined spoil disposal. Ala. Mar. Resour. Bull. *10,* 1–8 (1974)

Mayer, L.M.: Chemical water sampling in lakes and sediments with dialysis bags. Limnol. Oceanogr. *21,* 909–912 (1976)

Mayes, R., McIntosh, A.: Use of aquatic macrophytes as indicators of trace metal contamination in fresh water lakes. In: Trace Substances in Environmental Health, Vol. IX. Hemphill, D.D. (ed.). Columbia: Univ. Missouri 1975, pp. 157–167

McCabe, L.J., Symons, J.M., Lee, R.D., Robeck, G.G.: Survey of community water supply systems. J. A.W.W.A. *62,* 670–687 (1970)

McCaull, J.: Building a shorter life. Environment *13,* 3–41 (1971)

McCave, I.N.: Mud in the North Sea. In: North Sea Science. Goldberg, E.D. (ed.). Cambridge Mass.: MIT Press 1973, pp. 75–100

McCrone, A., Koch, R.: Some geochemical properties of Hudson River sediments, Kingston to Manhattan. Hudson River Ecol. Symp. Hudson River Valley Comm. N.Y., 41–59 (1966)

McElroy, A.D., Chiu, S.Y., Nebgen, J.W., Aleti, A., Vandegrift, E.: Water pollution from non-point sources. Water Res. *9,* 675–681 (1975)

McGarry, M.G., Polprasert, C., Whitaker T., Luo, M.-H.: Heavy metal pollution and development of the Quae Yai River, Thailand. In: Miner and the Environment. Int. Symp. Proc. Paper and Discussion, London, England, June 4–7 1974. London, England: Publ. Inst. Min. Metal 1975, pp. 715–730

McIntire, F.R., Neufeld, R.D.: Microbial methylation of mercury: a survey. Water Pollut. Control. 465–470 (1975)

McIntosh, A.: Notes on the use of copper sulfate in ponds. Bull. Environ. Contam. Toxicol. *12,* 425–429 (1974)

McKaveney, J.P., Fassinger, W.P., Stivers, D.A.: Removal of heavy metals from water and brine using silicon alloys. Environ. Sci. Technol. *6,* 1109–1113 (1972)

McKelvey, V.E.: Uranium in phosphate rock. U.S. Geol. Surv. Prof. Paper *300,* 477–512 (1956)

McKim, J.M., Cristensen, G.M., Tucker, J.H., Lewis, M.J.: Effects of pollution on freshwater fish. J.W.P.C.F. *45*, 1370–1405 (1973)

McLaughlin, R.J.W.: Iron and titanium oxides in soil clays and silts. Geochim. Cosmochim. Acta *5*, 85–96 (1954)

McLellon, W.M., Vickers, D.H., Charba, J.F., Bengstrom, G.I.: Environmental impact assessment of a sanitary landfill in a higher water table area. Proc. 29th Ind. Waste Conf., Purdue Univ., Ext. Ser. *145*, 94–102 (1974)

McLerran, C.J., Holmes, C.W.: Deposition of zinc and cadmium by marine bacteria in estuarine sediments. Limnol. Oceanogr. *19*, 998–1001 (1974)

McMullen, E.D.: Methylation of mercury in natural sediments. Vanderbilt Univ. Environ. Water Resour. Eng. Technol. Rep. *32*, 173–221 (1973)

Meade, R.H.: Landward transport of bottom sediments in estuaries of the Atlantic coastal plain. J. Sediment. Petrol. *39*, 222–234 (1969)

Meade, R.H.: Transport and deposition of sediments in estuaries. Geol. Soc. Am. Mem. *133*, 91–120 (1972)

Measures, C.I., Burton, J.D.: Determination of selenium (IV) and total selenium in oceanic water. Abstr. 1978 Spring Meeting AGU, EOS *59*, 307 (1978)

Meenan, W.R., Smythe, L.E.: Occurrence of beryllium as a trace element in environmental materials. Environ. Sci. Technol. *1*, 839–844 (1967)

Meinke, W.W.: Ultimate contribution of neutron activation analysis. J. Radioanal. Chem. *15*, 419–433 (1973)

Menard, H.W.: The Marine Geology of the Pacific. New York: McGraw-Hill 1964

Merlini, M., Argentesi, F., Brazelli, A., Oregioni, B., Pozzi, G.: The effects of sublethal amounts of cadmium and mercury on the metabolism of zinc-65 by freshwater fish. Int. Symp. Radioecology Applied to Man and Environment, Rome, Publ. No. 722 Biol. Dir. Comm. Eur. Comm., 1327–1344 (1971)

Merlini, M., Cadario, G., Oregioni, B.: The unionid mussel as a biogeochemical indicator of metal pollution. In: Environmental Biogeochemistry and Geomicrobiology, Vol. 3. Krumbein, W.E. (ed.). Ann Arbor: Sci. Publ. 1978, pp. 955–965

Merrill, J.R., Lyden, E.F.X., Honda, M., Arnold, J.R.: The sedimentary geochemistry of the beryllium isotopes. Geochim. Cosmochim. Acta *18*, 108–129 (1960)

Meyer, C.F.: Polluted groundwater, some causes, effects, controls and monitoring. EPA Rep. 600/4/73-oob (1973)

Mihm, U., Botzenhart, K., Noeske, G.: Untersuchungen über den Schwermetallgehalt in Sedimenten einer Trinkwassertalsperre. Zentralbl. Bakt. Hyg. Abt. I, Orig. *B 172*, 205–210 (1976a)

Mihm, U., Botzenhart, K., Noeske, G.: Die Bestimmung einiger Metalle und des Phosphats im Sediment der Vorsperre zur Wahnbachtalsperre. Dtsch. Gewässerkundl. Mitt. *20*, 47–51 (1976b)

Mill, G.: Mercury levels of sediments in the Wabigoon River system 1970–1974. Ontario Ministry of Environment. Cit. Buffa (1976)

Miller, G.E., Grant, P.M., Kishore, R., Steinkruger, F.J., Rowland, F.S., Guinn, V.P.: Mercury concentrations in museum specimens of tuna and swordfish. Science *175*, 1121–1122 (1972)

Miller, R.J., Johnson, R.D., Williams, R.E., Wai, C.M., Wiese, A.C., Mitchell, J.E.: Heavy metal problem of Silver Valley, North Idaho. Abstr. Int. Conf. Heavy Metal. Environ., Toronto, C-64 (1975)

Mills, A.L., Colwell, R.R.: Microbiological effects of metal ions in Chesapeake Bay water and sediment. Bull. Environ. Contam. Toxicol. *18*, 99–103 (1977)

Mills, E.L., Oglesby, R.T.: Five trace elements and vitamin B_{12} in Cayuga Lake, New York. Proc. 14th Conf. Great Lakes Res. Int. Assoc. Great Lakes Res. 256–267 (1971)

Milner, H.B.: Sedimentary Petrography. London: G. Allen & Unwin 1962

Mink, L.L., Williams, R.E., Wallace, A.T.: Effect of industrial and domestic effluents on the water quality of the Coeur d'Alene River Basin. Idaho Bur. Mines Geol. Pam. *49*, 30 (1971)

Mink, L.L., Williams, R.E., Wallace, A.T.: Effect of early day mining operations on present day water quality. Ground Water *10*, 310–314 (1972)

Mitchell, J.E.: The heavy metal problem of Silver Valley, northern Idahon. Proc. Int. Conf. Heavy Met., Toronto, 1975, *II/2*, 465–485 (1977)

Mitchell, R.L.: Trace elements in soil. In: Chemistry of the Soil. Bear, F.E. (ed.). Reinhold 1964, pp. 320–368

Mitchell, R.L.: Trace elements in soils. Techn. Bull. Minist. Agric. Fish. Fd. *21*, 8–20 (1971)

Mitchell, R.L.: Trace elements in soils and factors that affect their availability. Geol. Soc. Am. Bull. *83*, 1069–1076 (1972)

Molnar, F.M., Rothe, P., Förstner, U., Stern, J., Ogorelec, B., Šercelj, A., Culiberg, M.: Lakes Bled and Bohinj: Origin, composition, and pollution of recent sediments. Geologija (Ljubljana) *21*, 93–164 (1978)

Molt, E.L.: Spurenelemente in Oberflächen- und Trinkwasser. Paper pres. Arbeitsgemeinschaft der Rheinwasserwerke, Düsseldorf, Nov. 1967. Cit. Haberer u. Normann (1972)

Momoyama, K., Kobayashi, T., Yoshitsugu, K., Takayama, S.: Sediment of the fishing grounds of Yamaguchi Prefecture in the Inland Sea of Japan. Yamaguchi-ken Naikai Suisan Shinkenjo Hokoku 5, 79 pp. (1975). Cit. CA *85*, 112442

Montalvo, J.G., Jr., McKown, M.M.: Environmental implications of sediment bulk analysis for trace metals in offshore well-drilling operations. In: U.S. Environmental Protection Agency, Proc. Conf. Environ. Aspects of Chemical Use in Well-Drilling Operations, Washington D.C., U.S. EPA-560/1-75-004, 357–385 (1975)

Moore, J.R.: Bottom sediment studies, Buzzard Bay, Mass. J. Sediment. Petrol. *33* 511–558 (1963)

Moore, R.M., Burton, J.D.: Concentrations of dissolved copper in the eastern Atlantic Ocean, $23°N$ to $47°N$. Nature (London) *264*, 241–243 (1976)

Morel, F., McDuff, R.E., Morgan, J.J.: Interactions and chemostasis in aquatic chemical systems: role of pH, pE, solubility, and complexation. In: Trace Metals and Metal-Organic Interactions in Natural Waters. Singer, P.C. (ed.). Ann Arbor Sci. Publ. 1973, pp. 157–200

Morel, F.M.M., Morgan, J.J.: A numerical method for computing equilibria in aqueous chemical systems. Environ. Sci. Technol. *6*, 53–67 (1972)

Morgan, J.J.: Chemical equilibria and kinetic properties of manganese in natural waters. In: Principles and Applications of Water Chemistry. Faust, S.J., Hunter, J.V. (eds.). New York: John Wiley and Sons 1967, pp. 561–624

Morgan, J.J., Stumm, W.: The role of multivalent metal oxides in limnological transformations, as exemplified by iron and manganese. 2nd Int. Conf. Water Pollut. Res. *6*, 1–16 (1964a)

Morgan, J.J., Stumm, W.: Colloid-chemical properties of manganese dioxide. J. Coll. Sci. *19*, 347–353 (1964b)

Moriarty, F., French, M.C.: Mercury in waterways that drain into the wash in Eastham, England. Water Res. *11*, 367–372 (1977)

Morita, Y.: Distribution of copper and zinc in various phases of the earth materials. J. Earth Sci. Nagoya Univ. *3*, 33–56 (1955)

Morris, A.W.: Trace metal variations in sea water of the Menai Straits caused by a bloom of Phaeocystis. Nature (London) *233*, 427–428 (1971)

Morris, A.W., Bale, A.J.: The accumulation of cadmium, copper, manganese and zinc by Fucus vesiculosus in the Bristol Channel. Estuarine Coastal. Mar. Sci. *3*, 153–163 (1975)

Morris, A.W., Foster, P.F.: The seasonal variation of dissolved organic carbon in the inshore waters of the Menai Straits in relation to primary production. Limnol. Oceanogr. *16*, 987–989 (1971)

Morris, O.P., Russel, G.: Effect of chelation on toxicity of copper. Mar. Pollut. Bull. *4*, 159–160 (1973)

Morris, R.L., Johnson, L.G., Ebert, D.W.: Pesticides and heavy metals in the aquatic environment. Health Lab. Sci. *9*, 145–151 (1972)

Morrison, G.H., Pierce, J.O.: Sampling, sample preparation, and storage for analysis. Geochem. Environment *1*, 90–97, Natl. Acad. Sci. Wash., D.C. (1974)

Mortensen, G.L.: Complexing of metals by soil organic matter. Soil Sci. Soc. Am. Proc. *27*, 179–186 (1963)

Mortimer, C.H.: Chemical exchanges between sediments and water in the Great Lakes – speculations on probable regulatory mechanisms. Limnol. Oceanogr. *16*, 387–404 (1971)

Mortimer, D.C., Kudo, A.: Interaction between aquatic plants and river bed sediments in the uptake of mercury from flowing water. Proc. Int. Conf. Transp. Persist. Chem. Aquatic Ecosyst. Ottawa (1974)

Mortland, M.M.: Clay-organic complexes and interactions. Adv. Agron. *22*, 75–117 (1970)

Motosato, Y.: Report of Working Group for Counter Measures on Bottom Sediments in Minamata Bay. Dep. Civ. Eng. Kumamoto Univ. 1972. Cit. Buffa (1976)

Mottola, H.A.: **Nitrilotriacetic acid as a chelating agent**. Toxicol. Environ. Chem Rev. *2*, 99–161 (1974)

Mount, D.I., Stephan, C.E.: A method for detecting cadmium poisoning in fish. J. Wildlife Manage. *31*. 168–172 (1967)

Mouret, M.: Utilisation des bactères ferro-oxydans pour l'extraction de l'uranium des minéraux pauvres. Bull. Signal. CNRS *340*, 30 (1969)

Moyer, B.R., Budinger, T.F.: Cadmium levels in the shoreline sediments of San Francisco Bay. In: Trace Substances in Environmental Health, Vol. 8. Hemphill, D.D. (ed.). Univ. Missouri 1974, pp. 127–135

Müller, D., Schleichert, U.: Release of oxygen-consuming and toxic substances from anaerobic sediments by whirling-up and aeration. In: Interactions Between Sediments and Fresh Water. Golterman, H.L. (ed.). The Hague: Junk Publ. 1977, pp. 415–422

Müller, G.: Methoden der Sedimentuntersuchungen. Stuttgart: Schweizerbart 1964

Müller, G.: Beziehungen zwischen Wasserkörper, Bodensediment und Organismen im Bodensee. Naturwissenschaften 54, 454–466 (1967)

Müller, G.: Schadstoffuntersuchungen an datierten Sedimentkernen aus dem Bodensee. II. Historische Entwicklung von Schwermetallen – Beziehung zur Entwicklung polycyclischer aromatischer Kohlenwasserstoffe. Z. Naturforsch. 32c, 913–919 (1977a)

Müller, G.: Schadstoffuntersuchungen an datierten Sedimentkernen aus dem Bodensee. III. Historische Entwicklung von N- und P-Verbindungen – Beziehung zur Entwicklung von Schwermetallen und polycyclischen aromatischen Kohlenwasserstoffen. Z. Naturforsch. 32c, 920–925 (1977b)

Müller, G., Förstner, U.: Heavy metals in sediments of the Rhine and Elbe Estuaries: Mobilisation or mixing effect? Environ. Geol. 1, 33–39 (1975)

Müller, G., Förstner, U.: Schwermetalle in den Sedimenten der Elbe bei Stade: Veränderungen seit 1973. Naturwissenschaften 63, 242–243 (1976a)

Müller, G., Förstner, U.: Experimental mobilization of copper and zinc from aquatic sediments by some polyphosphate substitutes in detergents. Z. Wasser Abwasserforsch. 9, 150–152 (1976b)

Müller, G., Gastner, M.: The "Karbonat-Bombe", a simple device for the determination of the carbonate content in sediments, soils and other materials, Neues Jahrb. Mineral. Monatsh. 10, 466–469 (1971)

Müller, G., Prosi, F.: Cadmium in Fischen des mittleren und unteren Neckars. Veränderungen seit 1973. Naturwissenschaften 64, 530 (1977)

Müller, G., Prosi, F.: Verteilung von Zink, Kupfer und Cadmium in verschiedenen Organen von Plötzen (Rutilus rutilus L.) aus Neckar und Elsenz. Z. Naturforsch. 33c, 7–14 (1978)

Müller, G., Grimmer, G., Böhnke, H.: Sedimentary record of heavy metals and polycyclic aromatic hydrocarbons in Lake Constance. Naturwissenschaften 64, 427–431 (1977)

Müller, G., Irion, G., Förstner, U.: Formation and diagenesis of inorganic Ca-Mg carbonates in the lacustrine environment. Naturwissenschaften 59, 158–164 (1972)

Müller, J., Kretzler, W., Hirner, A.: Zur Methodik von Schwebstoffuntersuchungen an Flußwässern. GWF Wasser Abwasser 117, 220–222 (1976)

Mun, A.I., Idrisova, R.A.: Distribution of Cu, Zn, Co and Ni in the water and sediments of reservoirs. Trudy Inst. Khim. Nauk. A.N. Kaz. SSR 16, 225–232 (1967)

Murakami, K., Takeishi, K.: Behavior of heavy metals and PCBs in dredging and treating of bottom sediment. In: Proc. 2nd U.S.-Japan Experts Meeting "Management of Bottom Sediments Containing Toxic Substances". Peterson, S.A., Randolph, K.K. (eds.). Corvallis, Or.: U.S. EPA 1977, pp. 107–126

Murakami, T., Kida, A., Nakai, M., Matsunaga, S.: Heavy metal elements in sea sediments. Eisei Kagaku (Japan) 21, 275–281 (1975)

Murozumi, J., Chow, T.J., Patterson, C.: Chemical concentrations of pollutant aerosols, terrestrial dusts and sea salts in Greenland and Antarctic snow strata. Geochim. Cosmochim. Acta 33, 1247–1294 (1969)

Murphy, T.P., Lean, D.R.S., Nalewajko, C.: Blue-green algae: Their excretion of iron-selective chelators enables them to dominate other algae. Science 192, 900–902 (1976)

Murray, R.C.: Recent sediments of three Wisconsin lakes. Geol. Soc. Am. Bull. 67, 833–841 (1956)

Murray, C.N., Meinke, S.: Influence of soluble sewage material on adsorption and desorption behaviour of Ag, Cd, Co, and Zn sediment-seawater system. J. Oceanogr. Soc. Jpn. 30, 216–221 (1974)

Muzzarelli, R., Isolati, A.: Methylmercury acetate removal from water by chromatography. Water Air Soil Pollut. 1 (1971–72)

Mysels, K.J.: Introduction to Colloid Chemistry. New York: Interscience Publ. 1959

Mytelka, A.I., Czachor, J.S., Guggino, W.B., Golub, H.: Heavy metals in wastewater and treatment plant effluents. J.W.P.C.F. 45, 1859–1864 (1973)

Naeve, E.: Sources of pollution in the Mediterranean and its effects on living resources and fishing. Rev. Int. Oceanogr. Med. 35-36, 5–20 (1974)

Nakajima, J., Yarita, I., Kobayashi, S., Ogura, H.: An application of the principal component analysis to an evaluation of the heavy-metal contamination of water-bottom deposits. Mizu Shori Gijutsu 17, 721–727 (1976)

Nakamura, M., Nakano, S., Tachikawa, M.: Sediment deposits in Lake Biwa. IX. Heavy metal elements present in the bottom sediments in the southern part (Nan-ko) of Lake Biwa. Shiga Daigaku Kyoiku Gakubu Kiyo, Shizenkagaku 24, 89–96 (1974)

Nakanishi, J.: The cause and channels of mercury pollution in western Ontario. Congr. Sci. Hum. Environ., Tokyo, Japan, 1975, pp. 43–52. Cit. Buffa (1976)

Nakhshina, Y.P.: Distribution of trace elements in sediments of the Kiev Reservoir. Hydrobiol. J. 6, 7–11 (1974)

Namminga, H., Wilhm, J.: Heavy metals in water, sediments, and chironomids. J.W.P.C.F. 49, 1725–1731 (1977)

Napier, E., Wood, R.G., Chamberg, L.A.: Bacterial oxidation of pyrite and production of solutions for ore leaching. Symp. Adv. Extract. Metall. 1967, Inst. Min. Metall., 942–957 (1968)

Nauke, M.: In: Pollution in coastal waters: an interim report on the results of the priority programme from 1966 to 1974 – A Report of the German Research Society. Caspers, H. (ed.). Boppard: Boldt-Verlag 1975

Navrot, J., Amiel, A.J.: Metallic contaminants in dust storm deposits in the Rehovot area. Isr. J. Earth Sci. 23, 9–11 (1974)

Navrot, J., Amiel, A.J., Kronfeld, J.: Patella vulgata: A biological monitor of coastal metal pollution – a preliminary study. Environ. Pollut. 7, 303–308 (1974)

Neeb, R.: Inverse Polarography and Voltammetry. Weinheim: Verlag Chemie 1969

Nehring, R.B.: Aquatic insects as biological monitors of heavy metal pollution. Bull. Environ. Contam. Toxicol. 5, 147–154 (1976)

Nehring, R.B., Goettl, J.P., Jr.: Acute toxicity of a zinc-polluted stream to four species of salmonids. Bull. Environ. Contam. Toxicol. 12, 464–469 (1974)

Neihoff, R.A., Loeb, G.I.: The surface charge of particulate matter in seawater. Limnol. Oceanogr. 17, 7–16 (1972)

Nesterova, I.I.: Chemical composition of the suspended and dissolved loads of the Ob River. Geochemistry 4, 424–431 (1960)

Netzer, A., Wilkinson, P., Beszedits, S.: Removal of trace metals from wastewater by treatment with lime and discarded automotive tires. Water Res. 8, 813–817 (1974)

Neufeld, R.D., Hermann, E.F.: Heavy metal removal by acclimated activated sludge. J.W.P.C.F. 47, 310–329 (1975)

Neuland, H., Schrimpff, E., Hermann, R.: Zur Änderung der Spurenmetallgehalte im fließenden Wasserkörper und in den Sedimenten entlang eines Flußabschnittes des Roten Mains in Abhängigkeit von Redoxpotential, pH und anderen Einflußgrößen. Catena 5, 19–31 (1978)

Neumayr, V., Matthess, G.: Schwermetalle in Grundwässern der Westküste Schleswig-Holsteins. Vom Wasser 48, 17–39 (1977)

Nickless, G., Stenner, R., Terrile, N.: Distribution of cadmium, lead and zinc in the Bristol Channel. Mar. Pollut. Bull. 3, 188–190 (1972)

Nicol, I., Garret, R., Webb, J.S.: The role of some statistical and mathematical models in the interpretation of regional geochemical data. Econ. Geol. 64, 204–220 (1969)

Nielsen, S.A., Nathan, A.: Heavy metal levels in New Zealand molluscs. N.Z. J. Mar. Freshwater Res. 9, 476–481 (1975)

Nightingale, H.I.: Lead, zinc and copper in soils of urban storm runoff retention basins. J.A.W.W.A. 67, 443–446 (1975)

Nikiforova, E.M., Smirnova, R.S.: Metal technophility and lead technogenic anomalies. Abstr. Int. Conf. Heavy Met. Environ. Toronto C-94/96 (1975)

Nilsson, R.: Removal of metals by chemical treatment of municipal waste water. Water Res. 5, 51–60 (1971)

Nipkow, F.: Vorläufige Mitteilungen über Untersuchungen des Schlammabsatzes im Zürichsee. Z. Hydrol. 1, 1–23 (1920)

Nisimura, H.: Hg in fish and sediments in Tokuyama Bay. Univ. Tokyo (1974). Cit. Buffa (1976)

Nissenbaum, A.: Distribution of several metals in chemical fractions of sediment core from the Sea of Okhotsk. Isr. J. Earth Sci. 21, 143–154 (1972)

Nissenbaum, A.: Trace elements in Dead Sea sediments. Isr. J. Earth Sci. 23, 111–116 (1974)

Nissenbaum, A., Swaine, D.J.: Organic matter-metal interactions in Recent sediments: the role of humic substances. Geochim. Cosmochim. Acta 40, 809–816 (1976)

Nissenbaum, A., Presley, B.J., Kaplan, I.R.: Early diagenesis in a reducing fjord, Saanich Inlet, British Columbia – I. Chemical and isotopic changes in major components of interstitial water. Geochim. Cosmochim. Acta 36, 1007–1027 (1972)

Noddack, I., Noddack, W.: Die Häufigkeit der Schwermetalle in Meerestieren. Arch. Zool. 32-A, 35 (1940)

Noel-Lambot, F.: Distribution of cadmium, zinc, and copper in the mussel Mytilus edulis. Existence of cadmium-binding proteins similar to metallothioneins. Experientia 32 324 (1976)

Norval, E., Butler, L.R.P.: Trace elements in the human context and their determination by atomic absorption spectrometry. CSIR (1974), Pretoria S.A.

Nowak, H., Preul, F.: Untersuchungen über Blei- und Zinkgehalte in Gewässern des Westharzes. Beih. Geol. Jahrb. *105*, 68 pp. (1971)

Nriagu, J.O.: Lead orthophosphates. – IV. Formation and stability in the environment. Geochim. Cosmochim. Acta *38*, 887–898 (1974)

Nürnberg, H.W.: Potentialities and applications of advanced polarographic and voltammetric methods in environmental research and surveillance of toxic metals. Electrochim. Acta *22*, 935–949 (1977)

Nürnberg, H.W., Valenta, P.: Polarography and voltammetry in marine chemistry. In: The Nature of Seawater. Goldberg, E.D. (ed.). Berlin: Dahlem Konferenzen 1975, pp. 87–136

Nürnberg, H.W., Stoeppler, M., Valenta, P.: On the accuracy and reliability of trace metal determinations in environmental matrix types by advanced polarographic and spectroscopic techniques. Thalassia Jugosl. *11*, 85–100 (1975)

Nürnberg, H.W., Valenta, P., Mart, L., Raspor, B., Sipos, L.: Application of polarography and voltammetry to marine and aquatic chemistry. II. The polarographic approach to the determination and speciation of toxic metals in the marine environment. Z. Anal. Chem. *282*, 357–367 (1976)

Nusch, E.A.: Beurteilung der biologischen Schadwirkung organischer Komplexbildner und Komplexe aufgrund toxikologischer Testverfahren mit Bakterien, Algen, Protozoen, niederen Metazoen und Fischen. Z. Wasser Abwasserforsch. *10*, 49–61 (1977)

O'Connor, J.T.: Removal of trace inorganic constituents by conventional water treatment processes. In: Proc. 16th Water Qual. Conf., Trace Metals in Water Supplies: Occurrence, Significance, and Control. Univ. Ill., Coll. Eng., 99–110 (1974)

O'Connor, J.T., Kester, D.R.: Adsorption of copper and cobalt from fresh water and marine systems. Geochim. Cosmochim. Acta *39*, 1531–1543 (1975)

Odén, S.: Some effects on the acidity of air and precipitation. Conf. Luften, Larmet och Vi, Jönköping *1*, 85–94 (1971)

O'Hara, J.: Cadmium uptake by fiddler crabs exposed to temperature and salinity stress. J. Fish. Res. Board Canada *30*, 846–848 (1973)

Ohle, W.: Der schwefelsaure Tonteich bei Reinbek. Monographie eines idiotrophen Weihers. Arch. Hydrobiol. *30*, 604–662 (1936)

Ohle, W.: Die Seen als Opfer der Abwasserkalamität. Ber. Abwassertechn. Ver. *7*, 268–276 (1956)

Ohtsuka, T.: Countermeasures for pollution in Tokyo Bay. In: Management of Bottom Sediments Containing Toxic Substances. Proc. 2nd U.S.-Japan Experts Meet., Tokyo 1976. Environ. Res. Lab. U.S. EPA, Corvallis, Or. 1977, pp. 20–61

Olausson, E.: Methods for the chemical analysis of sediments. FAO Fish. Tech. Pap. *137*, 201–211 (1975a)

Olausson, E.: Man-made effect on sediments from Kattegat and Skagerrak. Geol. Fören. Stockholm Förh. *97*, 3–12 (1975b)

Olausson, E., Gustafsson, O., Mellin, T., Svensson, R.: Current level of heavy metal pollution and eutrophication in the Baltic proper. Medd. Maringeol. Lab. Göteborg *9*, 28 pp. (1977)

Oliver, B.G.: Heavy metal levels of Ottawa and Rideau River sediment. Environ. Sci. Technol. *7*, 135–137 (1973)

Oliver, B.G., Agemian, H.: Heavy metals in Ottawa and Rideau River sediments. Can. Inland Waters Dir. Rep. Ser. *37*, 10 pp. (1974)

Oliver, B.G., Cosgrove, E.G.: The efficiency of heavy metal removal by a conventional activated sludge treatment plant. Water Res. *8*, 869–874 (1974)

Oliver, B.G., Milne, J.B., LaBarre, N.: Chloride and lead in urban snow. J.W.P.C.F. *46*, 766–771 (1974)

Olphen, H. van: Introduction to Clay Colloid Chemistry. New York: Wiley Interscience 1963

Olson, B.H., Cooper, R.C.: In situ methylation of mercury in estuarine sediments. Nature *252*, 682–683 (1974)

Olson, B.H., Cooper, R.C.: Comparison of aerobic and anaerobic methylation of mercury chloride by San Francisco Bay sediments. Water Res. *10*, 113–116 (1976)

Olson, G.F., Mount, D.I., Snarski, V.M., Thorslund, T.W.: Mercury residues in fathead minnows, Pimephales promelas Rafinesque, chronically exposed to methylmercury in water. Bull. Environ. Contam. Toxicol. *14*, 129–134 (1975)

Olson, M.: Time and space dependence of pollutant levels in aquatic biota, field studies. In: Proc. Int. Conf. Transp. Persist. Chem. Aquatic Ecosyst., Ottawa, *III*, 49–60 (1974)

Ong, H.L., Bisque, R.E.: Coagulation of humic colloids by metal ions. Soil Sci. *106*, 220–224 (1968)

Ong, H.L., Swanson, V.E., Bisque, R.E.: Natural organic acids as agents of chemical weathering. Geol. Surv. Res. C130–C137 (1970)

Onishi, H.: Arsenic-abundance in natural waters. In: Handbook of Geochemistry, 33-I. Wedepohl, K.H. (ed.). Berlin, Heidelberg, New York: Springer 1970

Onstott, E.I., et al.: Removal of chromate from cooling tower blowdown by reaction with electrochemically generated ferrous hydroxide. Environ. Sci. Technol. 7, 333 (1973)

Oporowska, K.: Investigations on the content of copper in the ponds of some regions of Poland. Acta Hydrobiol. 18, 139–152 (1976)

Oppenheimer, C.H.: Bacterial activity in sediments of shallow marine bays. Geochim. Cosmochim. Acta 19, 244–260 (1960)

Oosterbaan, N.: An introductory bibliography on dredging and the environment. Terra et Aqua 7, 13–17 (1974)

Orren, M.J.: Trace elements (copper, iron and manganese) of the coast of South Africa. Invest. Rep. Div. Fish. Union S. Afr. 59, 1–40 (1967)

Osterberg, C., Kečkeš, S.: The state of pollution of the Mediterranean Sea. Ambio 6, 321–326 (1977)

Otsuka, H., Furuta, M., Arakawa, Y.: Effect of dredging on heavy metal distribution in lake sediments. Aichi-ken Kogai Chosa Senta Shoho 1976, 57–63

Ouellet, M.: Augmentation récente de métaux lourds dans les sédiments de plusieurs lacs du Quebec et de l'Ontario. In: Proc. Int. Conf. Heavy Met. Environ., Toronto (1975)

Overhoff, H., Forth, W.: Biologisch essentielle Elemente („Spurenelemente"). Deut. Ärzteblatt 1978, 301–305

Owen, R.M., Ullman, W.J.: Trace element profiles and inter-element correlations in a sediment core from Alpena-Basin, Lake Huron. 20th Conf. Great Lakes Research. Int. Assoc. Great Lakes Res. (1977)

Page, A.L.: Fate and effects of trace elements in sewage sludge when applied to agricultural lands. A literature review study. Cincinnati/Ohio: Office of Res. and Dev., U.S. Environ. Prot. Agency, Natl. Environ. Res. Cent. (1974)

Papadopoulos, C., Kanias, G.D., Elli, M.K.: Zinc content in otoliths of mackerel from the Aegean. Mar. Pollut. Bull. 9, 106–108 (1978)

Papakostidis, G., Grimanis, A.P., Zafiropoulos, D., Griggs, G.B., Hopkins, T.S.: Heavy metals in sediments from the Athens sewage outfall. Mar. Pollut. Bull. 6, 136–138 (1975)

Parker, P.L., Gibbs, A., Lawler, R.: Cobalt, iron and manganese in a Texas bay. Publ. Inst Mar. Sci. Univ. Tex. 9, 28–52 (1963)

Parker, R.L: Composition of the earth's crust. In: Data of Geochemistry, 6th ed. U.S. Geol. Surv. Prof. Pap. 440-D, 19 pp. (1967)

Parks, G.A.: Aqueous surface chemistry of oxides and complex oxides minerals In: Equilibrium Concepts in Natural Water Systems. Adv. Chem. Ser. 67, Washington, D.C.: Am. Chem. Soc., 121–160 (1967)

Parr, P.D., Taylor, F.G., Jr., Beauchamp, J.J.: Sensitivity of tobacco to mechanical draft cooling tower drift. Atmos. Environ. 10, 421–423 (1976)

Parsons, A., Dugan, P.: Production of extracellular polysaccharide matrix by Zoogloea ramigera. Appl. Microbiol. 21, 657–661 (1971)

Parsons, J.D.: Effects of acid mine wastes on aquatic ecosystems. Water Air Soil Pollut. 7, 333–354 (1977)

Pasternak, K.: Bottom sediments of the polluted dam reservoir at Otmuchów. Acta Hydrobiol. 12, 377–390 (1969)

Pasternak, K.: The spreading of heavy metals in flowing waters in the region of occurrence of natural deposits and of the zinc and lead industry. Acta Hydrobiol. 15, 145–166 (1973)

Pasternak, K.: The accumulation of heavy metals in the bottom sediments of the river Biala Przemsza as an indicator of their spreading by water courses from the centre of the zinc and lead mining smelting industries. Acta Hydrobiol. 16, 51–63 (1974a)

Pasternak, K.: The influence of the pollution of a zinc plant at Miasteczki Slaskie on the content of micro-elements in the environment of surface waters. Acta Hydrobiol. 16 273–297 (1974b)

Pasternak, K., Antoniewicz, A.: Preliminary investigations on the content of some trace components in surface waters of southern Poland. Acta Hydrobiol. 12, 111–124 (1970)

Pasternak, K., Antoniewicz, A.: The variability of copper, zinc and manganese content on the water of some rivers, streams and carp ponds. Acta Hydrobiol. 13, 251–268 (1971)

Pasternak, K., Glinski, J.: Some trace elements in mineral soils of the bottom of ponds. Pol. Soil Sci. 2, 15–24 (1969)

Pasternak, K., Glinski, J.: Occurrence and cumulation of microcomponents in bottom sediments of dam reservoirs of southern Poland. Acta Hydrobiol. 14, 225–255 (1972)

Patchineelam, S.R.: Untersuchungen über die Hauptbindungsarten und die Mobilisierbarkeit von Schwermetallen in fluviatilen Sedimenten. Diss. Univ. Heidelberg, 136 pp. (1975)

Patchineelam, S.R., Förstner, U.: Bindungsformen von Schwermetallen in marinen Sedimenten. Senckenbergiana Mar. *9*, 75–104 (1977)

Patrick, F.M., Loutit, M.: Passage of metals in effluents through bacteria to higher organisms. Water Res. *10*, 333–335 (1976)

Patrick, W.H., Jr., Gambrell, R.P., Khalid, R.A.: Physicochemical factors regulating solubility and bioavailability of toxic metals in contaminated dredged sediments. J. Environ. Sci. Health *A-12*, 475–492 (1977)

Patterson, C.C.: Contaminated and natural lead environments of man. Arch. Environ. Health *11*, 344–360 (1965)

Patterson, C.C.: Lead aerosol pollution in the high Sierra overrides. Natural mechanisms which exclude Pb from a food chain. Science *184*, 989–992 (1974)

Patterson, C.C., Settle, D.M.: The reduction of orders of magnitude errors in lead analyses of biological materials and natural waters by evaluation and controlling the extent and sources of industrial lead contamination introduced during sample collecting, handling and analysis. Proc. 7th IMR Symp. LaFleur, P.D. (ed.). NBS Spec. Publ. *422* (1975)

Patterson, C.C., Settle, D., Glover, B.: Analysis of lead in polluted coastal seawater. Mar. Chem. *4*, 305–319 (1976a)

Patterson, C.C., Settle, D., Schaule, B., Burnett, M.: Transport of pollutant lead to the ocean and within ocean ecosystems. In: Marine Pollution Transfer. Windom, H.L., Duce, R.A. (eds.). Lexington: D.C. Heath 1976b, pp. 23–38

Patterson, J.B.: Metal toxicities arising from industry. Technol. Bull. Min. Agric. Fish. Food *21*, 193–207 (1971)

Patterson, J.W., Minear, R.A.: Physical-chemical methods of heavy metals removal. In: Heavy Metals in the Aquatic Environment. Krenkel, P.A. (ed.). Oxford: Pergamon Press 1975, pp. 261–272

Payer, H.D., Runkel, K.H., Schramel, P., Stengel, E., Bhumiratana, A., Soeder, C.J.: Environmental influences on the accumulation of lead, antimony, arsenic, selenium, bromine and tin in unicellular algae cultivated in Thailand and in Germany. Chemosphere *6*, 413–418 (1976)

Pearson, R.: Hard and soft acids and bases, HSAB, Part I, Fundamental principles. J. Chem. Educ. *45*, 581–587 (1968a)

Pearson, R.: Hard and soft acids and bases, HSAB, Part II, Underlying theories. J. Chem. Educ. *45*, 643–648 (1968b)

Peden, J.D., Crothers, J.H., Waterfall, C.E., Beasley, J.: Heavy metals in Somerset marine organisms. Mar. Pollut. Bull. *4*, 7–9 (1973)

Penrose, W.R., Black, R., Hayward, M.J.: Limited arsenic dispersion in sea water, sediments, and biota near a continuous source. J. Fish. Res. Board Can. *32*, 1275–1281 (1975)

Pentreath, R.J.: The accumulation and retention of Zn-65 and Mn-54 by the plaice (Pleuronectes platessa L.). J. Exp. Mar. Biol. Ecol. *12*, 1–18 (1973); dto: . . . Fe-59 and Co-58. J. Exp. Mar. Biol. Ecol. *12*, 315–326 (1973)

Perhac, R.M.: Distribution of Cd, Co, Cu, Fe, Mn, Ni, Pb and Zn in dissolved and particulate solids from two streams in Tennessee. J. Hydrol. *15*, 177–186 (1972)

Perhac, R.M.: Heavy metal distribution in bottom sediment and water in the Tennessee River – Loudon Lake reservoir system. Water Resour. Res. Cent. Univ. Tenn.: Knoxville, Res. Rep. *40*, (1974a)

Perhac, R.M.: Water transport of heavy metals in solution and by different sizes of particulate solids. U.S. Dept. Interior, Water Resour. Res. Rep. *23*, (1974b)

Perhac, R.M., Wheelan, C.J.: A comparison of water, suspended solid and bottom sediment analysis for geochemical prospecting in a northern Tennessee zinc district. J. Geochem. Explor. *1*, 47–53 (1972)

Perkins, E.J., Gilchrist, J.R.S., Abbott, O.J., Halcrow, W.: Trace metals in Solway Firth sediments. Mar. Pollut. Bull. *4*, 59–61 (1973)

Peterson, S.A.: Hydraulic dredging as a lake restoration technique. In: Proc. 2nd U.S.-Japan Experts Meeting Management of Bottom Sediments Containing Toxic Substances. Peterson, S.A., Randolph, K.K. (eds.). Corvallis, Or.: U.S. EPA 1977, pp. 202–228

Petr, T.: Bioturbation and exchange of chemicals in the mud-water interface: In. Interactions Between Sediments and Fresh Water. Golterman, H.L. (ed.). The Hague: Junk Publ. 1977, pp. 216–226

Petri, H., Grohmann, A.: Die gesundheitliche Bedeutung des Zinks als Umweltfaktor des Menschen speziell in der Trinkwasserversorgung. WaBoLu-Ber. 7/71, 22 pp. (1971)

Pettyjohn, W.A.: Pickling liquors, strip mines and ground-water pollution. Ground Water *13*, 4–10 (1975)

References

Pezzetta, J.M., Iskandar, I.K.: Sediment characteristics in the vicinity of the Pulliam power plant, Green Bay, Wisconsin. Environ. Geol. *1,* 155–165 (1975)

Pfister, R.M., Dugan, P.R., Frea, I.: Microparticulates: isolation from water and identification of associated chlorinated pesticides. Science *166,* 878–879 (1969)

Pheiffer, T.H.: Heavy metal analyses of bottom sediments in the Potomac River estuary. Anapolis Field Office, Tech. Rep. *49,* U.S. Environ. Prot. Agency (1972)

Phillip, A.T., Pettis, R.W., Harris, J.E., Fabris, G.J., Boar, P.L., Bone, K.M.: Trace metals in Melbourne rivers – an inter-laboratory analysis. Proc. R. Austr. Chem. Inst. *42,* 209–215 (1975)

Phillips, D.J.H.: The common mussel Mytilus edulis as an indicator of pollution by zinc, cadmium, lead and copper. I. Effects of environmental variables on uptake of metals. Mar. Biol. *38,* 59–69 (1976a)

Phillips, D.J.H.: The common mussel Mytilus edulis as an indicator of pollution by zinc, cadmium, lead and copper. II. Relationship of metals in the mussel to those discharged by industry. Mar. Biol. *38,* 71–80 (1976b)

Phillips, D.J.H.: The use of biological indicator organisms to monitor trace metal pollution in marine and estuarine environments – a review. Environ. Pollut. *13,* 281–317 (1977)

Phillips, J.: Chemical processes in estuaries. In: The Estuarine Environment. Barnes, R.S.K., Green, J. (eds.). London: Applied Science 1972, pp. 33–50

Pierce, A.P., Botbol, J.M., Learned, R.E.: Mercury content of rocks, soils and stream sediments. In: Mercury in the Environment. U.S. Geol. Surv. Pap. *713,* 14–16 (1972)

Pierrard, J.M.: Photochemical decomposition of lead halides from automobile exhaust. Environ. Sci. Technol. *3,* 48 (1969)

Pigg, J., Coleman, M.S., Roach, B.: Distribution of lead in the upper Deep Fork River system of Oklahoma. Proc. Okla. Acad. Sci. *55,* 19–24 (1975)

Pilipchuk, M.F., Volkov, I.I.: Behaviour of molybdenum in processes of sediment formation and diagenesis in the Black Sea. In: The Black Sea, Geology, Chemistry and Biology. Degens, E.T., Ross, D.A. (eds.). Am. Assoc. Petrol. Geol. Mem. *20,* 542–553 (1974)

Pillai, T.N.V., Desai, M.V.M., Mathew, E., Ganapathy, S., Ganguly, A.K.: Organic materials in the marine environment and the associated metallic elements. Curr. Sci. *40,* 75–81 (1971)

Piotrowicz, S.R., Ray, B.J., Hoffman, G.L., Duce, R.A.: Trace metal enrichment in the sea-surface microlayer. J. Geophys. Res. *77,* 5243–5254 (1973)

Piper, D.Z.: The distribution of Co, Cr, Cu, Fe, Mn, Ni and Zn in Framvaren, a Norwegian anoxic fjord. Geochim. Cosmochim. Acta *35,* 531–550 (1971)

Pita, F.W., Hyne, N.J.: The depositional environment of zinc, lead, and cadmium in reservoir sediments. Water Res. *9,* 701–706 (1975)

Plato, P.A., Jacobson, A.P.: Cesium-137 in Lake Michigan sediments: areal distribution and correlation with other man-made material. Environ. Pollut. *10,* 19–34 (1976)

Poldervaart, A.: Chemistry of the earth's crust. Geol. Soc. Am. Spec. Pap. *62,* 119–144 (1955)

Polikarpov, G.G.: Radioecology of aquatic organisms. Amsterdam: North Holland 1966

Poon, C.P.C., Sheih, J.M.S.: Nutrient profiles of bay sediment. J.W.P.C.F. *48,* 2007–2017 (1976)

Popova, T.P.: Coprecipitation of some microconstituents from natural waters with calcium carbonate. Geochemistry *12,* 1256–1261 (1961)

Postma, H.: Sediment transport and sedimentation in the estuarine environment In: Estuaries. Lauff, G.G. (ed.). AAAS *83,* 158–179 (1967)

Potter, L., Kidd, D., Standiford, D.: Mercury levels in Lake Powell: Bioamplification of mercury in man-made desert reservoir. Environ. Sci. Technol. *9,* 41–46 (1975)

Powers, M.C.: Adjustment of land derived clays to the marine environment. J. Sediment. Petrol. *27,* 355–372 (1957)

Prakash, A.: NTA (nitrilotriacetic acid) – an ecological appraisal. Water Pollut. Control Direct. Can. Econ. Technol. Rev. Rep. EPS 3-WP-76-8 (1976)

Prakash, A., Jensen, A., Rashid, M.A.: Humic substances and aquatic productivity. In: Humic Substances, Their Structure and Function in the Biosphere. Povoledo, D.M., Golterman, H.L. (eds.). Wageningen: Pudoc 1975, pp. 259–268

Prater, B.E.: The metal content and dispersion characteristics of steelwork's effluents discharging to the Tees Estuary. Water Pollut. Control *74,* 63–78 (1975)

Pravdic, V.: Surface charge characterization of sea sediments. Limnol. Oceanogr. *15,* 230–233 (1970)

Premazzi, V.: Contenuto di metalli pesanti nei sedimenti del Lago Lugano. Annu. Rep. Lake Lugano. Ispra, Italy: Euratom 1973, pp. 96–101

Presley, B.J., Brooks, R.R., Kaplan, I.R.: Manganese and related elements in the interstitial water of marine sediments. Science *158,* 906–910 (1967)

Presley, B.J., Kolodny, Y., Nissenbaum, A., Kaplan, I.R.: Early diagenesis in a reducing fjord, Saanich Inlet, British Columbia. II. Trace element distribution in interstitial water and sediment. Geochim. Cosmochim. Acta *36*, 1073–1099 (1972)
Preston, A.: Artificial radioactivity in freshwater and estuarine systems. Proc. R. Soc. (London), Ser. B, *180*, 421–436 (1972)
Preston, A.: Heavy metals in British waters. Nature (London) *242*, 95–97 (1973)
Preston, A., Jefferies, D.F., Dutton, J., Harvey, B., Steele, A.K.: The concentration of selected heavy metals in sea water. Environ. Pollut. *3*, 69–82 (1972)
Pretorius, D.A.: The depositional environment of the Witwatersrand Goldfields: A chronological review of speculations and observations. Miner. Sci. Eng. *7*, 18–47 (1975)
Preuss, E., Kollmann, H.: Metallgehalte in Klärschlämmen. Naturwissenschaften *61*, 270–271 (1974)
Price, N.B.: Chemical diagenesis in sediment. WHOI-73-39 Tech. Rep. (Woods Hole, Mass.) (1973)
Price, N.B., Skei, J.M.: Areal and seasonal variations in the chemistry of suspended particulate matter in a deep water fjord. Estuar. Coast. Mar. Sci. *3*, 349–369 (1975)
Pritchard, D.W.: What is an estuary? Physical viewpoint. In: Estuaries. Lauff, G.H. (ed.). Am. Assoc. Adv. Sci. Publ. *83*, 3–5 (1967)
Pringle, B.H., Hissong, D.E., Katz, E.L., Mulawka, S.T.: Trace metal accumulation by estuarine molluscs. J. Sanit. Eng. Div. Am. Soc. Civ. Eng. *94*, 455–475 (1968)
Proctor, P.D., Kisvarsanyi, G., Garrison, E., Williams, A.: Heavy metal content of surface and ground waters of the Springfield-Joplin areas, Missouri. In: Trace Substances in Environmental Health, Vol. 7. Hemphill, D.D. (ed.). Columbia: Univ. Missouri 1973, pp. 63–73
Prosi, F.: Schwermetallbelastung in den Sedimenten der Elsenz und ihre Auswirkung auf limnische Organismen. Diss. Univ. Heidelberg 1977
Pryor, E.J.: Mineral Processing. Amsterdam: Elsevier 1965
Puel, D., Flateau, G., Laumond, F., Barelli, M., Aubert, M.: Etude de la toxicité des rivières de la province de Genes vis-à-vis de la biomasse marine. Rev. Int. Océanogr. Méd. *61–62*, 143–157 (1976)
Pustelnikov, O.S.: Geochemical features of suspended matter in connection with recent sedimentation processes in the Baltic Sea. In: 3rd Soviet-Swedish Symp. on the Pollution of the Baltic, Åkerblom, A. (ed.). Ambio Spec, Rep. *5*, 157–162 (1977)
Qasim, S.R., Burchinal, J.C.: Leaching from simulated landfills. J.W.P.C.F. *42*, 371–379 (1970)
Quentin, K.E.: Schadstoffe im Wasser als aktuelles Problem der Wasserversorgung. GWF Wasser Abwasser *113*, 377–381 (1972)
Quentin, K.E., Winkler, H.A.: Vorkommen und Nachweis von anorganischen Schadstoffen im Oberflächen- und Grundwasser. Zentralbl. Bakt. Hyg., Abt. Orig. B *158*, 514–523 (1974)
Qvarfort, U.: Tolkning av geokeiska anomalier inom kultur-påverkade områden. Kvartärgeol. Avdelningen, Uppsala Univ. *56*, 13 pp. (1974)
Rabe, F.W., Bauer, S.B.: Heavy metals in lakes of the Coeur d'Alène River Valley, Idaho. Northwest Sci. *51*, 183–197 (1977)
Ramamoorthy, S., Kushner, D.J.: Binding of mercuric and other heavy metal ions by microbial growth media. Microbial Ecol. *2*, 162–176 (1975)
Ramamoorthy, S., Rust, B.R.: Mercury sorption and desorption characteristics of some Ottawa River sediments. Can. J. Earth Sci. *13*, 530–536 (1976)
Ramamoorthy, S., Rust, B.R.: Heavy metal exchange processes in sediment-water systems. Environ. Geol. *2*, 165–172 (1978)
Rameau, J.Th.L.B.: Lead in environmental pollutant. In: Proc. Int. Symp. Environ. Health Aspects of Lead, Amsterdam (1972). Luxembourg: Commission of the European Communities 1973, pp. 189–197
Ramondetta, P.J., Harris, W.H.: Heavy metals distribution in Jamaica Bay sediments. Environ. Geol. *2*, 145–149 (1978)
Rankama, K.K.: Progress in Isotopic Geology. London: Interscience Publ. 1963, p. 471
Rankama, K.K., Sahama, T.G.: Geochemistry. Chicago: The University of Chicago Press 1950
Rashid, M.A.: Contribution of humic substances to the cation exchange capacity of different marine sediments. Marit. Sediments *5*, 44–50 (1969)
Rashid, M.A.: Role of humic acids of marine origin and their different molecular weight fractions in complexing di- and tri-valent metals. Soil. Sci. *111*, 298–305 (1971)
Rashid, M.A.: Adsorption of metals on sedimentary and peat humic acids. Chem. Geol. *13*, 115–123 (1974)
Rashid, M.A., King, L.H.: Molecular weight distribution measurements on humic and fulvic acid fractions from marine clays on the Scotian Shelf. Geochim. Cosmochim. Acta *33*, 147–151 (1969)

Rashid, M.A., King, L.H.: Chemical characteristics of fractionated humic acids associated with marine sediments. Chem. Geol. 7, 37–43 (1971)

Rashid, M.A., Leonard, J.L.: Modifications in the solubility and precipitation behaviour of various metals as a result of their interaction with sedimentary humic acid. Chem. Geol. 11, 89–97 (1973)

Rashid, M.A., Prakash, A.: Chemical characteristics of humic compounds isolated from some decomposed marine algae. J. Fish. Res. Board Can. 20, 55–60 (1972)

Rashid, M.A., Buckley, D.E., Robertson, K.R.: Interactions of a marine humic acid with clay minerals and a natural sediment. Geoderma 8, 11–27 (1972)

Rassbach, K.: Die Entwicklung der Flußwasseraufbereitung in Wiesbaden. In: Künstliche Grundwasseranreicherung am Rhein. Wiss. Ber. Stadtwerke Wiesbaden 2, 81–92 (1974)

Rathner, R., Ludmer, Z.: Separation of lithium by ion exchange chromatrography. Isr. J. Chem. 2, 21–24 (1964)

Ratkowsky, D.A., Thrower, S.J., Eustace, I.J., Olley, J.: A numerical study of the concentration of some heavy metals in Tasmania oysters. J. Fish. Res. Board Can. 31, 1165–1171 (1974)

Ratkowsky, D.A., Dix, T.G., Wilson, K.C.: Mercury in fish in the Derwent Estuary, Tasmania, and its relation to the position of the fish in the food chain. Aust. J. Mar. Freshwater Res. 26, 223–231 (1975)

Ravera, O., Premazzi, G.: A method to study the history of any persistent pollution in a lake by the concentration of Cs-137 from fallout. Proc. Int. Symp. Radioecol. Appl. Prot. Man Environ. Rome 1971, pp. 703–719

Ravera, O., Gommes, R., Muntau, H.: Cadmium distribution in aquatic environment and its effects on aquatic organisms. In: Problems of the Contamination of Man and his Environment by Mercury and Cadmium. Luxembourg: Eur. Comm. Coll. 1973, pp. 317–331

Rawson, J.: Solution of manganese dioxide by tannic acid. U.S. Geol. Surv. Prof. Pap. 475-C, C218–C219 (1963)

Raymont, J.E.G.: Some aspects of pollution in Southampton water. Proc. R. Soc. London B 180, 451–468 (1972)

Reay, P.F.: The accumulation of arsenic from arsenic-rich natural waters by aquatic plants. J. Appl. Ecol. 9, 557–565 (1972)

Reddy, C.N., Patrick, W.H., Jr.: Effect of redox potential on the stability of zinc and copper chelates in flooded soils. J. Soil Sci. Soc. Am. 41, 729–732 (1977a)

Reddy, C.N., Patrick, W.H., Jr.: Effect of redox potential and pH on the uptake of cadmium and lead by rice plants. J. Environ. Qual. 6, 259–262 (1977b)

Reddy, M.M.: A preliminary report: nutrients and metals transported by sediments within the Genesee River watershed, New York, U.S.A. In: Proc. Int. Symp. Interactions Between Sediments and Fresh Water, Amsterdam, 1976. Goltermann, H.L. (ed.). Wageningen/The Hague: Pudoc/Junk B.V. Publ. 1977, pp. 63–67

Reeburgh, W.S.: Observation of gases in Cheasapeake Bay sediments. Limnol. Oceanogr. 14, 368–375 (1969)

Reeder, S.W., Hitchon, B., Levinson, A.A.: Hydrogeochemistry of the surface waters of the Mackenzie River drainage basin, Canada. I. Factors controlling inorganic composition. Geochim. Cosmochim. Acta 36, 825–865 (1972)

Rehwoldt, R., Karimian-Teherani, D.: Uptake and effect of cadmium on Zebrafish. Bull. Environ. Contam. Toxicol. 13, 442–446 (1975)

Rehwoldt, R., Karimian-Teherani, D., Altmann, H.: Measurement and distribution of various heavy metals in the Danube River and Danube Canal aquatic communities in the vicinity of Vienna, Austria. Sci. Total Environ. 3, 341–348 (1975)

Reichenbach-Klinke, H.: Die Möglichkeiten toxischer Anreicherungen im Fisch. 1. Eur. Tierärztekongreß, 11.–18.9.1972 Wiesbaden, 134–137 (1972)

Reichert, J.K.: Beryllium, ein toxisches Element in der menschlichen Umgebung unter besonderer Berücksichtigung seines Vorkommens in Gewässern. Vom Wasser 40, 135–149 (1973)

Reichert, J.M., Haberer, K., Normann, S.: Untersuchungen über das Verhalten von Spurenelementen bei der Trinkwasseraufbereitung. Vom Wasser 39, 137–146 (1972)

Reimers, R.S., Krenkel, P.A., Eagle, M., Tragift, G.: Sorption phenomenon in the organics of bottom sediments. In: Heavy Metals in the Aquatic Environment. Krenkel, P.A. (ed.). Oxford: Pergamon Press 1975, pp. 117–136

Reinert, R.E., Stone, L.J., Willford, W.A.: Effect of temperature on accumulation of methylmercuric chloride and p,p'DDT by rainbow trout Salmo gairdneri. J. Fish. Res. Board. Can. 31, 1649–1652 (1974)

Reinhard, D., Förstner, U.: Metallanreicherungen in Sedimentkernen aus Stauhaltungen des mittleren Neckars. Neues Jahrb. Geol. Paläontol. Monatsh. 5, 301–320 (1976)

Renfro, W.C.: Transfer of ^{65}Zn from sediments by marine polychaete worms. Mar. Biol. *21,* 305–316 (1973)

Renfro, W.C., Fowler, S.W., Heyraud, M., La Rosa, J.: Relative importance of food and water in long-term zinc-65 accumulation by marine biota. J. Fish. Res. Board Can. *32,* 1339–1345 (1975)

Renzoni, A., Bacci, E., Falcia, L.: Mercury concentration in the water, sediments, and fauna of an area of the Tyrrhenian Coast. Rev. Int. Oceanogr. Méd. *31/32,* 17–45 (1973)

Reppert, R.T.: Aquatic life and the acid reaction. Proc. 5th Ann. Symp. Indus. Wastes. Md. Water Pollut. Control Comm., 27–49 (1964)

Reuter, J.H., Perdue, E.M.: Importance of heavy metal-organic matter interactions in natural waters. Geochim. Cosmochim. Acta *41,* 325–334 (1977)

Reynolds, A.: Kationenaustausch an Permutiten, insbesondere an Wasserstoff- und Schwermetallpermutiten. Kolloid-Beihefte *43,* 1–142 (1935)

Richardson, E.M., Epstein, E.: Retention of three insecticides on different size soil particles suspended in water. Soil Sci. Soc. Am. Proc. *35,* 884–891 (1971)

Richardson, E.W., Stobbe, E.D., Bernstein, S.: Ion exchange traps chromates for reuse. Environ. Sci. Technol. *2,* 1006 (1968)

Richert, J.G.: Les Eaux Souterraines Artificielles. Stockholm 1900. Cit. Trüb (1975)

Richins, R.T., Risser, A.C.: Total mercury in water, sediment, and selected aquatic organisms, Carson River, Nevada – 1972. Pestic. Monit. J. *9,* 44–54 (1975)

Richter, G.R., Washuettl.,J., Bancher, E., Altmann, H.: Bestimmung von Quecksilber und anderen Spurenelementen in Schlammproben des Neusiedlersees (Österreich). Ber. Österr. Studienges. Atomenergie *2357,* 25 pp. (1974)

Riley, J.P., with contributions by Robertson, D.E., Dutton, J.W.R., Mitchell, N.T., le B.Williams, P.J.: Analytical chemistry of sea water. In: Chemical Oceanography, 2nd ed., Vol. 3. Riley, J.P., Skirrow, G. (eds.). London, New York, San Francisco: Academic Press 1975, pp. 193–514

Riley, J.P., Chester, R. (eds.): Chemical Oceanography, Vol. 5, 402 pp.; Vol. 6, 414 pp. London, New York, San Francisco: Academic Press 1976

Riley, J.P., Roth, I.: The distribution of trace elements in some species of phytoplankton grown in culture. J. Mar. Biol. Assoc. U.K. *51,* 63–72 (1971)

Ritchie, J.A.: Arsenic and antimony in some New Zealand thermal water. N.Z. J. Sci. *4,* 218 (1961)

Ritchie, J.C., McHenry, R.J., Gill, A.C.: Dating recent reservoir sediments. Limnol. Oceanogr. *18,* 254–263 (1973)

Rittenhouse, G., Fulton, R.B. III, Grabowski, R., Bernard, J.L.: Minor elements in oil-field waters. Chem. Geol. *4,* 189–197 (1969)

Rivers, J.B., Pearson, J.E., Schultz, C.D.: Total and organic mercury in marine fish. Bull. Environ. Contam. Toxicol. *8,* 257–266 (1972)

Robbins, J.A., Edgington, D.N.: Determination of recent sedimentation in Lake Michigan using lead-210 and cesium-137. Geochim. Cosmochim. Acta *39,* 285–304 (1975)

Robbins, J.A., Edgington, D.N.: Major and minor elements in sediments of Southern Lake Michigan. Abstr. 20th Conf. Great Lakes Research. Int. Assoc. Great Lakes Res. (1977)

Robbins, J.A., Landström, E., Wahlgren, M.: Tributary inputs of soluble trace metals to Lake Michigan. Proc. 15th Conf. Great Lakes Res. Int. Assoc. Great Lakes Res., 270–290 (1972)

Robeck, G.G.: Purification of drinking water to remove pesticides and other poisonous chemicals: American practice. Special Subj. No. 10. Proc. 16th Water Qual. Conf. Univ. Ill. Coll. Eng. (1974a)

Robeck, G.G.: Purification of drinking water to remove pesticides and other poisonous chemicals: American practice. Proc. 16th Water Qual. Conf., Trace Metals in Water Supplies: Occurrence, Significance and Control. Univ. Ill. Coll. Eng., 139 pp. (1974b)

Robertson, D.E.: The distribution of cobalt in oceanic waters. Geochim. Cosmochim. Acta *34,* 553–567 (1970)

Robertson, D.E.: In: Pacific Northwest Laboratory Annual Report for 1970, Battelle Memorial Institute, Richland, Washington 1971. Cit. Brewer, P.G. (1975)

Robinson, A.R.: Sediment our greatest pollutant? In: Focus on Environmental Ecology. Tank, R.W. (ed.). New York: Oxford Univ. Press 1973, pp. 186–192

Robinson, K.: Thallium. In: United States Mineral Resources. Probst, D.A., Pratt, W.P. (eds.). U.S. Geol. Surv. Prof. Pap. *820,* 631 (1973)

Robitaille, D.R.: Sodium molybdate as a corrosion inhibitor in cooling tower water. Corros 76, Intl. Corrosion Forum on the Protection and Performance of Matter. Houston, Texas, Paper 93. Houston, Texas: Publ. NACE (1976)

Rogers, J.J.W., Adams, J.A.S.: Uranium – abundance in natural waters. In: Handbook of Geochemistry. Wedepohl, K.H. (ed.). Berlin, Heidelberg, New York: Springer 1970, 92-I/1–2

Rohatgi, N.K., Chen, K.Y.: Transport of trace metals by suspended particulates on mixing with seawater. J.W.P.C.F. *47*, 2298–2316 (1975)

Rohatgi, N.K., Chen, K.Y.: Fate of metals in wastewater discharge to ocean. J. Environ. Eng. Div. ASCE *102*, 675–685 (1976)

Rohde, G.: Sind bedenkliche Anreicherungen von Schwermetallen in Böden und Pflanzen nach fortgesetzten Einsatz von Müll- und Müllklärschlammkomposten möglich? ANS Mittl. *35*, 295–300 (1972)

Rohde, G.: Schwermetalle – ihre Bedeutung und mögliche Anreicherung im Boden durch Müll- und Müllklärschlammkomposte. ANS Mitt., 24 pp. (1974)

Rolfe, G.L., Jennett, J.C.: Environmental lead distribution in relation to automobile and mine and smelter sources. In: Heavy Metals in the Aquatic Environment. Krenkel, P.A. (ed.). Oxford: Pergamon Press 1975, pp. 231–241

Romano. R.R.: Fluvial transport of selected heavy metals in the Grand Calumet River system. In: Trace Substances in Environmental Health, Vol. 10. Hemphill, D.D. (ed.), 1976, pp. 207–216

Romeril, M.G.: The uptake and distribution of ^{65}Zn in oysters. Mar. Biol. *9*, 347–354 (1971)

Rona, E., Gilpatrick, L.O., Jeffrey, L.M.: Uranium determination in seawater. Trans. Am. Geophys. Union *37*, 697–701 (1956)

Rona, E., Hood, D.W., Muse, L., Buglio, B.: Activation analysis of manganese and zinc in seawater. Limnol. Oceanogr. *7*, 201–206 (1962)

Ronald, K., Tessaro, S.V., Uthe, J.F., Freeman, H.C., Frank, R.: Methylmercury poisoning in the harp seal (Pagophilus groenlandicus). Sci. Total Environ. *8*, 11–17 (1977)

Roschin, A.V., Timofeevskaya, L.A.: Chemical substances in the work environment: Some comparative aspects of USSR and US hygiene standards. Ambio *4*, 30–33 (1975)

Rosenbaum, J.B.: Minerals extraction and processing: new developments. Science *191*, 720–723 (1976)

Rosenqvist, I.Th.: Alternative sources for acidification of river water in Norway. Sci. Total Environ. *10*, 39–49 (1978)

Roskam, R.T.: Kopervergifting in zee. In: Koper in het Nederlandse Milieu. TNO Nieuws, 416–418 (1972)

Ros Vicent, J.C., Yangue, C.F., Parsi, P., Statham, G., Duursma, E.K.: The ease of release of some trace metals and radionuclides being sorbed for long periods by marine sediments. Bol. Inst. Esp. Oceangr. (1977) Cit. Duursma (1976)

Roth, I., Hornung, H.: Heavy metal concentrations in water, sediments, and fish from the Mediterranean coastal area, Israel. Environ. Sci. Technol. *11*, 265–269 (1977)

Rüb, F.: Technik der industriellen Abwasserbehandlung. Mainz: Krausskopf 1974

Ruch, R.R., Kennedy, E.J., Shimp, N.F.: Distribution of arsenic in unconsolidated sediments from southern Lake Michigan. Environ. Geol. Notes *37*, 1–16 (1970)

Ruch, R.R., Gluskoter, H.J., Shimp, N.F.: Occurrence and distribution of potentially volatile trace elements in coal. Environ. Geol. Notes *61*, 1–43 (1973)

Rühling, A., Tyler, G.: An ecological approach to the lead problem. Botariska Notiser *121*, 321 (1968)

Rump, H.H.: Mathematische Vorhersagemodelle für Pestizide und Schadstoffe in Gewässern der Niederrheinischen Bucht und der Nordeifel. Kölner Geogr. Arb. *34*, 122 (1976)

Rump, H.H.: Untersuchungen von Kupfer-, Zink- und Mangangehalten in kleinen Einzugsgebieten mit Hilfe statistischer Methoden. Deutsche Gewässerkdl. Mitt. *21*, 82–86 (1977)

Runnells, D.D.: Molybdenum in the environment: An interdisciplinary approach. Geol. Soc. Am. (1972)

Runnells, D.D.: Detection of molybdenum enrichment in the environment through comparative study of stream drainages, Central Colorado. In: Trace Substances in Environmental Health, Vol. 7. Hemphill, D.D. (ed.). Univ. Missouri 1973, pp. 99–104

Runnells, D.D., Brown, D., Lindberg, R.: Enrichment of molybdenum in the environment through comparative study of stream drainages, Central Colorado. Proc. 1st Ann. NSF Conf. Trace Contaminants, 1973. Fulkerson, W., Shults, W.D., van Hook, R.I. (eds.). Springfield, Va.: NTIS 1974, pp. 599–614

Ruohtula, M., Miettinen, J.K.: Retention and excretion of ^{203}Hg-labelled methylmercury in rainbow trout. Oikos *26*, 385–390 (1975)

Ruppert, D.F., Hopke, P.K., Clute, P., Metzger, W., Crowley, D.: Arsenic concentrations and distribution in Chautauqua Lake sediments. J. Radioanal. Chem. *23*, 159–169 (1974)

Russel, G.L., Morris, O.P.: Copper tolerance in the marine fouling alga Ectocarpus siliculosus. Nature (London) *228*, 288–289 (1970)

Rust, B.R., Waslenchuk, D.G.: The distribution and transport of bed sediments and persistent pollutants in the Ottawa River, Canada. Proc. Int. Conf. Transp. Persist. Chem. Aquatic Ecosyst., Ottawa *I*-25 (1974)

Rutherford, F., Church, T.: Use of silver and zinc to trace sewage sludge dispersal in coastal waters. In: Marine Chemistry in the Coastal Environment. Church, T.M. (ed.). Am. Chem. Soc. Symp. Ser. *18*, 440–452 (1975)

Rutherford, G.K.: Anthropogenic influences of sediment quality at a source. In: The Fluvial Transport of Sediment-Associated Nutriants and Contaminants. Shear, H., Watson, A.E.P. (eds.). Windsor: Int. Joint Comm., Great Lakes Regional Office 1977, pp. 95–104

Ryther, J.H., Menzel, D.W., Corwin, N.: Influence of the Amazon River outflow on the ecology of the western tropical Atlantic. I. Hydrography and nutrient chemistry. J. Mar. Res. *25*, 69–83 (1967)

Saager, R., Sinclair, A.J.: Factor analysis of stream sediment geochemical data from Mount Nansen Area, Yukon Territory, Canada. Mineral. Deposita *9*, 243–252 (1974)

Sackett, W., Arrhenius, G.O.S.: Distribution of aluminum species in the hydrosphere – I. Aluminum in the ocean. Geochim. Cosmochim. Acta *26*, 955–968 (1962)

Saelen, O.H.: Some features of the hydrography of Norwegian fjords. In: Estuaries. Lauff, G.H. (ed.). Am. Assoc. Adv. Sci. *83*, 63–70 (1967)

Salánki, J., Varanka, I.: Effect of copper and lead compounds on the activity of the fresh-water mussel. Ann. Biol. Tihany *43*, 21–27 (1976)

Salomons, W., De Groot, A.J.: Pollution history of trace metals in sediments, as affected by the Rhine River. In: Environmental Biogeochemistry, Vol. 1. Krumbein, W.E. (ed.). Ann Arbor Sci. Publ. 1978, pp. 149–162

Salomons, W., Mook, W.G.: Processes affecting trace metals in Lake IJssel. Abstr. 10th Int. Congr. on Sedimentology. Jerusalem 1978, pp. 569–570

Salomons, W., Hofman, P., Boelens, R., Mook, W.G.: The oxygen isotopic composition of the fraction less than 2 microns (clay fraction) in recent sediments from Western Europe. Mar. Geol. *18*, M23–M28 (1975)

Sames, C.-W.: Die Zukunft der Metalle. Frankfurt: Suhrkamp 1971

Sameshima, T.: Dredging of contaminated bed sediment in Japan. In: Proc. 2nd U.S.-Japan Experts Meet. Management of Bottom Sediments Containing Toxic Substances. Peterson, S.A., Randolph, K.K. (eds.). Corvallis, Or.: U.S. EPA 1977, pp. 1–19

Sanchez, I., Lee, G.F.: Sorption of copper on Lake Monona sediments – effect of NTA on copper release from sediments. Water Resour. Res. *7*, 587–593 (1973)

Särkkä, J., Hattula, M.L., Janatuinen, J., Paasivirta, J.: Mercury in sediments of Lake Päijänne, Finland. Bull. Environ. Contam. Toxicol. *20*, 21–27 (1978)

Saroma, A.M.: Heavy metal distribution in bottom sediments of lagoons and bays in Hokkaido. Chikyu Kaguku *25*, 149–159 (1972)

Sartor, J.D., Boyd, G.B., Agardy, F.J.: Water pollution aspects of street surface contaminants. J.W.P.C.F. *46*, 458–466 (1974)

Sasaki, M., Oka, H., Inoue, S., Kikuchi, T.: Mercury in river water from the River Muka. Kitami Kogyo Tanki Daigaku Kenkyu Hokoku *7*, 131–137 (1975)

Satake, M., Asano, T., Yamamoto, K., Yonekubo, T., Nagaosa, Y.: Distribution of heavy metals in Lake Biwa. Fukui Daigaku Kogakubu Kenkyu *23*, 109–114 (1975)

Satoh, E.: A method for disposing of waste water at dredged material reclamation sites. In: Proc. 2nd U.S.-Japan Experts Meeting Mangement of Bottom Sediments Containing Toxic Substances. Peterson, S.A., Randolph, K.K. (eds.). Corvallis, Or.: U.S. EPA 1977, pp. 169–190

Saward, D., Stirling, A., Topping, G.: Experimental studies on the effects of copper on a marine food chain. Mar. Biol. *29*, 351–361 (1975)

Saxby, J.D.: Metal-organic chemistry of the geochemical cycle. Rev. Pure Appl. Chem. *19*, 131–150 (1969)

Saxby, J.D.: Diagenesis of metal-organic complexes in sediments: formation of metal sulphides from cysteine complexes. Chem. Geol. *12*, 241–288 (1973)

Sayler, G.S., Colwell, R.R.: Partitioning of mercury and polychlorinated biphenyl by oil, water and suspended sediment. Environ. Sci. Technol. *10*, 1142–1145 (1976)

Sayler, G.S., Nelson, J.D., Jr., Colwell, R.R.: Role of bacteria in bioaccumulation of mercury in the oyster Crassostrea virginica. Appl. Microbiol. *30*, 91–96 (1975)

Sayles, F.L., Mangelsdorf, P.C., Jr., Wilson, T.R.S., Hume, D.N.: A sampler for the in situ collection of marine sedimentary pore waters. Deep Sea Res. *23*, 259–264 (1976)

Schafer, H.A.: Characteristics of municipal wastewater discharges, 1975. South. Calif. Coastal Water Res. Project. El Segundo, Annu. Rep. 57–60 (1976)

Schafer, H.A., Bascom, W.: Sludge in Santa Monica Bay. In: South. California Coastal Water Research Project, El Segundo, Annu. Rep. 1976, pp. 77–82

Schalscha, E.N., Gonzalez, C., Vergara, I., Galindo, G., Schatz, A.: Effect of drying on volcanic ash soils in Chile. Soil. Sci. Soc. Am. Proc. 29, 481–482 (1965)

Schalscha, E.N., Appelt, H., Schatz, A.: Chelating as a weathering mechanism. 1. Effect of complexing agents on the solubilization of iron from minerals and granodiorite. Geochim. Cosmochim. Acta 31, 587–596 (1967)

Schatz, A., Cheronis, N.D., Schatz, V., Trewlawney, G.S.: Chelation (sequestration) as a biological weathering factor in pedogenesis. Pa. Acad. Sci. 28, 44–51 (1954)

Schatz, A., Schatz, V., Martin, J.J.: Chelation as a biochemical weathering factor. Geol. Soc. Am. Bull. 68, 1792–1793 (1957)

Schaule, B., Patterson, C.: The occurrence of lead in the northeast Pacific and the effects of anthropogenic inputs. In: Proc. Int. Experts Discussion on Lead: Occurrence, Fate, and Pollution in the Marine Environment. Branica, M. (ed.). Oxford: Pergamon Press 1978

Scheelhase, F.: Erzeugung künstlichen Grundwassers aus Mainwasser. Geol. Rundsch. 3, 133–137 (1912)

Scheffer, F., Schachtschabel, P.: Lehrbuch der Bodenkunde. Stuttgart: Enke 1966

Scheffer, F., Welte, E., Ludwieg, F.: Zur Frage der Eisenoxydhydrate im Boden. Chem. Erde 19, 51–64 (1958)

Scheider, W.A., Adamski, J., Paylor, M.: Reclamation of acidified lakes near Sudbury, Ontario. Ont. Mines. Environ. Rep., 129 pp. (1975)

Scheiner, D.: Determination of ammonia and Kjeldahl nitrogen by indophenol method. Water Res. 10, 31–36 (1976)

Schell, W.R.: Sedimentation rates and mean residence times of Pb and ^{210}Pb in Lake Washington, Puget Sound estuaries and a coastal region. Abstr. Meet. Am. Soc. Limnol. Oceanogr. (1974)

Schell, W.R., Nevissi, A.: Heavy metals from waste disposal in central Puget Sound. Abstr. Int. Conf. Heavy Met. Environ., Toronto, C-43 (1975)

Schell, W.R., Nevissi, A.: Heavy metals from waste disposal in central Puget Sound. Environ. Sci. Technol. 11, 887–893 (1977)

Schindler, J.E., Alberts, J.J.: Behavior of mercury, chromium, and cadmium in aquatic systems. Rept. U.S. EPA 600/3-77-023, 70 pp. (1977)

Schindler, J.E., Alberts, J.J., Honick, K.R.: A preliminary investigation of organic-inorganic associations in stagnating system. Limnol. Oceanogr. 17, 952–957 (1972)

Schindler, P.W., Furst, B., Dick, R., Wolf, P.V.: Ligand properties of surface silanol groups. 1. Surface complex formation with Fe^{3+}, Cu^{2+}, Cd^{2+}, and Pb^{2+}. J. Colloid Interface Sci. 55, 469–475 (1976)

Schleichert, U.: Schwermetallgehalte der Schwebstoffe des Rheins bei Koblenz im Jahresablauf – eine gewässerkundliche Interpretation. Dtsch. Gewässerkundl. Mitt. 19, 150–157 (1975)

Schleichert, U., Hellmann, H.: Auftreten und Herkunft von Zink in Gewässern. Literaturber. 1972/3. Bundesanst. Gewässerkde. Koblenz (1973)

Schmidt, R.C.: Adsorption of Cu, Pb and Zn on some common rock forming minerals and its effect on lake sediments. Unpubl. Ph.D. Thesis, McGill Univ. Montreal, Quebec (1956). Cit. Coker and Nichol (1975)

Schmidt, U., Huber, F.: Methylation of organolead and lead (II) compounds to $(CH_3)_4$ Pb by microorganisms. Nature (London) 259, 157–158 (1976)

Schmoll, G., Förstner, U.: Chemical associations of heavy metals in lacustrine sediments: I. Calcareous lake sediments from different climatic zones. N. Jb. Miner. Abh. (1979, in press)

Schneider, H.: Zur Frage der Belastbarkeit des Rheins mit radioaktiven Nukliden. IV. Mitt.: Sorption von Radionukliden durch den planktonischen Anteil der Schwebstoffe des Rheins. GWF Wasser Abwasser 110, 624–652 (1969)

Schneider, H.: Zur Frage der Belastbarkeit des Rheins mit radioaktiven Nukliden. VI. Mitt.: Sorption von Radionukliden durch ausgewählte Minerale. GWF Wasser Abwasser 111, 21–26 (1970)

Schneider, H., Block, W.: Zur Frage der Belastbarkeit des Rheins mit radioaktiven Nukliden. III. Mitt.: Sorption von Radionukliden durch Sedimente des Rheins. GWF Wasser Abwasser 109, 1410–1415 (1968)

Schneider, W.: Geochemie und Hydrochemie des Flußgebietes der Diemel. Unpubl. Dr. Thesis, Univ. Bochum/Germany (1976)

Schnitzer, M., Wright, J.R.: Extractions of organic matter from podsolic soils by means of dilute inorganic acids. Can. J. Soil Sci. 37, 89–95 (1957)

Schofield, C.L.: Acid precipitation: effects on fish. Ambio 5, 228–230 (1976)

Schottel, J., Mandal, A., Toth, K., Clark, D., Silver, S.: Mercury and mercurial resistance determined by plasmids in Escherichia coli and Pseudomonas aeruginosa. Proc. Int. Conf. Transp. Persist. Chem. Aquatic Ecosyst. Ottawa (1974)
Schöttle, M., Friedman, G.: Effect of man's activities on distribution of trace elements in sub-bottom sediments of Lake George, New York. Sedimentology 21, 473–478 (1974)
Schöttler, U.: Untersuchungen zur Bestimmung der Reinigungswirkung von Böden am Beispiel von Schwermetallionen (Zn, Cu und Pb). Geol. Mitt. 12, 61–76 (1972)
Schöttler, U.: Die Reinigungswirkung von Böden am Beispiel von Schwermetallen. Z. Dtsch. Geol. Ges. 124, 555–566 (1973)
Schöttler, U.: Das Verhalten von Schwermetallen bei der Langsamsandfiltration. Z. Dtsch. Geol. Ges. 126, 373–384 (1975)
Schöttler, U.: Ausbreitung und Eliminierung von Spurenmetallen bei Infiltration und Untergrundpassage – Literaturstudie. Veröff. Inst. f. Wasserforschung, Dortmund 27, 99 pp. (1977a)
Schöttler, U.: Natürliche Filtrationssysteme und Schwermetalle, aufgezeigt am Beispiel von Quecksilber und Cadmium. Vom Wasser 49, 295–313 (1977b)
Schore, G.: Electronic equipment and ion exchange for use in activated treatment systems. 27th Purdue Industrial Waste Conf., Lafayette, Indiana, May 1972, pp. 312–334
Schroeder, H.A., Nason, A.P., Tipton, I.H., Balassa, J.J.: Essential trace metals in man: zinc. Relation to environmental cadmium. J. Chronic Dis. 20, 179–210 (1967)
Schroll, E.: Progress in the knowledge of indicator elements. Abstr. 2nd Symp. Origin and Distribution of the Elements, Paris-Unesco, p. 52 (1977)
Schroll, E., Krachsberger, H., Dolezel, P.: Hydrogeochemische Untersuchungen des Donauwassers in Österreich in den Jahren 1971 und 1972. Arch. Hydrobiol. Suppl. 44, 492–514 (1975)
Schultz, C.D., Crear, D., Pearson, J.E., Rivers, J.B., Hylin, J.W.: Total and organic mercury in the Pacific blue marlin. Bull. Environ. Contam. Toxicol. 15, 230–234 (1976)
Schulz-Baldes, M.: Lead uptake from sea water and food, and lead loss in the common mussel Mytilus edulis. Mar. Biol. 25, 177–193 (1974)
Schulz-Baldes, M., Lewin, R.A.: Lead uptake in two marine phytoplankton organisms. Biol. Bull. 150, 118–127 (1976)
Schumacher, A.: Quantitative Aspekte zwischen Stärke der Tubificidenbesiedlung und Schichtdicke der Oxidationszone in den Süßwasserwatten der Unterelbe. Arch. Fisch. Wiss. 14, 88–151 (1963)
Schutz, D.F., Turekian, K.K.: The investigation of the geographical and vertical distribution of several trace elements in seawater using neutron activation analysis. Geochim. Cosmochim. Acta 29, 259–313 (1965a)
Schutz, D.F., Turekian, K.K.: The distribution of cobalt, nickel, and silver in ocean water profiles around Pacific Antarctica. J. Geophys. Res. 70, 5519–5528 (1965b)
Schweiger, G.: Die toxikologische Einwirkung von Schwermetallsalzen auf Fische und Fischnährtiere. Inaug. Diss. Naturw. Fak. München (1956)
Schweisfurth, R.: Manganoxydierende Mikroorganismen in Trinkwasserversorgungsanlagen. GWF Wasser Abwasser 113, 562–572 (1972)
Schwertmann, U.: Inhibitory effect of soil organic matter on the crystallisation of amorphous ferric hydroxide. Nature (London) 212, 645–646 (1966)
Schwille, F.: Die chemischen Zusammenhänge zwischen Oberflächenwasser und Grundwasser im Moseltal zwischen Trier und Koblenz. Mitt. Dtsch. Gewässerkundl. Jahrb. 38, 73 pp. (1973)
Schwuger, M.J., Smolka, H.G.: Sodium-aluminum-silicates in the washing process. 1. Physicochemical aspects of phosphate substitution in detergents. Colloid Polymer Sci. 254, 1062–1069 (1976)
Schwuger, M.J., Smolka, H.G., Kurzendörfer, C.P.: Zur Verwendung von Na-Al-Silikaten in Waschmitteln, Teil IV: Modelluntersuchungen zum Ionenaustausch von Schwermetallen im Bereich geringer Ionenkonzentrationen. Tenside Deterg. 13, 305–312 (1976)
Sclater, F.R., Boyle, E.A., Edmond, J.M.: On the marine geochemistry of nickel. Earth Planet. Sci. Lett. 31, 119–128 (1976)
Scolari, G., Vernet, J.P.: Premiers résultats de la pollution par le mercure et autres métaux lourds dans les sédiments du bassin du Rhône et du Léman. Bull. ARPEA 71, 21–57 (1975)
Scott, D.M., Mercury concentration of white muscle in relation to age, growth and condition in four species of fishes from Clay Lake, Ontario. J. Fish. Res. Board Can. 31, 1723–1729 (1974)
Scott, R.C., Baker, F.B.: Ground-water sources containing high concentrations of radium. U.S. Geol. Surv. Prof. Paper 424-D, 357–359 (1961)
Scott, R.C., Baker, F.B.: Data on uranium and radium in ground water in the United States, 1954–1959. U.S. Geol. Surv. Prof. Paper 426, 115 pp. (1962)

Scott, W., Miner, D.H.: Sedimentation in Winona Lake and Tippecanoe Lake, Kosciusko Country Indiana. July 31, 1930 to July 3, 1935. Proc. Indiana Acad. Sci. *45*, 275–286 (1936)

Scrudato, R.J., Estes, E.L.: Clay-lead sorption relations. Environ. Geol. *1*, 167–170 (1975)

Seagle, S.M., Ehlman, A.J.: Manganese, zinc and copper in water, sediments and mussels in north-central Texas reservoirs. In: Trace Substances in Environmental Health, Vol. 8 Hemphill, D.D. (ed.). Columbia, Missouri: Univ. Missouri, 1974, pp. 101–106

Seeliger, U., Edwards, P.: Correlation coefficients and concentration factors of copper and lead in seawater and benthic algae. Mar. Pollut. Bull. *8*, 16–19 (1977)

Segall, H.J., Wood, J.M.: Mechanisms for the detoxification of methylmercury by selenium salts. U.S. Proc. Nat. Acad. Sci. (1975). Cit. Wood, J.M. (1977)

Segar, D.A., Cantillo, A.Y.: Trace metals in the New York Bight. Proc. Symp. Middle-Atlantic Continental Shelf and the New York Bight. Lawrence, Kansas, Spec. Symp. *2*, 171–178 (1976)

Segar, D.A., Pellenbarg, R.E.: Trace metals in carbonate and organic rich sediments. Mar. Pullut. Bull. *4*, 138–142 (1973)

Segar, D.A., Pellenbarg, R.E.: Trace metals in carbonate and organic rich sediments. Mar. Pollut. marine animals. II. Molluscs. J. Mar. Biol. Assoc. U.K. *51*, 131–136 (1971)

Seibold, E.: In: Brinkmann, R.: Lehrbuch der algemeinen Geologie, Vol. I. Stuttgart: Enke 1964

Semkin, R.G.: A limnological study of Sudbury area lakes. M. Sc. Thesis, McMaster Univ. Hamilton, Ontario (1975)

Semkin, R.G., Kramer, J.R.: Sediment geochemistry of Sudbury area lakes. Can. Mineral. *14*, 73–90 (1976)

Semmler, W.: Die Halden – ein hydrologisches Problem. Schlägel *9*, 694–698 (1958)

Seydel, I.S.: Distribution and circulation of arsenic through water, organisms and sediments of Lake Michigan. Arch. Hydrobiol. *71*, 17–30 (1972)

Shakman, R.A.: Nutritional influences on the toxicity of environmental pollutants. Arch. Environ. Health *28*, 105–113 (1974)

Shannon, E.E., Fowlie, P.J.A., Rush, R.J.: A study of nitrilotriacetic acid (NTA) degradation in a receiving stream. Technol. Dev. Rept. EPS 4-WP-74-7. Environmental Protection Serv., Canada, 1974

Shapiro, J.: Natural coloring substances of water and their relation to inorganic components. Geol. Soc. Am. Annu. Meet., New York, 148A (1963)

Shapiro, J.: Effect of yellow organic acids on iron and other metals in water. J. Am. Water Works Assoc. *56*, 161–179 (1964)

Shapiro, M.A.: Heavy metals in Westernport Bay. Proc. Int. Conf. Heavy Met. Environ., Toronto, 1975, *II*/1, 309–319 (1977)

Shaw, T.L., Brown, V.M.: The toxicity of some forms of copper to rainbow trout. Water Res. Arch. *8*, 377–382 (1974)

Shelton, R.G.J.: Sludge dumping in the Thames Estuary. Mar. Pollut. Bull. *2*, 24–27 (1971)

Shelton, R.G.J.: Some effects of dumped, solid wastes on marine life and fisheries. In: North Sea Science. Goldberg, E.D. (ed.). Cambridge, Mass.: MIT-Press 1973

Shen, Y.S.: Study of arsenic removal from drinking water. J. A.W.W.A. *65*, 543–548 (1973)

Sheppard, C.R., Bellamy, D.J.: Pollution of the Mediterranean around Naples. Mar. Pollut. Bull. *5*, 42–44 (1974)

Sheppard, J.C., Funk, W.H.: Trees as environmental sensors monitoring long-term heavy metal contamination of Spokane River, Idaho. Environ. Sci. Technol. *9*, 638–642 (1975)

Shibahara, M., Yamazaki, R., Nishida, K., Suzuki, J., Suzuki, H., Nishida, H., Tada, F.: Heavy metals in rivers of the Toyama Prefecture. Eisei Kaguku *21*, 173–182 (1975)

Shimizu, M., Kajihara, T., Suyama, I., Hiyama, Y.: Uptake of Co-58 by mussel Mytilus edulis. J. Radiat. Res. *12*, 17–28 (1971)

Shimp, N.F., Schleicher, J.A., Ruch, R.R., Heck, D.B., Leland, H.V.: Trace element and organic carbon accumulation in the most recent sediments of southern Lake Michigan. Environ. Geol. Notes *41*, 25 (1971)

Shiomi, M.T.: Great Lakes precipitation chemistry. I. Lake Ontario Basin. Proc. 16th Great Lakes Research, 581–602 (1973)

Sholkovitz, E.R.: Flocculation of dissolved organic and inorganic matter during the mixing of river water and seawater. Geochim. Cosmochim. Acta *40*, 831–845 (1976)

Sholkovitz, E.R., Boyle, E.A., Price, N.B.: The removal of dissolved humic acids and iron during estuarine mixing. Earth Planet. Sci. Lett. *40*, 130–136 (1978)

Shukla, S.S., Syers, J.K., Armstrong, D.E.: Arsenic interference in the determination of inorganic phosphate in lake sediments. J. Environ. Qual. *1*, 292–295 (1972)

Shuman, M.S., Haynie, C.L., Smock, L.A.: Modes of metal transport above and below waste discharge on the Haw River, North Carolina. Environ. Sci. Technol. *12*, 1066–1069 (1978)

Shuster, K.A.: Leachate damage assessment: Case studies. EPA/530 – SW-509, SW-514, SW-517, U.S. EPA, Cincinnati, Ohio (1976)
Sias, M., Wilhm, J.: Distribution of copper, lead and zinc in the sediment and water of Lake Carl Blackwell. Proc. Okla. Acad. Sci. *55*, 38–41 (1975)
Sibley, T.H., Morgan, J.J.: Equilibrium speciation of trace metals in fresh water: sea water mixtures. Proc. Int. Conf. Heavy Met. Environ., Toronto 1975, *1*, 319–338 (1977)
Siccama, T.G., Porter, E.: Lead in a Connecticut salt marsh. Biol. Sci. *22*, 232–234 (1972)
Sidle, R.C., Hook, J.E., Kardos, L.T.: Accumulation of heavy metals in soils from extended wastewater irrigation. J.W.P.C.F. *49*, 311–318 (1977)
Siebert, G., Werner, H.: Bergverkippung und Grundwasserbeeinflussung am Niederrehin. GWF Wasser Abwasser *111*, 520 (1970)
Sigworth, E.A., Smith, S.B.: Adsorption of inorganic compounds by activated carbon. J.A.W.W.A. *64*, 386–392 (1972)
Silker, W.B.: Variations in elemental concentrations in the Columbia River. Limnol. Oceanogr. *9*, 540–545 (1964)
Sillén, L.G.: The physical chemistry of seawater. In: Oceanography. Sears, M. (ed.). Washington: Am. Assoc, Adv. Sci, 1961, pp. . 549–581
Sillén, L.G.: Stability constants of metal-ion complexes. Sec. I: Inorganic ligands. Chem. Soc. London Spec. Publ. No. 17 (1964)
Silvey, W.D.: Occurrence of selected minor elements in the waters of California. U.S. Geol. Surv. Water Supply Pap. *1535-L*, 25 pp. (1967)
Simon, W.G.: Untersuchungsergebnisse an Grundproben aus dem Gebiet der Elbe zwischen Scheelenkuhlen und Cuxhaven und ihre Ausdeutung hinsichtlich der Sandwanderung. Mitt. Wasser- und Schiffahrtsdir. Hamburg. Mitt. Geol. Landesamt, Hamburg *11*, 153 pp. (1953)
Simone, R.E. De, Penley, M.W., Charbonneau, L., Smith, S.G., Wood, J.M., Hill, H.A.O., Pratt, J.M., Ridsdale, S., Williams, R.J.P.: The kinetics and mechanisms of cobalamin-dependent methyl and ethyl transfer to mercuric ion. Biochim. Biophys. Acta *304*, 851–863 (1973)
Simons, T.J.: Development of numerical models of Lake Ontario. Part. 2. Proc. 15th Conf. Great Lakes Res., 655–672 (1972)
Sims, R.R., Presley, B.J.: Heavy metal concentrations in organisms from an actively dredged Texas Bay. Bull. Environ. Contam. Toxicol. *16*, 520–526 (1976)
Singer, P.C.: Trace metals in water supplies. Occurance, significance and control. Proc. 16th Water Qual. Conf. Univ. Ill., Coll. Eng. (1974)
Singer, P.C.: Influence of dissolved organics on the distribution, transport, and fate of heavy metals in aquatic systems. In: Fate of Pollutants in the Air and Water Environment. Part. I. Suffet, I.H. (ed.). New York, 1977, pp. 155–182
Singer, P.C., Stumm, W.: Acidic mine drainage: The rate-determining step. Science *167*, 1121–1123 (1970)
Sinley, J.R., Goettl, J.P., Jr., Davies, P.H.: The effects of zinc on Rainbow Trout (Salmo gairdneri) in hard and soft water. Bull. Environ. Contam. Toxicol. *12*, 193–201 (1974)
Sivisankara, P.K., Thomas, C.C., Sondel, J.A., Hyche, C.M.: Determination of mercury in biological and environmental samples by neutron activation analysis. Anal. Chem. *43*, 1419–1425 (1971)
Skei, J.M., Price, N.B., Calvert, S.E., Holtedahl, H.: The distribution of heavy metals in sediments of Sörfjord, West Norway. Water Air Soil Pollut. *1*, 452–461 (1972)
Skei, J.M., Price, N.B., Calvert, S.E.: Particulate metals in waters of Sörfjord, West Norway. Ambio *2*, 122–124 (1973)
Skei, J.M., Saunders, M., Price, N.B.: Mercury in plankton from a polluted Norwegian fjord. Mar. Pollut. Bull. *7*, 34–36 (1976)
Skinner, B.J., White, D.E., Rose, H.J., Mays, R.E.: Sulfides associated with the Salton Sea geothermal brine. Econ. Geol. *62*, 316–342 (1967)
Skoch, E.J., Turk, J.M.: Fluctuations in the level of mercury in sediment collected from the island area of Lake Erie, 1964–1968. Proc. Conf. Great Lakes Res. *15*, 291–297 (1972)
Skougstad, M.W., Horr, C.A.: Occurrence of strontium in natural waters. U.S. Geol. Surv. Circ. *420*, 6 pp. (1960)
Slatt, R.M.: Geochemistry of bottom sediments, Conception Bay, southeastern Newfoundland. Can. J. Earth Sci. *11*, 768–784 (1974)
Slowey, F.J., Hood, D.W.: Copper, manganese and zinc concentrations in Gulf of Mexico waters. Geochim. Cosmochim. Acta *35*, 121–138 (1971)
Slowey, F.J., Jeffrey, L.M., Wood, D.W.: Evidence for organic complexed copper in sea water. Nature *214*, 377–378 (1967)
Sly, P.G.: Some influences of dredging in the Great Lakes. In: Interactions Between Sediments and Fresh Water. Golterman, H.L. (ed.). The Hague: Junk Publ. 1977, pp. 435–443

Sly, P.G., Thomas, R.L.: Review of geological research as it relates to an understanding of Great Lakes limnology. J. Fish. Res. Board Can. *31*, 795–825 (1974)
Smith, A.Y.: Heavy metal (Zn, Pb, Cu) content of stream sediments of part of Westmoreland County, New Brunswick. Geol. Surv. Canada Pap. *59-12*, 13 pp. (1960)
Smith, E.G., Berkes, F., Spence, J.A.: Mercury levels in fish in the La Grande River area, North Quebec, Bull. Environ. Contam. Toxicol. *13*, 673–677 (1975)
Smith, E.M., Daly, A.R.: The past, present and future prospects of burning municipal sewage sludge along with mixed municipal refuse. Proc. Natl. Conf. Municipal Sludge Manage. Disposal. Rockville, Maryland (1975)
Smith, I.C.: Control of mercury in sediments. Environ. Prot. Agency Dep. EPA-R2-72-043, 53 (1972)
Smith, J.D., Burton, J.D.: The occurrence and distribution of tin with particular reference to marine environments. Geochim. Cosmochim. Acta *36*, 621–633 (1972)
Smith, J.D., Nicholson, R.A., Moore, P.J.: Mercury in water of the tidal Thames. Nature (London) *232*, 393–394 (1971)
Smith, J.D., Nicholson, R.A., Moore, P.J.: Mercury in sediments from the Thames Estuary. Environ. Pollut. *4*, 153–157 (1973)
Smith, P.A., Moore, J.R.: The distribution of trace metals in the surficial sediments surrounding Keweenaw Point, Upper Michigan. Proc. 15th Conf. Great Lakes Res., 383–400 (1972)
Smith, R.A.H., Bradshaw, A.D.: Reclamation of toxic metalliferrous wastes using tolerant populations of grass. Nature (London) *227*, 376–378 (1970)
Smith, R.W., Frey, D.G.: Acid mine pollution on lake biology. Water Resour. Center, Indiana Univ. Rept. to U.S. EPA, Water Pollut. Control Res. Ser. 18050 EEC, 131 pp. (1971)
Smith, W.H.: Lead and mercury burden of urban woody plants. Science *176*, 1237–1239 (1972)
Snoeyink, V.L., Weber, V.I.: The surface chemistry of active carbon, a discussion of structure and surface functional groups. Environ. Sci. Technol. *1*, 228–232 (1967)
Snyder, S.L., Vigo, T.L.: Removal of mercury from aqueous solutions by n-(2-aminoethyl)aminodeoxycellulose cotton. Environ. Sci. Technol. *8*, 944–946 (1974)
Solbé, J.F. de L.G., Cooper, V.A.: Studies on the toxicity of copper sulphate to stone loach Noemacheilus barbatulus (L.) in hard water. Water Res. *10*, 523–527 (1976)
Solohub, J.T., Klovan, J.E.: Evaluation of grain-size parameters in lacustrine environments. J. Sediment. Petrol. *40*, 81–101 (1970)
Solomon, A.M., Kroener, D.F.: Suburban replacement of rural land uses reflected in the pollen rain of northeastern New Jersey. N.J. Acad. Sci. Bull. *16*, 30–44 (1971)
Somasundaram, P., Agar, G.E.: The zero point of charge of calcite. J. Colloid. Interface Sci. *24*, 433–438 (1967)
Sommer, S.W., Pyzik, A.J.: Geochemistry of middle Chesapeake Bay sediments from Upper Cretaceous to Present. Chesapeake Sci. *15*, 39–44 (1974)
Sommers, L.E., Floyd, M.: Microbial transformation of mercury in aquatic environments. Water Resour. Res. Rept., Purdue Univ. *54*, 80 pp. (1974)
Sontheimer, H.: Die Wasserqualität im Rheineinzugsgebiet in den Jahren 1971–1972. IAWR. 3rd Workshop Düsseldorf, 1973, pp. 33–46
Soong, K.-L.: Versuche zur adsorptiven Bindung von Schwermetall-Ionen an künstlichen Tongemischen. Unpubl. Diss., Univ. Heidelberg, 1974
Spalding, R.F., Sackett, W.M.: Uranium in runoff from the Gulf of Mexico distributive province: anomalous concentrations. Science *175*, 629–631 (1972)
Spangler, W.J., Spigarelli, J.L., Rose, J.M., Flippin, R.S., Miller, H.H.: Degradation of methylmercury by bacteria isolated from environmental samples. Appl. Mircobiol. *25*, 488–493 (1973)
Spence, J.A.: Mercury levels in fish in the La Grande River area, northern Quebec. Bull. Environ. Contam. Toxicol. *13*, 673–677 (1975)
Spencer, D.W., Brewer, P.G.: The distribution of copper, zinc and nickel in seawater of the Gulf of Maine and the Sargasso Sea. Geochim. Cosmochim. Acta *33*, 325–339 (1969)
Sprague, J.: Promising anti-pollutant: Chelating agent NTA protects fish from copper and zinc. Nature (London) *220*, 1345–1346 (1968)
Sprengel, C.: Über Pflanzenhumus, Humussäure und humussaure Salze. Kastners Arch. Naturlehre *8*, 145–220 (1826)
Steele, K.S., Wagner, G.H.: Trace metal relationships in bottom sediments of a freshwater stream – the Buffalo River, Arkansas. J. Sediment Petrol. *45*, 310–319 (1975)
Steelink, C.: What is humic acid? J. Chem. Educ. *40*, 379–384 (1963)
Steeman Nielsen, E., Wium-Anderson, S.: Copper ions as poison in the sea and in freshwater. Mar. Biol. *6*, 93–97 (1970)

Steinmann, P.: Fischvergiftungen und Wasserstoffionen. Z. Hydrol. *5* (1928)
Steinmann, P.: Toxikologie der Fische, 1927. In: Handbuch Binnenfischerei, Vol. VI (1928)
Stenner, R.D., Nickless, G.: Distribution of some heavy metals in organisms in Hardangerfjord and Skjerstadfjord, Norway. Water Air Soil Pollut. *3,* 279–291 (1974a)
Stenner, R.D., Nickless, G.: Absorption of cadmium, copper and zinc by dog whelks in the Bristol Channel. Nature (London) *247,* 198–199 (1974b)
Stenner, R.D., Nickless, G.: Heavy metals in organisms of the Atlantic coast of south-west Spain and Portugal. Mar. Pollut. Bull. *6,* 89–92 (1975)
Stenström, T., Lönsjö, H.: Cadmium availability to wheat: A study with radioactive tracers under field conditions. Ambio *3,* 87–90 (1974)
Stenström, T., Vahter, M.: Cadmium and lead in Swedish commercial fertilizers. Ambio *3,* 91–92 (1974)
Štern, J., Förstner, U.: Heavy metal distribution in the sediment of the Sava Basin in Slovenia. Geologija *19,* 259–274 (1976)
Stevens, J.D., Brown, B.E.: Occurrence of heavy metals in the blue shark Prionace glauca and selected pelagic fauna in the northeast Atlantic Ocean. Mar. Biol. *26,* 287–293 (1974)
Stevenson, F.J., Butler, J.H.A.: Chemistry of humic acids and related pigments. In: Organic Geochemistry, Methods and Results. Eglinton, G., Murphy, M.J.T. (eds.). Berlin, Heidelberg, New York: Springer 1969, pp. 534–557
Stewart, J., Schulz-Baldes, M.: Long-term lead accumulation in abalone (Haliotis sp.) fed on lead-treated brown algae (Egregia laevigata). Mar. Biol. *36,* 19–24 (1976)
Stiefel, R., Förstner, U.: Metallanreicherungen im Altrheinhafen von Mannheim – Auswirkungen der Neckarverschmutzung auf den Rhein. Jahresber. Mitt. Oberrheinischen Geol. Ver. *57,* 77–85 (1975)
Stiff, M.J.: Copper/bicarbonate equilibria in solutions of bicarbonate ion at concentrations similar to those found in natural waters. Water Res. *5,* 171–176 (1971)
Stiles, W.: Trace Elements in Plants and Animals. Cambridge Univ. Press 1946
Štirn, J., Avcin, A., Cencelj, J., Dorer, M., Gomiscek, S., Kveder, S., Malej, A., Meischner, D., Nozina, I., Paul, J., Tušnik, P.: Pollution problems of the Adriatic Sea – an interdisciplinary approach. Rev. Int. Océanogr. Méd. *35/36,* 21–78 (1974)
Stock, A., Cucuel, F.: Die Verbreitung des Quecksilbers. Naturwissenschaften *22,* 390–393 (1934)
Stockner, J.G., Benson, W.W.: The succession of diatom assemblages in the recent sediments of Lake Washington. Limnol. Oceanogr. *12,* 513–532 (1967)
Stöfen, D.: Blei als Umweltgift. Die verdeckte Bleivergiftung, ein Massenphänomen? Eschwege: G.E. Schroeder 1974a
Stöfen, D.: Grundheitliche Höchstwerte für Schadstoffe im Trinkwasser. GWF Wasser Abwasser *115,* 67–71 (1974b)
Stoffers, P., Summerhayes, G., Förstner, U., Patchineelam, S.R.: Copper and other heavy metal contamination in sediments from the New Bedford Harbor, Mass.: A preliminary note. Environ. Sci. Technol. *11,* 819–821 (1977)
Stokes, P.M.: Adaption of green algae to high levels of copper and nickel in aquatic environment. Proc. Int. Conf. Heavy Met. Environm., Toronto, Ont., *II/1,* 137–154 (1975)
Stokes, P.M., Hutchinson, T.C., Krauter, K.: Heavy metal tolerance in algae isolated from contaminated lakes near Sudbury, Ontario. Can. J. Bot. *51,* 2155–2168 (1973)
Strahler, A.N., Strahler, A.H.: Environmental Geoscience: Interaction between Natural Systems and Man. Santa Barbara/California: Hamilton Publishing Co. 1973
Strohal, P., Huljev, D., Lulic, S., Picer, M.: Antimony in the coastal marine environment, north Adriatic. Estuarine Coastl. Mar. Sci. *3,* 119–123 (1975)
Stumm, W.: Investigation on the corrosive behaviour of water. J. San. Eng. Div. ASCE, 27–45 (1960)
Stumm, W.: Die Rolle der Komplexbildung in natürlichen Gewässern und allfällige Beziehungen zur Eutrophierung. Gewässerschutz Wasser Abwasser *8,* 57–87 (1972)
Stumm, W.: Natürliche Gewässer weiterhin stark gefährdet. Neue Zürcher Zeitung, Beil. Forschung und Technik, No. 252, October 27, 1976
Stumm, W.: Die Beeinträchtigung aquatischer Ökosysteme durch die Zivilisation. Naturwissenschaften *64,* 157–165 (1977)
Stumm, W., Baccini, P.: Man-made chemical perturbation of lakes. In: Lakes – Chemistry, Geology, Physics. Lerman, A. (ed.). Berlin, Heidelberg, New York: Springer 1978, pp. 91–126
Stumm, W., Bilinski, H.: Trace metals in natural waters: Difficulties of interpretation arising from our ignorance on their speciation. In: Advances on Water Pollution Research. Jenkins, S.H. (ed.). Proc. 6th Int. Conf. Jerusalem. New York: Pergamon Press 1972, pp. 39–52

Stumm, W., Brauner, P.A.: Chemical speciation. In: Chemical Oceanography. Ch. 3. Riley, J.P., Skirrow, G. (eds.). New York: Academic Press 1975, pp. 173–239
Stumm, W., Lee, G.F.: The chemistry of aqueous iron. Schweiz. Z. Hydrol. 21 295–319 (1960)
Stumm, W., Lee, G.F.: Oxygenation of ferrous iron. Ind. Eng. Chem. 53, 143–146 (1961)
Stumm, W., Morgan, J.J.: Aquatic Chemistry. New York: Wiley 1970
Stumm, W., Hohl, H., Calang, F.: Interaction of metal ions with hydrous oxide surfaces. Croat. Chem. Acta 48, 491–504 (1976)
Suarez, D., Langmuir, D.: Heavy metal relationships in a Pennsylvania soil. Geochim. Cosmochim. Acta 40, 589–598 (1976)
Subramanian, V.: Experimental modelling of inter-elemental relationship in natural ferromanganese materials. Can. Mineral. 14, 32–39 (1976)
Subramanian, V., D'Anglejan, B.: Water chemistry of the St. Lawrence Estuary. J. Hydrol. 29, 341–354 (1976)
Subramanian, V., Gibbs, R.J.: Trace metal adsorption on clay minerals. Abstr. Annu. Meet. Geol. Soc. Amer. 1972
Suess, E.: Interaction of organic compounds with calcium carbonate. I. Association phenomena and geochemical implications. Geochim. Cosmochim. Acta 34, 157–168 (1970)
Suess, E.: Interaction of organic compounds with calcium carbonate – II. Organo-carbonate association in Recent sediments. Geochim. Cosmochim. Acta 37, 2435–2447 (1973)
Suess, E.: How can we distinguish between natural and anthropogenic materials in sediments and can we predict the effects of man's additions: UNESCO Workshop Biogeochem. Estuarine Sediment. Melreux/Belgium 1976, pp. 224–237 (1978)
Suess, E., Erlenkeuser, H.: History of metal pollution and carbon input in Baltic Sea sediments. Meyniana (Kiel) 27, 63–75 (1975)
Suffet, I.H., McGuire, M.J., Josephson, J., Ember, L.R.: Cleanup: that old magic works again! Environ. Sci. Technol. 12, 1138–1149 (1978)
Sugawara, K., Okabe, S.: Geochemistry of molybdenum in natural water. J. Earth Sci. Nagoya Univ. 8, 93–107 (1960)
Sugawara, K., Naito, H., Yamada, S.: Geochemistry of vanadium in natural waters. J. Earth Sci. Nagoya Univ. 4, 44–61 (1956)
Sugawara, K., Okabe, S., Tanaka, M.: Geochemistry of molybdenum in natural water. J. Earth Sci. Nagoya Univ. 9, 114–128 (1961)
Sugiyama, S., Hashimoto, M., Chiba, N., Hayase, Y., Yokoyama, T., Aikawa, H , Odaka, Y., Miyamoto, S.: The influence of L-cysteine on toxicity of methyl mercuric chloride (I). J. Med. Soc. Toho 22, 78–85 (1975)
Sumino, K., Hayakawa, K., Shibata, T., Kitamura, S.: Heavy metals in normal Japanese tissues. Arch. Environ. Health 30, 487–494 (1975)
Summerhayes, C.P., Ellis, J.P., Stoffers, P., Briggs, S.R., Fitzgerald, M.G.: Fine-grained sediment and industrial waste distribution and dispersal in New Bedford Harbor and Western Buzzard Bay, Massachusetts. Rep. 76-115. Woods Hole Oceanogr. Inst., 110 pp. (1977)
Sundberg, M.W., Meares, C.F., Goodwin, D.A., Diamanti, D.A.: Chelating agents for the binding of metal ions to macromolecules. Nature (London) 250, 587–588 (1974)
Sustar, J.E., Wakeman, T.H.: Dredging conditions influencing the uptake of heavy metals by organisms. In: Proc. 2nd U.S.-Japan Experts Meeting Management of Bottom Sediments Containing Toxic Substances. Peterson, S.A., Randolph, K.K. (eds.). Corvallis, Or.: U.S. EPA 1977, pp. 246–252
Sutton, J.A., Corrick, J.D.: Leaching of copper sulfide minerals with selected autotrophic bacteria. U.S. Bureau of Mines, Washington, D.C. Rep. Invest. 6423 (1964)
Suzuki, M., Yamada, T., Miyazaki, T., Kawazoe, K.: Accumulation of cadmium in the sediment of Tama River. Seisan Kankyu 27, 108–112 (1975)
Suzuki, M., Yamada, T., Kawazoe, K.: Sorption of cadmium by the sediment of Tama River. Seisan Kankyu 28, 92–95 (1976)
Suzuki, T., Miyama, T., Toyama, C.: The chemical form and bodily distribution of mercury in marine fish. Bull. Environ. Contam. Toxicol. 10, 247–355 (1973)
Swaine, D.J., Mitchell, R.L.: Trace-element distribution in soil profiles. J. Soil Sci. 11, 347–368 (1960)
Swanson, C., Wing, R.E., Doane, W.M., Russel, C.R.: Mercury removal from waste water with starch xanthate-cationic polymer complex. Environ. Sci. Technol. 7, 614–618 (1973)
Swanson, V.E., Frist, L., Rader, R.F., Jr., Huffman, C., Jr.: Metal sorption by northwest Florida humate. U.S. Geol. Surv. Prof. Pap 550-C, 174–177 (1966)
Sweeney, R., Foley, R., Merckel, C., Wyeth, R. Impacts of the deposition of dredged spoils on Lake Erie sediment quality and associated biota. J. Great Lakes Res. 1, 162–170 (1975)

Swisher, R.D., Taulli, T.A., Malec, E.J.: Biodegradation of NTA metal chelates in river water. In: Trace Metals and Metal-Organic Interactions in Natural Waters. Singer, P.C. (ed.). Ann Arbor: Ann Arbor Sci. Publ. 1973, pp. 237–263

Syers, J.K., Iskandar, I.K., Keeney, D.R.: Distribution and background levels of mercury in sediment cores from selected Wisconsin lakes. Water Air Soil Pollut. *2*, 105–118 (1973)

Sylvester, R.O., DeWalle, F.B.: Character and significance of highway runoff waters. Wash. State Highway Dept. Res. Progr. Rept. No. Y-1441 (1972)

Symader, W.: Multivariate Nährstoffuntersuchungen zu Vorhersagezwecken in Fließgewässern am Nordrand der Eifel. Kölner Geogr. Arb. *34*, 154 (1976)

Symader, W., Thomas, W.: Interpretation of average heavy metal pollution in flowing waters and by means of hierarchical grouping analysis using two different error indices. Catena *5*, 131–144 (1978)

Symader, W., Hermann, R., Rump, H., Richartz, H.: Der Summenparameter und ein anderer Weg zur Vereinfachung der Gewässergüteüberwachung. Dtsch. Gewässerkdl. Mitt. *21*, 118–120 (1977)

Szalay, A., Szilagyi, A.: The association of vanadium with humic acids. Geochim. Cosmochim. Acta *31*, 1–6 (1967)

Szebellédy, J., Literáthy, P.: Micropollutants over the Hungarian reach of the Danube River. 2nd Conf. Water Qual. Technol. Budapest (1970)

Szucs, F.K., Oostdam, B.L.: A comparative study of heavy metals in sediments of ocean dumpsites. Abstr. Int. Conf. Heavy Metl. Environ. C-52/54 (1975)

Tabata, K.: Studies on the toxicity of heavy metals to aquatic animals and the factors to decrease the toxicity. Bull. Tokai Reg. Fish Res. Lab. *58*, 203–261 (1969)

Takeuchi, T.: Distribution of mercury in the environment of Minamata Bay and the inland Ariake Sea. In: Environmental Mercury Contamination. Hartung, R., Dinman, B.D. (eds). Ann Arbor: Ann Arbor Sci. Publ. Inc. 1972, pp. 79–81

Talbot, V.W., Magee, R.J., Hussain, M.: Distribution of heavy metals in Port Philip Bay. Mar. Pollut. Bull. *7*, 53–55 (1976a)

Talbot, V.W., Magee, R.J., Hussain, M.: Cadmium in Port Philip Bay mussels. Mar. Pollut. Bull. *7*, 84–86 (1976b)

Tamm, C.O.: Acid precipitation: biological effects in soil and forest vegetation. Ambio *5*, 235–238 (1976)

Tanner, J.T., Friedman, M.H., Lincoln, D.N.: Mercury content of common foods determined by neutron activation analysis. Science *177*, 1102–1103 (1972)

Tardiff, R.G.: Health standards for metals in drinking water. Nat. Environ. Res. Cent. EPA, Cincinnati, Ohio, 113–139 (1972)

Tatekawa, M., Nakamura, M., Nakano, S.: The pollution of the Lake Biwa in the light of the distribution of heavy metals. Proc. Intern. Congr. on the Human Environment, Kyoto, Japan, 1975, pp. 402–407

Tatsumoto, M.T., Patterson, C.C.: Concentration of common lead in some Atlantic and Mediterranean waters and in snow. Nature (London) *199*, 350–352 (1963a)

Tatsumoto, M.T., Patterson, C.C.: The concentration of common lead in sea water. In: Earth Science and Meteoritics. Geiss, J., Goldberg, E.D. (eds). Amsterdam: North Holland Publ. 1963b

Taylor, D.: The natural distribution of trace metals in sediments from a coastal environment, Tor Bay, England. Estuarine Coastal Mar. Sci. *2*, 417–424 (1974)

Taylor, D.: Distribution of heavy metals in the sediment of an unpolluted estuarine environment. Sci. Total Environ. *6*, 259–264 (1976)

Taylor, J.K., Alvarez, R., Paulson, R.A., Rains, T.C., Rook, H.L.: Interactions of nitrilotriacetic acid with suspended and bottom material. Proj. No. EPA-WQO-16020 GFR. July, 1971

Taylor, S.R.: Abundance of chemical elements in the continental crust – a new table. Geochim. Cosmochim. Acta *28*, 1273–1285 (1964)

Temple, K.L., LeRoux, N.M.: Syngenesis of sulfide ores: sulfate-reducing bacteria and copper toxicity. Econ. Geol. *59*, 271–278 (1964)

Terwindt, J.H.J.: Mud transport in the Dutch delta area and along the adjacent coast line. Neth. J. Sea Res. *3*, 505–531 (1967)

Tessenow, U., Baynes, Y.: Redox-dependent accumulation of Fe and Mn in a littoral sediment supporting Isoetes lacustris. Naturwissenschaften *62*, 342 (1975)

Thayer, P.S., Kensler, C.J.: Current status of the environmental and human safety aspects of nitrilotriacetic acid (NTA). CRC Crit. Rev. Environ. Control *3*, 375–404 (1973)

Theis, T.L., Singer, P.C.: The stabilization of ferrous ions by organic compounds in natural waters. In: Trace Metals and Metal-Organic Interactions in Natural Waters. Singer, P.C. (ed.). Ann Arbor: Ann Arbor Science Publ. 1973, pp. 303–320

Theis, T.L., Singer, P.C.: Complexation of iron (II) by organic matter and its effect on iron (II) oxygenation. Environ. Sci. Technol. *8*, 569 (1974)

Thenard, P.: Observations sur le mémoire de M. Friedel. C.R. *70*, 1412–1414 (1870)

Theobold, P.K., Jr., Lakin, H.W., Hawkins, D.E.: The precipitation of aluminium, iron and manganese at the junction of Deer Creek with the Snake River in Summit County, Colo. Geochim. Cosmochim. Acta *27*, 121–132 (1963)

Thiem, A.: Die künstliche Erzeugung von Grundwasser. J. Gas Wasser *12*, (1898)

Thiem, L., Badorek, D., O'Connor, J.: Removal of mercury from drinking water using activated carbon. J. A.W.W.A. *68*, 447–451 (1976)

Thom, N.S.: Nitrilotriacetic acid: Literature survey. Water Res. *5*, 391–399 (1971)

Thomann, R.V., Szumski, D.S., DiToro, D.M., O'Connor, D.J.: A food chain model of cadmium in western Lake Erie. Water Res. *8*, 841–849 (1974)

Thomas, R.L.: A note on the relationship of grain size, clay content, quartz, and organic carbon in some Lake Erie and Lake Ontario sediments. J. Sediment. Petrol. *39*, 803–809 (1969a)

Thomas, R.L.: The qualitative distribution of feldspars in surficial bottom sediments from Lake Ontario. Proc. 12th Conf. Great Lakes Res., 364–379 (1969b)

Thomas, R.L.: The distribution of mercury in the sediment of Lake Ontario. Can. J. Earth Sci. *9*, 636–651 (1972)

Thomas, R.L.: The distribution of mercury in the surficial sediments of Lake Huron. Can. J. Earth Sci. *10*, 194–204 (1973)

Thomas, R.L.: The distribution and transport of mercury in the sediments of the Laurentian Great Lakes system. Proc. Int. Conf. Persist. Chem. Aquatic Ecosyst. *I*/1–16 (1974)

Thomas, R.L., Jaquet, J.-M.: Mercury in the surficial sediments of Lake Erie. J. Fish Res. Board Can. *33*, 404–412 (1976)

Thomas, R.L., Kemp, A.L.W., Lewis, C.F.M.: Distribution, composition and characteristics of the surficial sediments of Lake Ontario. J. Sediment Petrol. *42*, 66–82 (1972)

Thomas, R.L., Jaquet, J.-M., Kemp, A.L.W., Lewis, C.F.M.: Surficial sediments of Lake Erie. J. Fish. Res. Board. Can. *33*, 385–403 (1976)

Thomas, R.L., Jaquet, J.-M., Mudroch, A.: Sedimentation processes and associated changes in surface sediment trace metal concentrations in Lake St. Clair, 1970–1974. Proc. Int. Conf. Heavy Met. Environ., Toronto *II*/2, 691–708 (1977)

Thompson, J.A.J.: Copper in marine waters – effects of mining wastes. Proc. Int. Conf. Heavy Met. Environ., Toronto 1975, *II*/1, 273–284 (1977)

Thompson, J.A.J., McComas, F.T.: Copper and zinc levels in submerged mine tailings at Britannia Beach. B.C. Fish. Res. Board Can. Tech. Rep *473*, 33 pp. (1974)

Thompson, J.A.J., Paton, D.W.: Chemical delineation of a submerged mine tailing plume in Rupert and Holberg Inlets. B.C. Tech. Rep. Fish. Mar. Surv. Can. *506*, 24 pp. (1975)

Thompson, J.E., Duthie, J.R.: The biodegradability and treatment of NTA. J. Water Pollut. Control. Fed. *40*, 303–319 (1968)

Thomson, J., Turekian, K.K., McCaffrey, R.J.: The accumulation of metals in and release from sediments of Long Island Sound. In: Estuarine Research. Cronin, L.E. (ed.). New York: Academic Press 1975, pp. 28–44

Thorell, L.: Pollutants from Swedish municipal and industrial outlets into the Baltic. In: 3rd Soviet-Swedish Symp. on the Pollution of the Baltic. Åkerblom, A. (ed.). Ambio Spec. Rep. *5*, 213–218 (1977)

Thornton, I.: Geochemical parameters in the assessment of estuarine pollution. In: The Ecology of Resource – Degradation and Renewal. Chadwick, M.J., Goodman, G.T. (eds.). Oxford, London, Edinburgh, Melbourne: Blackwell 1975, pp. 157–169

Thornton, I.: Some aspects of environmental geochemistry in Britain. In: Proc. Int. Conf. Heavy Met. Environ., Toronto *II*/1, 17–33 (1977a)

Thornton, I.: Biogeochemical studies on molybdenum in the United Kingdom. In: Molybdenum in the Environment, Vol. 2. Chappell, W.R., Petersen, K.K. (eds.). New York: Marcel Dekker Inc., 1977b, pp. 341–369

Thornton, I., Webb, J.S.: Geochemical reconnaissance and the detection of trace element disorders in animals. In: Trace Metabolism in Animals. Mills, C.F. (ed.). Proc. WAAP/IBP Int. Symp. London: Livingstone 1970

Thornton, I., Webb, J.S.: Environmental geochemistry: Some recent studies in the United Kingdom. In: Trace Substances in Environmental Health. Hemphill, D.D. (ed.), Vol. VII. Univ. Missouri 1973, pp. 89–98

Thornton, I., Webb, J.S.: Trace elements in soils and surface waters contaminated by past metalliferous mining in parts of England. In: Trace Substances in Environmental Health. Hemphill, D.D. (ed.), Vol. IX. Columbia, Miss.: Univ. Missouri 1975, pp. 77–88

Thornton, I., Webb, J.S.: Application in the water industry of regional geochemical maps of England and Wales. J. Inst. Water Eng. Sci. *31*, 11–25 (1977)
Thornton, I., Watling, H., Darracott, A.: Geochemical studies in several rivers and estuaries for oyster rearing. Sci. Total Environ. *4*, 325–345 (1975)
Thorp, V.J., Lake, P.S.: Pollution of a Tasmanian river by mine effluents. II. Distribution of macroinvertebrates. Int. Rev. Ges. Hydrobiol. *58*, 885–892 (1973)
Thrower, S.J., Eustace, I.J.: Heavy metal accumulation in oysters grown in Tasmanian waters. Food Technol. Aust. *25*, 546–554 (1973)
Thurman, M.E.: Statistical study of content of molybdenum in stream sediments adjacent to a molybdenum mill. In: Natl. Sci. Found. Res. Appl. Natl. Needs (NSF-RANN) Trace Contam. Conf. Proc. Pacific Grove, Calif., Berkeley, Calif.: Univ. Calif. 1974, pp. 196–204
Timperley, M.H., Allan, R.J.: The formation and detection of metal dispersion haloes in organic lake sediments. J. Geochem. Exploration *3*, 167–190 (1974)
Tobschall, H.J., Göpel, C., Rast, U.: Geochemistry of organic gels and organic sediments from selected lakes in northern Norway. Abstr. 10th Int. Congr. on Sedimentology. Jerusalem 1978, pp. 682–683
Toerien, D.F., Hyman, K.L., Bruwer, M.J.: A preliminary trophic status classification of some South African impoundments. Water SA *1*, 15–23 (1975)
Tölg, G.: Zur Frage systematischer Fehler in der Spurenanalyse der Elemente. J. Vom Wasser *40*, 181–206 (1973)
Tonomura, K., Kanzaki, F.: The reductive decomposition of organic mercurials by cell free extract of a mercury-resistent pseudomonas. Biochem. Biophys. Acta *184*, 227–229 (1969)
Torma, A.E., Subramanian, K.N.: Selective bacterial leaching of lead suphilde. Int. J. Miner. Process. *1*, 125–134 (1974)
Tornabene, T.G., Edwards, H.W.: Microbial uptake of lead. Science *176*, 1334–1335 (1972)
Townsend, D.R., Kudo, A., Sayeed, H., Miller, D.R.: Mercury transport by bed sediment movement. Proc. Int. Conf. Transp. Persist. Chem. Aquatic Ecosyst., Ottawa, I-47-51 (1974)
Tratnyek, J.P.: Waste wool as a scavenger for mercury pollution in water. Water Pollut. Control Res. Ser. U.S. Environ. Prot. Ag., Washington, 1972. Cit. Wolery and Walters (1974)
Trefry, J.H.: The transport of heavy metals by the Mississippi River and their fate in the Gulf of Mexico. Diss. Texas A & M Univ., Dallas, 223 pp. (1977)
Trefry, J.H., Presley, B.J.: Heavy metal transport from the Mississippi River to the Gulf of Mexico. In: Marine Pollutant Transfer. Windom, H.L., Duce, R.A. (eds.). Lexington Books 1976a, pp.39–76
Trefry, J.H., Presley, B.J.: Heavy metals in sediments from San Antonio Bay and the northwest Gulf of Mexico. Environ. Geol. *1*, 283–294 (1976b)
Trostell, L.J., Wynne, D.J.: Determination of quartz (free silica) in refractory clays. J. Am. Ceram. Soc. *23*, 18–22 (1940)
Troup, B.N., Bricker, O.P.: Processes affecting the transport of materials from continents to oceans. In: Marine Chemistry in the Coastal Environment. Church, T.M. (ed). Am. Chem. Soc. Symp. Ser. *18*, 133–151 (1975)
Trudinger, P.A.: Microbes, metals and minerals. Miner. Sci. Eng. *3*, 13–25 (1971)
Trudinger, P.A., Bubela, B.: Microorganisms and the natural environment. Miner. Deposita *2*, 147–157 (1967)
Trueb, E.: Überblick über die Technik der künstlichen Anreicherung des Grundwassers, insbesondere in der Schweiz. Gas Wasser Abwasser *55*, 805–815 (1975)
Turekian, K.K.: Trace elements in seawater and other natural waters. Annu. Rep. AEC Contract AT-2912, Publ. Yale Univ., 59 pp. (1966)
Turekian, K.K.: The oceans, streams and atmosphere. In: Handbook of Geochemistry, Vol. I. Wedepohl, K.H. (ed.). Berlin, Heidelberg, New York: Springer 1969, pp. 297–323
Turekian, K.K.: Rivers, tributaries, and estuaries. In: Impingement of man on the oceans. Wood, D.W. (ed.). New York: Wiley-Interscience 1971, pp. 9–73
Turekian, K.K.: Fate of metals in estuaries. Chem. Eng. Prog. Symp. Ser. *71*, 26 (1975)
Turekian, K.K.: The fate of metals in the oceans. Geochim. Cosmochim. Acta *41*, 1139–1144 (1977)
Turekian, K.K., Johnson, D.G.: The barium distribution in sea water. Geochim. Cosmochim. Acta *30*, 1153–1174 (1966)
Turekian, K.K., Kleinkopf, M.D.: Estimates of the average abundance of Cu, Mn, Pb, Ti, Ni, and Cr in surface waters of Maine. Bull. Geol. Soc. Am. *67*, 1129–1132 (1956)
Turekian, K.K., Rona, P.A.: Eastern Atlantic fracture zones as potential disposal sites of radioactive waste. Environ. Geol. *2*, 59–62 (1977)

Turekian, K.K., Scott, M.: Concentrations of Cr, Ag, Mo, Ni, Co and Mn in suspended material in streams. Environ. Sci. Technol. *1*, 940–942 (1967)

Turekian, K.K., Wedepohl, K.H.: Distribution of the elements in some major units of the earth's crust. Bull. Geol. Soc. Am. *72*, 175–192 (1961)

Turekian, K.K., Harriss, R.C., Johnson, D.G.: The variations of Si, Cl, Na, Ca, Sr. Ba, Co, Ag in the Neuse River. Limnol. Oceanogr. *12*, 702–706 (1967)

Turner, R.R., Lindberg, S.E.: Behavior and transport of mercury in river-reservoir system downstream of inactive chloralkali plant. Environ. Sci. Technol. *12*, 918–923 (1978)

Turney, W.G.: Mercury pollution – Michigan's action program. J. W.P.C.F. *43*, 1427–1438 (1971)

Turney, W.G.: The mercury pollution problem in Michigan. In: Environmental Mercury Contamination. Hartung, R., Dinman, B.D. (eds.). Ann Arbor Sci. Publ. 1972

Twenkofel, W., McKelvey, V.: Sediments in freshwater lakes. Bull. Am. Assoc. Petrol. Geol. *25*, 826–849 (1941)

Tyler, P.A., Buckney, R.T.: Pollution of a Tasmanian river by mine effluents. I. Chemical evidence. Int. Rev. Gesamten Hydrobiol. *58*, 873–883 (1973)

Tzur, Y.: Interstitial diffusion and advection of solute in accumulating sediments. J. Geophys. Res. *76*, 4208–4211 (1971)

Ueda, T., Nakamura, R., Suzuki, Y.: Comparison of ^{115}Cd accumulation from sediments and sea water by polychaete worms. Bull. Japan Soc Sci. Fish. *42*, 299–306 (1976)

Ududov, P.A., Parilov, Y.U.S.: Certain regularities of migration of metals in natural waters. Geochemistry *8*, 763–769 (1961)

Ui, J., Kitamura, S.: Mercury in the Adriatic. Mar. Pollut. Bull. *2*, 56–58 (1971)

Underdal, B., Hastein, T.: Mercury in fish and water from a river and fjord in the Kragerø region, South Norway. Oikos *22*, 101–105 (1971)

Underwood, E.J.: Trace elements in human and animal nutrition. 3rd ed. New York: Academic Press 1971

Uthe, J.H., Bligh, E.G.: Preliminary survey of heavy metal contamination of Canadian freshwater fish. J. Fish. Res. Board Can. *28*, 786–788 (1971)

Uthe, J.F., Atton, F.M., Royer, L.M.: Uptake of mercury by caged rainbow trout (Salmo gairdneri) in the South Saskatchewan River. J. Fish. Res. Board Can. *30*, 643–650 (1973)

Vahrenkamp, H.: Metalle in Lebensprozessen. Chemie Unserer Zeit *7*, 97–105 (1973)

Valenta, P., Mart, L., Nürnberg, H.W., Stoeppler, M.: Voltammetrische simultane Spurenanalyse toxischer Metalle in Meerwasser, Binnengewässern, Trink- und Brauchwasser Vom Wasser *48*, 89–110 (1977)

Vallee, B.L., Ulmer, D.D.: Biochemical effects of mercury, cadmium, and lead. Annu. Rev. Biochem. *41*, 91–128 (1972)

Vallentyne, J.R.: The algal bowl: Lakes and mar. Dep. Environ. Fish. Mar. Serv , Ottawa, Misc. Spec. Publ. *22* (1974)

Van As, D., Fourie, H.O., Vleggaar, C.M.: Trace element concentrations in marine organisms from the Cape West Coast. S. Afr. J. Sci. *71*, 151–154 (1975)

Vandeginste, B.G., Salemink, P., Duinker, J.C.: Auto- and crosscorrelograms of particulate trace metals in the Rhine Estuary, Southern Bight and Dutch Wadden Sea. Neth. J. Sea Res. *10*, 59–70 (1976)

Van der Walt, S.R., Eldik, R. van, Potgieter, H.G.J.: The recovery of Fe, Mn, and Al from a mine water effluent. Water Res. *9*, 865–868 (1975)

Veeh, H.H., Burnett, W.C., Soutar, A.: Contemporary phosphorite on the continental margin of Peru. Science *181*, 845–847 (1973)

Veeh, H.H., Calvert, S.E., Price, N.B.: Accumulation of uranium in sediments and phosphorites on the South West African Shelf. Mar. Chem. *2*, 189–202 (1974)

Veen, J. van: Onderzoekingen in die Hoofden: S'Gravenhage, 1936. Cit. Simon (1953)

Venugopal, B., Luckey, T.D.: Toxicology of non-radioactive heavy metals and their salts. In: Heavy Metal Toxicity, Safety and Hormology. Luckey, T.D., Venugopal, B., Hutcheson, D. (eds.). Stuttgart: Thieme 1975, pp. 4–73

Veres, E.M., Hasty, R.A.: Mercury in bottom sediments of Rietvlei Reservoir. S.A. J. Sci. *72*, 86–87 (1976)

Vermeulen, A.J.: Acid precipitation in The Netherlands. Environ. Sci. Technol. *12*, 1016–1021 (1978)

Vernberg, F.J., Vernberg, W.B.: Pollution and physiology of marine organisms. London, New York, San Francisco: Academic Press 1974

Vernet, J.-P., Johnstone, L.M.: Heavy metal pollutants in the sediments of the Rhône River. Proc. Int. Conf. Transp. Persist. Chem. Aquatic Ecosyst. Ottawa, 1974

Vernet, J.-P., Ribordy, E.: Teneurs en métaux lourds des sédiments du Rhône Valaisan et de ses affluents. Murithienne 92, 9–20 (1975)

Vernet, J.-P., Thomas, R.L.: Levels of mercury in the sediments of some Swiss lakes including Lake Geneva and the Rhône River. Eclogae Geol. Helv. 65/2, 293–306 (1972a)

Vernet, J.-P., Thomas, R.L.: Le mercure dans l'environnement et le rôle de la géologie sédimentaire. Bull. B.R.G.M. 2, 43–61 (1972b)

Vernet, J.-P., Thomas, R.L.: The occurrence and distribution of mercury in the sediments of the Petit Lac (western Lake Geneva). Eclogae Geol. Helv. 65, 307–317 (1972c)

Vernet, J.-P., Chappuis, A., Favarger, P.-Y., Davaud, E.: Teneur en mercure des poissons du Léman, campagne 1975. Bull. A.R.P.E.A. 13, 23–44 (1976)

Vernet, J.-P., Rapin, F., Scolari, G.: Heavy metal content of lake and river sediments in Switzerland. In: Proc. Int. Symp. on Interactions Between Sediments and Freshwater, Amsterdam, 1976. Golterman, H.L. (ed.). The Hague: Junk B.V. Publ. 1977a, pp. 390–397

Vernet, J.-P., Rapin, F., Faverger, P.Y., Fernex, F.: Contamination des sédiments marins (Côte d'Azur) par les métaux lourds (Hg et Cd). Rev. Int. Oceanogr. Méd. 47, 91–95 (1977b)

Villa, O., Jr., Johnson, P.G.: Distribution of metals in Baltimore Harbour sediments. Annapolis Field Office Tech. Rep. 59, EPA-903/9-74-012, 71 pp. (1974)

Villiers, P.R. De: The chemical composition of the water of the Orange River at Viooldrif, Cape Province, Rep. S. Afr. Dep. Manwere Ann. Geol. Opname 1, 197 (1962)

Vinogradov, A.P.: Khimicheskii elementarnyi sostav organismov morya. Akad. Nauk. SSR, Biokhim. Lab. Trudy 3, 63–278 (1935)

Vinogradov, A.P.: Migration of molybdenum in the zone of weathering. Geokhimia 2, 120–126 (1957)

Vivian, C.M.G., Massie, K.S.: Trace metals in waters and sediments of the River Tawe, South Wales in relation to local sources. Environ. Pollut. 14, 47–61 (1977)

Vogt, G.: Schwermetallanreicherung in Fischen des Rheins. Paper pres. Bayer. Naturschutzakademie, Munich, Oct. 21, 1974

Vogt, G., Kittelberger, F.: Studie zur Aufnahme und Anreicherung von Schwermetallen in typischen Algenassoziationen des Rheines zwischen Germersheim und Gernsheim. In: Fisch und Umwelt, Vol. 3. Reichenbach-Klinke, H. (ed.). Stuttgart, New York: G. Fischer 1977, pp. 15–17

Volkov, I.I., Fomina, L.S.: Influence of organic material and processes of sulfide formation on distribution of some trace elements in deepwater sediments of Black Sea. In: The Black Sea, Geology, Chemistry and Biology. Degens, E.T., Ross, D.A. (eds.). Am. Assoc. Petrol. Geol., Mem. 20, 456–476 (1974)

Vonk, J.W., Sijepesteijn, A.K.: Studies on the methylation of mercuric chloride by pure cultures of bacteria and fungi. Antonie van Leeuwenhoek 39, 505–513 (1973)

Vuorela, I.: The indication of farming in pollen diagrams from southern Finland. Acta Bot. Fenn. 87, 1–140 (1970)

Wagner, G.: Die Untersuchung von Sinkstoffen aus Bodenseezuflüssen. Schweiz. Z. Hydrol. 38, 191–205 (1976)

Wagner, G.: FeS-Konkretionen im Bodensee. Int. Rev. Ges. Hydrobiol. 56, 265–272 (1971)

Wagner, K.H.: Die toxischen Inhaltsstoffe in Siedlungsabfällen und deren Aufbereitungsprodukten – Müll, Müllkompost, Müllklärschlammkompost und Klärschlamm. Fortschr. Ber. VDI Z. 15/10, 62 pp. (1977)

Waldhauer, R., Matte, A., Tucker, R.E.: Lead and copper in the waters of Raritan und lower New York Bays. Mar. Pollut. 9, 38–42 (1978)

Walker, G., Rainbow, P.S., Foster, P., Crisp, D.J.: Barnacles: Possible indicators of zinc pollution? Mar. Biol. 30, 57–65 (1975)

Walker, J.D., Colwell, R.R.: Mercury-reistant bacteria and petroleum degradation. Appl. Microbiol. 27, 285–287 (1974)

Walker, W.: Where have all the toxic chemicals gone? Ground Water 11, 11–14 (1973)

Walkley, A., Black, I.A.: An examination of the Degtjareff method for determining soil organic matter, and a proposed modification of the chromic acid titration method. Soil. Sci. 37, 29–38 (1934)

Wallace, G.T., Jr., Duce, R.A.: Concentration of particulate trace metals and particulate organic carbon in marine surface waters by a bubble flotation mechanism. Mar. Chem. 3, 157–181 (1975)

Wallace, G.T., Jr., Hoffman, G.L., Duce, R.A.: The influence of organic matter and atmospheric deposition on the particulate trace metal concentration of northwest Atlantic surface seawater. Mar. Chem. 5, 143–170 (1977)

Walter, C.M., June, F.C., Brown, H.G.: Mercury in fish, sediments, and water in Lake Oahe, South Dakota. J.W.P.C.F. 45, 2203–2210 (1973)

Walter, C.M., Brown, H.G., Hensley, C.P.: Distribution of total mercury in the fishes of Lake Oahe. Water Res. *8*, 413–418 (1974)

Walters, L.J., Jr.: Man's input of mercury, chromium, and nickel to Lake Erie sediments. Abstr. 20th Conf. Great Lakes Res. Int. Assoc. Great Lakes Res. (1977)

Walters, L.J., Jr., Herdendorf, C.E., Charlesworth, L.J., Anders, H.K., Jackson, W.B., Skoch, E.J., Webb, D.K., Kovacik, T.L., Sikes, C.S.: Mercury contamination and its relation to other physicochemical parameters in the western basin of Lake Erie..Proc. 15th Conf. Great Lakes Res. 306–316 (1972)

Walters, L.J., Kovacik, T.L., Herdendorf, C.E.: Mercury occurrence in sediment cores from western Lake Erie. Ohio J. Sci. *74*, 1–19 (1974a)

Walters, L.J., Wolery, T.J., Myser, R.D.: Occurrence of As, Cd, Co, Cr, Cu, Fe, Hg, Ni, Sb, and Zn in Lake Erie sediments. Proc. 17th Conf. Great Lakes Res. 219–234 (1974b)

Ward, N.I., Brooks, R.R., Reeves, R.D.: Copper, cadmium, lead and zinc in soils, stream sediments, waters, and natural vegetation around the Tui Mine, Te Aroha, New Zealand. N.Z. J. Sci. *19*, 81–89 (1976)

Ward, N.I., Brooks, R.R., Roberts, E.: Silver in soils, stream sediments, water, and vegetation near a silver mine and treatment plant at Maratoto, New Zealand. Environ. Pollut. *13*, 269–280 (1977)

Warner, T.B.: Mixing model prediction of fluoride distribution in Chesapeake Bay. J. Geophys. Res. *77*, 2728–2732 (1972)

Waslenchuk, D.G.: Mercury in fluvial bed sediments subsequent to contamination. Environ. Geol. *1*, 131–136 (1976)

Watling, H.R., Watling, R.J.: Trace metals in oysters from Knysna Estuary. Mar. Pollut. Bull. *7*, 45–48 (1976)

Watson, G.M.: Rum Jungle environmental studies. Summary Rep., Aust. At. Energy Commun. E *366*, 21 pp. (1975)

Wauschkuhn, A.: Rezente Sulfidbildung in vulkanischen Seen auf Hokkaido. Geol. Rundsch. *62*, 774–785 (1973)

Weatherley, A.H.: Ecology of zinc pollution of a river/lake system. Abstr. Int. Conf. Heavy Metals Environ., Toronto *C-308* (1975)

Weatherley, A.H., Dawson, P.: Zinc pollution in a freshwater system: analysis and proposed solutions. Search *4*, 471–476 (1973)

Webb, J.S., Nichol, I., Thornton, I.: The broadening scope of regional geochemical reconnaissance. 23rd Int. Geol. Congr. *6*, 131–147 (1968)

Webber, L.R., Beauchamp, E.G.: Heavy metals in corn grown on waste amended soils. Proc. Int. Conf. Heavy Met. Environ., Toronto, 1975, *II/1*, 443–452 (1977)

Webber, M.D., Corneau, D.G.M.: Metal extractability from sludge-soil mixtures. Proc. Int. Conf. Heavy Met. Environ., Toronto, 1975, *I*, 205–225 (1977)

Weber, H.: Vergiftungsversuche mit Kupfersulfat. Z. Hydrol. *4*, 105–126 (1934)

Wedepohl, K.H.: Composition and abundance of common sedimentary rocks, Chap. 8. In: Handbook of Geochemistry. Wedepohl, K.H. (ed.) Berlin, Heidelberg, New York: Springer 1969, pp. 250–271

Wedepohl, K.H.: Environmental influences on the chemical composition of shales and clays. Phys. Chem. Earth *8*, 307–333 (1970)

Wedepohl, K.H.: Zinc – abundance in natural waters and in the atmosphere. In: Handbook of Geochemistry. Wedepohl, K.H. (ed). Berlin, Heidelberg, New York: Springer 1972

Wedepohl, K.H.: Copper – abundance in natural waters and in the atmosphere. In: Handbook of Geochemistry. Wedepohl, K.H. (ed.). Berlin, Heidelberg, New York: Springer 1974a

Wedepohl, K.H.: Lead – abundance in natural waters and in the atmosphere. In: Handbook of Geochemistry. Wedepohl, K.H. (ed.). Berlin, Heidelberg, New York: Springer 1974b

Weeks, J.D.: Physical-chemical methods of heavy metals removal. Discussion of a paper given by J.W. Patterson and R.A. Minear. In: Heavy Metals in the Aquatic Enrivonment. Krenkel, P.A. (ed.). Oxford: Pergamon Press 1975, pp. 273–276

Weichart, G.: Pollution of the North Sea. Ambio *2*, 99–106 (1973a)

Weichart, G.: Verschmutzung der Nordsee. Naturwissenschaften *60*, 469–472 (1973b)

Weiden, C.H. van der, Arnoldus, M.J.H L., Meurs, C.T.: Desorption of metals from suspended material in the Rhine Estuary. Neth. J. Sea Res. *11*, 130–145 (1977)

Weiler, R.R., Chawla, V.K.: Dissolved mineral quality of Great Lake waters. Proc. 12th Conf. Great Lakes Res. Int. Assoc. Great Lakes Res. 301–818 (1969)

Weiss, A., Amstutz, G.C.: Ion-exchange reactions on clay minerals and cation selective membrane properties as possible mechanisms of economic metal concentration. Miner. Deposita *1*, 60–66 (1966)

Weiss, H.V., Koide, M., Goldberg, E.D.: Mercury in a Greenland ice sheet: evidence of recent input by man. Science *174*, 692–693 (1971)
Weissberg, B.G., Zobel, M.G.: Geothermal mercury pollution in New Zealand. Bull. Environ. Contam. Toxicol. *9*, 148–155 (1973)
Welsh, P., Denny, P.: Waterplants and the recycling of heavy metals in an English lake. In: Trace Substances in Environmental Health, Vol. X. Hemphill, D.D. (ed.). Columbia Univ. Missouri 1976, pp. 217–223
Welte, D.: Organic geochemistry of carbon. In: Handbook of Geochemistry. Wedepohl, K.H. (ed.). Berlin, Heidelberg, New York: Springer 1969
Wentink, G.R., Etzel, J.A.: Removal of metal ions by soils. J.W.P.C.F. *44*, 1561–1567 (1972)
Wentsel, R.S., Berry, J.W.: Cadmium and lead levels in Palestine Lake, Palestine, Indiana. Proc. Indiana Acad. Sci. *84*, 481–490 (1974)
Wentz, D.A.: Effect of mine drainage on the quality of stream in Colorado 1971–1972. Colorado Water Conservation Bd. Col. Water Resour. Circ. *21* (1974)
Wentz, D.A., Lee, G.F.: Sedimentary phosphorous in lake cores – observation on depositional patterns in Lake Mendota. Environ. Sci. Technol. *3*, 754–757 (1969)
Westöö, G.: Determination of methylmercury compounds in foodstuffs. I. Methylmercury compounds in fish, identification and determination. Acta Chem. Scand. *20*, 2131–2137 (1966)
Wheeler, R., Dunning, C.: Trace-element geochemistry of piston cores from western Michigan coastal lakes. Can. Mineral. *14*, 21–31 (1976)
Whipple, W., Jr., Hunter, J.V.: Nonpoint sources and planning for water pollution control. J.W.P.C.F. *49*, 15–23 (1977)
Whipple, W., Jr., Hunter, J.V., Yu, S.L.: Effects of storm frequency on pollution from urban runoff. J.W.P.C.F. *49*, 2243–2248 (1977)
Whitby, L.M., Hutchinson, T.C.: Heavy-metal pollution in the Sudbury mining and smelting region of Canada, II. Soil toxicity tests. **Environ. Conserv. *1*,** 191–200 (1974)
Whitby, L.M., Stokes, P.M., Hutchinson, T.C., Myslik, G.: Ecological consequence of acidic and heavy-metal discharges from the Sudbury smelters. Can. Mineral. *14*, 47–57 (1976)
White, A., Handler, P., Smith, E.L.: Principles of Biochemistry, 3rd. ed. New York: McGraw-Hill 1967
White, D.E., Hem, J.D., Waring, G.A.: Chemical composition of subsurface waters. In: Data of Geochemistry, 6th ed. U.S. Geol. Surv. Prof. Pap. *440-F*, 67 pp. (1963)
Whiteside, M.C.: Paleoecological studies of Potato Lake and its environment. Ecology *46*, 807–811 (1965)
Whitley, L.S.: The resistance of tubificid worms to three common pollutons. Hydrobiologia *32*, 193–205 (1968)
Widman, M.U., Epstein, M.M.: Polymer film overlay system for mercury contaminated sediments. Water Pollut. Control Res. Ser. U.S. EPA Wash. (1972). Cit. Wolery and Walters (1974)
Wiersma, J.H., Lee, G.F.: Selenium in lake sediments – analytical procedure and preliminary results. Environ. Sci. Technol. *5*, 1203–1206 (1971)
Wilber, W.G., Hunter, J.V.: Heavy metals in urban runoff. Proc. Southeastern Regional Conf. on Non-Point Sources and Water Pollut. (1975). Cit. Whipple and Hunter (1977)
Willey, J.D.: Geochemistry and environmental implications of the surficial sediments in northern Placentia Bay, Newfoundland. Can. J. Earth Sci. *13*, 1393–1410 (1976)
Williams, D.R.: The metals of life. The solution chemistry of metal ions in biological systems, Chap. 2. London: Van Nostrand Reinhold Company 1971
Williams, C.H., David, D.J.: Aust. J. Soil Sci. *11* (1973). Lit. cit. by Stenström, T., Vahter, M.: Ambio *3*, 91–92 (1974)
Williams, J.D.H., Syers, J.K., Harris, R.F., Armstrong, D.E.: Fractionation of inorganic phosphate in calcareous lake sediments. Soil Sci. Soc. Am. Proc. *35*, 250–255 (1971)
Williams, J.S., Joyce, J.C., Monk, J.T.: Stream-velocity effects on the heavy metal concentrations. J. Am. Water Works Assoc. *65*, 275–279 (1973)
Williams, L.G., Coffee, G.L.: Mercury monitoring technique using an organic substrate. J.W.P.C.F. *47*, 354–361 (1975)
Williams, P.M., Chan, K.S.: Distribution and speciation of iron in natural waters. J. Fish. Res. Board Can. *23*, 575–593 (1966)
Williams, R.J.P.: Heavy metals in biological systems. Endeavour *24*, 96–100 (1967)
Wilson, A.L.: Concentrations of trace metals in river waters, a review. Tech. Rep. No. 16. Water Res. Cent., Medmenhan Lab. and Stevenage Lab., U.K. (1976)
Winchester, J.W., Duce, R.A.: The air-water interface. In: Fate of Pollutants in the Air and Water Environments. Part I. Suffet, I.H. (ed.). New York, London, Sydney, Toronto: Wiley 1977, pp. 22–47

Windom, H.L.: Environmental aspects of dredging in estuaries. ASCE J. Waterways Harbors Coastal Eng. Div. 98, 475–505 (1972)

Windom, H.L.: Investigations of changes in heavy metal concentrations resulting from maintenance dredging of Mobile Bay ship channel, Mobile Bay, Alabama. Rep. U.S. Army Corps of Engineers, Mobile Distr., 46 pp. (1973)

Windom, H.L.: Heavy metal fluxes through salt-marsh estuaries. In: Estuarine Research. Cronin, L.E. (ed.)., Vol. I. New York: Academic Press 1975, pp. 137–152

Windom, H.L.: Environmental aspects of dredging in the coastal zone. CRC Crit. Rev. Environ. Control 5, 91–109 (1976)

Windom, H.L., Beck, K.C., Smith, R.: Transport of trace elements to the Atlantic Ocean by three southeastern streams. Southeast Geol. 12, 169–181 (1971)

Windom, H.L., Stickney, R., Smith, R., White, D., Taylor, F.: Arsenic, cadmium, copper, mercury, and zinc in some species of north Atlantic fishes. J. Fish. Res. Board Can. 30, 275–279 (1973)

Windom, H.L., Taylor, F., Stickney, R.: Mercury in North Atlantic plankton. J. Cons. Int. Expl. Mer. 35, 18–21 (1973)

Windom, H.L., Gardner, W.S., Dunstan, W.M., Paffenhofer, G.A.: Cadmium and mercury transfer in a coastal marine ecosystem. In: Marine Pollutant Transfer. Windom, H.L., Duce, R.A. (eds.). Lexington: Heath & Co. 1976, pp. 135–157

Wisseman, R.W., Cook, S.F., Jr.: Heavy metal accumulation in the sediments of a Washington lake. Bull. Environ. Contam. Toxicol. 18, 77–82 (1977)

Wittmann, G.T.W., Förstner, U.: Metal enrichment of sediments in inland waters – the Hartbeespoort Dam. Water SA 1, 76–82 (1975)

Wittmann, G.T.W., Förstner, U.: Heavy metal enrichment in mine dranage. I. The Rustenburg platinum mining area. S. Afr. J. Sci. 72, 242–246 (1976a)

Wittmann, G.T.W., Förstner, U.: Heavy metal enrichment in mine drainage. II. The Witwatersrand Goldfields. S. Afr. J. Sci. 72, 365–370 (1976b)

Wittmann, G.T.W., Förstner, U.: Metal enrichment of sediments in inland waters – the Jukskei and Hennops River drainage system. Water SA 2, 67–72 (1976c)

Wittmann, G.T.W., Förstner, U.: Heavy metal enrichment in mine drainage. III. The Klerksdorp, West Wits and Evander Goldfields. S. Afr. J. Sci. 73, 53–57 (1977a)

Wittmann, G.T.W., Förstner, U.: Heavy metal enrichment in mine drainage. IV. The Orange Free State Goldfield. S. Afr. J. Sci. 73, 374–378 (1977b)

Wixson, B.G., Bolter, E., Gale, N.L., Jennett, J.C., Purushothaman, K.: The lead industry as a source of trace metals in the environment. In: Cycling and Control of Metals. Natl. Environ. Res. Center (EPA) Cincinnati 1973, pp. 11–19

Wolery, T.J., Walters, L.J.: Pollutant mercury and sedimentation in the western basin of Lake Erie. Proc. 17th Conf. Great Lakes Res., 235–249 (1974)

Wolf, P. de: Mercury content of mussels from west European coasts. Mar. Pollut. Bull. 6, 61–63 (1975)

Wolfe, D.A., Cross, F.A., Jennings, C.D.: The flux of Mn, Fe, and Zn in an estuarine ecosystem. In: Radioactive Contamination of the Marine Environment. Vienna: I.A.E.A., 1973, pp. 159–175

Wolfenden, P.J., Lewin, J.: Distribution of metal pollutants in flood-plain sediments. Catena 4, 309–317 (1977)

Wolfenden, P.J., Lewin, J.: Distribution of metal pollutants in active stream sediments. Catena 5, 67–78 (1978)

Wollast, R.: Transport et accumulation des pollutants dans l'estuaire de l'Escaut. In: Project Mer. Vol. 10, L'Estuaire de l'Escaut. Nihoui, J.C.J., Wollast, R. (eds.). Progr. Polit. Scient. Bruxelles, 1976, pp. 191–218

Wollast, R., DeBroeu, F.: Study of the behaviour of dissolved silica in the estuary of the Scheldt. Geochim. Cosmochim. Acta 35, 613–620 (1971)

Wollast, R., Billen, G., MacKenzie, F.T: Behaviour of mercury in natural systems and its global cycle. In: Ecological Toxicology Research. McIntyre, A.D., Mills, C.F. (eds.). New York: Plenum Publishing Corporation 1971

Wollast, R., Billen, G., Longueville, M.T.: Accumulation et évolution du mercure dans les boues de la Sambre. In: Modèle Mathématique de la Pollution des Eaux Intérieures. Rapport de Synthèse No. 1, C.I.P.S. Bruxelles (1972)

Wong, P.T.S., Chau, Y.K., Luxon, P.L.: Methylation of lead in the environment. Nature (London) 253, 263–264 (1975)

Wood, D.K., Tchobanoglous, G.: Trace elements in biological waste treatment. J.W.P.C.F. 47, 1933–1945 (1975)

Wood, J.M.: A progress report on mercury. Environment 14, 33–39 (1972)

Wood, J.M.: Metabolic cycles for toxic elements in aqueous systems. Rev. Int. Océanogr. Méd. 31/32, 7–17 (1973)
Wood, J.M.: Biological cycles for toxic elements in the environment. Science 183, 1049–1052 (1974)
Wood, J.M.: Biological cycles for elements in the environment. Naturwissenschaften 62, 357–364 (1975)
Wood, J.M., Kennedy, F.S., Rosen, C.G.: Synthesis of methylmercury compounds by extracts of a methanogenic bacterium. Nature (London) 220, 173–174 (1968)
Wood, J.M., Segall, H.J., Ridley, W.P., Cheh, A., Chudyk, W., Thayer, J.S.: Metabolic cycles for toxic elements in the environment. Proc. Int. Conf. Heavy Metals in the Environment, Toronto 1975, 1, 49–68 (1977)
Wood, L.W.: Role of oligochaetes in the circulation of water and solutes across the mud-water interface. Verh. Int. Ver. Limnol. 19, 1530–1533 (1975)
Wood, P.C., et al.: In: Out of Sight, out of Mind. Vol. IV, Appendix D, 1973/74, pp. 57–67
Woodcock, J.T.: Copper waste dump leaching. Proc. Austral. Inst. Min. Metal. 224, 47–66 (1967)
Woolson, E.A., Kearney, P.C.: Consistence and reactions of carbon cacodylic acid in soils. Environ. Sci. Technol. 7, 47–50 (1973)
Wright, D.A.: Heavy metals in animals from the north east coast. Mar. Pollut. Bull. 7, 36–38 (1976)
Wright, J.R., Schnitzer, M.: Metallo-organic interactions associated with podzolization. Soil Sci. Soc. Am. Proc. 27, 171–176 (1963)
Wright, R.F., Galloway, J.N.: Acid precipitation and acidification of lakes. A comparison of northern Europe and eastern North America. Abstr. SIL-Congr., Kopenhagen, 301 (1977)
Wright, R.F., Gjessing, E.T.: Changes in the chemical composition of lakes. Ambio 4, 220–223 (1976)
Wright, R.F., Snekvik, E.: Acid precipitation: Chemistry and fish populations in 700 lakes in southern-most Norway. Abstr. SIL-Congr., Kopenhagen, 300 (1977)
Yamada, M., Tonomura, K.: Formation of methylmercury compounds from inorganic mercury by Clostridium cochlearium. J. Ferment Technol. 50, 159–166 (1972)
Yeaple, D.S., Feick, G., Horne, R.A.: Dredging of mercury-contaminated sediments. Proc. 4th Ann. Offshore Techn. Conf. Houston, Texas, Vol. I, Pap. 1584, 695–702 (1972)
Yim, W.W.S.: Heavy metal accumulation in estuarine sediments in historical mining, Cornwall. Mar. Pollut. Bull. 7, 147–150 (1976)
Yoshida, T., Ikegaki, Y.: A study on the behavior of mercury-contaminated sediments in Minamata Bay. In: Management of Bottom Sediments Containing Toxic Substances. Proc. 2nd U.S.-Japan Experts Meet., Oct. 1976, Tokyo. Environ. Res. Lab., U.S. EPA, Corvallis, Or., 1977, pp. 127–143
Yoshimura, S.: Kata-numa, a very strong acid-water lake of volcano Kata-numa, Miyagi Prefecture. Jpn. Arch. Hydrobiol. 26, 197–202 (1934)
Yostida, H.: Transference mechanism of mercury in marine environment (1967). Cit. Faberström and Jernelöv (1972)
Young, D.R., Johnson, J.N., Soutar, A., Isaacs, J.D.: Mercury concentrations in dated varved marine sediments collected of southern California. Nature (London) 244, 273–274 (1973)
Young, D.R., McDermott, D.J., Heesen, T.C., Jan, T.K.: Pollutant inputs and distributions of southern California. In: Marine Chemistry in the Coastal Environment. Church, T.M. (ed.). Am. Chem. Soc. Symp. Ser. 18, 424–439 (1975)
Young, R.H.F.: Pollution effects on surface and ground waters. J.W.P.C.F. 46, 1419–1429 (1974)
Zajicek, O.T., Pojasek, R.B.: Fulvic acid and aquatic manganese transport. Water Resour. Res. 12, 305–308 (1976)
Zaldivar, R.: Arsenic contamination of drinking water and food-stuffs causing endemic chronic poisoning. Beitr. Pathol. 151, 384–400 (1974)
Zauke, G.-P.: Mercury in benthic invertebrates of the Elbe estuary. Helgol. Wiss. Meeresunters. 29, 358–374 (1977)
Zemanski, G.M.: Removal of trace metals during conventional water treatment. J. Am. Water Works Assoc. 66, 606–609 (1974)
Zenz, D.R., Peterson, J.R., Brooman, D.L., LeUging, C.: Environmental impacts of land application of sludge. J.W.P.C.F. 48, 2332–2342 (1976)
Zievers, J.F., Novotny, C.J.: Curtailing pollution from metal finishing. Environ. Sci. Technol. 7, 209–213 (1973)
Zink-Nielsen, I., Krogh, O.: Orienterende undersøgelser af sedimentet fra Lille Belt for indhold af tungmetaller. In: Sediment Undersøgelser, Miljøstyrelsen Baeltprojektet, 45–50 (1976)
Zirino, A., Healy, M.L.: Voltammetric measurement of zinc in the northeastern tropical Pacific Ocean. Limnol. Oceanogr. 16, 773–778 (1971)

Zirino, A., Healy, M.L.: pH-controlled differential voltammetry of certain trace transition elements in natural water. Environ. Sci. Technol. 6, 243–249 (1972)

Zirino, A., Yamamoto, S.: A pH-dependent model for the chemical speciation of copper, zinc, cadmium and lead in sea water. Limnol. Oceanogr. 17, 661–671 (1972)

Zitko, V.: Structure activity relationships and the toxicity of trace elements to aquatic biota. In: Toxicity to Biota of Metal Forms in Natural Water. Andrew, R.W., Hodson, P.V., Konasewich, D.E. (eds.). Int. Joint Comm., Windsor, Ontario, 1976, pp. 9–43

Zitko, V., Carson, W.V.: Release of heavy metals from sediments by nitrilotriacetic acid (NTA). Chemosphere 3, 113–118 (1972)

Zitko, V., Carson, W.G.: A mechanism of the effects of water hardness on the lethality of heavy metals to fish. Chemosphere 5, 299–303 (1976)

Zitko, V., Carson, W.V., Carson, W.G.: Thallium, occurance in the environment and toxicity to fish. Bull. Environ. Contam. Toxicol. 13, 23–30 (1975)

Zoller, W.H., Gladney, E.S., Duce, R.A.: Atmospheric concentrations and sources of trace metals at the South Pole. Science 183, 198–200 (1974)

Züllig, H.: Sedimente als Ausdruck des Zustandes eines Gewässers. Schweiz. Z. Hydrol. 18, 7–143 (1956)

Subject Index

Aachen (D) 47, 357
Aare R. (CH) 169
Aberystwyth (GB) 33, 96
acceptors (Lewis acids) 6
acid mine drainage 33 ff., 82, 214, 258 ff., 344, 345
– precipitation 260 ff.
activated carbon 334 ff., 347 ff.
– sludge 345
Acushnet R. (Mass.) 182, 191
adaption 192, 294
Adirondack Mts. (N.Y.) 261
Adriatic Sea 184, 185
adsorption 207 ff., 357
– biologic 294
aeration 255, 330, 331, 353
aerosols 51
Aetna, Mt. (Colo.) 32
Afon Goch R. (GB) 289
age dependency 282 ff., 312
Agfardlikavsa (Greenland) 190
aging, ferric oxide 358
Agricola L. (CAN) 217
agriculture 48, 49
algae 18, 277, 285 ff.
– brown 288 ff.
– filamentous 291
algicides 149
alkaline waste 227
Allegheny R. (Pa.) 94, 95
Aller R. (D) 100, 159, 160, 162
alloys 365
Alpnach L. (CH) 313
alum system 347 ff.
aluminum, water 87
amalgam 36, 365
Amazon R. 91, 92, 123, 124, 201, 240, 241, 245
amended soils 354, 355
Amiata, Mt. (I) 158, 168
amino acids 280
amphibole 202
amphoteric 350, 351
analysis, sediment 113 ff., 238 ff.
– water 68 ff., 75 ff., 328

anchovy 320
Anglesey Is. (GB) 97, 289
annual cycles 107 ff.
anodic stripping voltametry 75, 76, 83, 328
anomaly 119, 120
antagonism 273
antimony
– air 49, 50
– groundwater 357
– PC-treatment 347 ff.
– rocks 134
– surface water 87, 93
– thermal water 357
arid regions 131, 232, 324
Arkansas R. (Ks.) 94, 95
arsenic
– air 49, 50
– algae 291
– cement production 52
– coal 52
– decarbonation 335
– drinking water 30, 340
– fish 310, 321
– health hazard 14, 15
– higher plants 292
– mosses 292
– mussels 321
– oil 52
– PC-treatment 332, 347 ff.
– rocks 134
– seaweed 321
– sediment 63, 64, 65, 148 ff., 175
– species in water 27
– surface water 87, 93, 94 ff., 104, 251, 252
– tolerance level 28
– zooplankton 321
artificial recharge 324, 327 ff.
Athens (GR) 185
atmosphere, fallout 49 ff., 149, 150, 197, 201
autotrophic 286
availability 272, 353 ff.
Avala, Mt. (YU) 157, 168, 170

background
– freshwater 87, 90 ff.

background
- organisms 282
- seawater 83ff., 87
- sediment 131ff., 146ff.
Back R. (Md.) 179
bacteria 236, 253, 265ff., 315
- leaching 266, 267
Baja California 52, 85, 176, 177
Balagtas R. (PI) 171
Baltic Sea 176, 188, 189
Baltimore Harbor 179, 269
bank filtration 324ff.
barge disposal 60, 61, 186, 358, 359
barium
- drinking water 30
- rocks 134
- surface water 87
barnacles 304
barrier concept 198, 199
battery 365
- cracking 166
- production 166
Beaulieu Estuary (GB) 238
Beldany, L. (Poland) 152
benthic food web 320, 321
- organisms 272, 313ff., 318
Berlin (D) 337ff.
beryllium
- health hazard 15
- PC-treatment 332, 347ff.
- rocks 134
- surface water 87, 95
- tolerance limits 28
bicarbonate 73, 74, 205
biologic half-life 287, 310
- waste treatment 341, 345ff.
biomagnification 318ff.
- benthic food web 321, 322
- via food 318, 319
- - water 318, 319
biomass 1, 286
bio-monitoring 65ff.
biotite 202
bioturbation 182, 256
birds 18, 152
bismuth, PC-treatment 347ff.
bitumen 221
bivalves 297ff.
Biwa, L. (J) 145
Björken, L. (S) 153
black marlin 308
Black Sea 319
bladder wrack 288ff.
blooming 285, 286
Blyth R. (GB) 226

bonding strength 222
Bougainville (PNG) 171
boron
- drinking water 340
- rocks 134
- surface water 87, 94, 95
- spring water 93
- tolerance limit 28
Bornholm Basin (Baltic) 188, 189
Bothnian Bay (Baltic) 176
brass 365
brickworks 53
Bridgeport (Conn.) 242
Bristol Channel (GB) 97, 175, 176
British Columbia, fjords 177, 190
brown trout 306, 308, 311
Buffalo R. (Ark.) 167
Buzzards Bay (Mass.) 179, 182

cable cover 365
cadmium
- air 49, 50
- bank filtrate 326
- batteries 59
- cement production 52
- coal 52
- decarbonation 335
- diatom, plankton 288
- dredging 258
- drinking water 30, 340
- groundwater 326, 357, 358
- health hazard 15, 21, 22
- mineral water 357
- mosses 292
- mussels 302, 303
- oil 52
- organ distribution 307
- PC-treatment 332, 347ff.
- phosphates 229
- pigment 59
- plastics 59
- rocks 44, 134
- sediment 148ff., 160, 161, 165, 175, 186, 191, 195
- sewage sludge 44, 352
- soil 58, 352, 353
- species in water 27, 78
- spring water 93
- surface water 83, 84, 87, 93, 94ff., 104
- waste water 89, 90, 181, 352, 354
- weight correlation 312
Cadiz Bay (E) 289
carbonates 134, 135, 205, 206
Cardigan Bay (GB) 176
Carnon R. (GB) 63, 97, 170

carp 282
catalysis 216, 260
catfish 315 ff.
cathode cell 153
cation exchange 207 ff., 246, 325, 353 ff.
Cayuga Basin (N.Y.) 166
cement production 52
cementation 337, 344, 345
cesium, dating 112
chalcopyrite 262
charcoal 260
Chesapeake Bay 179, 249, 250, 296
Ch'ien Chen R. (RC) 171
Cleveland (Ohio) 150
chlor-alkali electrolysis 40, 151 ff., 174
chlorinated hydrocarbons 269
chlorinity 236, 237
chlorophyll 9
chromium
– bank filtrate 326
– coal 52
– decarbonation 355
– drinking water 30, 340, 352, 354
– groundwater 326, 357
– macroalgae 290
– mineral water 357
– oil 52
– PC-treatment 332, 347 ff.
– phosphates 229
– poisoning 25, 26
– rocks 44, 134
– sediment 148 ff., 161, 175, 177, 191, 195
– sewage sludge 44, 352
– soil 58, 352
– species in water 27
– spring water 93
– surface water 87, 93, 94 ff.
– waste water 181
Cincinnati (Ohio) 347
circulation 192
cirripedia 304
citrate 279
Cladophora see filamentous algae
Clay Lake (CAN) 153, 164
Clearwater L. (CAN) 261, 262
Clyde Estuary (GB) 175, 237
coal residue 155, 189
coagulation 223 ff., 249, 250
coatings 230 ff.
– carbonate 234
– hydroxide 230, 231
– organic 232, 233
– phosphate 234
cobalt
– air 49, 50

– coal 52
– decarbonation 335
– groundwater 357
– oil 52
– mussels 301
– PC-treatment 332, 347 ff.
– phosphates 229
– pyrite 219
– rocks 134
– sediment 161, 186
– species in water 78
– spring water 93
– surface water 87, 95 ff.
– tolerance limits 28
– zooplankton 11, 296
Coeur d'Alène District 34, 152, 158
colloidal species 8, 82, 217
Colne R. (GB) 176
Colorado R. 94, 95, 320
combustion 52, 53
compartment 271, 272
complexing 235, 277, 325
– anionic 250
– humic substances 222, 223
– inorganic 248
– organic 262 ff.
– stability 223
conservative elements 129, 130, 236, 248
Constance, L. (Bodensee, D) 91, 106, 122, 147, 148, 159 ff. 243 ff.
Conway Bay (GB) 176, 252
copper
– air 49, 50
– artificial recharge 327, 328
– bank filtrate 326
– biota 11, 273, 274, 278, 290
– coal 52
– decarbonation 335
– drinking water 30, 340
– groundwater 326, 357
– health hazard 15, 24
– mineral water 357
– minerals 202, 219, 229
– oil 52
– PC-treatment 332, 347 ff.
– rocks 44, 134
– sediment 148 ff., 160, 161, 162, 165, 175, 177, 186, 191, 195
– sewage sludge 44, 352
– soil 58, 352, 353
– species sediment 182, 241, 242, 328, 329
– – water 78, 328, 329
– spring water 93
– surface water 83, 85, 87, 93, 94 ff., 103
– thermal water 357

copper
- tolerance limit   28
- waste water   89, 90, 181, 352, 354
coprecipitation   202, 234
- carbonate   227 ff.
- ferromanganese oxide   213 ff.
- iron sulfide   218, 219
Core Sound (N.C.)   296
Corpus Christi Bay (Tex.)   179, 191
corrosion   56, 326, 337 ff.
cosmetics   352
crabs   304
crustaceans   304 ff., 318
crustal rocks   4, 44, 147
crystallization   201, 202, 215
currents, tidal   254
Cuyahoga R. (Ohio)   94, 95, 150
cyanidation   38

Daphnia   275
Danube R.
- Austria   92, 93, 169, 170
- Germany   74, 98 ff., 159 ff., 311, 321
- Hungary   168, 170
dating   112, 146, 189
Dead Sea   240, 241
decarbonation   333 ff.
deep-sea clay   134, 135
d-electrons   7
deficiency, nutrition   12
de-icing salt   47
Delaware Bay   179
desorption   247 ff.
detergents   27, 44, 229, 230, 263, 264
detoxification   277, 278
detrital minerals   201, 202, 234, 239, 241, 246
Detroit R.   153, 154, 164
Derwent Estuary (Tasmania)   174, 175, 191, 302, 303, 315, 320
diagenesis   131, 182, 246
diatoms   286, 287
Diemel R. (D)   161, 162
diffusion   250, 254
digested sludge   346, 347
Dnepr R. (SU)   103, 220
dogfish   309
Don R. (SU)   103, 220, 221
donors (Lewis bases)   6, 14
Dovey R. (GB)   96, 97
dredging   48, 156, 182, 238, 239, 256 ff.
drinking water   29, 30, 324 ff.
Dryden (CAN)   153, 164
dumping   60, 61, 172, 178, 180 ff., 358, 359

EDTA   276 ff., 353, 354
- algae   287
- green oyster   278, 279
Eh-pH diagram   81, 206
Ekoln, L. (S)   152
Elbe R. (D)   88, 98 ff., 159 ff., 168, 186, 194 ff., 302, 305, 313, 320, 359
electric double layer   207
electrolysis   40, 337, 344, 345
electronegativity   5, 272
electroplating   27, 40, 149, 151, 340 ff.
Elsenz R. (D)   291, 305
Ems R. (D)   41
enzymes   10, 15, 17, 265, 266, 268
equilibrium   200, 209
- constant   204
- models   77 ff.
Erie, L.   129, 130, 143, 148 ff., 154, 155, 226, 227, 252
erosion   48, 159, 182
Erzgebirge (DDR)   34
essential elements   9 ff., 271
estuaries   191 ff., 247 ff.
- mixing processes   193 ff., 199, 200, 214, 215
- sediments   235
eutrophication   250, 251
evaporation   197, 228
exchange capacity   209 ff.
exoskeleton   84, 304
extraction, chemical   238 ff.

feeding habits   295
Feldsee (D)   250, 251
feldspar   201, 202
ferric hydroxide   214 ff., 230, 232 ff.
- system   347 ff.
ferromanganese deposits   217, 251, 252
fertilizer   27, 149, 229, 230, 365
Firth of Clyde (GB)   173, 174 ff.
Firth of Forth (GB)   175, 176
fish   306 ff.
- arsenic   310
- cadmium   306, 307
- mercury   308 ff.
- sex dependence   284
fishkill   254, 255
fjords   189, 190, 214
flagellate   285
flocculation   192, 223 ff., 235, 236, 330, 332
flounder   308
food chain   1, 17, 18, 318 ff.
Fos, Gulf de (F)   184
fossil fuel   52, 151
Foundry Cove (N.Y.)   227

Four Cantons, L. (Vierwaldstätter-See, CH) 108, 152, 313
fracture zone 86, 199
Framvaren Fj. (N) 190
Frankfurt (D) 357
Fraser R. (CAN) 178
freshwater fish 310ff.
Freundlich isotherm 209, 211, 212
fulvic acid 222
fungicides 57, 151, 174, 365
Fuoss effect 223, 224

galena 262
gammarids 305
Gannel R. (GB) 170, 176
gas bubbles 88, 254
gastropods 302
Genesee R. (N.Y.) 166
Gulf of Genova (Med.) 184
geothermal 54, 86, 158
German Bight (North Sea) 186ff., 242, 243
germicides 365
Gironde R. (F) 176, 193
goethite 214, 262
gold, water 87
goldfish 315, 316
Göteborg (S) 327
Gotland Basin (Baltic) 188
gradient, vertical 252ff.
grain size 121ff.
granites 134, 135
gravel overlays 155, 156
Great Bay (N.H.) 179, 236, 237
Greenland ice 50, 51
greigite 218
groundwater 197, 214, 326, 355ff.
guppies 315, 316
Gurk R. (A) 168
gyre, oceanic 199
gyttja 225

halibut 309
Hackett R. (CAN) 217
haddock 309
Haifa Bay (IL) 185
Halifax Bay (AUS) 174
halocline 188
Hardangerfjord (N) 190, 295, 320
hardness, water 263, 264, 274ff.
– toxicity 276
Hartbeespoort Dam (RSA) 142, 143
Hawaii 89
Hayle Estuary (GB) 176
hemocynin 304
hemoglobin 9, 11

herbicides 27, 49, 149
Hobart (Tasmania) 174, 175
homeostatic control 271
HSAB principle (Pearson) 5, 7, 14, 16
Hudson R. (N.Y.) 41, 163, 166, 181
Humber Estuary (GB) 175, 176, 186
humic acid 221ff.
humins 222
Huron L. 144
hydration 210
hydrocerussite 207
hydrolization 210
hydrolysis 215, 217, 236
hydrophilic colloids 223, 224
hydrotroilite 218, 219
Hyperion Plant (Los Angeles) 177, 248
hypochlorite 156
hypolimnion 108, 250

Ide Fjord (S) 190
IJssel Sea (NL) 227
Illinois R. 166
immobilization 200, 325, 327, 328
incorporation 224ff., 279
– modes 294
indicator organism 65ff., 296ff.
industry 38ff.
– battery 41, 166, 365
– electronic 41, 42
– food 42
– paint 41, 169, 365
– pulp and paper 39, 188
– shipping 172
– steel 39, 41
– textile 39, 41
infiltration 325ff.
interstitial water 250ff.
ion activity 204
– competition 275
– exchange 264, 265, 336, 337, 344, 345
– flotation 337
ionization 212
Irish Sea 88, 175, 176
iron
– bank filtrate 326
– biota 11, 290
– decarbonation 335
– drinking water 340
– groundwater 326, 357
– mineral water 357
– oxides 80, 230ff., 327
– phosphate 229
– soil 58
– species in water 78
– sulfide 80, 112, 218, 219

iron
- surface water 82, 87, 94 ff.
irrigation 94, 354
iso-electric point 208
isopods 305, 321
Itai-itai disease 21, 22, 104, 171

Japan, water 104
jewelry 365
Jintsu R. (J) 21, 22, 104, 171
Johannesburg (RSA) 341, 342
Jönköping (S) 151
Jubilee Creek (Ill.) 163, 164

Kamniska R. (YU) 168
Kansas R. 44, 229
Kaoshiung Hr. (RC) 174
kerogen 221
kidney 21, 283, 306, 307
Kohachiya R. (J) 171
Kura R. (SU) 104

La Grande R. (CAN) 31
lake trout 283, 311, 312
land application 48, 352 ff.
- spreading 327 ff.
landfill 47, 355 ff.
Landsort Deep (Baltic) 188, 189
Langmuir isotherm 209
largemouth bass 311
lead
- air 49, 50
- algae 287, 290, 291 ff.
- alkyls 52, 169
- bank filtrate 326
- cement production 52
- ceramics 23
- coal 52
- dating 112
- decarbonation 335
- drinking water 30, 340
- groundwater 326, 257
- health hazard 15, 16, 23, 24
- macrophytes 323
- mineral water 357
- minerals 202
- mosses 292
- mussels 302, 303
- nekton 323
- oil 52
- PC-treatment 323, 347 ff.
- phosphate 229
- pipes 23, 338
- rocks 44, 134
- sediment 148 ff., 160, 161, 175, 177, 186, 191, 195

- sewage sludge 44, 352
- soil 58, 352, 353
- species in water 27, 78
- spring water 93
- surface water 83, 87, 88, 93, 94 ff.
- thermal water 357
- tolerance limit 28
- tubificids 323
- waste water 60, 89, 90, 180 ff., 352, 354
leather tanning 41, 161
leeches 320, 321
Leine R. (D) 291, 292
lethal concentration 302
lignite 53
Ligurian Sea (I) 184
lime system 331, 347 ff.
lithium
- decarbonation 335
- rocks 134
- surface water 87
lithogenous 200, 201, 246
liver 14, 21, 283, 306
Liverpool Bay 88, 175, 176
Long Island Sound (Conn.) 179, 250, 254
Los Angeles 176, 177
- Harbor 240 ff.
- River 163, 164, 248
lysimeter 357, 358

Maas R. (NL) 186
mackerel 308
mackinawite 218
macroalgae 289 ff., 291
- lead 291
macrophytae 228
Maggiore, L. (I) 152, 292
magnetite 214
Main R. (D) 99 ff., 156, 159 ff., 258
major ions, water 72, 73
Malawi, L. 226
Mala Panew R. (PL) 169
manganese
- bank filtrate 326
- biota 11, 290
- decarbonation 335
- drinking water 340
- groundwater 326, 357
- mineral water 357
- oxide species 214
- phosphate 229
- rocks 134
- sediment 175
- sewage sludge 352
- soil 58, 352, 353
- species in water 78, 80

Subject Index

- spring water 93
- surface water 87, 93, 95 ff.
- thermal water 357
- waste water 352
Maraetai, L. (NZ) 31
marcasite 259
Marmorilik (Greenland) 310
mass balance 199
Maumee R. (Ind.) 166
median toxicity limit 275 ff.
Mediterranean Sea 184, 185
Meggen (D) 119, 344
Menai Straits (GB) 289
mercury
- air 49, 50
- alkyl 267, 268
- battery 59
- cement 52
- coal 52
- crustaceans 305
- decarbonation 335
- drinking water 30
- electrolysis 59
- food chain 17, 18
- freshwater fish 310, 311
- fungicides 57, 59
- geochemical cycle 32
- groundwater 357, 358
- health hazard 16, 18 ff.
- marlin 308, 309
- methylation 267, 268
- mussels 302, 303
- oil 52
- organ distribution 301, 308
- PC-treatment 332, 347 ff.
- phytoplankton 288
- pigments 59
- pore water 257, 258
- non predatory fish 311, 320, 231, 323
- predatory fish 311, 320, 321, 323
- restoration 151 ff.
- rocks 134
- sediment 148 ff., 152, 160, 164, 168, 173, 186, 191
- soil 58
- species in water 27, 78
- surface water 78, 87, 88, 93, 94 ff., 104
- tolerance limit 28
- waste water 343, 344, 354
- weight correlation 283
- zooplankton 296
Mersey Estuary (GB) 175, 176
metabolism 10 ff., 265, 266, 315
metals
- analysis 61 ff., 113 ff.

- classification 6
- densities 3
- earth's crust 4, 44, 147
- organics 224, 225
- properties 4
- species in water 7, 8, 26 27, 75 ff.
- world's consumption 58
methylation 267 ff.
- arsenic 269, 270
- lead 269, 270
- mercury 267, 268
- selenium 269, 270
Mexico, Gulf of 48, 179
Michigan, L. 53, 92, 93, 143, 147 ff., 198, 199, 251, 252, 291
microbial activity 265 ff., 324 ff.
microphytae 228
microzone (pH) 214
Mid-Atlantic Ridge 199
Minamata (J) 19, 20, 172, 173
mineral water 356
mining effluents 33 ff., 344, 345
mink 318
Miramichi R. (CAN) 263
Mississippi R. 167, 258
Miya R. (J) 171
Mobile Bay (Ala.) 242, 254, 258
mobilization 255, 256, 283, 313
- bacteria 315
- fish 315 ff.
- sludge worms 317, 318
Molonglo R. (AUS) 35, 145
molybdenum
- biota 11
- groundwater 357
- health hazard 16
- PC-treatment 347 ff.
- pyrite 219
- rocks 134
- stream sediments 63, 64, 175
- surface water 87, 95
- thermal water 357
Monongahela R. (Pa.) 94, 95
Monterey Bay (Cal.) 89, 285, 288
morids 308
Moscow R. (SU) 165, 169
Mosel R. (D) 99 ff.
Moste Dam (YU) 152
Muka R. (J) 171
Mullica R. (N.J.) 236, 237
multimedia filtration 331
Murderkill R. (Va.) 166
muscovite 202
mussel watch 66, 302
mussels 66, 297 ff., 303

mussels
- cobalt distribution 301
- - excretion 300, 301
- seasonal variation 298, 299
- spawning period 297
- zinc 299

Nagoya Hr. (J) 173
Naples (I) 184
Narragansett Bay (Mass.) 179
Neckar R. (D) 41, 43, 92, 93, 98ff., 159ff., 164, 168, 169, 305, 306, 312, 325, 326
Rio Negro (BR) 92
Nereis diversicolor see ragworm
neutralization 344, 345
New Bedford Hr. (Mass.) 182
Newfoundland (CAN) 179
Newport Bay (Cal.) 178
Newport Estuary (N.C.) 250
New York 39
New York Bight 52, 60, 178ff.
New Zealand 36, 158
Niagara R. (N.Y.) 110, 166
nickel
- air 49, 50
- bank filtrate 326
- coal 52
- decarbonation 335
- groundwater 326, 357
- macroalgae 290
- mineral water 357
- oil 52
- PC-treatment 332, 347ff.
- phosphate 229
- pyrite 219
- rocks 44, 147
- sediment 148ff., 161, 175, 177, 186
- sewage sludge 44, 352
- soil 58, 352, 353
- species in water 78
- surface water 86, 87, 93, 94ff.
- tolerance limit 28
- waste water 181, 352, 354
nitrate 86
non-conservative effects 199, 200, 236ff.
non-essential elements 12
North Sea 88, 175, 176, 185ff., 302
Nova Scotia 179
NTA 262ff., 276ff.
- rainbow trout 277, 278
nuclear waste 198, 199
nutrients 112, 117, 252, 256, 258

Oahe, L. (S. Dak.) 152
Ob R. (SU) 220
Odda (N) 55
Ohio R. 94, 95
oil 52, 355
- drilling 56
- field brines 356
- spill 166
Oker R. (D) 162, 354, 357
Okhotsk, Sea of (SU) 240, 241
olivine 201, 202
Ontario, L. 51, 73, 74, 110, 128, 129, 144, 146, 264, 291
Orbe R. (CH) 169
ore genesis 220
organ distribution 283ff., 301, 307, 308
organic carbon 222
organo-metal compounds 41, 42, 246
Orly (F) 152
Oslo Fjord (N) 186
osmosis 282
- reverse 337, 344, 345
Oswego R. (N.Y.) 164
Ottawa R. (CAN) 126, 127, 164, 313
otter 318
outfall plume 88, 89
oxidation 273, 326, 327ff., 330, 331
- number 6
oyster 279, 283, 284, 285, 297
- age-weight dependence 283
- sediment dependency 315
- toxicity to embryos 302
Oyster Bay (N.Y.) 296

Pacific oyster 283
paints 365
Pamlico R. (N.C.) 179
paper mill 39, 151ff., 188
pE diagram 80
pedogenesis 58, 220, 352
pelagic sediment 134, 197, 238
Peoria L. (Ill.) 166
perch 310, 312
Perch L. (CAN) 226
periodic table 3
pesticides 321, 361, 362
pH 13, 33, 37, 54, 78, 81, 94, 202, 205, 206, 208, 236, 248, 258ff., 279, 280, 353
phase concentration 244, 245
phenole 355
phenylmercury 153, 267ff.
phosphate 48, 229, 230
phosphorus 188, 353
photography 27, 352, 365
physico-chemical treatment (PCT) 330ff., 347ff.
phytoplankton 66, 67, 84, 109, 286ff., 288
pickling 40, 341

Pickwick Res. (Tenn.) 152
pigment 27, 41, 169, 365
pike 282, 311, 312
pilot plant 330
plagioclase 202
plaice 310
plumbogummite 207
pollen, dating 112, 129, 146
polychaetes 313ff.
polyethylene 156, 157
polymer film 155
polyphosphate 229, 230
Poole Hr. (GB) 175, 176
porbeagle 309
porosity 254
Port Phillip Bay (AUS) 174
Potamogeton sp. 292ff.
Potomac R. (Va.) 166, 179
precipitates, minerals 203ff., 235, 325
pre-purification 330, 331
primary consumer 295, 296
– seasonal variation 296
– producer 286
productivity, biologic 85, 86
pseudomorphosis 202, 229
Puget Sound (Wash.) 55, 178
pycnocline 190
pyrite 81, 259, 260, 262
pyrolusite 262
pyromorphite 207
pyroxene 201, 202

Qishon R. (IL) 170
quartz 201, 202
– correction method 128

radioactivity 38
– isotopes 198
– nuclides 198, 249
radium 357
ragworm 280, 281, 313ff.
– sediment dependency 314
rainbow trout 273, 278
– NTA 277, 278
Raritan Bay (N.J.) 179
recycling 363ff.
Red R. (GB) 170
redox effects 79ff., 232, 233
reducing conditions 253ff., 326
reef carbonate 238
refractory chemicals 1
relative pollution potential 89, 360
remobilization 197, 248, 325, 327, 328
removal rate 197
reprocessing 363ff.

residence time 197ff.
residual organics 221, 239, 244
respiration 294
restoration, lake 155, 156
Restronguet Cr. (GB) 176 191, 281, 304, 313ff.
reuse 363ff.
Rheidol R. (GB) 96, 97
Rhine R. (D) 98ff.
– bank filtration 325, 326
– fish 311, 312
– fishkill 254
– groundwater 357
– major ions 73
– organic pollutants 98, 99
– PC-treatment 330, 332
– sediment 122ff., 159ff., 167, 168, 240ff., 360, 361
– suspended solids 138, 139, 279
– water 98ff.
Rhine Estuary (NL) 88, 186, 193ff., 237, 238
Rhône (CH) 168, 169
Rietvlei Dam (RSA) 152
Rio Tinto Estuary (E) 191, 289
roach 282, 283, 306, 312
rock composition 147
– debris 200
Rotorua, L. (NZ) 31, 291
Rotterdam Hr. (NL) 125, 186
Ruhr R. (D) 99ff., 200, 291
– lead 292
– mosses 292

Saanich Inlet (CAN) 190
Sado R. (P) 300
Saguenay Fjord (CAN) 190
Sajo R. (H) 170
salinity 71, 72, 192, 247ff. 280
Salzach R. (A) 170
Sambre R. (B) 168
sampling
– sediments 114, 115
– water 68
sand filter 327
sandstone, composition 134, 135
San Francisco Bay 178
Sangchris, L. (Ill.) 54, 152
Santa Monica Canyon 177
Saoseo, L. (CH) 151
sapropel 225
Sargasso Sea 84ff., 310
Sarnia (CAN) 153, 154
Saronikos Gulf (GR) 185
scavenger 217, 222, 253
Scheldt R. (B) 176
Schuylkill R. (Pa.) 94, 95, 156

scrap metal 157, 364, 365
sea grass 289
seasonal variations 104 ff., 297 ff.
Seattle (Wash.) 150
sediment
– analysis 113 ff.
– chemical forms 238 ff., 328, 329
– digestion 118
– sampling 114, 115
– storage 115, 116
seed dressing 20, 152
Seine River (F) 168
selectivity 210, 212, 213
– extraction 238 ff.
– zeolites 265
selenium
– air 49, 50
– cement production 52
– coal 52
– drinking water 30, 340
– health hazard 16
– marlin 283
– oil 52
– PC-treatment 332, 347 ff.
– species in water 27
– surface water 87, 93
– tolerance limit 28
Seneca R. (N.Y.) 166
sequestering 236
sessility 283
Severn Estuary (GB) 175, 176, 308, 320
sewage
– disposal 355 ff., 358, 359
– effluent 151
– outfall 176 ff., 358
– sludge 44, 352 ff.
sex relation 284
shale composition 134, 135
shark 309
ship-bottom paint 56
Shiranui Sea (J) 172
shrimp 318
siderite 81
silica 86, 155, 260
sill 189, 190
silver
– air 49, 50
– drinking water 30, 340
– health hazard 16
– PC-treatment 347 ff.
– photochemisty 44
– rocks 44, 134
– sediment 175, 177, 191
– sewage sludge 44
– species in water 27, 78

– surface water 87, 93, 95, 99
– tolerance limit 28
smelter emission 27, 54, 55, 140, 151, 175, 178, 190, 260, 261
soil 58, 361, 362
– extract 353 ff.
sole 309
Solent (GB) 320
Solimoes R. (BR) 92
solubility 203 ff., 220, 247, 280
Solway Firth (GB) 175, 176
Sörfjord (N) 190, 191, 295, 296
sorption 234, 325
– carbonate 227 ff.
– clay minerals 210 ff.
– humic substances 222, 223
– hydrous oxides 213 ff.
– past history 212, 213
– radionuclides 212, 213
speciation
– sediment 238 ff.
– water 75 ff.
specific ion exchange 212
– surface area 209, 328
sphalerite 262
spoil heaps 47
spring water 92, 93
– – Fe/Mn-oxide 214
stabilization 220
stationary fish 283
statistic analysis 119
St. Clair R. (US/CAN) 40, 144, 153, 164, 310
St. Lawrence R. 164, 179
stickleback 273, 274
Stockholm (S) 189
Stola R. (PL) 169
strontium
– rocks 134
– surface water 87, 93
substitution
– lattice 211
– product 363
Sudburry (CAN) 54, 140, 142, 260, 261
sulfide 181, 205, 206, 252, 253, 259, 260, 349, 359
sulfosalts 259
sulfur 206
Superior, L. (CAN) 144, 157
superphosphate 229, 230
surface complexing 207, 208, 216, 217
– preparation 341, 342
Susquehanna R. (Pa.) 107, 163
Swansea Bay (GB) 175, 176
synergism 273
Sztola R. (PL) 34

Tacoma smelter (Wash.) 55, 149, 178
tailings 260
Takahara R. (J) 171
Tama R. (J) 170, 171
Tamar R. (GB) 34, 170, 283, 297
tanning industry 149, 161
Tay R. (GB) 168
Taylor Mts. (Alaska) 158, 164
technophility index 58, 361
Tees R. (GB) 41, 252
Tel Aviv 185
temperature 273, 285, 353
temporary elimination 325
Tennessee R. 94, 164, 166, 167
thallium
– health hazard 16
– groundwater 358
– occurrence 16
– PC-treatment 347ff.
Thames Estuary (GB) 88, 175, 176, 186
thermal fracturing 199
– water 356
thermocline 214
thiobacilli 266, 267
thiosulfate 279
tidal current 192
tin
– groundwater 357
– mineral water 357
– PC-treatment 347ff.
– soil 58
– surface water 87
– thermal water 357
– tolerance limit 28
titanium
– PC-treatment 347ff.
– production 181, 182, 186
– surface water 87, 93
Toce R. (I) 168
Togane mine (J) 104
Tokuyama Bay (J) 173
Tokyo Bay (J) 173
tolerance limit 27ff.
Tomogonops R. (CAN) 262, 263
Tor Bay (GB) 175, 176
toxicity 12ff., 271ff.
transport, organics 220
– river 104ff.
trickling filter 345
trophic level 1, 17, 272, 286, 296
– trout 283
Trummen, L. (S) 151, 152
Tsurumi R. (J) 171
tubificid worms 280, 316, 318ff., 321, 322, 323

Tuma L., (CH) 151
tunicate 225
tungsten
– groundwater 357
– rocks 134
turbulent exchange 254
Tuscany coast (I) 184
Tyrrhenian Sea (I) 83, 125, 184

ultrafiltration 75, 77
Ume R. (S) 168, 169
upwelling water 83, 84
uranium
– groundwater 357
– rocks 134
– surface water 87
urban storm water 44ff.
usage index 364ff.
US, mining effluents 34, 35
USSR, river water 102ff.

valence 6, 210
vanadium
– air 49, 50
– coal 52
– decarbonation 335
– oil 52
– PC-treatment 346ff.
– porphyrin 50
– rocks 134
– surface water 87, 93, 95
Van der Waals bonding 207, 208
Vänern, L. (S) 152
Vättern, L. (S) 151ff.
Växjö (S) 152
vegetable crops 353ff.
Vermillion R. (Ill.) 291
vitamin B_{12} 6, 11
volcanic activity 86
Volga R. (SU) 220

Wabigoon R. (CAN) 153, 154ff.
Wada R. (J) 171
Wadden Sea (NL) 193
Wahnbach Res. (D) 162
Waikato R. (NZ) 31, 292
Wanapitai R. (CAN) 164
Washington, L. (Wash.) 109, 144, 148ff., 198, 199
waste dumping 60, 61, 198, 358, 359
– wool 155
water discharge 104ff., 159
– quality criteria 26ff.
– standard 350, 351
waves 254

weathering 31, 220, 358
weight dependency 282, 283
Weschnitz R. (D) 161
Weser R. (D) 74, 98ff., 159ff., 186
Westernport Bay (AUS) 174
whirling-up 254, 255
white fish 311
White's Point (Cal.) 89
Wiesbaden (D) 330, 332
willemite 207
Windermere, L. (GB) 109, 147
Winnipeg, L. (CAN) 130, 152
Wisconsin, lakes 144, 148, 149
− rivers 164
Witwatersrand (RSA) 36, 342
Wupper R. (D) 168
Wyandotte (Mich.) 153, 154

yellow acids 222
Yodo R. (J) 171
Ystwyth R. (GB) 96, 97
Yukon R. (Alaska) 91, 92, 201, 240, 241, 245

zebrafish 284
zeolite 210, 264, 265
zero effluent system 342, 343
− point of charge 208
zinc
− air 49, 50
− algae 289, 291ff.
− bank filtrate 326
− cement production 52
− coal 52
− decarbonation 335
− drinking water 30, 340
− equivalent 355
− groundwater 326, 357
− health hazard 11, 17
− infiltration 327, 328
− mineral water 357
− minerals 202
− mussels 299, 302, 303
− oil 52
− PC-treatment 332, 347ff.
− phosphate 229
− refinery 190
− remobilization 327, 328
− rocks 44, 134
− sediment 148ff., 160, 161, 165, 175, 177, 186, 191, 195
− sewage sludge 44, 352
− soil 58, 352, 353
− species in water 78, 328, 329
− sources 59
− spring water 93
− surface water 86, 87, 93, 94ff.
− thermal water 357
− tolerance limit 28
− trout 273
− waste water 89, 90, 181, 352, 354
zooplankton
− metal variation 295, 296

D. M. Gates

Biophysical Ecology

1979.
(Springer Advanced Texts in Life Sciences)
ISBN 3-540-90414-X
In preparation

Biophysical Ecology presents a analytical treatment of the processes by which a plant or animal interacts with its environment. These processes include energy exchange, gas exchange, and chemical kinetics. A plant or animal interacts with its environment by processes which originate outside the organism, flow through the surface of the organism, and ultimately elicit responses through metabolic processes within the organism.

Energy exchange between an organism and the environment involves radiation, convection and evaporation or latent heat. Radiative exchange includes short-wave radiation such as solar, skylight, reflected sunlight, and long-wave radiation from thermal emission by the ground surface, vegetation, clouds, and the atmosphere. A detailed description is given of the geographical, seasonal, and diurnal dependencies of short and long wave length radiation. The spectral properties of organisms and the environment are described Analytical models of whole leaf photsynthesis and respiration are presented here for the first time.

Biophysical Ecology will be indispensable to advanced students in all life sciences; it is designed to provide an understanding of the ecology or organisms based on fundamental scientific principles.

Springer-Verlag
Berlin
Heidelberg
New York

Lakes: Chemistry, Geology, Physics

Editor: A. Lerman

1978. 206 figures, 61 tables. XI, 363 pages.
ISBN 3-540-90322-4

Contents
Introduction – *A. Lerman:* Man's Limnetic Drive. – *R. A. Ragotzkie:* Heat Budgets of Lakes. – *G. T. Csanady:* Water Circulation and Dispersal Mechanisms. – *P. G. Sly:* Sedimentary Processes in Lakes. – *W. Stumm, P. Baccini:* Man-Made Chemical Perturbation of Lakes. – *M. A. Barnes, W. A. Barnes:* Organic Compounds in Lake Sediments. – *S. Krishnaswami, D. Lal:* Radionuclide Limnochronology. – *F. Jones, C. J. Bowser:* The Mineralogy and Related Chemistry of Lake Sediments. – *P. H. Eugster, L. A. Hardie:* Saline Lakes. – *K. Kelts, K. J. Hsü:* Freshwater Carbonate Sedimentation. – *F. J. Pearson, T. B. Coplen:* Stable Isotope Studies of Lakes. – *D. M. Imboden, A. Lerman:* Chemical Models of Lakes.

Springer-Verlag
Berlin
Heidelberg
New York

Here is a multidisciplinary survey of fundamentals and recent developments in the fields of chemical, geological, and physical limnology. Emphasis is on inorganic and those biogeochemical processes that can be considered in terms of the macroscopic mechanisms controlling the evolution of lake systems. The book will be greatly appreciated by practicing scientists, engineers, and graduate students in the fields of limnology and related environmental sciences.